Halides of the Transition Elements

Halides of the First Row Transition Metals

Halides of the Transition Elements

is a series of three volumes consisting of
the present title together with the following

Halides of the Lanthanides and Actinides
by D. Brown

and

Halides of the Second and Third Row Transition Metals
by J. H. Canterford and R. Colton

Halides of the Transition Elements

Halides of the First Row Transition Metals

R. Colton
and
J. H. Canterford

Department of Inorganic Chemistry,
University of Melbourne,
Parkville, Victoria.

WILEY-INTERSCIENCE
a division of John Wiley & Sons Ltd.
London New York Sydney Toronto

Library of Congress catalog card No. 73-90912

SBN 471 16625 1

Printed in Great Britain at the Pitman Press, Bath

Preface

The halide chemistry of the first row transition elements is a popular field of research, but although the halide chemistries of several individual elements and of certain groups of compounds have been well reviewed previously, this is the first occasion that an attempt has been made to examine the complete halogen chemistry of these metals together.

In contrast to the Lanthanides, Actinides and the elements of the second and third transition series which are discussed in other volumes in this series, the information on the simple halides of the first row metals is very meagre and it is necessary in this case to discuss in detail the chemistry of the adducts formed by the halides in order to appreciate their true chemical behaviour. However, while the nature of the compounds discussed in this book is somewhat different, we have retained the style used in the previous volumes and placed considerable emphasis on physical measurements and structural determinations.

A major difficulty in writing a book of this type is the extraordinary rate at which material is being published. The main text includes the majority of all papers available prior to 1968 and we have included addenda at the end of each chapter to include information available to us till January 1969. In this short period some 600 relevant papers have been published. Whilst we have attempted to be as comprehensive as possible it is obvious that there will inevitably be omissions in the text as a result of the enormity of the subject under discussion.

Again we thank some of our colleagues for numerous helpful discussions, the Chemistry Department librarian, Miss I. M. Rennie, and her assistant Miss R. V. Bunevicius, for their help in obtaining many journals and our publishers for allowing us to insert the addenda at the galley proof stage. Special thanks are offered to Dr. T. A. O'Donnell who carefully read the complete manuscript and made many helpful suggestions.

University of Melbourne,
Victoria, Australia.
January 1969.

R.C.
J.H.C.

Abbreviations

The following list of ligand abbreviations will be used throughout the text. All other symbols and units are those in normal use.

py = pyridine
2-CH_3py = 2-methylpyridine = α-picoline
3-CH_3py = 3-methylpyridine = β-picoline
4-CH_3py = 4-methylpyridine = γ-picoline
2,6-lut = 2,6-lutidine = 2,6-dimethylpyridine
2,4,6-coll = 2,4,6-collidine = 2,4,6-trimethylpyridine
an = aniline
o-tol = *o*-toluidine
m-tol = *m*-toluidine
p-tol = *p*-toluidine
Q = quinoline
en = ethylenediamine
bipy = bipyridyl
phen = 1,10-phenanthroline

All temperatures are in °C unless specifically denoted °K. No distinction is, in general, made between a true melting point and a decomposition temperature: a wide temperature range normally implies decomposition.

Contents

Chapter 1

General Survey

It is interesting to observe the way in which the interests of Inorganic Chemists in the halide compounds of the first-row transition metals have evolved over the past fifteen years. As an example, some years ago a major area of investigation was the existence, or otherwise, of tetrahedrally coordinated nickel. The work of Gill and Nyholm[1] proved conclusively that such a stereochemistry can exist for nickel. Over the years there has been considerable effort applied to explain the magnetic and spectral properties of the halide compounds of the first-row transition metals in terms of first crystal-field theory and, more recently, ligand-field theory. At present a major interest is the synthesis and characterization of five coordinate complexes, and the modification of ligand-field theory to satisfactorily describe these compounds.

It is essential that a greater understanding of the bonding in halo compounds be developed and it is likely that this requirement, together with recent advances in experimental techniques, will increase the importance of the application of nuclear quadrupole resonance spectroscopy. It must be pointed out however, that the theoretical interpretation of NQR data is still not well developed.

In this chapter we will survey those features which are common to several of the elements of the first transition series. These topics include methods of preparation, spectral and magnetic properties and general chemistry, including the formation of adducts.

PREPARATIONS

This section comprises a general review of preparative methods for halogen compounds of the first-row transition metals. The examples of preparation cited are intended merely to illustrate general methods, and no attempt is made in this chapter to record all examples of a particular method, nor are highly specific preparations recorded. Such details will be found in the appropriate chapters.

Direct fluorination, although used frequently, is not as important in the preparation of binary fluorides of the first-row transition metals as it is with the second- and third-row metals. The main reason for this is that for most of the first-row transition metals, it is impossible to achieve a higher oxidation state with fluorine which is unavailable by other preparative methods.

Binary chlorides are in general more volatile than the corresponding fluorides and therefore direct chlorination of the metal is a practical method of preparing chlorides of most of the transition metals. Towards the right of the series, anhydrous binary chlorides are often prepared by dehydration of the appropriate hydrate in a stream of hydrogen chloride gas.

Complex fluorides of the first-row transition metals have been prepared in a number of ways which are not generally applicable to the compounds of the heavier transition metals. Such methods include reaction of the constituent fluorides in the melt, preparations in hydrofluoric acid, and preparations in methanol.

Chloro, bromo and iodo complexes are prepared either in the melt or in organic solvents. An interesting comparison with the second- and third-row elements is that the product of the reaction in the melt usually depends critically on the stoichiometric proportions of the constituent reactants. For example, reaction of titanium trichloride with alkali-metal chlorides can give four different products containing titanium(III), the actual product depending on the relative proportions of the halides used. With the heavier elements this condition does not seem to be so important. For example, reaction of niobium pentachloride and potassium chloride over quite a wide ratio of reactants yields only potassium hexachloroniobate(V).

Fluorides

Binary fluorides of the first-row transition metals have been prepared from the metals themselves using fluorine, hydrogen fluoride or halogen fluorides. In addition, as a direct consequence of the fewer oxide fluorides formed by these elements than by the heavier transition metals, fluorination of oxides has been used effectively in preparing binary fluorides.

Perhaps the most widely used preparative methods for binary fluorides of the first-row transition metals is halogen exchange between the corresponding chloride and anhydrous hydrogen fluoride, or by direct fluorination of a chloride of the metal, which naturally often gives concomitant oxidation.

Direct fluorination of the metal. The metals for which this reaction has been used are given in Table 1.1.

It will be noted that the method is only applicable to those elements towards the left of the Periodic Table that tend to form covalent volatile fluorides. Reaction with the remaining metals of the series tends to give

TABLE 1.1
Direct Fluorination of the Metal

Metal	Conditions	Product	References
Ti	Flow system, 200°	TiF_4	2
V	Flow system, 300°	VF_5	3
Cr	Flow system, 300–350°	CrF_5 or CrF_4[a]	4, 5
Mn	Fluidized bed, 600–700°	MnF_4	6

[a] Product depends on experimental conditions (see Chapter 4).

a protective layer of non-volatile ionic lower fluoride. This effect is well emphasized by the extensive use of nickel, and to a lesser extent of copper and iron, for apparatus for the handling of fluorine and volatile fluorides.

Direct fluorination of halides. Many of the difficulties associated with the method of preparing binary fluorides previously described, may be overcome by fluorination of either other halides or lower fluorides, as shown in Table 1.2.

TABLE 1.2
Direct Fluorination of Halides

Halide	Temperature	Product	References
$CrCl_3$ or CrF_3	500°	CrF_4	7
MnF_3	550°	MnF_4	8
$MnCl_2$	250–270°	MnF_3	9
FeF_2	550°	FeF_3	10
$FeCl_3$	Elevated	FeF_3	11
$CoCl_2$ or CoF_2	200–300°	CoF_3	10, 12, 13
$NiCl_2$	350–400°	NiF_2	13, 14

It is obvious from the table that good temperature control is required in the fluorination of manganese dichloride to give the trifluoride, to prevent the formation of manganese tetrafluoride impurity.

Use of halogen fluorides. A fairly general method of preparing binary fluorides is the action of the more reactive halogen fluorides, that is, chlorine trifluoride, bromine trifluoride and iodine pentafluoride, on the metal, a halide, or an oxide. The chlorine trifluoride reactions are almost invariably carried out in a flow system, but the bromine trifluoride and iodine pentafluoride reactions are usually performed under reflux conditions. Some typical examples of these types of reaction are given in Table 1.3.

TABLE 1.3
Preparations Using Halogen Fluorides

Reactant	Reagent	Conditions	Product	Reference
Ti	ClF_3	Flow system, 350°	TiF_4	15
V or VX_3	ClF_3	Flow system, elevated	VF_5	16
	BrF_3 or IF_5	Reflux	VF_5	16
$FeCl_3$	ClF_3	Flow system, 500°	FeF_3	17
	BrF_3	Reflux	FeF_3	18
$CoCl_2$	BrF_3	Reflux	CoF_3	18
NiO	ClF_3	Flow system, 25–180°	NiF_2	19

A major difficulty with the use of bromine trifluoride is the removal of the excess reactant and its decomposition products. In many cases this is not just a case of physical separation. There is a growing amount of evidence that the bromine trifluoride is involved in adduct formation and hence extremely difficult to remove. This difficulty also occurs with chlorine trifluoride and iodine pentafluoride although not to the same extent.

Anhydrous hydrogen fluoride. Use of anhydrous hydrogen fluoride is a common method of making binary fluorides, and indeed is one of the earliest methods of preparation of these compounds. The reagent can be used with a wide range of reactants of different chemical nature, at room temperature, in flow systems, or under conditions of high temperature and high pressure in a bomb, as shown in Table 1.4.

TABLE 1.4
Preparations Using Anhydrous Hydrogen Fluoride

Reactant	Conditions	Product	Reference
$TiCl_4$	Flow system	TiF_4	20
VCl_4	Room temperature	VF_4	21
VCl_3	Flow system, 600°	VF_3	22
CrO_3 or $K_2Cr_2O_7$	Room temperature	CrO_2F_2	23
$CrCl_3$	Flow system, 450–600°[a]	CrF_3	24
$MnCl_2$	Flow system, elevated	MnF_2	25
$FeCl_3$	Flow system, elevated	FeF_3	26
Co	Bomb, 180°	CoF_2	27
$CoCl_2$	Flow system, 300°	CoF_2	20
Ni	Bomb, 225°	NiF_2	27
$NiCl_2$	Flow system, >500°	NiF_2	28

[a] Product impure at high temperatures.

A major disadvantage of the use of this reagent is that the hydrogen fluoride must be very pure. The most common impurity is water which can cause hydrolysis. Production of high purity anhydrous hydrogen fluoride is quite difficult[29]. In addition, difficulties arise in flow system reactions because involatile surface fluoride can prevent complete fluorination. It is often necessary to remove the sample periodically from the apparatus, grind it, and then continue the hydrofluorination.

Thermal decomposition reactions. A very useful method of preparing binary fluorides of the elements to the left to the transition series is by the thermal decomposition of the ammonium salts of various complex fluorides, as shown in Table 1.5.

TABLE 1.5

Thermal Decomposition of Fluoro Complexes

Complex	Temperature	Product	Reference
$(NH_4)_2TiF_6$	150°	TiF_4	30
$(NH_4)_3VF_6$	500–600°	VF_3	31
$(NH_4)_3CrF_6$	450–500°	CrF_3	24
$(NH_4)_3CrF_6$	1100°	CrF_2	24
NH_4MnF_3	300°	MnF_2	32

This method can be of considerable importance, since the complex fluorides may be prepared in hydrofluoric acid or in methanol (see later), and thus provides a route to binary fluorides without the use of anhydrous hydrogen fluoride.

The thermal decomposition of ammonium hexafluorochromate(III) is interesting in that at comparatively low temperatures it produces chromium trifluoride but at higher temperatures chromium difluoride is formed. This is partly due to the reduction of the trifluoride by the liberated hydrogen.

Miscellaneous methods. A number of more specific fluorination reactions are given in Table 1.6.

Fluorination of vanadium pentaoxide in a flow system gives vanadium oxide trifluoride, which is volatile and sublimes from the reaction zone. In a closed system however, complete fluorination to vanadium pentafluoride occurs. Frequently such factors must be considered in detail before embarking on the preparation of a specific fluoride.

TABLE 1.6

Miscellaneous Preparative Methods for Fluorides

Reactants	Conditions	Product	Reference
TiO_2 and F_2	Flow system, >350°	TiF_4	2
V_2O_5 and F_2	Bomb, 200–475°	VF_5	33
V_2O_5 and F_2	Flow system, 450°	VOF_3	34
VF_3, H_2 and HF	Flow system, 1150°	VF_2	35
CrF_3 and Cr	Bomb, 1000°	CrF_2	24
Co_2O_3 and F_2	Flow system, 150–300°	CoF_3	36
NiO and F_2	Flow system, 375°	NiF_2	37
$NiF_2 \cdot 4H_2O$ and HF	Flow system, 350–400°	NiF_2	38

Chlorides

Direct chlorination of the metal is a fairly widely used method of preparation, particularly for the higher oxidation state covalent chlorides. The anhydrous chlorides in lower oxidation states, particularly of those elements near the middle of the first transition series, are often prepared from hydrates which are readily isolated from aqueous solutions. It is advisable however, that chlorides prepared by dehydration of hydrates be resublimed in a stream of hydrogen chloride, or in some cases chlorine, to remove oxide impurities formed by partial hydrolysis in the dehydration step. It is often convenient to carry out both reactions simultaneously.

Direct chlorination of metals and oxides. A representative selection of these reactions is given in Table 1.7. Invariably these reactions are carried out in a flow system.

TABLE 1.7

Direct Chlorination of Metals and Oxides

Reactants	Temperature	Product	References
TiO_2 and C	400–600°	$TiCl_4$	39, 40
V	200–500°	VCl_4	41, 42
V_2O_5 and C	800°	VCl_4	43
Cr	960–1000°	$CrCl_3$	44
Cr_2O_3 and C	800°	$CrCl_3$	45
Fe	300–350°	$FeCl_3$	46, 47
NiO and C	>300°	$NiCl_2$	48

The role of carbon is probably to reduce the oxides to the metal, or a carbide, which is then chlorinated. This is to be contrasted to the situation in the second- and third-row elements, where carbon is often used to minimize formation of volatile oxide chlorides.

Carbon tetrachloride with oxides. Carbon tetrachloride is frequently used to chlorinate metal oxides, usually in a flow system arrangement because of the high temperatures required. This is to be contrasted with the comparatively low temperatures required to chlorinate oxides of the second- and third-row transition elements, which makes sealed-tube reactions feasible. Some typical reactions involving carbon tetrachloride in a flow system are given in Table 1.8.

TABLE 1.8
Preparations with Carbon Tetrachloride

Oxide	Temperature	Product	Reference
TiO_2	600°	$TiCl_4$	49
V_2O_3	Red heat	VCl_3	50
Fe_2O_3	500°	$FeCl_3$	49
Co_3O_4	310–900°	$CoCl_2$	51

Hydrogen chloride. Hydrogen chloride reacts at high temperatures with a number of transition metals to produce binary chlorides. It also is extensively used in the dehydration of aquo compounds, since it reacts with any oxide impurity formed during dehydration. A selection of such reactions is given in Table 1.9.

TABLE 1.9
Preparations Using Hydrogen Chloride

Reactants	Temperature	Product	References
TiO_2 and C	1250°	$TiCl_4$	52
V	950°	VCl_2	53
Cr	900°	$CrCl_2$	54, 55
Fe_2O_3	300–1000°	$FeCl_3$	56
$NiCl_2 \cdot 6H_2O$	400–550°	$NiCl_2$	57, 58
Cu	800–1000°	CuCl	59
$CuCl_2 \cdot 2H_2O$	150°	$CuCl_2$	60

Thionyl chloride. Thionyl chloride has been used to chlorinate oxides and other salts to produce binary chlorides, but it has been most extensively used to dehydrate hydrates prepared from aqueous solutions. In most cases, the preparations are carried out under reflux conditions. Some examples are given in Table 1.10.

TABLE 1.10
Preparations Using Thionyl Chloride

Reactant	Product	References
VO_2	VCl_3	61
V_2O_3[a]	VCl_3	62
CrO_3	$CrCl_3$	63
$MCl_3 \cdot 6H_2O$ (M = Cr, Fe)	MCl_3	46, 64
M salt (M = Mn, Co, Ni, Cu)[b]	MCl_2	65
$MCl_2 \cdot 6H_2O$ (M = Co, Ni)	MCl_2	28, 64
$CuCl_2 \cdot 2H_2O$	$CuCl_2$	64

[a] Sealed tube reaction at 200°.
[b] Suitable salts are acetate, carbonate, formate and nitrate.

Lower chlorides. The principal methods of preparing lower chlorides of those elements towards the left of the first transition series, are reduction of higher chlorides with the metal itself or with hydrogen, and thermal decomposition methods. Some examples are given in Table 1.11.

TABLE 1.11
Preparation of Lower Chlorides

Reactants	Conditions	Product	References
$TiCl_4$ and H_2[a]	Flow system, >500°	$TiCl_3$	66, 67
$TiCl_4$ and Ti	Flow system, 900–950°	$TiCl_3$	68
$TiCl_4$ and Al	Sealed tube, 150°	$TiCl_3$	69
$TiCl_3$	Disproportionation, 475°	$TiCl_2$ (+$TiCl_4$)	67
VCl_4	Decomposition, 150–170°	VCl_3	70
VCl_3 and H_2	Flow system, 500°	VCl_2	70
VCl_3	Disproportionation, >400°	VCl_2 (+VCl_4)	70
$CrCl_3$ and Cr	Bomb, 900°	$CrCl_2$	71
$CuCl_2$	Decomposition, >300°	CuCl	72

[a] In this reaction the $TiCl_3$ must be rapidly removed from the hot zone to prevent disproportionation.

Miscellaneous methods. The dichlorides of manganese, cobalt and nickel are quite readily prepared by interaction of the metal with chlorine in ethanol or ether[73]. Another useful method of preparing these chlorides, and also copper dichloride, is to react the appropriate acetate with acetyl chloride in benzene or acetic acid[74,75].

Oxide Chlorides

Comparatively few oxide chlorides of the first-row transition metals are known, but it is true to say that although there is no general method of preparation which has been applied to a number of elements, practically every possible type of reaction has been investigated for at least one element. Table 1.12 shows the wide range of techniques used.

TABLE 1.12

Preparation of Oxide Chlorides

Reactants	Conditions	Product	Reference
$TiCl_4$ and O_3	At b.p. of $TiCl_4$	$TiOCl_2$	76
$TiCl_4$, Cl_2O and O_2	Room temperature	$TiOCl_2$	77
$TiCl_3$ and TiO_2	Sealed tube, 650–750°	TiOCl	78
V_2O_5, C and Cl_2	Flow system, 300–400°	$VOCl_3$	79
V_2O_5 and C_5Cl_8	Reflux	$VOCl_3$	80
V_2O_5 and $AlCl_3$	Sealed tube, 400°	$VOCl_3$	81
$VOCl_3$, Cl_2O and O_2	Room temperature	VO_2Cl	82
$VOCl_3$ and O_3	At b.p. of $VOCl_3$	VO_2Cl	82
V_2O_5 and VCl_3	Sealed tube, 600°	$VOCl_2$	20
$VOCl_3$ and VOCl	Thermal gradient, 450–250°	$VOCl_2$	79
$VOCl_2$	Decomposition in N_2 at 300°	VOCl	79
V_2O_5 and VCl_3	Thermal gradient, 720–620°	VOCl	20
CrO_3 or $M_2Cr_2O_7$ and MCl	Interaction in c. H_2SO_4	CrO_2Cl_2	83
Cr_2O_3 and $CrCl_3$	Thermal gradient, 1040–840°	CrOCl	84
Fe_2O_3 and HCl	Flow system, 300°	FeOCl	56
Fe_2O_3 and CCl_4	Bomb, 300°	FeOCl	85

Bromides

In general the methods of preparation of bromides closely parallel those used for preparing chlorides of the transition metals. However, direct interaction of the elements is of more general applicability because of the greater ease of carrying out sealed-tube reactions with bromine.

Bromination of the metal. Both bromine and hydrogen bromide have been used to prepare binary bromides as shown in Table 1.13.

One of the most interesting methods of preparing the dibromides, and indeed the dichlorides and diiodides, of iron through to copper is the direct interaction of the elements in ether or alcohol. This is an extremely useful preparative method[94] but there appears to have been little use made of it. It is interesting to observe that the thermal stability of chromium tribromide governs the products obtained by reacting chromium metal with bromine and hydrogen bromide under similar conditions. That is, it can be isolated in the presence of bromine but not with hydrogen bromide.

Halogen exchange. The classical method of preparing bromides by halogen exchange is by using hydrogen bromide, although in fact this method has not been widely used for the elements of the first transition series. Recently however, Druce, Lappert and Riley[95] have reported halogen-exchange reactions using boron tribromide as the brominating agent. The advantages of this very powerful brominating agent are that all the products derived from it are volatile, and the bromination reactions proceed at room temperature. For the first-row transition metals the reactions between boron tribromide and titanium tetrachloride, iron trichloride and copper dichloride have been studied, and in each case the corresponding bromide was obtained.

TABLE 1.13

Preparation of Bromides Using Bromine and Hydrogen Bromide

Reactants	Conditions	Product	Reference
Ti and Br_2	Flow system, 300–600°	$TiBr_4$	86
V and Br_2	Sealed tube, 400°	VBr_3	70
V and Br_2	Flow system, 550°	VBr_3	87
Cr and Br_2	Flow system, 750°	$CrBr_3$	88
Cr and HBr	Flow system, 750°	$CrBr_2$	89
Fe and Br_2	Flow system, 200°	$FeBr_3 + FeBr_2$	90
Fe_2O_3 and HBr	Flow system, 200–325°	$FeBr_2$	91
Co and HBr	Flow system, red heat	$CoBr_2$	92
Ni and Br_2	Ether, room temperature	$NiBr_2$	20
Cu and Br_2	Sealed tube, 300°	$CuBr_2$	93

Lower bromides. A number of miscellaneous reactions for the preparation of lower bromides have been reported, as shown in Table 1.14.

TABLE 1.14

Preparation of Lower Bromides

Reactants	Conditions	Product	Reference
$TiBr_4$ and H_2	Flow system, 750°	$TiBr_3$	96
$TiBr_4$ and $TiBr_2$	Sealed tube, <350°	$TiBr_3$	97
$TiBr_4$ and Ti	Sealed tube, 300°	$TiBr_3$	96
$TiBr_4$ and Ti	Sealed tube, 800°	$TiBr_2$	98
$TiBr_3$	Disproportionation >350°	$TiBr_2$	99
VBr_3 and H_2	Flow system, 450°	VBr_2	70
$FeBr_3$	Decomposition at 120°	$FeBr_2$	100
$CuBr_2$	Decomposition at 150°	CuBr	93

Miscellaneous methods. Acetyl bromide, in benzene or acetic acid, reacts with the acetates of manganese, cobalt, nickel and copper to give the corresponding bromides[74,75].

Anhydrous bromides have also been prepared by dissolving the carbonate of the transition metal in hydrobromic acid, evaporating the solution to dryness, and finally dehydrating the product in a stream of hydrogen bromide at elevated temperatures[101,102].

Oxide Bromides

Few oxide bromides of the first transition series elements are known. Titanium tetrabromide reacts with ozone and chlorine monoxide to give titanium oxide dibromide[103], and bromination of vanadium oxides gives vanadium oxide tribromide or vanadium oxide dibromide, depending on the conditions[103,104].

Iodides

There are two principal methods of preparing binary iodides of the transition metals, namely direct interaction of the elements, usually in a sealed tube, or alternatively, interaction of aluminium triiodide with the appropriate metal oxide. This latter method is particularly effective because of the large driving force due to the high lattice energy of aluminium oxide compared with that of the reactants.

Direct iodination of the metal. This method has been reported for most of the transition metals as shown in Table 1.15.

TABLE 1.15
Direct Iodination of the Metal

Metal	Conditions	Product	Reference
Ti[a]	Sealed tube, 150–250°	TiI_4	105
Ti[b]	Sealed tube, 750°	TiI_3	106
V[a]	Sealed tube, 300°	VI_3	107
Cr[b]	Sealed tube, 200–225°	CrI_3	108
Mn	Reaction in ether at 25°	MnI_2	109
Fe	Thermal gradient, 530–180°	FeI_2	110
Cu	Thermal gradient, 450–100°	CuI	111

[a] Excess iodine must be used to prevent formation of lower iodides.
[b] Stoichiometric quantities of metal and iodine necessary to give pure product.

Aluminium triiodide. Chaigneau[112–114] has made extensive use of this reagent in preparing binary iodides using both oxides and sulphides of the transition metals, as shown in Table 1.16. The reactions are invariably carried out in a sealed tube.

TABLE 1.16

Preparation of Iodides Using Aluminium Triiodide

Reactant	Temperature	Product
TiO_2	230°	TiI_4
V_2O_3	330°	VI_3
Cr_2S_3	350°	CrI_3
MnO_2	230°	MnI_2
MnS_2	230°	MnI_2
Fe_2O_3	300°	FeI_2
Co_3O_4	230°	CoI_2
CoS_2	230°	CoI_2
NiO	230°	NiI_2
CuS	230°	CuI

Miscellaneous methods. A variety of specific reactions for the preparation of particular iodides have been reported and some of these are given in Table 1.17.

TABLE 1.17

Miscellaneous Methods of Preparing Iodides

Reactants	Conditions	Product	Reference
TiI_4 and Ti	Thermal gradient, 600–300°	TiI_3	115
TiI_4 and Ti	Sealed tube, >800°	TiI_2	116
TiI_3	Disproportionation at 350°	TiI_2	117
VI_3	Decomposition at 400°	VI_2	118
$CoCO_3$ and aq. HI	Evaporate and dehydrate in HI at 200°	CoI_2	119
$NiCl_2$ and NaI	Interaction in ethanol at room temperature	NiI_2	120

Complex Fluorides

Usually complex fluorides are prepared by use of two general methods, either interaction in a melt, or fluorination using either fluorine or bromine trifluoride on the appropriate mixture of transition-metal salt and alkali-metal chloride. This last reaction is of considerable interest, since the final oxidation state achieved by the transition metal can depend solely on the stoichiometric ratio of the cation and the transition metal. As a specific example, fluorination of caesium tetrachlorocobaltate(II) gives caesium hexafluorocobaltate(IV), but under similar conditions fluorination of caesium hexacyanocobaltate(III) gives caesium hexafluorocobaltate(III). It would appear therefore, that factors such as the lattice energy of the resultant product can over-ride the oxidizing effect of fluorine in determining the oxidation state of the transition metal in the final product.

Preparation in the melt. A large number of complex fluorides of the transition metals have been prepared by direct interaction of the metal fluoride with alkali-metal fluorides, alkali-metal hydrogen difluorides, or alkaline-earth metal fluorides. The reactions are carried out under vacuum or in an atmosphere of hydrogen fluoride or inert gas. Some examples of these reactions are given in Table 1.18.

TABLE 1.18
Preparation of Complex Fluorides by Direct Interaction in the Melt

Reactants[a]	Product	References
TiF_4 and MF[b]	M_2TiF_6	121
MnF_2 and MF	M_2MnF_4	122
MnF_2 and MF	$MMnF_3$	122
FeF_3 and BaF_2	$Ba_3(FeF_6)_2$	123
FeF_3 and BaF_2	$BaFeF_5$	123
FeF_2 and BaF_2	Ba_2FeF_6	124
FeF_2 and BaF_2	$BaFeF_4$	124
FeF_2 and MF	$MFeF_3$	125, 126
CoF_2 and BaF_2	Ba_2CoF_6	127
CoF_2 and MF or MHF_2	M_2CoF_4	128
CoF_2 and MF	$MCoF_3$	128
NiF_2 and BaF_2	Ba_2NiF_6	127
NiF_2 and MF	M_2NiF_4	129
NiF_2 and MF or MHF_2	$MNiF_3$	129
CuF_2 and BaF_2	Ba_2CuF_6	127
CuF_2 and MF or MHF_2	M_2CuF_4	128, 130
CuF_2 and MF or MHF_2	$MCuF_3$	128, 130

[a] M = alkali metal.
[b] In a flow system at 350–400°.

Several interesting features are apparent in the Table. The proportions of alkali-metal fluoride and transition-metal fluoride are critical in determining the stoichiometry of the complex fluoride formed, although they do not effect the oxidation state of the metal. Thus, for example, with manganese difluoride either the tetrafluoromanganate(II) or the trifluoromanganate(II) ions are formed, according to the proportions of the reactants used. A second interesting feature is the stabilizing influence of the larger alkaline-earth cations on the salts containing anions with a large negative charge, such as the hexafluorocuprate(II) anion.

2

A series of compounds of the type $MTVF_6$, where M is an alkali metal and T a divalent transition metal, have been prepared by Babel and coworkers[131] by the interaction of stoichiometric amounts of alkali-metal fluoride or alkali-metal hydrogen difluoride, the transition-metal difluoride and vanadium trifluoride. Similar complexes were also prepared from chromium, iron and cobalt trifluorides[131].

Peacock[132,133] has prepared a number of hexafluorometallates(III) by reactions in the melt of the type

$$K_2MnF_5 \cdot H_2O + KHF_2 \xrightarrow{400^\circ} K_3MnF_6$$

The resultant potassium fluoride is leached out with formamide. This reaction has been used to prepare hexafluorometallates(III) of vanadium, chromium, manganese and iron.

Preparations with fluorine. Some examples of this type of reaction are given in Table 1.19.

TABLE 1.19

Preparation of Complex Fluorides Using Fluorine

Reactants[a]	Conditions	Product	References
VCl_3 and KCl	200–230°	K_2VF_6	134
$CrCl_3$ and MCl	350°	M_2CrF_6	134, 135
$MnCl_2$ and KCl	375–400°	K_2MnF_6	134
FeF_3 and KCl	100–375°	K_3FeF_6	136
Cs_2CoCl_4	350°	Cs_2CoF_6	137
$M_3Co(CN)_6$	300–350°	M_3CoF_6	138
$NiCl_2$ and MCl	Elevated	M_2NiF_6	136
$NiCl_2$ and KCl	Elevated	K_3NiF_6	139
$CuCl_2$ and KCl	250°	K_3CuF_6	136

[a] M = alkali metal.

The most important feature of this Table is the stabilization of certain oxidation states that are inaccessible in binary fluorides. Thus merely by the use of suitable stoichiometric proportions of alkali-metal and transition metal, complex salts of cobalt(IV), nickel(III), nickel(IV) and copper(III) may be prepared relatively easily.

Preparations with bromine trifluoride. Bromine trifluoride has not been widely used to prepare complex fluorides of the first-row transition metals. Those reactions which have been reported use a range of starting materials, as shown in Table 1.20.

TABLE 1.20
Preparation of Complex Fluorides with Bromine Trifluoride

Reactants	Product	Reference
TiF_4 and MF	M_2TiF_6	140
VCl_3 and MCl	MVF_6	22
$K_2Cr_2O_7$	$KCrOF_4$	141
CrF_4 and MCl	M_2CrF_6	5
CrF_4 and MCl	$MCrF_5$	5
$KMnO_4$ and KCl	K_2MnF_6	141

For the reasons stated when dealing with the preparation of binary fluorides with this reagent, there is sometimes considerable difficulty in removing the last traces of bromine from the product.

Reactions in hydrofluoric acid. Hydrofluoric acid is not commonly used to prepare complex fluorides of the transition metals. The few reactions which have been reported are given in Table 1.21.

TABLE 1.21
Reactions in Hydrofluoric Acid

Reactants	Product	Reference
TiO_2 and MF	M_2TiF_6	142
V_2O_5 and KI	K_3VF_6	143
$M_2Cr_2O_7$	$MCrO_3F$	144
MnO_2 and NH_4F	$(NH_4)_2MnF_5$	145
MnF_2 and TlF	$TlMnF_3$	146
FeF_3 and MCl	M_3FeF_6	143

Miscellaneous methods. Haendler and coworkers[147,148] have made extensive use of methanol as a reaction medium for the preparation of a wide range of complex fluorides of the first-row transition metals. The procedure is to add an alkali-metal fluoride or ammonium fluoride to a methanolic solution of the appropriate transition-metal bromide. The method has proved very useful in preparing trifluorometallates(II).

Machin and coworkers[149] have described the preparation of the potassium trifluorometallates(II) by mixing boiling aqueous solutions of potassium fluoride and the appropriate transition-metal difluoride. The difluoride is most conveniently prepared in solution by dissolving the freshly precipitated carbonate in the calculated amount of dilute hydrofluoric acid. It was found that, to achieve maximum purity and yield, the solutions must be saturated, the free acid concentration must be kept to a minimum, and the final solution must not be boiled for more than about ten seconds.

Complex Chlorides

As with complex fluorides, there are two general methods of preparation of complex chlorides. In this case the methods are reaction of the constituent chlorides in either the melt or in non-aqueous solvents. The solvents in the latter method may be inert, as with chloroform, or those that can actually act as a chlorinating agent, such as thionyl chloride.

As was the case with the complex fluorides for reactions in the melt, the stoichiometry is important in fixing the nature of the resultant complex chloride. Thus by merely altering the relative proportions, the interaction of alkali-metal chloride and chromium dichloride give either a tetrachlorochromate(II) or a trichlorochromate(II). However, particularly with copper, a new phenomenon appears, in that isomerism occurs for some complex anions. Thus the trichlorocuprates(II) may either contain dimeric Cu_2Cl_6 units, with a square-planar arrangement about the copper atom, or alternatively, may be polymeric, with octahedral coordination about the metal. The size of the cation is critical in determining which structure will be favoured.

Although octahedral coordination is favoured in complex chlorides, there are numerous examples of square-planar and tetrahedral coordination.

Reactions in the melt. Some examples of the preparation of complex chlorides by the interaction of alkali-metal chlorides and the appropriate transition-metal chloride in the molten state are given in Table 1.22. This is the most common, and in some cases the only reported method of preparing complex chlorides.

In the table, M is alkali metal and in most cases the reactions proceed with sodium, potassium, rubidium and caesium chlorides, although in a few cases the sodium salts cannot be isolated. It is noteworthy that in some cases the larger alkali-metal cations are required to isolate certain large anionic species, for example $[Ti_2Cl_9]^{3-}$. This is also true of anions of the type $CoCl_5^{3-}$ which actually contain the tetrahedral tetrachlorometallate(II) anion, together with free chloride ions.

Preparations in hydrochloric acid. Complex chlorides are not usually prepared in hydrochloric acid, and in some cases it is necessary to saturate the solution with hydrogen chloride gas to minimize hydrolysis. Salts prepared in this manner include hexachlorotitanates(IV)[171], oxopentachlorovanadates(IV)[172], oxotetrachlorovanadates(IV)[156], trioxochlorochromates(VI)[173] and tetrachloromanganates(II)[174].

Reactions in organic solvents. A large number of complex chlorides have been prepared in solvents such as chloroform and alcohol by direct interaction of the transition-metal chloride and the chloride of the cation. For solubility reasons the cations used are normally substituted quaternary ammonium, phosphonium or arsonium species, or large organic

TABLE 1.22

Preparation of Complex Chlorides in the Melt

Reactants	Product	References	Reactants	Product	References
$TiCl_4$ and MCl[a]	M_2TiCl_6	150, 151	$MnCl_2$ and MCl	M_2MnCl_4	162, 163
$TiCl_3$ and MCl	M_3TiCl_6	152, 153	$MnCl_2$ and MCl	$MMnCl_3$	162, 163
$TiCl_3$ and MCl	M_2TiCl_5	153	$FeCl_3$ and CsCl[b]	$Cs_3Fe_2Cl_9$	164
$TiCl_3$ and CsCl	$Cs_3Ti_2Cl_9$	154	$FeCl_3$ and MCl	$MFeCl_4$	165
$TiCl_3$ and MCl	$MTiCl_4$	152, 155	$FeCl_2$ and MCl[c]	M_3FeCl_5	166
$VOCl_3$ and MCl	M_2VOCl_4	156	$FeCl_2$ and MCl	M_2FeCl_4	166, 167
VCl_3 and MCl	M_3VCl_6	157	$FeCl_2$ and MCl	$MFeCl_3$	166, 167
VCl_3 and MCl	$M_3V_2Cl_9$	157	$CoCl_2$ and MCl[c]	M_3CoCl_5	168
VCl_2 and MCl	$MVCl_3$	158	$CoCl_2$ and MCl	M_2CoCl_4	168
$CrCl_3$ and MCl	M_3CrCl_6	159	$CoCl_2$ and MCl	$MCoCl_3$	168, 169
$CrCl_3$ and MCl	$M_3Cr_2Cl_9$	160	$NiCl_2$ and CsCl[d]	Cs_3NiCl_5	170
$CrCl_2$ and MCl	M_2CrCl_4	161	$NiCl_2$ and MCl	$MNiCl_3$	28
$CrCl_2$ and MCl	$MCrCl_3$	161			

[a] Salts thermally unstable and difficult to obtain pure.
[b] Two compounds isolated, only one of which is dimeric.
[c] M = rubidium and caesium.
[d] Rapid quenching necessary.

cations derived from amines. This method has been applied to the preparation of complex chlorides of all of the first-row transition metals, and a few typical examples are hexachlorotitanates(III)[175], hexachlorovanadates(IV)[176], tetrachloromanganates(II)[1,177] and tetrachloronickelates(II)[1,177].

Reactions in thionyl chloride. Adams and coworkers[178] have made an extensive study of the preparation of complex chlorides of the transition metals using thionyl chloride as the reaction medium. The nature of the solvent makes it unnecessary for the transition-metal chloride to be anhydrous. In addition, thionyl chloride will convert some complex oxide chlorides to the fully chlorinated anion. Thus oxotetrachlorovanadates(IV) are converted to hexachlorovanadates(IV) by thionyl chloride.

Complex Bromides and Iodides

Complex bromides have been reported for all the first-series transition metals with the exception of chromium, but complex iodides have only been prepared for elements to the right of manganese in the Periodic Table.

The reduced stability of the bromides, relative to the complex chlorides, is reflected in the fact that only the titanium compounds have been prepared in the melt[179]. The remaining complex bromides and all the complex iodides have been prepared by interaction in alcohol of quaternary halides and the appropriate transition-metal halide[1,180–182].

STRUCTURES

Binary Halides

The majority of the binary halides of the first-row transition metals fall into four main structural types, namely, the rutile, cadmium dichloride, cadmium diiodide and bismuth triiodide types. Little is known of the structures of the trifluorides, except that invariably they contain the metal in octahedral coordination. The dihalides of chromium and copper always show marked distortion from the ideal geometry, and this phenomenon has been ascribed to the Jahn–Teller effect.

The rutile structure. The difluorides of vanadium, manganese, iron, nickel and cobalt[35,183] adopt the tetragonal rutile structure shown in Figure 1.1. The unit-cell dimensions of each of these compounds are given in Table 1.23.

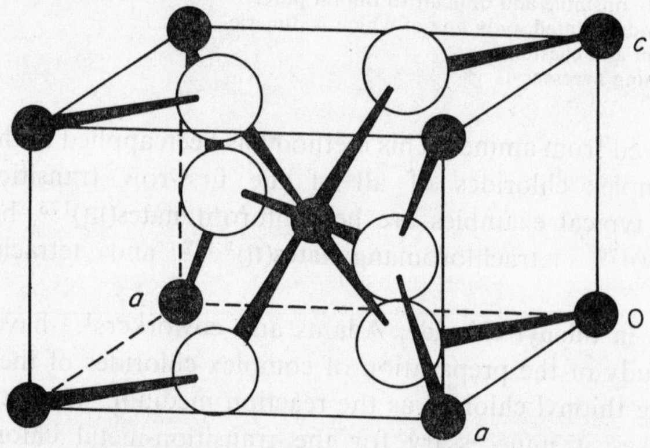

Figure 1.1 The rutile structure. (Reproduced by permission from C. Billy and H. M. Haendler, *J. Am. Chem. Soc.*, **79**, 1049 (1957))

Chromium difluoride and copper difluoride are both distorted from the ideal rutile structure as shown in Figure 9.1 (page 486). The compounds are both isomorphous and isostructural with each other.

The cadmium dichloride and cadmium diiodide structures. The cadmium dichloride structure is based on a cubic close-packed arrangement of the anions with half of the octahedral holes occupied by cations. This structure is adopted by the dichlorides of manganese, iron, cobalt and nickel, and also nickel diiodide[184–186]. The cadmium diiodide structure is based on hexagonal close-packing of the anions. The compounds which are known to adopt this structure are the dibromides and diiodides of titanium, vanadium, manganese, iron and cobalt[98,187–189]. There is some controversy over the structures of titanium dichloride and vanadium dichloride and this may be an indication of polymorphism in these compounds. All the dichlorides adopt the cadmium dichloride structure and all the dibromides and diiodides, with the exception of nickel diiodide, adopt the cadmium

TABLE 1.23

Unit Cell Dimensions of the Rutile-type Difluorides

Compound	*a* (Å)	*c* (Å)	Reference
VF_2	4.80	3.24	35
MnF_2	4.873	3.309	183
FeF_2	4.696	3.309	183
CoF_2	4.695	3.179	183
NiF_2	4.650	3.083	183

diiodide structure. The reason for the anomaly with nickel diiodide is not clear. The unit-cell dimensions of all of the dichlorides are $a = 3.5$ to 3.8 Å and $c = 17.4$ to 17.6 Å. The unit-cell dimensions of nickel diiodide are given as $a = 3.88$ and $c = 19.6$ Å[186]. Obviously it would be expected that the iodide would have a larger unit cell than the chlorides, but it is interesting to note that if the value of c is divided by three (which is the relationship between the two types of structure) the dimensions of nickel diiodide fall into the range expected by comparison with the other diiodides.

Since both the cadmium dichloride and cadmium diiodide structures are based on close-packed arrays of halide ions, there is only a small energy difference between the alternative structures. Thus the normal form of iron dichloride is the cadmium dichloride-type structure, but if the compound is sublimed and then rapidly condensed, a polymorph with the cadmium diiodide structure can be obtained[190]. Similarly, a pressure of about 2 Kbar produces a cadmium dichloride to cadmium diiodide transformation in iron dichloride[191].

Chromium and copper dihalides. Jack and Maitland [192] and Tracey and coworkers[54,193,194] have investigated the structures of chromium difluoride, and the other dihalides, respectively, by single-crystal X-ray diffraction techniques. The dichloride, dibromide and diiodide all consist of chains of CrX_6 octahedra which are formed by the sharing of short edges in the distorted octahedra, as may be seen for chromium dichloride in Figure 4.9 (page 185). It is important to note that if the octahedra were not distorted the structures would become regular three dimensional networks. The relative packing in the four chromium dihalides is shown in Figure 1.2.

Copper dichloride and dibromide consist of chains of copper atoms which are bridged by two halide atoms, to give square-planar coordination about the copper. The chains are stacked in such a manner that the resultant coordination about each copper is of the form of a tetragonally elongated octahedron.

Trihalides. The trifluorides of the first-row transition metals all have octahedrally coordinated metal atoms[10,26,192,195–197]. Only manganese trifluoride has been studied by single-crystal techniques[197], and for this compound there are three different metal-fluorine bond lengths. For the other trifluorides the octahedra appear to be regular.

The trichlorides, tribromides and triiodides of the transition metals under discussion have structures based on MX_6 octahedra. Most of the compounds are considered to have the bismuth triiodide-type structure. However, it is difficult to present a satisfactory discussion of the structures of these compounds because of the common occurrence of polymorphism and confusion in the literature as to the correct structure. The structure of chromium trichloride is an example of where there is confusion in the literature (page 171). Polymorphism is apparent in many of the trihalides, the extreme case being titanium trichloride. Natta and coworkers[198] have shown that there are three modifications of this compound, although all are based on octahedrally coordinated titanium.

Complex Halides

In this section only general comments on the stereochemistry of the complex halide anions will be made, since the overall crystal structures are often dominated by the nature of the cation.

Complex fluorides of the first transition series invariably contain the metal in an octahedral environment, although the empirical formulae of the anions may range from MF_6^{n-} to MF_3^-. In marked contrast, the other halide ions cause the transition metals to display a wide range of stereochemistries. To the left of the Periodic Table octahedral coordination of the metal is the most common stereochemistry. Although there is some evidence for possible tetrahedral coordination in halide anions of titanium, vanadium and chromium in lower oxidation states, this has not so

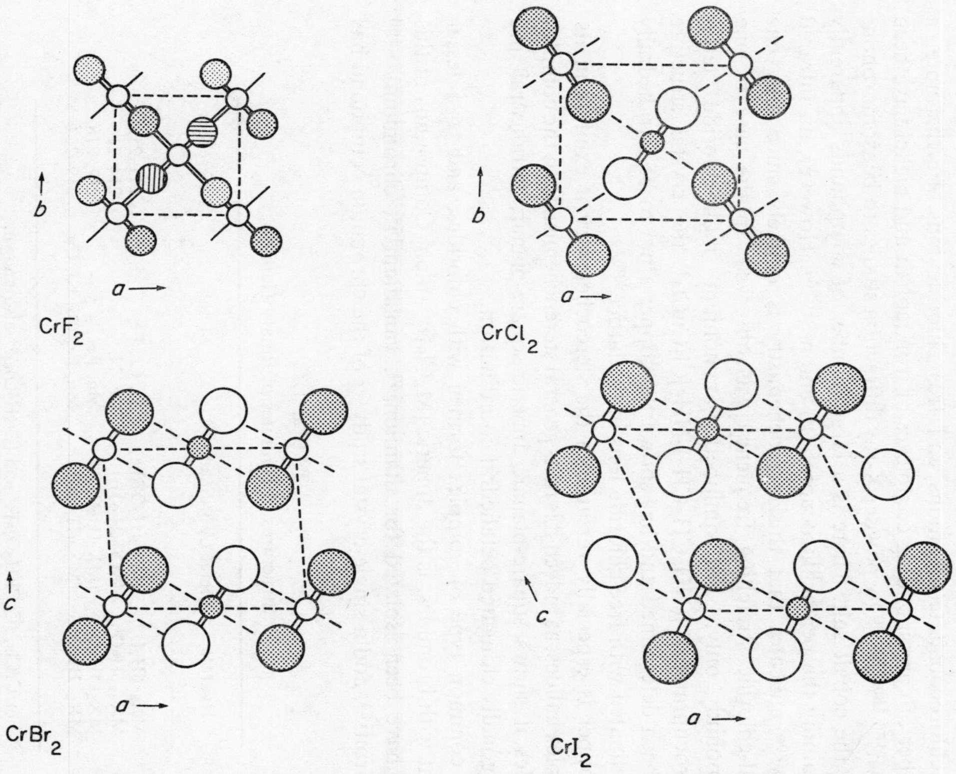

Figure 1.2 A comparison of the structures of the chromium dihalides. (Reproduced by permission from J. W. Tracey, N. W. Gregory and E. C. Lingafelter, *Acta Cryst.*, **15**, 672 (1962))

far been confirmed by crystal structure determinations. For each element beyond manganese some tetrahedral halide anions are known, and the tendency is for this stereochemistry to increase in importance. It may be noted at this point, that the compound caesium pentachlorocobaltate(II), Cs_3CoCl_5, has been shown[199] to contain equal numbers of tetrahedral tetrachlorocobaltate(II) anions and free chloride ions. Furthermore, a number of corresponding complexes of iron, cobalt and nickel have been shown[166,169,170,200] by powder X-ray diffraction studies to be isomorphous with the cobalt salt. There is a large number of compounds apparently containing the pentachlorocuprate(II) anion[201–204]. However, an infrared study[205] indicates that this stereochemistry is unlikely since only one metal-chlorine stretching frequency was observed. In the case of one compound only, bis(2-aminoethyl)ammonium monochloride tetrachlorocuprate(II) $[(NH_3CH_2CH_2)_2NH_2]Cl[CuCl_4]$, the crystal structure has been determined and has shown the copper atom to be tetrahedrally coordinated with free chloride ions in the lattice[206].

Copper is especially versatile in the stereochemistry it exhibits in its complex halides, as indeed it is in its general stereochemistry. In the complex halides it shows square-planar, dimeric square-planar, tetrahedral and tetragonally elongated octahedral coordination.

A common type of complex formed with chlorides, and to a lesser extent with bromides, is the dimeric $[M_2Cl_9]^{3-}$ anion. Compounds of this type have been isolated for titanium[154], vanadium[157], chromium[154,160] and iron[164], and a single-crystal study[154] of the chromium compound has

TABLE 1.24

Structures of Hydrates of Binary Halides

Hydrate	Metal Coordination	Metal	References
$MX_3\cdot6H_2O$[a]	$[MX_2(H_2O)_4]^+$	Ti, Cr, Fe	211–213
$MX_2\cdot6H_2O$[b]	$[MX_2(H_2O)_4]$	Co, Ni	214–216
$MX_2\cdot4H_2O$	$[MX_2(H_2O)_4]$	Mn, Fe	217, 218
$MX_2\cdot2H_2O$[c]	$[MX_4(H_2O)_2]_n$	Mn, Fe, Co, Ni, Cu	219–222

[a] For $CrCl_3$, $[CrCl(H_2O)_5]^{2+}$ and $[Cr(H_2O)_6]^{3+}$ also known.
[b] $CoI_2\cdot6H_2O$ and $NiI_2\cdot6H_2O$ have $[M(H_2O)_6]^{2+}$ cations[223,224].
[c] Polymeric with bridging halogens.

shown that the anion consists of two octahedra sharing a face, as shown in Figure 4.14 (page 198). Powder X-ray diffraction studies and infrared spectra indicate that the complexes for the other metals have the same structure.

Complexes of the type $[MX_3L]^-$, where M is cobalt or nickel, X is chloride, bromide or iodide, and L is a neutral ligand such as triphenylphosphine, have been isolated and in each case have been shown to have a distorted tetrahedral configuration[207–210].

Hydrates

There are four principal types of structure for the hydrates of the binary halides of the first-row transition metals as shown in Table 1.24.

With the exception of manganese dichloride tetrahydrate, all of the dihalotetraaquo octahedra have a *trans* configuration.

INFRARED AND ELECTRONIC SPECTRA

The metal-halogen stretching modes have proved to be of considerable diagnostic value in determining the stereochemistry of both complex halides and adducts of the binary halides. Clark[225] has shown that it is possible to correlate the position of the metal-halogen stretching frequency with the oxidation state, mass and coordination number of the central metal atom, and the stereochemistry of the complex or adduct. Thus Clark and Williams[226] were able to tabulate the structures of the pyridine adducts of the transition metals on the basis of the positions of the metal-halogen and metal-nitrogen stretching frequencies. Similarly, Adams and Lock[205] used infrared spectroscopy to distinguish between tetrahedral and square-planar tetrachlorocuprate(II) groups.

Adams[227] has made a very extensive survey of metal-ligand vibrations.

Electronic spectra have proved, in general, to be very useful for distinguishing the stereochemistry of the transition metal-halide compounds, although this approach is not definitive in d^1 and d^9 systems. For example, examination of the electronic spectra of nickel(II) compounds readily distinguishes between octahedral, square-planar, tetrahedral and five-coordinate structures, especially if complemented with magnetic measurements. There has also been much work reported on the theoretical interpretation of both spin-allowed and spin-forbidden transitions, particularly with manganese(II) and cobalt(II) compounds.

MAGNETIC PROPERTIES

From the chemist's point of view, the magnetic properties of the halides and halide complexes of the first-row transition metals are remarkably simple. Certain copper complexes show spin-spin interaction and a number of complex fluorides show antiferromagnetic behaviour at comparatively high temperatures. Apart from these examples the halides usually show close to spin-only values of the magnetic moment, and the data can safely be used as being diagnostic of oxidation state and stereochemistry.

A great amount of detailed study of magnetic properties at very low temperatures has also been done from the physicist's point of view. Almost all the compounds become antiferromagnetic at low temperature and a great deal of elegant work, using techniques including neutron-diffraction, nuclear magnetic resonance and specific heat measurements has been carried out.

It has not been our intention to discuss the details of the magnetic data which have been obtained, and in general only a summary of the major conclusions are given for most compounds. Konig[228] has made an excellent survey of all the magnetic and ESR data available for most of the compounds discussed in this book.

ADDUCTS

The chlorides, bromides and iodides of the first-row transition elements react with a wide range of donor ligands, to give both substitution and addition compounds. Little work has been reported on the preparation and characterization of adducts and substitution compounds of the fluorides of these elements. Most of the reactions reported are with the tetrafluorides of titanium[229–233] and vanadium[21]. There has been no systematic study of the chemistry of the remaining fluorides.

Donor ligands that have been reacted with the chlorides, bromides and iodides of the first-row transition metals range from such simple molecules as ammonia[234] and acetonitrile[235–237] to complex amines and arsines such as 1,4,8,11-tetraazaundecane[238] and methyl-bis(3-propanedimethylarsino)arsine[239]. Even with such simple donor molecules as acetonitrile, quite interesting chemical properties have been observed. For example, the diiodides of manganese, cobalt and nickel form complexes of the type $MI_2(CH_3CN)_3$ which have been shown[235] to be ionic and should be formulated as $[M(CH_3CN)_6^{2+}][MI_4^{2-}]$. Copper dichloride reacts readily with acetonitrile[237], and depending on the experimental conditions, two adducts may be isolated, namely $Cu_2Cl_4(CH_3CN)_2$ and $Cu_3Cl_6(CH_3CN)_2$. Willett and Rundle[237] have examined both of these adducts by single-crystal X-ray diffraction studies, which revealed copper in a tetragonally distorted octahedral environment (Figures 9.7 and 9.8, pages 500, 501).

In general, much of the work reported on the characterization of the adducts of the transition-metal halides has been in relation to the factors which influence a given stereochemical arrangement. As one would expect, steric hindrance is one of the most important features which determines stereochemistry. A typical series of papers on this subject is that by

Livingstone and his group[240–244], in which a variety of ligands, containing both sulphur and nitrogen donor sites, have been reacted with a wide range of transition-metal halides. For a detailed account of the chemistry of adducts with chelating ligands, the reader is referred to an excellent series of review articles edited by Dwyer and Mellor[245].

As mentioned in the introduction, five-coordinate complexes of the elements under consideration have become quite well established, and a number of these compounds have been studied by single-crystal X-ray diffraction techniques. Two groups have been particularly interested in the preparation and characterization of these complexes. Fowles and his coworkers have prepared a number of adducts of the trihalides of titanium, vanadium and chromium with trimethylamine, diethyl sulphide, etc., while Sacconi and his associates have studied the reactions between the dihalides of the elements from manganese to copper, particularly cobalt and nickel, and complex tridentate and tetradentate ligands such as bis(2-dimethylaminoethyl)sulphide, bis(2-dimethylaminoethyl)methylamine and tris(2-dimethylaminoethyl)amine. Five-coordinate complexes are also known with phosphines and arsines. Some of the compounds, which have been shown by single-crystal X-ray diffraction techniques to contain the metal in a five-coordinate environment, are given in Table 1.25.

TABLE 1.25

Some Five Coordinate Complexes

Compound	Reference
$TiBr_3[(CH_3)_3N]_2$	246
$CrCl_3[(CH_3)_3N]_2$	247
FeCl(salen); salen = *N,N′*-bis(salicylideneimine)	248
$FeCl(detc)_2$; detc = diethyldithiocarbamato	249
$CoCl_2(dienMe_5)$; $dienMe_5$ = bis(2-dimethylaminoethyl)methylamine	250
$CoCl_2(Et_4dien)$; Et_4dien = 1,1,7,7-tetramethyldiethylenetriamine	251
$CoCl_2$(paphy); paphy = 1,3-bis(2′-pyridyl)-2,3-diaza-1-propene	252
$CoBr_2(trenMe_6)$; $trenMe_6$ = tris(2-dimethylaminoethyl)amine	253
$CoBr_2[(C_6H_5)_2PH]_3$	254
$NiBr_2$(TAS); TAS = methyl-bis(3-propanedimethylarsino)arsine	255
$CuCl_2dmp\cdot H_2O$; dmp = 2,9-dimethyl-1,10-phenanthroline	256
$[CuBr(BPE + CH_3OH)]ClO_4$; BPE = 1,2 bis(2′-pyridylmethyleneamino)ethane	257
$[CuI(bipy)_2]ClO_4$; bipy = bipyridyl	258

The structures range from pure trigonal-bipyramidal to pure square-pyramidal. However, in general the actual structures lie between these two, that is, they may be considered as either distorted square-pyramidal or distorted trigonal-bipyramidal.

The preparation and characterization of a number of five-coordinate complexes has led to the appearance in the literature of several detailed energy diagrams to account for the observed electronic absorption spectra. Both specific and general treatments have been reported[259–262].

Probably the most studied group of donor molecules is pyridine and related ligands. The dichlorides, dibromides and diiodides of chromium, manganese, iron, cobalt, nickel and copper react readily with these ligands to give, in general, adducts of the types MX_2L_4 and MX_2L_2[226,263–271]. Complexes of the types MX_2L, MX_2L_3 and MX_2L_5 however, are also known. The stereochemistry of the adduct formed depends on experimental conditions such as the relative amounts of reactants and also on steric factors.

Of the vast number of adducts of these ligands known, only about ten have had their structures determined by single-crystal X-ray diffraction techniques. All of the adducts of the type MX_2L_4 that have been studied have *trans* octahedral structures[272,273]. Complexes of the general formula MX_2L_2 have four different types of structure:

(a) polymeric *trans* octahedral, e.g., $CoCl_2$(pyridine)$_2$
(b) tetrahedral, e.g., $CoCl_2$(*p*-toluidine)$_2$
(c) dimeric distorted square-pyramidal, e.g., $CuCl_2$(4-methylpyridine)$_2$
(d) *cis* square-planar, e.g., $CuCl_2$(1,8-naphthyridine)$_2$.

The five-coordinate structure of $CuCl_2$(4-methylpyridine)$_2$ is thought to be caused by steric hindrance. The structure of $CuCl_2$(1,8-naphthyridine)$_2$ is unusual, as it appears to be the only adduct of this type with a *cis* configuration.

The structure of the remaining adducts have been deduced from X-ray powder diffraction data, infrared, visible and ultraviolet spectroscopy and magnetic and conductance studies. Clark and Williams[226] have shown that the metal-halogen and metal-nitrogen stretching frequencies are particularly useful in distinguishing between octahedral monomeric, octahedral polymeric and tetrahedral configurations. Goodgame and Hayward[268] have examined the infrared spectra of a number of 4-methylpyridine adducts of the dihalides of manganese, cobalt, nickel and copper, and report that there are a number of useful ligand vibrations which enable one to distinguish between octahedral and tetrahedral stereochemistries.

The electronic spectra of the majority of the pyridine, substituted pyridine and related adducts of the transition-metal dihalides have been investigated in detail, particularly cobalt and nickel[271,278–281]. The results have been interpreted on the basis of the microsymmetry of the central metal atom. For example, Ludwig and Wittman[281] have shown that for adducts of the type NiI_2L_2, the nickel atom may have C_{2v} or D_{2h} symmetry. Ligands with substituents in the α position of the pyridine ring have D_{2h} symmetry, whereas those without α substitution have C_{2v} symmetry.

One of the most important general studies on pyridine adducts is that by Gill and her coworkers[263]. In a detailed examination of the modes of vibration of the pyridine ring coordinated to a metal atom, it was found that there were only very small shifts in the frequencies on coordination. This led to the suggestion that the electron density on the pyridine ring remains almost constant on coordination, indicating that back-donation plays an important part in the bonding of the pyridine ring to the metal atom.

The stoichiometry, structure and stability of a number of pyridine-type adducts, particularly with cobalt and nickel, have been discussed in terms of π-acceptor properties, steric hindrance, basicity of the ligand, etc.[269–271,282–289].

The thermal decomposition of a number of adducts of the dichlorides, dibromides and diiodides of manganese to nickel with pyridine and related ligands has been studied by differential thermal analysis and differential scanning calorimetry[290–295]. In general, the decomposition reactions involve a number of steps and Beech and coworkers[293–295] have been able to calculate heats of decomposition for the various steps in a number of cases. The relative values of the heats of decomposition have been related to the nature of the central metal atom and of the ligand, and to the stereochemistry of the adduct. One particularly interesting intermediate that has been isolated from a number of decomposition reactions is the adduct of empirical formula $MX_2L_{2/3}$. Electronic absorption spectra demonstrate clearly that these complexes contain the metal atom in an octahedral environment indicating that the compounds are polymeric with halogen bridges.

Another group of donor ligands that have been reacted with the first-row transition metal halides are phosphines and arsines. Booth[296] has reviewed this interesting section of transition-metal halide chemistry, literature up to about 1963 being covered. Since these ligands range from unidentates, such as triphenylphosphine, to tetradentates, such as tris(*o*-diphenylphosphinophenyl)arsine[297], it is not suprising that a variety of stoichiometries and stereochemistries have been reported for this type of adduct.

Adducts containing four-, five- and six-coordinate metal atoms are known, and for the four-coordinated complexes, both tetrahedral and square-planar stereochemistries have been reported. Venanzi and coworkers[292–298] and Hayter and Humiec[299] have observed square planar-tetrahedral isomerism with a number of diphenylalkylphosphine adducts of nickel dihalides. It was also found that while diphenylalkylphosphines may be square-planar or tetrahedral, trialkyl- and phenyldialkylphosphine-complexes are invariably square-planar, and triphenylphosphine adducts are tetrahedral. In addition, it was found that the change in structure of the diphenylalkylphosphine complexes occurs with nickel dibromide. That is, the chloro adducts are normally planar, the iodo complexes tetrahedral, while the bromo derivatives are often obtained in both isomeric forms. The tendency to form a square-planar complex rather a tetrahedral complex, and vice versa, has been shown to depend on the ligand-field strength of both the ligand and the halogen, and on solubility effects, in addition to crystal packing requirements.

O-phenylenebis(dimethylarsine) is perhaps the most interesting of all the phosphine and arsine ligands studied. Nyholm and his coworkers have examined the reactions of this ligand with a wide range of transition-metal halides and two very interesting properties of the ligand which have emerged are its ability to stabilize high oxidation states and high coordination numbers. This bidentate arsine reacts readily with titanium tetrachloride[300,301] to give a yellow-orange complex, which has been shown by single-crystal X-ray diffraction techniques to contain eight-coordinate titanium. This was the first example of a first-row transition metal with this stereochemistry. There are now at least nine known eight-coordinate adducts of this type with the chlorides and bromides of titanium, zirconium, hafnium, vanadium, niobium and tantalum[302]. *O*-phenylenebis(dimethylarsine) is able to stabilize halo complexes of, for example, iron(IV), nickel(IV) and nickel(III)[296]. For iron(IV), the arsine adduct is the only halo complex of this oxidation state.

REFERENCES

1. N. S. Gill and R. S. Nyholm, *J. Chem. Soc.*, **1959,** 3997.
2. H. M. Haendler, S. F. Bartram, R. S. Becker, W. J. Bernard and S. W. Bukata, *J. Am. Chem. Soc.*, **76,** 2177 (1954).
3. J. H. Canterford and T. A. O'Donnell, *Inorg. Chem.*, **5,** 1442 (1966).
4. T. A. O'Donnell and D. F. Stewart, *Inorg. Chem.*, **5,** 1434 (1966).
5. H. C. Clark and Y. N. Sadana, *Can. J. Chem.*, **42,** 50 (1964).
6. H. W. Roesky, O. Glemser and K. H. Hellberg, *Chem. Ber.*, **98,** 2046 (1965).
7. H. von Wartenberg, *Z. anorg. allgem. Chem.*, **247,** 135 (1941).
8. R. Hoppe, W. Dahne and W. Klemm, *Ann. Chim. (Paris)*, **658,** 1 (1962).
9. H. J. Emeleus and G. L. Hunt, *J. Chem. Soc.*, **1964,** 396.
10. M. A. Hepworth, K. H. Jack, R. D. Peacock and G. J. Westland, *Acta Cryst.*, **10,** 63 (1957).
11. K. Knox and D. W. Mitchell, *J. Inorg. Nucl. Chem.*, **21,** 253 (1961).
12. W. B. Burford, R. D. Fowler, J. M. Hamilton, H. C. Anderson, C. E. Weber and R. G. Sweet, *Ind. Eng. Chem.*, **39,** 321 (1947).

13. H. F. Priest, *Inorg. Syn.*, **3,** 173 (1950).
14. M. Faber, R. T. Meyer and J. L. Margrave, *J. Phys. Chem.*, **62,** 883 (1958).
15. R. W. Murray and H. M. Haendler, *J. Inorg. Nucl. Chem.*, **14,** 135 (1960).
16. N. S. Nikolaev and V. F. Sukhoverkhov, *Bul. Inst. Politeh. Iasi* [*NS*], **3,** 61 (1957).
17. D. B. Shinn, D. S. Crocket and H. M. Haendler, *Inorg. Chem.*, **5,** 1927 (1966).
18. A. G. Sharpe and H. J. Emeleus, *J. Chem. Soc.*, **1948,** 2135.
19. R. L. Farrar and H. A. Smith, *J. Phys. Chem.*, **59,** 763 (1955).
20. G. Braur, *Handbook of Preparative Inorganic Chemistry*, Academic Press, New York, 1963.
21. R. G. Cavell and H. C. Clark, *J. Chem. Soc.*, **1962,** 2692.
22. H. J. Emeleus and V. Gutmann, *J. Chem. Soc.*, **1949,** 2979.
23. A. Engelbrecht and A. V. Grosse, *J. Am. Chem. Soc.*, **74,** 5262 (1952).
24. B. J. Sturm, *Inorg. Chem.*, **1,** 665 (1962).
25. R. D. Fowler, H. C. Anderson, J. M. Hamilton, W. B. Burford, A. Spadetti, S. B. Bitterlich and I. Litant, *Ind. Eng. Chem.*, **39,** 343 (1947).
26. R. de Pape, *Compt. Rend.*, **260,** 4527 (1965).
27. E. L. Muetterties and J. E. Castle, *J. Inorg. Nucl. Chem.*, **18,** 148 (1961).
28. P. Allamagny, *Bull. Soc. Chim. France*, **1960,** 1099.
29. M. E. Runner, G. Balog and M. Kilpatrick, *J. Am. Chem. Soc.*, **78,** 5183 (1956).
30. S. Hartmann, *Z. anorg. allgem. Chem.*, **155,** 355 (1926).
31. B. J. Sturm and C. W. Sheridan, *Inorg. Syn.*, **7,** 87 (1963).
32. P. Nuka, *Z. anorg. allgem. Chem.*, **180,** 235 (1929).
33. A. Smalc, *Monatsh. Chem.*, **98,** 163 (1967).
34. L. E. Trevorrow, *J. Phys. Chem.*, **62,** 362 (1958).
35. J. W. Stout and W. O. J. Boo, *J. Appl. Phys.*, **37,** 966 (1966).
36. E. T. McBee, B. W. Hotten, C. R. Evans, A. A. Alberts, Z. D. Welch, W. B. Ligett, R. C. Shreyer and K. W. Krantz, *Ind. Eng. Chem.*, **39,** 310 (1947).
37. H. M. Haendler, W. L. Patterson and W. J. Bernard, *J. Am. Chem. Soc.*, **74,** 3167 (1952).
38. E. Catalano and J. W. Stout, *J. Chem. Phys.*, **23,** 1284 (1955).
39. A. V. Pamfilov and E. G. Shtardel, *J. Gen. Chem. USSR*, **7,** 258 (1937).
40. G. V. Seryakov, S. A. Vaks, V. V. Zheltova and E. P. Strashun, *Russ. J. Inorg. Chem.*, **12,** 3 (1967).
41. M. W. Duckworth, G. W. A. Fowles and R. A. Hoodless, *J. Chem. Soc.*, **1963,** 5665.
42. J. H. Simons and M. G. Powell, *J. Am. Chem. Soc.*, **67,** 75 (1945).
43. H. Oppermann, *Z. Chem.*, **2,** 376 (1962).
44. B. Morosin and A. Narath, *J. Chem. Phys.*, **40,** 1958 (1964).
45. C. T. Anderson, *J. Am. Chem. Soc.*, **59,** 488 (1937).
46. S. S. Todd and J. P. Coughlin, *J. Am. Chem. Soc.*, **73,** 4184 (1951).
47. S. Blairs and R. A. J. Shelton, *J. Inorg. Nucl. Chem.*, **28,** 1855 (1966).
48. Y. I. Ivashentsev and G. G. Bodunova, *Tr. Tomskogo Gos. Univ., Ser. Khim.*, **154,** 63 (1962).
49. Y. I. Ivashentsev, *Dokl. 7-i Nauch. Konf. Posvyashchen 40-letiya Velika. Oktyabr. Sots. Revolyutsii, Tomsk. Univ.*, **1947,** 157.

50. S. Bodforss, K. J. Karlsson and H. Sjodin, *Z. anorg. allgem. Chem.*, **221,** 382 (1935).
51. Y. I. Ivashentsev, *Tr. Tomskogo Gos. Univ., Ser. Khim.*, **157,** 77 (1963).
52. *British Patent*, 801, 424 (1958).
53. J. Villadsen, *Acta Chem. Scand.*, **13,** 2146 (1959)
54. J. W. Tracey, N. W. Gregory, E. C. Lingafelter, J. D. Dunitz, H. C. Mez, R. E. Scheringer, H. Y. Yakel and M. K. Wilkinson, *Acta Cryst.*, **14,** 927 (1961).
55. L. L. Handy and N. W. Gregory, *J. Chem. Phys.*, **19,** 1314 (1951).
56. H. Schafer, *Z. anorg. allgem. Chem.*, **259,** 53 (1949).
57. D. E. Milligan, M. E. Jacox and J. D. McKinley, *J. Chem. Phys.*, **42,** 902 (1965).
58. J. W. Johnson, D. Cubicciotti and C. M. Kelly, *J. Phys. Chem.*, **62,** 1107 (1958).
59. L. Brewer and N. L. Lofgren, *J. Am. Chem. Soc.*, **72,** 3038 (1950).
60 J. C. Fanning and H. B. Jonassen, *J. Inorg. Nucl. Chem.*, **25,** 29 (1963).
61. H. J. Seifert, *Z. anorg. allgem. Chem.*, **317,** 123 (1962).
62. H. Hecht, J. Jander and H. Schlapmann, *Z. anorg. allgem. Chem.*, **254,** 255 (1947).
63. E. Uhlemann and W. Fischbach, *Z. Chem.*, **3,** 470 (1963).
64. A. R. Pray, *Inorg. Syn.*, **5,** 153 (1957).
65. D. Khristov, S. Karaivanov and V. Kolushki, *God. Sofiskyia Univ., Khim. Fak.*, **55,** 49 (1960–61).
66. J. M. Sherfey, *Inorg. Syn.*, **6,** 57 (1960).
67. W. F. Krieve, S. P. Vango and D. M. Mason, *J. Chem. Phys.*, **25,** 519 (1960).
68. K. Funaki and K. Uchimura, *J. Chem. Soc. Japan*, **59,** 14 (1956).
69. Y. Ogawa, Y. Hisamatsu and K. Kawamura, *Nippon Kogyo Zasshi*, **73,** 565 (1957).
70. R. E. McCarley and J. W. Roddy, *Inorg. Chem.*, **3,** 60 (1964).
71. J. D. Corbett, R. J. Clark and T. F. Mundy, *J. Inorg. Nucl. Chem.*, **25,** 1287 (1963).
72. S. A. Shchukarev and M. A. Oranskaya, *Zh. Obschei Khim.*, **24,** 1926 (1954).
73 R. C. Osthoff and R. C. West, *J. Am. Chem. Soc.*, **76,** 4732 (1954).
74. G. W. Watt, P. S. Gentile and E. P. Helvenston, *J. Am. Chem. Soc.*, **77,** 2752 (1955).
75. H. D. Hardt, *Z. anorg. allgem. Chem.*, **301,** 87 (1959).
76. K. Dehnicke, *Angew. Chem., Intern. Ed. Engl.*, **2,** 325 (1963).
77. K. Dehnicke, *Z. anorg. allgem. Chem.*, **309,** 266 (1961).
78. H. Schafer, F. Wartenpfuhl and E. Weise, *Z. anorg. allgem. Chem.*, **295,** 268 (1958).
79. H. Oppermann, *Z. anorg. allgem. Chem.*, **351,** 113 (1967).
80. A. B. Bardawil, F. H. Collier and S. Y. Tyree, *J. Less-Common Metals*, **9,** 20, (1965).
81. R. B. Johannesen, *Inorg. Syn.*, **6,** 119 (1960).
82. K. Dehnicke, *Chem. Ber.*, **97,** 3354 (1964).
83. W. H. Hartford and M. Darrin, *Chem. Rev.*, **58,** 1 (1958).
84. H. Schafer and F. Wartenpfuhl, *Z. anorg. allgem. Chem.*, **308,** 282 (1961).
85. H. Schafer, *Z. anorg. allgem. Chem.*, **264,** 249 (1951).

86. S. A. Shchukarev, I. V. Vasil'kova and D. V. Korol'kov, *Russ. J. Inorg. Chem.*, **8,** 1006 (1963).
87. F. Ephraim and E. Ammann, *Helv. Chim. Acta*, **16,** 1273 (1933).
88. R. J. Sime and N. W. Gregory, *J. Am. Chem. Soc.*, **82,** 93 (1960).
89. R. J. Sime and N. W. Gregory, *J. Am. Chem. Soc.*, **82,** 800 (1960).
90. R. J. Sime and N. W. Gregory, *J. Phys. Chem.*, **64,** 86 (1960).
91. J. D. Christian and N. W. Gregory, *J. Phys. Chem.*, **71,** 1583 (1967).
92. R. C. Schoonmaker, A. H. Friedman and R. F. Porter, *J. Chem. Phys.*, **31,** 1586 (1959).
93. R. R. Hammer and N. W. Gregory, *J. Phys. Chem.*, **68,** 314 (1964).
94. G. Winter, *private communication.*
95. P. M. Druce, M. F. Lappert and P. N. K. Riley, *Chem. Commun.*, **1967,** 486.
96. R. C. Young and W. C. Schumb, *J. Am. Chem. Soc.*, **52,** 4233 (1930).
97. K. Funaki, K. Uchimura and Y. Kuniya, *Kogyo Kagaku Zasshi.*, **64,** 1914 (1961).
98. P. Ehrlich, W. Gutsche and H. J. Seifert, *Z. anorg. allgem. Chem.*, **312,** 80 (1961).
99. E. H. Hall and J. M. Blocher, *J. Phys. Chem.*, **63,** 1525 (1959).
100. N. W. Gregory and B. A. Thackrey, *J. Am. Chem. Soc.*, **72,** 3176 (1950).
101. T. J. Wydeven and N. W. Gregory, *J. Phys. Chem.*, **68,** 3249 (1964).
102. W. B. Hadley and J. W. Stout, *J. Chem. Phys.*, **39,** 2205 (1963).
103. K. Dehnicke, *Chem. Ber.*, **98,** 290 (1965).
104. F. A. Miller and W. K. Baur, *Spectrochim. Acta*, **17,** 112 (1961).
105. K. Funaki, K. Uchimura and H. Matsunayer, *Kogyo Kagaku Zasshi*, **64,** 129 (1961).
106. H. G. Schnering, *Naturwissenschaften*, **53,** 359 (1966).
107. T. A. Tolmacheva, V. M. Tsintsius and L. V. Andrianova, *Russ. J. Inorg. Chem.*, **8,** 281 (1963).
108. N. W. Gregory and L. L. Handy, *Inorg. Syn.*, **5,** 128 (1957).
109. J. W. Mellor, *A Comprehensive Treatise on Inorganic and Theoretical Chemistry*, Vol. 12, Longmans, London, p. 384.
110. H. Schafer and W. J. Hones, *Z. anorg. allgem. Chem.*, **288,** 62 (1956).
111. J. B. Wagner and C. Wagner, *J. Chem. Phys.*, **26,** 1597 (1957).
112. M. Chaigneau and M. Chastagnier, *Bull. Soc. Chim. France*, **1958,** 1192.
113. M. Chaigneau, *Bull. Soc. Chem. France*, **1957,** 886.
114. M. Chaigneau, *Compt. Rend.*, **242,** 263 (1956).
115. R. F. Rolsten and H. H. Sisler, *J. Am. Chem. Soc.*, **79,** 5891 (1957).
116. T. R. Ingraham and L. M. Pidgeon, *Can. J. Chem.*, **30,** 694 (1952).
117. J. D. Fast, *Rev. Trav. Chim.*, **58,** 174 (1939).
118. T. A. Tolmacheva, V. M. Tsintsius and E. E. Yudovich, *Russ. J. Inorg. Chem.*, **11,** 249 (1966).
119. W. E. Hatfield and J. T. Yoke, *Inorg. Chem.*, **1,** 463 (1962).
120. A. B. P. Lever, *J. Inorg. Nucl. Chem.*, **27,** 149 (1965).
121. *U.S. Patent*, 2, 816, 816 (1957).
122. I. N. Belyaev and O. Y. Revina, *Russ. J. Inorg. Chem.*, **11,** 1041 (1966).
123. J. Ravez, J. Viollet, R. de Pape and P. Hagenmuller, *Bull. Soc. Chem. France*, **1967,** 1325.
124. R. de Pape and J. Ravez, *Bull. Soc. Chim. France*, **1966,** 3283.

125. M. Kestigian, F. D. Leipzig, W. J. Croft and R. Guidoboni, *Inorg. Chem.*, **5,** 1462 (1966).
126. G. K. Wertheim, H. J. Guggenheim, H. J. Williams and D. N. E. Buchanan, *Phys. Rev.*, **158,** 446 (1967).
127. H. G. Schnering, *Z. anorg. allgem. Chem.*, **353,** 1 (1967).
128. W. Rudorff, G. Linke and D. Babel, *Z. anorg. allgem. Chem.*, **320,** 150 (1963).
129. W. Rudorff, J. Kaendler and D. Babel, *Z. anorg. allgem. Chem.*, **317,** 261 (1962).
130. O. Schmitz-Dumont and D. Grimm, *Z. anorg. allgem. Chem.*, **355,** 280 (1967).
131. D. Babel, G. Pausenwang and W. Viebahn, *Z. Naturforsch.*, **22b,** 1219 (1967).
132. R. D. Peacock, *J. Chem. Soc.*, **1957,** 4684.
133. R. D. Peacock, in *Progress in Inorganic Chemistry* (Ed. F. A. Cotton), Vol. 2, Interscience, New York, 1960.
134. E. Huss and W. Klemm, *Z. anorg. allgem. Chem.*, **262,** 25 (1950).
135. D. H. Brown, K. R. Dixon, R. D. W. Kemmitt and D. W. A. Sharp, *J. Chem. Soc.*, **1965,** 1559.
136. W. Klemm and E. Huss, *Z. anorg. allgem. Chem.*, **258,** 221 (1949).
137. R. Hoppe, *Rec. Trav. Chim.*, **75,** 569 (1956).
138. M. D. Meyers and F. A. Cotton, *J. Am. Chem. Soc.*, **82,** 5027 (1960).
139. H. Bode and E. Voss, *Z. anorg. allgem. Chem.*, **290,** 1 (1957).
140. A. G. Sharpe, *J. Chem. Soc.*, **1950,** 2907.
141. A. G. Sharpe and A. A. Woolf, *J. Chem. Soc.*, **1951,** 798.
142. B. Cox and A. G. Sharpe, *J. Chem. Soc.*, **1953,** 1783.
143. R. S. Nyholm and A. G. Sharpe, *J. Chem. Soc.*, **1952,** 3579.
144. H. Stammreich, O. Sala and D. Bassi, *Spectrochim. Acta*, **19,** 593 (1963).
145. R. Dingle, *Inorg. Chem.*, **4,** 1287 (1965).
146. D. E. Eastman and M. W. Shafer, *J. Appl. Phys.*, **38,** 1274 (1967).
147. H. M. Haendler, F. A. Johnson and D. S. Crocket, *J. Am. Chem. Soc.*, **80,** 2662 (1958).
148. D. S. Crocket and H. M. Haendler, *J. Am. Chem. Soc.*, **82,** 4158 (1960).
149. D. J. Machin, R. L. Martin and R. S. Nyholm, *J. Chem. Soc.*, **1963,** 1490.
150. L. A. Tsiorkina and M. A. Smirnov, *Russ. J. Inorg. Chem.*, **4,** 65 (1959).
151. R. J. Lister and S. N. Flengas, *Can. J. Chem.*, **41,** 1548 (1963).
152. R. V. Chernov and Y. K. Delmarskii, *Zh. Neorg. Khim.*, **6,** 2749 (1961).
153. P. Ehrlich, G. Kupa and K. Blankenstein, *Z. anorg. allgem. Chem.*, **299,** 213 (1959).
154. G. J. Wessel and D. J. W. Ijdo, *Acta Cryst.*, **10,** 466 (1957).
155. B. F. Markov and R. V. Chernov, *Ukr. Khim. Zh.*, **25,** 279 (1959).
156. I. S. Morozov and A. I. Morozov, *Izv. Akad. Nauk SSSR., Neorg. Mater.*, **3,** 1039 (1967).
157. S. A. Shchukarev and I. L. Perfilova, *Russ. J. Inorg. Chem.*, **8,** 1100 (1963).
158. H. J. Seifert and P. Ehrlich, *Z. anorg. allgem. Chem.*, **302,** 284 (1959).
159. S. A. Shchukarev, I. V. Vasil'kova, A. I. Efimov and B. Z. Pitirimov, *Russ. J. Inorg. Chem.*, **11,** 268 (1966).
160. I. V. Vasil'kova, A. I. Efimov and B. Z. Pitirimov, *Russ. J. Inorg. Chem.*, **9,** 493 (1964).

161. H. J. Seifert and K. Klatyk, *Z. anorg. allgem. Chem.*, **334,** 113 (1964).
162. H. J. Seifert and F. W. Koknat, *Z. anorg. allgem. Chem.*, **341,** 269 (1965).
163. B. F. Markov and R. V. Chernov, *Ukr. Khim. Zh.*, **24,** 139 (1958).
164. A. P. Ginsberg and M. B. Robin, *Inorg. Chem.*, **2,** 817 (1963).
165. C. M. Cook and W. E. Dunn, *J. Phys. Chem.*, **65,** 1505 (1961).
166. H. J. Seifert and K. Klatyk, *Z. anorg. allgem. Chem.*, **342,** 1, (1966).
167. N. V. Galitsky, V. I. Borodin and A. I. Lystov, *Ukr. Khim. Zh.*, **32,** 695 (1966).
168. H. J. Seifert, *Z. anorg. allgem. Chem.*, **307,** 133 (1960).
169. A. Engberg and H. Soling, *Acta Chem. Scand.*, **21,** 168 (1967).
170. E. Iberson, R. Gut and D. M. Gruen, *J. Phys. Chem.*, **66,** 65 (1962).
171. G. W. A. Fowles and D. Nicholls, *J. Inorg. Nucl. Chem.*, **18,** 130 (1961).
172. R. A. D. Wentworth and T. S. Piper, *J. Chem. Phys.*, **41,** 3884 (1964).
173. L. Helmholz and W. R. Foster, *J. Am. Chem. Soc.*, **72,** 4971 (1950).
174. H. P. de la Garanderie, *Compt. Rend.*, **255,** 2585 (1962).
175. G. W. A. Fowles and B. J. Russ, *J. Chem. Soc.*, **1967,** A, 517.
176. G. W. A. Fowles and R. A. Walton, *J. Inorg. Nucl. Chem.*, **27,** 735 (1965).
177. P. Pauling, *Inorg. Chem.*, **5,** 1498 (1966).
178. D. M. Adams, J. Chatt, J. M. Davidson and J. Gerratt, *J. Chem. Soc.*, **1963,** 2189.
179. I. I. Kozhina and D. V. Korol'kov, *Zh. Strukt. Khim.*, **6,** 97 (1965).
180. P. Ros, *Rec. Trav. Chim.*, **82,** 823 (1963).
181. A. Sabatini and L. Sacconi, *J. Am. Chem. Soc.*, **86,** 17 (1964).
182. F. A. Cotton, D. M. L. Goodgame and M. Goodgame, *J. Am. Chem. Soc.*, **83,** 4690 (1961).
183. W. H. Baur, *Acta Cryst.*, **11,** 488 (1958).
184. A. Ferrari, A. Braibantini and G. Bigliardi, *Acta Cryst.*, **16,** 846 (1963).
185. L. G. van Uitert, H. J. Williams, R. C. Sherwood and J. J. Rubin, *J. Appl. Phys.*, **36,** 1029 (1965).
186. D. Weigel, *Bull. Soc. Chim. France*, **1963,** 2087.
187. W. Klemm and L. Grimm, *Z. anorg. allgem. Chem.*, **249,** 198 (1941).
188. A. Ferrari and F. Giorgi, *Atti Acad. Lincei*, **9,** 1134 (1929).
189. A. Ferrari and F. Giorgi, *Atti Acad. Lincei*, **10,** 522 (1929).
190. R. O. MacLaren and N. W. Gregory, *J. Am. Chem. Soc.*, **76,** 2874 (1954).
191. A. Narath and J. E. Schirber, *J. Appl. Phys.*, **37,** 1124 (1966).
192. K. H. Jack and R. Maitland, *Proc. Chem. Soc.*, **1957,** 232.
193. J. W. Tracey, N. W. Gregory, J. M. Stewart and E. C. Lingafelter, *Acta Cryst.*, **15,** 460 (1962).
194. J. W. Tracey, N. W. Gregory and E. C. Lingafelter, *Acta Cryst.*, **15,** 672 (1962).
195. P. Ehrlich and G. Pietzka, *Z. anorg. allgem. Chem.*, **275,** 121 (1954).
196. K. H. Jack and V. Gutmann, *Acta Cryst.*, **4,** 246 (1951).
197. M. A. Hepworth and K. H. Jack, *Acta Cryst.*, **10,** 345 (1957).
198. G. Natta, P. Corradini and G. Allegra, *J. Polymer Sci.*, **51,** 399 (1961).
199. B. N. Figgis, M. Gerloch and R. Mason, *Acta Cryst.*, **17,** 506 (1964).
200. R. V. van Stapele, H. G. Beljers, P. F. Bongers and H. Zijlstra, *J. Chem. Phys.*, **44,** 3719 (1966).
201. W. E. Hatfield and T. S. Piper, *Inorg. Chem.*, **3,** 841 (1964).
202. G. C. Allen and N. S. Hush, *Inorg. Chem.*, **6,** 4 (1967).

203. M. Mori and S. Fujiwara, *Bull. Chem. Soc. Japan*, **36** 1636 (1963).
204. M. Mori, *Bull. Chem. Soc. Japan*, **34,** 1249 (1961).
205. D. M. Adams and P. J. Lock, *J. Chem. Soc.*, **1967,** A, 620.
206. B. Zaslow and G. L. Ferguson, *Chem. Commun.*, **1967,** 822.
207. F. A. Cotton, O. D. Faut and D. M. L. Goodgame, *J. Am. Chem. Soc.*, **83,** 344 (1961).
208. J. Bradbury, K. P. Forest, R. H. Nuttall and D. W. A. Sharp, *Spectrochim. Acta*, **23,** 2701 (1967).
209. D. M. L. Goodgame and M. Goodgame, *Inorg. Chem.*, **4,** 139 (1965).
210. M. Goodgame and F. A. Cotton, *J. Am. Chem. Soc.*, **84,** 1543 (1962).
211. H. L. Schlafer and H. P. Fritz, *Spectrochim. Acta*, **23,** 1409 (1967).
212. M. D. Lind, *J. Chem. Phys.*, **47,** 990 (1967).
213. I. G. Dance and H. C. Freeman, *Inorg. Chem.*, **4,** 1555 (1965).
214. J. Mizuno, *J. Phys. Soc. Japan*, **15,** 1412 (1960).
215. J. Mizuno, *J. Phys. Soc. Japan*, **16,** 1574 (1961).
216. A. Trutia and M. Musa, *Phys. Status Solidi*, **8,** 663 (1965).
217. A. Zalkin, J. D. Forrester and D. H. Templeton, *Inorg. Chem.*, **3,** 529 (1964).
218. B. R. Penfold and J. A. Grigor, *Acta Cryst.*, **12,** 850 (1959).
219. B. Morosin and E. J. Graebner, *J. Chem. Phys.*, **42,** 898 (1965).
220. B. Morosin, *J. Chem. Phys.*, **47,** 417 (1967).
221. D. Harker, *Z. Krist.*, **93,** 136 (1936).
222. B. Morosin, *J. Chem. Phys.*, **44,** 252 (1966).
223. S. A. Shchukarev, E. V. Stroganov, S. N. Andreev and O. F. Purvinskii, *Zh. Strukt. Khim.*, **4,** 63 (1963).
224. M. Gaudin-Louer and D. Weigel, *Compt. Rend.*, **264B,** 895 (1967).
225. R. J. H. Clark, *Spectrochim. Acta.*, **21,** 955 (1965).
226. R. J. H. Clark and C. S. Williams, *Inorg. Chem.*, **4,** 350 (1965).
227. D. M. Adams, *Metal-Ligand and Related Vibrations*, Arnold, London, 1967.
228. E. Konig, *Landolt-Bornstein*, Vol. 2, *Magnetic Properties of Coordination and Organometallic Transition Metal Compounds*, Springer-Verlag, New York, 1966.
229. G. S. Rao, *Z. anorg. allgem. Chem.*, **304,** 351 (1960).
230. E. L. Muetterties, *J. Am. Chem. Soc.*, **82,** 1082 (1960).
231. D. S. Dyer and R. O. Ragsdale, *Chem. Commun.*, **1966,** 601.
232. D. S. Dyer and R. O. Ragsdale, *Inorg. Chem.*, **6,** 8 (1967).
233. R. J. H. Clark and W. Errington, *J. Chem. Soc.*, **1967,** A, 258.
234. R. J. H. Clark and C. S. Williams, *J. Chem. Soc.*, **1966,** A, 1425.
235. B. J. Hathaway and D. G. Holah, *J. Chem. Soc.*, **1964,** 2400.
236. B. J. Hathaway and D. G. Holah, *J. Chem. Soc.*, **1964,** 2408.
237. R. D. Willett and R. E. Rundle, *J. Chem. Phys.*, **40,** 838 (1964).
238. B. Bosnich, R. D. Gillard, E. D. McKenzie and G. A. Webb, *J. Chem. Soc.*, **1966,** A, 1331.
239. G. A. Barclay and R. S. Nyholm, *Chem. Ind. (London)*, **1953,** 378.
240. L. F. Lindoy, S. E. Livingstone and T. N. Lockyer, *Australian J. Chem.*, **19,** 1391 (1966).
241. P. S. K. Chia, S. E. Livingstone and T. N. Lockyer, *Australian J. Chem.*, **19** 1835 (1966).

242. P. S. K. Chia, S. E. Livingstone and T. N. Lockyer, *Australian J. Chem.*, **20,** 239 (1967).
243. L. F. Lindoy, S. E. Livingstone and T. N. Lockyer, *Australian J. Chem.*, **20,** 471 (1967).
244. P. S. K. Chia and S. E. Livingstone, *Australian J. Chem.*, **21,** 339 (1968).
245. F. P. Dwyer and D. P. Mellor, *Chelating Agents and Metal Chelates*, Academic Press, New York, 1964.
246. B. J. Russ and J. S. Wood, *Chem. Commun.*, **1966,** 745.
247. G. W. A. Fowles, P. T. Greene and J. S. Wood, *Chem. Commun.*, **1967,** 971.
248. M. Gerloch and F. E. Mabbs, *J. Chem. Soc.*, **1967,** A, 1598.
249. B. F. Hoskins, R. L. Martin and A. H. White, *Nature*, **211,** 627 (1966).
250. M. Di Varia and P. L. Orioli, *Chem. Commun.*, **1965,** 590.
251. Z. Dori, R. Eisenberg and H. B. Gray, *Inorg. Chem.*, **6,** 483 (1967).
252. M. Gerloch, *J. Chem. Soc.*, **1966,** A, 1317.
253. M. Di Varia and P. L. Orioli, *Inorg. Chem.*, **6,** 955 (1967).
254. J. A. Bertrand and D. L. Plymale, *Inorg. Chem.*, **5,** 879 (1966).
255. G. A. Mair, H. M. Powell and D. E. Henn, *Proc. Chem. Soc.*, **1960,** 415.
256. H. S. Preston and C. H. L. Kennard, *Chem. Commun.*, **1967,** 1167.
257. B. F. Hoskins and F. D. Whillans, *Chem. Commun.*, **1966,** 798.
258. G. A. Barclay, B. F. Hoskins and C. H. L. Kennard, *J. Chem. Soc.*, **1963,** 5691.
259. M. Ciampolini, *Inorg. Chem.*, **5,** 35 (1966).
260. N. J. Norgett, J. H. M. Thornley and L. M. Venanzi, *J. Chem. Soc.*, **1967,** A, 540.
261. J. S. Wood, *Inorg. Chem.*, **7,** 852 (1968).
262. Z. Dori and H. B. Gray, *Inorg. Chem.*, **7,** 889 (1968).
263. N. S. Gill, R. H. Nuttall, D. E. Scaife and D. W. A. Sharp, *J. Inorg. Nucl. Chem.*, **18,** 79 (1961).
264. I. S. Ahuja, D. H. Brown, R. H. Nuttall and D. W. A. Sharp, *J. Inorg. Nucl. Chem.*, **27,** 1105 (1965).
265. N. S. Gill, G. A. Barclay, T. I. Christie and P. J. Pauling, *J. Inorg. Nucl. Chem.*, **18,** 88 (1961).
266. N. S. Gill and H. J. Kingdon, *Australian J. Chem.*, **19,** 2197 (1966).
267. C. E. Frank and L. B. Rogers, *Inorg. Chem.*, **5,** 615 (1966).
268. M. Goodgame and P. J. Hayward, *J. Chem. Soc.*, **1966,** A, 632.
269. D. P. Graddon, K. B. Heng and E. C. Watton, *Australian J. Chem.*, **21,** 121 (1968).
270. D. P. Graddon and E. C. Watton, *Australian J. Chem.*, **18,** 507 (1965).
271. S. M. Nelson and T. M. Shepard, *J. Chem. Soc.*, **1965,** 3276.
272. M. A. Porai-Koshits and A. S. Antsishkina, *Dokl. Akad. Nauk SSSR*, **92,** 333 (1953).
273. A. S. Antsishkina and M. A. Porai-Koshits, *Kristallografiya*, **3,** 676 (1958).
274. J. D. Dunitz, *Acta Cryst.*, **10,** 307 (1957).
275. G. B. Boki, T. I. Malinovskii and A. V. Ablov, *Kritsallografiya*, **1,** 49 (1956).
276. V. F. Duckworth, D. P. Graddon, N. C. Stephenson and E. C. Watton, *Inorg. Nucl. Chem. Letters*, **3,** 557 (1967).
277. K. Emmerson, paper given at American Crystallographic Society Meeting, Tuscon, 1968.

278. A. B. P. Lever and S. M. Nelson, *J. Chem. Soc.*, **1966,** A, 859.
279. D. M. L. Goodgame, M. Goodgame, M. A. Hitchman and M. J. Weeks, *J. Chem. Soc.*, **1966,** A, 1769.
280. A. B. P. Lever, S. M. Nelson and T. M. Shepard, *Inorg. Chem.*, **4,** 810 (1965).
281. W. Ludwig and G. Wittman, *Helv. Chim. Acta*, **47,** 1265 (1964).
282. S. Buffagni, L. M. Vallarino and J. V. Quagliano, *Inorg. Chem.*, **3,** 480 (1964).
283. L. M. Vallarino, W. E. Hill and J. V. Quagliano, *Inorg. Chem.*, **4,** 1598 (1965).
284. H. C. A. King, E. Koros and S. M. Nelson, *J. Chem. Soc.*, **1963,** 5449.
285. S. Buffagni, L. M. Vallarino and J. V. Quagliano, *Inorg. Chem.*, **3,** 671 (1964).
286. W. Libus and I. Uruska, *Inorg. Chem.*, **5,** 256 (1966).
287. J. deO. Cabral, H. C. A. King, S. M. Nelson, T. M. Shepard and E. Koros, *J. Chem. Soc.*, **1966,** A, 1348.
288. S. M. Nelson and T. M. Shepard, *J. Chem. Soc.*, **1965,** 3284.
289. H. C. A. King, E. Koros and S. M. Nelson, *J. Chem. Soc.*, **1964,** 4832.
290. J. R. Allan, D. H. Brown, R. H. Nuttall and D. W. A. Sharp, *J. Inorg. Nucl. Chem.*, **27,** 1865 (1965).
291. J. R. Allan, D. H. Brown, R. H. Nuttall and D. W. A. Sharp, *J. Inorg. Nucl. Chem.*, **27,** 1529 (1965).
292. J. R. Allan, D. H. Brown, R. H. Nuttall and D. W. A. Sharp, *J. Inorg. Nucl. Chem.*, **26,** 1895 (1964).
293 G. Beech, C. T. Mortimer and E. G. Tyler, *J. Chem. Soc.*, **1967,** A, 925.
294. G. Beech, S. J. Ashcroft and C. T. Mortimer, *J. Chem. Soc.*, **1967,** A, 929.
295. G. Beech, C. T. Mortimer and E. G. Tyler, *J. Chem. Soc.*, **1967,** A, 1111.
296. G. Booth, *Advances in Inorganic Chemistry and Radiochemistry* (Ed. H. J. Emeleus and A. G. Sharpe), Vol. 6, Academic Press, London, 1964.
297. L. M. Venanzi, *Angew. Chem., Intern. Ed. Engl.*, **3,** 453 (1964).
298. M. C. Browning, J. R. Mellor, D. J. Morgan, S. A. J. Pratt, L. E. Sutton and L. M. Venanzi, *J. Chem. Soc.*, **1962,** 693.
299. R. G. Hayter and F. S. Humiec, *Inorg. Chem.*, **4,** 1701 (1965).
300. R. J. H. Clark, R. S. Nyholm, P. Pauling and G. B. Robertson, *Nature*, **192,** 222 (1961).
301. R. J. H. Clark, J. Lewis and R. S. Nyholm, *J. Chem. Soc.*, **1962,** 2460.
302. R. J. H. Clark, W. Errington, J. Lewis and R. S. Nyholm, *J. Chem. Soc.*, **1966,** A, 989.

Chapter 2

Titanium

More starkly than for any other first-row transition element, the chemistry of titanium is dominated by the stability of its maximum oxidation state. Indeed, extreme difficulty has been experienced in preparing an authentic sample of titanium trifluoride, although as expected, the ease of preparation of lower-valent halides increases with increasing mass of the halogen.

Much of the interest in titanium halides has stemmed from their potential use in the preparation of white pigments and titanium metal.

Clark[1] has very recently reviewed the chemistry of titanium, including the halides.

HALIDES AND OXIDE HALIDES

The known halides and oxide halides of titanium are shown in Table 2.1.

TABLE 2.1
Halides and Oxide Halides of Titanium

Oxidation state	Fluoride	Chloride	Bromide	Iodide
IV	TiF_4, $TiOF_2$	$TiCl_4$, $TiOCl_2$	$TiBr_4$, $TiOBr_2$	TiI_4, $TiOI_2$
III	TiF_3, $TiOF$	$TiCl_3$, $TiOCl$	$TiBr_3$	TiI_3
II		$TiCl_2$	$TiBr_2$	TiI_2

Oxidation State IV

Titanium tetrafluoride. Titanium tetrafluoride is best prepared by the direct fluorination of titanium sponge at 200° in a flow system[2,3]. At this temperature titanium tetrafluoride is so volatile that a protective layer of fluoride is not formed on the remaining sponge and the reaction proceeds to completion.

Alternative methods of preparation include direct fluorination of titanium dioxide at above 350° in a flow system[2], fluorination of titanium metal in a bomb at elevated temperatures[4], the action[5] of chlorine trifluoride on titanium metal in a flow system at 350°, and the thermal decomposition of ammonium hexafluorotitanate(IV)[6] at 150°. Aynsley and coworkers[7] have found that the interaction of titanium metal and nitryl fluoride gives a mixture of the tetrafluoride and titanium dioxide, and on gentle heating the titanium tetrafluoride is readily separated by fractional sublimation. However, this is not a convenient or efficient method of preparing titanium tetrafluoride since the nitryl fluoride must first be prepared.

One of the first methods of preparing titanium tetrafluoride was the interaction of titanium tetrachloride and anhydrous hydrogen fluoride[8], but extremely pure hydrogen fluoride is required otherwise oxide fluorides and other contaminants are formed.

Titanium tetrafluoride is a white, extremely hygroscopic solid, which can be readily purified by vacuum sublimation. It is soluble in aqueous hydrogen fluoride to form the hexafluorotitanate(IV) ion. Elegant ^{19}F NMR studies of the hydrolysis of titanium tetrafluoride have recently been reported[9,10]. It has been unequivocally demonstrated that numerous species, both monomeric and polymeric, are produced in water.

Hall and coworkers[11] have measured the sublimation pressure of titanium tetrafluoride over a wide temperature range and the equation expressing the variation of sublimation pressure with temperature is

$$\log p_{\text{atm}} = 16.631 - 2.567 \log T - 5{,}332/T$$

The derived thermodynamic data from this study and also from fluorine bomb calorimetry by Hubbard and coworkers[4], are given in Table 2.2.

TABLE 2.2
Thermodynamic Properties of Titanium Tetrafluoride

Property	Value	Reference
ΔH_f° (kcal mole^{-1})	−394.2	4
ΔG_f° (kcal mole^{-1})	−372.7	4
ΔS_f° (cal deg^{-1} mole^{-1})	72.21	4
ΔH_{sub} (kcal mole^{-1})	22.87	11
ΔS_{sub} (cal deg^{-1} mole^{-1})	41.55	11

The structure of solid titanium tetrafluoride is not known, but Euler and Westrum[12] have suggested that it is intermediate between the monomeric molecular lattice-type of silicon tetrafluoride, and the infinitely polymeric coordination-type of zirconium tetrafluoride. The infrared spectrum of titanium tetrafluoride has been recorded several times[13–15] and strong absorptions were observed at 500, 590 and 750 cm^{-1}.

Titanium tetrafluoride forms adducts with nitriles, amines, arsenic ligands, etc., and also undergoes substitution reactions[13–29]. The most important study of titanium tetrafluoride adducts is the ^{19}F NMR study by Dyer and Ragsdale[20], who unequivocally showed that π-bonding occurs between the fluorine atoms and the titanium atom in the nitrogen-base adducts. Clark and Errington[27] have recorded titanium-fluorine stretching and bending modes in several titanium tetrafluoride adducts. Bands occur at about 630, 570 and 275 cm^{-1}.

Titanium trifluoride chloride. This yellow solid has been prepared by Vorres and Dutton[30] by the chlorination of titanium trifluoride. No experimental conditions for the preparation were reported. It sublimes more readily than titanium tetrafluoride and hydrolyses very readily in air to give a product reported to be titanium oxide difluoride (see below).

Titanium difluoride dichloride. This compound is prepared[31] by passing chlorine monofluoride through titanium tetrachloride at 0°. The difluoride dichloride precipitates as a very hydroscopic solid. It is very soluble in dry polar solvents and cryoscopic studies indicate it to be monomeric[31]. It is however, thermally unstable, giving titanium tetrafluoride and titanium tetrachloride above 125°. Its infrared spectrum shows absorptions at 286, 303, 444, 489, 578 and 614 cm^{-1} and this has been interpreted as evidence for C_{2v} symmetry for the molecule.

Titanium trifluoride bromide. Titanium trifluoride bromide has been prepared[32] by the action of bromine trifluoride on a solution of titanium tetrabromide in carbon tetrachloride at −10°. It is a very hygroscopic golden-brown solid. Its infrared spectrum in the solid phase has been interpreted on the basis of C_s symmetry, and the structure shown below was postulated.

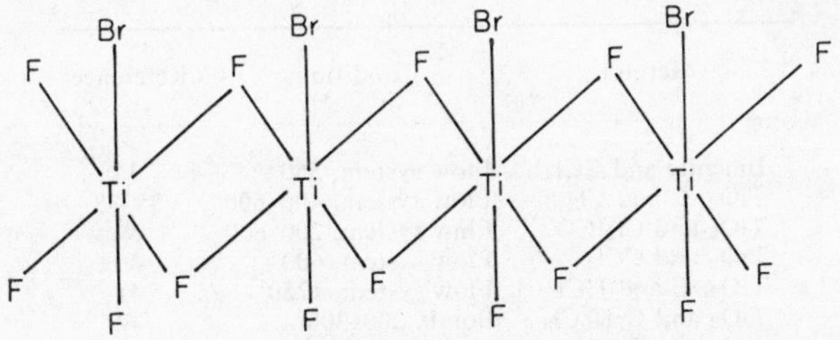

Titanium trifluoride bromide undergoes an interesting reaction with chlorine nitrate to give a compound which could be formulated as $TiF_3(NO_3)\cdot N_2O_5$[32]. The infrared spectrum of this compound shows the presence of the NO_2^+ ion, and together with its chemical properties, suggests that it should be formulated as $NO_2^+[TiF_3 \cdot (NO_3)_2]^-$. This compound loses nitrogen dioxide when heated to about 60°, and at 80° a white hydroscopic solid, Ti_2OF_6, is formed. The infrared spectrum of this compound shows that it contains the titanyl cation, TiO^{2+}, and should be formulated as $[TiO^{2+}][TiF_6^{2-}]$.

Titanium oxide difluoride. It has been claimed that the partial hydrolysis of titanium tetrafluoride, the hydrolysis of titanium trifluoride chloride and the dissolution of titanium dioxide in 50% hydrofluoric acid, all yield titanium oxide difluoride as a white solid[30,33]. X-ray powder diffraction data appeared to show a random distribution of oxygen and fluorine atoms in an octahedral environment about each titanium atom[33]. However, a subsequent and more detailed study of the properties of the material prepared by the above methods showed conclusively that the product obtained is in fact $TiO(OH)F$[31].

Authentic titanium oxide difluoride has been prepared as a yellow solid by the interaction of titanium difluoride dichloride and chlorine monoxide[31] in a polar solvent at 4°. The compound was shown by X-ray powder diffraction techniques to be cubic with $a = 3.71$ Å.

No chemical properties of titanium oxide difluoride have been reported.

Titanium tetrachloride. This compound is of great technological importance in the extraction and purification of titanium from mineral sources, and in the preparation of white pigments, etc. It is probably the most studied titanium compound, partly because of its economic importance and reactivity, and also because its volatility is suited to a wide variety of physico-chemical techniques.

Some methods of preparation of titanium tetrachloride are given in Table 2.3.

TABLE 2.3

Preparation of Titanium Tetrachloride

Method	Conditions	References
Ilmenite and CCl_4	Flow system, 550°	34
TiO_2, C and Cl_2	Flow system, 400–600°	35–38
TiO_2 and Cl_2/CO	Flow system, 200–600°	39
TiO_2 and CCl_4	Flow system, 600°	40
TiO_2, C and HCl	Flow system, 1250°	41
TiO_2 and $C_7H_5Cl_3$	Bomb, 200–300°	42
TiO_2 and C_5Cl_8	Reflux	43
TiN and HCl[a]	Flow system, 1300°	44

[a] $TiCl_2$ and $TiCl_3$ also formed.

Pure titanium tetrachloride is a colourless liquid, but it is often slightly yellow because of dissolved chlorine and iron trichloride. It is very readily hydrolysed giving titanium dioxide ultimately.

The vapour pressure of solid and liquid titanium tetrachloride has been measured several times and the results are expressed by the following equations.

Solid: $\log p_{mm} = 9.639 - 2453/T$ (-60 to $-23°$) (reference 45)

Liquid: $\log p_{mm} = 7.683 - 1964/T$ (-23 to $136°$) (reference 45)

$\log p_{mm} = 7.644 - 1948/T$ (-20 to $84°$) (references 46, 47)

$\log p_{mm} = 8.002 - 2078/T$ (40 to $84°$) (reference 48)

$\log p_{mm} = 7.669 - 1961/T$ (85 to $139.5°$) (reference 49)

The known thermodynamic data for titanium tetrachloride are given in Table 2.4.

TABLE 2.4

Thermodynamic Data for Titanium Tetrachloride

Property	Value	References
M.p.	−30°	50
	−24°	51
	−23.4°	52
	−23.22°	45
B.p.	135.9 ± 0.1°	45, 47, 52
	136.5 ± 0.1°	48, 53–55
ΔH°_{form} (liquid) (kcal mole^{-1})	−190.0	56
	−190.3	57
	−191.45	58
	−192.1	59
	−192.5	60
	−192.9	61
ΔH_{form} (gas) (kcal mole^{-1})	181.6	56
	−182.4	59, 62
	−182.9	60
	−183.0	61
ΔS°_{form} (cal deg^{-1} mole^{-1})	47.09	50
ΔH_{fus} (kcal mole^{-1})	22.33	50
ΔH_{vap} (kcal mole^{-1})	8.28	53
	8.65	48
	8.96	47, 63
	8.97	49
	9.6	60
ΔS_{fus} (cal deg^{-1} mole^{-1})	9.01	50
ΔS_{vap} (cal deg^{-1} mole^{-1})	21.12	48
Transition temperature	21.07°	47
	22.0°	49

Heat capacities and energy contents have been measured over the temperature range 0–300°K[64] and specific heats and entropies have been measured over the range 100–500°K[65].

Luchinskii[45] reported that the density of solid titanium tetrachloride at the triple point was 2.0279 g/cm³, while that of the liquid at the same temperature was 1.7993 g/cm³.

Brand and Sackmann[66] have determined the structure of solid titanium tetrachloride at −32° by X-ray diffraction studies. They found that it adopts the molecular $SnBr_4$ type lattice, with a regular tetrahedral configuration about each titanium atom, although each chlorine is crystallographically different in the lattice. The lattice is monoclinic with $a = 9.70$, $b = 6.48$, $c = 9.75$ Å and $\beta = 102.7°$. The ^{35}Cl NQR spectrum of titanium tetrachloride has been observed[67–69]. Four absorptions were observed in most cases in agreement with the crystal data, although in one early investigation[69] only three lines were observed because of experimental limitations.

Electron-diffraction studies[70–72] on the vapour of titanium tetrachloride have shown that the molecule is tetrahedral with an average titanium-chlorine distance of 2.18 Å.

The infrared and Raman spectra of titanium tetrachloride in the liquid and vapour phases, and also in solution in non-polar solvents, have been observed on numerous occasions and very good agreement on the positions of the absorptions has been obtained by the various groups[64,65,73–77]. In all cases the spectra were interpreted on the basis of T_d symmetry and Dove, Creighton and Woodward[75] gave the following assignments (cm⁻¹):

ν_1	ν_2	ν_3	ν_4
389	120	485	140

The chemistry of titanium tetrachloride is dominated by two major types of reaction, namely, addition of simple donor molecules to form, usually, octahedral complexes, and its ready participation in substitution reactions, for example, with the alkoxide ion. It is not our intention to discuss all adducts and reactions of titanium tetrachloride in detail, but reviews of these compounds are available[78–80].

The adducts of titanium tetrachloride are usually soluble in a variety of organic solvents and their constitution has been studied by a wide range of physico-chemical techniques, including infrared and NMR spectroscopy, measurement of colligative properties and X-ray crystallography.

One of the most interesting types of adduct is that formed with *o*-phenylenebis(dimethylarsine), D, which has the empirical formula

$TiCl_4 \cdot 2D$[81,82]. The compound was found to be a non-electrolyte suggesting that the titanium atom is eight-coordinate. This was confirmed by X-ray diffraction studies. The structure of the complex is shown in Figure 2.1, and consists of a distorted dodecahedron surrounding the titanium atom. The titanium-chlorine and titanium-arsenic bond distances are 2.46 and

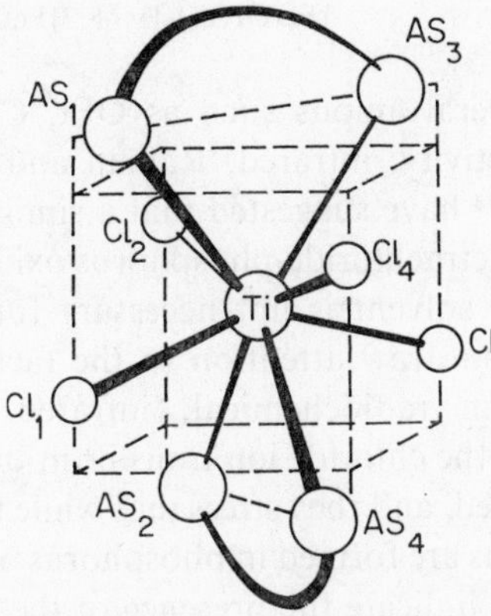

Figure 2.1 The structure of $TiCl_4 \cdot 2$Diarsine. As–As denotes *o*-phenylenebis(dimethylarsine). (Reproduced by permission from R. J. H. Clark, J. Lewis and R. S. Nyholm, *J. Chem. Soc.*, **1962**, 2460)

2.71 Å respectively. Unfortunately no details of the symmetry or dimensions of the unit cell have been reported.

Clark and coworkers[83] have recently reported that titanium tetrachloride reacts with *o*-phenylenebis(dimethylphosphine) to give a 1:2 adduct. This adduct has been shown to contain an eight-coordinate titanium atom.

The titanium tetrachloride-phosphorus oxide trichloride system is undoubtedly the most intensively studied adduct system of titanium tetrachloride[51,84–101]. There is complete unanimity that both 1:1 and 1:2 adducts are formed.

There has been some controversy over the mode of formation of the phosphorus oxide trichloride adducts. Gutmann[102,103] has suggested that phosphorus oxide trichloride ionizes to give free chloride ions:

$$POCl_3 \rightleftharpoons POCl_2^+ + Cl^-$$

In the completely analogous reaction of iron trichloride with triethyl phosphate, where the tetrachloroferrate(III) anion cannot form by ionization of the solvent (excess triethyl phosphate), Meek and Drago have shown by spectrophotometry that the tetrachloroferrate(III) anion is in fact formed[104]. Thus iron trichloride itself must undergo some dissociation to provide the additional chloride ion. They further showed that the following general mechanism applies:

$$FeCl_3 + Y_3PO \rightarrow [FeCl_3 \cdot OPY_3] \rightarrow [FeCl_{(3-x)}(OPY_3)_{(1+x)}]^{x+} + xFeCl_4^-$$
$$\rightleftharpoons$$
$$[Fe(OPY_3)_3]^{3+} + 3[FeCl_4^-]$$

where Y is one of several anions such as OEt, Cl, etc. This scheme is consistent with conductivity, infrared, Raman and other measurements, and Meek and Drago[104] have suggested that a similar mode of formation applies to the titanium tetrachloride-phosphorus oxide trichloride reaction. Thus ionization of the solvent is not necessary for complex formation. Meek and Drago[104] also draw attention to the fact that there is a large body of evidence from radiochemical, infrared, Raman and X-ray diffraction studies, that the chloride ion transfer in oxide chloride solvents has been over emphasized, and they stress that while there is good evidence that chlorometallate ions are formed in phosphorus oxide trichloride, there is as yet no evidence to indicate the presence of the $POCl_2^+$ ion.

Both phosphorus oxide trichloride adducts of titanium tetrachloride have been isolated. $TiCl_4 \cdot POCl_3$ melts at 105.7° and $TiCl_4 \cdot 2POCl_3$ at 105.0°[51,84,98,100]. Voitovich[86] showed that in dilute solutions in nitrobenzene, the 1:2 adduct is partially dissociated into the 1:1 complex and free phosphorus oxide trichloride.

Although, as mentioned above, it has been suggested that in solution in excess phosphorus oxide trichloride the adducts may be ionic in nature[99,102,103], there is no doubt that in the solid state both complexes are true adducts with the oxygen atom acting as the donor. For both adducts there is a marked decrease in the phosphorus-oxygen stretching frequency in the infrared spectrum[87,101], consistent with the formation of a titanium-oxygen bond. A similar conclusion was deduced from a ^{31}P NMR study of $TiCl_4 \cdot 2POCl_3$[99].

Branden and coworkers[95,96,105] have studied the structure of the adducts by X-ray diffraction studies, and fully confirmed the conclusions deduced from the infrared and NMR studies. Branden and Lindqvist[95] showed that there are in fact at least three different modifications of the 1:1 adduct. They showed that modification I is formed when the compound is recrystallized from carbon tetrachloride solutions. Modifications II and III were observed during zone-melting experiments, but no details were reported for the preparation of these polymorphs in a pure state. The lattice parameters of the three modifications are given in Table 2.5.

Oxotetrachlorotitanates(IV). Several oxotetrachlorotitanates(IV) have been prepared by Feltz[427,428] by the partial hydrolysis of hexachlorotitanates(IV) or titanium tetrachloride in organic solvents in the presence of quaternary ammonium chloride. Little is known of their chemical constitution and although Feltz discussed possible structures on the basis of their infrared spectra, Beattie and Fawcett[429] have pointed out that it is not possible to determine the structure of these complexes on the infrared data alone.

Oxotetrachlorotitanate(IV) hydrates have been prepared by Morozov and Toptygina[430] by the partial hydrolysis of hexachlorotitanates(IV) in hydrochloric acid solution. Dehydration to form the anhydrous compounds requires temperatures of the order of 150–200°, strongly suggesting that the water molecules are within the coordination sphere.

Hexabromotitanates(IV). The hexabromotitanate(IV) salts are not so well characterized as the corresponding chloro compounds, but in general, are prepared by the same techniques, as shown in Table 2.20.

TABLE 2.20
Preparation of Hexabromotitanates(IV)

Method	Conditions	References
MBr and $TiBr_4$	Sealed tube, elevated temp.	259, 431
MBr and $TiBr_4$	Interaction in molten $SbCl_3$	432
NH_4Br and $TiBr_4$	Conc. HBr saturated with HBr gas	433
$TiBr_4$ and amine	Conc. HBr saturated with HBr gas	411, 433
$TiBr_4$ and $(C_2H_5)_2NH_2Br$	Mix chloroform solutions	139

Like the corresponding chloro derivatives, the hexabromotitanates(IV) are readily hydrolysed by moist air, although they are stable in dry air. At elevated temperatures they dissociate in vacuum, but they may be melted without much decomposition in a sealed capillary. The melting points of the rubidium and caesium salts are 672° and 720° respectively[259]. The reported heats of formation are[259]

Compound	From constituent halides	From elements
Rb_2TiBr_6	−23.7 kcal mole^{-1}	−385.3 kcal mole^{-1}
Cs_2TiBr_6	−27.6 kcal mole^{-1}	−392.9 kcal mole^{-1}

The hexabromotitanates(IV) are dark-red crystalline solids. X-ray powder diffraction measurements show that they have the cubic K_2PtCl_6 type structure; the unit-cell parameters are given in Table 2.21.

TABLE 2.21
Lattice Parameters of Some Hexabromotitanates(IV)

Compound	*a* (Å)	Reference
K_2TiBr_6	10.35	432
	10.41	433
Rb_2TiBr_6	10.39	431
	10.46	432
Cs_2TiBr_6	10.57	431
	10.66	432

Oxidation State III

Hexafluorotitanates(III). Hexafluorotitanates(III) have been prepared by the interaction of titanium trichloride with a saturated solution of ammonium fluoride[434], and by electrolysis of an alkali-metal chloride melt containing an alkali-metal hexafluorotitanate(IV) under an argon atmosphere[435,436]. Alternatively, the potassium salt has been prepared by the interaction of titanium dioxide, potassium fluoride and potassium hydrogen fluoride, in the correct proportions, at 500° in an inert atmosphere[437] and also by the hydrogen reduction of potassium hexafluorotitanate(IV)[438]. In this latter method a compound of approximate composition K_2TiF_5 is formed. On careful sublimation of this product at about 800°, potassium tetrafluorotitanate(III) and potassium hexafluorotitanate(III) may be separated, but a more convenient method of separation is based on the insolubility of the tetrafluorotitanate(III) and the solubility of the hexafluorotitanate(III) salt in water[438].

The magnetic properties of the hexafluorotitanates(III) have only been investigated at room temperature[436]. Moments ranging from 1.70–2.09 BM were recorded. The moments greater than that expected for one unpaired electron could not be explained.

The absorption spectra[436] of the hexafluorotitanates(III) show two bands in the 15,000–16,000 and 18,900 cm^{-1} regions.

Hexachlorotitanates(III). The hexachlorotitantes(III) are invariably prepared by the interaction of the alkali-metal chloride and titanium trichloride in the melt at about 750–800°[439–450]. Pyridinium hexachlorotitanate(III) may be prepared by the direct interaction of pyridinium chloride and titanium trichloride in a sealed tube at 150°[451] or by interaction in chloroform[451,452].

The alkali-metal hexachlorotitanates(III) are considerably more stable than the corresponding hexachlorotitanates(IV) and they melt without appreciable decomposition. The melting points of some hexachlorotitanates(III) are:

Compound	Melting points	References
Na_3TiCl_6	553°	446
K_3TiCl_6	760°	444, 447, 448, 453
Rb_3TiCl_6	794°	447
Cs_3TiCl_6	770°	447

The pyridinium salt is orthorohombic with $a = 16.03$, $b = 16.83$ and $c = 20.83$ Å[451,452]. This compound has a room temperature magnetic moment of 1.66 BM[452].

The absorption spectrum of the hexachlorotitanate(III) ion in alkali-metal chloride melts shows a split band with peaks at 10,000 and 13,000 cm^{-1} attributed to the $E_g \leftarrow T_{2g}$ transition[454].

Pentachlorotitanates(III). Potassium pentachlorotitanate(III) has been reported as one of the products of the interaction of potassium chloride and titanium trichloride in the melt[446,453]. Ehrlich and coworkers[446] report that it melts at 605°, whereas Kamenitskii and his associates[453] found that it undergoes a phase transformation at 580°, and then decomposes to give a melt and potassium hexachlorotitanate(III). Morozov and coworkers[455,456] have prepared a number of pentachlorotitanate(III) hydrates by the interaction of titanium trichloride hexahydrate and alkali-metal chloride in hydrochloric acid solution. These compounds lose water at temperatures varying from 110 to 270° to give the anhydrous compounds. The same workers[456] report that the compounds formed by dehydration have quite different properties (melting point, X-ray data, etc.) from those of the same formula prepared directly in the melt. It is not clear whether these differences are due to dimorphism or the presence of hydrolysis products.

Caesium μ-trichlorohexachlorodititanate(III). This interesting compound may be prepared by heating a 3:2 mixture of caesium chloride and titanium trichloride in a sealed tube at about 700°[457]. X-ray powder diffraction patterns show it to be isomorphous with the corresponding chromium compound, for which a complete structural analysis has been carried out (Fig. 4.14, page 418). The lattice is hexagonal with $a = 7.32$ and $c = 17.97$ Å[457]. The dinuclear anion consists of two $TiCl_6$ octahedra sharing a face. The absorption spectrum[458] shows a band assigned to the $E_g \leftarrow T_{2g}$ transition at 13,400 cm^{-1}, but rather surprisingly, there appears to have been no investigation of the magnetic properties of this compound.

Tetrachlorotitanates(III). Alkali-metal tetrachlorotitanates(III) are invariably prepared in the melt[439,442,443,447,450]. Fowles and Russ[451] isolated the compound $(C_2H_5)_4NTiCl_4 \cdot 2CH_3CN$ by the interaction of the adduct $TiCl_3 \cdot 3CH_3CN$ with $(C_2H_5)_4NCl$ in acetonitrile. Thermal decomposition of this compound gave $(C_2H_5)_4NTiCl_4$.

The reported[447] melting points for the potassium, rubidium and caesium tetrachlorotitanates(III) are 654°, 800° and 781° respectively.

In alkali-metal chloride melts the tetrachlorotitanate(III) ion shows absorptions at 8,000 and 6,000 cm^{-1} attributed to the $T_{2g} \leftarrow E_g$ transition[454]. Fowles and Russ[451] have recorded the absorption spectrum of the tetraethylammonium salt and concluded that the titanium was not in a tetrahedral configuration, but is in fact in an octahedral environment. Absorptions were observed at 18,000, 14,300 and 12,300 cm^{-1} with the band at the highest energy being of a charge-transfer nature. They suggested a polymeric chain structure with halogen bridges such as

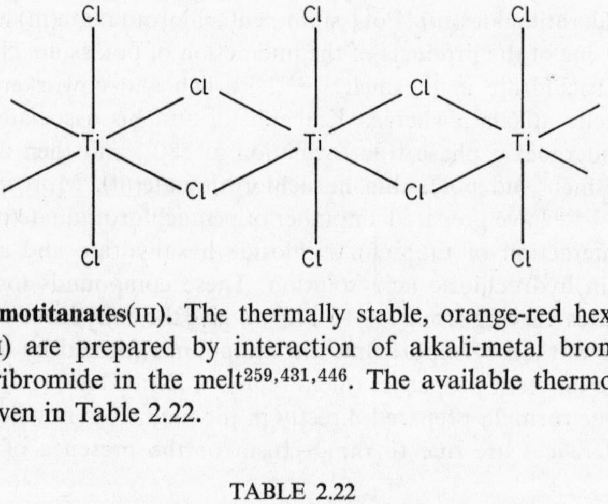

Hexabromotitanates(III). The thermally stable, orange-red hexabromotitanates(III) are prepared by interaction of alkali-metal bromide and titanium tribromide in the melt[259,431,446]. The available thermodynamic data are given in Table 2.22.

TABLE 2.22
Thermodynamic Data for Hexabromotitanates(III)

Property	Compound	Value	Reference
M.p.	K_3TiBr_6	662°	446
	Rb_3TiBr_6	664°	259
	Cs_3TiBr_6	670°	259
ΔH°_{form} (kcal mole^{-1})[a]	Rb_3TiBr_6	−11.8	362
	Cs_3TiBr_6	−14.4	362
ΔH°_{form} (kcal mole^{-1})[b]	Rb_3TiBr_6	−455.1	362
	Cs_3TiBr_6	−463.1	362

[a] From constituent bromides.
[b] From elements.

Kozhina and Korol'kov[431] indexed the X-ray powder diffraction patterns of the rubidium and caesium salts on the basis of hexagonal unit cells of space group $P6_3mc$. For the rubidium salt $a = 7.26$ and $c = 11.64$ Å, while for caesium hexabromotitanate(III) $a = 7.60$ and $c = 12.04$ Å.

μ-Tribromohexabromodititanates(III). Rubidium and caesium μ-tribromohexabromodititanates(III) are stable brown crystalline solids which are prepared in the melt[259,431]. Thermodynamic and crystallographic data for these compounds are given in Table 2.23.

TABLE 2.23
Properties of Rubidium and Caesium μ-tribromohexabromodititanates(III)

Property	$Rb_3Ti_2Br_9$	$Cs_3Ti_2Br_9$	Reference
M.p.	647	697	259
ΔH°_{form} (kcal mole^{-1})^a	−18.2	−29.4	362
ΔH°_{form} (kcal mole^{-1})^b	−606.4	−623.0	362
Lattice Symmetry	Hexagonal	Hexagonal	431
a	7.40 Å	7.58 Å	431
c	18.36 Å	18.51 Å	431

[a] From constituent bromides.
[b] From elements.

Oxidation State II

Few complex halides of titanium(II) are known, but those that are known have invariably been prepared by interaction of the constituent halides in the melt, under an inert atmosphere. The known properties of these compounds are summarized in Table 2.24.

TABLE 2.24

Properties of Complex Halides of Titanium(II)

Compound	M.P.	Other properties	Reference
K_3TiCl_5	766		459
Na_2TiCl_4		μ_{eff}(RT) = 2.4 BM	334
K_2TiCl_4	671		372
Rb_2TiCl_4	732		460
Cs_2TiCl_4	747		460
$KTiCl_3$	762		372
$RbTiCl_3$	852		460
$CsTiCl_3$	932		460
$RbTiBr_3$	840		259
		Hexagonal, $a = 7.46$, $c = 6.08$ Å	431
$CsTiBr_3$	855		259
		Hexagonal, $a = 7.64$, $c = 6.16$ Å	431

ADDENDA

Halides and Oxide Halides

Downing and Ragsdale[461] have prepared a number of titanium tetrafluoride adducts with substituted pyridines. ^{19}F NMR measurements on the 1:2 complexes with 2-chloropyridine and 2-bromopyridine indicate *cis* configurations. In addition, it was found that the stoichiometry of the adduct formed depended upon the pK_a of the ligand, strong bases favoring 1:1 complexes and weak bases 1:2 complexes.

The reaction between titanium tetrafluoride and 8-quinolinol (oxH) has been investigated in detail[462], both 1:1 and 1:2 adducts being isolated. Differential thermal analysis and thermogravimetric analysis indicate that at approximately 200° the 1:2 complex looses hydrogen fluoride to give difluoro-8-quinolinolatotitanium(IV), TiF_2ox_2. This latter compound can be sublimed at 300° and is stable to 400°. When heated with two moles of 8-quinolinol in a sealed tube at 160°, 8-quinolinolium fluoride and fluorotri-8-quinolinolatotitanium(IV) are produced. Some properties of the compounds derived from 8-quinolinol are given in Table 2.25. The

TABLE 2.25

Titanium Tetrafluoride Derivatives of 8-quinolinol

Compound	Properties
$TiF_4 \cdot oxH$	Yellow, $\nu_{Ti-F} = 610, 550$ cm^{-1}
$TiF_4 \cdot 2oxH$	Pink
TiF_2ox_2	Orange, $\nu_{Ti-F} = 627, 626$ cm^{-1}, *cis* octahedral
$TiFox_3$	Brown, $\nu_{Ti-F} = 609$ cm^{-1}, non-electrolyte, probably seven coordinate

mass spectrum of $TiF_4 \cdot oxH$ has been examined[463] at 180° and 240°. The major fragments are TiF_3ox^+ and $TiF_2ox_2^+$ at 180° and 240° respectively.

A polarographic investigation of titanium tetrachloride in dimethylsulphoxide and dimethylformamide has been reported[464] and Dahl and Johansen[465] have discussed the electronic structure of the same halide on the basis of a semi-quantitative LCAO–MO calculation.

The hydrolysis of titanium tetrachloride in non-aqueous solvents such as chloroform and carbon tetrachloride leads to a variety of complex oxide chlorides[466]. Products isolated include $Ti_3O_2Cl_8 \cdot 4H_2O$ and $Ti_6O_5Cl_{14} \cdot 13H_2O$. The infrared spectra of these compounds indicate the presence of —Ti—O—Ti—O— chains. Nabivanits and Kudritskaya[467,468] and also Davies and Long[469] have studied the hydrolysis of titanium tetrachloride in hydrochloric acid. The latter workers[469] report that the hexachlorotitanate(IV) anion is not formed, as shown by Raman spectroscopy. For a solution of titanium tetrachloride in hydrochloric acid with $[Cl^-]/[Ti] = 10.6$, the Raman spectrum showed lines at 343, 261 and 155 cm^{-1}. For a solution with $[Cl^-]/[Ti] = 7.3$ the lines occurred at 345, 259 and 161 cm^{-1} respectively[469]. The Raman spectra suggest that oxide chloride complexes of the type $TiO_2Cl_4^{4-}$ are formed[469]. However, it would appear that the hexachlorotitanate(IV) anion is formed in *fuming* hydrochloric acid[470]. Adams and Newton[470] observed three Raman-active modes for titanium tetrachloride in fuming hydrochloric acid which they assigned as follows on the basis of O_h symmetry:

$$\nu_1 = 347 \qquad \nu_2 = 275 \qquad \nu_5 = 171 \text{ cm}^{-1}$$

These frequencies are consistent with those for solid hexachlorotitanates(IV) (Table 2.33, page 90).

The titanium tetrachloride–phosphorus oxide trichloride system has been investigated again[471] and the formation of 1:1 and 1:2 adducts confirmed. Ryan and Rogers[472] have recorded the chlorine NQR spectra of both adducts, and in each case observed resonance frequencies for each of the crystallographically non-equivalent chlorine atoms bound to the phosphorus atoms. Smitskamp and coworkers[473] have studied the Raman spectrum of $TiCl_4 \cdot 2POCl_3$ in the solid and molten states, as well as in solution. The spectra clearly show that in solution, and also in the molten state, partial dissociation takes place.

Further infrared studies have been reported on titanium tetrachloride adducts with oxygen donor ligands. For example, Shvets and coworkers[474] have isolated 1:1 and 1:2 adducts with esters of β-phosphino carbonyl compounds. For the 1:1 compounds, the ligand acts as a chelate, being coordinated to the titanium atom by the P=O and C=O groups. In the 1:2 complexes, coordination is via the P=O group. Susz and associates[475,476] report that the carbonyl frequencies of aromatic carboxylic acids and *p*-substituted benzoyl chlorides are lowered by ca 150 cm^{-1} on formation of 1:1 titanium tetrachloride adducts.

Frazer and Rimmer[462] have published further work on the reaction between titanium tetrachloride and 8-quinolinol. The reported reaction scheme is on the next page.

The titanium–chlorine stretching frequencies of several of these complexes were reported[462]. Studd and Swallow[477] have published the full crystallographic details for $TiCl_2ox_2$ (Figure 2.5, page 48). $TiCl_2ox_2$ is hydrolytically stable and there is no appreciable chloride-ion exchange with tetraethylammonium chloride[478]. However $TiCl_2acac_2$ undergoes fairly rapid exchange[478], $t_{1/2}$ being 11 hr for a 0.1M solution in chloroform at 25°.

Titanium tetrachloride forms 1:1 and 1:2 adducts with dimethylglyoxime and the two isomers of benzil monoxime[479]. The thermal decomposition of these complexes was examined[479].

Raman spectroscopic studies[469] of the titanium tetrachloride-tri-n-butyl phosphate system indicate the formation of 1:1 and 1:2 adducts.

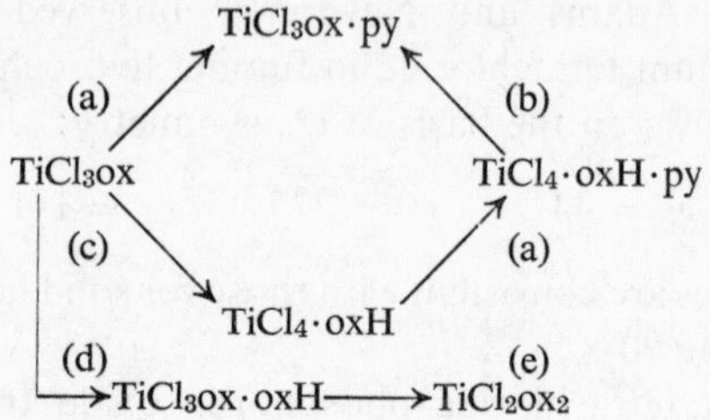

oxH = 8-quinolinol, ox = 8-quinolinolato, py = pyridine, (a) = py, (b) = reflux in $CHCl_3$, (c) = HCl, (d) = oxH, (e) = heat at 200°.

The spectra indicate that the 1:1 compound is of C_{2v} symmetry while the 1:2 complex has a *cis* octahedral configuration.

Titanium tetrahalides (Cl, Br, I) form 1:1 adducts with 3,3-dimethylacetylacetone which appear to have *cis* octahedral structures[480]. Both adducts and substitution compounds have been isolated with the tetrahalides and acetoacetanilide[481].

Rivest and coworkers[482] have prepared a number of 1:1 and 1:2 adducts of titanium tetrachloride with cyclic thioethers. Infrared spectra indicate that the 1:1 compounds are dimeric with halogen bridges, and that the 1:2 compounds have the expected *cis* octahedral structure.

Kyker and Schram[483] have reported the preparation of a very interesting and rather unusual 2:3 adduct of titanium tetrachloride with bis(dimethylamino)chloroborane, $[(CH_3)_2N]_2BCl$. The yellow crystalline solid may be made by direct interaction of the constituents, or by reaction with tetrakis(dimethylamino)diborane(4). It is thermally unstable in the gas phase, decomposing to its constituents. Spectroscopic studies (infrared, ultraviolet and ^1H NMR) indicate that in benzene and dichloromethane solutions, decomposition occurs to give titanium tetrachloride, bis(dimethylamino)chloroborane and several unidentified intermediate complexes. A detailed examination[484] of the infrared spectrum of the 2:3 complex in the solid state clearly showed that the chloroborane ligand utilizes both nitrogen atoms to bridge between two titanium tetrachloride groups. Three possible structures for the complex were suggested[483] on the basis of its physical properties. Unfortunately the infrared study[484] did not allow an unambiguous decision on the structure.

Titanium tetrachloride also reacts[483] with bis(dimethylamino)methylborane and tris(dimethylamino)borane. The titanium-containing products are $[(CH_3)_2N]_2TiCl_2$ and $(CH_3)_2NTiCl_3$ from the latter and $(CH_3)_2NTiCl_3$ from the former. The infrared spectral assignments for the trichloro complex were reported[483].

Titanium tetrachloride reacts readily with trimethylsilicon azide in dichloromethane to give $TiCl_2(N_3)_2$[485]. A detailed infrared examination of this complex indicates it to be an azide bridged polymer. The infrared spectrum was assigned as shown in Table 2.26.

TABLE 2.26

Infrared Spectrum (cm^{-1}) of $TiCl_2(N_3)_2$

Mode	Frequency	Mode	Frequency
νN_3 asym	2132	γ/δ ring (B_{2u})	488
νN_3 sym	1240, 1229	γ/δ ring (B_{3u})	455
γN_3	669, 657	$\nu TiCl_2$ asym	415
δN_3	563	$\nu TiCl_2$ sym	372

Haase and Hoppe[486] have reported that $TiCl_2(OC_2H_5)_2$ has a monoclinic unit cell with $a = 5.91$, $b = 10.91$, $c = 13.90$ Å and $\beta = 94.7°$. There are four molecules in the cell of space group $P2_1/c$. The structure of this complex is essentially the same as that of the corresponding phenoxy derivative (Figure 2.8, page 52). That is, the compound is dimeric with oxygen bridging groups, the titanium atoms being in a trigonal-bipyramidal environment. The important internuclear dimensions are given in Table 2.27.

TABLE 2.27

Internuclear Dimensions in $[TiCl_2(OC_2H_5)_2]_2$

Distance	Value (Å)	Angle	Value
Ti—Cl	2.19 and 2.20	Ti—O—Ti	109.0°
Ti—O	1.96 (bridge)	Cl—Ti—Cl	110.5°
	1.77 (terminal)		

The preparation and characterization of a number of other adducts of titanium tetrachloride have been reported[482,487–497]. Donor ligands include acrylonitrile, acetylacetone, substituted dioxans, phenols and thioxans.

Zavaritskaya and Podgaiskaya[498] have investigated the thermal decomposition of titanium dioxide dichloride while Fowles and co-workers[499] have prepared and characterized a number of adducts. Complexes of the type $TiOCl_2L_2$ are formed with acetonitrile, trimethylamine, tetrahydrofuran and tetrahydropyran. Adducts of the type $TiOCl_2L_n$ are formed with pyridine ($n = 2.5$), ethyleneglycol dimethyl ether ($n = 1$) and 1,4-dioxan ($n = 1.5$). The dioxan adduct may be used as an intermediate for the preparation of $TiOCl_2$(bipy) and $TiOCl_2(2\text{-}CH_3py)$. The trimethylamine adduct is monomeric in boiling acetonitrile, pyridine and chloroform, has a titanium–oxygen double bond stretching frequency of 976 cm^{-1} and has an orthorhombic unit cell with $a = 9.89$, $b = 10.81$ and $c = 13.50$ Å. It is isostructural with $VOCl_2 \cdot 2(CH_3)_3N$ which is known[500] to be trigonal-bipyramidal. Of the remaining complexes, only that of 2-methylpyridine has a terminal oxygen atom, as shown by infrared spectroscopy.

Further details on the preparation of titanium tetrabromide by halogen exchange between titanium tetrachloride and hydrogen bromide have been reported[501]. Titanium tetrabromide may also be prepared by refluxing the dioxide in boron tribromide[502]. Several adducts and substitution complexes of titanium tetrabromide have been investigated[462,480–482,490,495].

The preparation of titanium tetraiodide by the interaction of the elements in a sealed tube at 400–425° and its subsequent purification have been discussed by Lowry and Fay[503]. 8-Quinolinol has been reacted with titanium tetraiodide to give adducts and substitution compounds which are similar to the corresponding fluoro, chloro and bromo derivatives[462].

The $TiCl_3$-HCl-H_2O system has been examined by Karpinskaya and Andreev[504]. Over the range 0.1–10M HCl the species present were reported to be $Ti(H_2O)_6^{3+}$, $TiCl(H_2O)_5^{2+}$ and $TiCl_2(H_2O)_4^{+}$. For the HCl range 12–19M, $TiCl_5(H_2O)^{2-}$ is the predominant species. The polarographic properties of titanium trichloride in dimethylsulphoxide and dimethylformamide have been investigated[464].

There has been a continuing interest in the preparation and characterization of adducts of the trichloride. A number of these adducts are of the 1:2 stoichiometry and magnetic, spectral and X-ray diffraction studies show that both five- and six-coordinate complexes have been isolated.

For example, Chuchalin and coworkers[505] report that $TiCl_3(Bu_3PO_4)_2$ has an irregular trigonal-bipyramidal configuration, while Fowles and associates[506,507] have shown that the 1:2 adducts with dimethyl sulphide and tetrahydrothiophen are dimers, probably with a direct metal-metal bond. Wood[508] has published a paper on ligand-field theory to account for the electronic spectrum of the five-coordinate species $TiX_3 \cdot 2(CH_3)_3N$ (X = Cl, Br). The electronic spectra of $TiX_3 \cdot 2(CH_3)_3N$ (X = Cl, Br) and of $TiX_3 \cdot$ 2(2-methylpyridine) (X = Cl, Br, I) have been assigned[507] on the basis of the above ligand-field treatment. Some properties of a number of adducts and substitution complexes of titanium trichloride are noted in Table 2.28.

TABLE 2.28

Complexes of Titanium Trichloride

Compound	Properties	Reference
$TiCl_3L_3$	L = acetonitrile, μ = 1.71 BM	509
$TiCl_3L_3$	L = ethylenediamine, μ = 0.93 BM	509
$TiCl_3L_3$	L = 4-methylpyridine, μ = 1.74 BM	510
$TiCl_3L_2$	L = dimethyl sulphide, brown, μ = 1.03 BM	506
$TiCl_3L_2$	L = tetrahydrothiophen, grey, μ = 1.12 BM	506
$TiCl_3L_2$	L = nicotinamide, μ = 2.08 BM	509
$TiCl_3L$	L = dioxan, μ = 1.77 BM	510
$TiCl_3L$	L = thioxan, μ = 1.70 BM	510
$TiCl_2$(acac)	acac = acetylacetonato, red, m.p. = 270°, μ = 1.70 BM, ν_{Ti-O} = 468 cm^{-1}, tetrahedral	493
TiCl(acac)$_2$	acac = acetylacetonato, red-violet, m.p. = 213–214°, μ = 1.39 BM, ν_{Ti-O} = 452 cm^{-1}, dimer, octahedral	493

TABLE 2.29

Titanium Tribromide–Bipyridyl Complexes

Formula	Structure	μ (BM)
$TiBr_3 \cdot$bipy	$[TiBr_3 \cdot bipy]_2$	1.88
$TiBr_3 \cdot$2bipy	$[TiBr_2(bipy)_2]^+Br^-$	
$TiBr_3 \cdot 2bipy \cdot CH_3CN$	$[TiBr_2(bipy)_2]^+Br^- \cdot CH_3CH$	2.06
$TiBr_3 \cdot 2bipy \cdot CHCl_3$	$[TiBr_2(bipy)_2]^+Br^- \cdot CHCl_3$	1.85
$2TiBr_3 \cdot 3bipy$	$[TiBr_2(bipy)_2]^+[TiBr_4(bipy)]^-$	1.72

The reaction between titanium tribromide and bipyridyl (bipy) has been shown to give a number of complexes depending upon experimental conditions[511]. The stoichiometry, suggested structures and magnetic moments of the various compounds are given in Table 2.29.

The preparation and characterization of a number of other titanium tribromide adducts have been reported[490,506,507,510]. The magnetic properties of several of these complexes are given in Table 2.30.

TABLE 2.30
Magnetic Properties of Titanium Tribromide Adducts

Compound	L	μ (BM)	Reference
$TiBr_3L_4$	Ethylenediamine	1.58	510
$TiBr_3L_2$	Trimethylamine	1.71	510
$TiBr_3L_2$	Dimethyl sulphide	1.67	506
$TiBr_3L_2$	Tetrahydrothiophen	1.72	506

Titanium triiodide may be prepared by reducing the tetraiodide with aluminium powder at 280° in a sealed tube[512]. Fowles and co-workers[506,507,512] have made the first detailed study of the chemistry of titanium triiodide. It was found to react with nitrogen, oxygen and sulphur donor ligands to give adducts which have been characterized by the normal physico-chemical techniques. Some properties of a number of these red-black adducts are given in Table 2.31.

TABLE 2.31
Adducts of Titanium Triiodide

Formula	L	Structure	μ (BM)
TiI_3L_2	Trimethylamine	Monomeric	
TiI_3L_3	Pyridine	Monomeric	1.69
TiI_3L_2	2-methylpyridine	$[TiI_3L_2]_2$	1.65
TiI_3L_3	4-methylpyridine	Monomeric	1.89
TiI_3L_4	Methyl cyanide	$[TiI_2L_4]I$	1.62
$(TiI_3)_2L_3$	Bipyridyl	$[TiI_2L_2]^+[TiI_4L]^-$	1.39
TiI_3L_2	1,10-phenanthroline	$[TiI_2L_2]I$	1.39
TiI_3L_3	Tetrahydrofuran	Monomeric	1.64
TiI_3L_3	Dioxan	Monomeric	1.64
TiI_3L_2	Dimethyl sulphide	$[TiI_3L_2]_2$	1.68

The heat of formation of titanium dichloride has been reported[513] to be -121.3 kcal $mole^{-1}$, while the titanium–chlorine stretching frequency occurs[514] at about 290 cm^{-1}. It reacts readily with methyl cyanide under reflux conditions[514] to give black $TiCl_2(CH_3CN)_2$. The same adduct, which may be used as a starting material for other complexes, is also formed by the reaction between the ligand and sodium tetrachlorotitanate(II)[514]. Some properties of several of the titanium dichloride adducts are given in Table 2.32. Fowles and coworkers[514] suggested that these compounds were halogen-bridged polymers and they discussed possible structures.

TABLE 2.32

Adducts of Titanium Dichloride

Compound	Colour	μ (BM)	ν_{Ti-Cl} (cm^{-1})
$TiCl_2$(methyl cyanide)$_2$	Black	1.07	300, 260
$TiCl_2$(1,10-phenanthroline)	Dark blue	1.11	336, 317, 298
$TiCl_2$(bipyridyl)	Dark blue	1.09	350, 308
$TiCl_2$(pyridine)$_2$	Black	1.14	
$TiCl_2$(tetrahydrofuran)	Black	1.21	
$TiCl_2$(tetrahydropyran)	Black	0.99	

Titanium dibromide reacts with a number of nitrogen donor molecules to give adducts similar to those described above for the dichloride[514].

Complex Halides and Oxide Halides

Potassium hexafluorotitanate(IV) has been prepared[515] by the reaction between aqueous solutions of titanium sulphate and potassium fluoride at 70°. The hydrolysis of a number of hexafluorotitanates(IV) has been studied[516]. In alkaline media a number of complex oxide fluorides are formed.

Ehrlich and coworkers[517] have reinvestigated the preparation of hexachlorotitanates(IV) by the interaction of the constituent chlorides in a bomb at 550–750°. This method is not suitable for the preparation of the lithium and sodium salts. Quaternary ammonium hexachlorotitanates(IV) have been prepared by the interaction of the appropriate chlorides in thionyl chloride[518] and dichloromethane[519]. Impure hexachlorotitanates(IV) have been prepared[520] by the action of carbon tetrachloride on mixtures of alkali-metal titanates and titanium dioxide in a bomb at 400°.

A new determination[521] of the unit cell of caesium hexachlorotitanate(IV) gives $a = 10.23$ Å.

The infrared and Raman spectra of several hexachlorotitanates(IV) have been reported, as shown in Table 2.33.

Tetraethylammonium pentachlorotitanate(IV) has been prepared[518] by the interaction of a 1:1 mixture of the constituent chlorides in dichloromethane. The far infrared spectrum of this yellow complex suggests D_{3h} symmetry. The following fundamentals were assigned.

$$\nu_3 = 346 \qquad \nu_4 = 170 \qquad \nu_5 = 385 \qquad \nu_6 = 212\ \text{cm}^{-1}$$

By reaction of a 2:1 mixture of titanium tetrachloride and tetraethylammonium chloride in dichloromethane, yellow tetraethylammonium

TABLE 2.33

Infrared and Raman Spectra (cm^{-1}) of Some Hexachlorotitanates(IV)

Mode	Description	$(NH_4)_2TiCl_6$[470]	$(NMe_4)_2TiCl_6$[518]	$(NEt_4)_2TiCl_6$[518]	$(NEt)_2TiCl_6$[519]
ν_1	a_{1g}	331			320
ν_2	e_g	284			271
ν_3	t_{1u}	330	324	321	316
ν_4	t_{1u}	193	183	182	183
ν_5	t_{2g}	194			173

μ-trichlorohexachlorodititanate(IV) has been prepared[518]. The infrared spectrum (cm^{-1}) of this compound was assigned as follows:

Terminal Ti—Cl stretch	416, 379
Bridging Ti—Cl stretch	268, 230
Terminal bend	188, 171
Bridge bending or lattice mode	74, 57

Fowles, Lewis and Walton[499] have prepared yellow tetramethylammonium oxopentachlorotitanate(IV) by refluxing a mixture of the quaternary ammonium chloride and titanium oxide dichloride in dichloromethane. The complex has a metal-oxygen stretching frequency of 968 cm^{-1} suggesting the presence of a terminal oxygen group. The titanium-chlorine stretching frequency occurs at 325 cm^{-1}. The interaction of titanium oxide dichloride and the chlorides of the cations tetraethylammonium, tetraphenylarsonium, di-n-butylammonium and n-butylammonium, in methyl cyanide or methyl cyanide—dichloromethane mixtures led to the formation of the corresponding oxotetrachlorotitanates(IV)[499]. Pyridinium oxotetrachlorotitanate(IV) has been reported[522] to be one of the products of the hydrolysis of titanium tetrachloride in pyridine-chloroform solutions. Infrared spectra[499] indicate that the tetraethylammonium and tetraphenylarsonium salts contain terminal oxygen groups. For the tetraethylammonium complex this has been shown to be correct by a single crystal study by Haase and Hoppe[523]. The unit cell is tetragonal with $a = 9.90$ and $c = 12.34$ Å, contains two formula units and is of space group $P4_2nm$. The anion is in the form of a tetragonal pyramid with titanium-chlorine distances of 2.32 and 2.34 Å. The titanium-oxygen distance is 1.79 Å while the oxygen-titanium-chlorine angle is 102°. The titanium-oxygen stretching frequencies of the n-butylammonium and di-n-butylammonium oxotetrachlorotitanates(IV) occur at 815 and 890 cm^{-1} respectively, indicating a bridging environment[499].

Wendling and de Lavillandre[521] have prepared the monohydrates of rubidium and caesium peroxotetrachlorotitanate(IV) by the interaction of the constituent chlorides in hydrochloric acid-hydrogen peroxide solutions. Both complexes are orthorhombic, the unit-cell parameters (Å) being:

	a	b	c
$Rb_2Ti(O_2)Cl_4 \cdot H_2O$	14.32	13.74	10.053
$Cs_2Ti(O_2)Cl_4 \cdot H_2O$	14.65	14.282	10.29

The relationship between the orthorhombic cell and the simple cubic cell of the hexachlorotitanates(IV) was also discussed[521]. The electronic spectrum of the hydrated peroxotetrachlorotitanate(IV) anion has been investigated by Wendling[524].

The infrared and Raman spectra of tetraethylammonium hexabromotitanate(IV), prepared by interaction of the appropriate bromides in dichloromethane, have been recorded and the following assignments were reported[519].

$$\nu_1 = 192 \qquad \nu_3 = 244 \qquad \nu_4 = 119 \qquad \nu_5 = 115 \text{ cm}^{-1}$$

Clark and coworkers[519] have prepared several mixed hexahalotitanates(IV) of the tetraethylammonium cation by reacting the appropriate halides in dichloromethane solution. The orange to brown crystalline solids are extremely hygroscopic. The infrared and Raman absorptions, shown in Table 2.34, were assigned on the basis of a *cis* octahedral configuration for the anion.

Burkert, Fritz and Stefaniak[525] have reported the far infrared and ^{19}F NMR spectra and the magnetic moment of K_2NaTiF_6, which they prepared by adding a 2:1 mole ratio of potassium and sodium fluoride to titanium trichloride in hydrogen fluoride. These studies show that the hexafluorotitanate(III) anion is of D_{4h} symmetry, that is, it is a tetragonally distorted octahedron. The room temperature magnetic moment is 1.79 BM and the titanium–fluorine vibrational modes occur at 552, 515, 506, 294, 235 and 226 cm^{-1}.

The melting points of rubidium and caesium hexachlorotitanate(III) are reported[526] to be 779° and 760° respectively. The magnetic moment of pyridinium hexachlorotitanate(III) has been measured over a wide temperature range[510], the magnetic moment falling from 1.58 BM at 301°K to 1.21 BM at 77°K. The behaviour conforms to the Curie-Weiss law, the Weiss constant being 80°.

Rubidium and caesium μ-trichlorohexachloroditıtanate(III) have been prepared[526,527] by the interaction of the constituent chlorides in sealed tubes at elevated temperatures. The melting points of the two complexes

TABLE 2.34
Infrared (IR) and Raman (R) Spectra (cm^{-1}) of $TiX_4Y_2^{2-}$

Anion	Method	ν_{Ti-Cl}	ν_{Ti-Br}	ν_{Ti-I}
$TiCl_4Br_2^{2-}$	R	316, 300	212	
	IR	344, 312, 283	240	
$TiBr_4Cl_2^{2-}$	R	318	214	
	IR	344, 313	240	
$TiCl_4I_2^{2-}$	IR	331, 280		192 ?

are 671° and 705° respectively[526]. The magnetic behaviour of the caesium salt does not obey the Curie-Weiss law, the effective magnetic moment being about 1.3 BM at room temperature. Single-crystal electronic spectra for this complex have been recorded[527], the single absorption occurring at 13,300 and 12,600 cm^{-1} for parallel and perpendicular polarization respectively.

The magnetic moment of $(C_2H_5)_4NTiCl_4 \cdot 2CH_3CH$[510] falls from 1.76 BM at 300°K to 1.53 BM at 77°K, the Curie-Weiss law being obeyed with a Weiss constant of 30°.

Tetra- and trichlorotitanates(II) have been prepared[513] by fusing the appropriate amounts of the constituent chlorides in an inert atmosphere. The heats of formation (kcal mole^{-1}) of these complexes are:

$RbTiCl_3$	−246.6	Rb_2TiCl_4	−353.5
$CsTiCl_3$	−245.7	Rb_2TiCl_4	−354.4

The titanium–chlorine stretching frequencies in sodium tetrachlorotitanate(II) occur[514] at 300 and 260 cm^{-1}.

REFERENCES

1. R. J. H. Clark, *The Chemistry of Titanium and Vanadium*, Elsevier, Amsterdam, 1968.
2. H. M. Haendler, S. F. Bartram, R. S. Becker, W. J. Bernard and S. W. Bukata, *J. Am. Chem. Soc.*, **76,** 2177 (1954).
3. M. J. Steindler, D. V. Steidl and R. K. Steunenberg, *U.S.A.E.C. Report*, ANL 6002 (1959).
4. E. Greenberg, J. L. Settle and W. N. Hubbard, *J. Phys. Chem.*, **66,** 1345 (1962).
5. R. W. Murray and H. M. Haendler, *J. Inorg. Nucl. Chem.*, **14,** 135 (1960).
6. S. Hartmann, *Z. anorg. allgem. Chem.*, **155,** 355 (1926).
7. E. E. Aynsley, G. Hetherington and P. L. Robinson, *J. Chem. Soc.*, **1954,** 1119.
8. G. Brauer, *Handbook of Preparative Inorganic Chemistry*, Vol. 1, Academic Press, New York, 1963.
9. Y. A. Buslaev, D. S. Dyer and R. O. Ragsdale, *Inorg. Chem.*, **6,** 2208 (1967).
10. Y. A. Buslaev and V. A. Shcherbakov, *Dokl. Akad. Nauk SSSR*, **170,** 845 (1966).
11. E. H. Hall, J. M. Blocher and I. E. Campbell, *J. Electrochem. Soc.*, **105,** 275 (1958).
12. R. D. Euler and E. F. Westrum, *J. Phys. Chem.*, **65,** 132 (1961).
13. G. S. Rao, *Z. anorg. allgem. Chem.*, **304,** 176 (1960).
14. G. S. Rao, *Z. anorg. allgem. Chem.*, **304,** 351 (1960).
15. G. S. Rao, *Z. Naturforsch.*, **14b,** 689 (1959).
16. H. J. Emeleus and G. S. Rao, *J. Chem. Soc.*, **1958,** 4245.
17. R. C. Aggarwal and D. S. Bhusri, *Indian J. Chem.*, **4,** 19 (1966).
18 E. L. Muetterties, *J. Am. Chem. Soc.*, **82,** 1082 (1960).
19. D. S. Dyer and R. O. Ragsdale, *J. Phys. Chem.*, **71,** 2309 (1967).
20. D. S. Dyer and R. O. Ragsdale, *Inorg. Chem.*, **6,** 8 (1967).
21. D. S. Dyer and R. O. Ragsdale, *Chem. Commun.*, **1966,** 601.
22. D. S. Dyer and R. O. Ragsdale, *J. Am. Chem. Soc.*, **89,** 1528 (1967).
23. C. J. Liebenberg and F. W. G. Schoning, *J. South African Chem. Institute*, **20,** 57 (1967).
24. R. J. H. Clark, W. Errington, J. Lewis and R. S. Nyholm, *J. Chem. Soc.*, **1966,** A, 989.
25. G. A. Barclay, I. K. Gregor, M. J. Lambert and S. B. Wild, *Australian J. Chem.*, **20,** 1571 (1967).
26. I. Douek, M J. Frazer, Z. Goffer, M. Goldstein, B. Rimmer and H. A. Willis, *Spectrochim. Acta*, **23A,** 373 (1967).
27. R. J. H. Clark and W. Errington, *J. Chem. Soc.*, **1967,** A, 258.
28. R. C. Fay and R. N. Lowrey, *Inorg. Chem.*, **6,** 1512 (1967).
29. F. E. Dickinson, E. W. Gowling and F. F. Bentley, *Inorg. Chem.*, **6,** 1099 (1967).
30. K. S. Vorres and F. B. Dutton, *J. Am. Chem. Soc.*, **77,** 2019 (1955).
31. K. Dehnicke, *Naturwissenschaften*, **52,** 660 (1965).
32. G. Lange and K. Dehnicke, *Naturwissenschften*, **53,** 38 (1966).
33. K. S. Vorres and J. Donohue, *Acta Cryst.*, **8,** 25 (1955).
34. A. P. Buntin and Y. I. Ivashentsev, *Uchenye Zapiski Tomsk., Gos. Univ. im V. V. Kuibysheva*, **1959,** 137.

35. Y. Saeki and K. Funaki, *Kogyo Kagaku Zasshi*, **60,** 403 (1957).
36. A. V. Pamfilov and E. G. Shtandel, *J. Gen. Chem. USSR.*, **7,** 258 (1937).
37. J. Rakosnick, *Hutnicke Listy*, **9,** 268 (1954).
38. G. V. Seryakov, S. A. Vaks, V. V. Zheltova and E. P. Strashun, *Russ. J. Inorg. Chem.*, **12,** 3 (1967).
39. G. Nakazawa and Y. Okahara, *Nippon Kogyo Kaishi*, **74,** 183 (1958).
40. Y. I. Ivashentsev, *Dokl. 7–i Nauch. Konf. Posvyashchen 40–letiya Velikoi Oktyabr. Sots. Revolyutsii, Tomsk. Univ.*, **1957,** 157.
41. *British Patent*, 801, 424 (1958).
42. R. C. Schreyer, *J. Am. Chem. Soc.*, **80,** 3483 (1958).
43. A. B. Bardawil, F. N. Collier and S. Y. Tyree, *J. Less-Common Metals*, **9,** 20 (1965).
44. H. Schafer and W. Fuhr, *Z. anorg. allgem. Chem.*, **319,** 52 (1962).
45. G. P. Luchinskii, *Russ. J. Phys. Chem.*, **40,** 318 (1966).
46. K. Arii, *Science Repts. Tokyo Imp. Univ., First Ser.*, **22,** 182 (1933).
47. K. Arii, *Bull. Inst. Phys. Chem. Research (Tokyo)*, **8,** 714 (1929).
48. H. Schafer and F. Zeppernick, *Z. anorg. allgem. Chem.*, **272,** 274 (1953).
49. G. V. Seryakov, S. A. Vaks and L. S. Sidorina, *Titan. i Ego Splavy, Akad. Nauk SSSR Inst. Met.*, **1961,** 220.
50. W. M. Latimer, *J. Am. Chem. Soc.*, **44,** 90 (1922).
51. B. A. Voitovich and A. S. Barabanova, *Ukr. Khim. Zh.*, **29,** 1264 (1963).
52. Y. Saeki and K. Funaki, *Nippon Kagaku Zasshi*, **78,** 754 (1957).
53. F. P. Pike and C. T. Foster, *J. Chem. Eng. Data*, **4,** 305 (1959).
54. W. C. Schumb and R. F. Sundstrom, *J. Am. Chem. Soc.*, **55,** 596 (1933).
55. T. A. Zavaritskaya and N. I. Delarova, *Tr., Vses. Nauchn.-Issled. Alyumin.-Magnievyi Inst.*, **1966,** 96.
56. D. Altman, M. Farber and D. M. Mason, *J. Chem. Phys.*, **25,** 531 (1956).
57. P. Gross, C. Hayman and D. L. Levi, *Trans. Faraday Soc.*, **51,** 626 (1955).
58. P. Gross, C. Hayman and D. L. Levi, *Trans. Faraday Soc.*, **53,** 1601 (1957).
59. B. S. Sanderson and G. W. MacWood, *J. Phys. Chem.*, **60,** 316 (1956).
60. M. Farber and A. J. Darnell, *J. Chem. Phys.*, **23,** 1460 (1955).
61. G. B. Skinner and R. A. Ruehrwein, *J. Phys. Chem.*, **59,** 113 (1955).
62. W. H. Johnson, R. A. Nelson and E. J. Prosen, *J. Res. Nat. Bur. Std.*, **62,** 49 (1959).
63. N. N. Ruban and V. D. Ponomarev, *Tr. Inst. Met. i Obogashch., Akad. Nauk Kaz. SSR*, **4,** 19 (1962).
64. R. C. Herman, *J. Chem. Phys.*, **6,** 406 (1938).
65. N. J. Hawkins and D. R. Carpenter, *J. Chem. Phys.*, **23,** 1700 (1955).
66. P. Brand and H. Sackmann, *Z. anorg. allgem. Chem.*, **321,** 262 (1963).
67. A. H. Reddoch, *J. Chem. Phys.*, **35,** 1085 (1961).
68. H. G. Dehmelt, *J. Chem. Phys.*, **21,** 380 (1953).
69. R. P. Hamlen and W. S. Koski, *J. Chem. Phys.*, **25,** 360 (1956).
70. M. Kimura, K. Kimura, M. Aoki and S. Shibata, *Bull. Chem. Soc. Japan*, **29,** 95 (1956).
71. M. W. Lister and L. E. Sutton, *Trans. Faraday Soc.*, **37,** 393 (1941).
72. R. Wierl, *Ann. Physik.*, **8,** 521 (1931).
73. B. Moszynska, *Bull. Acad. Polon. Sci., Classe III*, **5,** 819 (1957).
74. B. Moszynska, *Bull. Acad. Polon, Sci., Ser. Sci., Math. Astron. Phys.*, **7,** 455 (1959).

75. M. F. A. Dove, J. A. Creighton and L. A. Woodward, *Spectrochim. Acta*, **18,** 267 (1962).
76. Y. Morino and H. Uehara, *J. Chem. Phys.*, **45,** 4543 (1966).
77. M. E. P. Rumpf, *Compt. Rend.*, **202,** 950 (1936).
78. I. Shiihara, W. T. Schwartz and H. W. Post, *Chem. Rev.*, **61,** 1 (1961).
79. A. K. Anagnostopoulos, *Chim. Chronika* (*Athens*), **31,** 159 (1966).
80. D. C. Bradley, *Progress in Inorganic Chemistry*, Vol. 2, Interscience, New York, 1960.
81. R. J. H. Clark, J. Lewis, R. S. Nyholm, P. Pauling and G. B. Robertson, *Nature*, **192,** 222 (1961).
82. R. J. H. Clark, J. Lewis and R. S. Nyholm, *J. Chem. Soc.*, **1962,** 2460.
83. R. J. H. Clark, R. H. U. Negrotti and R. S. Nyholm, *Chem. Commun.*, **1966,** 486.
84. B. F. Markov, B. A. Voitovich and A. S. Barabanova, *Russ. J. Inorg. Chem.*, **6,** 616 (1961).
85. V. Gutmann, *Z. anorg. allgem. Chem.*, **269,** 279 (1952).
86. B. A. Voitovich, *Russ. J. Inorg. Chem.*, **5,** 965 (1960).
87. J. C. Sheldon and S. Y. Tyree, *J. Am. Chem. Soc.*, **80,** 4775 (1958).
88. W. L. Groeneveld, J. W. van Spronsen and H. W. Kourvenhoven, *Rec. Trav. Chim.*, **72,** 950 (1953).
89. V. Gutmann, *Z. anorg. allgem. Chem.*, **270,** 179 (1952).
90 V. Gutmann and F. Mairinger, *Monatsh. Chem.*, **89,** 724 (1958).
91. B. A. Voitovich, A. S. Barabanova, V. P. Klochkov and E. V. Sharkina, *Ukr. Khim. Zh.*, **32,** 167 (1966).
92. D. I. Tsekhovol'skaya, *Russ. J. Inorg. Chem.*, **9,** 755 (1964).
93. D. S. Payne, *Rec. Trav. Chim.*, **75,** 620 (1956).
94. I. Lindqvist, *Acta Chem. Scand.*, **12,** 135 (1958).
95. C. I. Branden and I. Lindqvist, *Acta Chem. Scand.*, **14,** 726 (1960).
96. C. I. Branden, *Acta Chem. Scand.*, **16,** 1806 (1962).
97. B. A. Voitovich, *Titan. i Ego Splavy, Akad. Nauk SSSR, Inst. Met.*, **1961,** 188.
98. A. S. Barabanova and B. A. Voitovich, *Ukr. Khim. Zh.*, **31,** 352 (1965).
99. M. Becke-Goehring and A. Slawisch, *Z. anorg. allgem. Chem.*, **346,** 295 (1966).
100. B. A. Voitovich and N. F. Lozovskaya, *Ukr. Khim. Zh.*, **31,** 1136 (1965).
101. J. C. Sheldon and S. Y. Tyree, *J. Am. Chem. Soc.*, **81,** 2290 (1959).
102. V. Gutmann, *Rec. Trav. Chim.*, **75,** 605 (1956).
103. V. Gutmann, *J. Phys. Chem.*, **63,** 378 (1959).
104. D. W. Meek and R. S. Drago, *J. Am. Chem. Soc.*, **83,** 4322 (1961).
105. C. I. Branden, *Acta Chem. Scand.*, **17,** 759 (1963).
106. G. R. Badgley and R. L. Livingstone, *J. Am. Chem. Soc.*, **76,** 261 (1954).
107. Y. B. Kletenik and O. A. Osipov, *J. Gen. Chem. USSR*, **31,** 651 (1961).
108. Y. A. Lysenko and O. A. Osipov, *Zh. Neorg. Khim.*, **6,** 1656 (1961).
109. R. Aubin and R. Rivest, *Can. J. Chem.*, **36,** 915 (1958).
110. Y. A. Lysenko, *Zh. Obshch. Khim.*, **26,** 2963 (1956).
111. Y. A. Lysenko, O. A. Osipov and E. K. Akopov, *Zh. Neorg. Khim.*, **1,** 536 (1956).
112. Y. A. Lysenko and O. A. Osipov, *Zh. Obshch. Khim.*, **24,** 53 (1954).
113. O. A. Osipov and E. E. Kravtsov, *Stornik Statei Obshch. Khim., Akad. Nauk SSSR.*, **1,** 216 (1953).

114. O. A. Osipov, Y. A. Lysenko and E. K. Akopov, *J. Gen. Chem. USSR*, **25,** 233 (1955).
115. Y. A. Lysenko, O. A. Osipov and N. N. Feodos'ev, *Zh. Fiz. Khim.*, **28,** 700 (1954).
116. Y. N. Vol'nov, P. M. Glezer and I. Y. Raikina, *Stornik Statei Obshch. Khim., Akad. Nauk SSSR.*, **2,** 976 (1953).
117. O. A. Osipov and V. Suchkov, *Zh. Obshch. Khim.*, **22,** 1132 (1952).
118. M. F. Lappert, *J. Chem. Soc.*, **1962,** 542.
119. S. C. Jain and R. Rivest, *Can. J. Chem.*, **42,** 1079 (1964).
120. E. Rivet, R. Aubin and R. Rivest, *Can. J. Chem.*, **39,** 2343 (1961).
121. D. C. Bradley, D. C. Hancock and W. Wardlaw, *J. Chem. Soc.*, **1952,** 2773.
122. Y. A. Lysenko, O. A. Osipov and E. E. Kravtsov, *Russ. J. Inorg. Chem.*, **8,** 337 (1963).
123 V I. Gaivoronskii and O. A. Osipov, *Zh. Obshch. Khim.*, **33,** 2901 (1963).
124. B. Mori, J. Gohring, D. Cassimatis and B. P. Susz, *Helv. Chim. Acta*, **45,** 77 (1962).
125. O. A. Osipov and A. D. Semenov, *Zh. Obshch. Khim.*, **25,** 2059 (1955).
126. J. Gohring and B. P. Susz, *Helv. Chim. Acta*, **49,** 486 (1966).
127. Y. A. Lysenko, *Zh. Obshch. Khim.*, **37,** 11 (1967).
128. L. Brun, *Acta Cryst.*, **20,** 739 (1966).
129. B. P. Susz and A. Lachavanne, *Helv. Chim. Acta*, **41,** 634 (1958).
130. J. Gohring, G. P. Rossetti and B. P. Susz, *Helv. Chim. Acta*, **46,** 2639 (1963).
131. G. P. Rossetti, *Helv. Chim. Acta*, **47,** 2053 (1964).
132. G. P. Rossetti and B. P. Susz, *Helv. Chim. Acta*, **47,** 289 (1964).
133. G. P. Rossetti and B. P. Susz, *Helv. Chim. Acta*, **49,** 1899 (1966).
134. R. Weber and B. P. Susz, *Helv. Chim. Acta*, **50,** 2226 (1967).
135. I. R. Beattie and G. J. Leigh, *J. Inorg. Nucl. Chem.*, **23,** 55 (1961).
136. W. R. Trost, *Can. J. Chem.*, **30,** 842 (1952).
137. W. R. Trost, *Can. J. Chem.*, **30,** 835 (1952).
138. G. W. A. Fowles and R. A. Hoodless, *J. Chem. Soc.*, **1963,** 33.
139. B. J. Brisdon, T. E. Lester and R. A. Walton, *Spectrochim. Acta*, **23A,** 1969 (1967).
140. G. W. A. Fowles and R. A. Walton, *J. Chem. Soc.*, **1964,** 4330.
141. I. R. Beattie and M. Webster, *J. Chem. Soc.*, **1964,** 3507.
142. G. W. A. Fowles and R. A. Walton, *J. Less-Common Metals*, **5,** 510 (1963).
143. T. N. Sumarakova and D. S. Sakenova, *J. Gen. Chem. USSR*, **34,** 2717 (1964).
144. P. Dunn, *Australian J. Chem.*, **13,** 225 (1960).
145. O. C. Dermer and W. C. Fernelius, *Z. anorg. allgem. Chem.*, **221,** 83 (1934).
146. R. J. H. Clark, *J. Chem. Soc.*, **1963,** 1377.
147. K. S. Boustany, K. Bernauer and A. Jacot-Guillarmod, *Helv. Chim. Acta*, **50,** 1120 (1967).
148. S. C. Jain and R. Rivest, *J. Inorg. Nucl. Chem.*, **29,** 2787 (1967).
149. I. K. Shelomov, I. A. Kozlov and O. A. Osipov, *J. Gen. Chem. USSR*, **36,** 1950 (1966).
150. I. R. Beattie and T. Gilson, *J. Chem. Soc.*, **1965,** 6595.
151. A. K. Anagnostopoulos, *Chim. Chronika* (*Athens*), **31,** 141 (1966).
152. V. T. Panyushkin, E. S. Kagan, A. D. Garnovskii, O. A. Osipov and I. M. Semenova, *Zh. Obshsch. Khim.*, **37,** 1566 (1967).

153. W. Gerrard, M. F. Lappert and J. W. Wallis, *J. Chem. Soc.*, **1960,** 2141.
154. G. W. A. Fowles and F. H. Pollard, *J. Chem. Soc.*, **1953,** 2588.
155. J. Cueilleron and M. Charret, *Bull. Soc. Chim. France*, **1956,** 802.
156. M. Antler and A. W. Lanbengayer, *J. Am. Chem. Soc.*, **77,** 5250 (1955).
157. H. J. Coerver and C. Curran, *J. Am. Chem. Soc.*, **80,** 3522 (1958).
158. T. Karantassis, *Compt. Rend.*, **194,** 461 (1932).
159. G. W. A. Fowles and R. A. Walton, *J. Chem. Soc.*, **1964,** 2840.
160. S. C. Jain and R. Rivest, *Can. J. Chem.*, **41,** 2130 (1963).
161. R. Perrot and C. Devin, *Compt. Rend.*, **246,** 772 (1958).
162. R. J. Kern, *J. Inorg. Nucl. Chem.*, **25,** 5 (1963).
163. M. Kubota and S. R. Schulze, *Inorg. Chem.*, **3,** 853 (1964).
164. B. J. Hathaway and D. G. Holah, *J. Chem. Soc.*, **1965,** 537.
165. S. C. Jain and R. Rivest, *Inorg. Chem.*, **6,** 467 (1967).
166. E. N. Kharlamova, E. N. Gur'yanova and N. A. Slovokhotova, *J. Gen. Chem. USSR*, **37,** 284 (1967).
167. J. Chatt, R. L. Richards and D. J. Newman, *J. Chem. Soc.*, **1968,** A, 126.
168. R. Rivest, *Can. J. Chem.*, **40,** 2234 (1962).
169. E. A. Ionova and V. A. Golovnya, *Russ. J. Inorg. Chem.*, **11,** 74 (1966).
170. Y. A. Kharitonov, Y. A. Buslaev and E. A. Ionova, *Russ. J. Inorg. Chem.*, **11,** 1284 (1966).
171. D. Cassimatis, P. Gagnaux and B. P. Susz, *Helv. Chim. Acta*, **43,** 424 (1960).
172. D. Cassimatis and B. P. Susz, *Helv. Chim. Acta*, **44,** 943 (1961).
173. B. P. Susz and D. Cassimatis, *Helv. Chim. Acta*, **44,** 395 (1961).
174. T. N. Sumarokova, Y. Nevskaya and E. Yarmukhamedova, *Zh. Obshch. Khim.*, **30,** 1705 (1960).
175. E. C. Alyea and E. G. Torrible, *Can. J. Chem.*, **43,** 3468 (1965).
176. P. M. Hamilton, R. McBeth, W. Bekebrede and H. H. Sisler, *J. Am. Chem. Soc.*, **75,** 2881 (1953).
177. D. Schwartz and P. Reski, *J. Inorg. Nucl. Chem.*, **27,** 747 (1965).
178. J. B. Ott, J. R. Goates, R. J. Jensen and N. F. Mangelson, *J. Inorg. Nucl. Chem.*, **27,** 2005 (1965).
179. J. R. Goates, J. B. Ott, N. F. Mangelson and R. J. Jensen, *J. Phys. Chem.*, **68,** 2617 (1964).
180. P. Ehrlich and W. Siebert, *Z. anorg. allgem. Chem.*, **303,** 96 (1960).
181. J. Archambault and R. Rivest, *Can. J. Chem.*, **36,** 1461 (1958).
182. S. J. Kuhn and J. S. McIntyre, *Can. J. Chem.*, **43,** 375 (1965).
183. R. F. Kempa and W. H. Lee, *Z. anorg. allgem. Chem.*, **311,** 140 (1961).
184. D. Neubauer, J. Weiss and M. Becke-Goehring, *Z. Naturforsch.*, **14b,** 284 (1959).
185. T. N. Sumarakova and Y. Nevskaya, *Zh. Obshch. Khim.*, **27,** 3375 (1957).
186. N. A. Pushin, *Zh. Obshch. Khim.*, **18,** 1599 (1948).
187. R. Holtje, *Z. anorg. allgem. Chem.*, **190,** 241 (1930).
188. J. R. Partington and A. L. Whynes, *J. Chem. Soc.*, **1948,** 1952.
189. D. Nicholls and R. Swindells, *J. Chem. Soc.*, **1964,** 4204.
190. A. H. Norbury and A. I. P. Sinha, *J. Chem. Soc.*, **1966,** A, 1814.
191. W. L. Groeneveld, *Rec. Trav. Chim.*, **71,** 1152 (1952).
192. O. A. Osipov, V. M. Artemova, V. A. Kogan and Y. A. Lysenko, *J. Gen. Chem. USSR*, **32,** 1354 (1962).

193. V. N. Startsev, E. I. Krylov and Y. A. Koz'min, *Russ. J. Inorg. Chem.*, **10,** 1285 (1965).
194. J. Archambault and R. Rivest, *Can. J. Chem.*, **38,** 1331 (1960).
195. A. D. Westland and L. Westland, *Can. J. Chem.*, **43,** 426 (1965).
196. S. C. Jain and R. Rivest, *Can. J. Chem.*, **43,** 787 (1965).
197. A. V. Leshchenko, V. T. Panyushkin, A. D. Garnovskii and O. A. Osipov, *Russ. J. Inorg. Chem.*, **11,** 1155 (1966).
198. I. Pavlik, *Coll. Czech. Chem. Commun.*, **30,** 3052 (1965).
199. Y. Dutt and R. P. Singh, *Indian J. Chem.*, **4,** 424 (1966).
200. R. C. Paul, R. Prakash and S. S. Sandhu, *Indian J. Chem.*, **4,** 426 (1966).
201. A. Slawisch and M. Becke-Goehring, *Z. Naturforsch.*, **21b,** 589 (1966).
202. J. Chatt and R. C. Hayter, *J. Chem. Soc.*, **1963,** 1343.
203. G. B. L. Smith, *Chem. Rev.*, **23,** 165 (1938).
204. R. J. H. Clark, *J. Chem. Soc.*, **1965,** 5699.
205. R. J. H. Clark, M. L. Greenfield and R. S. Nyholm, *J. Chem. Soc.*, **1966,** A, 1254.
206. F. W. G. Schoning and C. J. Liebenberg, *J. South African Chem. Institute*, **20,** 61 (1967).
207. M. J. Frazer and Z. Goffer, *J. Chem. Soc.*, **1966,** A, 544.
208. I. Pavlik and K. Handlir, *Coll. Czech, Chem. Commun.*, **31,** 1958 (1966).
209. I. Pavlik, K. Handlir and V. Dvorak, *Coll. Czech. Chem. Commun.*, **30,** 3052 (1965).
210. R. J. Kern, *J. Inorg. Nucl. Chem.*, **24,** 1105 (1962).
211. E. E. Aynsley, N. N. Greenwood and J. B. Leach, *Chem. Ind.* (*London*), **1966,** 379.
212. R. A. Walton, *Inorg. Chem.*, **5,** 643 (1966).
213. R. J. H. Clark and W. Errington, *Inorg. Chem.*, **5,** 650 (1966).
214. M. F. Lappert and G. Srivastava, *J. Chem. Soc.*, **1966**, A, 210.
215. K. M. Ismailov, O. A. Osipov, A. D. Garnovskii, O. E. Kashireninov and N. L. Chikina, *Dokl. Akad. Nauk Azerb. SSR*, **21,** 34 (1965).
216. K. Baker and G. W. A. Fowles, *Proc. Chem. Soc.*, **1964**, 362.
217. A. D. Garnovskii, V. I. Minkin, O. A. Osipov, V. T. Panyushkin, L. K. Tsaeva and M. I. Knyazhanskii, *Russ. J. Inorg. Chem.*, **12,** 1288 (1967).
218. W. Wieker and A. R. Grimmer, *Z. Naturforsch.*, **22b,** 1220 (1967).
219. R. J. H. Clark and R. H. U. Negrotti, *Chem. Ind.* (*London*), **1968,** 154.
220. K. H. Thiele and K. Jacob, *Z. anorg. allgem. Chem.*, **356,** 195 (1968).
221. R. J. H. Clark, *Spectrochim. Acta*, **21,** 955 (1965).
222. S. Minami and T. Ishino, *Kogyo Kagaku Zasshi*, **61,** 66 (1958).
223. I. D. Varma and R. C. Mehrotra, *J. Prakt. Chem.*, **8,** 64 (1959).
224. A. N. Nesmeyanov, R. K. Friedlina and O. V. Nogina, *Izvt. Akad. Nauk SSSR, Otd. Khim. Nauk*, **1952,** 1037.
225. D. Behar and H. Feilchenfeld, *J. Organometal. Chem.*, **4,** 278 (1965).
226. K. A. Andrianov and A. I. Petrashko, *Dokl. Akad. Nauk SSSR*, **131,** 561 (1960).
227. K. A. Andrianov and N. A. Kurasheva, *Dokl. Akad. Nauk SSSR*, **131,** 825 (1960).
228. R. T. Cowdell and G. W. A. Fowles, *J. Chem. Soc.*, **1960,** 2522.
229. V. S. V. Nayar and R. D. Peacock, *J. Chem. Soc.*, **1964,** 2827.
230. E. Hayek, J. Puschmann and A. Czaloun, *Monatsh. Chem.*, **85,** 359 (1954).

231. V. Gutmann and A. M. Meller, *Monatsh. Chem.*, **92,** 470 (1961).
232. R. C. Mehrotra and R. A. Misra, *Indian J. Chem.*, **3,** 500 (1965).
233. D. Schwartz and W. Cross, *J. Chem. Eng. Data*, **8,** 463 (1963).
234. D. M. Puri and R. C. Mehrotra, *J. Less-Common Metals*, **3,** 247 (1961).
235. R. T. Cowdell, G. W. A. Fowles and R. A. Walton, *J. Less-Common Metals*, **5,** 386 (1963).
236. N. M. Cullinane and S. J. Chard, *Nature*, **164,** 710 (1949).
237. G. P. Luchinskii and E. S. Al'Aman, *Z. anorg. allgem. Chem.*, **225,** 321 (1935).
238. O. V. Nogina, R. K. Freidlina and A. N. Nesmeyanov, *Izvt. Akad. Nauk SSSR, Otd. Khim. Nauk*, **1950,** 327.
239. H. Funk, A. Schlegel and K. Zimmermann, *J. Prakt. Chem.*, **3,** 320 (1956).
240. G. S. Shvindlerman, T. G. Golenko and L. I. Red'kina, *Izvt. Akad. Nauk SSSR, Ser. Khim.*, **1966,** 158.
241. M. Cox, J. Lewis and R. S. Nyholm, *J. Chem. Soc.*, **1964,** 6113.
242. R. N. Kapoor, K. C. Pande and R. C. Mehrotra, *J. Indian Chem. Soc.*, **35,** 157 (1958).
243. K. L. Jaura, H. S. Banga and R. L. Kaushik, *J. Indian Chem. Soc.*, **39,** 531 (1962).
244. A. G. Swallow and B. F. Studd, *Chem. Commun.*, **1967,** 1197.
245. A. Rosenheim, W. Loewenstemm and L. Singer, *Ber.*, **36,** 1833 (1903).
246. W. Dilthey, *Ber.*, **37,** 588 (1904).
247. K. Waterpaugh and C. N. Caughlan, *Inorg. Chem.*, **6,** 963 (1967).
248. K. Watenpaugh and C. N. Caughlan, *Inorg. Chem.*, **5,** 1782 (1966).
249. K. Dehnicke, *Angew. Chem. Intern. Ed. Engl.*, **2,** 325 (1963).
250. K. Dehnicke, *Z. anorg. allgem. Chem.*, **309,** 266 (1961).
251. P. Ehrlich and W. Engel, *Z. anorg. allgem. Chem.*, **317,** 21 (1962).
252. P. Ehrlich and W. Engel, *Naturwissenschaften*, **48,** 716 (1961).
253. *German Patent*, 1,141,626 (1962).
254. K. Dehnicke, *Chem. Ber.*, **98,** 290 (1965).
255. D. I. Tsekhovol'skaya and B. M. Merenkova, *Russ. J. Phys. Chem.*, **40,** 269 (1966).
256. R. A. Nelson, W. H. Johnson and E. J. Prosen, *J. Res. Nat. Bur. Std.*, **62,** 67 (1959).
257. H. L. Schlafer and H. H. Schmidtke, *Z. Physik. Chem. (Frankfurt)*, **11,** 297 (1957).
258. J. K. Keavney and N. O. Smith, *J. Phys. Chem.*, **64,** 737 (1960).
259. S. A. Shchukarev, I. V. Vasil'kova and D. V. Korol'kov, *Russ. J. Inorg. Chem.*, **8,** 1006 (1963).
260. P. Ehrlich, W. Gutsche and H. J. Seifert, *Z. anorg. allgem. Chem.*, **312,** 80 (1961).
261. K. Funaki, K. Uchimura and Y. Kuniya, *Kogyo Kagaku Zasshi*, **64,** 1914 (1961).
262. G. P. Baxter and A. Q. Butler, *J. Am. Chem. Soc.*, **50,** 408 (1928).
263. J. C. Olsen and E. P. Ryan, *J. Am. Chem. Soc.*, **54,** 2215 (1932).
264. R. C. Young, *Inorg. Syn.*, **2,** 114 (1946).
265. P. M. Druce, M. F. Lappert and P. N. K. Riley, *Chem. Commun.*, **1967,** 486.
266. K. H. Gayer and G. Tennenhouse, *Can. J. Chem.*, **37,** 1373 (1959).

267. J. M. Blocher, R. F. Rolsten and I. E. Campbell, *J. Electrochem. Soc.*, **104,** 553 (1957).
268. E. H. Hall, J. M. Blocher and I. E. Campbell, *J. Electrochem. Soc.*, **105,** 271 (1958).
269. H. Kato and M. Abe, *Nippon Kagaku Zasshi*, **76,** 1182 (1955).
270. F. A. Miller and G. L. Carlson, *Spectrochim. Acta*, **16,** 6 (1960).
271. R. F. Rolsten and H. H. Sisler, *J. Am. Chem. Soc.*, **79,** 5891 (1957).
272. A. A. Boni, *J. Electrochem. Soc.*, **113,** 1089 (1966).
273. O. Hassel and H. Kringstad, *Z. Physik. Chem.*, **B15,** 274 (1932).
274. P. Brand and J. Schmidt, *Z. anorg. allgem. Chem.*, **348,** 257 (1966).
275. S. Prasad and R. C. Srivastava, *J. Indian Chem. Soc.*, **39,** 11 (1962).
276. R. F. Rolsten and H. H. Sisler, *J. Am. Chem. Soc.*, **79,** 1068 (1957).
277. R. F. Rolsten and H. H. Sisler, *J. Am. Chem. Soc.*, **79,** 1819 (1957).
278. F. Seel and H. Massat, *Z. anorg. allgem. Chem.*, **280,** 186 (1955).
279. P. J. Hendra and D. B. Powell, *J. Chem. Soc.*, **1960,** 5105.
280. G. W. A. Fowles and D. Nicholls, *J. Chem. Soc.*, **1959,** 990.
281. I. D. Varma and R. C. Mehrotra, *J. Less-Common Metals*, **1,** 263 (1959).
282. I. R. Krichevskii, G. F. Ivanovskii and E. K. Safronov, *Russ. J. Phys. Chem.*, **39,** 1436 (1965).
283. J. D. Fast, *Z. anorg. allgem. Chem.*, **241,** 42 (1939).
284. J. D. Fast, *Rec. Trav. Chim.*, **58,** 174 (1939).
285. K. Funaki, K. Uchimura and H. Matsunaga, *Kogyo Kagaku Zasshi*, **64,** 129 (1961).
286. J. M. Blocher and I. E. Campbell, *J. Am. Chem. Soc.*, **69,** 2100 (1947).
287. *U.S. Patent*, 2,904,397 (1959).
288. K. Dehnicke, *Z. anorg. allgem. Chem.*, **338,** 279 (1965).
289. M. Chaigneau, *Compt. Rend.*, **242,** 263 (1956).
290. P. P. Bhatnagar and R. A. Sharma, *J. Proc. Inst. Chemists* (*India*), **29,** 97 (1957).
291. M. Chaigneau, *Bull. Soc. Chim. France*, **1957,** 886.
292. T. R. Ingraham and L. M. Pidgeon, *Can. J. Chem.*, **30,** 694 (1952).
293. P. Ehrlich and G. Pietzka, *Z. anorg. allgem. Chem.*, **275,** 121 (1954).
294. J. M. Blocher and E. H. Hall, *J. Phys. Chem.*, **63,** 127 (1959).
295. S. Siegel, *Acta Cryst.*, **9,** 684 (1956).
296. B. L. Chamberland and A. W. Sleight, *Solid State Commun.*, **5,** 765 (1967).
297. R. D. Beyer and D. M. Mason, *Ind. Eng. Chem., Process Design Develop.*, **2,** 78 (1963).
298. *French Patent*, 1,252,168 (1960).
299. J. Ambroz, L. Ambroz and S. Dvorak, *Chem. prumysl*, **10,** 23 (1960).
300. Y. Ogawa, Y. Hisamatsu and K. Kawamura, *Nippon Kogyo Zasshi*, **73,** 565 (1957).
301. R. Schmidt, *Ber*, **58B,** 400 (1925).
302. W. F. Krieve, S. P. Vango and D. M. Mason, *J. Chem. Phys.*, **25,** 519 (1956).
303. G. Natta, P. Corradini, I. W. Bassi and L. Porri, *Atti Acad. Naz. Lincei, Rend., Classe Sci., Fis., Mat. Nat.*, **24,** 121 (1958).
304. J. M. Sherfey, *Inorg. Syn.*, **6,** 57 (1960).
305. C. Starr, F. Bitter and A. R. Kaufmann, *Phys. Rev.*, **58,** 977 (1940).
306. G. Pregaglia, G. Mazzanti and D. Morero, *Ann. Chim.* (*Rome*), **49,** 1784 (1959).

307. M. Farber, A. J. Darnell and F. Brown, *J. Chem. Phys.*, **23,** 1556 (1955).
308. M. Farber and A. J. Darnell, *J. Phys. Chem.*, **59,** 156 (1955).
309. H. Hartmann and G. Rinck, *Z. Physik. Chem.* (*Frankfurt*), **11,** 213 (1957).
310. O. Ruff and F. Neumann, *Z. anorg. allgem. Chem.*, **128,** 81 (1923).
311. T. R. Ingraham, K. W. Downes and P. Marier, *Inorg. Syn.*, **6,** 52 (1960).
312. T. R. Ingraham, K. W. Downes and P. Marier, *Can. J. Chem.*, **35,** 850 (1957).
313. L. V. Biryukova, *Zh. Vses. Khim. Obshch. im D. I. Mendeleeva*, **7,** 119 (1962).
314. L. V. Biryukova and Y. G. Saksonov, *Russ. J. Inorg. Chem.*, **5,** 477 (1960).
315. K. Funaki and K. Uchimura, *Bull. Tokyo Inst. Technol. Ser. B*, 191 (1960).
316. V. A. Ryabov, G. N. Zviadadze, O. V. Al'tshuler and D. M. Chizhikov, *Trudy Inst. Met. im A.A. Baikova*, **1957,** 85.
317. F. Meyer, A. Bauer and R. Schmidt, *Ber.*, **56B,** 1908 (1923).
318. K. Funaki and K. Uchimura, *J. Chem. Soc. Japan*, **59,** 14 (1956).
319. T. Ishino, H. Tamura and O. Nakagawa, *Kogyo Kagaku Zasshi*, **64,** 1344 (1961).
320. M. Billy and P. Brasseur, *Compt. Rend.*, **200,** 1765 (1935).
321. G. Natta, P. Corradini and G. Allegra, *J. Polymer Sci.*, **51,** 399 (1961).
322. H. Schafer and E. Sibbing, *Angew. Chem.*, **69,** 479 (1957).
323. B. S. Sanderson and G. E. MacWood, *J. Phys. Chem.*, **60,** 314 (1956).
324. H. Schafer, G. Breil and G. Pfeffer, *Z. anorg. allgem. Chem.*, **276,** 325 (1954).
325. S. F. Belov and S. I. Skylarenko, *Tsvet. Met.*, **31,** 37 (1958).
326. D. G. Clifton and G. E. MacWood, *J. Phys. Chem.*, **60,** 309 (1956).
327. R. B. Head, *Australian J. Chem.*, **13,** 332 (1960).
328. W. H. Johnson, A. A. Gilliland and E. J. Prosen, *J. Res. Nat. Bur. Std.*, **64A,** 515 (1960).
329. D. G. Clifton, *Diss. Abs.*, **15,** 987 (1955).
330. B. S. Sanderson, *Diss. Abs.*, **15,** 992 (1955).
331. W. Klemm and E. Krose, *Z. anorg. allgem. Chem.*, **253,** 218 (1947).
332. J. W. Reed and G. E. MacWood, 133rd Meeting American Chemical Society, San Franscisco 1958.
333. J. W. Reed, *Diss. Abs.*, **17,** 1479 (1957).
334. J. Lewis, D. J. Machin, I. E. Newnham and R. S. Nyholm, *J. Chem. Soc.*, **1962,** 2036.
335. W. Klemm and E. Krose, *Z. anorg. allgem. Chem.*, **253,** 209 (1947).
336. S. Ogawa, *J. Phys. Soc. Japan*, **15,** 1901 (1960).
337. E. Konig, Landolt-Bornstein, *Magnetic Properties of Coordination and Organometallic Transition Metal Compounds*, Springer-Verlag, Vol. 2, New York, 1966.
338. C. Dijkgraaf, *Nature*, **201,** 1121 (1964).
339. M. W. Duckworth, G. W. A. Fowles and R. A. Hoodless, *J. Chem. Soc.*, **1963,** 5665.
340. A. F. Reid and P. C. Wailes, *J. Organometal. Chem.*, **2,** 329 (1964).
341. R. J. H. Clark, J. Lewis, D. J. Machin and R. S. Nyholm, *J. Chem. Soc.*, **1963,** 379.
342. H. L. Schlafer and R. Goetz, *Z. Physik. Chem.* (*Frankfurt*), **41,** 97 (1964).
343. H. Hartmann, H. L. Schlafer and K. H. Hansen, *Z. anorg. allgem. Chem.*, **284,** 153 (1956).

344. S. Prasad and K. S. Devi, *Indian J. Chem.*, **4,** 543 (1966).
345. H. L. Schlafer and W. Schroeder, *Z. anorg. allgem. Chem.*, **347,** 45 (1966).
346. H. L. Schlafer and W. Schroeder, *Z. anorg. allgem. Chem.*, **347,** 59 (1966).
347. G. J. Sutton, *Australian J. Chem.*, **12,** 122 (1959).
348. G. W. A. Fowles, R. A. Hoodless and R. A. Walton, *J. Chem. Soc.*, **1963,** 5873.
349. R. J. H. Clark and M. L. Greenfield, *J. Chem. Soc.*, **1967,** A, 409.
350. W. Giggenbach and C. H. Brubaker, *Inorg. Chem.*, **7,** 129 (1968).
351. L. D. Calvert and C. M. Pleass, *Can. J. Chem.*, **40,** 1473 (1962).
352. G. W. A. Fowles, P. T. Greene and J. S. Wood, *Chem. Commun.*, **1967,** 971.
353. H. Hartmann and H. L. Schlafer, *Z. Physik. Chem. (Leipzig)*, **197,** 116 (1951).
354. H. L. Schlafer and H. P. Fritz, *Spectrochim. Acta*, **23A,** 1409 (1967).
355. H. Schafer, F. Wartenpfuhl and E. Weise, *Z. anorg. allgem. Chem.*, **295,** 268 (1958).
356. H. Schafer, E. Weise and F. Wartenpfuhl, *Angew. Chem.*, **69,** 479 (1957).
357. R. C. Young and W. C. Schumb, *J. Am. Chem. Soc.*, **52,** 4233 (1930).
358. R. C. Young and W. M. Leaders, *Inorg. Syn.*, **2,** 116 (1946).
359. R. F. Rolsten and H. H. Sisler, *J. Phys. Chem.*, **62,** 1024 (1958).
360. E. H. Hall and J. M. Blocher, *J. Phys. Chem.*, **63,** 1525 (1959).
361. E. H. Hall and J. M. Blocher, *J. Electrochem. Soc.*, **105,** 40 (1958).
362. S. A. Shchukarev, I. V. Vasil'kova and D. V. Korol'kov, *Russ. J. Inorg. Chem.*, **9,** 980 (1964).
363. G. W. A. Fowles, P. T. Greene and T. E. Lester, *J. Inorg. Nucl. Chem.*, **29,** 2365 (1967).
364. B. J. Russ and J. S. Wood, *Chem. Commun.*, **1966,** 745.
365. H. G. Schnering, *Naturwissenschaften,* **53,** 359 (1966).
366. A. Herczog and L. M. Pidgeon, *Can. J. Chem.*, **34,** 1687 (1956).
367. V. Gutmann, H. Nowotny and G. Ofner, *Z. anorg. allgem. Chem.*, **278,** 78 (1955).
368. E. L. Gal'perin and R. A. Sandler, *Soviet Phys - Crystallography*, **7,** 169 (1962).
369. H. Tadenuma, H. Ikeda and K. Fujita, *Kogyo Kagaku Zasshi*, **59,** 356 (1956).
370. P. Ehrlich, H. J. Hein and H. Kuknl, *Z. anorg. allgem. Chem.*, **292,** 139 (1957).
371. N. C. Baenziger and R. E. Rundle, *Acta Cryst.*, **1,** 274 (1948).
372. P. Ehrlich and H. Kuknl, *Z. anorg. allgem. Chem.*, **292,** 146 (1957).
373. K. Komarek and P. Herasymenko, *J. Electrochem. Soc.*, **105,** 216 (1958).
374. M. Faber and A. J. Darnell, *J. Chem. Phys.*, **25,** 526 (1956).
375. D. G. Clifton and G. E. MacWood, *J. Phys. Chem.*, **60,** 311 (1956).
376. W. Klemm and L. Grimm, *Z. anorg. allgem. Chem.*, **249,** 209 (1942).
377. W. Klemm and L. Grimm, *Z. anorg. allgem. Chem.*, **249,** 198 (1942).
378. B. Cox and A. G. Sharpe, *J. Chem. Soc.*, **1953,** 1783.
379. B. Cox, *J. Chem. Soc.*, **1954,** 3251.
380. S. Siegel, *Acta Cryst.*, **5,** 683 (1952).
381. C. J. Janz, C. Solomons, J. H. Gardner, J. Goodkin and C. T. Brown, *J. Phys. Chem.*, **62,** 823 (1958).
382. R. D. W. Kemmitt and D. W. A. Sharp, *J. Chem. Soc.*, **1961,** 2496.

383. S. Aleonard, *Compt. Rend.*, **260,** 1977 (1965).
384. K. Aotani, *Kogyo Kagaku Zasshi*, **62,** 1368 (1959).
385. *U.S. Patent*, 2,816,816 (1957).
386. A. G. Sharpe, *J. Chem. Soc.*, **1950,** 2907.
387. H. M. Haendler, F. A. Johnson and D. S. Crocket, *J. Am. Chem. Soc.*, **80,** 2662 (1958).
388. P. A. W. Dean and D. F. Evans, *J. Chem. Soc.*, **1967,** A, 698.
389. D. H. Brown, K. R. Dixon, R. D. W. Kemmitt and D. W. A. Sharp, *J. Chem. Soc.*, **1965,** 1559.
390. C. Cipriani, *Periodico Mineral* (*Rome*), **24,** 361 (1956).
391. B. Cox, *J. Chem. Soc.*, **1956,** 876.
392. J. A Ibers and C. H. Holm, *Acta Cryst.*, **10,** 139 (1957).
393. D. H. Brown, K. R. Dixon, C. M. Livingston, R. H. Nuttall and D. W. A. Sharp, *J. Chem. Soc.*, **1967,** A, 100.
394. R. D. Peacock and D. W. A. Sharp, *J. Chem. Soc.*, **1959,** 2762.
395. V. G. Bamburov, N. V. Demenev and V. M. Polyakova, *Izv. Sibirsk. Otd. Akad. Nauk SSSR.*, **1962,** 73.
396. R. Weiss, J. Fischer and B. Chevrier, *Compt. Rend.*, **260,** 3401 (1965).
397. R. Weiss, J. Fischer and B. Chevrier, *Acta Cryst.*, **20,** 534 (1966).
398. R. Weiss, J. Fischer and G. Keib, *Compt. Rend.*, **259,** 1125 (1964).
399. J. Fischer, G. Keib and R. Weiss, *Acta Cryst.*, **22,** 338 (1967).
400. A. Decian, J. Fischer and R. Weiss, *Acta Cryst.*, **22,** 340 (1967).
401. P. Bouy, *Ann. Chim.* (*Paris*), **4,** 853 (1959).
402. A. A. Woolf, *J. Chem. Soc.*, **1950,** 1053.
403. J. A. Chandler and R. S. Drago, *J. Inorg. Nucl. Chem.*, **21,** 283 (1961).
404. J. A. Chandler, J. E. Wuller and R. S. Drago, *Inorg. Chem.*, **1,** 65 (1962).
405. J. E. Wuller, *Diss. Abs.*, **23,** 2688 (1963).
406. L. A. Tsiorkina and M. A. Smirnov, *Russ. J. Inorg. Chem.*, **4,** 65 (1959).
407. R. L. Lister and S. N. Flengas, *J. Electrochem. Soc.*, **111,** 343 (1964).
408. R. L. Lister and S. N. Flengas, *Can. J. Chem.*, **41,** 1548 (1963).
409. J. H. Mui and S. N. Flengas, *Can. J. Chem.*, **40,** 997 (1962).
410. I. S. Morozov and G. M. Toptygina, *Russ. J. Inorg. Chem.*, **5,** 42 (1960).
411. G. W. A. Fowles and D. Nicholls, *J. Inorg. Nucl. Chem.*, **18,** 130 (1961).
412. P. Ehrlich and E. Framm, *Z. Naturforsch.*, **9b,** 326 (1954).
413. K. F. Guenther, *Inorg. Chem.*, **3,** 923 (1964).
414. S. S. Sandhu, B. S. Chakkal and G. S. Sandhu, *J. Indian Chem. Soc.*, **37,** 329 (1960).
415. K. Goyal, R. C. Paul and S. S. Sandhu, *J. Chem. Soc.*, **1959,** 322.
416. R. C. Paul, K. Chander and G. Singh, *J. Indian Chem. Soc.*, **35,** 869 (1958).
417. D. Schwartz and D. Naegli, *Inorg. Nucl. Chem. Letters*, **2,** 149 (1966).
418. D. M. Adams, J. Chatt, J. M. Davidson and J. Gerratt, *J. Chem. Soc.*, **1963,** 2189.
419. T. C. Waddington and S. N. Nabi, *Proc. Pakistan Sci. Conf.*, **12,** Pt. 3, C7 (1960).
420. S. N. Flengas and T. R. Ingraham, *Can. J. Chem.*, **38,** 813 (1960).
421. J. A. Bland and S. N. Flengas, *Can. J. Phys.*, **39,** 941 (1961).
422. F. Schossberger, *Ind. Eng. Chem.*, **51,** 669 (1959).
423. G. Engel, *Centr. Mineral. Geol.*, **1934A,** 285.
424. E. Wendling, *Bull. Soc. Chim. France*, **1967**, 5.

425. J. Wernet, *Z. anorg. allgem. Chem.*, **272,** 279 (1953).
426. R. A. Walton and B. J. Brisdon, *Spectrochim. Acta*, **23A,** 2222 (1967).
427. A. Feltz, *Z. anorg. allgem. Chem.*, **334,** 242 (1964).
428. A. Feltz, *Z. anorg. allgem. Chem.*, **338,** 155 (1965).
429. I. R. Beattie and V. Fawcett, *J. Chem. Soc.*, **1967,** A, 1583.
430. I. S. Morozov and G. M. Toptygina, *Russ. J. Inorg. Chem.*, **5,** 1218 (1960).
431. I. I. Kozhina and D. V. Korol'kov, *Zh. Strukt. Khim.*, **6,** 97 (1965).
432. K. F. Guenther, *Inorg. Chem.*, **3,** 1788 (1964).
433. J. Jander, H. Machatzke and D. Mecke, *Z. anorg. allgem. Chem.*, **294,** 181 (1958).
434. D. Negoiu, *Acap. Rep. Populare Romine Studii Cercetari Chim.*, **11,** 61 (1963).
435. N. F. H. Bright and J. G. Wurm, *Can. J. Chem.*, **36,** 615 (1958).
436. H. E. Bedon, S. M. Horner and S. Y. Tyree, *Inorg. Chem.*, **3,** 647 (1964).
437. *U.S. Patent*, 2,672,399 (1954).
438. P. Ehrlich and F. Pietzka, *Naturwissenschaften*, **40,** 509 (1953).
439. R. V. Chernov and Y. K. Delimarskii, *Fiz. Khim. Rasplavlen. Solei i Shlakov, Akad. Nauk SSSR, Ural'sk. Filial Inst. Elektrokhim., Tr. Vses. Soveshch., Sverdlovsk*, **1960,** 146.
440. Y. K. Delimarskii and R. V. Chernov, *Dopovidi Akad. Nauk Ukr. RSR*, **1961,** 1508.
441. M. V. Kamenetskii and L. I. Shevlyakova, *Izv. Vyss. Uchebn. Zavedenii, Tsvetn. Met.*, **5,** 89 (1962).
442. R. V. Chernov and Y. K. Delimarskii, *Zh. Neorg. Khim.*, **6,** 2749 (1961).
443. B. F. Markov and R. V. Chernov, *Ukr. Khim. Zh.*, **27,** 34 (1961).
444. M. V. Kamenetskii, *Nauch.-Tekh. Inform. Byull. Leningrad. Politekh. Inst.*, **1957,** 3.
445. M. V. Kamenetskii, A. A. Kostyukov and A. N. Popov, *Izv. Vyss. Uchebn. Zavedenni, Tsvetn. Met.*, **3,** 119 (1960).
446. P. Ehrlich, G. Kupa and K. Blankenstein, *Z. anorg. allgem. Chem.*, **299,** 213 (1959).
447. B. F. Markov and R. V. Chernov, *Ukr. Khim. Zh.*, **25,** 279 (1959).
448. M. V. Kamenetskii, *Tsvetn. Met.*, **31,** 39 (1958).
449. Y. K. Delimarskii and R. V. Chernov, *Dopovidi Akad. Nauk Ukr. RSR*, **1960,** 795.
450. B. N. Podzorov, I. V. Vasil'kova and R. A. Sandler, *Russ. J. Inorg. Chem.*, **12,** 16 (1967).
451. G. W. A. Fowles and B. J. Russ, *J. Chem. Soc.*, **1967,** A, 517.
452. B. J. Russ and G. W. A. Fowles, *Chem. Commun.*, **1966,** 19.
453. M. V. Kamenetskii, A. A. Kostyukov and T. C. Hsiao, *Fiz Khim. Rasplavlen. Solei i Shlakov, Akad. Nauk SSSR, Ural'sk. Filial Inst. Elektrokhim., Tr. Vses. Soveshch., Sverdlovsk*, **1960,** 54.
454. D. M. Gruen and R. L. McBeth, *Nature*, **194,** 468 (1962).
455. I. S. Morozov, G. M. Toptygina and N. P. Lipatova, *Russ. J. Inorg. Chem.*, **6,** 1279 (1961).
456. I. S. Morozov, G. M. Toptygina and N. P. Lipatova, *Russ. J. Inorg. Chem.*, **6,** 1282 (1961).
457. G. J. Wessel and D. J. W. Ijdo, *Acta Cryst.*, **10,** 466 (1957).
458. C. Dijkgraaf, J. P. C. van Heel and J. P. G. Rousseau, *Nature*, **211,** 185 (1966).

459. Hsi Chung Li and D. M. Chizhikov, *Izv. Akad. Nauk SSSR, Otd. Tekh. Nauk, Met. i Toplivo*, **1961,** 22.
460. P. Ehrlich and R. Schmitt, *Z. anorg. allgem. Chem.*, **308,** 91 (1961).
461. J. W. Downing and R. O. Ragsdale, *Inorg. Chem.*, **7,** 1675 (1968).
462. M. J. Frazer and B. Rimmer, *J. Chem. Soc.*, **1968,** A, 69.
463. M. J. Frazer, W. E. Newton, B. Rimmer and J. R. Majer, *Chem. Commun.*, **1968,** 1336.
464. V. Gutmann and M. Michlmayr, *Monatsh. Chem.*, **99,** 316 (1968).
465. J. P. Dahl and H. Johansen, *Theoret. Chim. Acta*, **11,** 26 (1968).
466. E. A. Ionova, Y. A. Buslaev and Y. Y. Kharitonov, *Izv. Akad. Nauk SSSR, Neorg. Mater.*, **1968,** 71.
467. B. I. Nabivanets and L. N. Kudritskaya, *Russ. J. Inorg. Chem.*, **12,** 616 (1967).
468. B. I. Nabivanets and L. N. Kudritskaya, *Russ. J. Inorg. Chem.*, **12,** 789 (1967).
469. J. E. D. Davies and D. A. Long, *J. Chem. Soc.*, **1968,** A, 2560.
470. D. M. Adams and D. C. Newton, *J. Chem. Soc.*, **1968,** A, 2262.
471. B. A. Voitovich and E. V. Zvagolskaya, *Ukr. Khim. Zh.*, **33,** 1258 (1967).
472. M. T. Rogers and J. A. Ryan, *J. Phys. Chem.*, **72,** 1340 (1968).
473. C. C. Smitskamp, K. Olie and H. Gerding, *Z. anorg. allgem. Chem.*, **359,** 318 (1968).
474. A. A. Shvets, O. A. Osipov and A. M. Shakirova, *J. Gen. Chem. USSR*, **37,** 2588 (1967).
475. B. P. Susz and B. Petitpierre, *Helv. Chim. Acta*, **50,** 392 (1967).
476. B. P. Susz and J. C. Jaccard, *Helv. Chim. Acta*, **50,** 97 (1967).
477. B. F. Studd and A. G. Swallow, *J. Chem. Soc.*, **1968,** A, 1961.
478. J. M. Bull, M. J. Frazer and J. Measures, *Chem. Commun.*, **1968,** 1310.
479. J. Charalambous and M. J. Frazer, *J. Chem. Soc.*, **1968,** A, 2361.
480. A. L. Allred and D. W. Thompson, *Inorg. Chem.*, **7,** 1196 (1968).
481. D. N. Sen and P. Umapathy, *J. Indian Chem. Soc.*, **45,** 810 (1968).
482. R. Rivest, H. S. Ahuja and S. C. Jain, *J. Inorg. Nucl. Chem.*, **30,** 2459 (1968).
483. G. S. Kyker and E. P. Schram, *J. Am. Chem. Soc.*, **90,** 3672 (1968).
484. G. S. Kyker and E. P. Schram, *J. Am. Chem. Soc.*, **90,** 3678 (1968).
485. N. Wieberg and K. H. Schmid, *Chem. Ber.*, **100,** 748 (1967).
486. W. Haase and H. Hoppe, *Acta Cryst.*, **24B,** 281 (1968).
487. K. Andra, *Chem. Ber.*, **101,** 1013 (1968).
488. L. V. Orlova, A. D. Garnovskii, O. A. Osipov and O. A. Raevskii, *J. Gen. Chem. USSR*, **37,** 1704 (1967).
489. M. R. Farona and G. R. Tompkin, *Spectrochim. Acta*, **24A,** 788 (1968).
490. K. L. Baker and G. W. A. Fowles, *J. Chem. Soc.*, **1968,** A, 801.
491. D. Gervais, M. Basso-Bert, J. Labarre and F. Gallais, *Compt. Rend.*, **266C,** 1183 (1968).
492. G. H. Dahl and B. P. Block, *Inorg. Chem.*, **6,** 1439 (1967).
493. J. J. Salzmann, *Helv. Chim. Acta*, **51,** 601 (1968).
494. E. Lindner, R. Lehner and H. Scheer, *Chem. Ber.*, **100,** 1331 (1967).
495. S. C. Jain and R. Rivest, *Can. J. Chem.*, **45,** 139 (1967).
496. A. D. Garnovskii, O. A. Osipov, K. M. Ismailov and N. D. Chikina, *Russ. J. Inorg. Chem.*, **12,** 80 (1967).

497. U. A. Nevskaya, A. K. Nurmakova and T. N. Sumarokova, *Izv. Akad. Nauk Kaz. SSR, Ser. Khim.*, **18,** 20 (1968).
498. T. A. Zavaritskaya and M. N. Podgaiskaya, *Zh. Priklad. Khim.*, **41,** 948 (1968).
499. G. W. A. Fowles, D. F. Lewis and R. A. Walton, *J. Chem. Soc.*, **1968,** A, 1468.
500. J. E. Drake, J. Vekris and J. S. Wood, *J. Chem. Soc.*, **1968,** A, 1000.
501. R. B. Johannesen and C. L. Gordon, *Inorg. Syn.*, **9,** 46 (1967).
502. M. F. Lappert and B. Prokai, *J. Chem. Soc.*, **1967,** A, 124.
503. R. N. Lowry and R. C. Fay, *Inorg. Syn.*, **10,** 1 (1967).
504. N. M. Karpinkskaya and S. N. Andreev, *Russ. J. Inorg. Chem.*, **13,** 25 (1968).
505. L. K. Chuchalin, L. K. Peshchevitskii, I. A. Kuzin, Z. A. Grankina and E. V. Kulepov, *Zh. Strukt. Khim.*, **9,** 213 (1968).
506. G. W. A. Fowles, T. E. Lester and R. A. Walton, *J. Chem. Soc.*, **1968,** A, 198.
507. P. C. Crouch, G. W. A. Fowles and R. A. Walton, *J. Chem. Soc.*, **1968,** A, 2172.
508. J. S. Wood, *Inorg. Chem.*, **7,** 852 (1968).
509. G. D. McDonald, M. Thompson and E. M. Larsen, *Inorg. Chem.*, **7,** 648 (1968).
510. D. J. Machin, K. S. Murray and R. A. Walton, *J. Chem. Soc.*, **1968,** A, 195.
511. G. W. A. Fowles and T. E. Lester, *J. Chem. Soc.*, **1968,** A, 1180.
512. G. W. A. Fowles, T. E. Lester and B. J. Russ, *J. Chem. Soc.*, **1968,** A, 805.
513. D. V. Korol'kov and V. D. Zahlarghevskaya, *Russ. J. Inorg. Chem.*, **12,** 1561 (1967).
514. G. W. A. Fowles, T. E. Lester and R. A. Walton, *J. Chem. Soc.*, **1968,** A, 1081.
515. G. A. Lopatkina and N. S. Masalovich, *Zh. Priklad. Khim.*, **40,** 2632 (1967).
516. A. A. Lastochkina, I. A. Sheka and L. A. Malinko, *Izv. Sib. Otd. Akad. Nauk SSSR, Ser. Khim. Nauk*, **1968,** 37.
517. P. Ehrlich, H. Kueknl and G. Mueller, *Z. anorg. allgem. Chem.*, **357,** 172 (1968).
518. J. A. Creighton and J. H. S. Green, *J. Chem. Soc.*, **1968,** A, 808.
519. R. J. H. Clark, L. Maresca and R. J. Puddephatt, *Inorg. Chem.*, **7,** 1603 (1968).
520. S. M. Horner, W. W. Horner, W. B. Robertson, F. N. Collier and S. Y. Tyree, *J. Less-Common Metals*, **15,** 29 (1968).
521. E. Wendling and J. de Lavillandre, *Bull. Soc. Chim. France*, **1967,** 2142.
522. Y. Y. Kharitonov, E. A. Ionova and Y. A. Buslaev, *Izv. Akad. Nauk SSSR, Neorg. Mater.*, **4,** 720 (1968).
523. W. Hasse and H. Hoppe, *Acta Cryst.*, **24B,** 282 (1968).
524. E. Wendling, *Rev. Chim. Minerale*, **4,** 425 (1967).
525. P. B. Burkert, H. P. Fritz and G. Stefaniak, *Z. Naturforsch.*, **23b,** 872 (1968).
526. D. V. Korol'kov and G. N. Kudryashova, *Zh. Neorg. Khim.*, **13,** 1626 (1968).
527. R. Saillant and R. A. D. Wentworth, *Inorg. Chem.*, **7,** 1606 (1968).

Chapter 3

Vanadium

Proceeding from titanium to vanadium, there is a marked decrease in the stability of the maximum oxidation state. For vanadium, the only binary pentahalide is the fluoride, but oxygen stabilizes vanadium(V) to such an extent that even vanadium oxide tribromide is known. Vanadium(III) is not a strong reducing agent in halide systems, but vanadium(III) halides are rapidly oxidized to higher states even by air.

The chemistry of the vanadium halides has recently been reviewed by several authors[1–4].

HALIDES AND OXIDE HALIDES

The known halides and oxide halides of vanadium are given in Tables 3.1 and 3.2.

TABLE 3.1
Halides of Vanadium

Oxidation state	Fluoride	Chloride	Bromide	Iodide
V	VF_5			
IV	VF_4	VCl_4	VBr_4	
III	VF_3	VCl_3	VBr_3	VI_3
II	VF_2	VCl_2	VBr_2	VI_2

TABLE 3.2
Oxide Halides of Vanadium

Oxidation state	Fluoride	Chloride	Bromide	Iodide
V	VOF_3	$VOCl_3$	$VOBr_3$	
	VO_2F	VO_2Cl		
IV	VOF_2	$VOCl_2$	$VOBr_2$	VOI_2
III	VOF	$VOCl$		

As expected, oxide halides are formed predominantly in the higher oxidation states and, as mentioned above, exceptional stability is conferred on vanadium(v).

Oxidation State V

Vanadium pentafluoride. Undoubtedly vanadium pentafluoride is best prepared by direct fluorination of vanadium metal at about 300° in a flow system[5–8]. It may also be prepared by the action of halogen fluorides such as chlorine trifluoride, bromine trifluoride and iodine pentafluoride on vanadium metal or other vandium halides[6,9–11]. The use of halogen fluorides in the preparation of vanadium pentafluoride is to be avoided if possible. It has been found[12] that material prepared in this way is often contaminated with free halogen and even halogen fluoride itself. The removal of this contamination is relatively difficult and requires several purification procedures. With bromine trifluoride in particular, relatively stable adducts are formed. Naturally, the presence of quite small amounts of halogen and halogen fluorides can cause erroneous results when precise physical measurements are being made or when chemical reactivity is being studied.

Vanadium pentafluoride has also been produced by the action of excess fluorine on vanadium pentaoxide in a bomb at 200–475° and under a pressure of 10–50 atmospheres[13], and has been identified as an intermediate product in the reaction between vanadium pentaoxide and nitrosyl fluoride[14]. Vanadium pentafluoride was first reported by Ruff and Lickfett[15] as resulting from the thermal disproportionation of vanadium tetrafluoride. Emeleus and Gutmann[6] showed that the product described as vanadium pentafluoride by Ruff and Lickfett was in fact vanadium oxide trifluoride, formed by attack on the glass apparatus. However, vanadium tetrafluoride does disproportionate to give vanadium pentafluoride and vanadium trifluoride, but this method of preparing vanadium pentafluoride requires very careful preparation and handling of the vanadium tetrafluoride.

Although vanadium pentafluoride is normally described as a yellow, low-melting solid[6–8], it is in fact a colourless compound when rigorously purified[16]. The yellow colour is probably due to traces of vanadium oxide trifluoride or vanadium pentaoxide. It is extremely readily hydrolysed by moist air. At 25°, 3.3 moles of vandium pentafluoride will dissolve in 1000 gm of anhydrous hydrogen fluoride[17]. The Raman spectrum of vanadium pentafluoride in anhydrous hydrogen fluoride does *not* show the presence of the hexafluorovanadate(v) anion, indicating that vanadium pentafluoride is a weak acid in anhydrous hydrogen fluoride.

Vanadium pentafluoride is very volatile and the variation of vapour pressure with temperature has been investigated several times. The following equations were deduced from the experimental observations.

Solid: $\log p_{mm} = 11.049 - 2608/T$ (−20 to 190°) (reference 7)

Liquid: $\log p_{mm} = 10.430 - 3423/T$ (20−45°) (reference 7)

$\log p_{mm} = 11.764 - 3387/T - 171{,}910/T^2$ (24–58°) (reference 8)

$\log p_{mm} = 1.002 + 3654/T - 979{,}350/T^2$ (58–86°) (reference 8)

The agreement between the two vapour pressure equations for similar temperature ranges is quite good.

The known thermodynamic data for vanadium pentafluoride are summarized in Table 3.3.

TABLE 3.3
Thermodynamic Data for Vanadium Pentafluoride

Property	Value	References
M.p.	19.0°	8
	19.5°	7
B.p.	47.9°	8
	48.3°	7
ΔH°_{form} (kcal mole^{-1})	−35.2	18
ΔH_{fus} (kcal mole^{-1})	11.94	7
ΔH_{vap} (kcal mole^{-1})	10.62	7, 8

Solid vanadium pentafluoride is known to have an orthorhombic unit cell[19] and preliminary X-ray diffraction data[20] give $a = 5.4$, $b = 7.5$ and $c = 16.6$ Å. The results indicate that vanadium pentafluoride has a similar structure to molybdenum oxide tetrafluoride[19], that is, it has a chain structure with *cis* bridging groups, giving the vanadium atoms octahedral coordination.

As with the other transition-metal pentafluorides[21], vanadium pentafluoride melts to give a dense, very viscous liquid. The density varies with temperature according to the equation[22]

$$\rho = 2.483 - 0.00349(T - 25) \text{ g cm}^{-3}$$

while the viscosity is given by[22]

$$\eta = 124 - 7.2(T - 25) \text{ cp}$$

over the temperature range 25–32°.

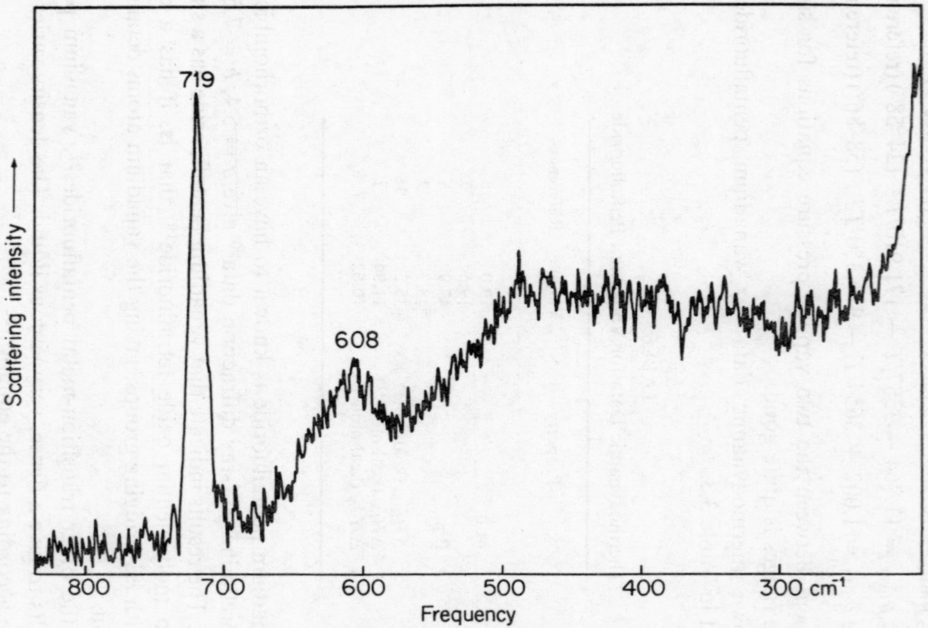

Figure 3.1 Raman spectrum of gaseous vanadium pentafluoride. (Reproduced by permission from H. H. Claassen and H. Selig, *J. Chem. Phys.*, **44**, 4039 (1966))

Liquid vanadium pentafluoride has a Trouton constant[7] of 33.1, indicative of considerable association. The presence of polymeric species in liquid vanadium pentafluoride has been elegantly demonstrated by Claassen and Selig[16] by using Raman spectroscopy. By comparing the Raman spectra of gaseous vanadium pentafluoride, which is known from vapour density measurements to be monomeric[7,8], and of liquid vanadium pentafluoride as a function of temperature, they showed that as the temperature increased, the fraction of monomer also increased. Figures 3.1,

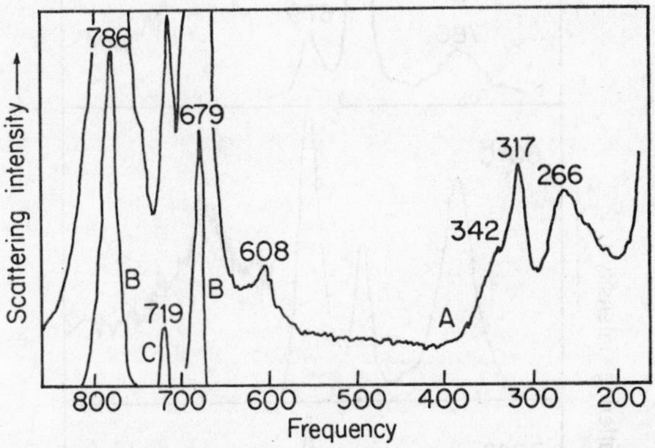

Figure 3.2 Raman spectrum of liquid vanadium pentafluoride at room temperature. (Reproduced by permission from H. H. Claassen and H. Selig, *J. Chem. Phys.*, **44**, 4039 (1966))

3.2 and 3.3 show the Raman spectra of gaseous and liquid vanadium pentafluoride at several temperatures and clearly demonstrate that although there is some monomer at room temperature, the predominant species at this temperature are polymeric.

At 25° vanadium pentafluoride has a specific conductivity[7] of 2.4×10^{-4}, indicating that partial dissociation occurs in the liquid state. The conductivity is said to be due to a small amount of the dissociation reaction

$$2VF_5 \rightleftharpoons VF_4^+ + VF_6^-$$

Some support for this proposal is in the fact that potassium hexafluorovanadate(V) may be isolated from the interaction of potassium fluoride and vanadium pentafluoride[6].

As stated earlier, density measurements indicate that vanadium pentafluoride is monomeric in the vapour phase[7,8]. The infrared spectrum of the vapour has been recorded several times. Blanchard[10] compared the spectrum of vanadium pentafluoride with that of other trigonal-bipyramidal compounds and concluded that vanadium pentafluoride has this geometry. Cavell and Clark[23], in a more detailed study, could not positively distinguish between a trigonal bipyramid and a square pyramid, although they favoured the former. The problem was solved by a

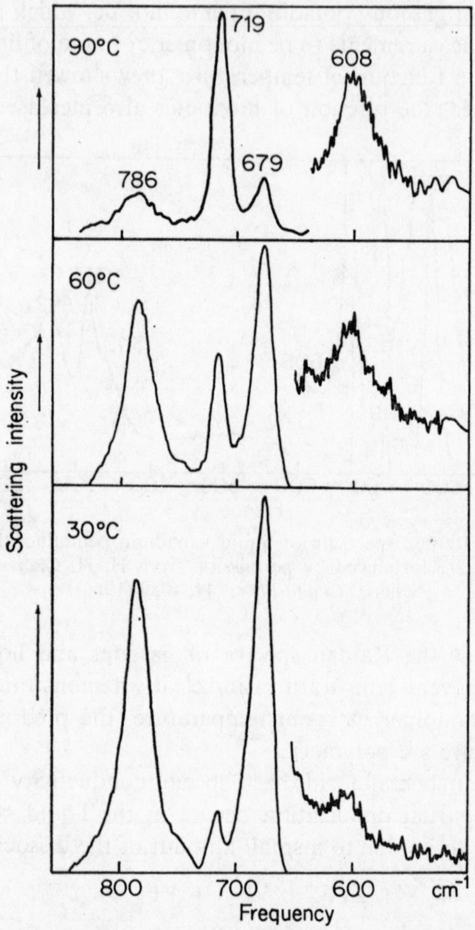

Figure 3.3 Raman spectrum of liquid VF_5 between 600 and 850 cm^{-1} as a function of temperature. (Reproduced by permission from H. H. Claassen and H. Selig, *J. Chem. Phys.*, **44**, 4039 (1966))

combination of infrared and Raman spectroscopy by Claassen and Selig[16] who confirmed the D_{3h} symmetry. The following fundamental frequencies were assigned: ν_1 719, ν_2 608, ν_3 784, ν_4 331, ν_5 810, ν_6 282, ν_7 200 and ν_8 350 cm^{-1}. Claassen and Selig[16] carried out a modified Urey-Bradley potential function calculation, to obtain the various force constants. They calculated an average vanadium-fluorine bond distance of 1.80 Å. At about the same time as this study, Romanov and Spiridonov[24] observed the electron-diffraction pattern of vanadium pentafluoride vapour and also concluded that the molecule is a trigonal bipyramid. These workers estimated the mean vanadium-fluorine distance to be 1.71 Å.

Vanadium pentafluoride is an extremely reactive compound and this is exemplified by the fact that it can attack the inert polymer Kel-F[5,8], presumably by chlorine abstraction from the polymer. It is both a powerful fluorinating and oxidizing agent and it also readily forms adducts.

The reactions of vanadium pentafluoride with covalent chlorides involve halogen-exchange and oxidation-reduction reactions, the latter resulting from the instability of other vanadium(V) halide species. The reactions often occur at very low temperatures and in some cases with explosive violence. They are very complex, involving a number of successive reactions[5,7,12]. Canterford and O'Donnell[12] have shown that these reactions may be rationalized on the basis of complete halogen exchange producing vanadium tetrachloride and the appropriate fluoride. As soon as some vanadium tetrachloride is produced, it itself undergoes further reactions. One of the most important of these is the reaction with more vanadium pentafluoride to give a mixed halide of vanadium(IV)[12].

Vanadium pentafluoride is readily reduced to the tetrafluoride by a variety of reductants[5,17,25]. Probably the most efficient and experimentally convenient reagent is phosphorus trifluoride, since the reaction proceeds at temperature below room temperature and the oxidation product is volatile. Selig and Frlec[17] found that the reduction of vanadium pentafluoride with hydrazinium fluoride, $N_2H_6F_2$, proceeded via the unstable complex $N_2H_6(VF_6)_2$, which readily decomposes on standing to give vanadium tetrafluoride. Vanadium pentafluoride is reduced also by ammonia, pyridine and ethylenediamine, but in these cases adducts of vanadium tetrafluoride are formed. The adducts are $VF_4 \cdot NH_3$, $VF_4 \cdot C_5H_6N$ and $VF_4 \cdot (en)_3$ (en = ethylenediamine) respectively; the first two adducts may also be made by direct interaction of the constituents[11]. It is interesting to note that in the corresponding reactions of niobium pentafluoride with these reagents, adduct formation occurs *without* reduction[26]. This is consistent with the known stability of niobium(V)[5,27].

Vanadium pentafluoride reacts readily with sulphur dioxide and sulphur trioxide at room temperature[28] to give vanadium oxide trifluoride in both cases. Thionyl fluoride is formed in the sulphur dioxide reaction, whereas sulphur trioxide gives pyrosulphuryl fluoride, $S_2O_5F_2$.

Nitryl fluoride reacts with vanadium pentafluoride at −78° to give

$NO_2^+[VF_6^-]$[28], which is also formed by the action of liquid nitryl fluoride on vanadium tetrafluoride in the presence of iodine pentafluoride[11], and from the action of nitryl fluoride on vanadium pentaoxide[14]. The complex is thermally unstable and the dissociation pressure over the temperature range 25–90° is given by the expression[28]

$$\log p_{mm} = 9.726 - 2725/T$$

Similarly, Clark and Emeleus[28] observed that nitrosyl fluoride reacted with vanadium pentafluoride to give $NO^+[VF_6^-]$. This complex is also thermally unstable, the dissociation pressure over the temperature range 25–90° being given by the expression[28]

$$\log p_{mm} = 13.45 - 4048/T$$

$NO^+[VF_6^-]$ may also be made by refluxing a mixture of nitrosyl chloride, vanadium pentaoxide and bromine trifluoride[29].

Moody and Selig[30] have found that vanadium pentafluoride reacts with xenon hexafluoride and xenon oxide tetrafluoride to form $2XeF_6 \cdot VF_5$ and $2XeOF_4 \cdot VF_5$ respectively. Both complexes are completely dissociated in the vapour phase, as indicated by infrared and molecular weight measurements.

Vanadium oxide trifluoride. Whereas the high pressure reaction between vanadium pentaoxide and fluorine gives the pentafluoride[13], vanadium oxide trifluoride is most conveniently prepared by fluorination of vanadium pentaoxide at about 450° in a flow system[13,31,32]. Other, but less efficient methods, include the action of bromine trifluoride or chlorine trifluoride on vanadium pentaoxide, bromine trifluoride on vanadium oxide trichloride, and the interaction of vanadium pentafluoride with sulphur dioxide and sulphur trioxide[6,10,28]. Vanadium oxide trifluoride has also been isolated as an intermediate product in the nitrosyl fluoride and vanadium pentaoxide reaction[14]. It is a yellow solid which vaporizes as the monomer[32,33]. A vapour pressure of one atmosphere is reached at 110°[32].

The infrared spectrum of gaseous vanadium oxide trifluoride has been recorded by Blanchard[10] and also by Selig and Claassen[33]. The latter workers concluded that vanadium oxide trifluoride has a similar stereochemistry to the other vanadium oxide trihalides and assigned the following fundamentals on the basis of C_{3v} symmetry: ν_1 1058, ν_2 721.5, ν_3 258, ν_4 806, ν_5 308 and ν_6 204 cm^{-1}.

Virtually no chemistry of vanadium oxide trifluoride has been investigated, but the adducts $2VOF_3 \cdot 3IOF_3$ and $VOF_3 \cdot 2SeF_4$ have been prepared by indirect methods[34,35].

Vanadium dioxide fluoride. Weidlein and Dehnicke[36] prepared vanadium dioxide fluoride by passing a 1:1 mixture of fluorine and nitrogen over vanadium dioxide chloride at 75–80°. It is a crystalline solid which is insoluble in non-polar solvents but instantly hydrolysed by water. It is thermally unstable, decomposing above 300° to give vanadium oxide trifluoride and vanadium pentaoxide.

The infrared spectrum of vanadium dioxide fluoride shows bands at 1135, 1024, 875 and 558 cm^{-1}. Weidlein and Dehnicke[36] suggested that the infrared spectrum indicates an ionic formulation, $[VO_2^+]F^-$. This suggestion is supported by the compound's insolubility in non-polar solvents and by its reaction with antimony pentafluoride. A series of compounds can be isolated from the reaction with antimony pentafluoride; one of the first products is $VO_2^+[Sb_3F_{16}^-]$, containing the known trimeric antimony(v) anion, and on heating the reaction mixture to about 250° the ultimate product is $VO_2^+[SbF_6^-]$.

Vanadium oxide trichloride. The principal methods of preparing this well-known compound are summarized in Table 3.4.

Vanadium oxide trichloride is a lemon-yellow liquid at room temperature and it is extremely sensitive to moisture, giving red fumes in air. The red fumes are probably the pentaoxide. The vapour pressure of liquid vandium oxide trichloride has been measured several times and the following equations were deduced from the experimental data.

$$\log p_{\text{mm}} = 7.69 - 1920/T \quad (15.4\text{–}125°) \quad \text{(reference 40)}$$

$$\log p_{\text{mm}} = 7.37 - 1794/T \quad (18\text{–}90°) \quad \text{(reference 52)}$$

$$\log p_{\text{mm}} = 7.70 - 1921/T \quad (19\text{–}99°) \quad \text{(reference 48)}$$

$$\log p_{\text{mm}} = 7.60 - 1888/T \quad (58\text{–}124°) \quad \text{(reference 53)}$$

$$\log p_{\text{mm}} = 7.802 - 1968/T \quad (35\text{–}120°) \quad \text{(reference 41)}$$

It is clear by inspection that in general there is good agreement between the various equations.

Thermodynamic data for vanadium oxide trichloride derived from the vapour pressure and other measurements are given in Table 3.5.

The infrared and Raman spectra of gaseous vanadium oxide trichloride have been interpreted on the basis of C_{3v} symmetry[38,58]. Very good agreement was obtained between the two groups, and the fundamentals assigned by Miller and Cousins[38] are given in Table 3.6.

Palmer[47] has recorded the electron-diffraction pattern of vanadium oxide trichloride and concluded that the molecule possesses C_{3v} symmetry with the dimensions shown in Table 3.7.

TABLE 3.4
Preparation of Vanadium Oxide Trichloride

Method	Conditions	References
V_2O_5, C and Cl_2	Flow system, 300–400°	37–41
V_2O_5 and CH_3COCl	Reflux	42
V_2O_5 and C_5Cl_8	Reflux	43
V_2O_5 and $AlCl_3$	Sealed tube, 400°	44
V_2O_5 and trichloromethylbenzene	Sealed tube, 200–400° stoichiometric amounts	45
V_2O_3, C and Cl_2	Flow system, 200°	46–48
$VOCl_2$ decomposition	Greater than 300°	49
VCl_4 and O_2	Flow system, 300°	50
VCl_3 and O_2	Flow system, 200°	49, 51

TABLE 3.5
Thermodynamic Data for Vanadium Oxide Trichloride

Property	Value	References
M.p.	−79.5°	38, 54
	−78.9°	48, 49
	−77°	39, 55
B.p.	125.6	40
	126.7 ± 0.1	39, 52, 53
	127	38, 41, 49
	127.2	46, 48
ΔH°_{form} (kcal mole^{-1})	−177.2	56
ΔH_{fus} (kcal mole^{-1})	2.29	57
ΔH_{vap} (kcal mole^{-1})	8.21	52
	8.64	53
	8.7	54
	8.8	40, 48
ΔS_{fus} (cal deg^{-1} mole^{-1})	11.7	57

TABLE 3.6
Fundamental Vibrational Assignments for Vanadium Oxide Trichloride

Mode	Description	Value (cm^{-1})
ν_1	V═O stretch	1035
ν_2	V—Cl stretch	408
ν_3	VCl_3 deformation	165
ν_4	V—Cl stretch	504
ν_5	VCl_3 deformation	249
ν_6	Cl_3—V—O rock	129

The binary systems $VOCl_3$—MCl_4, where M is carbon, germanium, silicon, tin, titanium and vanadium, have been studied in detail by Oppermann[57].

Vanadium oxide trichloride forms a number of adducts with donor molecules such as phosphorus oxide trichloride[59–61], dioxan[62], dimethylformamide[63] and alkyl nitriles[64]. It also undergoes substitution reactions particularly with phenols and alcohols[64–67].

TABLE 3.7

Internuclear Dimensions of Vanadium Oxide Trichloride

Distance		Angle	
V—O	1.56 Å	Cl—V—Cl	111.3°
V—Cl	2.12 Å	Cl—V—O	108.2°

Vanadium dioxide chloride. This compound may be prepared by passing ozone through boiling vanadium oxide trichloride[68,69] or by passing chlorine monoxide, diluted with oxygen, through vanadium oxide trichloride at room temperature[41,69]. It is a deep orange hygroscopic crystalline solid. It is thermally unstable above about 150° forming vanadium oxide trichloride and vanadium pentaoxide[41,69]. The equilibrium decomposition pressure of vanadium oxide trichloride over solid vanadium dioxide chloride in the temperature range 100–175° has been measured by Oppermann[41] and is given by the equation

$$\log p_{\text{mm}} = 12.076 - 4154/T$$

The heat and entropy of decomposition are 19.0 kcal mole^{-1} and 41.8 cal deg^{-1} mole^{-1} respectively[41], while the heat of formation of vanadium dioxide chloride has been reported to be −183 kcal mole^{-1}.

Dehnicke[69] has recorded the infrared spectrum of vanadium dioxide chloride and it shows absorptions at 990, 855, 500 and 455 cm^{-1}.

Very little is known of the chemistry of vanadium dioxide chloride except that it reacts with carbon tetrachloride at room temperature to give vanadium oxide trichloride, and with pyridine the adduct $2VO_2Cl \cdot 3C_5H_5N$ is formed[69].

Vanadium oxide tribromide. Vanadium oxide tribromide has been prepared by brominating either a mixture of vanadium pentaoxide and carbon in a flow system at red heat[70] or, alternatively, a vanadium trioxide-carbon mixture at 200° in a flow system[71]. It is a deep red liquid which is readily hydrolysed by moist air. It melts at $-59°$ and boils at 170°. Vanadium oxide tribromide is thermally unstable, and slowly decomposes at room temperature even in the dark. Heating causes rapid decomposition[71].

Miller and Baer[71] have examined the infrared spectrum of liquid vanadium oxide tribromide and assigned the observed spectrum on the basis of C_{3v} symmetry. The fundamentals and their assignments are given in Table 3.8.

TABLE 3.8
Fundamental Vibrational Assignments for Vanadium Oxide Tribromide

Mode	Description	Value (cm^{-1})
ν_1	V═O stretch	1025
ν_2	V—Br stretch	271
ν_3	VBr_3 deformation	120
ν_4	V—Br stretch	400
ν_5	Br_3—V—O rock	212
ν_6	VBr_3 deformation	83

Nothing is known of the chemistry of vanadium oxide tribromide except that Dehnicke[68] found that it reacts with ozone at room temperature to give a compound of empirical formula $V_4O_9Br_2$.

Oxidation State IV

Vanadium tetrafluoride. Vanadium tetrafluoride was first prepared by Ruff and Lickfett[15] by the interaction of vanadium tetrachloride and liquid anhydrous hydrogen fluoride. They described their product as an easily hydrolysed brown solid which disproportionated above 325° to vanadium pentafluoride and vanadium trifluoride. Cavell and Clark[11] have made a detailed study of the preparation and properties of vanadium tetrafluoride and they found that pure vanadium tetrafluoride is a lime-green solid which on exposure to moist air turns brown. They therefore concluded that the product prepared by Ruff and Lickfett had suffered surface hydrolysis. Cavell and Clark[11] found that the interaction of vanadium tetrachloride and anhydrous hydrogen fluoride is most efficiently carried out in an inert solvent such as trifluorochloromethane.

Vanadium tetrafluoride may also be prepared by the fluorination of vanadium metal in a flow system at the relatively low temperature of 200°; the vanadium pentafluoride also formed in this reaction is readily removed by sublimation[11]. This method usually gives a more massive product than the previously described method which is less susceptible to surface hydrolysis. Kolditz and coworkers[72] found that vanadium trichloride is converted to the tetrafluoride by heating it to 150° with fluorine in an atmosphere of argon. These workers[72] also reported that they obtained a light green form of vanadium tetrafluoride, but they concluded that the two reported forms are different polymorphs of the tetrafluoride. As noted previously, vanadium tetrafluoride is readily produced by the reduction of vanadium pentafluoride with phosphorus trifluoride[5,25] and with hydrazinium fluoride[17]. This latter method is of interest since it gave the first sample of crystalline vanadium tetrafluoride.

Cavell and Clark[11] found that Ruff and Lickfett's[15] temperature for the disproportionation of vanadium tetrafluoride was too high. They found that it disproportionates slowly even at room temperature; at 100–120° *in vacuo* it is rapidly decomposed. More recently, however, Selig and Frlec[17] found that in Kel-F apparatus the disproportionation begins at about 150° and they suggested that the lower temperature reported by Cavell and Clark[11] was possibly due to partial hydrolysis. Indeed, Cavell and Clark themselves reported that it is possible to sublime vanadium tetrafluoride slowly at about 100° in vacuum with only a small amount of decomposition.

The heat of formation of vanadium tetrafluoride has been determined[18] to be -321 kcal mole^{-1}, and by using the heats of formation of the tri-, tetra- and pentafluorides, the heat of disproportionation of vanadium tetrafluoride has been calculated[73] to be 29 kcal mole^{-1}.

Cavell and Clark[11] indexed the X-ray powder diffraction pattern of vanadium tetrafluoride on the basis of a hexagonal cell with $a = 5.38$ and $c = 5.16$ Å. However no firm conclusions as to the positions of the atoms in the unit cell could be deduced, no doubt partially due to the inherent instability of the compound. On the basis of its chemical and physical properties it does not appear that vanadium tetrafluoride is of the same structural type as either zirconium tetrafluoride or the molecular tetrafluorides such as that of silicon. It was suggested[11] that the structure may consist of a condensed system, containing VF_6 octahedra each sharing two edges with other octahedra. Such a structure is consistent with the propensity of vanadium to achieve octahedral coordination and the intermediate volatility of the compound between that of vanadium pentafluoride and vanadium trifluoride.

Selig and Frlec[17] attempted to carry out a single-crystal X-ray diffraction study of vanadium tetrafluoride. They found an orthorhombic (pseudo-hexagonal) unit cell with $a = 9.32$, $b = 5.35$ and $c = 5.16$ Å. However they were not able to deduce the structure from their data.

The magnetic susceptibility of vanadium tetrafluoride has been measured over a wide temperature range[11]. The Curie-Weiss law is obeyed with a Weiss constant of 198° and a magnetic moment of 2.17 BM. A detailed interpretation of this anomalously high magnetic moment must await the determination of the structure of the compound.

The infrared spectrum of vanadium tetrafluoride has been recorded by Cavell and Clark[23] and absorptions were observed at 1025, 837, 780 and 530 cm^{-1}. No assignment of these bands is possible at this stage.

Vanadium tetrafluoride forms 1:1 adducts with selenium tetrafluoride, ammonia and pyridine[11], the last two also being formed indirectly from vanadium pentafluoride[26].

Vanadium fluoride chlorides. Kolditz and coworkers[72] found that the adduct $VCl_4 \cdot PCl_5$, when dissolved in arsenic trichloride, reacts quite readily with arsenic trifluoride to produce phosphorus pentafluoride and mixed halides of vanadium(IV). They were able to isolate vanadium trifluoride chloride, VF_3Cl, and divanadium pentafluoride trichloride, $V_2F_5Cl_3$, depending on the amount of arsenic trifluoride added. It has also been found[12] that the mixed halides are formed in a number of halogen-exchange reactions of vanadium pentafluoride.

Both mixed halides are brown powders which are very moisture sensitive. Moreover, they are thermally unstable, and they decompose even under water-pump vacuum. The mixed halides are sparingly soluble in non-polar solvents and tend to react slowly with polar solvents such as acetonitrile.

Vanadium oxide difluoride. Ruff and Lickfett[15] claimed to have prepared vanadium oxide difluoride as a yellow solid, by passing anhydrous hydrogen fluoride over a compound reported as vanadium oxide dibromide at 600–700°. Very few properties of the compound were given and there has been no confirmation of this preparative method. The only other report of vanadium oxide difluoride is that it may be formed by the interaction between vanadium dioxide and aqueous solutions containing more than 70% hydrogen fluoride[74].

Several hydrates of vanadium oxide difluoride are known. These are prepared by the electrolytic reduction of vanadium pentaoxide in aqueous hydrogen fluoride[75] or by the dissolution of vanadium dioxide in aqueous hydrogen fluoride[74]. A vanadium-oxygen stretching frequency of 984 cm^{-1} has been reported for one of the hydrates of vanadium oxide difluoride[76,77].

Vanadium tetrachloride. Vanadium tetrachloride is most conveniently prepared by direct chlorination of vanadium metal in a flow system[40,78–83] at 200–500°. In some cases ferrovanadium has been used and the iron trichloride formed has been separated by sublimation. Other methods of preparation include the chlorination of vanadium trichloride[51,83], the disproportionation of vanadium trichloride[49,83–85], the interaction of stoichiometric amounts of vanadium pentaoxide and trichloromethylbenzene at 200–300° in a sealed tube[45] and also chlorination of a mixture of vanadium pentaoxide and carbon[39,40] at 800°. If a lower temperature is used in this last reaction, considerable quantities of vanadium oxide trichloride are also formed.

Vanadium tetrachloride is a deep red-brown liquid which is extremely readily hydrolysed and fumes copiously in moist air. Care must be taken in purification since vanadium oxide trichloride is normally formed as a by-product and its removal is relatively difficult because of its similar volatility. Vanadium tetrachloride is thermally unstable, decomposing readily when refluxed in vacuum to give vanadium trichloride and chlorine[79,86].

The vapour pressure of vanadium tetrachloride has been measured twice and the results are expressed by the following equations:

$$\log p_{\text{mm}} = 7.581 - 1998/T \text{ (30–80°)} \quad \text{(reference 79)}$$

$$\log p_{\text{mm}} = 7.62 - 2020/T \text{ (30–153°)} \quad \text{(reference 40)}$$

It is clear that the agreement between the independent sets of data is quite good.

Reported thermodynamic data for vanadium tetrachloride derived from the vapour pressure measurements and from other sources are summarized in Table 3.8.

The remarkably wide range of melting point and boiling point recorded for vanadium tetrachloride is probably best attributed to impurities such as vanadium oxide trichloride.

The electron-diffraction pattern of vanadium tetrachloride vapour has been examined by three groups of workers and all agree that in the gas phase vanadium tetrachloride has a tetrahedral configuration. The earliest measurement by Lipscomb and Whittaker[90] gave an average vanadium-chlorine bond distance of 2.03 Å. More recent measurements by Spiridonov and Romanov[84] and by Morino and Uehara[91] give an average vanadium-chlorine bond length of 2.14 Å. Spiridonov and Romanov[84] report that Jahn-Teller distortion of the tetrahedron was either absent or could not be detected by this method. However, Morino and Uehara[91], who also examined the infrared spectrum of vanadium tetrachloride, made a detailed study of the vibronic interactions in vanadium tetrachloride and found that these interactions suggested, but did not confirm, the existence of Jahn-Teller distortion.

The existence of Jahn-Teller distortion in vanadium tetrachloride has been discussed in theoretical terms by Ballhausen and coworkers[92,93]. From calculations, Ballhausen and Liehr[92] suggest that below 200°K, vanadium tetrachloride assumes a static distortion in which two of the tetrahedral angles are closed by 6° and four of the angles are opened by 3°. Above 200°K the molecule undergoes a pseudo-rotation in the sense

TABLE 3.9
Thermodynamic Data for Vanadium Tetrachloride

Property	Value	References
M.p.	−109°	83
	−28°	49, 55, 82
	−25.7°	79
	−20.5°	39
B.p.	145.9°	87
	148.5°	83
	152.0°	49
	153.0°	39
	153.7°	40
ΔH°_{form} (kcal mole^{-1})	−136.2	88
	−145	89
ΔH_{fus} (kcal mole^{-1})	2.3	57
ΔH_{vap} (kcal mole^{-1})	9.9	40, 87
ΔH_{dec} (kcal mole^{-1})	13.8	79
ΔS_{fus} (cal deg^{-1} mole^{-1})	9.1	57
ΔS_{vap} (cal deg^{-1} mole^{-1})	23.7	87

that each chlorine atom is displaced from the regular tetrahedral position and rotates around the tetrahedral bond with a radius of 0.11 Å. Ballhausen and Liehr[92] point out however, that as yet there is no unambiguous experimental evidence for the presence of Jahn-Teller distortion in this compound. (However see Addenda.)

Dove, Creighton and Woodward[82] have made a detailed study of the Raman spectrum of a carbon tetrachloride solution of vanadium tetrachloride and of the infrared spectra of the liquid, saturated vapour and carbon tetrachloride solution. By comparison with the spectrum of titanium tetrachloride, they assigned the observed absorptions on the basis of tetrahedral symmetry. Some difficulty was encountered in making the following assignments: ν_1 383, ν_2 128, ν_3 475 and ν_4 128 cm^{-1}. On the basis of Urey-Bradley force field calculations, Creighton and coworkers[94] later gave ν_2 129 and ν_4 105 cm^{-1}. Other assignments of the fundamental frequencies of vanadium tetrachloride have been given by Morino and Uehara[91] and by Grubb and Belford[95].

The electronic absorption spectrum of vanadium tetrachloride has been examined by several workers in the near infrared, visible and ultraviolet regions[95–99]. Penella and Taylor[85] give the following assignments for the visible and ultraviolet spectrum:

24,800 cm^{-1}	$^2E \leftarrow T_1$
33,900 cm^{-1}	$^3T_2 \leftarrow T_1$
45,400 cm^{-1}	$^2E \leftarrow {}^2T_2$
50,000 cm^{-1}	$^3T_2 \leftarrow {}^2T_2$

Alderdice[99] gives the frequencies for the first two of the above assignments as 25,000 and 34,100 cm^{-1} respectively.

Blakenship and coworkers[96–98] have reported that gaseous vanadium tetrachloride shows a broad band with a maximum at approximately 9000 cm^{-1} in the near infrared region. This band has been resolved into three components at 6600, 7850 and 9250 cm^{-1}. It was suggested that the fine-structure of this broad band was attributable to Jahn-Teller splitting of the excited state. A detailed examination of this broad band was also reported to support the view that vanadium tetrachloride is a normal liquid, that is, non-associated. Similar bands to those reported above have been observed for solutions of vanadium tetrachloride in carbon tetrachloride[100].

The magnetic properties of vanadium tetrachloride have been examined by Clark and Machin[100] and by several other workers[101]. The compound obeys the Curie Law almost exactly with a magnetic moment of 1.61 BM[100]. However, there may be some inconsistency here since using this result and the value of $10D_q$ derived from the absorption spectrum, a value of 270 cm^{-1} is found for the spin-orbit coupling constant which is *greater* that the value of 250 cm^{-1} for the free ion. Usually the free ion spin-orbit coupling constant is decreased on compound formation.

Vanadium tetrachloride forms a number of adducts with a variety of donor molecules which include amines, nitrogen and sulphur heterocycles, arsines, etc.[61,62,78,102–113]. The majority of the adducts appear to be normal octahedral complexes, although there are exceptions such as the eight-coordinate $VCl_4 \cdot 2D$ [D = *o*-phenylenebis(dimethylarsine)][114]. In general, little is known about these compounds. Clark[105] has shown that the octahedral adducts exhibit a broad vanadium-chlorine stretching frequency at about 360 cm^{-1} with a shoulder at 385 cm^{-1} and a much weaker band at about 310 cm^{-1}.

As expected, vanadium tetrachloride undergoes a number of substitution reactions, particularly with ammonia, amines and alcohols, forming compounds such as $VCl(NH_3)_3$ and $VCl_2(CH_3CN)_2$[115–119].

Vanadium oxide dichloride. Interaction of vanadium pentaoxide and vanadium trichloride in a sealed tube at 600° yields the oxide dichloride, the product being collected at one end of the tube at room temperature[83]. Alternative methods include the interaction of vanadium oxide trichloride with vanadium oxide chloride[41] over the temperature gradient of 450–250°, and the passage of hydrogen chloride gas over vanadium trichloride hexahydrate[49] at 160°.

Vanadium oxide dichloride is a green crystalline solid which is thermally unstable above about 300°, decomposing to vanadium oxide trichloride and vanadium oxide chloride[41,49,120]. Oppermann[41] reported that the equilibrium vapour pressure of vanadium oxide trichloride over solid vanadium oxide dichloride in the temperature range 250–400° is given by the expression

$$\log p_{\text{mm}} = 10.780 - 5192/T$$

The heat and entropy of decomposition are 23.75 kcal mole^{-1} and 36.2 cal deg^{-1} mole^{-1} respectively. The heat of formation of vanadium oxide dichloride is reported[41] to be -165 kcal mole^{-1}.

With donor molecules such as bipyridyl, tertiary phosphines and arsines, pyridine, amines, etc., vanadium oxide dichloride forms a number of adducts which may be prepared directly or indirectly[103,105,116,121–125]. Indirect

TABLE 3.10

Properties of Some Adducts of Vanadium Oxide Dichloride

Compound	μ (BM)	$\nu_{V=O}$ (cm^{-1})
$VOCl_2.2N(CH_3)_3$	1.73	990
$VOCl_2.2NH(CH_3)_2$	1.68	1000
$VOCl_2.2S(CH_3)_2$	1.77	995
$VOCl_2.2S(C_2H_5)_2$	1.78	1010

methods of preparing these adducts include the aerial oxidation of the corresponding vanadium trichloride adduct, partial hydrolysis of a vanadium tetrachloride adduct and preparation from vanadium oxide trichloride.

Baker and coworkers[125] have prepared four vanadium oxide dichloride adducts which, from their magnetic properties, infrared and absorption spectra, and in one case X-ray powder diffraction pattern, have been shown to be five-coordinate. Some properties of these particular adducts are given in Table 3.10.

The trimethylamine complex is isomorphous with $VCl_3 \cdot 2N(CH_3)_3$ which is known to have a trigonal-bipyramidal structure.

A hydrate, $VOCl_2 \cdot xH_2O$, may be prepared as a syrupy liquid by evaporating a mixture of vanadium pentaoxide, ethanol, hydrochloric acid and water until all the pentaoxide has reacted[75]. The vanadium-oxygen stretching frequency occurs[76,77] at 990 cm^{-1}. A hydrochloric acid solution of vanadium oxide dichloride is readily prepared by reducing vanadium pentaoxide in hydrochloric acid with mercury[126].

Vanadium tetrabromide. McCarley and Roddy[127] have reported that vanadium tribromide and bromine react together in a sealed tube at 325°. Condensation of the product at −78° resulted in the isolation of a purplish solid which was shown to be vanadium tetrabromide. This is stable at −45° but decomposes slowly above −23° giving vanadium tribromide and bromine.

Shchukarev and coworkers[128,129], and McCarley and Roddy[127] have proposed the formation of vanadium tetrabromide to account for the observation that vanadium tribromide may be transported by a stream of bromine using an inert carrier gas. These workers, as well as Spiridonov and Romanov[84], have shown that vanadium tribromide disproportionates at relatively high temperatures in the presence of an inert carrier gas to give vanadium tetrabromide vapour.

The heat and entropy of formation of vanadium tetrabromide are −94 kcal $mole^{-1}$ and 44.5 cal deg^{-1} $mole^{-1}$ respectively [128].

In the vapour phase, vanadium tetrabromide is intensely violet and monomeric. From electron-diffraction measurements Spiridonov and Romanov[84] deduced an average vanadium-bromine bond distance of 2.30 Å based on a tetrahedral configuration.

Vanadium oxide dibromide. Dehnicke[130] has prepared vanadium oxide dibromide by bromination of vanadium trioxide at 600° in a flow system. It shows infrared absorptions at 871, 360 and 293 cm^{-1}.

A hydrate, $VOBr_2 \cdot xH_2O$, which has been prepared[75] in a manner analogous to that described for the corresponding vanadium oxide dichloride complex, shows a vanadium-oxygen double bond stretching frequency[76,77] at 996 cm^{-1}.

Vanadium oxide diiodide. Morette[131] has briefly reported the preparation of vanadium oxide diiodide by dissolving vanadium pentaoxide in hydriodic acid. No properties of this compound were given.

Oxidation State III

Vanadium trifluoride. The major methods of preparing vanadium trifluoride are summarized in Table 3.11.

Sturm and Sheridan[136] made a study of the preparation of vanadium trifluoride via vanadium chlorides and the trihydrate of vanadium trifluoride and concluded that none of these methods is satisfactory. They found that the best method of preparing vanadium trifluoride is the thermal decomposition of ammonium hexafluorovanadate(III), $(NH_4)_3VF_6$, at 500–600° in an inert atmosphere. The major difficulty in preparing pure

TABLE 3.11
Preparation of Vanadium Trifluoride

Method	Conditions	References
VN and NF_3	Bomb, 400°	132
VCl_3 and anhydrous HF	Flow system, 600°	6
VCl_2 and anhydrous HF	Flow system, 700°	6
V and anhydrous HF	Bomb, 225°	133
VF_4 disproportionation	125°	11
$VF_3.3H_2O$ and anhydrous HF or inert gas	Flow system, 300°	134, 135

vanadium trifluoride is its involatile nature, so that in several of the preparative methods only a surface coating of the trifluoride is formed on the substrate.

Vanadium trifluoride is a yellow-green solid, insoluble in water and melting at approximately 1400° without decomposition[136]. Siderov and coworkers[137] used mass-spectrometric techniques to determine the vapour pressure of vanadium trifluoride over the temperature range 650–920°. The results are expressed by the equation

$$\log p_{\text{mm}} = 12.357 - 15{,}603/T$$

Jack and Gutmann[138] indexed the X-ray powder diffraction pattern of vanadium trifluoride on the basis of a rhombohedral unit cell with $a = 3.73$ Å and $\alpha = 57.9°$. The lattice consists of regular VF_6 octahedra in an approximately hexagonal close-packed array. The vanadium-fluorine bond distance is 1.95 Å.

Vanadium trifluoride in the solid phase shows a strong broad band at 540 cm^{-1} in its infrared spectrum which may be considered to be the ν_3 frequency for the VF_6 octahedral unit[23].

Clark[139] has recorded the diffuse-reflectance spectrum of vanadium trifluoride and observed bands at 12,000, 14,700 and 22,500 cm^{-1}. He has assigned the bands at 12,000 and 14,700 cm^{-1} to the $^3T_{2g} \leftarrow {}^3T_{1g}(F)$

transition and has suggested that the symmetry of the metal is not exactly octahedral.

At room temperature, vanadium trifluoride has an effective magnetic moment of 2.55 BM[140]. From a ^{19}F NMR study, Saraswati[141] deduced a Weiss constant of 34°. Wollan and coworkers[142] have shown from a neutron-diffraction study that there is no evidence for any magnetic ordering in vanadium trifluoride at temperatures as low as 4.2°K.

Vanadium trifluoride trihydrate. Vanadium trifluoride trihydrate, $VF_3 \cdot 3H_2O$, may be prepared by the electrolytic reduction of vanadium pentaoxide in aqueous hydrogen fluoride[134]. Maak and coworkers[143] have shown that it is isostructural with the β-form of aluminium trifluoride trihydrate.

Ballhausen and Winther[144] have recorded the diffuse-reflectance spectrum of vanadium trifluoride trihydrate and observed absorptions at 11,500, 16,200 and 24,500 cm^{-1}. These bands were assigned as:

11,500 cm^{-1}	$^1E_{1g}, {}^1T_{2g} \leftarrow {}^3T_{1g}(F)$
16,200 cm^{-1}	$^3T_{2g} \leftarrow {}^3T_{1g}(F)$
24,500 cm^{-1}	$^3T_{1g}(P) \leftarrow {}^3T_{1g}(F)$

Vanadium oxide fluoride. Black, crystalline vanadium oxide fluoride has been prepared[145] by interacting the stoichiometric amounts of vanadium trifluoride and vanadium trioxide in a press at 60–65 Kbar pressure and at a temperature of about 1000°. The compound is antiferromagnetic and has the tetragonal rutile-type structure[145]. The unit-cell dimensions were found to vary slightly from sample to sample, apparently due to the formation of non-stoichiometric compounds.

Vanadium trichloride. Vanadium trichloride may be prepared by a variety of methods and a number of these depend on the thermal instability and ease of reduction of vanadium tetrachloride. The methods using sulphur monochloride, carbon disulphide and hydrogen sulphide require special purification techniques to remove sulphur contamination. Vanadium trichloride hexahydrate *cannot* be dehydrated in a stream of dry hydrogen chloride gas without some hydrolysis occurring[120]. The methods of preparation of vanadium trichloride are given in Table 3.12.

Vanadium trichloride is a violet, very hygroscopic crystalline solid which is soluble in acidified water. In the absence of air, the vanadium trichloride may be recovered from the acidified solution as the hexahydrate. Vanadium trichloride is thermally unstable, disproportionating above about 400° to the tetrachloride and dichloride[49,120,149,150]. The equilibrium dissociation pressure of vanadium tetrachloride over solid vanadium

TABLE 3.12
Preparation of Vanadium Trichloride

Method	Conditions	References
V and ICl	Reflux	83, 146
$VCl_3.6H_2O$ and $SOCl_2$	Reflux	111
VCl_4 and H_2S	Flow system, room temperature	147
VCl_4 and HI	Interaction in CCl_4	148
VCl_4 and CS_2	Sealed tube, 250°	39, 149
VCl_4 decomposition	150–170° in CO_2, N_2 or H_2 stream or in vacuum	80, 83, 86, 150, 151
V_2O_3 and $SOCl_2$	Sealed tube, 200°	83, 152
VO_2 and $SOCl_2$	Reflux	111
V_2O_5 and CCl_4	Flow system, red heat	153
V_2O_5 and S_2Cl_2	Reflux	51, 78, 83, 154
V_2O_5 and S_2Cl_2	Sealed tube, 250–300°	155
V_2O_5 and C_5Cl_8	Reflux	43

trichloride has been measured several times and the results are expressed by the following equations:

$$\log p_{\mathrm{mm}} = 9.84 - 7801/T \quad (352\text{–}567°) \quad \text{(reference 86)}$$

$$\log p_{\mathrm{atm}} = 8.70 - 8350/T \quad (425\text{–}655°) \quad \text{(reference 150)}$$

$$\log p_{\mathrm{mm}} = 11.449 - 8237/T \ (550\text{–}675°) \quad \text{(reference 149)}$$

By correcting for the dissociation pressure of vanadium tetrachloride, McCarley and Roddy[86] report that the true vapour pressure of vanadium trichloride over the temperature range 352–567° is given by the expression

$$\log p_{\mathrm{mm}} = 11.20 - 9777/T$$

Thermodynamic data for vanadium trichloride from the vapour pressure measurements and from other sources are given in Table 3.13.

Shomate[51] has measured the heat capacity of vanadium trichloride over a wide temperature range and recorded a sharp peak in the heat capacity at 104.9°K.

X-ray powder diffraction studies have shown that vanadium trichloride has a hexagonal layer lattice-type structure[157,158], similar to that of iron trichloride. The unit-cell dimensions have been reported as[157] $a = 6.01$ and $c = 17.34$ Å and as[158] $a = 6.045$ and $c = 17.45$ Å.

The magnetic susceptibility of vanadium trichloride has been measured at three temperatures by Klemm and Hoschek[159] and from room temperature to about 14°K by Starr and coworkers[160]. Good agreement was obtained between the independent measurements. The compound obeys the Curie-Weiss law to about 30°K, with a Weiss constant of 30°. The magnetic moment is 2.85 BM. Rather surprisingly, there appears to be no magnetic anomaly associated with the heat capacity maximum at about 105°K.

The diffuse-reflectance spectrum of vanadium trichloride has been recorded by Clark[139]. A relatively sharp band was observed at 12,500 cm^{-1} and was assigned to a $^3T_{2g} \leftarrow {}^3T_{1g}(F)$ transition. A more intense band, with the main peak at 17,300 cm^{-1} and a shoulder at 19,600 cm^{-1} was also observed. Clark suggested that the 17,300 cm^{-1} band is associated with charge-transfer while the shoulder is the $^3T_{1g}(P) \leftarrow {}^3T_{1g}(F)$ transition.

TABLE 3.13
Thermodynamic Data for Vanadium Trichloride

Property	Value	References
ΔH°_{form} (kcal $mole^{-1}$)	−143	89, 156
ΔH_{sub} (kcal $mole^{-1}$)	44.7	86
ΔH_{diss} (kcal $mole^{-1}$)	36.0	86
	37.7	149
	38.0	150
ΔS_{sub} (cal deg^{-1} $mole^{-1}$)	38.0	86
ΔS_{diss} (cal deg^{-1} $mole^{-1}$)	32	86
	39.2	149
	39.6	150

Vanadium trichloride forms a number of adducts with donor molecules such as nitriles[78,105,161–163], alcohols[116,164], amines[153,165,166], ethers[124,154,162,167] and tertiary arsines[104]. Magnetic and spectral studies indicate that the majority of these adducts are six-coordinate. However, vanadium trichloride forms a number of adducts in which the vanadium atom is five-coordinate as shown by crystallographic, magnetic and spectral studies.

$VCl_3 \cdot 2N(CH_3)_2$ is probably the best established five-coordinate compound. It was prepared by Fowles and Pleass[168] by heating the product of the reaction between vanadium tetrachloride and trimethylamine to 100°. The red-violet crystalline product sublimed from the hot zone. It has a room temperature magnetic moment of 2.7 BM[168]. Calvert and Pleass[169] have shown that this complex has an orthorhombic unit cell with $a = 9.838$, $b = 10.14$ and $c = 13.14$ Å. The space group is *Pnma* and there are eight molecules in the unit cell. Although it was not possible to unambiguously demonstrate the five-coordinate nature of the vanadium atom directly, it was shown that the cell parameters are almost identical with those of the corresponding TiX_3 adducts[169]. Fowles and coworkers[170] and Russ and Wood[170] have recently shown by single-crystal X-ray diffraction studies that the adducts $CrCl_3 \cdot 2N(CH_3)_3$ and $TiBr_3 \cdot 2N(CH_3)_3$ are five coordinate and pointed out the similarity of the cell characteristics of the vanadium trichloride adduct. Furthermore, Beattie and Gilson[172] have examined the infrared spectrum of $VCl_3 \cdot 2N(CH_3)_3$ and consider the observed spectrum to be consistent with a monomeric five-coordinate structure, probably of D_{3h} symmetry with a planar VCl_3 unit. Thus the compound is apparently structurally identical with the corresponding chromium trichloride and titanium tribromide adducts.

Duckworth and coworkers have prepared $VCl_3 \cdot 2S(CH_3)_3$, $VCl_3 \cdot 2S(C_2H_5)_2$, and $VCl_3 \cdot 2C_4H_8S$ in addition to the trimethylamine adduct. From dipole moment, infrared, visible and ultraviolet spectroscopic studies they concluded that the central metal atom in these adducts also has a *trans* trigonal-bipyramidal configuration.

Issleib and Bohn[175] have prepared a number of adducts of the general type $VCl_3 \cdot 2L$ where L is a phosphine or phosphine oxide. Molecular weight, conductivities and magnetic susceptibilities suggest that the compounds are also five-coordinate.

The reactions between vanadium trichloride and pyridine, bipyridyl and *o*-phenanthroline (py, bipy and phen respectively) have been recently investigated by Fowles and Greene[176]. With pyridine, the green complex

TABLE 3.14
Properties of Some Vanadium Trichloride Adducts

Compound	Colour	Probable Structure
$VCl_3(bipy)_2$	Mustard yellow	$[VCl_2(bipy)_2]Cl$
$VCl_3(phen)_2$	Mustard yellow	$[VCl_2(phen)_2]Cl$
$VCl_3(bipy)$	Yellow-brown	$[VCl_3(bipy)]_2$
$2VCl_3.3bipy$	Golden brown	$[VCl_2(bipy)_2][VCl_4(bipy)]$

$VCl_3(py)_3$ was found, but with the other ligands the reaction product isolated depended critically on the experimental conditions. The complexes shown in Table 3.14 were characterized by their magnetic, spectral and conductance properties.

When vanadium trichloride is reacted with methanol or ethanol in the presence of the corresponding lithium alkoxide, complete replacement of the chlorine atoms takes place to give the alkoxide $V(OR)_3$[164]. Various workers have studied the reaction between vanadium trichloride and liquid ammonia[80,177,178]. It would appear that initially an adduct of composition $VCl_3 \cdot 6NH_3$ is formed, but solvolysis occurs and the final product is probably $VCl_2(NH)_2 \cdot 4NH_3$[177].

McCarley and coworkers[158] report that vanadium trichloride may be transported in bromine vapour at 365–400° and suggest that transportation is via the formation and subsequent decomposition of a compound which is possibly VCl_3Br.

Vanadium trichloride hexahydrate. This hydrate may be prepared by dissolving vanadium trichloride in concentrated hydrochloric acid saturated with hydrogen chloride gas[179], or by dissolving vanadium trioxide in concentrated hydrochloric acid[49,120]. It is a bright green crystalline solid, and is reported to decompose to the green tetrahydrate on heating to 85–90° in a stream of hydrogen chloride[179]. Heating to higher temperatures causes partial hydrolysis[49,120] and at 160° vanadium oxide dichloride is formed. Thermogravimetric analysis in an inert atmosphere suggests that at 130° a trihydrate is formed and that the product on heating to higher temperatures includes vanadium oxide dichloride and vanadium oxide chloride.

On the basis of its absorption spectrum, Horner and Tyree[179] suggested that vanadium trichloride hexahydrate and vanadium trichloride tetrahydrate have four water molecules and two chlorine atoms directly bound to the metal atom. Thus the compounds should be formulated as $[VCl_2(H_2O)_4]Cl \cdot 2H_2O$ and $[VCl_2(H_2O)_4]Cl$ respectively.

Vanadium oxide chloride. This compound may be prepared either by heating vanadium oxide dichloride in a nitrogen stream[41,83,120] at 300°, or by the interaction of vanadium pentaoxide and vanadium trichloride[83,180] in a sealed tube over a temperature gradient of 720–620°, or by reducing vanadium oxide trichloride with hydrogen[36,149] at 600°. In the thermal gradient method, the di- tri- and tetrachlorides and vanadium oxide trichloride are formed as minor by-products. The two latter compounds are removed by vacuum distillation while the other two are readily removed by dimethylformamide. McCarley and Roddy[37] have shown that the temperature of the hydrogen reduction of vanadium oxide trichloride must be strictly controlled. Above 650° and below 750°, vanadium trioxide and vanadium dichloride, as well as vanadium oxide trichloride are formed, whilst at 850° the product consists of vanadium trioxide and a trace of vanadium metal. The formation of these products results from the thermal instability of vanadium oxide chloride itself[49,83,149].

Oppermann[149] proposes that the decomposition of vanadium oxide chloride above 700° can be expressed by the equation

$$7VOCl \rightarrow 2V_2O_5 + 2VCl_2 + VOCl_3$$

and he gives the equilibrium decomposition pressure of vanadium oxide trichloride over solid vanadium oxide chloride for the temperature range 700–850° as

$$\log p_{mm} = 11.878 - 10{,}216/T$$

The heat and entropy of decomposition are 46.72 kcal mole^{-1} and 41.75 cal deg^{-1} mole^{-1} respectively. Values for the heat of formation of vanadium oxide chloride have been reported as[149] −144 kcal mole^{-1} and as[180] −149 kcal mole^{-1}.

Vanadium oxide chloride is a brown crystalline paramagnetic solid which is isomorphous with titanium and iron oxide chlorides[37,120,180]. Quite good agreement has been obtained between the three groups of workers for the cell parameters of the orthorhombic lattice. Schafer and Wartenpfuhl[180] give $a = 3.78$, $b = 3.30$ and $c = 7.91$ Å.

Vanadium tribromide. Vanadium tribromide may be prepared directly from the elements in a flow system[80,83,181] at about 550° or in a sealed tube[86] at 400°, or by the interaction of vanadium pentaoxide and carbon tetrabromide[129] in a sealed tube at 350°. It is readily purified by sublimation in a bromine stream[86].

Vanadium tribromide is a black crystalline solid which gives a violet vapour. It is thermally unstable, both decomposing to vanadium dibromide, and bromine and disproportionating to vanadium tetrabromide and vanadium dibromide[84,127–129]. For the disproportionation reaction, McCarley and Roddy[127] report that for the temperature range 371–532°, the equilibrium decomposition pressure of vanadium tetrabromide over solid vanadium tribromide is given by the expression

$$\log p_{mm} = 10.74 - 8240/T$$

For the same reaction and for the temperature range 490–650°, Shchukarev and coworkers[129] give the expression

$$\log p_{mm} = 9.125 - 8325/T$$

Partial explanation for the poor agreement between the two sets of data is the possibility of the other mode of decomposition of vanadium tribromide, and the thermal instability of vanadium tetrabromide itself. The heat and entropy of disproportionation are given as 38 kcal mole^{-1}

and 41.5 cal deg^{-1} mole^{-1} respectively[128,129]. For the decomposition reaction

$$VBr_3 \rightarrow VBr_2 + 1/2Br_2$$

McCarley and Roddy[127] give the expression

$$\log p_{mm} = 5.20 - 5070/T$$

for the equilibrium vapour pressure of bromine over the temperature range 371–532°. The heat and entropy of decomposition are 23.1 kcal mole^{-1} and 10.6 cal deg^{-1} mole^{-1}. The true vapour pressure of vanadium tribromide, corrected for the various decomposition reactions, is given by[86]

$$\log p_{mm} = 11.12 - 7470/T$$

for the temperature range 314–427°. The heat and entropy of sublimation are 43.3 kcal mole^{-1} and 37.7 cal cal deg^{-1} mole^{-1} respectively. The heat of formation has been reported as[182] −107 kcal mole^{-1} and as[128] −118 kcal mole^{-1} while the entropy of formation is calculated to be[128] 60.6 cal deg^{-1} mole^{-1}.

Vanadium tribromide, like the corresponding trichloride, is a layer lattice, the hexagonal unit cell having the dimensions[158] $a = 6.40$ and $c = 18.53$ Å. The magnetic susceptibility of vanadium tribromide has been measured at 20°, −78° and −183° by Klemm and Hoschek[159] who give effective magnetic moments at these temperatures of 2.67, 2.58 and 2.51 BM respectively.

Although the chemistry of vanadium tribromide has not been studied as extensively as that of vanadium trichloride, it appears to behave in a similar manner. It forms six-coordinate adducts with alkyl nitriles, tetrahydrofuran, pyridine, etc.[78,105,163,167,176] and five-coordinate complexes with trimethylamine, dimethyl sulphide, diethyl sulphide and tetrahydrothiophen[172–174]. Vanadium tribromide reacts with liquid ammonia to give $VBr_2 \cdot (NH_2) \cdot xNH_3$ after initial formation of $VBr_3 \cdot 6NH_3$[80,177,178,181].

Vanadium triiodide. Vanadium triiodide has been prepared by direct iodination of the metal[83,183,184] in a sealed tube at 120–300°, of vanadium diiodide in a sealed tube[183,185] at 300° and by direct interaction of aluminium triiodide and vanadium trioxide[186] in a sealed tube at 330°. It is readily purified by vacuum sublimation at 80–100°.

Vanadium triiodide is a dark brown crystalline solid which is very hygroscopic, being readily soluble in water to give a brown solution which gradually turns green. It is thermally unstable[183–186] decomposing to the diiodide and iodine above about 300° providing the iodine can be removed from the system. In a sealed system an equilibrium is established.

Tolmacheva and coworkers[184] report that the equilibrium dissociation pressure of iodine over solid vanadium triiodide in the temperature range 300–500° is given by the expression

$$\log p_{atm} = 5.0 - 4800/T$$

The heat and entropy of dissociation are 22 kcal mole^{-1} and 27 cal^{-1} deg^{-1} mole^{-1} respectively while the heat and entropy of formation are reported[184] to be -67 kcal mole^{-1} and 48.5 cal deg^{-1} mole^{-1} respectively.

Virtually no chemistry of vanadium triiodide has been reported.

Oxidation State II

Vanadium difluoride. Although Emeleus and Gutmann[6] reported that anhydrous hydrogen fluoride did not react with vanadium metal at red heat, and that vanadium dichloride was converted to vanadium trifluoride under similar conditions, Stout and Boo[187] report that the trifluoride is formed by the action of anhydrous hydrogen fluoride on vanadium metal at 1250°, and that controlled reduction of the trifluoride by a mixture of hydrogen and anhydrous hydrogen fluoride at 1150° gives vanadium difluoride.

The blue needle-shaped crystals were shown by X-ray powder diffraction measurements to be isomorphous with the difluorides of manganese, iron, cobalt and nickel. The structure is a tetragonal rutile-type one with $a = 4.80$ and $c = 3.24$ Å.

Heat capacity measurements of vanadium difluoride indicate a sharp maximum at 7.0°K associated with a long-range magnetic ordering. There are strong exchange interactions between neighbouring vanadium atoms which form one dimensional chains parallel to the c axis. Exchange forces between the vanadium atom and its eight nearest neighbour vanadium atoms in the (111) direction are relatively weak. Stout and Lau[188] have measured the magnetic susceptibility of vanadium difluroride and a Curie-Weiss law was found to hold between 77 and 300°K with a Weiss constant of about 85°. Below 77°K, the magnetic suceptibility parallel to the tetragonal axis is greater than that in the plane perpendicular to the axis, whereas at room temperature the difference is in the opposite direction.

Seifert and Gerstenberg[189] reported the preparation of vanadium difluoride tetrahydrate by the electrolytic reduction of vanadium pentaoxide in aqueous hydrogen fluoride under a nitrogen atmosphere. Heating to 100° gave the dihydrate, but even at 200° not all the water could be removed. As expected, the hydrates are very air sensitive.

Vanadium dichloride. Vanadium dichloride may be prepared by the hydrogen reduction of vanadium trichloride at 500° in a flow system[49,80,83,86,160,190], by the interaction of vanadium metal and hydrogen chloride gas at 950° in a flow system[191], and by the disproportionation of vanadium trichloride[49,83,120,149,150,192] at temperatures greater than 500°.

Vanadium dichloride is a light green crystalline solid. It is quite stable thermally, melting at approximately 1350°[49,120]. Oranskaya and Perfilova[192] report that the vapour pressure of vanadium dichloride in the temperature range 910–1100° is given by

$$\log p_{\text{atm}} = 5.725 - 9721/T$$

McCarley and Roddy[86] give the expression

$$\log p_{\text{mm}} = 8.713 - 9804/T$$

for the temperature range 490–631°. These workers report that the heat and entropy of sublimation of vanadium dichloride are 44.8 kcal mole^{-1} and 26.6 cal deg^{-1} mole^{-1} respectively, in close agreement with the values of 44 kcal mole^{-1} and 26.1 cal deg^{-1} mole^{-1} respectively given by the Russian workers[192]. The heat and entropy of formation calculated from the hydrogen reduction of vanadium dichloride have been given as -110 kcal mole^{-1} and 37 cal deg^{-1} mole^{-1} respectively[193].

Only X-ray powder diffraction data has been reported so far for vanadium dichloride and there appears to be some confusion over the structure, probably due to polymorphism. Klemm and Grimm[194] reported that vanadium dichloride was *not* isomorphous with titanium diiodide and several other dihalides, all of which have the cadmium diiodide structure. Villadsen[191] found that his sample of vanadium dichloride, prepared by the action of hydrogen chloride on vanadium metal at 950°, did have the hexagonal cadmium diiodide structure with $a = 3.60$ and $c = 5.83$ Å. The vanadium-chlorine bond distance was calculated to be 2.55 Å. Identical unit-cell dimensions were reported by Ehrlich and Seifert[120] for a sample prepared by the thermal disproportionation of vanadium trichloride at 800°. On the other hand, McCarley and Roddy[37] report a rhombohedral unit cell with $a = 6.20$ Å and $\alpha = 33.8°$. This may be transformed to a pseudo-hexagonal cell with $a = 3.60$ and $c = 17.50$ Å. It would seem that a detailed examination of vanadium dichloride using single-crystal X-ray diffraction techniques would be desirable.

The effective magnetic moment of vanadium dichloride at room temperature was given by Klemm and Hoschek[159] as 2.41 BM, considerably lower than the spin-only value. The magnetic moment falls to 1.49 BM at $-183°$. A later reinvestigation of the magnetic properties of vanadium dichloride by Starr and coworkers[160] showed that the Curie-Weiss law is obeyed between about 14°K and 300°K. A Weiss constant of 565° gave a magnetic moment of 4.15 BM, but of course this value obtained by the conventional method of calculation has no real significance[195].

Vanadium dichloride reacts with liquid ammonia to give $VCl_2 \cdot 6NH_3$ initially, which decomposes on heating to a variety of products[80]. With methylamine the adduct $VCl_2 \cdot 6NH_2CH_3$ is formed[165] while the adduct $VCl_2 \cdot 4C_5H_5N$ is also known[116].

The tetrahydrate $VCl_2 \cdot 4H_2O$ has been prepared by the electrolytic reduction of vanadium pentaoxide in non-oxidizing acids, followed by addition of hydrochloric acid and evaporation to dryness[196].

Vanadium dibromide. Vanadium dibromide is prepared as a light brown to red solid, which is thermally stable to at least 900° in vacuum, by the hydrogen reduction of vanadium tribromide at 450° in a flow system[80,83,86].

Over the temperature range 541–716° the vapour pressure of vanadium dibromide is given by the expression[86]

$$\log p_{\text{atm}} = 9.08 - 10{,}460/T$$

Tsintsius and Yudovich[197] give the expression

$$\log p_{\text{mm}} = 5.9 - 9830/T$$

for the temperature range 800–905°. The heat and entropy of sublimation of vanadium dibromide have been given as 45 kcal mole^{-1} and 27 cal deg^{-1} mole^{-1} respectively[86,197]. Shchukarev and coworkers[128] give the heat and entropy of formation as -83 kcal mole^{-1} and 36 cal deg^{-1} mole^{-1} respectively.

Vanadium dibromide has the hexagonal cadmium diiodide structure with $a = 3.77$ and $c = 6.18$ Å[194]. Like vanadium dichloride, the magnetic moment of vanadium dibromide is considerably lower than the spin-only value; at 20° the magnetic moment is 2.80 BM and at $-183°$ it is 1.82 BM[159].

Vanadium dibromide reacts with liquid ammonia in a manner similar to the reaction of vanadium dichloride, to give the adduct $VBr_2 \cdot 6NH_3$[80]. The hydrate $VBr_2 \cdot 6H_2O$ has been prepared in the same way as the tetrahydrate of vanadium dichloride[196].

Vanadium diiodide. This compound is prepared by the thermal decomposition of vanadium triiodide at about 400°, the iodine being removed from the system to prevent recombination of the products[83,183,185,186]. It is a dark violet solid which sublimes above 850° in vacuum, but it begins to decompose to the metal[185,197] above 1000°.

Tsintsius and Yudovich[197] found that the vapour pressure of vanadium diiodide over the temperature range 850–1016° is given by

$$\log p_{\text{atm}} = 2.56 - 5600/T$$

The heat and entropy of sublimation are 44 kcal $mole^{-1}$ and 28 cal deg^{-1} $mole^{-1}$ respectively. The heat and entropy of formation are −63 kcal $mole^{-1}$ and 35 cal deg^{-1} $mole^{-1}$ respectively[184].

Vanadium diiodide has the cadmium diiodide-type hexagonal lattice with $a = 4.00$ and $c = 6.67$ Å[194]. At 15° the magnetic susceptibility of vanadium diiodide is 4440×10^{-6} and this rises to 9400×10^{-6} at −183°[198].

COMPLEX HALIDES AND OXIDE HALIDES

No complex iodides of vanadium in any oxidation state have been prepared, and the only complex halides of vanadium(v) are hexafluorovanadate(v), oxopentafluorovanadate(v), oxotetrafluorovanadate(v) and oxopentachlorovanadate(v).

Oxidation State V

Hexafluorovanadates(v). Potassium hexafluorovanadate(v) may be prepared by direct interaction of vanadium pentafluoride and potassium fluoride at room temperature[6] and a range of other hexafluorovanadates(v) have been prepared by refluxing a mixture of vanadium trichloride, bromine trifluoride and the appropriate chloride[6].

The hexafluorovanadates(v) are, in general, white very hygroscopic solids which rapidly hydrolyse in moist air. The potassium salt is thermally stable in vacuum up to 300°, but above 330° it decomposes to its constituent fluorides[6].

The structures of a number of hexafluorovanadates(v) have been investigated by Cox[199], and later by Kemmitt and coworkers[200], using X-ray powder diffraction techniques. The symmetry and lattice parameters of these salts are given in Table 3.15.

TABLE 3.15

Symmetries and Lattice Parameters of Hexafluorovanadates(v)

Compound	Symmetry	Parameters	References
$LiVF_6$	Rhombohedral $LiSbF_6$ type	$a = 5.30$ Å, $\alpha = 56.3°$	200
$NaVF_6$	Rhombohedral $LiSbF_6$ type	$a = 5.63$ Å, $\alpha = 56.55°$	200
$AgVF_6$	Cubic CsCl type	$a = 9.52$ Å	199
	Tetragonal $KNbF_6$ type	$a = 4.90$ Å, $c = 9.42$ Å	200
KVF_6	Rhombohedral $KOsF_6$ type	$a = 4.92$ Å, $\alpha = 97.2°$	199, 200
$RbVF_6$	Rhombohedral $KOsF_6$ type	$a = 5.01$ Å, $\alpha = 97.0°$	199, 200
$TlVF_6$	Rhombohedral $KOsF_6$ type	$a = 5.10$ Å, $\alpha = 95.2°$	200
$CsVF_6$	Rhombohedral $KOsF_6$ type	$a = 5.24$ Å, $\alpha = 96.2°$	199, 200

The vanadium-fluorine stretching frequency in potassium hexafluorovanadate(V) occurs[201] at 715 cm^{-1}.

Oxopentafluorovanadates(V). Baker and Haendler[202] have prepared the ammonium and potassium salts of the oxopentafluorovanadate(V) anion by adding the univalent fluoride to a mixture of vanadium metal and bromine in methanol. The ammonium salt has also been prepared by adding ammonium fluoride to a mixture of vanadium metal and silver nitrate in methanol[203]. In the latter reaction, vanadium nitrate is initially formed and this reacts with the ammonium fluoride to give the oxofluoro anion.

X-ray powder diffraction studies indicate that both the ammonium and potassium salts are tetragonal[202], the unit-cell parameters being $a = 9.17$ and $c = 17.64$ Å for the ammonium salt and $a = 8.75$ and $c = 17.09$ Å for potassium oxopentafluorovanadate(V).

The infrared spectrum of potassium oxopentafluorovanadate(V) has been recorded by Kharitonov and Buslaev[204], who observed absorptions at 989, 630, 596 and 482 cm^{-1}. The absorption at 989 cm^{-1} was attributed to the vanadium-oxygen double bond stretching frequency and an octahedral configuration was postulated for the anion.

Oxotetrafluorovanadate(V). Glemser and coworkers[132] report that nitrosyl oxotetrafluorovanadate(V), $NO^+(VOF_4^-)$, may be prepared by the interaction of nitrosyl fluoride and vanadium oxide trifluoride, or by treating vanadium pentaoxide with nitrogen trifluoride at 400° in a bomb. The compound shows a nitrogen-oxygen stretching frequency at 2330 cm^{-1} and a vanadium-oxygen double bond stretching frequency at 1020 cm^{-1}.

Oxopentachlorovanadates(V). Wendling[205] has prepared rubidium and caesium oxopentachlorovanadate(V) and has shown that they have cubic K_2PtCl_6 type lattices. The unit-cell parameters are 9.85 Å and 10.14 Å for the rubidium and caesium salts respectively.

Oxidation State IV

Hexafluorovanadates(IV). Hexafluorovanadates(IV) have been prepared by fluorination of the corresponding pentafluorovanadate(III)[206,207]. The potassium salt has also been prepared by direct fluorination of a 2:1 mixture of potassium chloride and vanadium trichloride[208] at 200–230°, and by the interaction of potassium fluoride and vanadium tetrafluoride in selenium tetrafluoride[11]. The temperature at which direct fluorination is carried out is of importance because of the existence of several polymorphic forms of the hexafluorovanadates(IV). For example, fluorination[207] of potassium pentafluorovanadate(III) at 300° gives a trigonal modification of potassium hexafluorovanadate(IV), whereas the hexagonal form is obtained at 500°. The lattice symmetry and dimensions of a number of hexafluorovanadates(IV) are given in Table 3.16.

TABLE 3.16

Symmetry and Lattice Parameters of Hexafluorovanadates(IV)

Compound	Symmetry	Lattice Parameters	References
K_2VF_6	Trigonal K_2GeF_6 type	$a = 5.67$, $c = 4.64$ Å	11,206
	Hexagonal K_2MnF_6 type	$a = 5.79$, $c = 9.34$ Å	206
Rb_2VF_6	Trigonal K_2GeF_6 type	$a = 5.88$, $c = 4.78$ Å	206
	Hexagonal K_2MnF_6 type	$a = 5.95$, $c = 9.58$ Å	206
$CsVF_6$	Trigonal K_2GeF_6 type	$a = 6.17$, $c = 4.98$ Å	206
	Hexagonal K_2MnF_6 type	$a = 6.28$, $c = 10.08$ Å	206
	Cubic K_2PtCl_6 type	$a = 9.03$ Å	206

The magnetic moments of a number of hexafluorovanadates(IV) have been determined over a wide temperature range and the Curie-Weiss law is obeyed in each case, but with large Weiss constants as shown in Table 3.17.

TABLE 3.17

Magnetic Properties of Some Hexafluorovanadates(IV)

Compound	μ (BM)	θ°	Reference
K_2VF_6	2.01	78	206
	2.05	118	11
Rb_2VF_6	1.99	100	206
Cs_2VF_6	2.03	103	206

The variation of magnetic moment with temperature does not conform to the theory of Kotani. Cavell and Clark[11] have suggested the presence of antiferromagnetic interactions, probably via the tetragonal distortion of the octahedral coordination about the vanadium atom.

The vanadium-fluorine stretching frequency occurs at 583 cm^{-1} in potassium hexafluorovanadate(IV)[201].

Oxopentafluorovanadates(IV). Ammonium oxopentafluorovanadate(IV) has been prepared by adding ammonium hydrogen fluoride to a solution of vanadyl sulphate in aqueous hydrogen fluoride[75]. The sodium salt has been prepared by titrating a vanadyl solution against sodium fluoride[209]. The ammonium salt[76,77,204] shows infrared absorptions at 934 and

946 cm^{-1}. The electronic absorption spectrum of the ammonium salt has been recorded[210,211] and absorptions were observed at 13,160, 13,660, 14,100 and 18,200 cm^{-1}.

Oxotetrafluorovanadates(IV). Anhydrous oxotetrafluorovanadates(IV) are known[1] but there has been little recent work on these compounds. Markin and coworkers[212] have prepared blue ammonium oxotetrafluorovanadate(IV) monohydrate $(NH_4)_2VOF_4 \cdot H_2O$, by dissolving vanadium tetraoxide in excess hydrofluoric acid, followed by addition of ammonia solution to pH 3. The compound has a monoclinic unit cell with $a = 10.2$, $b = 9.60$, $c = 7.00$ Å and $\beta = 80.0°$. The space group is $P2_1/c$ and there are four molecular entities in the cell.

Hexachlorovanadates(IV). Salts of the hexachlorovanadate(IV) anion have been prepared by the interaction of vanadium tetrachloride, or the vanadium tetrachloride-acetonitrile adduct $VCl_4 \cdot 2CH_3CN$, and an amine hydrochloride in chloroform solution[213] and also by dissolving the corresponding oxotetrachlorovanadate(IV) in thionyl chloride[107]. The latter method has been used to prepare the pyridinium, quinolinium and isoquinolinium salts. In order to prepare caesium hexachlorovanadate(IV) it is necessary to use a 20–30% solution of thionyl chloride in nitromethane to overcome solubility difficulties[107].

Hexachlorovanadates(IV) are dark coloured, very hygroscopic solids. The magnetic moments of a number of hexachlorovanadates(IV) have been determined and are close to the spin-only value as shown in Table 3.18.

TABLE 3.18

Magnetic Properties of Some Hexachlorovanadates(IV)

Compound	μ (BM)	$\theta°$	Reference
Cs_2VCl_6	1.83	138	214
$[NH(C_2H_5)_3]_2VCl_6$	1.75[a]		213
$[NH_2(C_2H_5)_2]_2VCl_6$	1.74[a]		213
$[C_5H_6N]_2VCl_6$	1.76	145	214
$[CH_9N_8]_2VCl_6$	1.76	84	214

[a] Room temperature measurement only.

Kilty and Nicholls[107] have recorded the absorption spectrum of the hexachlorovanadate(IV) anion in acetonitrile solution and observed bands at 15,400 and 20,500 cm^{-1}. The first band was assigned to the $^2E_{1g} \leftarrow {}^2T_{2g}$ transition while the higher energy band is of charge-transfer nature.

Oxopentachlorovanadates(IV). Solutions of the oxopentachlorovanadate(IV) anion are prepared by reduction of vanadium pentaoxide in hydrochloric acid solution with the stoichiometric amount of oxalic acid[215]. The caesium salt, which melts at 509°, has very recently been prepared by fusing a 1:1 mixture of caesium oxotetrachlorovanadate(IV) and caesium chloride[216]. The electronic absorption and electron-spin resonance spectra of the oxopentachlorovanadate(IV) anion have been investigated[215,217].

Oxotetrachlorovanadates(IV). Morozov and Morozov[216] have prepared the potassium, rubidium, caesium and ammonium salts of the oxotetrachlorovanadate(IV) anion by heating a 1:1 mixture of vanadium oxide trichloride and the appropriate chloride in a sealed tube at 270–530°. The interaction of vanadium oxide trichloride and the chloride, in concentrated hydrochloric acid saturated with hydrogen chloride gas, gives the appropriate oxotetrachlorovanadate(IV) dihydrate[216]. These are readily dehydrated, although in the case of the caesium salt some chlorine is also lost. X-ray powder diffraction data indicate that with the exception of the caesium salt, the two methods give identical products.

Feltz[218] reacted the adduct $VOCl_2 \cdot 2C_4H_8O_2$ in acetonitrile solution with ammonium or quaternary ammonium chloride to give the corresponding oxotetrachlorovanadate(IV). In most cases the product was isolated as an acetonitrile adduct but the acetonitrile can be readily removed under vacuum. Reaction of $VOCl_2 \cdot 2C_4H_8O_2$ with the chlorides of potassium, rubidium and caesium in liquid sulphur dioxide also gives salts of the oxotetrachlorovanadate(IV) anion. The ethanol adduct $[(CH_3)_4N]_2VOCl_4 \cdot C_2H_5OH$ has been prepared by mixing vanadium oxide dichloride hydrate and tetramethylammonium chloride in ethanol[75].

Kilty and Nicolls[219] prepared a series of oxotetrachlorovanadate(IV) hydrates by adding the appropriate chloride to a hot solution of vanadium oxide dichloride in an ethanolic hydrogen chloride solution, followed by cooling to 0° while being saturated with hydrogen chloride gas. The hydrates could be dehydrated by standing over phosphorus pentaoxide or by heating in vacuum.

The magnetic properties of a number of oxotetrachlorovanadates(IV) have been investigated. Those which have been examined over a wide temperature range obey the Curie-Weiss law, with small values of the Weiss constant. The available magnetic data are summarized in Table 3.19.

The ESR spectrum of the oxotetrachlorovanadate(IV) anion in a mixture of acetic and hydrochloric acids shows an average g value of 1.969[220].

The infrared spectra of a number of oxotetrachlorovanadate(IV) salts have been recorded[218,219]. The vanadium-oxygen double bond stretching frequencies fall in the range 918–1012 cm^{-1}.

The absorption spectra of the oxotetrachlorovanadate(IV) dihydrates show a broad unresolved band at about 14,500 cm^{-1} which is thought to be due to $^2E_{1g} \leftarrow {}^2B_2$ and $^2B_1 \leftarrow {}^2B_2$ transitions. It has been suggested[219] that the dihydrates of the oxotetrachlorovanadate(IV) ion have a basic square-pyramidal structure with a weakly held water molecule *trans* to the vanadium-oxygen double bond. The anhydrous salts and also the monohydrate show absorptions at about 12,000 and 13,800 cm^{-1} assigned

TABLE 3.19
Magnetic Properties of Some Oxotetrachlorovanadates(IV)

Compound	μ (BM)	$\theta°$	Reference
$[C_5H_6N]_2VOCl_4$	1.76	10	214
	1.76[a]		219
$[C_5H_6N]_2VOCl_4 \cdot 2H_2O$	1.72	4	214
	1.74[a]		219
$[C_9H_8N]_2VOCl_4$	1.75	0	214
	1.77[a]		219
$[C_9H_8N]_2VOCl_4 \cdot 2H_2O$	1.72[a]		219
$[i\text{-}C_9H_8N]_2VOCl_4$	1.78[a]		219
$(NH_4)_2VOCl_4$	1.77[a]		219
$(NH_4)_2VOCl_4 \cdot 2H_2O$	1.73[a]		219
$Cs_2VOCl_4 \cdot H_2O$	1.73[a]		219

[a] Room temperature measurement only.

respectively to the $^2B_1 \leftarrow {}^2B_2$ and $^2E_g \leftarrow {}^2B_2$ transitions. Above 20,000 cm^{-1} there are charge-transfer bands although it was suggested that the $^2A_1 \leftarrow {}^2B_2$ transition occurs as a shoulder at approximately 24,000 cm^{-1} in some of the spectra[219].

Oxotrichlorovanadates(IV). Kilty and Nicholls[219] have prepared a number of both anhydrous and hydrated oxotrichlorovanadate(IV) salts using similar methods to those described for the preparation of oxotetrachlorovanadates(IV). The vanadium-oxygen double bond stretching frequencies occur in the range 973–1015 cm^{-1} and the room temperature magnetic moments are about 1.7 BM. It is suggested that the trihydrate compounds should be formulated as $M[VOCl_3(H_2O)_2] \cdot H_2O$ since the infrared spectra show that water is coordinated to the vanadium atom. Kilty and Nicholls also suggested that the anhydrous compounds have a pseudo-tetrahedral structure.

Oxidation State III

Hexafluorovanadates(III). The methods used to prepare hexafluorovanadates(III) are given in Table 3.20.

TABLE 3.20
Preparation of Hexafluorovanadates(III)

Method	Conditions	Reference
V and KHF_2	1200°	221
V_2O_3 and NH_4HF_2	250°	136
V_2O_5, KI and aqueous HF		140
$K_2VF_5 \cdot H_2O$ and KHF_2	Fuse	222
V, $N_2H_6F_2$ and aqueous HF	Evaporate to dryness	223

Babel and coworkers[224] have isolated a number of hexafluorovanadates(III) of the type $CsMVF_6$, where M is a divalent transition metal, by fusing mixtures of caesium hydrogen fluoride, transition metal difluoride and vanadium trifluoride at 600–1000° in an inert atmosphere. The compounds all have cubic lattices with the parameters shown in Table 3.21.

TABLE 3.21
Lattice Parameters of Some Hexafluorovanadates(III)

Compound	*a* (Å)	Compound	*a* (Å)
$CsNiVF_6$	10.36	$CsFeVF_6$	10.48
$CsCuVF_6$	10.39	$CsMnVF_6$	10.57
$CsCoVF_6$	10.42		

Potassium hexafluorovanadate(III) is a green compound which is reported to be cubic[222]. It melts at 1020° without decomposition[221]. The ammonium salt is reported[225,226] to be cubic and isomorphous with the α-form of ammonium hexafluoroferrate(III) with $a = 9.04$ Å. As expected, the ammonium salt is thermally unstable and decomposes at 500–600° in an inert atmosphere to give vanadium trifluoride[136]. Both the ammonium and potassium salts have room temperature magnetic moments of 2.79 BM[140]. Potassium hexafluorovanadate(III) has a vanadium-fluorine stretching frequency at[201] 511 cm^{-1}.

The reflectance spectrum of the hexafluorovanadate(III) anion has been recorded. The observed bands and their assignments are given in Table 3.22.

TABLE 3.22
Absorption Spectrum of the Hexafluorovanadate(III) ion

Absorption (cm^{-1})	Assignment	References
10,200		144
10,260	$^1E_{1g}, ^1T_{2g} \leftarrow ^3T_{1g}(F)$	162
14,800	$^3T_{2g} \leftarrow ^3T_{1g}(F)$	139, 144, 162
23,000		144
23,250	$^3T_{1g}(P) \leftarrow ^3T_{1g}(F)$	139
23,260		162

Pentafluorovanadates(III). Interaction of vanadium trifluoride and alkali-metal fluorides in 40% hydrofluoric acid leads to the formation of pentafluorovanadate(III)monohydrates[206]. These compounds have cubic lattices, with $a = 8.42$, 8.42 and 8.45 Å for the ammonium, rubidium and thallium salts respectively.

The magnetic moments of a number of pentafluorovanadate(III) monohydrates have been examined and the results are given in Table 3.23.

TABLE 3.23
Magnetic Properties of Pentafluorovanadate(III) Hydrates

Compound	μ (BM)	$\theta°$	Reference
$K_2VF_5 \cdot H_2O$	2.72[a]		162
	2.85	26	206
$Rb_2VF_5 \cdot H_2O$	2.86	45	206
$Cs_2VF_5 \cdot H_2O$	2.78	24	206

[a] Room temperature measurement only.

The absorption spectrum of the potassium salt has bands at about 10,000, 15,600 and 23,600 cm^{-1} and the assignments are identical to those given for similar bands for the hexafluorovanadate(III) ion[144,162]. This suggests that the compounds should be formulated as $[VF_5(H_2O)]^{2-}$.

The hydrates may be dehydrated by heating under high vacuum, and although no structural information is available it is probable that the resulting compounds contain condensed VF_6 octahedra[206].

Hexachlorovanadates(III). A number of alkali-metal hexachlorovanadates(III) have been prepared by fusing mixtures of vanadium trichloride and the appropriate chloride[227–230]. Pyridinium hexachlorovanadate(III) has been prepared by Fowles and Russ[231] by heating vanadium trichloride and pyridinium chloride in a sealed tube at 150°, and by the interaction in chloroform of pyridinium chloride and the adduct $VCl_3 \cdot 3CH_3CN$. The pyridinium salt shows absorption maxima at 11,400 and 18,000 cm^{-1} which have been assigned to the $^3T_{2g} \leftarrow {}^2T_{1g}(F)$ and $^3T_{1g}(P) \leftarrow {}^3T_{1g}(F)$ transitions[231].

As would be expected the stability of the alkali-metal hexachlorovanadates(III) increases with the size of the cation. Some thermodynamic data for these complexes are given in Table 3.24

TABLE 3.24

Thermodynamic Data for Alkali-metal Hexachlorovanadates(III)

Property	Compound	Value	References
M.p.	Na_3VCl_6	555°	228, 230
	K_3VCl_6	722°	228, 230
	Rb_3VCl_6	749°	228, 230
	Cs_3VCl_6	726°	228
ΔH°_{form} (kcal $mole^{-1}$)[a]	Na_3VCl_6	−6.1	227, 228, 230
	K_3VCl_6	−14.4	227, 228, 230
	Rb_3VCl_6	−19.7	227, 228, 230
	Cs_3VCl_6	−20.7	228

[a] From constituent chlorides.

Pentachlorovanadates(III). Interaction of alkali-metal chloride or ammonium chloride in concentrated hydrochloric acid with vanadium trichloride results in the formation of pentachlorovanadate(III) hydrates. The anhydrous salts may be prepared by heating the hydrate at 200–300° in a stream of hydrogen chloride gas[120]. Crayton and Thompson[232] reported that they obtained a green tetrahydrate $K_2VCl_5 \cdot 4H_2O$ by reacting hydrochloric acid solutions of vanadium trichloride and potassium chloride at −30° and saturating the solution with hydrogen chloride gas. The violet anhydrous salt was obtained by heating the tetrahydrate in argon or nitrogen at 100–125°. However, Horner and Tyree[179] reported that they could not repeat the work of Crayton and Thompson and they suggested on the basis of unspecified X-ray data that the green product prepared by Crayton and Thompson was a mixture of hydrated potassium tetrachlorovanadate(III) and potassium chloride. It is important to note that in the earlier work by Ehrlich and Seifert[120], diagrams of the X-ray powder diffraction patterns for the various pentachlorovanadates(III) were reported, as well as those of alkali-metal chloride, and these clearly showed that the compounds are *not* mixtures as suggested by Horner and Tyree. Whilst it is possible that Crayton and Thompson may have prepared an impure product, there appears to be little doubt that the pentachlorovanadates(III) do exist.

μ-Trichlorohexachlorodivanadates(III). Alkali-metal salts of this dinuclear anion are readily prepared by fusion of vanadium trichloride and the appropriate alkali-metal chloride in the correct stoichiometry[228–230]. Quaternary ammonium salts have also been prepared in thionyl chloride solution[233].

Thermodynamic data for these compounds are given in Table 3.25.

TABLE 3.25
Thermodynamic Data for Some μ-Trichlorohexachlorodivanadates(III)

Property	Compound	Value	References
M.p.	$K_3V_2Cl_9$	628°	228, 230
	$Rb_3V_2Cl_9$	688°	228, 230
	$Cs_3V_2Cl_9$	701°	228
ΔH°_{form} (kcal mole^{-1})[a]	$K_3V_2Cl_9$	−14.1	228
		−18.6	230
	$Rb_3V_2Cl_9$	−26.7	228
		−27.6	230
	$Cs_3V_2Cl_9$	−36.8	228

[a] From constituent chlorides.

Wessel and Ijdo[234] have shown by X-ray powder diffraction studies that the caesium salt is isomorphous with the corresponding chromium compound, for which complete structural data are available (page 197). For the hexagonal unit cell $a = 7.24$ and $c = 17.94$ Å.

The tetraethylammonium salt shows vanadium-chlorine stretching frequencies at[233] 335 and 296 cm^{-1} and its magnetic moment is 2.39 BM at room temperature, falling to 1.93 BM at 92°K[235].

Tetrachlorovanadates(III). Clark, Nyholm and Scaife[163] have observed that the interaction of the adduct $VCl_3 \cdot 3CH_3CN$ with the chloride of a large cation, resulted in the formation of salts of the general type $R^+[VCl_4 \cdot 2CH_3CN]^-$. Heating these complexes to 100° in vacuum gave the tetrachlorovanadate(III) salts. It was also found that the acetonitrile in the original reaction product $[VCl_4 \cdot 2CH_3CN]^-$ may be replaced with pyridine and *o*-phenanthroline[163]. Diffuse-reflectance spectra clearly show the distorted octahedral arrangement in this compound. Clark and coworkers[163] suggested that the tetrachlorovanadate(III) salts contained tetrahedrally coordinated vanadium.

Casey and Clark[235] have extended the earlier work on the preparation and properties of tetrachlorovanadates(III). From magnetic and spectral data they have shown that tetraphenylarsonium tetrachlorovanadate(III) is in fact tetrahedral. The magnetic moment is almost independent of temperature over a wide temperature range and has a value of 2.55 BM at room temperature. The diffuse-reflectance spectrum shows an absorption centred at about 9000 cm^{-1} attributed to the $^3T_{1g}(F) \leftarrow {}^3A_{2g}$ transition and a second band at approximately 15,000 cm^{-1} was assigned to a $^3T_{1g}(P) \leftarrow {}^3A_{2g}$ transition. The diffuse-reflectance spectrum of the tetraethylammonium salt could not be interpreted satisfactorily, but it suggested a very distorted octahedral arrangement about the vanadium atom. The magnetic moment of this compound fell from 2.30 BM at room temperature to 1.88 BM at 93°K and it was suggested that this compound has a polymeric structure, similar to that noted previously for the corresponding titanium(III) compound (page 80).

Hexabromovanadate(III). Pyridinium hexabromovanadate(III) has been prepared by the reaction between pyridinium bromide and the adduct $VBr_3 \cdot 3CH_3CN$ in chloroform solution[231]. The physico-chemical properties of the orange-brown solid are, as expected, very similar to those of the corresponding chloro derivative.

Tetrabromovanadates(III). Salts of the tetrabromovanadate(III) anion have been prepared in a manner similar to that described for the corresponding tetrachloro derivatives[163,235]. The magnetic and spectral data for the tetraethylammonium salt suggest that, unlike the corresponding chloro derivative, it contains tetrahedrally coordinated vanadium.

Oxidation State II

Trichlorovanadates(II). The potassium and caesium salts of the trichlorovanadate(II) anion have been prepared by fusing vanadium dichloride with the alkali-metal chloride under an inert atmosphere[236]. Potassium trichlorovanadate(II) melts at 946° while the caesium salt melts at 1084°. Both compounds are probably isostructural with hexagonal caesium trichloronickelate(II); for the potassium salt $a = 6.90$ and $c = 5.98$ Å while for the caesium salt $a = 7.23$ and $c = 6.03$ Å.

Several hydrates of the trichlorovanadate(II) anion have been prepared by electrolytic reduction of vanadium pentaoxide in a non-oxidizing acid, followed by addition of the appropriate chloride and evaporation to dryness[196].

ADDENDA

Halides and Oxide Halides

Mean square amplitudes of vibration and force constants have been calculated for vanadium pentafluoride[237] and vanadium oxide trifluoride, oxide trichloride and oxide tribromide[238,239].

Vanadium dioxide fluoride has been identified as one of the species present in the $VO_2^+ - F^- - H_2O$ system at high fluoride ion concentrations[240].

Vanadium oxide trichloride may be prepared by refluxing vanadium pentaoxide in thionyl chloride until all the oxide is dissolved[241]. The excess thionyl chloride is destroyed by adding sodium carbonate, and the product distilled off.

Vanadium oxide trichloride reacts[242] with triphenylphosphine in petroleum ether to give a red-brown 1:1 adduct which melts with partial decomposition at greater than 210°. The V=O stretching frequency occurs at 1,000 cm^{-1}.

Johannesen, Candela and Tsang[243] have carried out a very elegant and detailed examination of the magnetic properties of vanadium tetrachloride. Their study indicates the presence of a static Jahn–Teller distortion, the ground state being a mixture of 58 % flattened and 42 % elongated configurations. ESR measurements give $g_{\parallel} = 1.920$ and $g_{\perp} = 1.899$.

Vanadium tetrachloride reacts[244] with chlorazide, ClN_3, in non-aqueous solvents to give $VCl_4 \cdot N_3$. Molecular nitrogen is lost on standing to give vanadium(chlorimide)trichloride, $VCl_3 \cdot NCl$. Straehle and Baernighausen[244] have studied this compound by single crystal X-ray techniques. The unit cell, which is of space group $P\bar{1}$ and contains two molecules, has $a = 7.64$, $b = 7.14$, $c = 5.91$ Å, $\alpha = 112.4°$, $\beta = 94.9°$ and $\gamma = 107.8°$. The structure consists of Cl_3VNCl molecules linked together into dimers, the dimeric units being weakly associated to give layers along the (001) plane. Each vanadium atom is surrounded by four chlorine atoms and one nitrogen atom to give tetragonal pyramidal stereochemistry. In addition, there is a fifth chlorine atom at a very much greater distance so as to give in effect a very distorted octahedral configuration. The important internuclear dimensions (Å) are: V—V = 3.707, V—N = 1.64, N—Cl = 1.59, and V—Cl = 2.204, 2.298, 2.383, 2.464 and 2.600. The vanadium-nitrogen-chlorine bonding system is almost linear, and together with the vanadium-nitrogen and nitrogen-chlorine bond distances, indicates the presence of resonance.

Anhydrous vanadium oxide dichloride has been prepared relatively simply by treating an aqueous acidic solution of the oxide dichloride (page 125) with excess thionyl chloride and refluxing[245]. It has an orthorhombic unit cell[246] with $a = 3.842$, $b = 11.761$ and $c = 3.383$ Å.

Drake, Vekris and Wood[247] have shown by a three dimensional X-ray diffraction study that $VOCl_2 \cdot 2(CH_3)_3N$ is five coordinate, the symmetry being C_{2v}. The compound has an orthorhombic unit cell with $a = 9.64$, $b = 10.41$ and $c = 12.18$ Å. There are four formula units in the cell of space group *Pnma*. The important internuclear dimensions are given in Table 3.26. The electronic spectrum, which is almost identical in the solid

TABLE 3.26

Internuclear Dimensions in $VOCl_2 \cdot 2(CH_3)_3N$

Distance	Value (Å)	Angle	Value
V—Cl	2.25	O—V—N	88.0°, 92.3°
V—N	2.17	O—V—Cl	119.7
V—O	1.59	Cl—V—N	89.4°, 90.2°
		Cl—V—Cl	120.5°

state and in benzene, chloroform and dioxan solutions, has been interpreted on the basis of the crystallographic study[247].

Garvey and Ragsdale[248] have prepared and characterized a number of adducts of vanadium oxide dichloride with 4-substituted pyridine *N*-oxides. The adducts are of the type $VOCl_2L_2$. The decomposition temperatures and V═O stretching frequencies of several of these complexes are given in Table 3.27. Adducts of the type $VOCl_2L_2$ have also been prepared with hexamethylphosphoramide and *N,N,N′,N,′*-tetramethylurea[245]. The physico-chemical properties of these compounds indicate that they contain five-coordinate vanadium. Complexes with phosphines and phosphine oxides have also been reported[249].

Vanadium oxide dibromide may be conveniently prepared by refluxing vanadium pentaoxide in boron tribromide[250]. Adducts of the type $VOBr_2L_2$ with hexamethylphosphoramide and *N,N,N′*,N′-tetramethylurea have been prepared and characterized by du Preez and Sadie[245]. Spectra and conductance studies indicate that the complexes are five-coordinate. Selbin and Vigee[249] have prepared several vanadium oxide dibromide adducts with phosphines and phosphine oxides.

Anhydrous vanadium trifluoride has been prepared[251] by dehydrating the trihydrate at 150° and then heating in anhydrous hydrogen fluoride at 800° in a flow system.

Dichlorophosphoryltriphenylphosphonium chloride, $[(C_6H_5)_3PPCl_2O]Cl$, reacts[252] with vanadium trichloride to give the previously known triphenylphosphine oxide adduct, which has a vanadium–chlorine stretching frequency of 331 cm^{-1}. Wood[253] has published a paper on ligand-field theory to account for the electronic spectra of five coordinate

TABLE 3.27
Vanadium Oxide Dichloride–4-Substituted Pyridine N-oxide Adducts

Substituent	Decomposition temperature	$\nu_{V=O}$ (cm^{-1})
Methyl	192–194°	998
Chloro	202–203°	1004
Bromo	196°	1002
Nitrito	219–222°	1000

$VX_3 \cdot 2(CH_3)_3N$ (X = Cl, Br). The electronic spectra of these complexes and of the corresponding 2-methylpyridine adducts have been recorded by Crouch and associates[254]. The infrared spectra of the previously unreported 2-methylpyridine complexes are listed below.

$VCl_3 \cdot 2$(2-methylpyridine)	$\nu_{V-Cl} = 400\ cm^{-1}$	$\nu_{V-N} = 245\ cm^{-1}$
$VBr_3 \cdot 2$(2-methylpyridine)	$\nu_{V-Br} = 320\ cm^{-1}$	$\nu_{V-N} = 242\ cm^{-1}$

A re-examination[254] of the electronic spectra of the complexes $VX_3 \cdot 2(CH_3)_2S$ and $VX_3 \cdot 2C_4H_8S$ (X = Cl, Br) shows that the compounds are not five coordinate in the solid state (page 130) but are dimers with octahedrally coordinated vanadium. However, it is important to note[254] that the thioether compounds break down in non-polar solvents to give trigonal bipyramidal monomers.

Seifert and coworkers[255] have made a further study of the electrolytic reduction of vanadium pentaoxide in aqueous hydrogen fluoride. As well as the previously reported vanadium difluoride tetrahydrate, careful control of experimental conditions resulted in the precipitation of $V_2F_5 \cdot 7H_2O$. This mixed oxidation state hydrate is a grey crystalline solid which is rapidly oxidized in air. It has a room temperature magnetic moment of 3.35 BM and its X-ray diffraction pattern shows it to be isomorphous with $Fe_2F_5 \cdot 7H_2O$ (page 283). On heating to 200° in vacuo, an impure sample of the anhydrous compound, V_2F_5, was formed. This has a tetragonal unit cell, of the rutile type, with $a = 4.77$ and $c = 3.20$ Å – compare with vanadium difluoride (page 134). Vanadium difluoride tetrahydrate was found[255] to obey the Curie law with a magnetic moment of 3.85 BM.

Seifert and Auel[256,257] have studied the preparation and properties of a number of adducts of vanadium dichloride with oxygen and nitrogen donor ligands. Metastable solutions of vanadium(II), which were prepared by cathodic reduction of vanadium trichloride in methanol–hydrochloric acid solutions, were used in the preparation of these adducts. In addition, it was reported[256] that vanadium dichloride tetrahydrate and vanadium dichloride dihydrate are isostructural with the corresponding iron dichloride hydrates, the unit-cell parameters being:

	a	*b*	*c*	*β*
$VCl_2 \cdot 4H_2O$	5.97 Å	7.17 Å	8.49 Å	110.81°
$VCl_2 \cdot 2H_2O$	7.31 Å	8.78 Å	3.59 Å	97.69°

The room temperature magnetic moments of the two hydrates are 3.86 and 3.24 BM respectively[256].

Methanol adducts of vanadium dibromide and vanadium diiodide have been reported[256].

Complex Halides and Oxide Halides

Khalilova and Morozov[258] have prepared hydrated oxopentachlorovanadates(V) by saturating 2:1 mixtures of alkali-metal chloride and vanadium oxide trichloride in concentrated hydrochloric acid with hydrogen chloride gas. The anhydrous salts are obtained by heating the hydrates in a stream of hydrogen chloride gas.

The vanadium–oxygen double bond and axial vanadium–fluorine stretching frequencies in ammonium oxopentafluorovanadate(IV) are reported to occur at 940 and 735 cm^{-1} respectively[259].

Hydrated alkali-metal oxotetrachlorovanadates(IV) may be prepared by passing hydrogen chloride gas through concentrated hydrochloric acid solutions containing stoichiometric amounts of the appropriate chloride and vanadium tetrachloride[260]. The anhydrous salts may be prepared by the interaction of a suitable adduct of vanadium oxide dichloride such as $VOCl_2 \cdot 2C_4H_8O$ or $VOCl_2 \cdot 2CH_3CN$, with the alkali-metal chloride in liquid sulphur dioxide[261]. Substituted quaternary ammonium salts have been prepared in the analogous manner, except that methyl cyanide was used as the reaction medium[261]. The initial products obtained from this reaction are solvated, the pure compounds being obtained on heating.

The infrared spectra[261] of these compounds indicate the presence of terminal oxygen atoms. The positions of the vanadium-oxygen double bond stretching frequencies in the oxotetrachlorovanadates(IV) are given in Table 3.28. The presence of a terminal oxygen atom and hence a five-

TABLE 3.28
Infrared Spectra (cm^{-1}) of Some Oxotetrachlorovanadates(IV)

Cation	$\nu_{V=O}$	Cation	$\nu_{V=O}$	Cation	$\nu_{V=O}$
K^+	912	NH_4^+	929	$NH(CH_3)_3^+$	1018
Rb^+	952	$NH_3CH_3^+$	930, 960	$N(CH_3)_4^+$	1010
Cs^+	989	$NH_2(CH_3)_2^+$	970, 982	$N(C_2H_5)_4^+$	1013

coordinate structure for the tetraethylammonium salt has been confirmed by a crystallographic study[262]. The unit cell is tetragonal with $a = 10.12$ and $c = 12.68$ Å and is isostructural with that of the corresponding titanium complex, which is known[262] to have a tetragonal pyramidal structure.

Pyridinium oxotrichlorovanadate(IV) has been prepared from pyridinium hydrochloride and a suitable adduct of vanadium oxide dichloride in methyl cyanide[261].

Cousseins and Cretenet[251,263] have studied the vanadium trifluoride–alkali-metal fluoride systems, and as well as hexafluorovanadates(III) they isolated anions of the types VF_4^-, VF_5^{2-}, $V_3F_{11}^{2-}$ and $V_5F_{17}^{2-}$. Some properties of these complexes are given in Table 3.29. Mass spectrometric studies[264] of the vanadium trifluoride–sodium fluoride system revealed that sodium tetrafluorovanadate(III) molecules exist in the saturated vapour at high temperatures. Potassium and rubidium tetrafluorovanadate(III), prepared by direct interaction of stoichiometric amounts of the appropriate fluorides, have orthorhombic and tetragonal cells respectively[265], with the following parameters.

	a	b	c
KVF_4	7.596 Å	7.738 Å	12.28 Å
$RbVF_4$	7.596 Å		6.315 Å

Interaction of potassium chloride and vanadium trichloride in a sealed tube gives both hexachlorovanadate(III) and μ-trichlorohexachlorodivanadate(III)[266]. The caesium salt of the dinuclear anion has also been prepared by the analogous reaction[267]. The effective room temperature magnetic moment of this red complex was found to be 2.74 BM.

Ammonium tetrafluorovanadate(II) dihydrate results from the addition of ammonium fluoride to an aqueous solution of vanadium dichloride in the absence of air[255]. It obeys the Curie-Weiss law with a magnetic moment of 3.84 BM and a Weiss constant of 7°.

Seifert and coworkers[268] have prepared tetramethylammonium trichlorovanadate(II) from the appropriate chlorides in alcohol. The unit

TABLE 3.29
Properties of Some Fluorovanadates(III)

Compound	M.p.	Transition temperature	*a* (Å)*	Peritectic point and reaction
KVF_4		775°		840°, $KVF_4 \rightleftharpoons K_2V_3F_{11}$ + liquid
$RbVF_4$		115°		928°, $RbVF_4 \rightleftharpoons Rb_2V_5F_{17}$ + liquid
$CsVF_4$				885°, $CsVF_4 \rightleftharpoons Cs_2V_5F_{17}$ + liquid
$TlVF_4$				855°, $TlVF_4 \rightleftharpoons VF_3$ + liquid
K_2VF_5				795°, $K_2VF_5 \rightleftharpoons K_3VF_6$ + liquid
Rb_2VF_5				Decomposes in solid state at 690°
K_3VF_6	1175°	200°	8.70	
Rb_3VF_6	1128°	130, 258, 345°	9.07	
Cs_3VF_6	1075°	345°	9.47	
Tl_3VF_6	736°	115°	9.04	
$K_2V_3F_{11}$				860°, $K_2V_3F_{11} \rightleftharpoons K_2V_5F_{17}$ + liquid
$K_2V_5F_{17}$				920°, $K_2V_5F_{17} \rightleftharpoons VF_3$ + liquid
$Rb_2V_5F_{17}$				996°, $Rb_2V_5F_{17} \rightleftharpoons VF_3$ + liquid
$Cs_2V_5F_{17}$				1008°, $Cs_2V_5F_{17} \rightleftharpoons VF_3$ + liquid

* High temperature form.

cell is hexagonal with $a = 9.146$ and $c = 6.227$ Å and a room-temperature magnetic moment of 2.13 BM was reported. The heat of formation of potassium trichlorovanadate(II) from the chlorides and from the elements are −20.9 and −235.1 kcal mole^{-1} respectively[269].

REFERENCES

1. J. Selbin, *Chem. Rev.*, **65**, 153 (1965).
2. J. Selbin, *Coord. Chem. Rev.*, **1**, 293 (1966).
3. D. Nicholls, *Coord. Chem. Rev.*, **1**, 379 (1966).

4. R. J. H. Clark, *The Chemistry of Titanium and Vanadium*, Elsevier, Amsterdam, 1968.
5. J. H. Canterford and T. A. O'Donnell, *Inorg. Chem.*, **5,** 1442 (1966).
6. H. J. Emeleus and V. Gutmann, *J. Chem. Soc.*, **1949,** 2979.
7. H. C. Clark and H. J. Emeleus, *J. Chem. Soc.*, **1957,** 2119.
8. L. E. Trevorrow, J. Fischer and R. K. Steunenberg, *J. Am. Chem. Soc.*, **79,** 5167 (1957).
9. N. S. Nikolaev and V. F. Sukhoverkhov, *Bul. Inst. Politeh Iasi* [*NS*], **3,** 61 (1957).
10. S. Blanchard, *J. Chim. Phys.*, **61,** 747 (1964).
11. R. G. Cavell and H. C. Clark, *J. Chem. Soc.*, **1962,** 2692.
12. J. H. Canterford and T. A. O'Donnell, *Australian J. Chem.*, **21,** 1421 (1968).
13. A. Smalc, *Monatsh. Chem.*, **98,** 163 (1967).
14. E. E. Aynsley, G. Hetherington and P. L. Robinson, *J. Chem. Soc.*, **1954,** 1119.
15. O. Ruff and H. Lickfett, *Ber.*, **44,** 2539 (1911).
16. H. H. Claassen and H. Selig, *J. Chem. Phys.*, **44,** 4039 (1966).
17. H. Selig and B. Frlec, *J. Inorg. Nucl. Chem.*, **29,** 1887 (1967).
18. R. G. Cavell and H. C. Clark, *Trans. Faraday Soc.*, **59,** 2706 (1963).
19. A. J. Edwards, G. R. Jones and B. R. Steventon, *Chem. Commun.*, **1967,** 462.
20. A. J. Edwards, Private communication.
21. J. H. Canterford, R. Colton and T. A. O'Donnell, *Rev. Pure Appl. Chem.*, **17,** 123 (1967).
22. R. G. Cavell and H. C. Clark, *J. Chem. Soc.*, **1963,** 4261.
23. R. G. Cavell and H. C. Clark, *Inorg. Chem.*, **3,** 1789 (1964).
24. G. V. Romanov and V. P. Spiridonov, *Zh. Strukt. Khim.*, **7,** 882 (1966).
25. J. H. Canterford and T. A. O'Donnell, *Inorg. Chem.*, **6,** 541 (1967).
26. R. G. Cavell and H. C. Clark, *J. Inorg. Nucl. Chem.*, **17,** 257 (1961).
27. J. H. Canterford and R. Colton, *Halides of the Transition Elements, The Second and Third Row Transition Metals*, Wiley, London 1968.
28. H. C. Clark and H. J. Emeleus, *J. Chem. Soc.*, **1958,** 190.
29. A. G. Sharpe and A. A. Woolf, *J. Chem. Soc.*, **1951,** 798.
30. G. J. Moody and H. Selig, *J. Inorg. Nucl. Chem.*, **28,** 2429 (1966).
31. H. M. Haendler, S. F. Bartram, R. S. Becker, W. J. Bernard and S. W. Bukata, *J. Am. Chem. Soc.*, **76,** 2177 (1954).
32. L. E. Trevorrow, *J. Phys. Chem.*, **62,** 362 (1958).
33. H. Selig and H. H. Claassen, *J. Chem. Phys.*, **44,** 1404 (1966).
34. E. E. Aynsley, R. Nichols and P. L. Robinson, *J. Chem. Soc.*, **1953,** 623.
35. R. D. Peacock, *J. Chem. Soc.*, **1953,** 3617.
36. J. Weidlein and K. Dehnicke, *Z. anorg. allgem. Chem.*, **348,** 278 (1966).
37. R. E. McCarley and J. W. Roddy, *J. Inorg. Nucl. Chem.*, **15,** 293 (1960).
38. F. A. Miller and L. R. Cousins, *J. Chem. Phys.*, **26,** 329 (1957).
39. H. Oppermann, *Z. Chem.*, **2,** 376 (1962).
40. A. A. Sytnik, A. A. Furman and A. S. Kulyasova, *Russ. J. Inorg. Chem.*, **11,** 543 (1966).
41. H. Oppermann, *Z. anorg. allgem. Chem.*, **351,** 113 (1967).
42. A. Chretien and G. Oechsel, *Compt. Rend.*, **206,** 254 (1938).
43. A. B. Bardawil, F. N. Collier and S. Y. Tyree, *J. Less-Common Metals*, **9,** 20 (1965).

44. R. B. Johannesen, *Inorg. Syn.*, **6,** 119 (1960).
45. R. C. Schreyer, *J. Am. Chem. Soc.*, **80,** 3483 (1958).
46. F. E. Brown and F. A. Griffitts, *Inorg. Syn.*, **1,** 106 (1939).
47. K. J. Palmer, *J. Am. Chem. Soc.*, **60,** 2360 (1938).
48. A. V. Komandin and M. L. Vlodavets, *Zh. Fiz. Khim.*, **26,** 1291 (1952).
49. V. V. Pechkovskii and N. I. Vorob'ev, *Russ. J. Inorg. Chem.*, **10,** 779 (1965).
50. *British Patent*, 957,971 (1964).
51. H. Funk and W. Weiss, *Z. anorg. allgem. Chem.*, **295,** 327 (1958).
52. I. S. Morozov and D. Y. Toptygin, *Zh. Neorg. Khim.*, **1,** 2601 (1956).
53. N. N. Ruban and V. D. Ponomarev, *Tr. Inst. Met. i Obogashch., Akad. Nauk Kaz. SSR*, **5,** 34 (1962).
54. H. Flood, J. Gorrissen and R. Veimo, *J. Am. Chem. Soc.*, **59,** 2494 (1937).
55. A. Morette, *Compt. Rend.*, **202,** 1846 (1936).
56. I. L. Perfilova, I. V. Kozlova, S. A. Shchukarev and I. V. Vasil'kova, *Vestn. Leningr. Univ. Ser. Fiz. i Khim.*, **16,** 130 (1961).
57. H. Oppermann, *Z. Physik. Chem. (Leipzig)*, **236,** 161 (1967).
58. H. J. Eichhoff and F. Weigel, *Z. anorg. allgem. Chem.*, **275,** 267 (1954).
59. B. A. Voitovich, *Titan i Ego Splavy, Akad. Nauk. SSSR Inst. Met.*, **1961,** 88.
60. A. S. Barabanova and B. A. Voitovich, *Ukr. Khim. Zh.*, **31,** 352 (1965).
61. B. F. Barkov, B. A. Voitovich and A. S. Barabanova, *Ukr. Khim. Zh.*, **27,** 580 (1961).
62. D. Cozzi and S. Cecconi, *Ricerca Sci.*, **23,** 609 (1953).
63. C. Ringel and H. A. Lehmann, *Z. anorg. allgem. Chem.*, **353,** 158 (1967).
64. H. Funk, W. Weiss and M. Zeising, *Z. anorg. allgem. Chem.*, **296,** 36 (1958).
65. G. C. Kleinkopf and J. M. Shreeve, *Inorg. Chem.*, **3,** 607 (1964).
66. V. Gutmann and A. Meller, *Monatsh. Chem.*, **92,** 470 (1961).
67. S. Prasad and K. N. Upadhyaya, *J. Indian Chem. Soc.*, **37,** 543 (1960).
68. K. Dehnicke, *Angew. Chem. Intern. Ed. Engl.*, **2,** 325 (1963).
69. K. Dehnicke, *Chem. Ber.*, **97,** 3354 (1964).
70. F. G. Nunez and E. Figueroa, *Compt. Rend.*, **206,** 437 (1938).
71. F. A. Miller and W. K. Baer, *Spectrochim. Acta*, **17,** 112 (1961).
72. L. Kolditz, V. Neumann and G. Kilch, *Z. anorg. allgem. Chem.*, **325,** 275 (1963).
73. R. G. Cavell and H. C. Clark, *J. Chem. Soc.*, **1965,** 444.
74. Y. A. Buslaev and M. P. Gustyakova, *Izv. Akad. Nauk SSSR, Ser. Khim.*, **1963,** 1533.
75. J. Selbin and L. H. Holmes, *J. Inorg. Nucl. Chem.*, **24,** 1111 (1962).
76. J. Selbin, L. H. Holmes and S. P. McGlynn, *Chem. Ind. (London)*, **1961,** 746.
77. J. Selbin, L. H. Holmes and S. P. McGlynn, *J. Inorg. Nucl. Chem.*, **25,** 1359 (1962).
78. M. W. Duckworth, G. W. A. Fowles and R. A. Hoodless, *J. Chem. Soc.*, **1963,** 5665.
79. J. H. Simons and M. G. Powell, *J. Am. Chem. Soc.*, **67,** 75 (1945).
80. F. Ephraim and E. Ammann, *Helv. Chim. Acta*, **16,** 1273 (1933).
81. A. G. Wittaker and D. M. Yost, *J. Chem. Phys.*, **17,** 188 (1949).
82. M. F. A. Dove, J. A. Creighton and L. A. Woodward, *Spectrochim. Acta*, **18,** 267 (1962).
83. G. Brauer, *Handbook of Preparative Inorganic Chemistry*, Vol. 1, Academic Press, New York, 1963.

84. V. P. Spiridonov and G. V. Romanov, *Zhur. Strukt. Khim.*, **8,** 160 (1967).
85. F. Pennella and W. J. Taylor, *J. Mol. Spectroscopy*, **11,** 321 (1963).
86. R. E. McCarley and J. W. Roddy, *Inorg. Chem.*, **3,** 60 (1964).
87. S. A. Shchukarev, M. A. Oranskaya, T. A. Tolmacheva and A. K. Yakhkind, *Zh. Neorg. Khim.*, **1,** 30 (1956).
88. P. Gross and C. Hayman, *Trans. Faraday Soc.*, **60,** 45 (1964).
89. S. I. Sklyarenko, L. P. Ruzinov and Y. U. Samson, *Russ. J. Inorg. Chem.*, **7,** 1377 (1962).
90. W. N. Lipscomb and A. G. Whittaker, *J. Am. Chem. Soc.*, **67,** 2019 (1945).
91. Y. Morino and H. Uehara, *J. Chem. Phys.*, **45,** 4543 (1966).
92. C. J. Ballhausen and A. D. Liehr, *Acta Chem. Scand.*, **15,** 775 (1961).
93. C. J. Ballhausen and J. De Heer, *J. Chem. Phys.*, **43,** 4304 (1965).
94. J. A. Creighton, J. H. S. Green and W. Kynaston, *J. Chem. Soc.*, **1966,** A, 208.
95. E. L. Grubb and R. L. Belford, *J. Chem. Phys.*, **39,** 244 (1963).
96. F. A. Blankenship and R. L. Belford, *J. Chem. Phys.*, **36,** 633 (1962).
97. F. A. Blankenship and R. L. Belford, *J. Chem. Phys.*, **37,** 675 (1962).
98. E. L. Grubb, F. A. Blankenship and R. L. Belford, *J. Phys. Chem.*, **67,** 1562 (1963).
99. D. S. Alderdice, *J. Mol. Spectroscopy*, **15,** 509 (1965).
100. R. J. H. Clark and D. J. Machin, *J. Chem. Soc.*, **1963,** 4430.
101. E. Konig, Landolt-Bornstein, *Magnetic Properties of Coordination and Organometallic Transition Metal Compounds*, Springer-Verlag, New York, 1966.
102. W. L. Groeneveld, *Rec. Trav. Chim.*, **71,** 1152 (1952).
103. R. J. H. Clark, *J. Chem. Soc.*, **1963,** 1377.
104. R. J. H. Clark, M. L. Greenfield and R. S. Nyholm, *J. Chem. Soc.*, **1966,** A, 1254.
105. R. J. H. Clark, *Spectrochim. Acta*, **21,** 955 (1965).
106. B. E. Bridgland, G. W. A. Fowles and R. A. Walton, *J. Inorg. Nucl. Chem.*, **27,** 383 (1965).
107. P. A. Kilty and D. Nicholls, *J. Chem. Soc.*, **1965,** 4915.
108. S. Prasad and R. C. Srivastava, *Indian J. Chem.*, **3,** 87 (1965).
109. S. Prasad and R. C. Srivastava, *Z. anorg. allgem. Chem.*, **340,** 325 (1965).
110. S. Prasad and R. C. Srivastava, *Z. anorg. allgem. Chem.*, **337,** 221 (1965).
111. H. J. Seifert, *Z. anorg. allgem. Chem.*, **317,** 123 (1962).
112. E. E. Aynsley, N. N. Greenwood and J. B. Leach, *Chem. Ind.* (*London*), **1966,** 379.
113. R. J.H. Clark and W. Errington, *Inorg. Chem.*, **5,** 650 (1966).
114. R. J. H. Clark, J. Lewis and R. S. Nyholm, *J. Chem. Soc.*, **1962,** 2460.
115. G. W. A. Fowles and D. Nicholls, *J. Chem. Soc.*, **1958,** 1687.
116. H. Funk, G. Mohaupt and A. Paul, *Z. anorg. allgem. Chem.*, **302,** 119 (1959).
117. M. W. Duckworth and G. W. A. Fowles, *J. Less-Common Metals*, **4,** 338 (1958).
118. G. W. A. Fowles and C. M. Pleass, *J. Chem. Soc.*, **1957,** 1674.
119. J. C. W. Chien and C. R. Boss, *J. Am. Chem. Soc.*, **83,** 3767 (1961).
120. P. Ehrlich and H. J. Seifert, *Z. anorg. allgem. Chem.*, **301,** 282 (1959).
121. A. Feltz, *Z. anorg. allgem. Chem.*, **354,** 225 (1967).
122. S. M. Horner, S. Y. Tyree and D. L. Venezky, *Inorg. Chem.*, **1,** 844 (1962).

123. A. K. Majumdar, A. K. Mukherjee and R. G. Bhattacharya, *J. Inorg. Nucl. Chem.*, **26,** 386 (1964).
124. R. J. Kern, *J. Inorg. Nucl. Chem.*, **24,** 1105 (1962).
125. K. L. Baker, D. A. Edwards, G. W. A. Fowles and R. G. Williams, *J. Inorg. Nucl. Chem.*, **29,** 1881 (1967).
126. V. L. Zolotavin, *Zh. Obshch. Khim.*, **24,** 433 (1954).
127. R. E. McCarley and J. W. Roddy, *Inorg. Chem.*, **3,** 54 (1964).
128. S. A. Shchukarev, T. A. Tolmacheva and V. M. Tsintsius, *Russ. J. Inorg. Chem.*, **7,** 345 (1962).
129. S. A. Shchukarev, T. A. Tolmacheva and V. M. Tsintsius, *Russ. J. Inorg. Chem.*, **7,** 777 (1962).
130. K. Dehnicke, *Chem. Ber.*, **98,** 290 (1965).
131. A. Morette, *Bull. Soc. Chim. France*, **1957,** 551.
132. O. Glemser, J. Wegner and R. Mews, *Chem. Ber.*, **100,** 2474 (1967).
133. E. L. Muetterties and J. E. Castle, *J. Inorg. Nucl. Chem.*, **18,** 148 (1961).
134. G. Scagliarini and A. Airoldi, *Gazz. Chim. Ital.*, **55,** 44 (1925).
135. *U.S. Patent*, 2,743,161 (1956).
136. B. J. Sturm and C. W. Sheridan, *Inorg. Syn.*, **7,** 87 (1963).
137. L. N. Sidorov, M. Y. Denisov, P. A. Akishin and V. B. Shol'ts, *Russ. J. Phys. Chem.*, **40,** 620 (1966).
138. K. H. Jack and V. Gutmann, *Acta Cryst.*, **4,** 246 (1951).
139. R. J. H. Clark, *J. Chem. Soc.*, **1964,** 417.
140. R. S. Nyholm and A. G. Sharpe, *J. Chem. Soc.*, **1952,** 3579.
141. V. Saraswati, *J. Phys. Soc. Japan*, **23,** 647 (1967).
142. E. O. Wollan, H. R. Child, W. C. Koehler and M. K. Wilkinson, *Phys. Rev.*, **112,** 1132 (1958).
143. I. Maak, P. Eckerlin and A. Rabenau, *Naturwissenschaften*, **48,** 218 (1964).
144. C. J. Ballhausen and F. Winther, *Acta Chem. Scand.*, **13,** 1729 (1959).
145. B. L. Chamberland and A. W. Sleight, *Solid State Commun.*, **5,** 765 (1967).
146. V. Gutmann, *Monatsh. Chem.*, **81,** 1155 (1950).
147. *Hungarian Patent*, 149,600 (1962).
148. *Hungarian Patent*, 148,834 (1961).
149. H. Oppermann, *Z. anorg. allgem. Chem.*, **351,** 127 (1967).
150. M. A. Oranskaya, Y. S. Lebedev and I. L. Perfilova, *Russ. J. Inorg. Chem.*, **6,** 132 (1961).
151. C. H. Shomate, *J. Am. Chem. Soc.*, **69,** 220 (1947).
152. H. Hecht, J. Jander and H. Schlapmann, *Z. anorg. allgem. Chem.*, **254,** 255 (1947).
153. S. Bodforss, K. J. Karlsson and H. Sjodin, *Z. anorg. allgem. Chem.*, **221,** 382 (1935).
154. E. Kurras, *Naturwissenschaften*, **46,** 171 (1959).
155. H. Funk and C. Muller, *Z. anorg. allgem. Chem.*, **244,** 94 (1940).
156. S. A. Shchukarev, I. V. Vasil'kova, I. L. Perfilova and L. V. Chernykh, *Russ. J. Inorg. Chem.*, **7,** 779 (1962).
157. W. Klemm and E. Krose, *Z. anorg. allgem. Chem.*, **253,** 218 (1947).
158. R. E. McCarley, J. W. Roddy and K. O. Berry, *Inorg. Chem.*, **3,** 50 (1964).
159. W. Klemm and E. Hoschek, *Z. anorg. allgem. Chem.*, **226,** 359 (1936).
160. C. Starr, F. Bitter and A. R. Kaufmann, *Phys. Rev.*, **58,** 977 (1940).
161. R. J. Kern, *J. Inorg. Nucl. Chem.*, **25,** 5 (1963).

162. D. J. Machin and K. S. Murray, *J. Chem. Soc.*, **1967,** A, 1498.
163. R. J. H. Clark, R. S. Nyholm and D. E. Scaife, *J. Chem. Soc.*, **1966,** A, 1296.
164. D. C. Bradley and M. L. Mehta, *Can. J. Chem.*, **40,** 1710 (1962).
165. G. W. A. Fowles and P. G. Lanigan, *J. Less-Common Metals*, **6,** 396 (1964).
166. R. J. H. Clark and M. L. Greenfield, *J. Chem. Soc.*, **1967,** A, 409.
167. G. W. A. Fowles, P. T. Greene and T. E. Lester, *J. Inorg. Nucl. Chem.*, **29,** 2365 (1967).
168. G. W. A. Fowles and C. M. Pleass, *Chem. Ind.* (*London*), **1955,** 1743.
169. L. D. Calvert and C. M. Pleass, *Can. J. Chem.*, **40,** 1473 (1962).
170. G. W. A. Fowles, P. T. Greene and J. S. Wood, *Chem. Commun.*, **1967,** 971.
171. B. J. Russ and J. S. Wood, *Chem. Commun.*, **1966,** 745.
172. I. R. Beattie and T. Gilson, *J. Chem. Soc.*, **1965,** 6595.
173. M. W. Duckworth, G. W. A. Fowles and P. T. Greene, *J. Chem. Soc.*, **1967,** A, 1592.
174. M. W. Duckworth, G. W. A. Fowles and R. G. Williams, *Chem. Ind.* (*London*), **1962,** 1285.
175. K. Issleib and G. Bohn, *Z. anorg. allgem. Chem.*, **301,** 188 (1959).
176. G. W. A. Fowles and P. T. Greene, *J. Chem. Soc.*, **1967,** A, 1869.
177. G. W. A. Fowles, P. G. Lanigan and D. Nicholls, *Chem. Ind.* (*London*), **1961,** 1167.
178. H. Remy and I. May, *Naturwissenschaften*, **48,** 524 (1961).
179. S. M. Horner and S. Y. Tyree, *Inorg. Chem.*, **3,** 1173 (1964).
180. H. Schafer and F. Wartenpfuhl, *J. Less-Common Metals*, **3,** 29 (1961).
181. D. Nicholls, *J. Inorg. Nucl. Chem.*, **24,** 1001 (1962).
182. S. A. Shchukarev, I. V. Vasil'kova, M. A. Oranskaya, V. M. Tsintsius and N. A. Subbotina, *Vestn. Leningr. Univ., Ser. Fiz. i Khim.*, **16,** 125 (1961).
183. A. Morette, *Compt. Rend.*, **207,** 1218 (1938).
184. T. A. Tolmacheva, V. M. Tsintsius and L. V. Andrianova, *Russ. J. Inorg. Chem.*, **8,** 281 (1963).
185. T. A. Tolmacheva, V. M. Tsintsius and E. E. Yudovich, *Russ. J. Inorg. Chem.*, **11,** 249 (1966).
186. M. Chaigneau, *Bull. Soc. Chim. France*, **1957,** 886.
187. J. W. Stout and W. O. J. Boo, *J. Appl. Phys.*, **37,** 966 (1966).
188. J. W. Stout and H. Y. Lau, *J. Appl. Phys.*, **38,** 1472 (1967).
189. H. J. Seifert and B. Gerstenberg, *Angew. Chem.*, **73,** 657 (1961).
190. R. E. Young and M. E. Smith, *Inorg. Syn.*, **4,** 128 (1953).
191. J. Villadsen, *Acta Chem. Scand.*, **13,** 2146 (1959).
192. M. A. Oranskaya and I. L. Perfilova, *Russ. J. Inorg. Chem.*, **6,** 131 (1965).
193. S. A. Shchukarev, M. A. Oranskaya and T. A. Tolmacheva, *Russ. J. Inorg. Chem.*, **5,** 3 (1960).
194. W. Klemm and L. Grimm, *Z. anorg. allgem. Chem.*, **249,** 198 (1941).
195. B. N. Figgis and J. Lewis, *Modern Coordination Chemistry* (Ed. J. Lewis and R. G. Wilkins), Interscience, New York, 1964.
196. H. J. Seifert and B. Gerstenberg, *Z. anorg. allgem. Chem.*, **315,** 56 (1962).
197. V. M. Tsintsius and E. E. Yudovich, *Russ. J. Inorg. Chem.*, **9,** 555 (1964).

198. W. Klemm and L. Grimm, *Z. anorg. allgem. Chem.*, **249,** 209 (1942).
199. B. Cox, *J. Chem. Soc.*, **1956,** 876.
200. R. D. W. Kemmitt, D. R. Russell and D. W. A. Sharp, *J. Chem. Soc.*, **1963,** 4408.
201. R. D. Peacock and D. W. A. Sharp, *J. Chem. Soc.*, **1959,** 2762.
202. A. E. Baker and H. H. Haendler, *Inorg. Chem.*, **1,** 127 (1962).
203. W. G. Bottjer and H. H. Haendler, *Inorg. Chem.*, **4,** 913 (1965).
204. Y. Y. Kharitonov and Y. A. Buslaev, *Izv. Akad. Nauk SSSR, Ser. Khim.*, **1964,** 808.
205. E. Wendling, *Bull. Soc. Chim. France*, **1967,** 5.
206. W. Liebe, E. Weise and W. Klemm, *Z. anorg. allgem. Chem.*, **311,** 281 (1961).
207. R. Hoppe, *Rec. Trav. Chim.*, **75,** 569 (1956).
208. E. Huss and W. Klemm, *Z. anorg. allgem. Chem.*, **262,** 25 (1950).
209. S. Ahrland and B. Noren, *Acta Chem. Scand.*, **15,** 1595 (1958).
210. J. Selbin and T. R. Ortolano, *Inorg. Chem.*, **2,** 1315 (1963).
211. T. R. Ortolano, J. Selbin and S. P. McGlynn, *J. Chem. Phys.*, **41,** 262 (1964).
212. V. N. Markin, O. A. Lobaneva and S. S. Tolkachev, *Vestn. Leningr. Univ., Ser. Fiz. i Khim.*, **20,** 133 (1965).
213. G. W. A. Fowles and R. A. Walton, *J. Inorg. Nucl. Chem.*, **27,** 735 (1965).
214. D. J. Machin and K. S. Murray, *J. Chem. Soc.*, **1967,** A, 1330.
215. R. A. D. Wentworth and T. S. Piper, *J. Chem. Phys.*, **41,** 3884 (1964).
216. I. S. Morozov and A. I. Morozov, *Izv. Akad. Nauk SSSR, Neorg. Mater.*, **3,** 1039 (1967).
217. K. DeArmond, B. B. Garrett and H. S. Gutowsky, *J. Chem. Phys.*, **42,** 1019 (1965).
218. A. Feltz, *Z. Chem.*, **7,** 23 (1967).
219. P. A. Kilty and D. Nicholls, *J. Chem. Soc.*, **1966,** A, 1175.
220. H. Kon and N. E. Sharpless, *J. Phys. Chem.*, **70,** 105 (1966).
221. B. M. Wanklyn, *J. Inorg. Nucl. Chem.*, **27,** 481 (1965).
222. R. D. Peacock, *Progress in Inorganic Chemistry* (Ed. F. A. Cotton), Interscience, New York, 1960.
223. J. Slivnik, J. Pezdiic and B. Sedej, *Monatsh. Chem.*, **98,** 204 (1967).
224. D. Babel, G. Pausenwang and W. Viebahn, *Z. Naturforsch.*, **22b,** 1219 (1967).
225. H. Bode and E. Voss, *Z. anorg. allgem. Chem.*, **290,** 1 (1957).
226. L. Passerini and R. Pirani, *Gazz. Chim. Ital.*, **62,** 279 (1932).
227. A. I. Efimov, I. V. Vasil'kova, E. K. Smirnova, N. D. Zaitseva, T. S. Shemyakina and I. L. Perfilova, *Khim. Redkikh Elementov, Leningr. Gos. Univ.*, **1964,** 38.
228. S. A. Shchukarev, I. L. Perfilova and L. N. Garin, *Russ. J. Inorg. Chem.*, **11,** 774 (1966).
229. I. V. Vasil'kova and I. L. Perfilova, *Russ. J. Inorg. Chem.*, **10,** 1248 (1965).
230. S. A. Shchukarev and I. L. Perfilova, *Russ. J. Inorg. Chem.*, **8,** 1100 (1963).
231. G. W. A. Fowles and B. J. Russ, *J. Chem. Soc.*, **1967,** A, 517.
232. P. H. Crayton and W. A. Thompson, *J. Inorg. Nucl. Chem.*, **25,** 742 (1963).
233. D. M. Adams, J. Chatt, J. M. Davidson and J. Gerratt, *J. Chem. Soc.*, **1963,** 2189.
234. G. J. Wessel and D. J. Ijdo, *Acta Cryst.*, **10,** 466 (1957).
235. A. T. Casey and R. J. H. Clark, *Inorg. Chem.*, **7,** 1598 (1968).

236. H. J. Seifert and P. Ehrlich, *Z. anorg. allgem. Chem.*, **302,** 284 (1959).
237. A. Mueller, *Z. Chem.*, **7,** 35 (1967).
238. G. Nagarajan and A. Mueller, *Z. Physik. Chem.* (*Leipzig*), **237,** 297 (1968).
239. A. Mueller, B. Krebs, A. Fadini, O. Glemser, S. J. Cyvin, J. Brunvoll, B. N. Cyvin, I. Elvebredd, G. Hagen and B. Vizi, *Z. Naturforsch.*, **23a,** 1656 (1968).
240. A. A. Ivakin, *Russ. J. Inorg. Chem.*, **12,** 939 (1967).
241. S. K. Anand, R. K. Multani and B. D. Jain, *Chem. Ind.* (*London*), **1968,** 743.
242. K. Tarama, S. Yoshida, H. Kani and S. Osaka, *Bull. Chem. Soc. Japan*, **41,** 1271 (1968).
243. R. B. Johannesen, G. A. Candella and T. Tsang, *J. Chem. Phys.*, **48,** 5544 (1968).
244. J. Straehle and H. Baernighausen, *Z. anorg. allgem. Chem.*, **357,** 325 (1968).
245. J. G. H. du Preez and F. G. Sadie, *Inorg. Chim. Acta*, **1,** 202 (1967).
246. R. J. Sime, *Z. Krist.*, **124,** 238 (1967).
247. J. E. Drake, J. Vekris and J. S. Wood, *J. Chem. Soc.*, **1968,** A, 1000.
248. R. G. Garvey and R. O. Ragsdale, *J. Inorg. Nucl. Chem.*, **29,** 745 (1967).
249. J. Selbin and G. Vigee, *J. Inorg. Nucl. Chem.*, **30,** 1644 (1968).
250. M. F. Lappert and B. Prokai, *J. Chem. Soc.*, **1967,** A, 129.
251. J. C. Cousseins and J. C. Cretenet, *Compt. Rend.*, **265C,** 1464 (1967).
252. E. Lindner, R. Lehner and H. Scheer, *Chem. Ber.*, **100,** 1331 (1967).
253. J. S. Wood, *Inorg. Chem.*, **7,** 852 (1968).
254. P. C. Crouch, G. W. A. Fowles and R. A. Walton, *J. Chem. Soc.*, **1968,** A, 2172.
255. H. J. Seifert, H. W. Loh and K. Jungnickel, *Z. anorg. allgem. Chem.*, **360,** 62 (1968).
256. H. J. Seifert and T. Auel, *Z. anorg. allgem. Chem.*, **360,** 50 (1968).
257. H. J. Seifert and T. Auel, *J. Inorg. Nucl. Chem.*, **30,** 2081 (1968).
258. N. K. Khalilova and I. S. Morozov, *Russ. J. Inorg. Chem.*, **13,** 517 (1968).
259. B. N. Sathyanarayana and C. C. Patel, *J. Inorg. Nucl. Chem.*, **30,** 207 (1968).
260. N. K. Khalilova and I. S. Morozov, *Russ. J. Inorg. Chem.*, **13,** 700 (1968).
261. A. Feltz, *Z. anorg. allgem. Chem.*, **355,** 120 (1967).
262. W. Haase and H. Hoppe, *Acta Cryst.*, **24B,** 282 (1968).
263. J. C. Cretenet and J. C. Cousseins, *Compt. Rend.*, **267C,** 240 (1968).
264. L. N. Sidorov, U. M. Korenev, V. B. Shol'ts, P. A. Akishin and V. P. Frolov, *Russ. J. Phys. Chem.*, **41,** 371 (1967).
265. D. Babel, *Z. Naturforsch.*, **23a,** 1417 (1968).
266. A. V. Storonkin, I. V. Vasil'kova and U. A. Fedorov, *Russ. J. Phys. Chem.*, **41,** 357 (1967).
267. R. Saillant and R. A. D. Wentworth, *Inorg. Chem.*, **7,** 1606 (1968).
268. H. J. Seifert, H. Fink and E. Just, *Naturwissenschaften*, **55,** 297c (1968).
269. I. V. Vasil'kova, A. I. Efimov and E. K. Lupenko, *Vestn. Leningrad. Univ., Fiz. Khim.*, **23,** 73 (1968).

Chapter 4

Chromium

In proceeding from titanium, chromium is the first of the transition elements for which the lower oxidation states become important. Although for chromium the chemistry is most extensive for the trivalent state, the divalent state, which in succeeding elements becomes the dominant one, is already important.

Apart from the binary fluorides, the halide chemistry of chromium(V) and chromium(VI) is essentially that of oxide halides and except for chromium tetrafluoride and its associated complex fluorides, there is no halide chemistry of chromium(IV).

The chemistry of chromium(III) is dominated by the formation of octahedrally coordinated complexes, presumably due in part to the maximum crystal-field stabilization energy obtained by this stereochemistry for the d^3 configuration.

HALIDES AND OXIDE HALIDES

The known halides and oxide halides of chromium are listed in Tables 4.1 and 4.2.

TABLE 4.1
Halides of Chromium

Oxidation State	Fluoride	Chloride	Bromide	Iodide
VI	CrF_6			
V	CrF_5			
IV	CrF_4	$[CrCl_4]$		
III	CrF_3	$CrCl_3$	$CrBr_3$	CrI_3
II	CrF_2	$CrCl_2$	$CrBr_2$	CrI_2

TABLE 4.2
Oxide Halides of Chromium

Oxidation State	Fluoride	Chloride	Bromide
VI	$CrOF_4$, CrO_2F_2	CrO_2Cl_2	CrO_2Br_2
V	[$CrOF_3$]	$CrOCl_3$	
III		$CrOCl$	$CrOBr$

In addition, the mixed oxidation state compound Cr_2F_5 is known.

Oxidation State VI

Chromium hexafluoride. This compound, which is thermally unstable above $-100°$, has been prepared in small yields by Glemser and co-workers[1] by heating chromium metal and fluorine at 200 atmospheres pressure to 400° in a bomb. Addition of small amounts of manganese powder increases the yield of chromium hexafluoride. The lemon-yellow compound decomposes above $-100°$ in the absence of a large pressure of fluorine, to give the pentafluoride and fluorine. At $-180°$ it shows an intense chromium-fluorine stretching frequency[2] at 785 cm^{-1}.

Chromium oxide tetrafluoride. Chromium oxide tetrafluoride has been prepared by Edwards[3] as a by-product when chromium metal is heated in fluorine in a glass flow system. The oxygen may arise from oxide film on the metal, but is more likely to be derived from the glass apparatus. The separation of the chromium oxide tetrafluoride from the major products of the reaction is difficult and requires very careful fractional sublimation.

Chromium oxide tetrafluoride is a dark red solid which melts at 55° and is extremely readily hydrolysed. It has a monoclinic unit cell with $a = 12.3$, $b = 5.4$, $c = 7.3$ Å and $\beta = 104°$.

Chromium dioxide difluoride. The methods reported for preparing chromium dioxide difluoride are given in Table 4.3.

Chromium dioxide difluoride was first prepared by von Wartenberg[4] by the fluorination of chromium dioxide dichloride. Engelbrecht and Grosse[5] suggested that from its chemical properties, von Wartenberg's sample was not pure and contained free fluorine. Other possible impurities include chlorine trifluoride and related compounds. Engelbrecht and Grosse found that the most suitable method of preparing chromium dioxide difluoride was the interaction of chromium trioxide and anhydrous hydrogen fluoride at room temperature, since it is possible to decant the

TABLE 4.3

Preparation of Chromium Dioxide Difluoride

Method	Conditions	References
CrO_2Cl_2 and F_2	Flow system, 200°	4
CrO_3 and anhydrous HF	Room temperature	5
$K_2Cr_2O_7$ and ahydrous HF	Room temperature	6
Cr and NO_2F	Flow system, mild heat	7
CrO_3 and SeF_4	Reflux	8
CrO_3 and SF_4	5°	9, 10
CrO_3 and CoF_3	Bomb, 450°	11

excess hydrogen fluoride off at −78° and recrystallize the compound from anhydrous hydrogen fluoride.

Chromium dioxide difluoride is a dark violet-red solid, melting at 31.6° to give an orange-red liquid and a red-brown vapour. It is a very reactive compound, attacking glass and silica. Although it is stable in the dark, it very rapidly polymerizes to a white solid on exposure to sunlight[4,5].

The vapour pressure of solid chromium dioxide difluoride is given by the expression[5]

$$\log p_{\mathrm{mm}} = 6.252 + 955/T - 599{,}333/T^2$$

and for the liquid

$$\log p_{\mathrm{mm}} = 8.85 - 1785/T$$

Thermodynamic data for chromium dioxide difluoride are given in Table 4.4.

TABLE 4.4

Thermodynamic Data for Chromium Dioxide Difluoride

Property	Value	Reference
M.p.	30.5°	9
	31.6°	5
Quasi b.p.[a]	29.6	5
ΔH_{fus} (kcal mole^{-1})	5.6	5
ΔH_{sub} (kcal mole^{-1})	13.8	5
ΔH_{vap} (kcal mole^{-1})	8.2	5

[a] Temperature at which vapour pressure is 1 atmosphere.

Hobbs[12] has recorded the infrared spectrum of chromium dioxide difluoride and assigned the following fundamentals on the basis of

C_{2v} symmetry: ν_1 1006, ν_2 727, ν_3 304, ν_4 182, ν_5 422, ν_6 1016, ν_7 274, ν_8 789 and ν_9 259 cm^{-1}.

Flesch and Svec[11,13] have studied the reaction between chromium dioxide difluoride and chromium dioxide dichloride to give chromium dioxide chloride fluoride, CrO_2ClF. Infrared and mass-spectrometric studies confirmed the existence of this species as a true compound and it shows a characteristic absorption in its infrared spectrum at 750 cm^{-1}.

Chromium dioxide dichloride. Chromium dioxide dichloride is normally prepared by refluxing a mixture of alkali-metal chloride and concentrated sulphuric acid with chromium trioxide or an alkali-metal dichromate, and distilling the product from the mixture[12,14–17]. It has also been prepared by the interaction of chromium trioxide or a dichromate with thionyl chloride[18]. The preparation, purification and properties of chromium dioxide dichloride have been reviewed by Hartford and Darrin[19].

Chromium dioxide dichloride is a dark red liquid which freezes at −96.5° and boils at 117°. It fumes copiously in moist air and is sensitive to both light and elevated temperatures. At temperatures below 200° only chlorine is evolved, but above this temperature both chlorine and oxygen are released. The initial decomposition products are of the type $[CrO_2]_nCl_2$ where n is 3 or 4. On further heating the oxide Cr_5O_9 is formed and the final involatile product is chromium(III) oxide, Cr_2O_3[20].

Palmer[14] has recorded the electron-diffraction pattern of chromium dioxide dichloride and the dimensions given in Table 4.5 were deduced on the basis of a tetrahedral molecule.

TABLE 4.5

Internuclear Dimensions of Chromium Dioxide Dichloride

Distance	Value	Angle	Value
Cr—O	1.57 Å	O—Cr—O	105.1°
Cr—Cl	2.12 Å	Cl—Cr—Cl	113.3°
Cl—Cl	3.54 Å	Cl—Cr—O	109.6°
O—O	2.49 Å		

The infrared and Raman spectra of chromium dioxide dichloride vapour have been recorded several times[12,21,22] and there is quite good agreement between the various groups for the assignments of the fundamental frequencies. The assignments of Miller and coworkers[22] appear to be the most satisfactory, and they are given in Table 4.6.

The electronic spectra of chromium dioxide dichloride in the solid and gaseous phases and also in solution have been recorded several times[23–26]. Very recently, Dunn and Francis[26] have recorded the electronic spectrum of crystalline chromium dioxide dichloride at 1.7°K in the region

TABLE 4.6
Fundamental Frequencies of Chromium Dioxide Dichloride

Mode	Description	Value (cm^{-1})
ν_1	Cr—O symmetrical stretch	981
ν_2	Cr—Cl symmetrical stretch	465
ν_3	O—Cr—O bend	357
ν_4	O—Cr—Cl bend	140
ν_5	Torsion	224
ν_6	Cr—O asymmetrical stretch	995
ν_7	Rock	211
ν_8	Cr—Cl asymmetrical stretch	496
ν_9	Rock	287

6000–3800 Å. A very detailed examination of the fine structure of the observed bands was carried out.

Chromium dioxide dichloride reacts with peroxydisulphyldifluoride, $S_2O_6F_2$, at room temperature[27] to give $CrO_2(SO_3F)_2$ and with pyridine, the adducts $CrO_2Cl_2 \cdot C_5H_5N$ and $CrO_2Cl_2 \cdot 2C_5H_5N$ have been prepared[28].

Chromium dioxide dibromide. Zellner[29] has prepared chromium dioxide dibromide by the interaction of chromium dioxide dichloride and a tenfold excess of hydrogen bromide at −60° to −80°. Prepared in this manner the product is normally contaminated with some chromium dioxide dichloride. Krauss and Stark[30] have found that chromium dioxide dibromide is also formed by the reaction between chromium trioxide and hydrogen bromide in the presence of phosphorus pentaoxide.

Chromium dioxide dibromide is thermally unstable even below room temperature and very little is known of its physico-chemical properties. However, it has been reported[30] that its molecular weight in carbon tetrachloride indicates that it is a monomer.

Oxidation State V

Chromium pentafluoride. Chromium pentafluoride was first prepared in small amounts by von Wartenberg[4] by fluorinating chromium metal, chromium trichloride or chromium trifluoride at 350–500° in a flow system. Edwards[3] has reinvestigated the preparation and properties of chromium pentafluoride and found that in the direct fluorination of chromium metal, the yield is remarkably enhanced if the tube used for the preparation is sloped downwards in such a manner that the vapour of chromium pentafluoride can escape readily from the hot zone. The experimental arrangement has been described in detail by O'Donnell and Stewart[31]. Chromium pentafluoride is formed as a by-product in the preparation of chromium hexafluoride and also by the thermal decomposition of chromium hexafluoride[1], although it itself appears to be thermally unstable at reasonably elevated temperatures.

Chromium pentafluoride is a red solid which melts at 30° to give a dark red viscous liquid[3]. X-ray diffraction measurements indicate that it is isomorphous with technetium pentafluoride and the orthorhombic unit cell[3] has $a = 5.5$, $b = 7.4$ and $c = 16.3$ Å.

O'Donnell and Stewart[31] have shown that chromium pentafluoride is a powerful oxidant and fluorinating agent. They examined a series of comparative reactions with reagents such as phosphorus and arsenic trifluorides, phosphorus trichloride and carbon disulphide and showed that its reactivity is considerably greater than that of molybdenum and tungsten hexafluorides.

Chromium oxide trifluoride. All attempts to prepare pure chromium oxide trifluoride have failed[32–34]. Both chromium tetrafluoride and chromium trioxide have been used as starting materials and these have been reacted with bromine trifluoride, bromine pentafluoride and chlorine trifluoride. In all cases the products obtained contained a certain amount of residual halogen fluoride, probably in the form of a complex, which could not be removed. However, the magnetic moments of the products clearly demonstrated the presence of chromium(V)[32,33].

Chromium oxide trichloride. Krauss and coworkers[35–37] have prepared chromium oxide trichloride, a very dark red to black crystalline solid, by the interaction of chromium trioxide and thionyl chloride or sulphuryl chloride, chromium dioxide dichloride and thionyl chloride, and chromium dioxide dichloride and boron trichloride. The compound may be purified by sublimation at room temperature in a vacuum. It is light sensitive and on standing at room temperature gradually disproportionates to chromium dioxide dichloride and solid chromium(III) compounds.

Oxidation State IV

Chromium tetrafluoride. Chromium tetrafluoride is best prepared by the fluorination of chromium metal at about 350° in a flow system[4,32]. Chromium trifluoride or chromium trichloride may be used instead of chromium metal, but in these cases higher temperatures are required[4].

In all cases the fluorination apparatus is arranged in a horizontal position to minimize the amount of chromium pentafluoride formed. Any chromium pentafluoride and chromium dioxide difluoride formed as by-products may be removed by vacuum sublimation.

Although chromium tetrafluoride was originally described as a brown solid[4], Clark and Sadana[32] have shown that pure chromium tetrafluoride is a green solid which turns brown with the slightest contact with moisture. The exactly analogous property of vanadium tetrafluoride has already been referred to. Chromium tetrafluoride may be sublimed in vacuum at temperatures in excess of 100°, the sublimed material often having a bluish tinge[32].

X-ray analysis indicated that the product obtained by sublimation was amorphous, but because of its intermediate volatility it is probably similar to the tetrafluorides of titanium and vanadium, that is, it probably consists of chains of CrF_6 units joined by shared edges.

The magnetic susceptibility of chromium tetrafluoride has been measured over a wide temperature range. The Curie-Weiss law is obeyed with a Weiss constant of 78° and the magnetic moment is 3.02 BM[32].

Chromium tetrafluoride is surprisingly inert chemically[32]. No reaction occurs with ammonia, sulphur dioxide, sulphur trioxide, selenium tetrafluoride or even bromine trifluoride at room temperature. On refluxing with bromine trifluoride however, reduction occurs and the product obtained analysed to approximately $CrF_3 \cdot 0{\cdot}5BrF_3$. Similarly, refluxing with selenium tetrafluoride gave a mixture of compounds which could be separated by rapid removal of the selenium tetrafluoride at 80–90°. The final products obtained were pink $CrF_2 \cdot SeF_4$ and buff coloured $CrF_3 \cdot SeF_4$.

Chromium tetrachloride. Von Wartenberg[38] has obtained evidence from pressure measurements for the formation of chromium tetrachloride in the gas phase, by passing chlorine over chromium trichloride at 700°, while Galitsky and Guskov[39] observed the disproportionation of chromium trichloride vapour into chromium dichloride and chromium tetrachloride between 600° and 1000°. However, chromium tetrachloride has never been obtained in the solid state.

Oxidation State III

Chromium trifluoride. Most workers have prepared chromium trifluoride by passing anhydrous hydrogen fluoride over heated chromium trichloride in a flow system. For example, von Wartenberg[40] used a temperature of 1100°, while Knox and Mitchell[41] carried out the reaction at 850°. In a very extensive and thorough investigation of the lower fluorides of chromium, Sturm[42] has shown that the use of such high temperatures leads to the formation of an impure product. He has shown that chromium trifluoride disproportionates at elevated temperatures to chromium difluoride and chromium pentafluoride. In order to obtain a pure specimen of chromium trifluoride a temperature between 450° and 600° is required[42]. Consistent with this is the report by Muetterties and Castle[43] that the reaction between chromium metal and anhydrous hydrogen fluoride in a bomb at 300° gave chromium trifluoride, whereas at 900° chromium difluoride was the product. Sturm[42] suggested that the thermal decomposition of ammonium hexafluorochromate(III) at 450–500° in a flow system with helium as the carrier gas, is also a useful method of preparing pure chromium trifluoride. The system must be flushed with an inert gas to remove the hydrogen formed and nitrogen is to be avoided to prevent the formation of chromium nitride. Other less important methods of preparing chromium trifluoride are given in Table 4.7.

TABLE 4.7
Methods of Preparing Chromium Trifluoride

Method	Conditions	Reference
CrF_2 and anhydrous HF	Bomb, 400°	44
$CrCl_3$ and NF_3	Bomb, 480°	45
Cr_2O_3 and NF_3	Bomb, 430°	45
CrO_3, C and anhydrous HF	Flow system, heat	46

Chromium trifluoride is a green crystalline solid which melts in a closed system at 1404°, but disproportionates at temperatures greater than 600° in an open system which allows removal of the chromium pentafluoride formed[42].

Zmbov and Margrave[47] used mass-spectrometric techniques to study the vapour pressure of chromium trifluoride. The results showed it to be monomeric in the vapour phase, and for the temperature range 633–785° the vapour pressure of chromium trifluoride is given by the expression

$$\log p_{\text{atm}} = 7.20 - 12{,}640/T$$

The heat of sublimation of chromium trifluoride is 60 kcal mole^{-1} and the heat of formation has been reported as being[47] -265.2 kcal mole^{-1} and as[40] -266.1 kcal mole^{-1}.

The crystal structure of chromium trifluoride has been determined by X-ray diffraction techniques by Jack and Maitland[48] and also by Knox[49].

Good agreement was obtained between the two groups for dimensions of the rhombohedral unit cell at room temperature with $a = 5.264$ Å and $\alpha = 56.61°$. However, a study of the structure at various temperatures[49] showed that as the temperature was increased, a increased at a greater rate than did α. Knox[49] suggested that if there was a transition to a cubic modification before melting, it would only occur at a very high temperature. In agreement with this, Sturm[42] has shown that there is no phase transition below 1192°. The structure of chromium trifluoride consists of regular CrF_6 octahedra joined by shared corners. The chromium-fluorine bond distance is 1.90 Å and the Cr—F—Cr angle is 146° due to the irregular packing of the octahedra[48,49].

The magnetic properties of chromium trifluoride have received considerable attention. The Curie-Weiss law is obeyed above about −100°. Hansen and Griffel[50] report a magnetic moment of 3.85 BM and a Weiss constant of 124°. Other reported values of the Weiss constant are[51] 133° and[52] 80°. From a study of the magnetic properties of chromium trifluoride below 100°K, Hansen and Griffel[50,53,54] concluded that it is ferromagnetic. On the other hand, Wollan and coworkers[52] concluded from a neutron-diffraction study that the super lattice reflections of the neutron-diffraction patterns were characteristic of an antiferromagnetic structure.

As anticipated, an electron-spin resonance study of chromium trifluoride gave a g value of 2.00[55], indicating normal magnetic behaviour.

Clark[56] has observed the diffuse-reflectance spectrum of chromium trifluoride. Two bands at 16,100 and 22,900 cm^{-1} were assigned to the $^4T_{2g} \leftarrow {}^4A_{2g}$ and $^4T_{1g}(F) \leftarrow {}^4A_{2g}$ transitions respectively.

The only chemical reactions of chromium trifluoride which have been reported are based on its fluoride-acceptor properties to form complex fluorides which will be described in detail later in this chapter, and the formation of adducts with selenium tetrafluoride[32] and bromine trifluoride[33].

Chromium trifluoride trihydrate. This hydrate may be prepared by reducing chromium trioxide in 10% hydrofluoric acid with ethanol in the presence of glucose[57]. It is a green compound which is isomorphous with the α-form of aluminium trifluoride trihydrate with unit-cell dimensions $a = 9.36$ and $c = 9.45$ Å[58]. The diffuse-reflectance spectrum of chromium trifluoride trihydrate has been observed and the following assignments were made[59].

14,930 cm^{-1}	$^2E_{1g} \leftarrow {}^4A_{2g}$
16,420 cm^{-1}	$^4T_{2g} \leftarrow {}^4A_{2g}$
23,420 cm^{-1}	$^4T_{1g}(F) \leftarrow {}^4A_{2g}$
37,100 cm^{-1}	$^4T_{1g}(P) \leftarrow {}^4A_{2g}$

Chromium trichloride. Anhydrous chromium trichloride has been prepared in a number of ways using the metal, an oxide or a hydrated chloride as the starting material. The principal methods which have been used are given in Table 4.8.

TABLE 4.8
Preparation of Chromium Trichloride

Method	Conditions	References
Cr and Cl_2	Flow system, 960–1000°	60–62
Cr, Cl_2 and CCl_4	Flow system, 850–900°	63
Cr_2O_3, C and Cl_2	Flow system, 800°	64
$Cr_2O_3 \cdot xH_2O$ and CCl_4	Flow system, 630°	65
Cr_2O_3 and S_2Cl_2	Sealed tube, 410°	66
CrO_3, C and Cl_2	Flow system, 800°	67, 68
CrO_3 and $SOCl_2$	Reflux	68
CrO_3 and CCl_4	Reflux	68
$CrCl_3 \cdot 6H_2O$ and CCl_4	Flow system, 625°	53, 69
$CrCl_3 \cdot 6H_2O$, Cl_2 and CCl_4	Flow system, 600°	70
$CrCl_3 \cdot 6H_2O$ and $SOCl_2$	Reflux	62, 71, 72

It is important to note that it is not possible to dehydrate chromium trichloride hexahydrate by merely heating it, since some hydrolysis always occurs. Chromium trichloride is readily purified by sublimation in a stream of chlorine at about 850°, or in vacuum at slightly higher temperatures. However, sublimation in a chlorine stream is preferable since there is some evidence for the disproportionation of chromium trichloride at high temperatures in vacuum[39,73]. Sublimation of chromium trichloride is advisable following all the methods described in Table 4.8. At the high temperatures involved in this procedure, the rather inert violet form of chromium trichloride is invariably obtained. There have been reports of a more reactive, water soluble, peach coloured modification of chromium trichloride, resulting from low temperature methods of preparation[66,72]. Little is known of this modification however, and all further properties described in this section refer to the violet form.

Chromium trichloride melts[74] at 1150° and from vapour pressure measurements the heat and entropy of sublimation were found to be 59.5 kcal mole^{-1} and 48.4 cal deg^{-1} mole^{-1} respectively[39]. The determination of the true vapour pressure of chromium trichloride is complicated by concomitant disproportionation. The heat of formation has been reported as[73] -121.9 kcal mole^{-1} and as[40] -136.2 kcal mole^{-1}.

The structure of chromium trichloride is of considerable interest in connection with its magnetic properties which are discussed below. Wooster[61] reported that chromium trichloride has a hexagonal unit cell with $a = 6.02$ and $c = 17.3$ Å. Each chromium is surrounded by an octahedron of chlorine atoms, the octahedra themselves being cubic close packed. A more recent neutron-diffraction study[75] gave the hexagonal unit-cell parameters as $a = 5.952$ and $c = 17.47$ Å. On the other hand, a very detailed single crystal X-ray diffraction study[62] has shown that while the description as an infinite sandwich layer structure is correct, the true unit cell is of monoclinic symmetry, although this may be reduced to a psuedo-hexagonal cell. The parameters of the monoclinic cell are $a = 5.959$, $b = 10.321$, $c = 6.114$ Å and $\beta = 108.50°$. There are four molecules in the cell of space group $C2/m$. The chromium-chlorine bond distances are 2.340 and 2.347 Å. The dimensions of the pseudo-hexagonal unit cell are $a = 5.959$ and $c = 17.394$ Å[62]. It was also found[62] that chromium trichloride undergoes a phase transition near 240°K to give a rhombohedral unit cell of space group $R\bar{3}$ with $a = 5.942$ and $c = 17.333$ Å. The chromium-chlorine distances become 2.328 and 2.365 Å.

The magnetic properties of chromium trichloride have been examined in detail by a number of workers using a variety of techniques. The magnetic susceptibility has been measured both as a function of temperature and of field strength[50,53,54,76,77]. Above about 100°K the Curie-Weiss law is obeyed and the magnetic moment is close to the spin-only value as shown below.

μ (BM)	$\theta°$	References
3.69	31	50, 53
3.90	27	76, 77

There appears to be no magnetic anomaly at the temperature of the reported phase transition[77].

Specific heat measurements[53,67,76,78,79] show an anomaly at 16.8°K which has been interpreted as due to a magnetic transition. Neutron-diffraction studies confirm than an anomaly exists at this temperature[75]. There is however some controversy over the magnetic properties of chromium trichloride below 16.8°K. Cable and coworkers[75] consider that chromium trichloride becomes antiferromagnetic, with magnetic moments in each layer of metal atoms being aligned in a parallel fashion, but with adjacent layers aligned in the opposite direction. On the other hand, Hansen and coworkers[50,53,54] and Leech and Manuel[76,80] consider that chromium trichloride is ferromagnetic and that the similarity to true antiferromagnetic compounds is due to a two dimensional domain structure.

The spectroscopic splitting factor, g, for chromium trichloride is 1.997 at room temperature, but increases to 2.37 at 13.15°K[55,76,81].

The diffuse-reflectance spectrum of chromium trichloride shows absorptions at 13,500 and 18,900 cm^{-1} which have been assigned to the $^4T_{2g} \leftarrow {}^4A_{2g}$ and $^4T_{1g}(F) \leftarrow {}^4A_{2g}$ transitions[56,82].

Chromium trichloride forms a large number of adducts with a wide variety of donor molecules which include alkyl nitriles, ethers, amines, alcohols, phosphines, etc.[83–95].

Probably the most widely studied compounds are those formed with amines. These are normally of the general type $CrCl_3 \cdot nL$, where n may be 2 or 3. In some cases, one or more of the chlorine atoms may be displaced from the immediate coordination sphere of the metal[96–106]. Much of the work on these compounds, including kinetic and spectroscopic data, is discussed in detail in several review articles[107–109]. Although it is usually assumed that these complexes contain octahedrally coordinated chromium, only one crystal structure of this type, that is where at least one chlorine atom is covalently bound to the chromium atom, has been reported. Ooi and coworkers[110] have determined the structure of *trans* dichlorobis(ethylenediamine)chromium(III) chloride hydrochloride dihydrate $[Cr(en)_2Cl_2]Cl \cdot HCl \cdot 2H_2O$. The crystals are monoclinic with $a = 10.97$, $b = 7.88$, $c = 9.12$ Å and $\beta = 111.5°$. The space group is $P2_1/c$ and there are two molecules in the cell. The four ethylenediamine nitrogen atoms are almost coplanar at a distance of 2.12 Å from the chromium atom, while the two chlorine atoms lie above and below this plane at 2.33 Å. The two ethylenediamine molecules are in the gauche form. The non-coordinated chlorine ion, the water and hydrogen chloride molecules are arranged in a manner identical to that in the isomorphous cobalt salt, that is, they are held together by hydrogen bonding.

One of the most interesting adducts of chromium trichloride is that formed with trimethylamine[111–113] with empirical formula $CrCl_3 \cdot 2[(CH_3)_3N]$. Dipole moment, infrared, visible and ultraviolet spectra all indicated that the complex was monomeric with a *trans* trigonal-bipyramidal configuration. This geometry has been confirmed by Fowles and coworkers[112] by single-crystal X-ray diffraction techniques. The structure of the compound is shown in Figure 4.1. It should be noted that the molecule possesses only approximately C_{2v} symmetry because of the eclipsed configuration of the amine groups. The principal internuclear dimensions are given in Table 4.9.

With pyridine, the adduct $CrCl_3 \cdot 3C_5H_5N$ is very readily formed[86,96]. The infrared spectrum of the compound has been examined[114] and bands attributed to chromium-chlorine stretches were observed at 364, 341 and 307 cm^{-1}, while the chromium-nitrogen stretch occurs at 221 cm^{-1}. The diffuse-reflectance spectrum of $CrCl_3 \cdot 3C_5H_5N$ has also been reported[115].

Bands at 15,870 and 22,220 cm^{-1} were assigned to the $^4T_{2g} \leftarrow {}^4A_{2g}$ and $^4T_{1g}(F) \leftarrow {}^4A_{2g}$ transitions.

Complexes with other nitrogen heterocycles, such as bipyridyl and

TABLE 4.9
Internuclear Dimensions in $CrCl_3 \cdot 2[(CH_3)_3N]$

Distance	Value (Å)	Angle	Value
Cr—Cl	2.216	Cl—Cr—Cl(m)	124.2°
Cr—Cl(m)	2.244	Cl—Cr—Cl′	111.6°
Cr—N(1)	2.198	Cl(m)—Cr—N(1)	89.5°
Cr—N(2)	2.168	Cl(m)—Cr—N(2)	91.7°

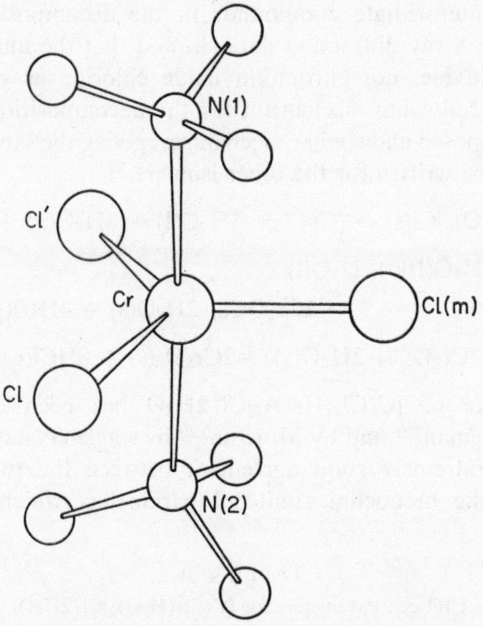

Figure 4.1 The structure of $CrCl_3 \cdot 2[(CH_3)_3N]$. (Reproduced by permission from G. W. A. Fowles, P. T. Greene and J. S. Wood, *Chem. Commun.*, **1967**, 971)

1,10-phenanthroline, are well known[116,117] and, as with the amine compounds, much work has been reported on spectroscopic and kinetic studies.

Howell and coworkers[118] have isolated the adduct $CrCl_3 \cdot QP$ [QP is tris(*o*-diphenylphosphinophenyl)phosphine] as a dichloromethane solvate. From spectral, conductivity and magnetic studies they concluded that only three of the phosphorus atoms of the potentially tetradentate ligand are coordinated to the metal.

Chromium trichloride hexahydrate. The hexahydrate of chromium trichloride exists in three isomeric forms. The normal dark green, commercially available form has the formula $[CrCl_2(H_2O)_4]Cl \cdot 2H_2O$. The remaining isomers $[Cr(H_2O)_6]Cl_3$ and $[CrCl(H_2O)_5]Cl_2 \cdot H_2O$, which are violet and pale green respectively, are far less common. They are normally prepared from the usual isomer[17].

The thermal decomposition of the three isomers has been studied several times[70,119,120], but the most reliable results appear to be those of Cathers and Wendlandt[120]. These workers found that the three isomers gave almost identical differential thermal analysis curves indicating the formation of the same intermediate compounds in the decomposition reactions. Analytical and X-ray diffraction data showed that the final product was chromium(III) oxide, not chromium oxide chloride as previously suggested[119]. The following mechanism for the decomposition of the violet isomer was proposed and similar mechanisms giving the same intermediate products may be written for the other isomers.

$$2[Cr(H_2O)_6]Cl(s) \rightarrow (CrCl_3)_2 \cdot 3H_2O(l) + 8H_2O(g) + H_2O(l)$$

$$2[(CrCl_3)_2 \cdot 3H_2O(l)] + H_2O(l) \rightarrow 2Cr_2OCl_4 \cdot 2H_2O(s) + 4HCl(g) + H_2O(g)$$

$$2Cr_2OCl_4 \cdot 2H_2O(s) \rightarrow 2Cr_2O_3(s) + 8HCl(g)$$

The structure of $[CrCl_2(H_2O)_4]Cl \cdot 2H_2O$ has been determined by Dance and Freeman[121] and by Morosin[122] by single-crystal X-ray diffraction studies. Extremely good agreement between the two groups was obtained for the monoclinic unit-cell parameters which are given in Table 4.10.

TABLE 4.10

Unit-cell Parameters for $[CrCl_2(H_2O)_4]Cl \cdot 2H_2O$

Parameter	Dance and Freeman[121]	Morosin[122]
a(Å)	12.053	12.055
b(Å)	6.840	6.832
c(Å)	11.640	11.648
β	94.16°	94.14°
Space group	*C*2/*c*	*C*2/*c*

There are four molecules in the unit cell. The structure consists of almost perfectly octahedral *trans* $[CrCl_2(H_2O)_4]$ units, chloride ions and water molecules. The dimensions of the octahedral units are shown in Figure 4.2 and the units of the structure are linked together by means of hydrogen-bonding as shown in Figure 4.3. Each water molecule coordinated to the metal atom forms a hydrogen bond to the free chloride ion and also to a free water molecule. In addition the chloride ion and a free water molecule participate in a slightly longer interaction (3.197 Å). Extremely good

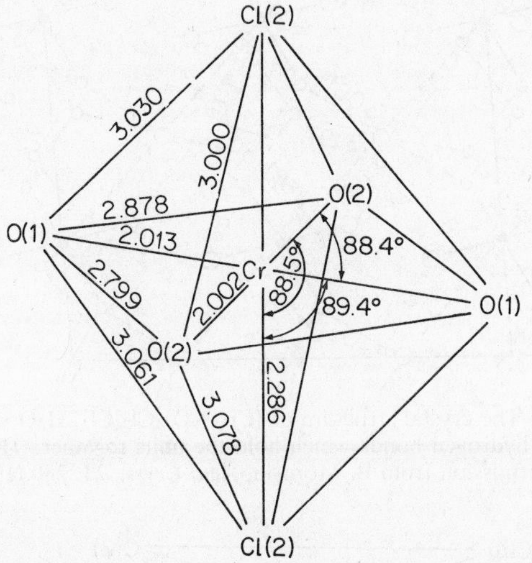

Figure 4.2 The octahedral coordination of the $[Cr(H_2O)_4Cl_2]^+$ ion. (Reproduced by permission from B. Morosin, *Acta Cryst.*, **21**, 280 (1966))

agreement was obtained between the two groups for the hydrogen bond distances, those of Morosin being 2.681, 2.707, 3.061, 3.112 and 3.197 Å. An interesting feature of the structure is the relative position of the one remaining hydrogen atom of the free water molecules. The non-hydrogenic environment of this particular water molecule is a somewhat distorted trigonal prism, as shown in Figure 4.4. Geometrical considerations suggest that the remaining hydrogen atom lies along the pseudo three-fold axis formed by the three near 100° 'a' angles, resulting in a tetrahedrally disposed set of hydrogen linkages about the oxygen atom.

$[Cr(H_2O)_6]Cl_3$ has a rhombohedral unit cell[123] with $a = 7.95$ Å and $\alpha = 97°$. There are two molecules in the unit cell of space group $R\bar{3}c$. The structure consists of regular $Cr(H_2O)_6^{3+}$ octahedra and chloride ions. The infrared spectrum of this isomer has been investigated[124] and bands at 800, 541 and 490 cm^{-1} were assigned to the H_2O rock, H_2O wag and Cr—O stretching modes respectively.

Chromium oxide chloride. Chromium oxide chloride is best prepared by the interaction of chromium(III) oxide and excess chromium trichloride

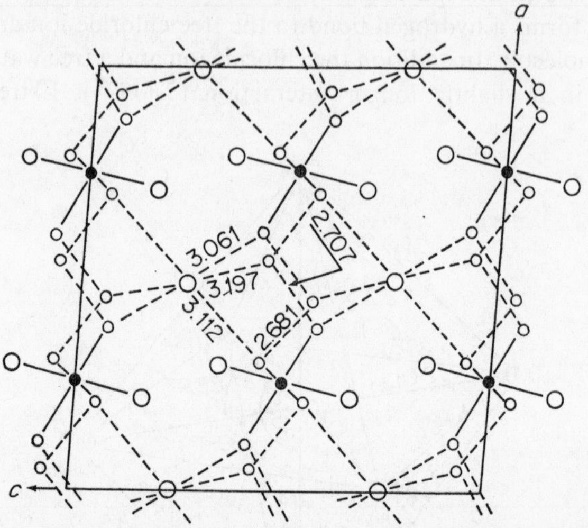

Figure 4.3 The crystal structure of $[Cr(H_2O)_4Cl_2]Cl\cdot 2H_2O$ showing the network of hydrogen bonds which hold the units together. (Reproduced by permission from B. Morosin, *Acta Cryst.*, **21**, 280 (1966))

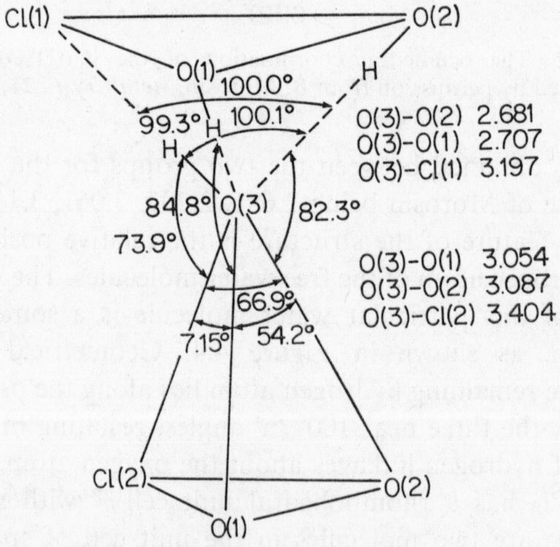

Figure 4.4 The nearest neighbour environment about the non-coordinated water molecule in $[Cr(H_2O)_4Cl_2]Cl\cdot 2H_2O$. (Reproduced by permission from B. Morosin, *Acta Cryst.*, **21**, 280 (1966))

over a thermal gradient of 1040–840° in a silica tube[125]. By heating the product with an aqueous chromium dichloride solution and then with warm dilute hydrofluoric acid, the excess chromium trichloride and the silicon dioxide, which is formed as a by-product, are removed. Less efficient methods of preparation include the interaction of chromium(III) oxide and silicon tetrachloride, and also chromium trichloride with water, bismuth trioxide, titanium dioxide or silicon dioxide[125]. It has also been reported that the passage of oxygen over chromium trichloride at 200–500° results in the formation of chromium oxide chloride[126], but Pechkovskii and coworkers[70] report that this reaction gives chromium(III) oxide at 350°. It could be that the product reported as chromium oxide chloride may have been a mixture of chromium trichloride and chromium(III) oxide.

Chromium oxide chloride is a green crystalline solid which is stable to hot water, and also to alkali, acids and air at room temperature. When heated in air it decomposes to chromium(III) oxide and chlorine, whereas heating in vacuum leads to the formation of chromium trichloride and chromium(III) oxide[125].

The heat and entropy of formation of chromium oxide chloride are -139 kcal mole^{-1} and 16.3 cal deg^{-1} mole^{-1} respectively. It has an orthorhombic unit cell[125] with $a = 3.84$, $b = 3.20$ and $c = 7.98$ Å. Chromium oxide chloride obeys the Curie-Weiss law with a magnetic moment of about 3.9 BM.

Chromium tribromide. Chromium tribromide is best prepared as a lustrous black crystalline solid by bromination of chromium metal at about 750° in a flow system[17,127]. In the presence of bromine vapour, the apparent vapour pressure over chromium tribromide rises very sharply at about 850°, indicating the formation of some chromium tetrabromide in the vapour phase. In vacuum, chromium tribromide decomposes to the dibromide and bromine at high temperatures.

Vapour pressure measurements[127] show that chromium tribromide is monomeric in the vapour phase, and the sublimation pressure over the temperature range 535–795° is given by the expression

$$\log p_{\text{mm}} = 12.82 - 12{,}380/T$$

It has been reported[128] that chromium tribromide has a hexagonal unit cell with $a = 6.26$ and $c = 18.2$ Å. In view of the confusion over the reported hexagonal symmetry of chromium trichloride, it is possible that a more rigorous investigation of the structure of chromium tribromide may reveal similar anomalies. Indeed, preliminary data collected by Morosin and Narath[62] indicate that at room temperature chromium tribromide has a rhombohedral cell of space group $R\bar{3}$, which is transformed to a monoclinic unit cell at about 150°.

The magnetic properties of chromium tribromide have been studied by a number of workers using a variety of techniques. Above about 100°K the Curie-Weiss law is obeyed as follows

μ (BM)	θ°	Reference
3.85	47	129
3.94	51	50

The spectroscopic splitting factor, g, has been measured over a wide temperature range and was found to be close to the free electron value[130].

Specific heat measurements indicate an anomaly at 32.53°K[131], while NMR measurements show that the Curie temperature also occurs at approximately this temperature[132,133]. Below this temperature, chromium tribromide acts as a ferromagnet[54,132–134].

Virtually no chemistry has been reported for chromium tribromide, but the small number of compounds that are known indicate that its chemistry is similar to that of chromium trichloride[84,95,118].

Chromium oxide bromide. A brief report indicates that this compound, having an orthorhombic unit cell with $a = 3.86$, $b = 3.23$ and $c = 8.36$ Å, results from the reaction between chromium tribromide and arsenic trioxide at 380° in a sealed tube[135].

Chromium triiodide. Chromium triiodide is prepared by the interaction of stoichiometric amounts of the elements at 200–225° in a sealed tube[136,137]. If higher temperatures, or excess chromium metal are used, then the product may contain some chromium diiodide[60,138]. As a result of the thermal instability of chromium triiodide, it is advisable to decompose the whole product to the diiodide and to subsequently react this with excess iodine at 500° in a sealed tube[138]. An alternative method of preparation is the reaction of aluminium triiodide with chromium trisulphide at 350° in a sealed tube[139].

Chromium triiodide is a black crystalline solid which does not dissolve readily in water. The equilibrium vapour pressure of iodine over heated chromium triiodide has been measured and is given by the expressions[138,140]

$$\log p_{\text{mm}} = 11.832 - 8936/T \quad (310\text{–}374°)$$

$$\log p_{\text{mm}} = 12.04 - 8649/T \quad (390\text{–}664°)$$

The heat of dissociation is 20.6 kcal mole^{-1} and the heat of formation has been reported[137,141] as -47.8 kcal mole^{-1} and as[142] -54.2 kcal mole^{-1} the former value being the more reliable.

Chromium triiodide is reported to have a hexagonal structure, isomorphous with chromium trichloride, with $a = 6.86$ and $c = 19.88$ Å[136,137,143]. Above about 80°K it obeys the Curie-Weiss law with a Weiss constant of 70° and a magnetic moment of 4.00 BM[50,136]. At 1.50°K the g value is 2.07[136]. Below 68°K chromium triiodide acts as a ferromagnet[132,136].

The Mixed Oxidation State Cr(III), Cr(II)

Sturm[42] studied the phase equilibria in the chromium trifluoride-chromium difluoride system and showed that a single intermediate compound existed for the composition range $CrF_{2.40}$ to $CrF_{2.45}$. The compound however, is best prepared by reacting equimolar amounts of chromium trifluoride and chromium difluoride in a sealed tube at 800–900°. A single-crystal study by Steinfink and Burns[144] has shown however that the compound has the ideal composition $CrF_{2.5}$. These workers draw the analogy with FeO, which although it has an ideal 1:1 composition and the rock salt structure, in fact exists only in the composition range $FeO_{1.06}$ to $FeO_{1.12}$.

$CrF_{2.5}$ has a monoclinic unit cell[144] with $a = 7.773$, $b = 7.540$, $c = 7.440$ and $\beta = 124.25°$. The space group is $C2/c$ and there are four molecules in the unit cell. The structure of this interesting compound is shown in Figure 4.5 and consists of two distinct types of CrF_6 octahedra, the shape and size of individual octahedra being determined by the oxidation state of the metal. The octahedra containing trivalent chromium are nearly regular with an average chromium-fluorine bond distance of 1.89 Å. The divalent metal octahedra are tetragonally elongated and the long chromium-fluorine bond directions are tilted at about 18.5° to the planar CrF_4 entity and directed towards the corners. The chromium(III) octahedra share corners with each other and also share corners with the chromium(II) octahedra, while the chromium(II) units share edges among themselves and corners with the chromium(III) octahedra. The interatomic dimensions are summarized in Table 4.11.

Osmond[145], from a theoretical treatment of the magnetic properties of $CrF_{2.5}$, concluded that the compound should be antiferromagnetic at low temperatures. He suggests a magnetic structure with equal numbers of ferromagnetically coupled cations of both valence states in (011) planes, and opposed spin directions in alternate planes.

Oxidation State II

Chromium difluoride. The compound was first reported by Poulenc[146,147] as being formed by passing anhydrous hydrogen fluoride over chromium metal at red-heat in a flow system, or by the interaction of anhydrous

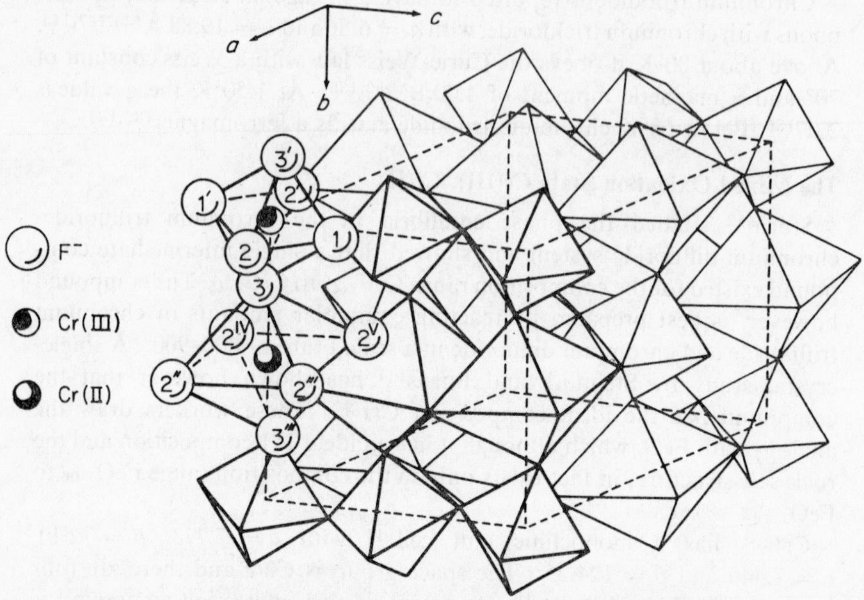

Figure 4.5 The structure of $CrF_{2.5}$. (Reproduced by permission from H. Steinfink and J. H. Burns, *Acta Cryst.*, **17**, 823 (1964))

TABLE 4.11
Interatomic Dimensions in $CrF_{2.5}$

Cr(III) Octahedra		Cr(II) Octahedra	
Cr—F(1)	1.894 Å	Cr—F(2″)	2.572 Å
Cr—F(2)	1.904 Å	Cr—F(3)	1.955 Å
Cr—F(3)	1.877 Å	Cr—F(2‴)	2.010 Å
F(2)—Cr—F(1)	90.28°	F(2‴)—Cr—F(3)	90.38°
F(2)—Cr—F(3)	90.34°	F(2‴)—Cr—F(2ᵛ)	71.58°
F(3)—Cr—F(1)	88.84°	F(2″)—Cr—F(3)	83.21°

hydrogen fluoride and chromium dichloride at room temperature. Sturm[42] has shown that neither of these methods of preparation are satisfactory; for the flow-system reaction, he found that a surface coating of chromium difluoride on the metal prevented complete attack and furthermore, he showed that the fluoride was a mixture of chromium difluoride, chromium trifluoride and the mixed oxidation state fluoride $CrF_{2.5}$.

He has also shown that the halogen-exchange reaction does not proceed at room temperature, while at 700° a mixture of chromium fluorides is obtained. In both of these reactions, Sturm showed that the formation of chromium trifluoride results from the *oxidation* of the difluoride by anhydrous hydrogen fluoride. This most unusual reaction has been confirmed by Glemser and coworkers[44] who reacted chromium difluoride and anhydrous hydrogen fluoride in a bomb at 400° and obtained chromium trifluoride. Another older method of preparing chromium difluoride is the hydrogen reduction of chromium trifluoride at elevated temperatures[148]. However, this method is unsatisfactory since further reduction to the metal is difficult to avoid.

Sturm[42] found that the most efficient methods of preparing very pure chromium difluoride were to reduce chromium trifluoride with the stoichiometric amount of chromium metal in an evacuated nickel bomb at 1000°, and the thermal decomposition of ammonium hexafluorochromate(III) at 1100° in a flow system. In this latter reaction, decomposition to chromium trifluoride commences at 450° and this is partially reduced by the hydrogen formed in the decomposition reaction. The remaining chromium trifluoride disproportionates to give chromium difluoride and volatile chromium pentafluoride which is removed from the reaction zone. Since ammonium hexafluorochromate(III) is readily prepared, it is obviously not profitable to attempt to reduce chromium trifluoride itself directly. A less useful method of preparing chromium difluoride is the reduction of chromium trichloride with the stoichiometric amount of tin difluoride at 1000°, the volatile tin tetrachloride and tin dichloride being readily removed[42]. However the product is normally contaminated with several per cent of chlorine. The oxidation of chromium metal with the fluorides of more noble metals such as tin difluoride, bismuth trifluoride and lead difluoride, also result in the formation of chromium difluoride[42].

Chromium difluoride is a blue-green solid which melts[42] at 894°. By using mass-spectrometric techniques, it has been shown that chromium difluoride is largely monomeric in the vapour phase[149]. The vapour pressure of chromium difluoride has been measured as a function of temperature and is given by the expressions[149]

$$\log p_{\mathrm{atm}} = 9.14 - 18{,}220/T \quad (839\text{–}894°)$$

$$\log p_{\mathrm{atm}} = 8.18 - 17{,}070/T \quad (894\text{–}1060°)$$

The heat of sublimation is 86.8 kcal mole^{-1} and the heat of formation has been given[149] as -181 kcal mole^{-1}.

Chromium difluoride has a distorted rutile structure[48], the monoclinic unit cell having $a = 4.732$, $b = 4.718$, $c = 3.505$ Å and $\beta = 96.52°$.

There are two molecules in the unit cell and the space group is $P2_1/n$. It is isomorphous with copper difluoride and the unit cell and its internuclear distances are shown in Figure 4.6.

The magnetic moment of chromium difluoride is 4.3 BM at room temperature and from a neutron-diffraction study over a wide temperature

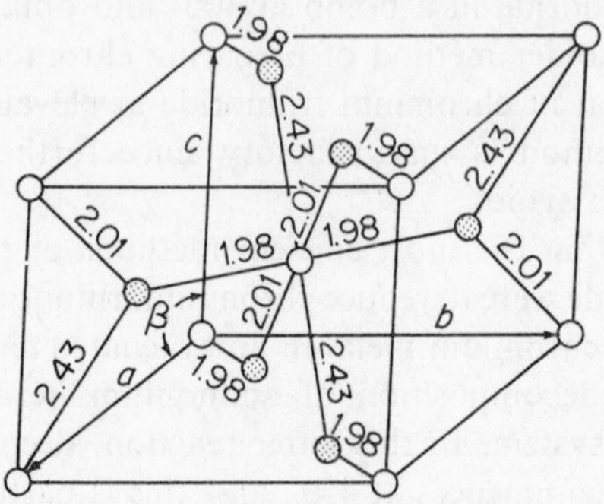

Figure 4.6 The structure of CrF_2. (Reproduced by permission from J. W. Cable, M. K. Wilkinson and E. O. Wollan, *Phys. Rev.*, **118**, 950 (1960))

range, a Neel temperature of 53°K was reported[150]. The neutron-diffraction patterns showed that the magnetic unit cell is identical with the crystallographic unit cell and that the magnetic axis is parallel to the direction of

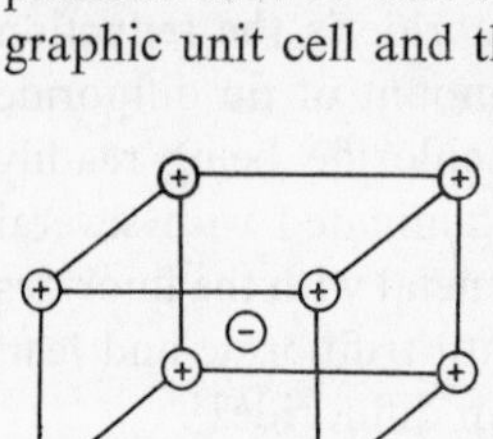

Figure 4.7 The magnetic structure of CrF_2. (Reproduced by permission from J. W. Cable, M. K. Wilkinson and E. O. Wollan, *Phys. Rev.*, **118**, 950 (1960))

the longest chromium-fluorine bonds. The magnetic cell is shown in Figure 4.7.

The optical spectrum of a single crystal of chromium difluoride at 77°K has been observed by Holloway and Kestigian[151]. Four weak bands were observed and assigned as follows:

16,970 cm^{-1}	$^3E_{1g} \leftarrow {}^5E_{1g}$
18,650 cm^{-1}	$^3T_{2g} \leftarrow {}^5E_{1g}$
19,160 cm^{-1}	$^3A_{1g} \leftarrow {}^5E_{1g}$
20,200 cm^{-1}	$^3A_{2g} \leftarrow {}^5E_{1g}$

Fackler and Holah[152] observed bands at 11,500 and 14,700 cm^{-1} in addition to the spin-forbidden bands noted above, and assigned the band at 11,500 cm^{-1} to the ${}^5T_{2g} \leftarrow {}^5E_{1g}$ transition.

Chromium dichloride. Chromium dichloride may be prepared in a number of ways, but all methods require the use of high temperatures. The most common preparation is the hydrogen reduction of chromium trichloride at about 500°. This reaction is best carried out with a mixture of hydrogen and anhydrous hydrogen chloride, to prevent reduction to the metal[17,67,70,153–155]. Chromium metal can be reacted with hydrogen chloride gas alone[17,154,156,157] at about 1200°, but with a mixture of chlorine and hydrogen chloride the same product is obtained at a lower temperature[157–159]. Finally, chromium dichloride has been prepared by the reduction of chromium trichloride with chromium metal[60] in a bomb at about 900°.

Chromium dichloride is a white, very hygroscopic solid which may be readily purified by sublimation in vacuum. It is readily soluble in water to form a blue solution. The melting point has been reported as[70] 806°, as[158] 820° and as[60,160] 824°. The heat of formation has been determined[155] as −50.4 kcal $mole^{-1}$. The mass spectrum of the vapour of chromium dichloride over the temperature range 440–700° shows that although some dimeric species are present, chromium dichloride vapour exists principally as a monomer[161].

The structure of chromium dichloride has been examined by a number of groups by both single-crystal and powder X-ray diffraction techniques. Quite good agreement has been obtained for the orthorhombic unit-cell parameters which are given in Table 4.12.

TABLE 4.12
Unit Cell Parameters of Chromium Dichloride

a (Å)	*b* (Å)	*c* (Å)	*Z*	Space Group	Reference
6.65	5.99	3.48			156
6.624	5.974	3.488	2	*Pnnm*	157
6.64	5.98	3.48	2	*Pnnm*	154
6.638	5.984	3.476	2	*Pnnm*	150

As shown in Figure 4.8, each chromium atom is at the centre of a plane of four equidistant chlorine atoms, with the two remaining chlorines completing the octahedron at somewhat longer distances, almost at right angles to the planar $CrCl_4$ unit. The elongated and slightly distorted

$CrCl_6$ octahedra share their shortest edges to form densly packed chains parallel to the *c* axis, as shown in Figure 4.9. A comparison of the bond angles and internuclear distances is given in Table 4.13, the notation used being that shown in Figure 4.8.

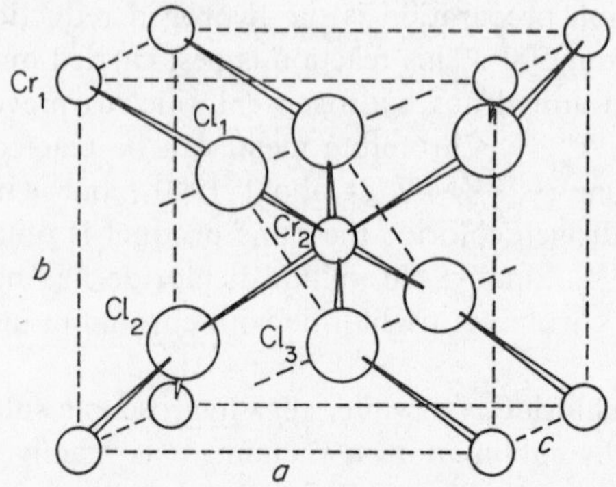

Figure 4.8 The unit cell of $CrCl_2$. (Reproduced by permission from J. W. Tracey and co-workers, *Acta Cryst.*, **14**, 927 (1961))

TABLE 4.13
Internuclear Dimensions in Chromium Dichloride

Dimension	Oswalt[157]	Tracy *et al*[154]	Cable *et al*[150]
Cr(2)—Cl(1)	2.375 Å	2.395 Å	2.40 Å
Cr(2)—Cl(2)	2.910 Å	2.915 Å	2.89 Å
Cl(1)—Cr(2)—Cl(3)	85.6°	86.2°	
Cl(1)—Cr(2)—Cl(2)	90.0°	90.1°	

Heat capacity measurements on chromium dichloride[67,155,162] show a maximum at 16.06°K indicating a magnetic transition. Neutron-diffraction studies as a function of temperature, show that at low temperatures chromium dichloride becomes antiferromagnetic and a Neel temperature of 20°K has been reported[150]. Starr and coworkers[77] found that chromium dichloride obeys the Curie-Weiss law above about 40°K with a Weiss constant of 149° and a magnetic moment of 5.13 BM. Cable and co-workers[150], using neutron-diffraction techniques, found that the magnetic unit cell had the parameters *b* and *c* double those of the crystallographic unit cell. The magnetic unit cell consists of ferromagnetic (001) planes with adjacent planes having anti-parallel spins as shown in Figure 4.10.

As with chromium difluoride, the axis of magnetization lies parallel to the longest chromium-halogen bond.

Clark[56] has recorded the diffuse-reflectance spectrum of chromium dichloride and observed an intense band at 11,300 cm^{-1} assigned to the $^5T_{2g} \leftarrow {}^5E_{1g}$ transition. Fackler and Holah[152] give a value of 11,600 cm^{-1}

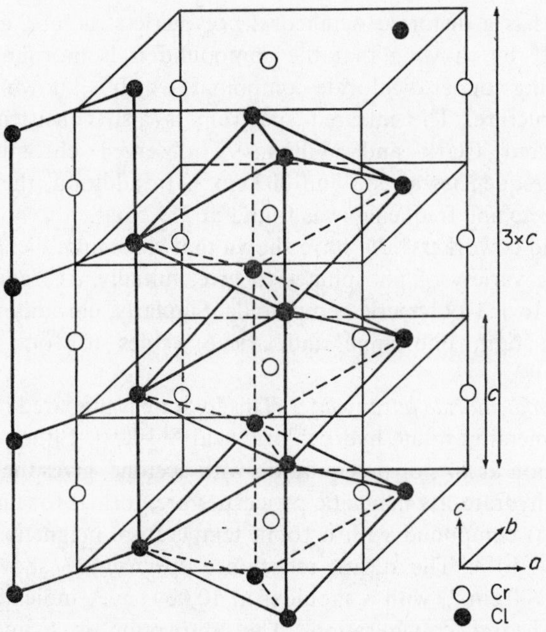

Figure 4.9 The crystal structure of $CrCl_2$. (Reproduced by permission from H. R. Oswalt, *Helv. Chim. Acta*, **44**, 1049 (1961))

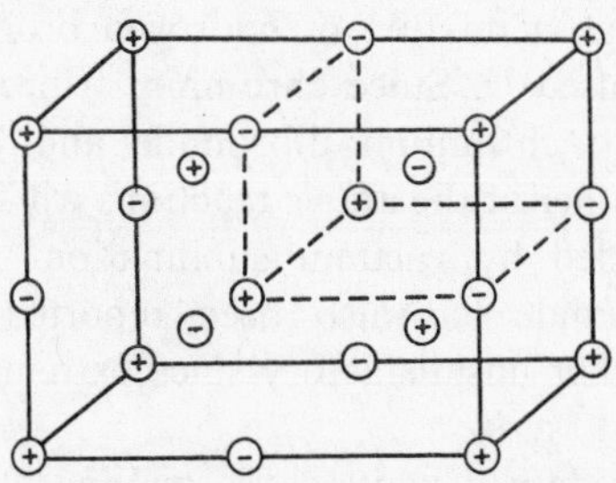

Figure 4.10 The magnetic structure of $CrCl_2$. (Reproduced by permission from J. W. Cable, M. K. Wilkinson and E. O. Wollan, *Phys. Rev.*, **118**, 950 (1960))

for this transition, as well as a shoulder at 8,350 cm^{-1} and weak spin-forbidden bands at 16,100, 17,400 and 19,000 cm^{-1}. These workers suggested that Clark's spectrum shows the presence of chromium(III) as an impurity. The absorption spectrum of chromium dichloride vapour has also been reported[163].

Chromium dichloride reacts readily with pyridine in alcohol[164,165] to give green $CrCl_2 \cdot 2C_5H_5N$. Magnetic and spectral data suggested that the compound has a distorted octahedral polymeric structure, and this was confirmed[166] by showing that the compound is isomorphous with the corresponding copper dichloride compound, which is known to have this type of structure. The infrared spectrum is also consistent with this structure, and Clark and Williams[114] observed chromium-chlorine stretching frequencies at 382 and 303 cm^{-1}. In addition, the chromium-nitrogen stretching frequency was found at 219 cm^{-1}.

Issleib and coworkers[95,167] have shown that chromium dichloride reacts with a wide variety of phosphines to give, initially, 1:2 adducts which decompose to 1:1 polymeric compounds. Similarly, chromium dichloride reacts with both phosphine and arsine oxides to form stable 1:2 adducts[167–169].

Chromium dichloride tetrahydrate. This hydrate is prepared by dissolving chromium metal in dilute hydrochloric acid[152,170,171]. Either evaporation of the solution at 80°, or precipitation with acetone, gives the blue hygroscopic tetrahydrate. Its magnetic properties are normal for an octahedral chromium(II) compound with a room temperature magnetic moment of 4.95 BM[152,171,172]. The diffuse-reflectance spectrum[152] shows a strong band at 14,700 cm^{-1}, with a shoulder at 10,000 cm^{-1}, indicative of a distorted octahedral configuration. The absorption is assigned to the $^5T_{2g} \leftarrow {}^5E_{1g}$ transition.

Chromium dibromide. Chromium dibromide may be prepared by passing hydrogen bromide gas over chromium metal at about 750° in a flow system[161,173], although a mixture of hydrogen bromide and bromine at 890° has also been used[174]. Since chromium tribromide decomposes at high temperatures to chromium dibromide and bromine[127], any tribromide formed in either of the above reactions will be decomposed when the product is purified by vacuum sublimation. Hydrogen reduction of chromium tribromide has also been reported to give chromium dibromide, but is rather unsatisfactory since extremely pure hydrogen is required[17].

Chromium dibromide is a white solid, melting[160] at 842°. It dissolves in air-free water to give a blue solution. The mass spectrum of chromium dibromide in the temperature range 440–700° shows that although some dimeric species are present, the compound is predominantly monomeric in the vapour phase[161].

The vapour pressure of chromium dibromide in the temperature range 564–810° is given by the expression[173]

$$\log p_{\text{mm}} = 11.06 - 12{,}050/T$$

From a single-crystal X-ray diffraction study[175] it has been shown that chromium dibromide has a monoclinic unit cell with $a = 7.114$, $b = 3.649$, $c = 6.217$ Å and $\beta = 93.9°$. The space group is $C2/m$ and there are two molecular entities in the cell. As shown in Figure 4.11 each chromium

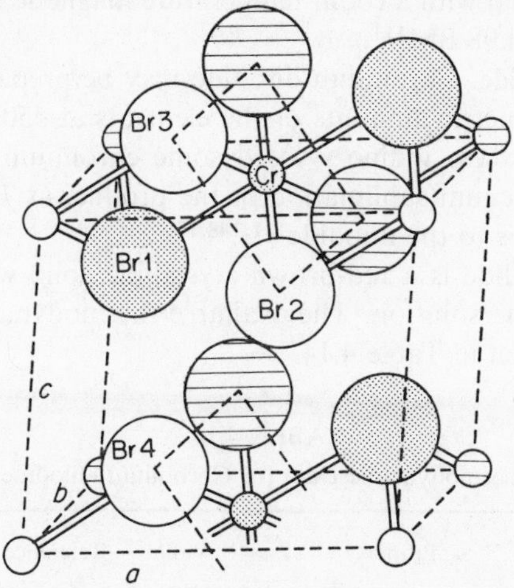

Figure 4.11 The structure of $CrBr_2$. (Reproduced by permission from J. W. Tracey, N. W. Gregory and E. C. Lingafelter, *Acta Cryst.*, **15**, 672 (1962))

atom is at the centre of a slightly irregular plane of four bromine atoms at 2.54 Å from the chromium atom, the remaining two bromine atoms being on an axis nearly perpendicular to the plane at 3.00 Å. The Br(1)—Cr—Br(3) angle is 88.1° while that of Br(2)—Cr—Br(3) is 88.4°. Each bromine atom is shared by three chromium atoms. The distorted $CrBr_6$ octahedra are linked by sharing their shortest edges to give densely packed planar 'ribbons' parallel to the *b* axis; they also share four long edges forming a layer structure.

Chromium dibromide reacts with pyridine to give $CrBr_2 \cdot 2C_5H_5N$. Absorption spectra and magnetic studies[164,165] suggest that the chromium atom has a markedly distorted octahedral stereochemistry.

Scaife[169] has shown that two isomers exist for the compound $CrBr_2 \cdot 2[(C_6H_5)_3PO]$. On the basis of spectral and magnetic studies he concluded that the green isomer has a tetragonally-distorted octahedral structure, while the yellow isomer has the form of a flattened tetrahedron. 1:2 adducts are also known with triphenylarsine oxide[168] and aliphatic phosphines[95].

Chromium dibromide hexahydrate. This hydrate is prepared by dissolving chromium metal in 40% hydrobromic acid and either evaporating *in vacuo* or precipitating with acetone[152,170,171]. The diffuse-reflectance spectrum of this dark blue compound shows absorptions at 10,500 and 14,900 cm^{-1} indicative of a distorted octahedral configuration[152]. It is magnetically normal with a room-temperature magnetic moment reported as 4.90 BM[152] or 4.98 BM[171].

Chromium diiodide. Chromium diiodide may be prepared by the interaction of stoichiometric amounts of the elements at 550–700° in a sealed tube[60,176,177]. If excess iodine is used some chromium triiodide is also formed, but on vacuum sublimation of the product at 700° any triiodide present decomposes to the diiodide[137,138,178].

Chromium diiodide is a red-brown crystalline solid which dissolves in water to give a blue solution. The available thermodynamic data for this compound are given in Table 4.14.

TABLE 4.14
Thermodynamic Data for Chromium Diiodide

Property	Value	Reference
M.p.	868°	60
ΔH°_{form} (kcal $mole^{-1}$)	−37.8	141
	−38.5	179
ΔH_{sub} (kcal $mole^{-1}$)	71.4	180
ΔS_{sub} (cal deg^{-1} $mole^{-1}$)	54.8	180

Chromium diiodide has a monoclinic unit cell with $a = 7.545$, $b = 3.929$, $c = 7.505$ Å and $\beta = 115.6°$. There are two molecules in the unit cell and the space group is $C2/m$[178]. The structure and packing arrangement are essentially the same as those described for chromium dibromide, as shown in Figure 4.12. The chromium-iodine bond distances are 2.740 and 3.237 Å.

Very little chemistry of chromium diiodide has been reported, but adducts with pyridine[165] and triphenylphosphine oxide[169] are known.

Chromium diiodide hydrate. It has been reported that chromium metal dissolves in hydriodic acid to give a solution from which a bluish hydrate of chromium diiodide may be isolated[152,170,171]. There is some controversy

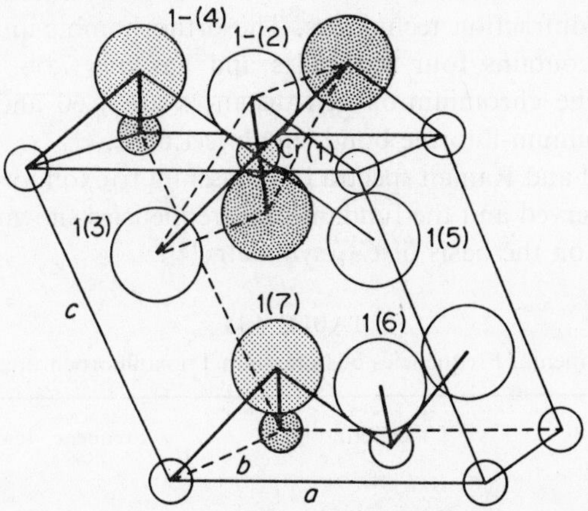

Figure 4.12 The structure of CrI_2. (Reproduced by permission from J. W. Tracey, N. W. Gregory, J. M. Stewart and E. C. Lingafelter, *Acta Cryst.*, **15**, 460 (1962))

as to whether there are five or six molecules of water in this hydrate. Earnshaw and coworkers[171] have shown that it is possible to change the water content of the solid by pumping.

COMPLEX HALIDES AND OXIDE HALIDES

Rather surprisingly, no complex bromides or iodides of chromium have been reported. Both titanium and vanadium form complex bromides but complex iodides are not formed by first-row transition metals before manganese.

As with the binary halides, there is little reported chemistry of chromium(V) and chromium(IV) and the surprising feature of the lower oxidation state compounds is that although a wide variety are known, few have been studied more than superficially.

Oxidation State VI

Trioxofluorochromates(VI). Salts of the trioxofluorochromate(VI) anion may be prepared by reacting the appropriate dichromate with 40% hydrofluoric acid[181,182]. They are red, easily hydrolysed solids.

The potassium and caesium salts have the tetragonal scheelite structure[183,184] with $a = 5.46$ and $c = 12.89$ Å for the potassium compound and $a = 5.72$ and $c = 14.50$ Å for the caesium salt. It is considered that the trioxofluorochromate(VI) anion is very close to a regular tetrahedron with the oxygen and fluorine atoms arranged in a statistical manner in the lattice. Hanic[185] determined the structure of the ammonium salt by single-crystal X-ray diffraction techniques. The orthorhombic unit cell of space group *Pbnm* contains four molecules and has $a = 7.56$, $b = 9.10$ and $c = 6.02$ Å. The chromium-oxygen distances are 1.60 and 1.62 Å (×2) while the chromium-fluorine bond length is 1.68 Å.

The infrared and Raman spectra of potassium trioxofluorochromate(VI) have been observed and the fundamental frequencies shown in Table 4.15 were assigned on the basis of C_{3v} symmetry[181].

TABLE 4.15

Fundamental Frequencies of Potassium Trioxofluorochromate(VI)

Mode	Description	Frequency (cm^{-1})
ν_1	Cr—O symmetrical stretch	911
ν_2	Cr—F stretch	637
ν_3	(O—Cr—O; O—Cr—F)	338
ν_4	Cr—O asymmetrical stretch	955
ν_5	O—Cr—O	370
ν_6	O—Cr—F	261

Trioxochlorochromates(VI). Orange crystals of potassium trioxochloro-chromate(VI) separate on cooling a solution in which potassium dichromate and hot concentrated hydrochloric acid have been reacted[17,186–188]. Potassium trioxochlorochromate(VI) may also be prepared[17,189] by the interaction of potassium chromate and chromium dioxide dichloride at 90–100°. The anion is readily hydrolysed in water, but is stable in dilute hydrochloric acid.

X-ray powder diffraction data for potassium trioxochlorochromate(VI) show that it has a monoclinic unit cell[187,190] with $a = 7.79$, $b = 7.50$, $c = 7.80$ Å and $\beta = 91.33°$. The space group is $P2_1/c$ and there are four molecules in the cell. Within experimental error the trioxochloro-chromate(VI) anion has C_{3v} symmetry. The chromium-chlorine distance is 2.16 Å and the chromium-oxygen bond lengths are 1.528, 1.529 and 1.539 Å. The ammonium salt has a completely analogous structure[191] with $a = 7.77$, $b = 7.72$, $c = 7.96$ Å and $\beta = 90.45°$.

The infrared and Raman spectra of the trioxochlorochromate(VI) anion have been recorded several times[186,192,193] and the observed bands have been assigned on the basis of C_{3v} symmetry. The assignments of

Stammreich and coworkers[186] are given below, the description of the modes being analogous to those for the trioxofluorochromate(VI) ion: ν_1 907, ν_2 438, ν_3 295, ν_4 954, ν_5 365 and ν_6 209 cm^{-1}.

Oxidation State V

Oxotetrafluorochromates(V). Sharpe and Woolf[34] prepared the silver and potassium salts of the oxotetrafluorochromate(V) anion by refluxing the corresponding dichromate with bromine trifluoride, and subsequently removing the solvent. The potassium salt may also be prepared by the action of chlorine trifluoride on potassium chromate[33], and by refluxing a mixture of potassium chloride and chromium trioxide with bromine trifluoride[34]. The compounds are purple and very moisture sensitive.

The X-ray powder diffraction pattern of potassium oxotetrafluorochromate(V) has been indexed[33] on the basis of an orthorhombic unit cell with $a = 13.24$, $b = 10.31$ and $c = 8.32$ Å. At room temperature this compound has an effective magnetic moment of 1.76 BM[194].

Oxopentachlorochromates(V). Krauss and coworkers[36] have prepared several oxopentachlorochromates(V) by reacting chromium trioxide or chromium dioxide dichloride with an organic base in thionyl chloride. Alkali-metal oxopentachlorochromates(V) may be prepared by dissolving chromium trioxide in anhydrous acetic acid saturated with hydrogen chloride and adding to this a solution of the alkali-metal chloride in hydrochloric acid, which is also saturated with hydrogen chloride[195].

Caesium oxopentachlorochromate(V) has the K_2PtCl_6 type cubic structure[195,196] with $a = 10.19$ Å. The chromium-oxygen double bond stretching frequency in the potassium salt[193] occurs as 952 cm^{-1} and at 925 cm^{-1} in the caesium salt[195]. In the latter compound the chromium-chlorine stretch has been observed at 336 cm^{-1}. The magnetic susceptibilities of potassium and caesium oxopentachlorochromate(V) have been measured over a wide temperature range, the Curie-Weiss law being obeyed by each compound.

Compound	μ (BM)	$\theta°$	Reference
K_2CrOCl_5	1.92	14	36
Cs_2CrOCl_5	1.80	14	195

The electron-spin resonance spectrum of the oxopentachlorochromate(V) anion has been observed by Hare and coworkers[197] who found a single line with $g = 1.986$, and by Kon and Sharpless[198] who found $g_{\parallel} = 2.008$ and $g_{\perp} = 1.977$.

Oxotetrachlorochromates(V). Reduction of chromium trioxide in dilute acetic acid with hydrogen chloride gas, followed by addition of the calculated quantity of organic base in acetic acid, leads to the formation of salts of the oxotetrachlorochromate(V) anion[199]. Alternatively, these salts can be isolated by the interaction of the same materials in thionyl chloride[36]. Kon[199] has measured the spectroscopic splitting factor of a number of oxotetrachlorochromate(V) salts in anhydrous acetic acid and also of the solids. The values in the solid state vary according to the nature of the organic cation, and the results have been interpreted on the basis of square-pyramidal (C_{4v}) symmetry. The isoquinolinium compound obeys the Curie-Weiss law[36] and has a magnetic moment of 1.82 BM with a Weiss constant of 7°.

Oxidation State IV

Hexafluorochromates(IV). Alkali-metal salts of this anion are most conveniently prepared by fluorinating the appropriate mixture of chromium trichloride and alkali-metal chloride at about 350° in a flow system[200,201]. The potassium and caesium salts have also been prepared by heating mixtures of the alkali-metal chloride, chromium tetrafluoride and bromine trifluoride[32] at about 130°. This method suffers from the disadvantage that it is difficult to remove all traces of bromine trifluoride from the product.

The hexafluorochromates(IV) are pink hygroscopic solids and, as with many other hexafluorometallates(IV), they tend to be polymorphic. The known lattice symmetries and parameters are given in Table 4.16.

TABLE 4.16
Symmetries and Lattice Parameters of Hexafluorochromates(IV)

Compound	Symmetry	Parameters	Reference
K_2CrF_6	Cubic	$a = 8.104$ Å	32
		$a = 8.15$ Å	202
	Hexagonal	$a = 5.69$, $c = 9.34$ Å	202
Rb_2CrF_6	Cubic	$a = 8.51$ Å	202
	Hexagonal	$a = 5.94$, $c = 9.67$ Å	202
Cs_2CrF_6	Cubic	$a = 8.916$ Å	32
		$a = 9.00$ Å	202

The potassium salt, which decomposes above 300° to form chromium pentafluoride and a chromium(III) compound[201], shows a chromium-fluorine stretching frequency[203] at 556 cm^{-1}.

Pentafluorochromates(IV). Alkali-metal salts of this anion have been prepared by the interaction of stoichiometric amounts of alkali-metal chloride and chromium tetrafluoride in bromine trifluoride, the residual bromine trifluoride being removed at 160° in vacuum[32]. The compounds have magnetic moments of about 3.15 BM at room temperature and are considered to contain condensed CrF_6 octahedral units[32]. The known[32] unit-cell symmetries and parameters of the pentafluorochromates(IV) are given in Table 4.17.

TABLE 4.17

Symmetries and Lattice Parameters of Pentafluorochromates(IV)

Compound	Symmetry	Parameters
$KCrF_5$	Hexagonal	$a = 8.739$, $c = 5.226$ Å
$RbCrF_5$	Hexagonal	$a = 6.985$, $c = 12.12$ Å
$CsCrF_5$	Cubic	$a = 8.107$ Å

Oxidation State III

Hexafluorochromates(III). The methods of preparing some salts of the hexafluorochromate(III) anion are summarized in Table 4.18.

TABLE 4.18

Preparation of Hexafluorochromates(III)

Method	Conditions	Reference
KCl, $CrCl_3$ and F_2	Flow system, heat	204
$K_2CrF_5 \cdot H_2O$ and KHF_2	Fuse at 400°	205
$CrF_3 \cdot 3H_2O$ and NH_4HF_2	Fuse at 200°	42
CrO_3 and NH_4HF_2	Fuse at 200°	42
$(NH_4)_2Cr_2O_7$ and NH_4HF_2	Fuse at 200°	42
Cr, hydrazinium fluoride and aqueous hydrogen fluoride	Evaporate to dryness[a]	206

[a] The product of this reaction is $(N_2H_5)_3CrF_6$.

The addition of an aqueous solution of chromium trichloride to an aqueous ammonium fluoride solution at 90° results in the precipitation of a mixture of the hexafluoro and pentafluoroaquochromate(III) salts. When this mixture is heated with ammonium fluoride at 140° pure green ammonium hexafluorochromate(III) is produced[59]. Fusion of chromium trifluoride with a 2:1 mixture of potassium and sodium hydrogen difluorides results in the formation of NaK_2CrF_6 which, as will be described below, has been shown to be a true compound and not merely a mixture[41].

Recently Babel and coworkers[207] have prepared a number of hexafluorochromates(III) by heating a mixture of alkali-metal hydrogen difluoride, transition-metal difluoride and chromium trifluoride in an inert atmosphere at 600–1000°.

The X-ray powder diffraction data for a number of hexafluorochromates(III) are given in Table 4.19.

TABLE 4.19
Symmetries and Parameters of Some Hexafluorochromates(III)

Compound	Symmetry	Parameters	References
K_3CrF_6	Cubic	$a = 8.54$ Å	204
	Tetragonal	$a = 8.56, c = 8.62$ Å	205
$(NH_4)_3CrF_6$	Cubic	$a = 9.01$ Å	204, 208
NaK_2CrF_6	Cubic	$a = 8.266$ Å	41
$KNiCrF_6$	Cubic	$a = 10.18$ Å	207
$RbCuCrF_6$	Cubic	$a = 10.17$ Å	207
$RbNiCrF_6$	Cubic	$a = 10.21$ Å	207
$CsNiCrF_6$	Cubic	$a = 10.28$ Å	207
$CsCuCrF_6$	Cubic	$a = 10.32$ Å	207
$CsCoCrF_6$	Cubic	$a = 10.35$ Å	207
$CsFeCrF_6$	Cubic	$a = 10.40$ Å	207
$CoMnCrF_6$	Cubic	$a = 10.49$ Å	207

Knox and Mitchell[41] have shown from single-crystal X-ray diffraction studies that NaK_2CrF_6 belongs to the *Fm*3*m* space group and that the potassium and fluorine atoms are in an essentially cubic close packed array with the small chromium and sodium atoms in octahedral holes. The structural arrangement, which is shown in Figure 4.13, is thus related to that of the perovskites. The chromium-fluorine bond distance is 1.933 Å.

Potassium hexafluorochromate(III) shows two chromium-fluorine stretching frequencies[203] at 535 and 522 cm^{-1} and at room temperature the hydrazinium salt has a magnetic moment of 4.3 BM[206].

The diffuse-reflectance spectrum of ammonium hexafluorochromate(III) shows four absorptions which have been assigned as shown below[59].

$$15{,}670 \text{ cm}^{-1} \quad {}^2E_{1g} \leftarrow {}^4A_{2g}$$
$$16{,}060 \text{ cm}^{-1} \quad {}^4T_{2g} \leftarrow {}^4A_{2g}$$
$$22{,}780 \text{ cm}^{-1} \quad {}^4T_{1g}(F) \leftarrow {}^4A_{2g}$$
$$35{,}100 \text{ cm}^{-1} \quad {}^4T_{1g}(P) \leftarrow {}^4A_{2g}$$

Pentafluorochromates(III). Pentafluoroaquochromates(III) may be isolated from saturated aqueous solutions of chromium trifluoride and the alkali-metal fluoride[209–211]. These salts may be dehydrated by heating

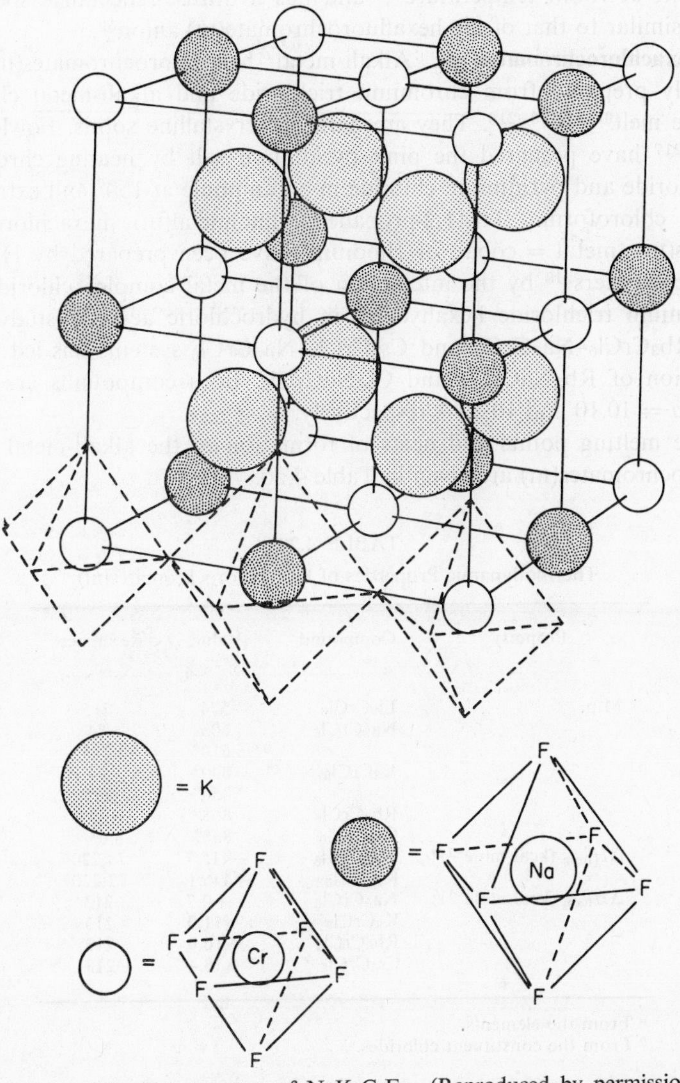

Figure 4.13 The structure of NaK_2CrF_6. (Reproduced by permission from K. Knox and D. W. Mitchell, *J. Inorg. Nucl. Chem.*, **21**, 253 (1961))

in vacuum but nothing is known of anhydrous pentafluorochromates(III)[212].

The pentafluoroaquochromate(III) salt has a magnetic moment of 3.8 BM at room temperature[194] and has a diffuse-reflectance spectrum very similar to that of the hexafluorochromate(III) anion[59].

Hexachlorochromates(III). Alkali-metal hexachlorochromates(III) are readily prepared from chromium trichloride and alkali-metal chloride in the melt[63,74,213–216]. They are dark red crystalline solids. Fowles and Russ[217] have prepared the pink pyridinium salt by heating chromium trichloride and pyridinium chloride in an ampoule at 150° and extracting with chloroform. Tris(1,2-propanediamine)metal(III) hexachlorochromates(III) [metal = cobalt or rhodium] have been prepared by Hatfield and coworkers[218] by the interaction of the metal complex chloride and chromium trichloride hexahydrate in hydrochloric acid. A study[219] of the Rb_3CrCl_6–Na_3CrCl_6 and Cs_3CrCl_6–Na_3CrCl_6 systems has led to the isolation of $Rb_2NaCrCl_6$ and $Cs_2NaCrCl_6$. Both compounds are cubic with a = 10.10 and 10.20 Å respectively.

The melting points and heats of formation of the alkali-metal hexachlorochromates(III) are given in Table 4.20.

TABLE 4.20
Thermodynamic Properties of Hexachlorochromates(III)

Property	Compound	Value	References
M.p.	Li_3CrCl_6	574°	215
	Na_3CrCl_6	603°	74
		610°	215
	K_3CrCl_6	840°	74
		850°	215
	Rb_3CrCl_6	868°	63
	Cs_3CrCl_6	838°	63
ΔH°_{form} (kcal mole^{-1})^a	Na_3CrCl_6	−415.7	73,220
	K_3CrCl_6	−446.1	73,220
ΔH°_{form} (kcal mole^{-1})^b	Na_3CrCl_6	+0.7	213
	K_3CrCl_6	−11.0	213
	Rb_3CrCl_6	−16.0	213
	Cs_3CrCl_6	−18.4	213

^{a} From the elements.
^{b} From the constituent chlorides.

The sodium salt undergoes polymorphic transitions at [74] 440° and at[216] 590°.

The magnetic susceptibility of tris(1,2-propanediamine)cobalt(III) hexachlorochromate(III), $[Co(pn)_3]CrCl_6$, has been measured over a wide temperature range and the Curie-Weiss law is obeyed with a Weiss constant of 2° and the magnetic moment is 3.87 BM[218]. The diffuse-reflectance spectrum of the corresponding complex rhodium hexachlorochromate(III) and also the pyridinium salt have been recorded[217,218] and the following assignments were made.

	$^4T_{2g} \leftarrow {}^4A_{2g}$	$^4T_{1g}(F) \leftarrow {}^4A_{2g}$
$(pyH)_3CrCl_6$	12,700 cm^{-1}	18,200 cm^{-1}
$[Rh(pn)_3]CrCl_6$	13,180 cm^{-1}	18,700 cm^{-1}

μ-Trichlorohexachlorodichromates(III). The dark blue alkali-metal salts of this dimeric anion are readily prepared by interaction of chromium trichloride and alkali-metal chloride in the appropriate stoichiometry in the melt[63,215,216,221]. Quaternary ammonium salts have been prepared by refluxing the base and chromium trichloride in thionyl chloride[222]. The tetraethylammonium salt has also been prepared by the reaction between tetraethylammonium chloride and chromium trichloride hexahydrate in acetonitrile solution[223].

Thermodynamic properties of a number of salts of the dimeric anion are given in Table 4.21.

TABLE 4.21

Thermodynamic Properties of Some μ-Trichlorohexachlorodichromates(III)

Property	Compound	Value	References
M.p.	$K_3Cr_2Cl_9$	813°	215
	$Rb_3Cr_2Cl_9$	874°	63
	$Cs_3Cr_2Cl_9$	894°	63
ΔH°_{form} (kcal mole^{-1})[a]	$K_3Cr_2Cl_9$	−568.0	73, 220
ΔH°_{form} (kcal mole^{-1})[b]	$K_3Cr_2Cl_9$	−12.7	213
	$Rb_3Cr_2Cl_9$	−25.8	213
	$Cs_3Cr_2Cl_9$	−34.7	213

[a] From the elements.
[b] From the constituent chlorides.

Wessel and Ijdo[221] have shown that the caesium salt has a hexagonal cell with $a = 7.22$ and $c = 17.93$ Å. The anion, whose structure is shown in Figure 4.14, is very similar to the corresponding tungsten complex except that the two metal atoms are *not* drawn together, indicating that negligible metal-metal bonding occurs. There is pyramidal distortion of the two $CrCl_6$ octahedra resulting in the angle α being 176° and not 180°.

The chromium-chlorine distances are 2.34 and 2.52 Å, while the chromium-chromium distance is 3.12 Å. The potassium salt is isomorphous with $Cs_3Cr_2Cl_9$ with $a = 6.88$ and $c = 17.52$ Å[216].

The magnetic properties of caesium μ-trichlorohexachlorodichromate(III) confirm that there is little or no interaction between the two chromium atoms in the anion. It has a magnetic moment of 3.82 BM at goom temperature[221]. Earnshaw and Lewis[223] studied the magnetic properties of the tetraethylammonium salt over a wide temperature range and

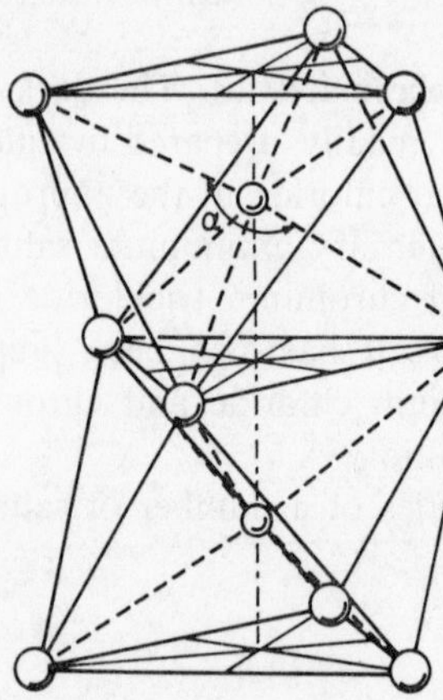

Figure 4.14 The structure of the $[Cr_2Cl_9]^{3-}$ anion. (Reproduced by permission from G. J. Wessel and D. W. J. Ijdo, *Acta Cryst.*, **10**, 466 (1957))

found that the Curie-Weiss law was obeyed with a Weiss constant of 12°, a magnetic moment of 3.94 BM and a *g* value of 2.08.

The diffuse-reflectance spectrum of caesium μ-trichlorohexachlorodichromate(III) has been recorded[82] and the absorptions were assigned as follows:

$$12{,}800\ \text{cm}^{-1} \quad {}^4T_{2g} \leftarrow {}^4A_{2g}$$
$$18{,}000\ \text{cm}^{-1} \quad {}^4T_{1g}(F) \leftarrow {}^4A_{2g}$$

Tetrachlorochromate(III). Machin and coworkers[224] have prepared the compound of empirical formula $PCrCl_8$ by the interaction of chromium dioxide dichloride with a slight excess of phosphorus pentachloride in a sealed tube at 140°. The blue compound has a magnetic moment of 3.59 BM at room temperature and it was suggested that it contains the tetrahedral $CrCl_4^-$ anion. However, it was also pointed out that it was not possible to eliminate the alternative formulation of a long chain octahedral polymeric anion. This type of anion is known for titanium(III) and vanadium(III). It is obvious that more work needs to be done on this interesting compound.

Oxidation State II

Tetrafluorochromate(II). Sodium tetrafluorochromate(II) has been prepared by the interaction of aqueous solutions of sodium fluoride and chromium dichloride or chromium sulphamate in the correct proportions[225,226]. The pale blue product is heated to about 120° in vacuum to remove the last traces of water. It is extremely hygroscopic and was reported initially as a dihydrate[226]. At room temperature the magnetic moment is 4.74 BM[226], and a Weiss constant of 0° suggests that there is little interaction between magnetic centres. The diffuse-reflectance spectrum shows a strong absorption at 15,000 cm^{-1}.

Trifluorochromates(II). These blue compounds are prepared by the interaction of chromium(II) acetate and an alkali-metal hydrogen difluoride in an inert solvent, or alternatively, chromium(II) acetate and alkali-metal fluoride in hydrofluoric acid[226–229]. The compounds are quite difficult to prepare pure.

Potassium trifluorochromate(II) has a distorted perovskite structure[227,228]. The tetragonal unit cell has the parameters $a = 4.274$ and $c = 4{\cdot}109$ Å. The structure consists of distorted CrF_6 octahedra, the distortion being a tetragonal compression rather than the more usual tetragonal elongation. The chromium-fluorine bond distances are 2.00 and 2.14 Å. The rubidium salt is similar with $a = 4.347$ and $c = 4.043$ Å[229].

Above 90°K potassium trifluorochromate(II) obeys the Curie law with a magnetic moment of 4.93 BM[226]. Below 90°K it becomes antiferromagnetic and a Neel temperature of 40°K has been reported[230]. Neutron-diffraction measurements confirmed the antiferromagnetic behaviour and the results were interpreted on the basis of an 'A' type magnetic structure, that is, ferromagnetic (001) sheets coupled antiferromagnetically along the $\langle 001 \rangle$ direction[231]. The magnetic unit cell has the dimensions $a = 8.544$ and $c = 7.966$ Å at 4.2°K.

The chromium-fluorine stretching frequency occurs at 481 cm^{-1} in potassium trifluorochromate(II)[203].

Pentachlorochromate(II). Regardless of the relative proportions of sodium chloride and chromium dichloride, the only compound obtained on fusion of the mixture is sodium pentachlorochromate(II)[158,232]. In marked contrast, the other alkali-metal chlorides react in the melt with chromium dichloride to give either tetrachlorochromates(II) or trichlorochromates(II), but under no circumstances do they give the pentachlorochromate(II) salts. This curious effect is probably related to the small ionic radius of the sodium ion, but apparently no comprehensive investigation has been made with other small cations.

Tetrachlorochromates(II). Alkali-metal tetrachlorochromates(II) are prepared by reactions in the melt of alkali-metal chloride and chromium dichloride in the correct proportions[158,159,233,234]. The melting points of the compounds are given in Table 4.22.

TABLE 4.22

Melting Points of Tetrachlorochromates(II)

Compound	Value	References
K_2CrCl_4	478°	158
	481°	234
	490°	233
Rb_2CrCl_4	551°	158
Cs_2CrCl_4	564°	158, 159, 234

The caesium salt is tetragonal with $a = 5.215$ and $c = 16.460$ Å[159].

Trichlorochromates(II). Like the preceding complexes, the trichlorochromates(II) are prepared by fusing together alkali-metal chloride and chromium dichloride in the correct ratios[158,159,233,234]. The melting points of these salts are listed in Table 4.23.

TABLE 4.23

Melting Points of Trichlorochromates(II)

Compound	Value	References
$KCrCl_3$	494°	158
	495°	234
	501°	233
$RbCrCl_3$	606°	158
$CsCrCl_3$	709°	158, 159

The caesium salt is isomorphous with the corresponding nickel compound and has a hexagonal cell with $a = 7.249$ and $c = 6.228$ Å[158,159].

ADDENDA

Halides and Oxide Halides

Kidd[235] has recorded the ^{17}O NMR spectrum of chromium dioxide dichloride and has related the NMR shift with the π-bond character of the chromium-oxygen linkage.

Opperman[236] has studied the reaction

$$2CrCl_3 + Cl_2 \rightleftharpoons 2CrCl_4$$

in the gas phase and has calculated thermodynamic data for the reaction. The sublimation pressure of chromium trichloride was also determined[236] over the temperature range 700–900°. The sublimation pressure equation is

$$\log p_{mm} = 11.887 - 11{,}500/T$$

The heat and entropy of sublimation are 52.7 kcal mole^{-1} and 41.2 cal deg^{-1} mole^{-1} respectively[236].

The fluorescence spectrum of chromium trichloride has been reported by Vierke and Hansen[237].

Chromium trichloride reacts with thiourea and substituted thioureas[238] to form octahedral adducts of the type $CrCl_3L_3$. Dubicki, Kakos and Winter[239] have prepared green $CrCl_2(OCH_3)\cdot 2CH_3OH$ by adding an equimolar amount of sodium methoxide in methanol to a methanolic solution of chromium trichloride. One molecule of the methanol may be

TABLE 4.24

Properties of Some Chromium(III) Chloro Alkoxides

Compound	M.p.	μ (BM)	ν_{Cr-O} (cm^{-1})	ν_{Cr-Cl} (cm^{-1})
$CrCl_2(OCH_3)\cdot CH_3OH$		3.66	550, 450	320
$CrCl_2(OCH_3)\cdot 2CH_3OH$	84–86°	3.75	560, 450	325, 275
$CrCl_2(OCH_3)\cdot CH_3OH\cdot CH_3COCH_3$	74–75°	3.80	550, 445	350
$CrCl_2(OCH_3)\cdot CH_3OH\cdot CH_3CN$	98–100°	3.78	555, 455	360, 345, 310
$CrCl_2(OCH_3)\cdot CH_3OH\cdot C_4H_8O_2$	56–58°	3.79	550, 450	350

replaced by one molecule of acetone, acetonitrile or dioxan. The magnetism and spectra of these compounds indicate that they are dimeric. The di-methanolate loses one molecule of methanol at 100° and for the resulting compound a tetrameric structure was proposed. Some properties of these chloro alkoxides are given in Table 4.24.

Wood[240] has reported a ligand-field theory treatment which he used to account for the electronic spectrum of trigonal bipyramidal CrX_3L_2 (L = $(CH_3)_3N$; X = Cl, Br). The adduct $CrCl_3L_3$, where L is tetrahydrofuran, has been used[241,242] as the starting material for the preparation of chromium organometallic compounds. The visible and near infrared spectra of the complexes $[CrX_2D_2]ClO_4$, where X = Cl, Br or I and D is *o*-phenylenebis(dimethylarsine), have been recorded and interpreted on the basis of a *trans* octahedral configuration[243]. Other adducts of chromium trichloride that have been investigated include those with acrylonitrile[244] and pyridine-2-aldehyde 2′-pyridylhydrazone (paphy)[245].

The thermal decomposition of $[CrCl_2(H_2O)_4]Cl \cdot 2H_2O$ has been reinvestigated[246], the results agreeing quite well with those of Cathers and Wendlandt[120] (page 174).

A further study on the detailed magnetic properties of chromium tribromide has been reported by Matricardi and coworkers[247].

The preparation and properties of adducts of the type $CrCl_2L_2$, where L is pyridine or acetonitrile, have been reported by Holah and Fackler[248]. The light blue-green adduct $CrCl_2(PMT)_2$, where PMT is pentamethylenetetrazole, decomposes at 160° and has a room-temperature magnetic moment of 4.55 BM[249]. Holah and Fackler[248] have also given further details on the preparation of chromium dibromide and diiodide hexahydrates by dissolving chromium metal in the appropriate hydrohalic acid.

Complex Halides and Oxide Halides

The electronic spectrum of the oxopentachlorochromate(V) anion has been investigated by Wendling[250].

Garton and Wanklyn[251] have described the preparation of K_3CrF_6, K_2NaCrF_6 and Cs_2KCrF_6 by the flux method. For potassium hexafluorochromate(III), X-ray powder diffraction data indicated[251] a tetragonal unit cell with a = 12.08 and c = 12.16 Å. The luminescence spectrum of the same salt has also been reported[237]. De Kozak and Cousseins[252] have investigated the CrF_3-TlF and CrF_3–RbF systems and their results are summarized below.

Tl_3CrF_6	M.p. = 785°, transformation temperature = 212°, cubic unit cell of high temperature form has a = 8.98 Å
$TlCrF_4$	Decomposes at 871°, tetragonal with a = 7.38 and c = 12.87 Å
Rb_3CrF_6	M.p. = 1255°, transformation temperature = 410°, cubic unit cell of high temperature form has a = 9.03 Å
$Rb_2Cr_5F_{17}$	Decomposes at 1135°, orthorhombic with a = 25.59, b = 7.40 and c = 14.69 Å
$Rb_7Cr_6F_{25}$	Decomposes at 906°

Rubidium tetrafluorochromate(III), prepared by direct interaction of the constituent fluorides in the melt, has a tetragonal unit cell[253] with $a = 7.438$ and $c = 6.442$ Å.

Alkali-metal hexachlorochromates(III) and μ-trichlorohexachlorodichromates(III) have been prepared by interaction of the appropriate chlorides in sealed tubes[254–257]. Some thermodynamic properties of several of these complexes are listed in Table 4.25. Galitskii and co-

TABLE 4.25

Thermodynamic Data for Some Chlorochromates(III)[256]

Compound	M.p.	ΔH_f^0 (kcal mole^{-1})^a
Na_3CrCl_6	613°	+0.7
K_3CrCl_6	820°	−11.0
Rb_3CrCl_6	834°	−16.0
Cs_3CrCl_6	812°	−18.4
$K_3Cr_2Cl_9$		−12.7
$Rb_3Cr_2Cl_9$		−25.8

^{a} From constituent chlorides.

workers[254] have measured the saturated vapour pressures of a number of chlorochromates(III). The constants for the vapour pressure equation

$$\log p_{mm} = A - B/T - C \log T$$

are given in Table 4.26. The heat and entropy of sublimation of these complexes fall in the ranges 205–234 kcal mole^{-1} and 155–180 cal deg^{-1} mole^{-1} respectively[254].

Adams and Morris[258] report chromium-chlorine frequencies of 315 and 200 cm^{-1} in the infrared spectrum of tris(1,2-propanediamine)-cobalt(III) hexachlorochromate(III).

Saillant and Wentworth[257] have prepared chloro, bromo and iodo derivatives of the dinuclear anion $Cr_2X_9^{3-}$ by heating stoichiometric amounts of the appropriate halides in sealed tubes. The magnetic moments for the complexes $K_3Cr_2Cl_9$, $Rb_3Cr_2Cl_9$, $Cs_3Cr_2Cl_9$, $Cs_3Cr_2Br_9$ and $Cs_3Cr_2I_9$ are 3.77, 3.77, 3.76, 3.81 and 3.88 BM respectively. The potassium and rubidium chloro derivatives have hexagonal unit cells:[257]

	a	*c*
$K_3Cr_2Cl_9$	6.84 Å	16.53 Å
$Rb_3Cr_2Cl_9$	6.86 Å	17.15 Å

Electronic spectra of the chloro and bromo salts were recorded and found to be consistent with octahedral chromium(III). Potassium μ-trichlorohexachlorodichromate(III) was observed[257] to react with hot pyridine to give $CrCl_3 \cdot 3C_5H_5N$.

Hardt and Streit[259] have prepared a number of tetrachloro, trichloro and tribromochromates(II) by adding a suitable acetate to a solution of

TABLE 4.26

Vapour Pressure Data for Some Chlorochromates(III)

Phase	Compound	A	B	C	Temp. (°K)
Solid	Li_3CrCl_6	12.135	10,700	0	298–1014
	K_3CrCl_6	11.000	10,500	0	298–1110
	Rb_3CrCl_6	11.125	11,000	0	298–1119
	Cs_3CrCl_6	11.960	11,640	0	298–1085
	$K_3Cr_2Cl_9$	11.330	10,600	0	298–1073
	$Rb_3Cr_2Cl_9$	11.800	11,410	0	298–1119
	$Cs_3Cr_2Cl_9$	12.305	12,195	0	298–1135
Liquid	Li_3CrCl_6	24.433	6,935	5.305	1014–1530
	Na_3CrCl_6	47.25	14,863	10.5	930–1260
	K_3CrCl_6	6.911	5,980	0	1110–1485
	Rb_3CrCl_6	6.575	6,000	0	1119–1560
	Cs_3CrCl_6	7.155	6,600	0	1085–1547
	$K_3Cr_2Cl_9$	16.140	7,250	2.5	1073–1370
	$Rb_3Cr_2Cl_9$	7.04	6,030	0	1119–1450
	$Cs_3Cr_2Cl_9$	7.063	6,180	0	1135–1481

chromium(II) in an acetyl halide-acetic acid mixture. Pentahalochromates(III) were prepared in an analogous manner using a chromium(III) solution.

The heat of formation of potassium trichlorochromate(II) from the chlorides and from the elements are -5.8 and -203.7 kcal mole^{-1} respectively[260]. For potassium tetrachlorochromate(II) the values are -1.5 and -303.6 kcal mole^{-1} respectively[260].

REFERENCES

1. O. Glemser, H. Roesky and K. H. Hellberg, *Angew. Chem. Intern. Ed. Eng.*, **2**, 266 (1963).
2. K. H. Hellberg, A. Muller and O. Glemser, *Z. Naturforsch.*, **21b**, 118 (1966).
3. A. J. Edwards, *Proc. Chem. Soc.*, **1963**, 205.
4. H. von Wartenberg, *Z. anorg. allgem. Chem.*, **247**, 135 (1941).
5. A. Engelbrecht and A. V. Grosse, *J. Am. Chem. Soc.*, **74**, 5262 (1952).
6. K. Wiechert, *Z. anorg. allgem. Chem.*, **261**, 310 (1950).

7. E. E. Aynsley, G. Hetherington and P. L. Robinson, *J. Chem. Soc.*, **1954,** 1119.
8. N. Bartlett and P. L. Robinson, *J. Chem. Soc.*, **1961,** 3549.
9. H. L. Krauss and F. Schwarzbach, *Chem. Ber.*, **94,** 1205 (1961).
10. W. T. Smith, *Angew. Chem. Intern. Ed. Eng.*, **1,** 467 (1962).
11. G. D. Flesch and H. J. Svec, *J. Am. Chem. Soc.*, **80,** 3189 (1958).
12. W. E. Hobbs, *J. Chem. Phys.*, **28,** 1220 (1958).
13. G. D. Flesch and H. J. Svec, *J. Am. Chem. Soc.*, **81,** 1787 (1959).
14. K. J. Palmer, *J. Am. Chem. Soc.*, **60,** 2360 (1938).
15. H. H. Sisler, *Inorg. Syn.*, **2,** 205 (1946).
16. A. A. Vakhrushev, *Uchenye Zapiski Udmurt, Gos. Pedagog. Inst.*, **1957,** 143.
17. G. Brauer, *Handbook of Preparative Inorganic Chemistry*, Academic Press, New York, 1963.
18. J. H. Freeman and C. E. C. Richards, *J. Inorg. Nucl. Chem.*, **7,** 287 (1958).
19. W. H. Hartford and M. Darrin, *Chem. Rev.*, **58,** 1 (1958).
20. S. Z. Markarov and A. A. Vakhrushev, *Izv. Akad. Nauk SSSR, Otd. Khim. Nauk*, **1960,** 1731.
21. H. Stammreich, K. Kawai and Y. Tavares, *Spectrochim. Acta*, **15,** 438 (1959).
22. F. A. Miller, G. L. Carlson and W. B. White, *Spectrochim. Acta*, **15,** 709 (1959).
23. R. von Ritschl, *Z. Physik.*, **42,** 172 (1927).
24. R. deL. Kronig, A. Schaafsma and P. K. Peerlkamp, *Z. Physik. Chem.*, **822,** 323 (1933).
25. P. K. Peerlkamp, *Physica*, **1,** 150 (1933).
26. T. M. Dunn and A. H. Francis, *J. Mol. Spectroscopy*, **25,** 86 (1968).
27. M. Lustig and G. H. Cady, *Inorg. Chem.*, **1,** 714 (1962).
28. J. Bernard and M. Camelot, *Compt. Rend.*, **258,** 5881 (1964).
29. H. Zellner, *Monatsh. Chem.*, **80,** 317 (1949).
30. H. L. Krauss and K. Stark, *Z. Naturforsch.*, **17b,** 1 (1962).
31. T. A. O'Donnell and D. F. Stewart, *Inorg. Chem.*, **5,** 1434 (1966).
32. H. C. Clark and Y. N. Sadana, *Can. J. Chem.*, **42,** 50 (1964).
33. H. C. Clark and Y. N. Sadana, *Can. J. Chem.*, **42,** 702 (1964).
34. A. G. Sharpe and A. A. Woolf, *J. Chem. Soc.*, **1951,** 798.
35. H. L. Krauss and G. Muenster, *Z. Naturforsch*, **17b,** 344 (1962).
36. H. L. Krauss, M. Leder and G. Muenster, *Chem. Ber.*, **96,** 3008 (1963).
37. R. B. Johannesen and H. L. Krauss, *Chem. Ber.*, **97,** 2094 (1964).
38. H. von Wartenberg, *Z. anorg. allgem. Chem.*, **250,** 122 (1942).
39. N. V. Galitsky and V. M. Guskov, *Izv. Vyss. Uchebn. Zavendenii, Tsvet. Met.*, **8,** 75 (1965).
40. H. von Wartenberg, *Z. anorg. allgem. Chem.*, **249,** 100 (1942).
41. K. Knox and D. W. Mitchell, *J. Inorg. Nucl. Chem.*, **21,** 253 (1961).
42. B. J. Sturm, *Inorg. Chem.*, **1,** 665 (1962).
43. E. L. Muetterties and J. E. Castle, *J. Inorg. Nucl. Chem.*, **18,** 148 (1961).
44. H. W. Roesky, O. Glemser and K. H. Hellberg, *Chem. Ber.*, **99,** 459 (1966).
45. O. Glemser, J. Wegner and R. Mews, *Chem. Ber.*, **100,** 2474 (1967).
46. V. Shcherbakova, *J. Appl. Chem. USSR*, **10,** 1405 (1937).
47. Z. F. Zmbov and J. L. Margrave, *J. Inorg. Nucl. Chem.*, **29,** 673 (1967).

48. K. H. Jack and R. Maitland, *Proc. Chem. Soc.*, **1957,** 232.
49. K. Knox, *Acta Cryst.*, **13,** 507 (1960).
50. W. N. Hansen and M. Griffel, *J. Chem. Phys.*, **30,** 913 (1959).
51. H. Bizette and B. Tsai, *Compt. Rend.*, **211,** 252 (1940).
52. E. O. Wollan, H. R. Child, W. C. Koehler and M. K. Wilkinson, *Phys. Rev.*, **112,** 1132 (1958).
53. W. N. Hansen and M. Griffel, *J. Chem. Phys.*, **28,** 902 (1958).
54. W. N. Hansen, *J. Appl. Phys.*, **30,** 304S (1959).
55. Y. Ting and D. Williams, *Phys. Rev.*, **82,** 507 (1951).
56. R. J. H. Clark, *J. Chem. Soc.*, **1964,** 417.
57. S. T. Talipov and V. E. Antipov, *Dokl. Akad. Nauk SSSR*, **12,** 27 (1949).
58. I. Maak, P. Eckerlin and A. Rabenau, *Naturwissenschaften*, **48,** 218 (1964).
59. H. L. Schlafer, H. Gausmann and H. U. Zander, *Inorg. Chem.*, **6,** 1528 (1967).
60. J. D. Corbett, R. J. Clark and T. F. Mundy, *J. Inorg. Nucl. Chem.*, **25,** 1287 (1963).
61. N. Wooster, *Z. Krist.*, **74,** 363 (1930).
62. B. Morosin and A. Narath, *J. Chem. Phys.*, **40,** 1958 (1964).
63 A. I. Efimov and B. Z. Pitirimov, *Russ. J. Inorg. Chem.*, **8,** 1042 (1963).
64. R. L. Annis, R. G. Banner, J. G. Kourilo, F. S. Mahne and T. S. Perrin, *Mines Mag. (Denver)*, **49,** 21 (1959).
65. A. Vavoulis, T. E. Austin and S. Y. Tyree, *Inorg. Syn.*, **6,** 129 (1960).
66. H. Funk and C. Muller, *Z. anorg. allgem. Chem.*, **244,** 94 (1940).
67. C. T. Anderson, *J. Am. Chem. Soc.*, **59,** 488 (1937).
68. E. Uhlemann and W. Fischbach, *Z. Chem.*, **3,** 470 (1963).
69. G. B. Hesig, B. Fawkes and R. Hedin, *Inorg. Syn.*, **2,** 193 (1946).
70. V. V. Pechkovskii, S. A. Amirova, N. I. Vorob'ev and T. V. Ostrovskaya, *Russ. J. Inorg. Chem.*, **9,** 1113 (1964).
71. D. Grdenic and M. Gjorgjevic, *Croat. Chem. Acta.*, **30,** 105 (1958).
72. A. R. Pray, *Inorg. Syn.*, **5,** 153 (1957).
73. S. A. Shchukarev, I. V. Vasil'kova, A. I. Efimov and B. Z. Pitirimov, *Russ. J. Inorg. Chem.*, **11,** 247 (1966).
74. B. G. Korshunov and B. Y. Raskin, *Russ. J. Inorg. Chem.*, **7,** 584 (1962).
75. J. W. Cable, M. K. Wilkinson and E. O. Wollan, *J. Phys. Chem. Solids*, **19,** 29 (1961).
76. J. W. Leech and A. J. Manuel, *Proc. Phys. Soc.*, **69B,** 210 (1956).
77. C. Starr, F. Bitter and A. R. Kaufmann, *Phys. Rev.*, **58,** 977 (1940).
78. H. Bizette and C. Terrier, *J. Phys. Radium*, **23,** 486 (1962).
79. O. N. Trapeznikova, L. V. Shubnikov and G. Muliatin, *Physik. Z. Sowjetunion*, **9,** 237 (1936).
80. J. W. Leech and A. J. Manuel, *Proc. Phys. Soc.*, **69B,** 220 (1956).
81. Y. Ting, L. D. Farringer and D. Williams, *Phys. Rev.*, **97,** 1037 (1955).
82. C. Dijkrgraaf, J. P. C. van Heel and J. P. G. Rousseau, *Nature*, **211,** 185 (1966).
83. R. J. Kern, *J. Inorg. Nucl. Chem.*, **25,** 5 (1963).
84. B. J. Hathaway and D. G. Holah, *J. Chem. Soc.*, **1965,** 537.
85. R. J. H. Clark, M. L. Greenfield and R. S. Nyholm, *J. Chem. Soc.*, **1966,** A, 1254.
86. *U.S. Patent*, 3,076,833 (1963).

87. M. J. Frazer, W. Gerrard and R. Twaits, *J. Inorg. Nucl. Chem.*, **25,** 637 (1963).
88. G. W. A. Fowles, P. T. Greene and T. E. Lester, *J. Inorg. Nucl. Chem.*, **29,** 2365 (1967).
89. F. A. Cotton and R. Francis, *J. Am. Chem. Soc.*, **82,** 2986 (1960).
90. F. A. Cotton, R. Francis and W. D. Horrocks, *J. Phys. Chem.*, **64,** 1534 (1960).
91. W. E. Bull, S. K. Madan and J. E. Willis, *Inorg. Chem.*, **2,** 303 (1963).
92. R. J. Kern, *J. Inorg. Nucl. Chem.*, **24,** 1105 (1962).
93. H. Herwig and H. Zeiss, *J. Org. Chem.*, **23,** 1404 (1958).
94. K. Issleib and G. Doll, *Z. anorg. allgem. Chem.*, **305,** 1 (1960).
95. K. Issleib and H. O. Frohlich, *Z. anorg. allgem. Chem.*, **298,** 84 (1958).
96. J. C. Taft and M. M. Jones, *J. Am. Chem. Soc.*, **82,** 4196 (1960).
97. R. J. H. Clark and M. L. Greenfield, *J. Chem. Soc.*, **1967,** A, 409.
98. B. Bosnich, R. D. Gillard, E. D. McKenzie and G. A. Webb, *J. Chem. Soc.*, **1966,** A, 1331.
99. O. Kling and H. L. Schlafer, *Z. anorg. allgem. Chem.*, **313,** 187 (1961).
100. H. L. Schlafer and O. Kling, *Z. anorg. allgem. Chem.*, **302,** 1 (1959).
101. H. L. Schlafer and O. Kling, *Z. anorg. allgem. Chem.*, **309,** 245 (1961).
102. M. Linhard and M. Weigel, *Z. anorg. allgem. Chem.*, **271,** 101 (1952).
103. M. Linhard and M. Weigel, *Z. anorg. allgem. Chem.*, **271,** 115 (1952).
104. C. L. Rollinson and R. C. White, *Inorg. Chem.*, **1,** 281 (1962).
105. T. B. Jackson and J. O. Edwards, *Inorg. Chem.*, **1,** 398 (1962).
106. G. Morgan and F. H. Burstall, *J. Chem. Soc.*, **1937,** 1649.
107. F. Basolo and R. G. Pearson, *Mechanisms of Inorganic Reactions*, 2nd ed., Wiley, New York, 1967.
108. J. H. Dunlop and R. D. Gillard, *Advances in Inorganic Chemistry and Radiochemistry*, **9,** 185 (1966).
109. *Modern Coordination Chemistry* (Ed. J. Lewis and R. G. Wilkins), Interscience, New York, 1960.
110. S. Ooi, Y. Komiyama and H. Kuroya, *Bull. Chem. Soc. Japan*, **33,** 354 (1960).
111. M. W. Duckworth, G. W. A. Fowles and P. T. Greene, *J. Chem. Soc.*, **1967,** A, 1592.
112. G. W. A. Fowles, P. T. Greene and J. S. Wood, *Chem. Commun.*, **1967,** 971.
113. G. W. A. Fowles and P. T. Greene, *Chem. Commun.*, **1966,** 784.
114. R. J. H. Clark and C. S. Williams, *Inorg. Chem.*, **4,** 350 (1965).
115. E. Konig and H. L. Schlafer, *Z. Physik. Chem. (Frankfurt)*, **26,** 371 (1960).
116. J. A. Broomhead and F. P. Dwyer, *Australian J. Chem.*, **14,** 250 (1961).
117. W. W. Brandt, F. P. Dwyer and E. A. Gyafas, *Chem. Rev.*, **54,** 960 (1954).
118. I. V. Howell, L. M. Venanzi and D. C. Goodall, *J. Chem. Soc.*, **1967,** A, 395.
119. D. Kiraly, K. Zalatnai and M. T. Beck, *J. Inorg. Nucl. Chem.*, **11,** 170 (1959).
120. R. E. Cathers and W. W. Wendlandt, *J. Inorg. Nucl. Chem.*, **27,** 1015 (1965).
121. I. G. Dance and H. C. Freeman, *Inorg. Chem.*, **4,** 1555 (1965).
122. B. Morosin, *Acta Cryst.*, **21,** 280 (1966).
123. K. R. Andress and C. Carpenter, *Z. Krist.*, **87,** 446 (1934).

124. I. Nakagawa and T. Shimanouchi, *Spectrochim. Acta*, **20,** 429 (1964).
125. H. Schafer and F. Wartenpfuhl, *Z. anorg. allgem. Chem.*, **308,** 282 (1961).
126. *U.S. Patent*, 3,134,640 (1964).
127. R. J. Sime and N. W. Gregory, *J. Am. Chem. Soc.*, **82,** 93 (1960).
128. H. Braekken, *Kgl. Norskii Videnskab. Selskab. Forh.*, **5,** No. 11 (1932).
129. I. Tsubokawa, *J. Phys. Soc. Japan*, **15,** 1664 (1960).
130. J. F. Dillon, *J. Appl. Phys.*, **33,** 1191 (1962).
131. L. D. Jennings and W. N. Hansen, *Phys. Rev.*, **139,** 1694 (1965).
132. A. Narath, *Phys. Rev.*, **140,** 854 (1965).
133. S. D. Senturia and G. B. Benedek, *Phys. Rev. Letters*, **17,** 475 (1966).
134. J. F. Dillon, *J. Phys. Soc. Japan*, **19,** 1662 (1964).
135. M. Danot and J. Rouxel, *Compt. Rend.*, **262C,** 1879 (1966).
136. J. F. Dillon and C. E. Olson, *J. Appl. Phys.*, **36,** 1259 (1965).
137. N. W. Gregory and L. L. Handy, *Inorg. Syn.*, **5,** 128 (1957).
138. L. L. Handy and N. W. Gregory, *J. Am. Chem. Soc.*, **72,** 5049 (1950).
139. M. Chaigneau and M. Chastagnier, *Bull. Soc. Chim. France*, **1958,** 1192.
140. L. L. Handy and N. W. Gregory, *J. Am. Chem. Soc.*, **74,** 2050 (1952).
141. N. W. Gregory and T. R. Burton, *J. Am. Chem. Soc.*, **75,** 6054 (1953).
142. *Selected Values of Chemical Thermodynamic Properties, Nat Bur. Std. (U.S.) Circ.* 500.
143. L. L. Handy and N. W. Gregory, *J. Am. Chem. Soc.*, **74,** 891 (1952).
144. H. Steinfink and J. H. Burns, *Acta Cryst.*, **17,** 823 (1964).
145. W. P. Osmond, *Proc. Phys. Soc.*, **87,** 767 (1966).
146. M. Poulenc, *Compt. Rend.*, **116,** 253 (1893).
147. M. Poulenc, *Ann. Chim. Phys.*, **2,** 60 (1894).
148. K. Jellinek and A. Rudat, *Z. anorg. allgem. Chem.*, **175,** 281 (1928).
149. R. A. Kent and J. L. Margrave, *J. Am. Chem. Soc.*, **87,** 3582 (1965).
150. J. W. Cable, M. K. Wilkinson and E. O. Wollan, *Phys. Rev.*, **118,** 950 (1960).
151. W. W. Holloway and M. Kestigian, *Spectrochim. Acta*, **22,** 1381 (1966).
152. J. P. Fackler and D. G. Holah, *Inorg. Chem.*, **4,** 954 (1965).
153. A. Burg, *Inorg. Syn.*, **3,** 150 (1950).
154. J. W. Tracey, N. W. Gregory, E. C. Lingafelter, J. D. Dunitz, H. C. Mez, R. E. Scheringer, H. Y. Yakel and M. K. Wilkinson, *Acta Cryst.*, **14,** 927 (1961).
155. J. W. Stout and R. C. Chisholm, *J. Chem. Phys.*, **36,** 979 (1962).
156. L. L. Handy and N. W. Gregory, *J. Chem. Phys.*, **19,** 1314 (1951).
157. H. R. Oswalt, *Helv. Chim. Acta*, **44,** 1049 (1961).
158. H. J. Seifert and K. Klatyk, *Z. anorg. allgem. Chem.*, **334,** 113 (1964).
159. H. J. Seifert and K. Klatyk, *Naturwissenschaften*, **49,** 539 (1962).
160. W. Fischer and R. Gewehn, *Z. anorg. allgem. Chem.*, **222,** 303 (1935).
161. R. C. Schoonmaker, A. H. Friedman and R. F. Porter, *J. Chem. Phys.*, **31,** 1586 (1959).
162. J. W. Stout and R. C. Chisholm, *J. Phys. Soc. Japan*, **17,** 522S (1962).
163. C. W. DeKock and D. M. Gruen, *J. Chem. Phys.*, **44,** 4387 (1966).
164. N. S. Gill, R. S. Nyholm, G. A. Barclay, T. I. Christie and P. J. Pauling, *J. Inorg. Nucl. Chem.*, **18,** 88 (1961).
165. D. G. Holah and J. P. Fackler, *Inorg. Chem.*, **4,** 1112 (1965).
166. H. Lux, L. Eberle and D. Sarre, *Chem. Ber.*, **97,** 503 (1964).

167. K. Issleib, A. Tzschach and H. O. Frohlich, *Z. anorg. allgem. Chem.*, **298,** 164 (1958).
168. L. F. Larkworthy and D. J. Phillips, *J. Inorg. Nucl. Chem.*, **29,** 2101 (1967).
169. D. E. Scaife, *Australian J. Chem.*, **20,** 845 (1967).
170. H. Lux and G. Illmann, *Chem. Ber.*, **91,** 2143 (1958).
171. A. Earnshaw, L. F. Larkworthy and K. S. Patel, *J. Chem. Soc.*, **1965,** 3267.
172. A. Earnshaw, L. F. Larkworthy and K. S. Patel, *Proc. Chem. Soc.*, **1963,** 281.
173. R. J. Sime and N. W. Gregory, *J. Am. Chem. Soc.*, **82,** 800 (1960).
174. H. Kueknl and W. Ernst, *Z. anorg. allgem. Chem.*, **317,** 84 (1962).
175. J. W. Tracey, N. W. Gregory and E. C. Lingafelter, *Acta Cryst.*, **15,** 672 (1962).
176. F. Hein and I. Wintner-Holder, *Z. anorg. allgem. Chem.*, **202,** 81 (1931).
177. F. Hein and G. Bahr, *Z. anorg. allgem. Chem.*, **251,** 241 (1943).
178. J. W. Tracey, N. W. Gregory, J. M. Stewart and E. C. Lingafelter, *Acta Cryst.*, **15,** 460 (1962).
179. C. Cerny and L. Bartovska, *Coll. Czech. Chem. Commun.*, **31,** 3031 (1966).
180. T. L. Allen, *J. Am. Chem. Soc.*, **78,** 5476 (1956).
181. H. Stammreich, O. Sala and D. Bassi, *Spectrochim. Acta*, **19,** 593 (1963).
182. R. Bogvad and A. H. Nielsen, *Acta Cryst.*, **4,** 77 (1951).
183. J. A. A. Ketelaar and E. Wegerif, *Rec. Trav. Chim.*, **57,** 1269 (1938).
184. J. A. A. Ketelaar and E. Wegerif, *Rec. Trav. Chim.*, **58,** 948 (1939).
185. F. Hanic, *Mat-Fyz. Casopis*, **5,** 231 (1955).
186. H. Stammreich, O. Sala and K. Kawai, *Spectrochim. Acta*, **17,** 226 (1961).
187. L. Helmholz and W. R. Foster, *J. Am. Chem. Soc.*, **72,** 4971 (1950).
188. G. P. Haight, D. C. Richardson and N. C. Coburn, *Inorg. Chem.*, **3,** 1777 (1964).
189. H. H. Sisler, *Inorg. Syn.*, **2,** 208 (1946).
190. S. Gawrych, *Roczniki Chem.*, **19,** 413 (1939).
191. F. Hanic and J. Madar, *Chem. Zvesti*, **10,** 82 (1956).
192. T. Dupuis, *Compt. Rend.*, **246,** 3332 (1958).
193. C. G. Barraclough, J. Lewis and R. S. Nyholm, *J. Chem. Soc.*, **1959,** 3552.
194. R. S. Nyholm and A. G. Sharpe, *J. Chem. Soc.*, **1952,** 3579.
195. D. Brown, *J. Chem. Soc.*, **1964,** 4944.
196. E. Wendling, *Bull. Soc. Chim. France*, **1967,** 5.
197. C. R. Hare, I. Bernal and H. B. Gray, *Inorg. Chem.*, **1,** 831 (1962).
198. H. Kon and N. E. Sharpless, *J. Chem. Phys.*, **42,** 906 (1965).
199. H. Kon, *J. Inorg. Nucl. Chem.*, **25,** 933 (1963).
200. D. H. Brown, K. R. Dixon, R. D. W. Kemmitt and D. W. A. Sharp, *J. Chem. Soc.*, **1965,** 1559.
201. E. Huss and W. Klemm, *Z. anorg. allgem. Chem.*, **262,** 25 (1950).
202. H. Bode and E. Voss, *Z. anorg. allgem. Chem.*, **286,** 136 (1956).
203. R. D. Peacock and D. W. A. Sharp, *J. Chem. Soc.*, **1959,** 2762.
204. H. Bode and E. Voss, *Z. anorg. allgem. Chem.*, **290,** 1 (1957).
205. R. D. Peacock, *J. Chem. Soc.*, **1957,** 4684.
206. J. Slivnivk, J. Pezdiic and B. Sedej, *Monatsh. Chem.*, **98,** 204 (1967).
207. D. Babel, G. Pausewang and W. Viebahn, *Z. Naturforsch.*, **22b,** 1219 (1967).
208. L. Passerini and R. Pirani, *Gaz. Chim. Ital.*, **62,** 289 (1932).

209. S. T. Talipov and T. I. Fedorova, *Trudy, Stredneaziat Univ., Khim. Nauk*, **1953,** 47.
210. S. T. Talipov and V. E. Antipov, *Trudy Inst. Khim., Akad. Nauk Uzbek. SSSR*, **3,** 206 (1952).
211. S. T. Talipov and E. L. Krukovskaya, *Trudy Sredneaziat Gos. Univ. im V.I. Lenina, Khim.*, **84,** 3 (1958).
212. R. D. Peacock, *Progress in Inorganic Chemistry* (ed. F. A. Cotton), Vol. 2, Interscience, New York, 1960.
213. S. A. Shchukarev, I. V. Vasil'kova, A. I. Efimov and B. Z. Pitirimov, *Russ. J. Inorg. Chem.*, **11,** 268 (1966).
214. S. N. Shkol'nikov and A. M. Volkov, *Izv. Vyssh. Uchebn. Zavedenii, Tsvetn. Met.*, **5,** 65 (1962).
215. I. V. Vasil'kova, A. I. Efimov and B. Z. Pitirimov, *Russ. J. Inorg. Chem.*, **9,** 493 (1964).
216. C. M. Cook, *J. Inorg. Nucl. Chem.*, **25,** 123 (1963).
217. G. W. A. Fowles and B. J. Russ, *J. Chem. Soc.*, **1967,** A, 517.
218. W. E. Hatfield, R. C. Fay, C. E. Pfluger and T. S. Piper, *J. Am. Chem. Soc.*, **85,** 265 (1963).
219. I. I. Kozhina and P. S. Shapkin, *Russ. J. Inorg. Chem.*, **12,** 1161 (1967).
220. I. V. Vasil'kova, A. I. Efimov and B. Z. Pitirimov, *Russ. J. Inorg. Chem.*, **9,** 417 (1964).
221. G. J. Wessel and D. J. W. Ijdo, *Acta Cryst.*, **10,** 466 (1957).
222. D. M. Adams, J. Chatt, J. M. Davidson and J. Gerratt, *J. Chem. Soc.*, **1963,** 2189.
223. A. Earnshaw and J. Lewis, *J. Chem. Soc.*, **1961,** 396.
224. D. J. Machin, D. F. C. Morris and E. L. Short, *J. Chem. Soc.*, **1964,** 4658.
225. A. J. Deyrup, *Inorg. Chem.*, **3,** 1645 (1964).
226. A. Earnshaw, L. F. Larkworthy and K. S. Patel, *J. Chem. Soc.*, **1966, A,** 363.
227. A. J. Edwards and R. D. Peacock, *J. Chem. Soc.*, **1959,** 4126.
228. K. Knox, *Acta Cryst.*, **14,** 583 (1961).
229. J. C. Cousseins and A. DeKozak, *Compt. Rend.*, **263C,** 1533 (1966).
230. S. Yoneyama and K. Hirakawa, *J. Phys. Soc. Japan*, **21,** 183 (1966).
231. V. Scatturin, L. Corliss, N. Elliott and J. Hastings, *Acta Cryst.*, **14,** 19 (1961).
232. J. C. Shiloff, *J. Phys. Chem.*, **64,** 1566 (1960).
233. S. N. Shkol'nikov and A. M. Volkov, *Izv. Vyss. Uchebn. Zavedenii, Tsvetn. Met.*, **7,** 82 (1964).
234. R. Gut and R. Gnehur, *Chimica*, **16,** 289 (1962).
235. R. G. Kidd, *Can. J. Chem.*, **45,** 605 (1967).
236. H. Opperman, *Z. anorg. allgem. Chem.*, **359,** 51 (1968).
237. G. Vierke and K. H. Hansen, *Z. Physik. Chem. (Frankfurt)*, **59,** 109 (1968).
238. E. Cervone, P. Cancellieri and C. Furlani, *J. Inorg. Nucl. Chem.*, **30,** 2431 (1968).
239. L. Dubicki, G. A. Kakos and G. Winter, *Australian J. Chem.*, **21,** 1461 (1968).
240. J. S. Wood, *Inorg. Chem.*, **7,** 852 (1968).

241. R. P. A. Sneeden and H. H. Zeiss, *J. Organometal. Chem.*, **13,** 369 (1968).
242. R. P. A. Sneeden and H. H. Zeiss, *J. Organometal. Chem.*, **13,** 377 (1968).
243. R. D. Feltham and W. Silverthorn, *Inorg. Chem.*, **7,** 1154 (1968).
244. M. F. Farons and G. R. Tompkins, *Spectrochim. Acta*, **24A,** 788 (1968).
245. F. Lions, I. G. Dance and J. Lewis, *J. Chem. Soc.*, **1967,** A, 565.
246. P. Lumme and K. Junkkarinen, *Suomen Kemistilehti.*, **41B,** 261 (1968).
247. V. R. Matricardi, W. G. Lehmann, N. Kitamura and J. Silcox, *J. Appl. Phys.*, **38,** 1297 (1968).
248. D. G. Holah and J. P. Fackler, *Inorg. Syn.*, **10,** 26 (1967).
249. D. M. Bowers and A. I. Popov, *Inorg. Chem.*, **7,** 1594 (1968).
250. E. Wendling, *Rev. Chim. Minerale*, **4,** 425 (1967).
251. G. Garton and B. M. Wanklyn, *J. Crystal Growth.*, **1,** 49 (1967).
252. A. de Kozak and J. C. Cousseins, *Compt. Rend.*, **267C,** 74 (1968).
253. D. Babel, *Z. Naturforsch.*, **23a,** 1417 (1968).
254. N. V. Galitskii, K. P. Minina and G. M. Borisovskaya, *Russ. J. Inorg. Chem.*, **12,** 1386 (1967).
255. A. V. Storonkin, I. V. Vasil'kova and U. A. Federov, *Russ. J. Phys. Chem.*, **41,** 357 (1967).
256. P. S. Shapkin, I. V. Vasil'kova and M. P. Susarev, *Russ. J. Inorg. Chem.*, **12,** 1498 (1967).
257. R. Saillant and R. A. D. Wentworth, *Inorg. Chem.*, **7,** 1606 (1968).
258. D. M. Adams and D. M. Morris, *J. Chem. Soc.*, **1968,** A, 694.
259. H. D. Hardt and G. Streit, *Naturwissenschaften*, **55,** 443 (1968).
260. I. V. Vasil'kova, A. I. Efimov and E. K. Lupenko, *Vestn. Leningrad. Univ., Fiz. Khim.*, **1968,** 73.

Chapter 5

Manganese

The halide chemistry of manganese is dominated by oxidation state II. As a direct consequence of its d^5 high-spin configuration, there is no Jahn-Teller distortion and no spin-allowed electronic transitions, so there has been considerable investigation of the magnetic and spectral properties to study the usually unobservable spin-forbidden transitions.

HALIDES AND OXIDE HALIDES

The known halides and oxide halides of manganese are given in Table 5.1.

TABLE 5.1
Halides and Oxide Halides of Manganese

Oxidation state	Fluoride	Chloride	Bromide	Iodide
VII	MnO_3F	[MnO_3Cl]		
IV	MnF_4			
III	MnF_3	[$MnCl_3$]		
II	MnF_2	$MnCl_2$	$MnBr_2$	MnI_2

The features to note in the table are the absence of binary halides above oxidation state IV and the general paucity of oxide halides.

Oxidation State VII

Manganese trioxide fluoride. Manganese trioxide fluoride is prepared by reacting potassium permanganate with anhydrous hydrogen fluoride[1,2], with iodine pentafluoride at 40°[3] or with fluorosulphonic acid[1]. The last method is reported to be the most satisfactory, but in each case the product needs to be purified by distillation under vacuum.

Manganese trioxide fluoride is a dark green solid which melts at −78°

to give a dark green liquid[1,3]. The vapour pressure of the liquid over the temperature range $-15°$ to $10°$ is given by the expression[1]

$$\log p_{\text{mm}} = 8.2 - 1007/T$$

A boiling point of 60° was obtained by extrapolation and the heat of vaporization was calculated to be 8.1 kcal mole^{-1}.

Manganese trioxide fluoride is stable at low temperatures, but at room temperature and above it decomposes explosively to manganese difluoride, manganese dioxide and oxygen[1]. As expected it is instantly hydrolysed by moisture.

In the vapour phase, manganese trioxide fluoride is monomeric and the microwave spectrum indicates almost perfect tetrahedral symmetry[4]. The manganese-fluorine and manganese-oxygen bond distances are 1.724 and 1.586 Å respectively while the oxygen-manganese-fluorine angle is 108.45°.

Manganese trioxide chloride. This highly unstable compound has been prepared by Michel and Doiwa[5] by passing anhydrous hydrogen chloride gas through a solution of potassium permanganate in concentrated sulphuric acid. The green-violet gas condenses at $-50°$ to give a very dark liquid. At room temperature it undergoes violent decomposition.

Oxidation State IV

Manganese tetrafluoride. This very hygroscopic blue solid may be prepared by fluorinating either manganese powder in a fluidized bed[6] at 600–700°, manganese trifluoride[7] at 550° or lithium hexafluoromanganate(IV)[8] at 550°.

At room temperature manganese tetrafluoride decomposes slowly to manganese trifluoride and fluorine[6,7] and thus it is expected to be an extremely powerful fluorinating agent. It obeys the Curie-Weiss law[7,8] with a magnetic moment of 3.84 BM and a Weiss constant of 10°.

Oxidation State III

Manganese trifluoride. The principal methods used to prepare manganese trifluoride are given in Table 5.2.

Manganese trifluoride is a red-purple solid which is thermally stable, but it is instantly hydrolysed by traces of moisture. In anhydrous hydrogen fluoride its solubility[15] is 0.164 gm/100 ml at 11.5°. Manganese trifluoride has been used extensively as a fluorinating agent in organic systems, both on the laboratory scale and in industry[10,11,16–18].

The vapour pressure of manganese trifluoride has been determined from mass-spectrometric measurements[19] and this showed that manganese trifluoride is monomeric in the vapour phase. The variation of vapour pressure in the temperature range 726–960° is given by the expression

$$\log p_{\text{atm}} = 7.68 - 14{,}250/T$$

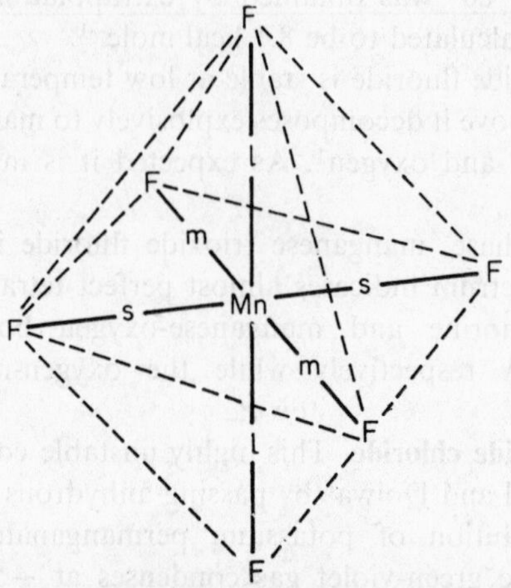

Figure 5.1 Dimensions of the MnF_6 octahedron in MnF_3: l = 2.1 Å, m = 1.9 Å, s = 1.8 Å. (Reproduced by permission from M. A. Hepworth and K. H. Jack, *Acta Cryst.*, **10**, 345 (1957))

TABLE 5.2
Preparation of Manganese Trifluoride

Method	Conditions	References
MnF_2 and F_2	Flow system, 250°	9, 10
$MnCl_2$ and F_2	Flow system, 270°	11
MnI_2 and F_2	Flow system, 250°	9
$Mn(IO_3)_2$ and BrF_3[a]	Reflux, evap. at 140°	12, 13
Mn oxides and F_2[b]	Flow system, 100–150°	14

[a] Best to finally fluorinate at 500° to remove traces of BrF_3[13].
[b] The product from this reaction may be contaminated with a little MnF_2.

The heat of sublimation and the heat of formation are 68 and −238 kcal mole^{-1} respectively[19].

The structure of manganese trifluoride has been determined by single-crystal X-ray diffraction studies[13,20]. The lattice is monoclinic with $a = 8.509$, $b = 5.037$, $c = 13.448$ Å and $\beta = 92.74°$. The space group is

$C2/c$ and there are two molecular units in the cell. The structure consists of distorted MnF_6 octahedra which share their corners. The distorted nature of the octahedra is illustrated in Figure 5.1. The average manganese-fluorine-manganese angle is 146°. The fluorine atom positions are mid-way between the close-packed hexagonal and the rhenium trioxide defective cubic close-packed arrangements.

Manganese trifluoride obeys the Curie-Weiss law[21] with a magnetic moment of 5.0 BM and a Weiss constant of 8°. At very low temperatures manganese trifluoride becomes antiferromagnetic and the Neel temperature has been reported as[22] 43° and as[21] 47°. Neutron-diffraction measurements[22] indicate that the antiferromagnetic structure consists of ferromagnetic layers in which the spins in alternate layers are oppositely orientated.

The diffuse-reflectance spectrum of manganese trifluoride has been recorded by Clark[23] and the following assignments were reported:

12,000 cm^{-1}	$^3T_{1g}(H) \leftarrow {}^5E_g$
19,000 cm^{-1}	$^5T_{2g} \leftarrow {}^5E_g$
19,100 cm^{-1}	$^3E_g(H) \leftarrow {}^5E_g$
23,200 cm^{-1}	$^3T_{2g}(H) \leftarrow {}^5E_g$

Manganese trichloride. This black, thermally unstable compound has been prepared by treating manganese(II) acetate with hydrogen chloride[24] at −100° and by the interaction of manganese dioxide and hydrogen chloride in anhydrous ethanol at −63°, followed by precipitation with carbon tetrachloride or ligroin[25]. Manganese trichloride thermally decomposes above −40°. However, it is possible to stabilize it by forming 1:3 adducts with amines[26]. These complexes are stable at room temperature, but are extremely moisture sensitive. With bipyridyl, 1,10-phenanthroline and acetylacetone, it is possible to carry out substitution reactions.

Oxidation State II

Manganese difluoride. Manganese difluoride is probably most conveniently prepared by the interaction of manganese carbonate and hydrofluoric acid[9,27]. Other methods of preparation include the reaction between manganese powder and anhydrous hydrogen fluoride at 180° in a bomb[28], the passage of anhydrous hydrogen fluoride over heated manganese dichloride[10] and the thermal decomposition of ammonium trifluoromanganate(II) at 300° in a stream of carbon monoxide[29]. All of these methods give a powder and there has been a considerable amount of work reported[30–33] on the preparation of single crystals of manganese difluoride suitable for spectral and magnetic studies.

Manganese difluoride is pink and only sparingly soluble in water. It melts[33–35] at about 920° and below the melting point it vaporizes as a linear monomer[36,37]. The sublimation pressure of manganese difluoride over the temperature range 881–920° is given by the expression[36]

$$\log p_{atm} = 8.70 - 15{,}960/T$$

The heat of sublimation is 76.1 kcal $mole^{-1}$.

The structure of manganese difluoride has been investigated by a number of workers, the interest arising because of its magnetic properties (see below). Both single-crystal and powder X-ray diffraction studies show that the compound has a distorted rutile structure of space group *P4/mnm*, the tetragonal unit cell containing two formula units. The structure consists of tetragonally elongated MnF_6 octahedra. The room temperature unit-cell parameters and manganese-fluorine bond distances are given in Table 5.3.

TABLE 5.3

Room Temperature Cell Parameters of Manganese Difluoride

Unit-cell parameters (Å)	Mn—F Distance (Å)	References
$a = 4.8734$, $c = 3.103$	2.14 (×2), 2.11 (×4)	33
$a = 4.8734$, $c = 3.3099$	2.14 (×2), 2.11 (×4)	38
$a = 4.873$, $c = 3.309$	2.13 (×2), 2.10 (×4)	39, 40
$a = 4.885$, $c = 3.293$		41

The variation of the lattice parameters of manganese difluoride with temperature has been investigated by Gibbons[42] and by Strong[43]. Both workers found a slight variation in the value of *a* with temperature, but *c* decreases quite markedly, particularly below the Neel temperature as shown in Figures 5.2 and 5.3.

The magnetic properties of manganese difluoride have been studied in great detail by a number of workers using a variety of techniques. Above about 80°K the Curie-Weiss law is obeyed and the reported values for the magnetic moment and Weiss constant are:

μ (BM)	$\theta°$	Reference
5.98	97	44
5.73	113	45

As these results imply, the spectroscopic splitting-factor is close to the free electron value[46,47].

At low temperatures manganese difluoride becomes antiferromagnetic and the reported values of the Neel temperature range from 66°K to 75°K[42,43,46–54]. The pressure dependence and the field-strength dependence of the Neel temperature have both been investigated[48,55]. Heat capacity

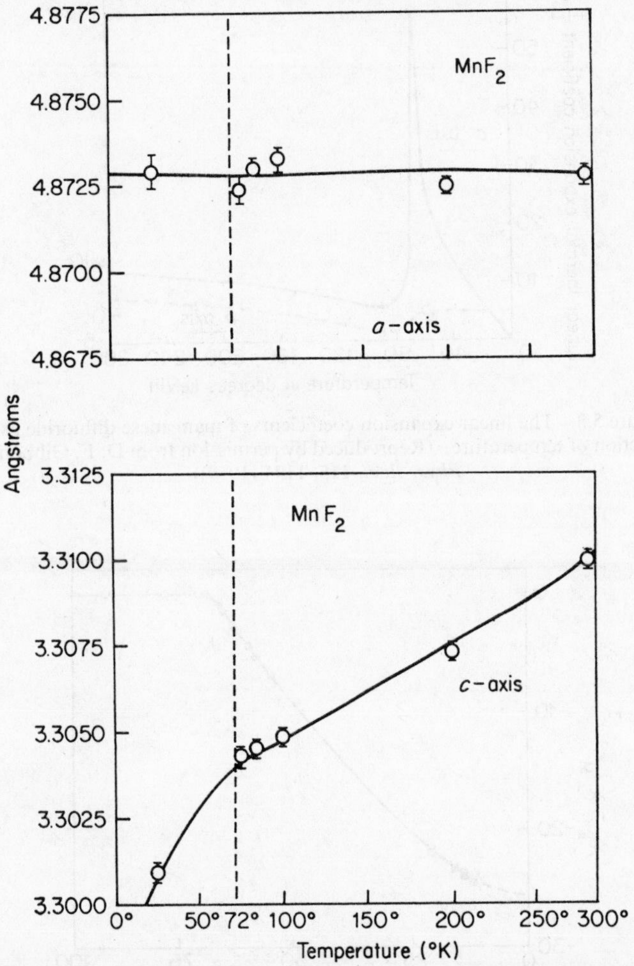

Figure 5.2 Lattice constants of MnF_2 as a function of temperature. (Reproduced by permission from S. L. Strong, *J. Phys. Chem. Solids*, **19,** 51 (1961))

measurements show a discontinuity at 66.5°K consistent with the measurements noted above[50,56]. The variation of the molar susceptibility both parallel to, and perpendicular to the tetragonal axis have been examined

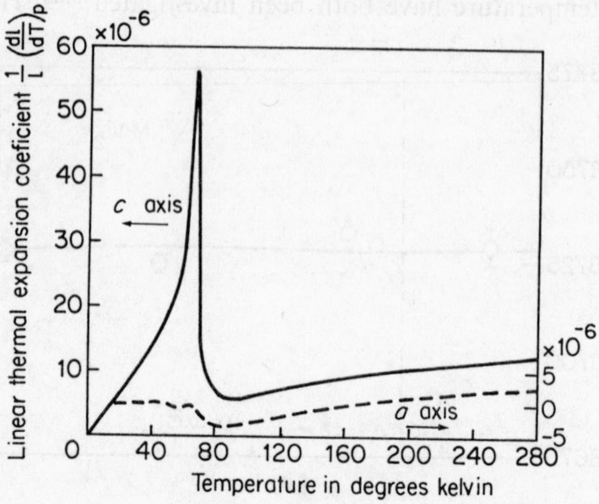

Figure 5.3 The linear expansion coefficients of manganese difluoride as a function of temperature. (Reproduced by permission from D. F. Gibbons, *Phys. Rev.*, **115,** 1194 (1959))

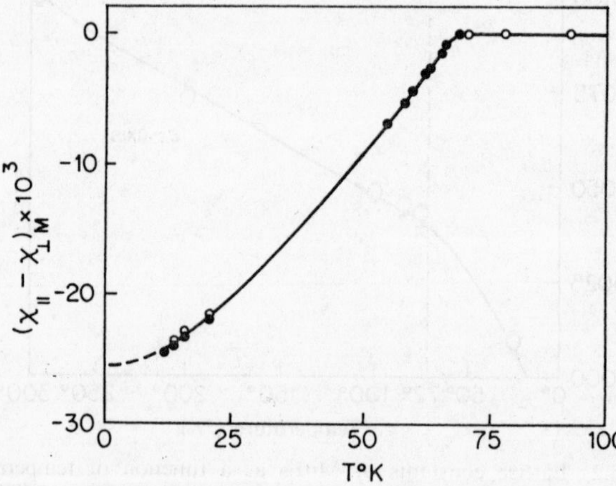

Figure 5.4 The magnetic anisotropy of MnF_2. The arrow indicates the temperature of the heat capacity maximum. (Reproduced by permission from J. W. Stout and M. Griffel, *Phys. Rev.*, **76,** 144 (1949))

in detail[49,54,57]. As can be seen from Figure 5.4, the magnetic anisotropy of manganese difluoride becomes quite large below the Neel temperature. At 20°K $(\chi_{\parallel} - \chi_{\perp})_M$ was found to be independent of field strength. Thermal conductivity measurements[58] also show an anomaly at 67°K. The magnetic structure which can account for the magnetic properties has been discussed by several groups[46,55,57]. The magnetic spins of the manganese atoms are aligned in the *c* axis direction, and the direction of the spin of the manganese atom at the centre of the tetragonal cell is opposite to that of the atoms at the corners of the cell, as shown in Figure 5.5.

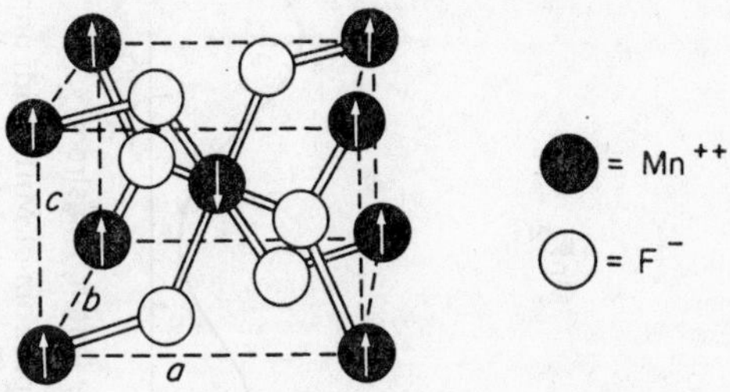

Figure 5.5 The magnetic structure of MnF_2. (Reproduced by permission from P. Heller, *Phys. Rev.*, **146**, 403 (1966))

As is the case for most compounds of manganese(II), the electronic absorption spectrum of manganese difluoride has been studied in great detail by numerous workers[59–68]. It is well known that the spectra of octahedral manganese(II) compounds are characterized by two features:

(a) the bands are of weak intensity, arising from the fact that the ground state is the only sextet state, so that all transitions are spin-forbidden, and

(b) some of the bands are extremely sharp because the differences of the energies of the ground and excited states are independent of the ligand-field strength.

The absorption spectrum of a single crystal of manganese difluoride at room temperature is shown in Figure 5.6. Most workers agree on the assignment of the transitions occurring below about 35,000 cm^{-1}, but

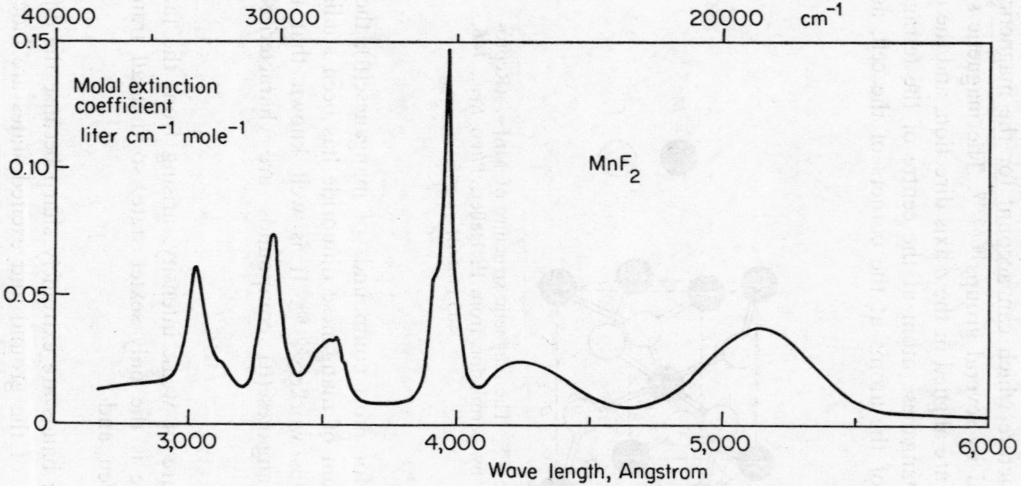

Figure 5.6 The electronic absorption spectrum of MnF_2 with unpolarized light propagated parallel to the *c* axis. (Reproduced by permission from J. W. Stout, *J. Chem. Phys.*, **31**, 709 (1959))

there is considerable disagreement concerning the assignment of the higher energy bands. The assignments of the lower energy bands are[60]

$$
\begin{array}{ll}
19{,}440\ \mathrm{cm^{-1}} & {}^4T_{1g}(G) \leftarrow {}^6A_{1g}(S) \\
23{,}500\ \mathrm{cm^{-1}} & {}^4T_{2g}(G) \leftarrow {}^6A_{1g}(S) \\
25{,}180\ \mathrm{cm^{-1}} & {}^4A_{1g}(G) \leftarrow {}^6A_{1g}(S) \\
25{,}500\ \mathrm{cm^{-1}} & {}^4E_{g}(G) \leftarrow {}^6A_{1g}(S) \\
\left.\begin{array}{l}28{,}120\ \mathrm{cm^{-1}} \\ 28{,}370\ \mathrm{cm^{-1}}\end{array}\right\} & {}^4T_{2g}(D) \leftarrow {}^6A_{1g}(S) \\
30{,}230\ \mathrm{cm^{-1}} & {}^4E_{g}(D) \leftarrow {}^6A_{1g}(S) \\
33{,}060\ \mathrm{cm}_{1g} & {}^4T_{1g}(P) \leftarrow {}^6A_{1g}(S)
\end{array}
$$

The Raman spectrum of manganese difluoride has been investigated by Porto and coworkers[69] and four Raman-active modes were observed. These are (cm^{-1})

$$
\begin{array}{llll}
B_{1g} & 61 & A_{1g} & 341 \\
E_g & 247 & B_{2g} & 476
\end{array}
$$

The infrared spectrum of manganese difluoride has been investigated both above and below the Neel temperature[70–72]. The modes and their origin, particularly below the Neel temperature, have been discussed in detail.

Manganese dichloride. Manganese dichloride is a pink crystalline solid[73] and the principal methods of preparation are given in Table 5.4.

TABLE 5.4
Preparation of Manganese Dichloride

Method	References
Mn and Cl_2 in ethanol	74
Dissolve Mn in conc. HCl, heat product to 580° in stream of HCl gas	75
$MnCl_2 \cdot 4H_2O$, dehydrate in vacuum at 200°[a]	76–79
$MnCl_2 \cdot 4H_2O$, heat in HCl gas at 350–400°	80, 81
MnO_2 and NH_4Cl in sealed tube at 650–700°	82
MnO_2 and acetyl chloride at room temperature	83
Mn acetate and acetyl chloride in benzene	84
Mn salt and thionyl chloride, reflux[b]	85

[a] Advisable to heat product in dry HCl or Cl_2 to avoid contamination by hydrolysis products.
[b] Suitable salts are the nitrate, carbonate, formate and acetate.

Manganese dichloride is very soluble in water, from which the hexa-, tetra- and dihydrates may be isolated, depending on the experimental conditions (see below).

The vapour pressure of anhydrous manganese dichloride over the temperature range 725–950° is given by the expression[86]

$$\log p_{\mathrm{mm}} = 8.559 - 8448/T$$

Mass-spectrometric studies[37,76] indicate that manganese dichloride vaporizes principally as a linear monomeric species.

Thermodynamic data derived from the vapour pressure study and from other sources are given in Table 5.5.

TABLE 5.5
Thermodynamic Properties of Manganese Dichloride

Property	Value	References
M.p.	652 ± 2°	86–88
B.p.	1190°	89
	1225°	86
$\Delta H^\circ_{\mathrm{form}}$ (kcal mole^{-1})	−115.2	80
	−111.6	90
ΔH_{vap} (kcal mole^{-1})	28.78	89
	35.6	86
ΔS_{vap} (cal deg^{-1} mole^{-1})	23.67	86

Manganese dichloride has a hexagonal lattice[91] of the cadmium dichloride type with $a = 3.711$ and $c = 17.59$ Å. Leroi and coworkers[92] have recorded the infrared spectrum of the saturated vapour of manganese dichloride at 800–1000° and have reported that the asymmetrical manganese-chlorine stretching frequency (ν_3) for the monomer occurs at 467 cm^{-1}.

Manganese dichloride obeys the Curie-Weiss law[78] above about 15°K. The magnetic moment is 5.73 BM and the Weiss constant is 3.3°. Heat capacity measurements[81,93] show an anomaly at about 2°K. More detailed studies[75,94,95] of the heat capacity, particularly by Murray, indicate that in fact the anomaly has a distinct doublet nature. Murray[95] found maxima of the specific heat at 1.81 and 1.96°K in the absence of a magnetic field. He showed that there is a divergence of the maxima on application of a magnetic field and at 7.26 Koe the maxima occur at 1.49 and 2.03°K. Murray points out that this divergence is contrary to present theoretical predictions and has suggested that at 1.96°K, in the absence of a magnetic field, manganese dichloride changes from being paramagnetic to antiferromagnetic, and that at 1.81°K a slightly different antiferromagnetic structure is generated.

The absorption spectrum of manganese dichloride has been recorded and discussed in detail by several groups[23,96–98] using both single-crystal and diffuse-reflectance techniques. As with manganese difluoride, there is general agreement on the assignment of the low energy bands, but there are some differences in the interpretation of the high energy transitions. The assignments given by Pappalardo[96] are

18,500 cm^{-1}	$^4T_{1g}(G) \leftarrow {}^6A_{1g}(S)$
22,000 cm^{-1}	$^4T_{2g}(G) \leftarrow {}^6A_{1g}(S)$
23,590 cm^{-1}	$^4A_{1g}(G) \leftarrow {}^6A_{1g}(S)$
23,825 cm^{-1}	$^4E_{g}(G) \leftarrow {}^6A_{1g}(S)$
28,065 cm^{-1}	$^4T_{2g}(D) \leftarrow {}^6A_{1g}(S)$
36,500 cm^{-1}	$^4T_{1g}(P) \leftarrow {}^6A_{1g}(S)$
38,400 cm^{-1}	$^4A_{2g}(F) \leftarrow {}^6A_{1g}(S)$
40,650 cm^{-1}	$^4T_{1g}(F) \leftarrow {}^6A_{1g}(S)$
42,370 cm^{-1}	$^4T_{2g}(F) \leftarrow {}^6A_{1g}(S)$

A considerable amount of work has been reported on the preparation and properties of manganese dichloride adducts with organic molecules in which a nitrogen atom acts as the donor atom. The organic compounds used include amines, pyridine, substituted pyridines, aniline, quinoline, *o*-phenanthroline and bipyridyl[99–126]. In the following discussion, and in the subsequent chapters, the above ligands will be abbreviated using widely accepted notations. These abbreviations are listed at the beginning of the book.

The adducts $MnCl_2(py)_2$ and $MnCl_2(py)_4$ are readily prepared by direct interaction of the constituents, in suitable ratios, in alcohol, or alternatively, by refluxing manganese dichloride in pyridine[108,109].

Very little is known about the properties of $MnCl_2(py)_4$ except that it decomposes[108] at 120° to give $MnCl_2(py)_2$. Libus and Uruska[120] have used the manganese dichloride-pyridine-chlorobenzene system to investigate the characteristics of the

$$MnCl_2(py)_4 \rightleftharpoons MnCl_2(py)_2 + 2py$$

equilibrium. The spectra of the solutions obtained suggested that $MnCl_2(py)_2$ exists in a tetrahedral configuration in pyridine-chlorobenzene solutions and the calculations made were based on this structure.

Initially, pink crystalline $MnCl_2(py)_2$ was considered[110] to have a tetrahedral stereochemistry; but more recent work, particularly infrared studies, has shown quite conclusively that the compound is polymeric, the manganese atom having an octahedral configuration. Zannetti and Serra[126] have shown by X-ray diffraction techniques that $MnCl_2(py)_2$

has a monoclinic unit cell, of space group $P2_1/n$ containing two formula units, with $a = 17.40$, $b = 8.75$, $c = 3.71$ Å and $\beta = 91°$.

The infrared spectrum of the adduct has been studied in great detail[101,104,111,118,127]. Gill and her coworkers[101] have shown that the spectrum of the pyridine molecule is not altered greatly on coordination and have concluded that the electron density over the pyridine ring remains almost constant. They suggested that back-donation plays an important part in the bonding of the pyridine ring to the manganese atom. Clark and Williams[104] report a strong manganese-chlorine stretching mode at 233 cm^{-1} and a manganese-nitrogen absorption at 212 cm^{-1}. These observations are in good agreement with the results of Frank and Rogers[118].

The magnetic moment of $MnCl_2(py)_2$ has been reported several times[106,110,113] with values ranging from 5.89 to 5.97 BM at room temperature. The diffuse-reflectance spectrum of the adduct has been recorded and interpreted on the basis of a *trans* octahedral polymer[112,113]. Konig and Schlafer[112] reported the following assignments.

16,670 cm^{-1}, 19,050 cm^{-1}	$^4T_{1g} \leftarrow {}^6A_{1g}$
20,400 cm^{-1}	$^4T_{2g} \leftarrow {}^6A_{1g}$
23,700 cm^{-1}	$^4A_{1g} + {}^4E_g \leftarrow {}^6A_{1g}$

Groups led by Sharp[108] and Beech[116] have studied the thermal decomposition of $MnCl_2(py)_2$. There is overall agreement as to the mode of decomposition, although naturally there are slight discrepancies with respect to the temperatures of the decomposition reactions in each case. $MnCl_2(py)_2$ decomposes according to the following equations:

$$MnCl_2(py)_2(s) \rightarrow MnCl_2(py)(s) + py(g) \quad (1)$$

$$MnCl_2(py)(s) \rightarrow MnCl_2(py)_{2/3}(s) + 1/3py(g) \quad (2)$$

$$MnCl_2(py)_{2/3}(s) \rightarrow MnCl_2(s) + 2/3py(g) \quad (3)$$

The data reported by Beech and coworkers are summarized in Table 5.6.

Infrared and absorption spectra indicate that both $MnCl_2(py)$ and $MnCl_2(py)_{2/3}$ contain manganese in an octahedral environment, the chlorine atoms acting as bridging groups[108].

The thermal decomposition of the adducts $MnCl_2(3\text{-}CH_3py)_2$ and $MnCl_2(4\text{-}CH_3py)_2$ have been shown[108,116] to follow the same route as that for $MnCl_2(py)_2$. The data shown in Tables 5.7 and 5.8, using the same notation as in Table 5.6, were obtained by Beech and coworkers[116].

Infrared absorption spectra of the decomposition products of both $MnCl_2(3\text{-}CH_3py)_2$ and $MnCl_2(4\text{-}CH_3py)_2$ indicate that they are polymeric

TABLE 5.6

Thermodynamic Data for the Thermal Decomposition of $MnCl_2(py)_2$

Reaction[a]	1	2	3
ΔH_{decomp} (kcal mole^{-1}	14.7	3.1	10.5
T_i (°K)	410	520	550
T_p (°K)	460	540	600
T_f (°K)	470	550	610

[a] T_i, T_p and T_f are initial, peak (i.e., when rate of change of ΔH greatest) and final temperatures of the decomposition reaction respectively.

TABLE 5.7

Thermodynamic Data for the Thermal Decomposition of $MnCl_2(3\text{-}CH_3py)_2$

Reaction	1	2	3
ΔH_{decomp} (kcal mole^{-1})	13.6	5.1	13.4
T_i (°K)	390	480	545
T_p (°K)	450	510	615
T_f (°K)	465	530	625

TABLE 5.8

Thermodynamic Data for the Thermal Decomposition of $MnCl_2(4\text{-}CH_3py)_2$

Reaction	1	2	3
ΔH_{decomp} (kcal mole^{-1})	15.1	5.4	14.0
T_i (°K)	380	500	565
T_p (°K)	440	540	615
T_f (°K)	455	560	630

with the manganese atoms in an octahedral environment[108]. The diffuse-reflectance spectra of $MnCl_2(3\text{-}CH_2py)_2$ and its decomposition products have been recorded[108] and the bands and their assignments are given in Table 5.9.

Some properties of several substituted pyridine complexes of manganese dichloride are given in Table 5.10.

Heating manganese dichloride directly with aniline or a substituted aniline, or alternatively, addition of an ethanolic solution of manganese dichloride to the ligand, results in the formation of pink manganese dichloride adducts[102,114]. The second preparative method normally gives

TABLE 5.9

Electronic Spectra (cm^{-1}) of $MnCl_2(3\text{-}CH_3py)_2$ and its Decomposition Products

Compound	$^4T_{1g} \leftarrow {}^6A_{1g}$	$^4T_{2g} \leftarrow {}^6A_{1g}$	$^4E_g, {}^4A_{1g} \leftarrow {}^6A_{1g}$
$MnCl_2(3\text{-}CH_3py)_2$	18,500	20,410	23,530
$MnCl_2(3\text{-}CH_3py)$	18,520	21,740	23,810
$MnCl_2(3\text{-}CH_3py)_{2/3}$	18,520	21,740	23,810

TABLE 5.10

Properties of Some $MnCl_2$-Substituted Pyridine Adducts

Compound	Properties	References
$MnCl_2(3\text{-}CH_3py)_4$	$\nu_{Mn\text{-}Cl} = 222\ cm^{-1}$	107
$MnCl_2(3\text{-}CH_3py)_2$	$\nu_{Mn\text{-}Cl} = 222, 204\ cm^{-1}$	107
$MnCl_2$(2,6-lut)	Decomposes at 300° to $MnCl_2$	108
$MnCl_2$(2,4,6-coll)	Decomposes at 350° to $MnCl_2$	108

an adduct which is solvated, but gentle heat to about 100° gives the non-solvated complex[102,114]. The manganese-chlorine stretching frequencies (cm^{-1}) for a number of aniline and substituted aniline adducts are shown below[102,114].

$MnCl_2(an)_2$	227	$MnCl_2(p\text{-tol})_2$	227
$MnCl_2(o\text{-tol})_2$	233	$MnCl_2(3,4\text{-xyl})_2$	233
$MnCl_2(m\text{-tol})_2$	227		

The diffuse-reflectance spectra[102,114] of these adducts have been interpreted on the basis of an octahedral polymeric configuration, and the reported bands and their assignments are given in Table 5.11.

Brown and coworkers[115] have shown that manganese dichloride reacts readily with quinoline to give the adduct $MnCl_2Q_2$ when the base is added to an ethanolic solution of the chloride, and when the hydrated chloride is refluxed in quinoline. The above workers showed that this adduct undergoes thermal decomposition reactions which are completely

TABLE 5.11
Electronic Spectra (cm^{-1}) of Some $MnCl_2$-Aromatic Amine Adducts

Compound	${}^4T_{1g} \leftarrow {}^6A_{1g}$	${}^4T_{2g} \leftarrow {}^6A_{1g}$	${}^4E_g, {}^4A_{1g} \leftarrow {}^6A_{1g}$
$MnCl_2(an)_2$	18,180	21,740	25,000
$MnCl_2(o\text{-tol})_2$	17,700	20,830	23,810
$MnCl_2(m\text{-tol})_2$	17,860	21,280	23,530
$MnCl_2(p\text{-tol})_2$	18,180		23,530
$MnCl_2(3,4\text{-xyl})_2$	17,700		23,260

analogous to those of the corresponding pyridine adduct. The decomposition reaction temperatures are[115]

$$MnCl_2Q_2 \xrightarrow{190^\circ} MnCl_2Q \xrightarrow{220^\circ} MnCl_2Q_{2/3} \xrightarrow{340^\circ} MnCl_2$$

The infrared and electronic absorption spectra of the intermediate products of decomposition clearly demonstrate octahedral coordination about the manganese atoms. The manganese-chlorine stretching frequencies (cm^{-1}) for the quinoline adducts are[127]

$MnCl_2Q_2$	241	$MnCl_2Q_{2/3}$	262	$MnCl_2Q$	240, 230

Frank and Rogers[118] report manganese-chlorine and manganese-nitrogen stretching frequencies of 246 and 196 cm^{-1} respectively for $MnCl_2Q_2$.

Perhaps one of the most interesting complexes of manganese dichloride is that formed with bis(2-dimethylaminomethyl)methylamine, (dienMe). Ciampolini and Speroni[125] prepared the adduct $MnCl_2$(dienMe) as a white solid by the interaction of a hot solution of manganese dichloride in butanol and the ligand. The compound has a room-temperature magnetic moment of 5.85 BM and is a non-conductor in chloroform and in nitrobenzene. Together with the diffuse-reflectance spectrum, the experimental data have been interpreted on the basis of the manganese atom being five coordinate, although the actual stereochemistry is unknown.

Manganese dichloride reacts with hydrazine[128] to form the complex $MnCl_2(N_2H_4)_2$. Ferrari and coworkers[129] have studied the compound by means of single-crystal X-ray diffraction techniques. The compound is monoclinic and consists of chains of octahedral $MnCl_2(N_2H_4)_4$ units, the hydrazine molecules acting as bridging groups. The manganese-nitrogen and manganese-chlorine bond distances are 2.266 and 2.570 Å respectively. An interesting feature of this structure is that the hydrazine groups are in the staggered form. Sacconi and Sabatini[128] have examined the infrared spectrum of $MnCl_2(N_2H_4)_2$ and reported the assignments given in Table 5.12.

TABLE 5.12
Infrared Spectrum (cm^{-1}) of $MnCl_2(N_2H_4)_2$

Mode	Frequency	Mode	Frequency
NH_2 stretch	3298, 3237, 3151	NH_2 twist	1156
		N-N stretch	960
NH_2 bend	1606, 1575	NH_2 asym. rock	590
		NH_2 sym. rock	518
NH_2 wag	1348, 1299	Mn-N stretch	343

TABLE 5.13
Infrared Absorptions (cm^{-1}) of $MnCl_2(tu)_4$

Mode	Frequency	Intensity
ν_{Mn-Cl}	187	Very strong, broad
	134	Medium, broad
ν_{Mn-s}	227	Medium
δ_{MnSC}	162	Weak
δ_{SMnS}	112	Weak

Manganese dichloride reacts readily with thiourea (tu) in alcohol to give the adduct $MnCl_2(tu)_4$[130–133]. The complex melts[130] at 193–195° and from X-ray powder diffraction patterns, has been shown to be isostructural with $NiCl_2(tu)_4$, for which a full structural analysis is available (Fig. 8.3, page 418). Thus the manganese compound has a *trans* octahedral configuration and its infrared spectrum[130,131] is consistent with this stereochemistry. The infrared spectral studies show conclusively that the thiourea molecules are bound to the manganese atom via the sulphur atoms. The most detailed infrared study is that by Adams and Cornell[131] whose assignments are given in Table 5.13.

Manganese dichloride reacts readily with phosphine and arsine oxides in solvents such as nitromethane and ethanol to give 1:2 adducts[127,134–137]. The adducts have been examined in detail using such techniques as conductance, infrared, visible and ultraviolet spectroscopy, magnetochemistry and X-ray diffraction All these techniques show unequivocally that the adducts are tetrahedral in the solid state. Some properties of several of these adducts are given in Table 5.14.

TABLE 5.14
Phosphine and Arsine Oxide Adducts of Manganese Dichloride

Compound	Properties	References
$MnCl_2[(C_6H_5)_3PO]_2$	Pale yellow, m.p. = 244°, μ = 5.92 BM, $\nu_{Mn\text{-}Cl}$ = 327, 317, 292 cm^{-1}	127, 137
$MnCl_2[(C_6H_5)_3AsO]_2$	Pale yellow, m.p. = 236°, μ = 6.03 BM, $\nu_{As\text{-}O}$ = 894 cm^{-1}, $\nu_{Mn\text{-}O}$ = 370 cm^{-1} $\nu_{Mn\text{-}Cl}$ = 311, 287 cm^{-1}	136, 137
$MnCl_2[CH_3(C_6H_5)_2AsO]_2$	$\nu_{As\text{-}O}$ = 865, 847 cm^{-1}, $\nu_{Mn\text{-}O}$ = 395 cm^{-1}, $\nu_{Mn\text{-}Cl}$ = 298, 272 cm^{-1}	136

A number of other adducts of manganese dichloride are known and some properties of a number of these are given in Table 5.15.

TABLE 5.15
Adducts of the Type $MnCl_2L_n$

Compound and Properties	References
L = CH_3CN, n = 2, pink, μ = 5.83 BM, octahedral	138
L = NH_2CH_2COOH, n = 2, isolated as a dihydrate, loses water at 35°, ferroelectric at room temperature, monoclinic with a = 9.96, b = 8.53, c = 6.86 Å, β = 107°, Z = 2, $P2_1$	139
L = $(CH_3)_2SO$(DMSO), n = 3, pale yellow, shown by infrared spectrum to be $[Mn(DMSO)_6^{2+}][MnCl_4^{2-}]$	140–143
L = ROH, n = 1,2,3,4; R = CH_3, C_2H_5, C_3H_7, C_4H_9	74, 144, 145
L = $(C_6H_5)_3P$, n = 2, m.p. = 232°, μ = 5.81 BM	146
L = $(C_6H_5)_3As$, n = 2, m.p. = 240°, μ = 5.99 BM	146
L = *N,N*-dimethylethylenediamine *N*-oxide, n = 1, tan, m.p. = 123°, μ = 5.86 BM	119
L = pyrazine (pyz), n = 2, decomposes above 300° to $MnCl_2$	117
L = 2-picolyamine, pink, μ = 5.91 BM, octahedral	122

Other donor ligands that have been reacted with manganese dichloride include ammonia, dioxan, sulphur monochloride and urea[74,147–157].

Manganese dichloride hexahydrate. This hydrate has been isolated[158,159] from aqueous solutions of manganese dichloride at low temperatures. At −2° it converts rapidly to the tetrahydrate and because of this, very little is known about its properties.

Manganese dichloride tetrahydrate. Manganese dichloride tetrahydrate is readily prepared from a saturated aqueous solution of manganese dichloride at room temperature[158,160–163]. Its heat of formation[164,165] is −404 kcal mole^{-1}. Borchardt and Daniels[77] have examined the thermal decomposition of manganese dichloride tetrahydrate by the differential thermal technique and their results are expressed by the following equations.

$$MnCl_2 \cdot 4H_2O(s) \xrightarrow{50^\circ} MnCl_2 \cdot 2H_2O(s) + 2H_2O(l)$$

$$MnCl_2 \cdot 2H_2O(s) \xrightarrow{135^\circ} MnCl_2 \cdot H_2O(s) + H_2O(g)$$

$$MnCl_2 \cdot H_2O(s) \xrightarrow{210^\circ} MnCl_2(s) + H_2O(g)$$

Delain[166] showed that manganese dichloride tetrahydrate has a monoclinic unit cell with $a = 11.9$, $b = 9.55$, $c = 6.15$ Å and $\beta = 111°$. Zalkin and coworkers[162] refined the structure of this compound using single-crystal X-ray diffraction techniques. These workers deduced a slightly different space group for the unit cell and gave the following parameters: $a = 11.186$, $b = 9.513$, $c = 6.186$ Å and $\beta = 99.74°$. The space group of this cell is $P2_1/n$ and it contains four molecular units. The lattice contains discrete octahedral groups with each manganese atom coordinated by two chlorine atoms and four water molecules. It is interesting to note that the two chlorine atoms have the *cis* configuration. The average manganese-chlorine and manganese-oxygen bond distances are 2.488 and 2.206 Å respectively. Originally it was concluded[162] that only half of the hydrogen atoms took part in hydrogen bonding, but subsequently, Baur[167] showed that all eight hydrogen atoms are involved.

The magnetic properties of manganese dichloride tetrahydrate have been examined in some detail[168]. The compound has a magnetic moment of 5.94 BM at room temperature[169] and at very low temperatures it becomes antiferromagnetic. The Neel temperature is 1.6°K[170–172] and a specific heat anomaly has also been reported at this temperature[173,174]. The spectroscopic splitting factor has been given[175] as 2.00. Spence and Nagarajan[176] have studied the structure of manganese dichloride tetrahydrate in the antiferromagnetic state by ^1H and ^{35}Cl NMR spectroscopy.

From their results they deduced that the magnetic cell was of space group $P2/_1'a$.

The H_2O rock, H_2O wag and manganese-oxygen stretching modes in manganese dichloride tetrahydrate have been observed[177] at 655, 560 and 395 cm^{-1} respectively.

Tsujikawa and Kanda[178] have investigated the temperature dependence, fine-structure and intensity of the ${}^4A_{1g} \leftarrow {}^6A_{1g}$ transition in manganese dichloride tetrahydrate which occurs at about 24,200 cm^{-1}.

Manganese dichloride dihydrate. Manganese dichloride dihydrate crystallizes out from a saturated aqueous solution of manganese dichloride at about 65° [160,163,179–181]. It rapidly absorbs water at room temperature.

The structure of manganese dichloride dihydrate was first investigated by Vainshtein[163] who found it to be isomorphous with the corresponding cobalt compound. Close agreement with Vainshteins's results has recently been reported by Morosin and Graebner[179]. The unit cell is monoclinic and the parameters are given in Table 5.16.

TABLE 5.16

Unit-cell Parameters of Manganese Dichloride Dihydrate

Parameter	Vainshtein[163]	Morosin and Graebner[179]
a (Å)	7.403	7.409
b (Å)	8.770	8.800
c (Å)	3.701	3.691
β	98.1°	98.67°
Z	2	2
Space group	$C2/m$	$C2/m$

The structure consists of polymeric chains of manganese and chlorine atoms in an arrangement that is nearly square-planar, the chains being parallel to the *c* axis and held together by hydrogen bonds. These bonds are formed between the chlorine atom of an adjacent polymeric chain and the hydrogen atoms of the water molecule completing the octahedral coordination about the manganese atom. The manganese-chlorine bond distances are 2.515 and 2.592 Å, the manganese-oxygen distance is 2.150 Å and the chlorine-manganese-chlorine angle is 87.45°.

Manganese dichloride dihydrate is reported[182] to become magnetically ordered at very low temperatures but no details are as yet available.

The electronic absorption spectrum of manganese dichloride dihydrate has been investigated by Lawson[180] and by Goode[181]. The spectra were recorded at 298, 77 and 4.2°K by both workers. In general, quite good agreement was obtained for the positions and assignments of the bands. Both workers have discussed the structure of the spectra in detail. The room temperature bands and their assignments reported by Lawson[180] are:

19,030 cm^{-1}	${}^4T_{1g}(G) \leftarrow {}^6A_{1g}(S)$
21,440 cm^{-1}	${}^4T_{2g}(G) \leftarrow {}^6A_{1g}(S)$
24,320 cm^{-1}	${}^4E_g$, ${}^4A_{1g}(G) \leftarrow {}^6A_{1g}(S)$
26,840 cm^{-1}	${}^4T_{2g}(D) \leftarrow {}^6A_{1g}(S)$
28,740 cm^{-1}	${}^4E_g(D) \leftarrow {}^6A_{1g}(S)$
30,740 cm^{-1}	${}^4T_{1g}(P) \leftarrow {}^6A_{1g}(S)$
39,170 cm^{-1}	${}^4A_{2g}(F) \leftarrow {}^6A_{1g}(S)$
40,215 cm^{-1}	${}^4T_{1g}(F) \leftarrow {}^6A_{1g}(S)$
43,070 cm^{-1}	${}^4T_{2g}(F) \leftarrow {}^6A_{1g}(S)$

Manganese dibromide. Manganese dibromide has been prepared[183] by dissolving manganese carbonate in hydrobromic acid, dehydrating at 100° and finally heating the product in hydrogen bromide gas at 725°. It has also been prepared by the interaction of manganese acetate and acetyl bromide in benzene[84], the product being dried at 200° in a nitrogen atmosphere.

Manganese dibromide is a pink hygroscopic solid which melts[184] at 695°. It is readily soluble in water from which several hydrates may be isolated. Schoonmaker and coworkers[76] have shown that manganese dibromide vaporizes mainly as a monomer, although dimeric species were observed.

Ferrari and Giorgi[185] have shown that manganese dibromide has a hexagonal unit cell with $a = 3.820$ and $c = 6.188$ Å.

Manganese dibromide has a magnetic moment of 5.82 BM at room temperature[168], but it does not obey either the Curie or the Curie-Weiss laws at normal temperatures[186]. Over the temperature range 46.4–77.3°K the magnetic susceptibility of manganese dibromide, measured by the inductance method[187], is given by the expression

$$\chi_M = 4.38/T + 4.7$$

The Neel temperature has been reported[188–190] as 2.16°K and there is also a specific heat maximum at this temperature[183,187]. The magnetic structure of manganese dibromide has been deduced by Koehler and coworkers[188] from neutron-diffraction measurements on a single crystal at 1.35°K.

The electronic absorption spectrum of manganese dibromide has been recorded, but there is some controversy[97,98] over the calculation of the various parameters from the assignments given[96].

Manganese dibromide reacts readily with pyridine to form initially $MnBr_2(py)_4$[108,191]. This adduct decomposes[108] at 140° to give pink $MnBr_2(py)_2$ which itself is thermally unstable[108]; $MnBr_2(py)$ is obtained at 210° while at 300° manganese dibromide is formed. Infrared[104] and magnetic studies[106,113] indicate that $MnBr_2(py)_2$ is a polymer, the octahedra about each manganese atom being bridged by bromine atoms. Clark and Williams[104] report a manganese-nitrogen stretching frequency of 212 cm^{-1}. As with the corresponding chloro compound, the pyridine ring vibrations are not greatly changed on coordination[101]. $MnBr_2(py)_2$ has a room temperature magnetic moment of 5.96 BM[106,113].

Both 3-methylpyridine and 4-methylpyridine react[108] with manganese dibromide, when refluxed in excess ligand, to form adducts of the type $MnBr_2L_4$. Allan and coworkers[108] have studied the thermal decomposition of these adducts and reported the following decomposition sequence:

$$MnBr_2L_4 \xrightarrow{T_1} MnBr_2L_2 \xrightarrow{T_2} MnBr_2L \xrightarrow{T_3} MnBr_2L_{2/3} \xrightarrow{T_4} MnBr_2$$

The temperatures for these reactions are

L = 3-methylpyridine: $T_1 = 180°$, $T_2 = 210°$, $T_3 = 260°$, $T_4 = 340°$

L = 4-methylpyridine: $T_1 = 180°$, $T_2 = 240°$, $T_3 = 280°$, $T_4 = 340°$

All of the intermediate products are considered to be octahedral polymers with the bromine atoms acting as the bridging groups. $MnBr_2(4\text{-}CH_3py)_4$ has a manganese-bromine stretching frequency[107] of 237 cm^{-1}.

Aniline and substituted anilines react with manganese dibromide under the same conditions noted above for manganese dichloride[102,114] to give, in general, adducts of the type $MnBr_2L_2$. Infrared, visible and ultraviolet spectra indicate that all of the complexes of this type are octahedral with bridging bromine atoms[102,114].

Manganese dibromide reacts with a large variety of other donor molecules and some properties of a number of these adducts are given in Table 5.17.

Manganese dibromide tetrahydrate. This hydrate may be prepared by evaporating a saturated aqueous solution of manganese dibromide at room temperature[197]. Specific heat and magnetic studies indicate that below about 2.15°K manganese dibromide tetrahydrate becomes antiferromagnetic[169–171,198]. A magnetic moment of 5.93 BM and a Weiss constant of 2° have been reported[168]. The temperature dependence, intensity and fine-structure of the ${}^4A_{1g} \leftarrow {}^6A_{1g}$ transition has been investigated[178] in some detail.

Manganese dibromide dihydrate. Evaporation of a saturated aqueous solution of manganese dibromide at about 65° produces manganese dibromide dihydrate[197]. The unit cell[199] has the parameters $a = 7.763$, $b = 8.990$ and $c = 3.881$ Å. The structure is completely analogous to that of manganese dichloride dihydrate; the manganese atom is surrounded by an approximately square plane of bromine atoms, each bromine being shared by two metal atoms. The water molecules fill the remaining octahedral positions[199].

TABLE 5.17
Adducts of the Type $MnBr_2L_n$

Compound and Properties	References
L = ROH, $n = 1,2$; R = C_2H_5, *i*-C_3H_7	144
L = 9,10-phenanthrenequinone, $n = 1$, violet, $\mu = 5.84$ BM	192
L = 1,2-chrysenequinone, $n = 1$, black	192
L = CH_3CN, $n = 2$, pink, μ 5.79 BM, octahedral	138
L = $(C_6H_5)_3PO$, $n = 2$, pale green, m.p. = 243°, $\mu = 6.02$ BM, $\nu_{Mn\text{-}Br} = 248, 229$ cm^{-1}, tetrahedral	127, 137
L = $(C_6H_5)_3AsO$, $n = 2$, pale yellow-green, m.p. = 214°, $\mu = 6.03$ BM, $\nu_{As\text{-}O} = 910, 880$ cm^{-1}, $\nu_{Mn\text{-}O} = 405, 393$ cm^{-1}, $\nu_{Mn\text{-}Br} = 230$ cm^{-1}, tetrahedral	127, 136, 137
L = $CH_3(C_6H_5)_2AsO$, $n = 2$, $\nu_{As\text{-}O} = 859$ cm^{-1}, $\nu_{Mn\text{-}O} = 397$ cm^{-1}, $\nu_{Mn\text{-}Br} = 236$ cm^{-1}	136
L = *N,N*-dimethylacetamide (DMA), $n = 2$, isolated as tetrahydrate, light yellow, $\mu = 5.90$ BM	154
L = C_5H_5NO, $n = 2$, $\mu = 5.62$ BM, tetrahedral	193
L = $(CH_3)_3NO$, $n = 2$, $\mu = 5.64$ BM, tetrahedral	193
L = *o*-phenylenebis(dimethylarsine) (D or diars), $n = 2$, white, $\mu = 5.95$ BM, monomeric, non-electrolyte	194
L = $(C_6H_5)_3P$, $n = 2$, yellow, m.p. = 229°, $\mu = 6.04$ BM	146
L = $(C_6H_5)_3As$, $n = 2$, cream, m.p. = 238°, $\mu = 5.93$ BM	146
L = 2-picolyamine, pink, $\mu = 5.93$ BM, octahedral	122
L = bis(2-dimethylaminoethyl)methylamine (dienMe), buff, five coordinate, ΔH°_{form} (from constituents) = -28.74 kcal mole^{-1}	125, 196
L = Schiff base formed between *N*-methyl-*o*-aminobenzaldehyde and *N,N*-diethylethylenediamine (MABen-NEt_2), $n = 1$, hygroscopic, non-electrolyte, $\mu = 5.86$ BM, five coordinate	196

Manganese diiodide. Manganese diiodide has been prepared by direct interaction of the elements in ether[200], and by the interaction of either the dioxide or disulphide with aluminium triiodide at 230° in a sealed tube[201,202].

Manganese diiodide is a pink crystalline solid melting[184] at 613°. It is readily soluble in water and from these solutions a number of hydrates may be isolated[200]. It has a hexagonal unit cell[203,204], Cable and coworkers[204] reporting parameters of $a = 4.146$ and $c = 6.289$ Å.

From room temperature to about 60°K the magnetic properties of manganese diiodide deviate only slightly from a Curie law[186] with a magnetic moment of 5.88 BM. Manganese diiodide becomes antiferromagnetic at low temperatures and a Neel temperature of 3.4°K has been reported[190,204]. A neutron-diffraction study by Cable and coworkers[204] has shown that the magnetic structure of manganese diiodide is of the helical type. In this structure the moments within the (307) plane are aligned antiferromagnetically but rotate by $2\pi/16$ in successive (307) planes.

As expected, manganese diiodide forms adducts with donor molecules such as pyridine, aniline, triphenylphosphine, triphenylarsine and their oxides etc.[101,102,106,121,122,137,140,146,194,205,206]. In general the properties of these adducts are similar to those previously described for manganese dichloride and manganese dibromide.

COMPLEX HALIDES

There are no reported complex oxide halides of manganese. Apart from the fluoro complexes, the complex halides of manganese are essentially those of oxidation state II. Although potassium hexachloromanganate(IV) has been prepared, it is thermally unstable at room temperature.

Oxidation State IV

Hexafluoromanganates(IV). These yellow hygroscopic salts have been prepared by two general methods, namely the direct fluorination of suitable mixtures of manganese and alkali-metal salts, and the electrolytic oxidation of manganese(II) or the reduction of manganese(VII) in 40% hydrofluoric acid. The details of various preparations are given in Table 5.18.

The structures of a number of hexafluoromanganates(IV) have been investigated by X-ray powder diffraction techniques, and in a number of cases polymorphism has been observed. The known lattice symmetries and parameters are given in Table 5.19.

The magnetic moments of a number of alkali- and alkaline-earth hexafluoromanganates(IV) have been determined[209,211,220] and they fall in the range 3.80–3.95 BM. The manganese-fluorine stretching frequency has been reported[221] as 625 cm^{-1} in potassium hexafluoromanganate(IV).

Pentafluoromanganates(IV). Hoppe and coworkers[7,211] have prepared a number of alkali-metal pentafluoromanganates(IV) by fluorinating the appropriate trifluoromanganate(II) at 450–500° in a flow system. The potassium salt is also reported[222] to result from refluxing potassium permanganate in selenium tetrafluoride and removing the excess selenium tetrafluoride under vacuum at 160°. The pentafluoromanganates(IV) are

TABLE 5.18
Preparation of Hexafluoromanganates(IV)

Method	Conditions	References
$MnCl_2$, KCl and F_2	Flow system, 375–400°	9, 207
$MnSO_4$, Ca salt and F_2[a]	Flow system, 500°	208–210
$MnSO_4$, Ba salt and F_2[a]	Flow system, 500°	208–210
$Li_2Mn(SO_4)_2$ and F_2	Flow system, 250–350°	211
$SrMnO_3$ and F_2	Flow system, 400°	208–210, 212
$BaMnO_4$ and F_2	Flow system, 400°	208–210, 212
$KMnO_4$, KHF_2 and reducing agent[b]	Reduction in 40% aq. HF	213, 214
MnF_2 and MF or MHF_2 (M = alkali metal)	Electrolytic oxidation in 40% aq. HF	215, 216
$KMnO_4$, KCl and BrF_3	Reflux	12
MnF_4, $NOBrF_4$ and BrF_3	Reflux	217

[a] Suitable salts are the carbonate, chloride and sulphate.
[b] Suitable reducing agents are hydrogen peroxide and ether.

TABLE 5.19
Lattice Parameters of Hexafluoromanganates(IV)

Compound	Symmetry	Parameters	References
Li_2MnF_6	Hexagonal	$a = 8.42$, $c = 4.59$ Å	211
Na_2MnF_6	Hexagonal	$a = 9.03$, $c = 5.15$ Å	215, 218
K_2MnF_6	Hexagonal	$a = 5.67$, $c = 9.35$ Å	219
		$a = 5.70$, $c = 9.30$ Å	207
	Trigonal	$a = 5.71$, $c = 4.65$ Å	219
Rb_2MnF_6 (70–100°)	Cubic	$a = 8.40$ Å	219
(<40°)	Hexagonal	$a = 5.86$, $c = 9.50$ Å	219
Cs_2MnF_6	Cubic	$a = 8.92$ Å	219
$MgMnF_6$	Hexagonal	$a = 5.01$, $c = 13.17$ Å	208, 209
$CaMnF_6$	Hexagonal	$a = 5.21$, $c = 14.13$ Å	208, 209
$SrMnF_6$	Hexagonal	$a = 7.02$, $c = 6.78$ Å	208, 209
$BaMnF_6$	Hexagonal	$a = 7.35$, $c = 7.09$ Å	208, 209, 212

readily hydrolysed by moist air. Their magnetic moments[211] are close to the spin-only value for three unpaired electrons. It is probable that the pentafluoromanganates(IV) contain condensed MnF_6 octahedral units.

Hexachloromanganate(IV). Potassium hexachloromanganate(IV) has been prepared by reduction of calcium permanganate with 40% hydrochloric acid followed by addition of potassium chloride[9]. It is a black, unstable solid which liberates chlorine even at room temperature in dry air.

Oxidation State III

Hexafluoromanganate(III). Peacock[221] has prepared potassium hexafluoromanganate(III) by fusing a mixture of potassium pentafluoroaquomanganate(III) with potassium hydrogen difluoride in an inert atmosphere. The excess potassium fluoride was leached out with formamide. The purple-blue complex has a tetragonal unit cell with $a = 8.75$ and $c = 8.30$ Å and has a room-temperature magnetic moment of 4.95 BM. A strong manganese-fluorine stretching frequency was observed[221] at 560 cm^{-1} with a weaker band at 617 cm^{-1}.

Pentafluoromanganates(III). Pink potassium pentafluoroaquomanganate(III) has been isolated in three ways, namely by the interaction of manganese dioxide and potassium hydrogen difluoride in hydrofluoric acid[214], by electrolytic oxidation of manganese difluoride in 40% hydrofluoric acid, followed by addition of a saturated solution of potassium hydrogen difluoride in hydrofluoric acid[216], and by the oxidation of manganese sulphate with potassium permanganate in hydrofluoric acid containing potassium fluoride[220]. The caesium salt of the pentafluoroaquomanganate(III) ion has been prepared[224] by dissolving manganese trifluoride in a solution of hydrofluoric acid containing caesium fluoride. Potassium pentafluoroaquomanganate(III) has a room-temperature magnetic moment[225] of 3.32 BM.

Anhydrous ammonium pentafluoromanganate(III) has been prepared[226,227] by adding a solution of hydrated manganese(III) oxide in hydrofluoric acid to a concentrated solution of ammonium fluoride in the same solvent. The complex has an orthorhombic unit cell[226] with $a = 6.20$, $b = 7.94$ and $c = 10.72$ Å. The space group is *Pnma* and there are four formula units in the cell. The structure consists of tetragonally elongated MnF_6 octahedra, the manganese-fluorine bond distances being 1.84 and 2.12 Å. The long manganese-fluorine bonds link octahedra into infinite, kinked chains.

The electronic absorption spectrum of a single crystal of ammonium pentafluoromanganate(III) has been investigated by Dingle[226] who reported the following bands and assignments.

12,750 cm^{-1}	$^5A_{1g}$ or $^5B_{1g} \leftarrow {}^5B_{1g}$
18,200 cm^{-1}	$^5B_{2g} \leftarrow {}^5B_{1g}$
21,000 cm^{-1}	$^5E_g \leftarrow {}^5B_{1g}$

Tetrafluoromanganates(III). Hydrogen reduction[7,211] of the appropriate pentafluoromanganate(IV) gave dark brown to violet alkali-metal tetrafluoromanganates(III) which obey the Curie-Weiss law with magnetic moments and Weiss constants as follows:

	μ (BM)	$\theta°$
$LiMnF_4$	4.7	10
$KMnF_4$	4.87	27
$RbMnF_4$	4.98	7

Hexachloromanganates(III). Hatfield and coworkers[228] have prepared $[Co(C_3H_{10}N_2)_3][MnCl_6]$, tris(1,2-propenediamine)cobalt(III) hexachloromanganate(III), by the interaction of hydrated manganese sulphate and the complex cobalt chloride in hydrochloric acid. The analogous rhodium compound was prepared in a similar manner[228]. The cobalt complex, which is dark brown, obeys the Curie law almost exactly with a magnetic moment of 4.91 BM. The rhodium complex shows an absorption at 17,540 cm^{-1} which has been assigned to the $^5T_{2g} \leftarrow {}^5E_g$ transition[228].

Pentachloromanganate(III). Tetraethylammonium pentachloromanganate(III) is recorded[229] to be formed by the reduction of manganese dioxide, suspended in carbon tetrachloride, with hydrogen chloride, followed by extraction with ether and addition of the ether extract to a solution of tetraethylammonium chloride in ethanol. The dark green solid obeys the Curie law with a magnetic moment of 5.00 BM[229].

Potassium pentachloromanganate(III) has been reported[9] as a brown solid, but very little is known of the physico-chemical properties of this compound.

Oxidation State II

Tetrafluoromanganates(II). Alkali-metal tetrafluoromanganates(II) are usually prepared by the interaction of manganese difluoride and the alkali-metal fluoride in the melt[34,35,230–232]. Barium tetrafluoromanganate(II) is also prepared in this manner[233]. Breed[234] has prepared single crystals of potassium tetrafluoromanganate(II) by fusing potassium trifluoromanganate(II) with potassium hydrogen difluoride at 900°. Care must be taken when fusing alkali-metal fluorides with manganese difluoride to use the stoichiometric proportions of the two fluorides, since the trifluoromanganates(II) are also prepared by this method and Chretien and Cousseins[232] have isolated $K_3Mn_2F_7$ from the potassium fluoride-manganese difluoride system. The heptafluoro compound has a tetragonal cell[232] with $a = 4.19$ and $c = 21.66$ Å.

Potassium tetrafluoromanganate(II) undergoes a polymorphic transformation at 787° and melts[35,230] at 795°. The melting points[34,35,230] of the rubidium and caesium salts are 796° and 725° respectively. Caesium tetrafluoromanganate(II) is reported[231,232] to be dimorphic but no other details are available.

Unit-cell parameters from X-ray powder diffraction investigations of the tetrafluoromanganates(II) are listed in Table 5.20.

TABLE 5.20
Unit-cell Parameters of Some Tetrafluoromanganates(II)

Compound	Symmetry	Parameters	References
K_2MnF_4	Tetragonal	$a = 4.20, c = 13.41$ Å	232, 234
Rb_2MnF_4	Tetragonal	$a = 4.25, c = 13.92$ Å	232
Cs_2MnF_4[a]	Tetragonal	$a = 4.31, c = 14.63$ Å	232
$BaMnF_4$	Orthorhombic	$a = 15.08, b = 4.22, c = 5.99$ Å	233

[a] High temperature form.

Breed[234] has shown from magnetic studies that potassium tetrafluoromanganate(II) becomes antiferromagnetic at about 75°K. Single-crystal magnetic susceptibility measurements indicate that even at room temperature potassium tetrafluoromanganate(II) is magnetically anisotropic. At this temperature $(\chi_{\perp} - \chi_{\parallel})_g$ is approximately 0.5×10^{-6} emu. The room temperature magnetic moment is 6.0 BM. The magnetic behaviour at low temperatures indicates that the magnetic spins are aligned along the *c* axis.

Trifluoromanganates(II). Because of the interest in the magnetic and spectral properties of the trifluoromanganates(II) much work has been reported on the preparation of samples of high purity in powder or crystalline form. Direct interaction in the melt[34,230,231,235,236] and preparation by mixing boiling solutions of manganese difluoride and alkali-metal chloride[29,211,237–239], in general give reasonably pure samples. For the preparation of single crystals further treatment is required[31,240]. Other less important methods of preparing trifluoromanganates(II) are given in Table 5.21.

TABLE 5.21
Preparation of Trifluoromanganates(II)

Method	Conditions	References
$MnCl_2$, KF and KCl	Melt	241
MnF_2 and NH_4F	100,000 psi pressure	242
$MnBr_2$ and KF or RbF	Interaction in methanol	243
MnF_2 and MF	Saturated aqueous solutions	244, 245
MnF_2, TlF and 40% HF	Evaporate, heat in anhydrous HF at 500–600°	246

The melting points of the sodium, potassium, rubidium, caesium and thallium trifluoromanganates(II) have been reported[34,35,230,246] as 762°, 1032°, 986°, 780° and 820° respectively. The potassium salt has a manganese-fluorine stretching frequency[221] of 407 cm^{-1}.

The crystal structures of the trifluoromanganates(II) have been extensively studied by both powder and single-crystal X-ray diffraction techniques. With the exception of the sodium and caesium salts, the trifluoromanganates(II) have the cubic perovskite structure at room temperature. The unit-cell dimensions at room temperature for several of these complexes are given in Table 5.22.

TABLE 5.22

Room Temperature Unit-cell Parameters of Some Trifluoromanganates(II)

Compound	Symmetry	Parameters	References
$KMnF_3$	Cubic	a = 4.182–4.194 Å	211, 232, 240, 243, 247–250
$RbMnF_3$	Cubic	a = 4.241–4.25 Å	211, 232, 243
$CsMnF_3$	Hexagonal	a = 6.213, c = 15.074 Å	236, 251
NH_4MnF_3	Cubic	a = 4.238 and 4.250 Å	211, 243
$TlMnF_3$	Cubic	a = 4.250 Å	245, 246

The perovskite trifluoromanganates(II) have the structure shown in Figure 5.7, consisting of a manganese atom at the centre of a cube, surrounded by six fluorine atoms at the centres of the cubic faces. The structure of hexagonal caesium trifluoromanganate(II) has been discussed in detail by Lee and coworkers[236,251]. An important feature of this structure, which is shown in Figure 5.8, is that there are two crystallographically different types of manganese atom. The structure is built up of six close-packed layers of caesium and fluorine atoms with the manganese atoms located in the fluorine octahedral holes between the layers. The atoms designated Csl have twelve fluorine neighbours arranged as in hexagonal close-packing. The Cs2 atoms also have twelve fluorine neighbours but these are arranged as in cubic close packing. One third of the manganese atoms, designated Mnl, occupy the centres of fluorine octahedra as in the perovskite structure. The remaining two thirds of the manganese atoms, Mn2, share one face and three corners with other octahedra. The Mn2 atoms are distorted out of their close packed positions to a distance 3.004 Å apart. The distortion of the structure does not affect greatly the various manganese-fluorine bond distances (average = 2.13 Å) but the fluorine-fluorine distances fall in the range 2.69–3.52 Å.

The structure of potassium trifluoromanganate(II), and, to a lesser

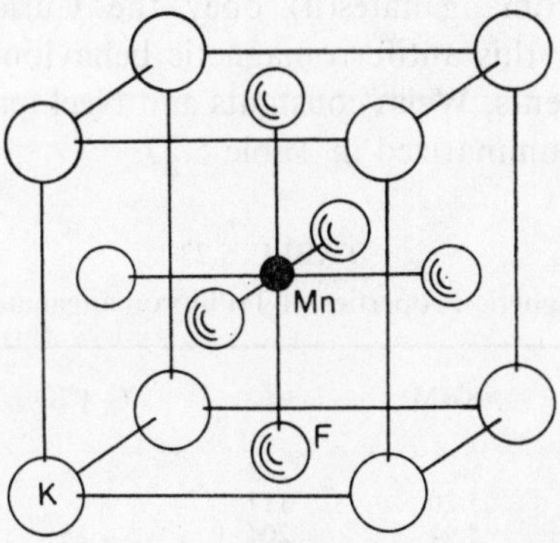

Figure 5.7 The unit cell of potassium trifluoromanganate(II) and other perovskite fluorides at room temperature. (Reproduced by permission from O. Beckman and K. Knox, *Phys. Rev.*, **121**, 376 (1961))

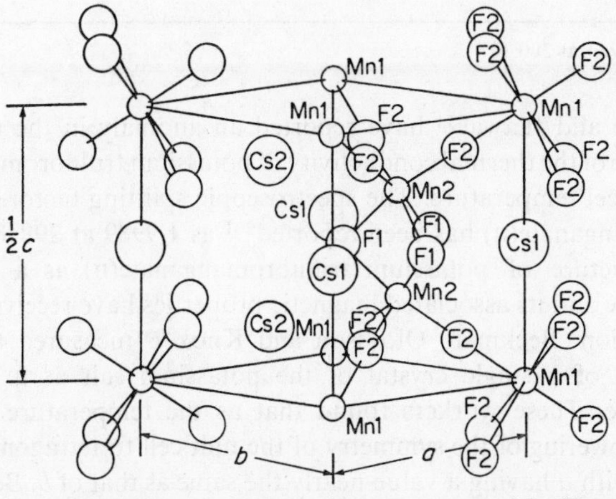

Figure 5.8 The unit cell of caesium trifluoromanganate(II). (Reproduced by permission from K. Lee, A. M. Portis and G. L. Witt, *Phys. Rev.*, **132**, 144 (1963))

extent, those of rubidium and thallium trifluoromanganate(II) have been investigated at low temperatures. A considerable amount of information has been gained from these studies, particularly in relation to the magnetic properties of the trifluoromanganates(II). This work will be discussed after the magnetic properties have been described.

The magnetic properties of the trifluoromanganates(II) have been studied by a number of workers using a variety of techniques. Above about 100°K the trifluoromanganates(II) obey the Curie-Weiss law but at temperatures below this antiferromagnetic behaviour becomes apparent. The magnetic moments, Weiss constants and Neel temperatures that have been reported are summarized in Table 5.23.

TABLE 5.23
Magnetic Properties of Trifluoromanganates(II)

Compound	μ (BM)	$\theta°$	T_N (°K)	References
$NaMnF_3$	5.20	117		239
$KMnF_3$	5.94	204	89	241
	6.04	202	95	241
	6.15	158	88	237
	4.86[a]		<80	238
$RbMnF_3$			82	251–253
$TlMnF_3$		125	85	245
		138	76	246

[a] μ_{eff} at 300°K.

Suemune and Ikawa[254] have reported an anomaly in the temperature dependence of the thermal conductivity of potassium trifluoromanganate(II) near the Neel temperature. The spectroscopic splitting factor for caesium trifluoromanganate(II) has been reported[251] as 1.9989 at 298°K.

The structure of potassium trifluoromanganate(II) as a function of temperature and its associated magnetic properties have received considerable attention. Beckman, Olovsson and Knox[255] measured the unit-cell dimensions of a single crystal of the potassium salt as a function of temperature. These workers found that as the temperature is lowered, there is a lowering of the symmetry of the unit cell to tetragonal or orthorhombic, with *a* having a value nearly the same as that of *b*. Below 200°K, the ratio *c*/*a* increases linearly with decreasing temperature and reaches a maximum value of 1.007 at the Neel temperature. A more detailed later investigation by Beckman and Knox[256] showed that at 184°K the lattice changes from cubic to orthorhombic. The orthorhombic unit cell is shown in Figure 5.9. The cell retains this symmetry at 95°K and 65°K, although as shown in Table 5.24, it was possible to index the X-ray data on the basis of a pseudo-tetragonal cell.

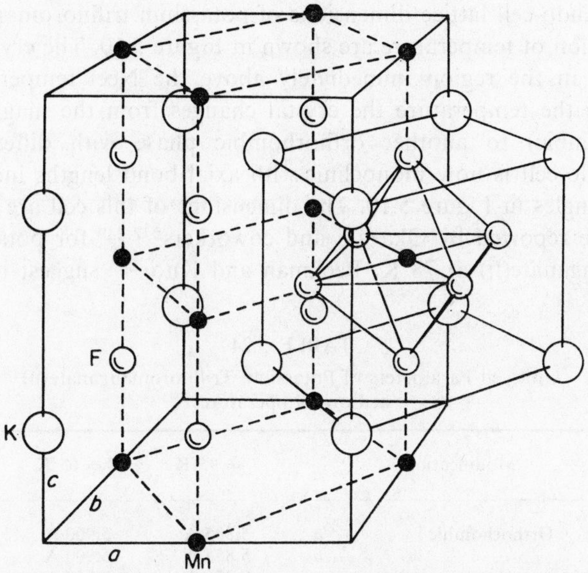

Figure 5.9 The relationship between the cubic and orthorhombic phases of potassium trifluoromanganate(II). (Reproduced by permission from O. Beckman and K. Knox, *Phys. Rev.*, **121,** 379 (1961))

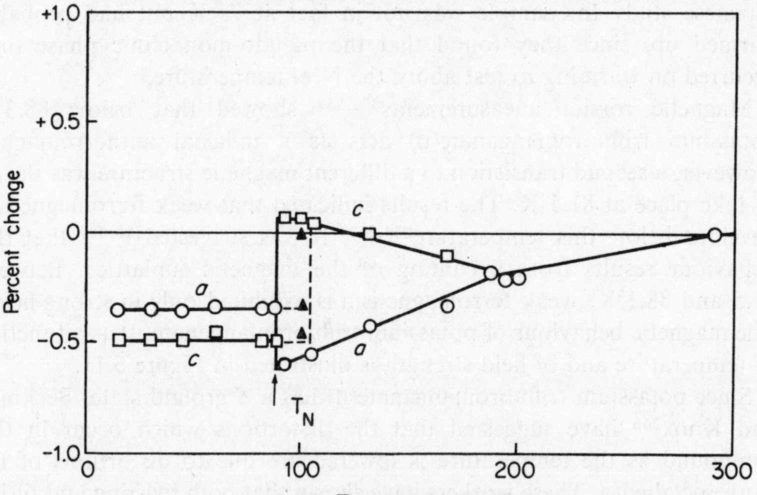

Figure 5.10 Pseudocell dimensions of potassium trifluoromanganate(II) as a function of temperature. (Reproduced by permission from O. Beckman and K. Knox, *Phys. Rev.*, **121,** 376 (1961))

The pseudo-cell lattice dimensions of potassium trifluoromanganate(II) as a function of temperature are shown in Figure 5.10. The crystal shows hysteresis in the region immediately above the Neel temperature. On increasing the temperature the crystal changes from the magnetic state (orthorhombic) to another orthorhombic phase with different axes. The pseudo cell is now monoclinic with axial bond lengths indicated by black triangles in Figure 5.10. The dimensions of this cell are consistent with those reported by Okazaki and coworkers[247,249] for potassium trifluoromanganate(II) at 78°K. Beckman and Knox[256] suggest that in the

TABLE 5.24
Unit-cell Parameters of Potassium Trifluoromanganate(II) at Low Temperatures

Modification		$T = 95°K$	$T = 65°K$
Orthorhombic	*a*	5.885 Å	5.900 Å
	b	5.855 Å	5.900 Å
	c	8.376 Å	8.330 Å
Pseudo Tetragonal	*a*	4.161 Å	4.172 Å
	b	4.188 Å	4.165 Å

Japanese study the sample was not in fact at 78°K but had probably warmed up, since they found that the pseudo-monoclinic phase only occurred on warming to just above the Neel temperature.

Magnetic torsion measurements[256,257] showed that below 88.3°K potassium trifluoromanganate(II) acts as a uniaxial antiferromagnet. However, a second transistion to a different magnetic structure was shown to take place at 81.5°K. The results indicated that weak ferromagnetism develops below this temperature[256,257]. It was suggested[256,257] that this behaviour results from a canting of the magnetic sublattice. Between 81.5 and 88.3°K, weak ferromagnetism is exhibited only in strong fields. The magnetic behaviour of potassium trifluoromanganate(II) as a function of temperature and of field strength is illustrated in Figure 5.11.

Since potassium trifluoromanganate(II) has a *S* ground state, Beckman and Knox[256] have suggested that the distortions which occur in this compound as the temperature is lowered are due to distortions of the MnF_6 octahedra. These workers have shown that both twisting and tilting of the MnF_6 octahedra occurs, and that the magnetic properties are closely related to the degree of twisting and tilting. At 95°K only tilting of the MnF_6 octahedra is apparent. The distortion of the MnF_6 octahedra at

65°K is marked, while at 84°K, the almost regular octahedra are merely tilted and rotated.

Neutron-diffraction measurements by Scatturin and coworkers[258] show that in potassium trifluoromanganate(II) the spin of the manganese atom is coupled antiferromagnetically with the spins of its six nearest neighbours. Pickart and coworkers[259] report that the sodium, rubidium and ammonium trifluoromanganates(II) also have this magnetic structure.

Teaney and coworkers[260] have shown that for rubidium trifluoromanganate(II) there is no distortion from cubic symmetry down to 20°K. They also report[260] a small anomaly in the thermal expansion of rubidium

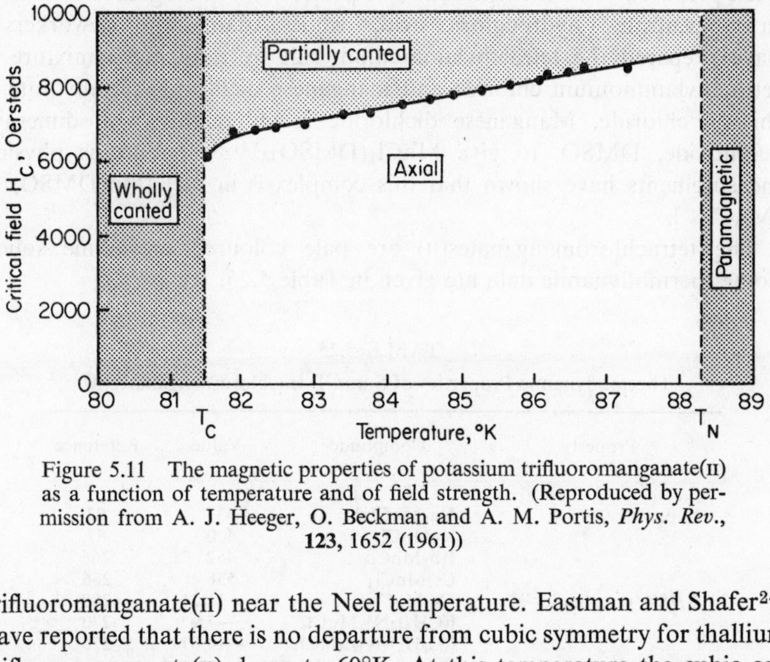

Figure 5.11 The magnetic properties of potassium trifluoromanganate(II) as a function of temperature and of field strength. (Reproduced by permission from A. J. Heeger, O. Beckman and A. M. Portis, *Phys. Rev.*, **123**, 1652 (1961))

trifluoromanganate(II) near the Neel temperature. Eastman and Shafer[246] have reported that there is no departure from cubic symmetry for thallium trifluoromanganate(II) down to 60°K. At this temperature the cubic cell has $a = 4.238$ Å[246].

The electronic absorption spectra of the potassium, rubidium and caesium trifluoromanganates(II) have created much interest[62,63,65,67,261–265]. As in the case of manganese difluoride, there is good agreement between the various groups on the positions and assignments of the bands below about 30,000 cm^{-1}; but above this frequency there is some controversy as to the origin and assignments of the bands[62]. In general terms, the spectra of the trifluoromanganates(II) are almost identical with that of manganese difluoride.

Tetrachloromanganates(II). Alkali-metal tetrachloromanganates(II) are invariably prepared by direct interaction of the constituent chlorides in the melt[87,266]. As expected, the correct stoichiometries of the two chlorides must be used, since other chloro complexes are known to form in the melt. Interaction of saturated aqueous solutions of the alkali-metal chloride and manganese dichloride invariably leads to the formation of hydrated salts[160,161,267,268]. Substituted ammonium, phosphonium and arsonium salts of the tetrachloromanganate(II) anion are prepared by mixing the appropriate chloride and manganese dichloride in a 2:1 in a solvent such as acetone or alcohol[269–277]. Pyridinium tetrachloromanganate(II) has been prepared[278] by the interaction of pyridine and manganese dichloride in concentrated hydrochloric acid, while Adams and coworkers[279] have prepared the tetraethylammonium salt by refluxing a mixture of tetraethylammonium chloride and manganese dichloride tetrahydrate in thionyl chloride. Manganese dichloride reacts readily with dimethylsulphoxide, DMSO, to give $MnCl_2(DMSO)_3$[141–143]. Various physical measurements have shown that this complex is in fact $[Mn(DMSO)_6^{2+}]$ $[MnCl_4^{2-}]$.

The tetrachloromanganates(II) are pale coloured crystalline solids. Some thermodynamic data are given in Table 5.25.

TABLE 5.25

Thermodynamic Properties of Some Tetrachloromanganates(II)

Property	Compound	Value	Reference
M.p.	Na_2MnCl_4[a]	437°	87
	K_2MnCl_4	490°	87
	Rb_2MnCl_4	462°	266
	Cs_2MnCl_4	538°	266
ΔH°_{form} (kcal mole^{-1})[b]	Cs_2MnCl_4	−11.21	280
	$[(CH_3)_4N]_2MnCl_4$	−7.92	270
	$[(C_2H_5)_4N]_2MnCl_4$	−14.07	270

[a] Peritectic point.
[b] From constituent chlorides.

The relatively few structural data at present available for the anhydrous tetrachloromanganates(II) are summarized in Table 5.26.

The infrared spectra of several tetrachloromanganates(II) have been recorded[279,282,283]. In general, good agreement was obtained between the three groups. The most detailed study is that by Sabatini and Sacconi[238] whose results are given in Table 5.27.

As expected, the magnetic moments of the tetrachloromanganates(II) fall in the range 5.8–5.95 BM at room temperature[271,272,275].

The electronic absorption spectrum of the tetrachloromanganate(II) anion has been recorded in solution[275,277,284,285] and by single-crystal

TABLE 5.26
Crystallographic Data for Some Tetrachloromanganates(II)

	Rb_2MnCl_4[87]	$(CH_3NH_3)_2MnCl_4$[281]	$(C_2H_5NH_3)_2MnCl_4$[281]	$[CH_3(C_6H_5)_3As]_2MnCl_4$[276]
Symmetry	Tetragonal	Tetragonal	Tetragonal	Cubic
a (Å)	5.051	7.29	7.28	15.63
c (Å)	16.18	19.40	22.10	
Z	2	4	4	4
Space group	*I4/mmm*	$P4_2/ncm$	$P4_2/ncm$	$P2_13$

TABLE 5.27
Infrared Frequencies (cm^{-1}) for Some Tetrachloromanganates(II)

Mode	Compound	Value
ν_3 (Mn—Cl stretch)	$[(CH_3)_4N]_2MnCl_4$	284*s*
	$[(C_2H_5)_4N]_2MnCl_4$	284*s*
ν_4 (Cl—Mn—Cl bend)	$[(CH_3)_4N]_2MnCl_4$	123*m*
	$[(C_2H_5)_4N]_2MnCl_4$	118*m*
ν_5 (Cl—Mn—Cl bend)	$[(CH_3)_4N]_2MnCl_4$	79*w*
	$[(C_2H_5)_4N]_2MnCl_4$	78*w*

s = strong, *m* = medium, *w* = weak.

techniques[286]. The results have been interpreted on the basis of T_d symmetry. Furlani and Furlani[277] gave the following bands and their assignments for a dimethylformamide solution containing 0.1 M excess chloride ion.

21,300 cm^{-1}	${}^4T_{1g}(G) \leftarrow {}^6A_{1g}(S)$
22,530 cm^{-1}	${}^4T_{2g}(G) \leftarrow {}^6A_{1g}(S)$
23,180 cm^{-1}, 23,400 cm^{-1}	${}^4A_{1g}, {}^4E_g(G) \leftarrow {}^6A_{1g}(S)$
26,600 cm^{-1}	${}^4T_{1g}(P) \leftarrow {}^6A_{1g}(S)$
27,300 cm^{-1}, 27,650 cm^{-1}	${}^4T_{2g}(D) \leftarrow {}^6A_{1g}(S)$
28,060 cm^{-1}	${}^4E_g(D) \leftarrow {}^6A_{1g}(S)$
35,600 cm^{-1}	${}^4A_{2g}(F) \leftarrow {}^6A_{1g}(S)$; ${}^4T_{1g}(F) \leftarrow {}^6A_{1g}(S)$
37,120 cm^{-1}	${}^4T_{2g}(F) \leftarrow {}^6A_{1g}(S)$

Tetrachloromanganate(II) dihydrates. The interaction of an alkali-metal chloride, or ammonium chloride, and manganese dichloride in the correct stoichiometries in water normally gives the appropriate tetrachloromanganate(II) dihydrate[161,267,287].

Greenberg and Walden[161] have made a detailed study of the manganese dichloride-ammonium chloride-water system. They isolate the compounds $6NH_4Cl \cdot MnCl_2 \cdot 2H_2O$ and $2NH_4Cl \cdot MnCl_2 \cdot 2H_2O$ and in addition showed the existence of a range of solid solutions between these two compounds. The latter salt has a tetragonal unit cell with $a = 7.514$ and

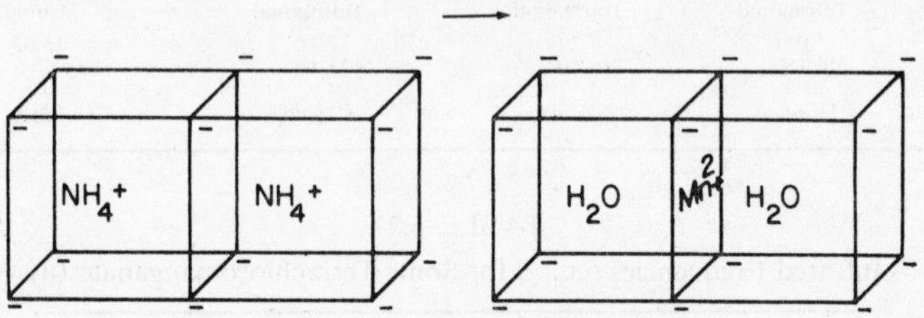

Figure 5.12 The relationship between the structures of ammonium chloride and ammonium tetrachloromanganate(II) dihydrate. (Reproduced by permission from A. L. Greenberg and G. H. Walden, *J. Chem. Phys.*, **8**, 645 (1940))

$c = 8.245$ Å, and its structure may be considered as two NH_4Cl unit cubes having a common face and with chlorines at the corners as shown in Figure 5.12. A manganese ion is placed at the common face in the centre of the four chlorine atoms, and water molecules are substituted for the ammonium ions in the body centres.

Jensen[267] has determined the crystal structures of rubidium and caesium tetrachloromanganate(II) dihydrate by single-crystal techniques. Both compounds are triclinic. The space group of the cell is $P\bar{1}$ and contains one molecule. The unit-cell parameters are given in Table 5.28, together with bond distances.

Each manganese atom is surrounded by four chlorines and two water molecules to give *trans* $[MnCl_4(H_2O)_2]$ octahedra.

Recent specific heat measurements[287] on rubidium tetrachloromanganate(II) dihydrate show an anomaly at 2.24 Å, indicative of a magnetic transition.

Goode[181] has recorded the absorption spectrum at several temperatures of ammonium tetrachloromanganate(II) dihydrate.

Trichloromanganates(II). Alkali-metal trichloromanganates(II) have been prepared by fusing 1:1 mixture of the appropriate chloride and manganese dichloride[87,266]. Alternatively, Kestigian and coworkers[288,289] have found that heating a 1:1 mixture of alkali-metal chloride and manganese dichloride tetrahydrate in an atmosphere of anhydrous hydrogen chloride, also gives the trichloromanganates(II). Evaporation of an aqueous solution of tetramethylammonium chloride and manganese dichloride gives the tetramethylammonium salt[290].

TABLE 5.28
Unit-cell Parameters and Bond Distances of $Rb_2MnCl_4 \cdot 2H_2O$ and $Cs_2MnCl_4 \cdot 2H_2O$

Parameter	$Rb_2MnCl_4 \cdot 2H_2O$	$Cs_2MnCl_4 \cdot 2H_2O$
a	5.66 Å	5.74 Å
b	6.48 Å	6.66 Å
c	7.01 Å	7.27 Å
α	66.7°	67.0°
β	87.7°	87.0°
γ	84.8°	84.3°
Mn—Cl	2.54 and 2.58 Å	2.54 Å
Mn—O	2.08 Å	2.13 Å

TABLE 5.29
Thermodynamic Properties of Some Trichloromanganates(II)

Property	Compound	Value	Reference
M.p.	$NaMnCl_3$	428°	87
	$KMnCl_3$	490°	87
		507°	288
	$RbMnCl_3$	552°	266
		571°	289
	$CsMnCl_3$	593°	266
		616°	289
ΔH°_{form} (kcal mole^{-1})[a]	$NaMnCl_3$	−0.71	280
	$KMnCl_3$	−3.63	280
		−4.4	291
	$RbMnCl_3$	−6.21	280
	$CsMnCl_3$	−9.78	280

[a] From constituent chlorides.

Some thermodynamic properties of the alkali-metal trichloromanganates(II) are listed in Table 5.29.

The unit-cell parameters of several trichloromanganates(II) are given in Table 5.30.

Only the structure of tetramethylammonium trichloromanganate(II) has been studied by single-crystal techniques[290]. Figure 5.13 shows that the structure consists of infinite linear chains of $MnCl_6$ octahedra joined by three bridging chlorine atoms. The tetramethylammonium ions are discrete but somewhat distorted. The manganese-chlorine bond distance

TABLE 5.30

Unit-cell Parameters of Some Trichloromanganates(II)

Parameter	$KMnCl_3$[288]	$RbMnCl_3$[87]	$CsMnCl_3$[289]	$(CH_3)_4NMnCl_3$[290]
Symmetry	Tetragonal	Hexagonal	Hexagonal	Hexagonal
a (Å)	10.024	7.165	7.288	9.151
c (Å)	9.972	17.815	27.44	6.494
Z		6	9	2
Space group		$P6_3/mmc$		$P6_3/m$

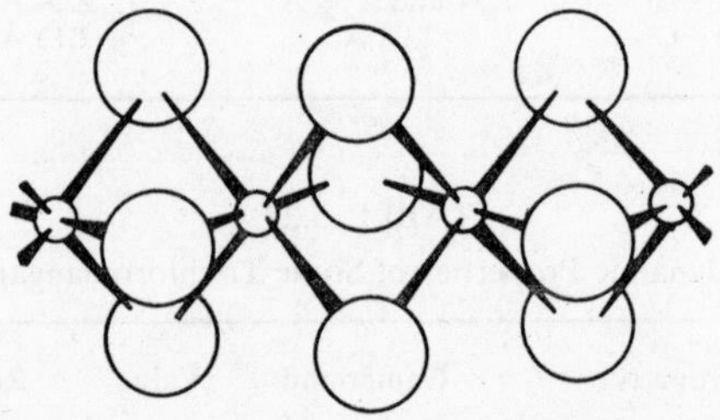

Figure 5.13 The linear chain of $MnCl_6$ octahedra in tetramethylammonium trichloromanganate(II). (Reproduced by permission from B. Morosin and E. J. Graebner, *Acta Cryst.*, **23**, 766 (1967))

is 2.560 Å and the chlorine-manganese-chlorine bond angles of 84.09° and 95.91° indicate slight trigonal elongation of the octahedron about the manganese atom.

Kedzie and coworkers[292] report that the alkali-metal trichloromanganates(II) become antiferromagnetic at low temperatures. The Neel temperatures reported were 100°, 86° and 69° for the potassium, rubidium and caesium salts respectively.

Lawson[286] has recorded the polarized spectrum of tetramethylammonium trichloromanganate(II). The spectrum is consistent with a trigonal distortion of the $MnCl_6$ octahedral, in agreement with the single-crystal structure determination[290].

Trichloromanganate(II) dihydrates. Jensen and coworkers[293] have prepared caesium trichloromanganate(II) dihydrate by evaporating a

1:1 mixture of caesium chloride and manganese dichloride in water at 20°. According to Jensen[267] the analogous reaction does not produce the corresponding rubidium salt, although Seifert and Koknat[87] reported that they obtained the rubidium salt by this method. Jensen[294] has reported that a mixture of α- and β-rubidium trichloromanganate(II) dihydrate is precipitated when a saturated solution of manganese dichloride tetrahydrate and rubidium chloride (5:1) in 8 M hydrochloric acid is cooled from 80° to 50°. A transformation of α → β takes place if this mixture is allowed to stay in the mother liquor for several months at 25°. On the other hand, the β-form is transformed into the α-modification if the temperature is kept at 0°. Suss[295] and Croft and coworkers[288] have reported the isolation of potassium trichloromanganate(II) dihydrate.

Jensen and coworkers[293,294] have studied the crystal structures of the caesium, and the α- and β-rubidium salts by single-crystal techniques. The unit-cell parameters are listed in Table 5.31.

TABLE 5.31

Unit-cell Parameters of Some Trichloromanganate(II) Dihydrates

Parameter	α-$RbMnCl_3 \cdot 2H_2O$	β-$RbMnCl_3 \cdot 2H_2O$	$CsMnCl_3 \cdot 2H_2O$
Symmetry	Orthorhombic	Triclinic	Orthorhombic
a	9.005 Å	6.65 Å	9.060 Å
b	7.055 Å	7.01 Å	7.285 Å
c	11.340 Å	9.03 Å	11.455 Å
α		92.3°	
β		109.4°	
γ		112.9°	
Z	4	2	4
Space group	*Pcca*	*P*1	*Pcca*

α-Rubidium trichloromanganate(II) dihydrate and caesium trichloromanganate(II) dihydrate are isomorphous and isostructural. In these compounds, and also in the β-rubidium salt, each manganese atom is surrounded by an octahedron of four chlorines and two water molecules. In the former two compounds, the water molecules are in a *cis* arrangement and the octahedra share corners to form a zig-zag chain. In β-rubidium trichloromanganate(II) dihydrate however, the water molecules are in the *trans* configuration. The octahedra are joined by sharing edges giving $[Mn_2Cl_6(H_2O)_4]$ groups. The structures of both forms of rubidium trichloromanganate(II) dihydrate are shown in Figure 5.14. The important bond distances of the rubidium and caesium salts are given in Table 5.32, the notation used being that given in Figure 5.14.

Goode[181] has investigated the single-crystal optical absorption spectrum of potassium trichloromanganate(II) dihydrate.

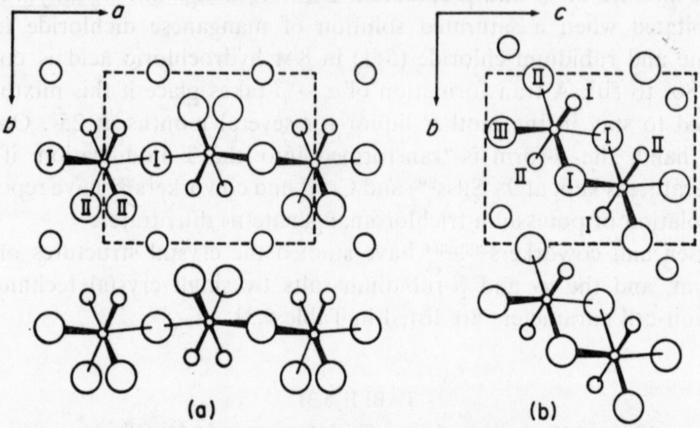

Figure 5.14 The structure of rubidium trichloromanganate(II) dihydrate. (a) α-form. (b) β-form. (Reproduced by permission from S. J. Jensen, *Acta Chem. Scand.*, **21**, 889 (1967))

TABLE 5.32
Bond Distances (Å) in Rubidium and Caesium Trichloromanganate(II) Dihydrates

Distance	α-$RbMnCl_3 \cdot 2H_2O$	β-$RbMnCl_3 \cdot 2H_2O$	$CsMnCl_3 \cdot 2H_2O$
Mn—Cl_I	2.549	2.625 and 2.543	2·57
Mn—Cl_{II}	2.531	2.488	2.50
Mn—Cl_{III}		2.501	
Mn—O_I	2.117	2.202	2.08
Mn—O_{II}		2.228	

Miscellaneous chloromanganates(II). A number of complex chloromanganates(II) have been isolated from the interaction of alkali-metal chloride and manganese dichloride[87,266,280]. The melting points of these compounds are given in Table 5.33.

$K_3Mn_2Cl_7$ has a tetragonal unit cell with $a = 5.027$ and $c = 25.325$ Å[87]. The heat of formation of Cs_3MnCl_5 from its constituent chlorides is reported[280] to be -9.81 kcal mole^{-1}.

Tetrabromomanganates(II). Substituted ammonium, phosphonium and arsonium tetrabromomanganates(II) have been prepared by methods similar to those for the corresponding chloro complexes[272,275–277,285]. The heat of formation of the tetraethylammonium compound from its constituents is reported[296] to be -10.12 kcal mole^{-1}. The magnetic moments of the complexes are in the range 5.8–5.9 BM[272,275].

The infrared spectra of several tetrabromomanganates(II) have been recorded[279,282,283]. Manganese-bromine stretching frequencies in the 220 cm^{-1} region were reported.

TABLE 5.33
Miscellaneous Chloromanganates(II)

Compound	M.p.	Reference	Compound	M.p.	Reference
$NaMn_2Cl_5$	448	87	$Rb_3Mn_2Cl_7$	475	266
Na_4MnCl_6	458^a	87	$CsMn_4Cl_9$	537	266
$K_3Mn_2Cl_7$	437^a	87	Cs_3MnCl_5	511	266
K_4MnCl_6	448^a	87			

^{a} Peritectic point.

The absorption spectrum of the tetrabromomanganate(II) ion has been recorded in solution[275,277,284,285]. The general features and assignments for the spectrum are similar to those for the tetrachloromanganate(II) anion.

Tetraiodomanganates(II). The methods of preparation, magnetic, spectral and other properties of the substituted ammonium, phosphonium and arsonium tetraiodomanganates(II) are completely analogous to those described for the corresponding chloro and bromo complexes[272,275,276,285]. The manganese-iodine stretching frequency[283] occurs at about 185 cm^{-1}.

Hathaway and Holah[138] reacted manganese diiodide with acetonitrile and isolated a yellow adduct $MnI_2(CH_3CN)_3$. This compound had a room-temperature magnetic moment of 5.86 BM. The diffuse-reflectance spectrum and conductance indicate that the compound should be formulated as $[Mn(CH_3CN)_6^{2+}][MnI_4^{2-}]$.

ADDENDA

Halides and Oxide Halides

The preparation of manganese trioxide chloride, manganese dioxide dichloride and manganese oxide trichloride by the reaction of manganese heptaoxide with chlorosulphonic acid has been described by Briggs[297].

The three compounds are liquids at room temperature but are explosively unstable. They are soluble in carbon tetrachloride to give dark coloured solutions, but undergo slow reaction with the solvent.

Further detailed studies on the magnetic[298–300], spectral[301–305] and far infrared[306] properties of manganese difluoride have been reported.

Mehra[307] has investigated the electronic spectrum of a single crystal of manganese dichloride at low temperatures and reports a previously unobserved band at 30,500 cm^{-1}. This band has also been observed by Foster and Gill[305] who in addition recorded the spectra of the dibromide and diiodide. The observed bands and their assignments are given in Table 5.34.

TABLE 5.34

Electronic Spectra (cm^{-1}) of Manganese Dihalides

Transition	$McCl_2$[307]	$MnCl_2$[305]	$MnBr_2$[305]	MnI_2[305]
${}^4T_{1g}(G) \leftarrow {}^6A_{1g}(S)$	18,500	18,900	18,800	18,180
${}^4T_{2g}(G) \leftarrow {}^6A_{1g}(S)$	22,000	22,220	21,930	21,110
	23,574			
${}^4A_{1g}(G), {}^4E_g(G) \leftarrow {}_6A_{1g}(S)$	23,800	23,870	23,420	22,420
${}^4T_{2g}(D) \leftarrow {}^6A_{1g}(S)$	27,000	27,030	26,740	25,900
${}^4E_g(D) \leftarrow {}^6A_{1g}(S)$	28,066	28,250	27,630	26,450
${}^4T_{1g}(P) \leftarrow {}^6A_{1g}(S)$	30,500	30,490	29,760	28,900
${}^4A_{2g}(F), {}^4T_{1g}(F) \leftarrow {}^6A_{1g}(S)$	38,400			
${}^4T_{2g}(F) \leftarrow {}^6A_{1g}(S)$	40,700			

A number of adducts of manganese dichloride have been prepared with ligands that have nitrogen, oxygen and sulphur donor atoms. Some properties of a number of these adducts are summarized in Table 5.35.

One interesting type of reaction that has been reported[319] is that between dichlorobis(pyridine)manganese(II), $MnCl_2(py)_2$, and thiourea (tu) in methanol. The white product obtained analysed as $MnCl_2(py)_2(tu)_4$; it was found to melt at 177°, to be a non-conductor in non-aqueous solvents and to have a room-temperature magnetic moment of 6.6 BM. It was suggested[319] that the manganese atom was in an eight-coordinate environment.

Other donor ligands that have been reacted with manganese dichloride include 4-methylpyridine[320], aniline[321], acrylonitrile[322], hydrazine[323] and imidazole[324].

There is an anomaly in the thermal expansion coefficient of manganese dichloride tetrahydrate at 1.622°K due to antiferromagnetic ordering[325]. Lumme and Raivio[326] have studied the thermal decomposition of this hydrate in both static air and dynamic nitrogen systems. The reported results are summarized in Table 5.36.

Di Vaira and Orioli[327] have shown that $MnBr_2(Me_6tren)$, where

TABLE 5.35

Adducts of the Type $MnCl_2L_n$

Compound and Properties	Reference
L = 3-methylpyridine, $n = 2$, white, $\mu = 6.13$ BM, octahedral	308
L = 3-methylpyridine, $n = 1$, brown, $\mu = 6.10$ BM, tetrahedral dimer	308
L = 4-methylpyridine, $n = 1$, pink, $\mu = 6.20$ BM, tetrahedral dimer	308
L = 4-vinylpyridine, $n = 2$, buff, $\mu = 6.05$ BM, octahedral	308
L = 4-vinylpyridine, $n = 1$, light brown, $\mu = 5.98$ BM, tetrahedral dimer	308
L = 2-pyridone (TP), $n = 1$, white, $\mu = 5.97$ BM, octahedral	309
L = 2,9-dimethyl-1,10-phenanthroline (dmp), $n = 1$, brown	310
L = 2,2′,2″-terpyridyl (terpy), $n = 1$, $\mu = 5.96$ BM, five-coordinate	311
L = pyridine-2-aldehyde 2′-pyridylhydrazone (paphy), $n = 1$, orange, $\mu =$ 5.94 BM	312
L = pentamethylenetetrazolme (PMT), $n = 1$, pale pink, m.p. > 300°, $\mu =$ 5.44 BM, $\nu_{Mn-Cl} = 206\ cm^{-1}$, octahedral	313
L = bis-[2-(2-pyridyl)ethyl]amine (dpea), $n = 1$, white, $\mu = 6.02$ BM, octahedral	314
L = di-2-pyridyl ketone (DPK), $n = 2$, $\mu = 6.1$ BM, *cis* octahedral	315
L = 1,4-dioxan, $n = 1$, white, $\mu = 5.88$ BM, $\nu_{Mn-Cl} = 250\ cm^{-1}$, octahedral	316
L = biuret, $n = 1$, white, $\mu = 5.90$ BM, octahedral	317
L = 1,2-bis(methylsulphonyl)ethane (BMSE), $n = 1$, isolated as dihydrate, m.p. = 208°, $\nu_{S-O} = 1000\ cm^{-1}$, octahedral	318

TABLE 5.36

Thermal Decomposition of Manganese Dichloride Tetrahydrate

Reaction	Temperature Range	
	Static Air	Dynamic Nitrogen
$MnCl_2 \cdot 4H_2O \rightarrow MnCl_2 \cdot H_2O + 3H_2O$	<50–175°	<50–191°
$MnCl_2 \cdot H_2O \rightarrow MnCl_2 + H_2O$	175–227°	191–228°
$MnCl_2 \rightarrow Mn + Cl_2$		650–950°

Me_6tren is tris(2-dimethylaminoethyl)amine, is isomorphous with the corresponding iron, cobalt, nickel and copper complexes. Thus the manganese atom is in a trigonal bipyramidal configuration and this is the first five-coordinate manganese(II) compound to be confirmed by X-ray analysis. The unit cell is cubic with $a = 12.216$ Å, is of space group

$P2_13$, and contains four formula units. The internuclear dimensions, using the notation in Figure 7.14 (page 360), are given in Table 5.37.

TABLE 5.37
Internuclear Dimensions in $[MnBr(Me_6tren)]^+$

Distance	Value (Å)	Angle	Value
Mn—Br(1)	2.491	N(1)—Mn—N(2)	80.7°
Mn—N(1)	2.19	N(2)—Mn—Br(1)	99.3°
Mn—N(2)	2.27	N(2)—Mn—N(2′)	117.4°

Both manganese dibromide and diiodide form adducts with a number of ligands that have been found to react with the dichloride. Some properties of these adducts are summarized in Table 5.38.

TABLE 5.38
Adducts of the Type MnX_2L_n

Compound and Properties	Reference
X = Br, L = 4-methylpyridine, $n = 2$, white, $\mu = 6.10$ BM, octahedral	308
X = Br, L = 4-methylpyridine, $n = 1$, light pink, $\mu = 6.20$ BM, tetrahedral dimer	308
X = Br, L = 4-vinylpyridine, $n = 2$, buff, $\mu = 6.28$ BM, octahedral	308
X = Br, L = 4-vinylpyridine, $n = 1$, light brown, $\mu = 6.00$ BM, tetrahedral dimer	308
X = Br, L = 2,2′,2″-terypyridyl (terpy), $n = 1$, $\mu = 5.98$ BM, five coordinate	311
X = Br, L = pyridine-2-aldehyde 2′-pyridylhydrazone (paphy), $n = 1$, orange, $\mu = 5.95$ BM	312
X = Br, L = pentamethylenetetrazole (PMT), $n = 1$, pale pink, m.p. > 300°, $\mu = 5.63$ BM, octahedral	313
X = Br, L = bis-[2-(2-pyridyl)ethyl]amine (dpea), $n = 1$, pale green, $\mu =$ 5.88 BM, five coordinate	314
X = Br, L = 1,4-dioxan, $n = 2$, white, $\mu = 5.85$ BM, octahedral	316
X = I, L = 2,2′,2″-terpyridyl (terpy), $n = 1$, $\mu = 5.94$ BM, five coordinate	311
X = I, L = pyridine-2-aldehyde 2′-pyridylhydrazone (paphy), $n = 1$, red	312

The Neel temperature of manganese dibromide tetrahydrate has been reported[328] as 2.13°K.

Pale red manganese dibromide dihydrate and black manganese diiodide monohydrate have been prepared[326] by dissolving the calculated amount of manganese carbonate in the appropriate hydrohalic acid.

The thermal decomposition of both of these hydrates has been studied in static air and dynamic nitrogen atmospheres[326]. The results are summarized in Table 5.39.

TABLE 5.39

Thermal Decomposition of Manganese Dihalide Hydrates

Reaction	Temperature Range	
	Static Air	Dynamic Nitrogen
$MnBr_2 \cdot 2H_2O \rightarrow MnBr_2 \cdot H_2O + H_2O$	70–137°	60–146°
$MnBr_2 \cdot H_2O \rightarrow MnBr_2 + H_2O$	137–197°	146–197°
$MnI_2 \cdot H_2O \rightarrow MnI_2 + H_2O$	85–170°	86–160°
$MnI_2 \rightarrow Mn + I_2$		560–943°

Complex Halides

Moews[329] has prepared a number of hexachloromanganates(IV) by adding a suitable permanganate to a hydrochloric acid solution of the appropriate chloride saturated with hydrogen chloride gas at 0°. The potassium salt has also been prepared by the reaction between manganese(III) acetate and potassium acetate in the presence of a large excess of acetyl chloride[330]. These complexes have face-centred cubic unit cells, of the K_2PtCl_6 type, with the following dimensions (Å)[329].

K_2MnCl_6	9.645	$(NH_4)_2MnCl_6$	9.80
Rb_2MnCl_6	9.82	$[(CH_3)_4N]_2MnCl_6$	12.70
Cs_2MnCl_6	10.17		

For the potassium salt, a manganese–chlorine bond distance of 2.28 Å was reported[329]. This salt has manganese–chlorine vibrations[331] at 358 and 200 cm^{-1} and obeys the Curie-Weiss law[332] with a magnetic moment of 3.90 BM and a Weiss constant of 30°. Moews[329] recorded the electronic spectrum of potassium hexachloromanganate(IV) but his results have more recently been challenged[333].

Adams and Morris[331] report infrared absorptions at 342 and 183 cm^{-1} in tris(1,2-propenediamine)cobalt(III) hexachloromanganate(III).

Davis, Fackler and Weeks[334] have examined the electronic spectra of several manganese(III) complex halides. The influence of the Jahn–Teller effect was discussed, and the spectra were assigned as shown in Table 5.40 on the basis of a $^5B_{1g}$ ground state for a complex of D_{4h} symmetry.

Potassium tetrafluoromanganate(II), prepared[335] by fusing a suitable mixture of manganese difluoride and potassium hydrogen difluoride at 850°, has a tetragonal unit cell with $a = 4.171$ and $c = 13.259$ Å. At 4.2°K the parameters are 5.581 and 13.242 Å respectively, while the magnetic moment at this temperature is 4.54 BM[335]. Copla and coworkers[336] have measured the heat capacity of potassium tetrafluoromanganate(II) at temperatures below 1°K.

There have been a number of further studies on the magnetic[337–341], spectral[302,303,342–345] and structural properties[336,340,346] of the alkali-metal trifluoromanganates(II). A new determination[347] of the melting

TABLE 5.40
Electronic Spectra (cm^{-1}) of Halomanganates(III)

Transition	K_2MnF_5	K_2MnCl_5	$[(C_2H_5)_4N]_2MnCl_5$	$[Co(pn)_3]MnCl_5$
$^5A_{1g} \leftarrow {}^5B_{1g}$	12,100	12,000	11,300	8,333
$^5B_{2g} \leftarrow {}^5B_{1g}$	18,750	18,500	16,200	17,540
$^5E_g \leftarrow {}^5B_{1g}$	21,400		23,530	22,400

point of potassium trifluoromanganate(II) gives a value of 1013°. Roscoe and Haendler[340] have prepared solid solutions of ammonium trifluoromanganate(II) and zincate(II) by adding an excess of bromine to the appropriate amounts of manganese and zinc suspended in methanol, followed by the appropriate amount of ammonium fluoride in a saturated methanol solution. Magnetic moments, X-ray diffraction patterns and decomposition temperatures were measured as a function of the mole per cent of ammonium trifluoromanganate(II). The thermal decomposition products resulting from the solid solutions were shown to be themselves solid solutions of the type $Mn_xZn_{1-x}F_3$. The magnetic moments, unit-cell parameters and decomposition temperatures are given in Table 5.41. Anomalies occur in the neutron elastic properties and ^{19}F NMR spectrum of potassium trifluoromanganate(II) in the region of the 184°K phase transition[346,347] (page 242).

Gill and Taylor[348] have given details for the preparation of tetraethylammonium tetrachloromanganate(II) and the corresponding bromo derivative. There has been considerable interest shown over the last year or so on the preparation and characterization of complexes derived from dithioacetylacetone. Furuhashi and coworkers[349] prepared yellow $Mn(C_5H_7S_2)_2Cl_4$ by passing hydrogen chloride and hydrogen sulphide through an ethanol solution containing acetylacetone and manganese dichloride. These workers suggested that the complex contained octahedral manganese(II), the ligand acting as a monodentate sulphur donor. However, a very detailed study by Heath, Martin and Stewart[350] on the preparation and characterization of a number of salts containing dithioacetylacetone showed quite clearly that this complex, as well as several

TABLE 5.41
The NH_4MnF_3-NH_4ZnF_3 System

Mole % of NH_4MnF_3	a (Å)	μ (BM)	T_{decomp}
0.0	4.120	0.00	210°
1.0		0.60	
12.5	4.134	1.90	
25.0	4.158	2.62	215°
50.0	4.178	3.77	225°
75.0	4.219	4.37	233°
87.5	4.232	4.69	
100.0	4.242	5.00	240°

others, should be formulated as 3,5-dimethyldithiolium tetrachlorometallate(II). Furuhashi and associates[351] have also prepared the corresponding bromo derivative which has a room-temperature magnetic moment of 6.02 BM. The chloro compound has a moment of 5.92 BM[349].

The infrared and Raman spectra of the tetrachloro-, tetrabromo- and tetraiodomanganates(II) have been investigated in detail[352,353]. The results are summarized in Table 5.42.

The unit-cell parameters of several halomanganates(II) have been reported[305]. The results are given below.

		a	c
α-Cs_2MnCl_4	Tetragonal	5.137 Å	16.93 Å
$(NH_3CH_3)_2MnCl_4$	Tetragonal	7.20 Å	19.37 Å
$(CH_3)_4NMnCl_3$	Hexagonal	9.13 Å	6.49 Å
$(CH_3)_4NMnBr_3$	Hexagonal	9.44 Å	6.76 Å

Jensen[354,355] has continued his studies on the structures of hydrated halomanganates(II). Potassium tetrachloromanganate(II), prepared from aqueous solutions containing a 2:1 mixture of potassium chloride and manganese dichloride tetrahydrate at 45°, has a tetragonal unit cell[355], containing two formula units and of space group *I4/mmm*, with $a = 7.415$ and $c = 8.220$ Å. The structure consists of potassium ions and *trans*

$[MnCl_4(H_2O)_2]^{2-}$ groups. The manganese–chlorine and manganese–oxygen bond distances are 2.531 and 2.175 Å respectively. The transition to antiferromagnetic behaviour of rubidium tetrachloromanganate(II) dihydrate below 2.24°K has been confirmed by ^1H, ^{35}Cl, ^{85}Rb and ^{87}Rb NMR studies[356].

Potassium trichloromanganate(II) dihydrate, isolated from aqueous solutions containing the stoichiometric amounts of the chlorides at 20°, has a triclinic unit cell[354] with $a = 6.49$, $b = 6.91$, $c = 9.91$ Å, $\alpha = 96.8°$, $\beta = 114.1°$ and $\gamma = 112.6°$. There are two formula units in the cell of

TABLE 5.42

Infrared (IR) and Raman (R) Spectra (cm^{-1}) of Tetrahalomanganates(II)

Complex[a]	Method	State of sample	ν_1	ν_2	ν_3	ν_4	Reference
K_2MnCl_4	R	$MnCl_2$-KCl melt	255				352
$(NHEt_3)_2MnCl_4$	R	CH_3CN solution	249				352
	R	Solid	256				352
	IR	Nujol mull			301, 278	120	352
$(NEt_4)_2MnCl_4$	R	Solid	258			116	353
	R	CH_3NO_2 solution	251				353
	IR	Solid			284	118	353
$(NEt_4)_2MnBr_4$	R	Solid	157				353
	R	CH_3NO_2 solution	156				353
	IR	Solid			221	84	353
$(NBu_4)_2MnBr_4$	R	$CHCl_3$ solution	195	65	226	81	352
	R	Solid	195				352
	IR	Nujol mull			221, 209	89	352
$(NEt_4)_2MnI_4$	R	Solid	108				352
	IR	Nujol mull			193, 188	56	352
$(NBu_4)_2MnI_4$	R	$CHCl_3$ solution	116	46		58	352
	R	Solid	107				
	R	Solid	111				353

[a] Et = Ethyl. Bu = *n*-Butyl.

space group $P\bar{1}$. This hydrate is isostructural with the β form of the corresponding rubidium salt (Figure 5.14, page 252). That is, the structure consists of potassium cations and *trans* $[Mn_2Cl_6(H_2O)_4]^{2-}$ anions. The bond distances, using the notation shown in Figure 5.14, are Mn—Cl_1 = 2.595 and 2.570 Å, Mn—Cl_2 = 2.490 Å, Mn—Cl_3 = 2.482 Å, Mn—O_1 = 2.182 Å and Mn—O_2 = 2.187 Å. Specific heat measurements show that the potassium salt has a magnetic transition at 2.70°K[357]. The electronic absorption spectrum of caesium trichloromanganate(II) dihydrate has been recorded and discussed by Le Paillier-Mall[358].

The electronic spectrum of the tetraiodomanganate(II) anion has been observed and assigned as for the corresponding chloro and bromo complexes[305].

REFERENCES

1. A. Engelbrecht and A. V. Grosse, *J. Am. Chem. Soc.*, **76**, 2042 (1954).
2. K. Wiechert, *Z. anorg. allgem. Chem.*, **261**, 310 (1950).
3. E. E. Aynsley, *J. Chem. Soc.*, **1958**, 2425.
4. A. Javan and A. Engelbrecht, *Phys. Rev.*, **96**, 649 (1954).
5. D. Michel and A. Doiwa, *Naturwissenschaften*, **53**, 129 (1966).
6. H. W. Roesky, O. Glemser and K. H. Hellberg, *Chem. Ber.*, **98**, 2046 (1965).
7. R. Hoppe, W. Dahne and W. Klemm, *Ann. Chim.* (*Paris*), **658**, 1 (1962).
8. R. Hoppe, W. Dahne and W. Klemm, *Naturwissenschaften*, **48**, 429 (1961).
9. G. Braur, *Handbook of Preparative Inorganic Chemistry*, 2nd ed., Academic Press, New York, 1963.
10. R. D. Fowler, H. C. Anderson, J. M. Hamilton, W. B. Burford, A. Spadetti, S. B. Bitterlich and I. Litant, *Ind. Eng. Chem.*, **39**, 343 (1947).
11. H. J. Emeleus and G. L. Hunt, *J. Chem. Soc.*, **1964**, 396.
12. A. G. Sharpe and A. A. Woolf, *J. Chem. Soc.*, **1951**, 798.
13. M. A. Hepworth and K. H. Jack, *Acta Cryst.*, **10**, 345 (1957).
14. E. E. Aynsley, R. D. Peacock and P. L. Robinson, *J. Chem. Soc.*, **1950**, 1622.
15. A. W. Jache and G. H. Cady, *J. Phys. Chem.*, **56**, 1106 (1952).
16. G. Fuller, M. Stacey, J. C. Tatlow and C. R. Thomas, *Tetrahedron*, **18**, 123 (1962).
17. M. Stacey and J. C. Tatlow, *Advances in Fluorine Chemistry*, Vol. 1, Butterworths, London, 1960.
18. R. A. Rausch, R. A. Davis and D. W. Osborne, *J. Org. Chem.*, **28**, 494 (1963).
19. Z. F. Zmbov and J. L. Margrave, *J. Inorg. Nucl. Chem.*, **29**, 673 (1967).
20. M. A. Hepworth, K. H. Jack and R. S. Nyholm, *Nature*, **179**, 211 (1957).
21. R. M. Bozorth and J. W. Nielsen, *Phys. Rev.*, **110**, 879 (1958).
22. E. O. Wollan, H. R. Child, W. C. Koehler and M. K. Wilkinson, *Phys. Rev.*, **112**, 1132 (1958).
23. R. J. H. Clark, *J. Chem. Soc.*, **1964**, 417.
24. A. Chretien and G. Varga, *Bull. Soc. Chim. France*, **3**, 2385 (1936).
25. J. H. Krepelka, *Coll. Czech. Chem. Commun.*, **7**, 105 (1935).
26. H. Funk and H. Kreis, *Z. anorg. allgem. Chem.*, **349**, 45 (1967).
27. A. Kurtenacker, W. Finger and F. Hey, *Z. anorg. allgem. Chem.*, **211**, 83 (1933).
28. E. L. Muetterties and J. E. Castle, *J. Inorg. Nucl. Chem.*, **18**, 148 (1961).
29. P. Nuka, *Z. anorg. allgem. Chem.*, **180**, 235 (1929).
30. N. N. Mikhailov and S. V. Petrov, *Kristallografiya*, **11**, 443 (1966).
31. K. Nassau, *J. Appl. Phys.*, **32**, 1820 (1961).
32. S. Legrand and M. Binard, *Bull. Soc. Chim. France*, **1964**, 1900.
33. M. Griffel and J. W. Stout, *J. Am. Chem. Soc.*, **72**, 4351 (1950).

34. I. N. Belyaev and O. Y. Revina, *Russ. J. Inorg. Chem.*, **11,** 1041 (1966).
35. I. N. Belyaev and O. Y. Revina, *Russ. J. Inorg. Chem.*, **11,** 772 (1966).
36. R. A. Kent, T. C. Ehlert and J. L. Margrave, *J. Am. Chem. Soc.*, **86,** 5090 (1964).
37. A. Buchler, J. L. Stauffer and W. Klemperer, *J. Chem. Phys.*, **40,** 3471 (1964).
38. J. W. Stout and S. A. Reed, *J. Am. Chem. Soc.*, **76,** 5279 (1954).
39. W. H. Baur, *Acta Cryst.*, **11,** 488 (1958).
40. W. H. Baur, *Naturwissenschaften*, **44,** 349 (1957).
41. A. E. van Arkel, *Rec. Trav. Chim.*, **45,** 437 (1926).
42. D. F. Gibbons, *Phys. Rev.*, **115,** 1194 (1959).
43. S. L. Strong, *J. Phys. Chem. Solids*, **19,** 51 (1961).
44. L. Corliss, Y. Delabarre and N. Elliot, *J. Chem. Phys.*, **18,** 1256 (1950).
45. H. Bizette and T. Belling, *Compt. Rend.*, **209,** 205 (1939).
46. R. A. Erickson, *Phys. Rev.*, **90,** 779 (1953).
47. F. M. Johnson and A. H. Nethercot, *Phys. Rev.*, **104,** 847 (1956).
48. D. N. Astrov, S. I. Novikova and M. P. Orlova, *Zh. Eksptl. i Theoret. Fiz.*, **37,** 1197 (1959).
49. S. Foner, *J. Phys. Radium*, **20,** 336 (1959).
50. J. W. Stout and H. E. Adams, *J. Am. Chem. Soc.*, **64,** 1535 (1942).
51. M. Griffel and J. W. Stout, *J. Chem. Phys.*, **18,** 1455 (1950).
52. H. A. Alperin, P. J. Brown, R. Nathans and S. J. Pickart, *Phys. Rev. Letters*, **8,** 237 (1962).
53. P. Heller and G. B. Benedek, *Phys. Rev. Letters*, **8,** 428 (1962).
54. H. Bizette and B. Tsai, *Compt. Rend.*, **238,** 1575 (1954).
55. P. Heller, *Phys. Rev.*, **146,** 403 (1966).
56. J. W. Stout and E. Catalano, *J. Chem. Phys.*, **23,** 2013 (1955).
57. J. W. Stout and M. Griffel, *Phys. Rev.*, **76,** 144 (1949).
58. G. A. Slack, *Phys. Rev.*, **122,** 1451 (1961).
59. D. M. Finlayson, I. S. Robertson, T. Smith and R. W. H. Stevenson, *Proc. Phys. Soc.*, **76,** 355 (1960).
60. J. W. Stout, *J. Chem. Phys.*, **31,** 709 (1959).
61. V. V. Eremenko and A. I. Belyaeva, *Soviet Phys.—Solid State*, **5,** 2106 (1964).
62. J. Ferguson, *Australian J. Chem.*, **21,** 307 (1968).
63. J. Ferguson, *Solid State Commun.*, **5,** 773 (1967).
64. P. G. Russell, D. S. McClure and J. W. Stout, *Phys. Rev. Letters*, **16,** 176 (1966).
65. R. Stevenson, *Phys. Rev.*, **152,** 531 (1966).
66. D. D. Sell, R. L. Green, W. M. Yen and A. L. Schawlow, *J. Appl. Phys.*, **37,** 1229 (1966).
67. R. Stevenson, *Can. J. Phys.*, **43,** 1732 (1965).
68. R. L. Green, D. D. Sell, W. M. Yen and A. L. Schawlow, *Phys. Rev. Letters*, **15,** 656 (1965).
69. S. P. S. Porto, P. A. Fleury and T. C. Damen, *Phys. Rev.*, **154,** 522 (1967).
70. P. A. Fleury, S. P. S. Porto, L. E. Cheesman and H. J. Guggenheim, *Phys. Rev. Letters*, **17,** 87 (1966).
71. D. Bloor and D. H. Martin, *Proc. Phys. Soc.*, **78,** 774 (1961).
72. G. Parisot, *Compt. Rend.*, **265B,** 1192 (1967).

73. N. V. Sidgwick, *The Chemical Elements and Their Compounds*, O.U.P., London, 1950.
74. R. C. Osthoff and R. C. West, *J. Am. Chem. Soc.*, **76,** 4732 (1954).
75. R. A. Butera and W. F. Giauque, *J. Chem. Phys.*, **40,** 2379 (1964).
76. R. C. Schoonmaker, A. H. Friedman and R. F. Porter, *J. Chem. Phys.*, **31,** 1586 (1959).
77. H. J. Borchardt and F. Daniels, *J. Phys. Chem.*, **61,** 917 (1957).
78. C. Starr, F. Bitter and A. R. Kaufmann, *Phys. Rev.*, **58,** 977 (1940).
79. R. C. Chisholm and J. W. Stout, *J. Chem. Phys.*, **36,** 972 (1962).
80. M. F. Koehler and J. P. Coughlin, *J. Phys. Chem.*, **63,** 605 (1959).
81. R. B. Murray and L. D. Roberts, *Phys. Rev.*, **100,** 1067 (1955).
82. N. F. Dashniani, *Trudy Inst. Priklad. Khim. i Elektrokhim., Akad. Nauk Gruzin SSR*, **1,** 111 (1959).
83. A. Chretien and G. Oechsel, *Compt. Rend.*, **206,** 254 (1938).
84. G. W. Watt, P. S. Gentile and E. P. Helvenston, *J. Am. Chem. Soc.*, **77,** 2752 (1955).
85. D. Khristov, S. Karaivanov and V. Kolushki, *Godishnik Sofiskyia Univ., Khim. Fak.*, **55,** 49 (1960–61).
86. H. Schafer, L. Bayer, G. Breil, K. Etzel and K. Krehl, *Z. anorg. allgem. Chem.*, **278,** 300 (1959).
87. H. J. Seifert and F. W. Koknat, *Z. anorg. allgem. Chem.*, **341,** 269 (1965).
88. J. D. Corbett, R. J. Clark and T. F. Mundy, *J. Inorg. Nucl. Chem.*, **25,** 1287 (1963).
89. C. G. Mair, *Bur. of Mines, Tech. Paper* 360 (1925).
90. A. Konneker and W. Biltz, *Z. anorg. allgem. Chem.*, **242,** 225 (1939).
91. A. Ferrari, A. Braibanti and G. Bigliardi, *Acta Cryst.*, **16,** 846 (1963).
92. G. E. Leroi, T. C. James, J. T. Hougen and W. Klemperer, *J. Chem. Phys.*, **36,** 2879 (1962).
93. M. O. Kostryakova, *Dokl. Akad. Nauk SSSR*, **96,** 959 (1954).
94. R. B. Murray, *Phys. Rev.*, **100,** 1071 (1955).
95. R. B. Murray, *Phys. Rev.*, **128,** 1570 (1962).
96. R. Pappalardo, *J. Chem. Phys.*, **31,** 1050 (1959).
97. J. W. Stout, *J. Chem. Phys.*, **33,** 303 (1960).
98. R. Pappalardo, *J. Chem. Phys.*, **33,** 613 (1960).
99. S. Prasad and K. P. Kacker, *J. Indian Chem. Soc.*, **35,** 719 (1958).
100. S. Prasad and V. R. Reddy, *J. Indian Chem. Soc.*, **35,** 722 (1958).
101. N. S. Gill, R. H. Nuttall, D. E. Scaife and D. W. A. Sharp, *J. Inorg. Nucl. Chem.*, **18,** 79 (1961).
102. I. S. Ahuja, D. H. Brown, R. H. Nuttall and D. W. A. Sharp, *J. Inorg. Nucl. Chem.*, **27,** 1105 (1965).
103. S. Prasad and P. J. Sarma, *J. Proc. Inst. Chemists (India)*, **37,** 246 (1965).
104. R. J. H. Clark and C. S. Williams, *Inorg. Chem.*, **4,** 350 (1965).
105. E. G. Cox, A. J. Shorter, W. Wardlaw and W. J. R. Way, *J. Chem. Soc.*, **1937,** 1556.
106. N. S. Gill, R. S. Nyholm, G. A. Barclay, T. I. Christie and P. J. Pauling, *J. Inorg. Nucl. Chem.*, **18,** 88 (1961).
107. N. S. Gill and H. J. Kingdon, *Australian J. Chem.*, **19,** 2197 (1966).
108. J. R. Allan, D. H. Brown, R. H. Nuttall and D. W. A. Sharp, *J. Inorg. Nucl. Chem.*, **27,** 1865 (1966).

109. W. S. Fyfe, *J. Chem. Soc.*, **1950,** 790.
110. D. P. Mellor and C. R. Coryell, *J. Am. Chem. Soc.*, **60,** 1786 (1938).
111. R. J. H. Clark and C. S. Williams, *Chem. Ind.* (*London*), **1964,** 1317.
112. E. Konig and H. L. Schlaffer, *Z. Physik. Chem.* (*Frankfurt*), **26,** 371 (1960).
113. R. W. Asmussen and H. Soling, *Acta Chem. Scand.*, **11,** 1331 (1957).
114. I. S. Ahuja, D. H. Brown, R. H. Nuttall and D. W. A. Sharp, *J. Inorg. Nucl. Chem.*, **27,** 1625 (1965).
115. D. H. Brown, R. H. Nuttall and D. W. A. Sharp, *J. Inorg. Nucl. Chem.*, **26,** 1151 (1964).
116. G. Beech, C. T. Mortimer and E. G. Tyler, *J. Chem. Soc.*, **1967,** A, 1111.
117. G. Beech and C. T. Mortimer, *J. Chem. Soc.*, **1967,** A, 1115.
118. C. W. Frank and L. B. Rogers, *Inorg. Chem.*, **5,** 615 (1966).
119. J. T. Summers and J. V. Quagliano, *Inorg. Chem.*, **3,** 1767 (1964).
120. W. Libus and I. Uruska, *Inorg. Chem.*, **5,** 256 (1966).
121. M. Goodgame and P. J. Hayward, *J. Chem. Soc.*, **1966,** A, 632.
122. G. J. Sutton, *Australian J. Chem.*, **14,** 550 (1961).
123. P. Pfeiffer and F. Tappermann, *Z. anorg. allgem. Chem.*, **215,** 273 (1933).
124. J. A. Broomhead and F. P. Dwyer, *Australian J. Chem.*, **14,** 250 (1961).
125. M. Ciampolini and G. P. Speroni, *Inorg. Chem.*, **5,** 45 (1966).
126. R. Zannetti and R. Serra, *Gazz. Ital. Chim.*, **90,** 328 (1960).
127. J. R. Allan, D. H. Brown, R. H. Nuttall and D. W. A. Sharp, *J. Inorg. Nucl. Chem.*, **27,** 1305 (1965).
128. L. Sacconi and A. Sabatini, *J. Inorg. Nucl. Chem.*, **25,** 1389 (1965).
129. A. Ferrari, A. Braibanti, G. Bigliardi and F. Dallavalle, *Z. Krist.*, **119,** 284 (1963).
130. C. D. Flint and M. Goodgame, *J. Chem. Soc.*, **1966,** A, 744.
131. D. M. Adams and J. B. Cornell, *J. Chem. Soc.*, **1967,** A, 884.
132. K. Swaminathan and H. M. N. H. Irving, *J. Inorg. Nucl. Chem.*, **26,** 1291 (1964).
133. A. Rosenheim and V. J. Meyer, *Z. anorg. allgem. Chem.*, **49,** 17 (1906).
134. D. J. Phillips and S. Y. Tyree, *J. Am. Chem. Soc.*, **83,** 1806 (1961).
135. D. M. L. Goodgame and F. A. Cotton, *J. Chem. Soc.*, **1961,** 2298.
136. G. A. Rodley, D. M. L. Goodgame and F. A. Cotton, *J. Chem. Soc.*, **1965,** 1499.
137. D. M. L. Goodgame and F. A. Cotton, *J. Chem. Soc.*, **1961,** 3735.
138. B. J. Hathaway and D. G. Holah, *J. Chem. Soc.*, **1964,** 2400.
139. R. Pepinsky, K. Vedam and Y. Okaya, *Phys. Rev.*, **110,** 1309 (1958).
140. F. A. Cotton and R. Francis, *J. Am. Chem. Soc.*, **82,** 2986 (1960).
141. R. S. Drago and D. Meek, *J. Phys. Chem.*, **65,** 1446 (1961).
142. F. A. Cotton, R. Francis and W. D. Horrocks, *J. Phys. Chem.*, **64,** 1534 (1960).
143. D. W. Meek, D. K. Staub and R. S. Drago, *J. Am. Chem. Soc.*, **82,** 6013 (1960).
144. J. G. F. Druce, *J. Chem. Soc.*, **1937,** 1407.
145. D. E. Zvyagintsev and A. Z. Chkhenkilo, *J. Gen. Chem. USSR*, **11,** 791 (1941).
146. L. Naldini, *Gazz. Chim. Ital.*, **90,** 1337 (1960).
147. J. R. Partington and A. L. Whynes, *J. Chem. Soc.*, **1948,** 1952.
148. R. J. Kern. *J. Inorg. Nucl. Chem.*, **25,** 5 (1963).

149. I. G. Druzhinin and K. Rysemendeev, *Izv. Vyssh. Uchebn. Zavedenii, Khim. i Khim. Tekhnol.*, **5,** 7 (1962).
150. R. W. Asmussen, *Z. anorg. allgem. Chem.*, **243,** 127 (1939).
151. R. Chand, G. S. Handard and K. Lal, *J. Indian Chem. Soc.*, **35,** 28 (1958).
152. W. W. Wendlandt and J. P. Smith, *Nature*, **201,** 73 (1964).
153. I. G. Druzhinin and K. Rysmendeev, *Izv. Akad. Nauk Kirg. SSR., Ser. Estestv. i Tekhn. Nauk*, **4,** 21 (1962).
154. W. E. Bull, S. K. Madan and J. E. Willis, *Inorg. Chem.*, **2,** 303 (1963).
155. I. S. Ahuja, *J. Inorg. Nucl. Chem.*, **29,** 2091 (1967).
156. M. Nardelli and I. Chierci, *J. Chem. Soc.*, **1960,** 1952.
157. R. Gomer and C. N. Tyson, *J. Am. Chem. Soc.*, **66,** 1331 (1944).
158. L. A. Ozerov, *Trudy Voronezh. Univ.*, **28,** 24 (1953).
159. J. W. Mellor, *A Comprehensive Treatise on Inorganic and Theoretical Chemistry*, Vol. 12, Longmans, London, 1932, p. 349.
160. H. Benrath, *Z. anorg. allgem. Chem.*, **220,** 145 (1934).
161. A. L. Greenberg and G. H. Walden, *J. Chem. Phys.*, **8,** 645 (1940).
162. A. Zalkin, J. D. Forrester and D. H. Templeton, *Inorg. Chem.*, **3,** 529 (1964).
163. B. K. Vainshtein, *Dokl. Akad. Nauk. SSSR*, **83,** 227 (1952).
164. A. Walkley, *J. Electrochem. Soc.*, **94,** 41 (1948).
165. A. Walkley, *J. Electrochem. Soc.*, **93,** 316 (1948).
166. C. Delain, *Compt. Rend.*, **238,** 1245 (1954).
167. W. H. Baur, *Inorg. Chem.*, **4,** 1840 (1965).
168. E. Konig, Landolt-Bornstein, *Magnetic Properties of Coordination and Organometallic Transition Metal Compounds*, Vol. 2, Springer-Verlag, New York, 1966.
169. H. M. Gijsman, N. J. Poulis and J. van den Handel, *Physica*, **25,** 954 (1959).
170. I. Tsujikawa, *J. Phys. Soc. Japan*, **13,** 315 (1958).
171. B. Bolger, *Bull. Inst. Intern. Froid Annexe*, **1955,** 244.
172. M. A. Lasheen, J. van den Brock and C. J. Gorter, *Physica*, **24,** 1061 (1958).
173. H. Forstat, G. O. Taylor and B. R. King, *J. Phys. Soc. Japan*, **15,** 528 (1960).
174. S. A. Friedberg and J. D. Wasscher, *Physica*, **19,** 1072 (1953).
175. F. W. Lancaster and W. Gordy, *J. Chem. Phys.*, **19,** 1181 (1951).
176. R. D. Spence and V. Nagarajan, *Phys. Rev.*, **149,** 191 (1966).
177. I. Nakagawa and T. Shimanouchi, *Spectrochim. Acta*, **20,** 429 (1964).
178. I. Tsujikawa and E. Kanda, *J. Phys. Soc. Japan*, **18,** 1382 (1963).
179. B. Morosin and E. J. Graebner, *J. Chem. Phys.*, **42,** 898 (1965).
180. K. E. Lawson, *J. Chem. Phys.*, **44,** 4159 (1966).
181. D. H. Goode, *J. Chem. Phys.*, **43,** 2830 (1965).
182. A. Narath, *Phys. Rev.*, **136,** 766 (1964).
183. W. B. Hadley and J. W. Stout, *J. Chem. Phys.*, **39,** 2205 (1963).
184. G. Devoto and A. Guzzi, *Gazz. Chim. Ital.*, **59,** 591 (1929).
185. A. Ferrari and F. Giorgi, *Atti Acad. Lincei*, **9,** 1134 (1929).
186. W. J. de Haas, B. H. Schultz and J. Koolhaas, *Physica*, **7,** 57 (1940).
187. J. W. Stout, W. B. Hadley and C. L. Brandt, *U.S. Dept. Com., Office Tech. Serv., PB Report* 143, 420 (1958).
188. W. C. Koehler, E. O. Wollan and M. K. Wilkinson, *Phys. Rev.*, **110,** 638 (1958).
189. R. Stevenson, *Can. J. Phys.*, **40,** 1385 (1962).

190. L. G. van Uitert, H. J. Williams, R. C. Sherwood and J. J. Rubin, *J. Appl. Phys.*, **36,** 1029 (1965).
191. H. Grossmann, *Ber.*, **37,** 559 (1904).
192. P. J. Crowley and H. M. Haendler, *Inorg. Chem.*, **1,** 904 (1962).
193. K. Issleib and A. Krebich, *Z. anorg. allgem. Chem.*, **313,** 338 (1961).
194. R. S. Nyholm and G. J. Sutton, *J. Chem. Soc.*, **1958,** 564.
195. P. Paoletti and M. Ciampolini, *Inorg. Chem.*, **6,** 64 (1967).
196. L. Sacconi, I. Bertini and R. Morassi, *Inorg. Chem.*, **6,** 1548 (1967).
197. J. W. Mellor, *A Comprehensive Treatise on Inorganic and Theoretical Chemistry*, Vol. 12, Longmans, London, 1932, p. 381.
198. D. G. Kapadnis and R. Hartmans, *Physica*, **22,** 181 (1956).
199. B. Morosin, *J. Chem. Phys.*, **47,** 417 (1967).
200. J. W. Mellor, *A Comprehensive Treatise on Inorganic and Theoretical Chemistry*, Vol. 12, Longmans, London, 1932, p. 384.
201. M. Chaigneau, *Bull. Soc. Chim. France*, **1957,** 886.
202. M. Chaigneau and M. Chastagnier, *Bull. Soc. Chim. France*, **1958,** 1192.
203. A Ferrari and F. Giorgi, *Atti Acad. Lincei*, **10,** 522 (1929).
204. J. W. Cable, M. K. Wilkinson, E. O. Wollan and W. C. Koehler, *Phys. Rev.*, **125,** 1860 (1962).
205. H. Rheinboldt, A. Luyken and H. Schmittmann, *J. Prakt. Chem.*, **149,** 30 (1937).
206. G. J. Sutton, *Australian J. Chem.*, **19,** 2059 (1966).
207. E. Huss and W. Klemm, *Z. anorg. allgem. Chem.*, **262,** 25 (1950).
208. R. Hoppe, *J. Inorg. Nucl. Chem.*, **8,** 435 (1958).
209. R. Hoppe and K. Blinne, *Z. anorg. allgem. Chem.*, **291,** 269 (1957).
210. R. Hoppe, *Chem. Cood. Cmpds., Symp., Rome*, **1957,** 437.
211. R. Hoppe, W. Liebe and W. Dahne, *Z. anorg. allgem. Chem.*, **307,** 276 (1961).
212. R. Hoppe, *Rec. Trav. Chim.*, **75,** 569 (1956).
213. H. Bode, H. Jenssen and F. Bandte, *Angew. Chem.*, **65,** 304 (1953).
214. W. G. Palmer, *Experimental Inorganic Chemistry*, Cambridge, London, 1962.
215. B. Cox, *J. Chem. Soc.*, **1954,** 3251.
216. B. Cox and A. G. Sharpe, *J. Chem. Soc.*, **1954,** 1798.
217. P. Bouy, *Ann. Chim. (Paris)*, **4,** 853 (1959).
218. D. H. Brown, K. R. Dixon, R. D. W. Kemmitt and D. W. A. Sharp, *J. Chem. Soc.*, **1965,** 1559.
219. H. Bode and W. Wendt, *Z. anorg. allgem. Chem.*, **269,** 165 (1952).
220. J. T. Grey, *J. Am. Chem. Soc.*, **68,** 605 (1946).
221. R. D. Peacock and D. W. A. Sharp, *J. Chem. Soc.*, **1959,** 2762.
222. R. D. Peacock, *J. Chem. Soc.*, **1953,** 3617.
223. R. D. Peacock, *J. Chem. Soc.*, **1957,** 4684.
224. I. G. Ryss and B. S. Vitokhnovskaya, *Zh. Neorg. Khim.*, **3,** 1185 (1958).
225. R. S. Nyholm and A. G. Sharpe, *J. Chem. Soc.*, **1952,** 3579.
226. R. Dingle, *Inorg. Chem.*, **4,** 1287 (1965).
227. J. W. Mellor, *A Comprehensive Treatise on Inorganic and Theoretical Chemistry*, Vol. 12, Longmans, London, 1932, p. 345.
228. W. E. Hatfield, R. C. Fay, C. E. Pfluger and T. S. Piper, *J. Am. Chem. Soc.*, **85,** 265 (1963).

229. N. S. Gill, *Chem. Ind. (London)*, **1961,** 989.
230. I. N. Belyaev and O. Y. Revina, *Russ. J. Inorg. Chem.*, **11,** 772 (1966).
231. J. C. Cousseins, *Rev. Chim. Minerale*, **1,** 573 (1964).
232. A. Chretien and J. C. Cousseins, *Compt. Rend.*, **259,** 4696 (1964).
233. J. C. Cousseins and M. Samouel, *Compt. Rend.*, **265C,** 1121 (1967).
234. D. J. Breed, *Phys. Letters*, **23,** 181 (1966).
235. I. N. Belyaev and O. Y. Revina, *Fiz-Khim. Analiz. Solvkh. Sistem.*, **1962,** 77.
236. A. Zalkin, K. Lee and D. H. Templeton, *J. Chem. Phys.*, **37,** 697 (1962).
237. K. Hirakawa, K. Hirakawa and T. Hashimoto, *J. Phys. Soc. Japan*, **15,** 2063 (1960).
238. D. J. Machin, R. L. Martin and R. S. Nyholm, *J. Chem. Soc.*, **1963,** 1490.
239. D. J. Machin and R. S. Nyholm, *J. Chem. Soc.*, **1963,** 1500.
240. K. Knox, *Acta Cryst.*, **14,** 583 (1961).
241. S. Ogawa, *J. Phys. Soc. Japan*, **14,** 1115 (1959).
242. D. S. Crocket and R. A. Grossman, *Inorg. Chem.*, **3,** 644 (1964).
243. D. S. Crocket and H. M. Haendler, *J. Am. Chem. Soc.*, **82,** 4158 (1960).
244. Y. P. Simonov, L. R. Batsonova and L. M. Kovba, *Zh. Neorg. Khim.*, **2,** 2410 (1957).
245. S. A. Kezhaev, A. G. Tutov and V. A. Bokov, *Soviet Phys.—Solid state*, **7,** 2325 (1966).
246. D. E. Eastman and M. W. Shafer, *J. Appl. Phys.*, **38,** 1274 (1967).
247. A. Okazaki, Y. Suemune and T. Fuchikami, *J. Phys. Soc. Japan*, **14,** 1823 (1959).
248. R. L. Martin, R. S. Nyholm and N. C. Stephenson, *Chem. Ind. (London)*, **1956,** 83.
249. A. Okazaki and Y. Suemune, *J. Phys. Soc. Japan*, **16,** 671 (1961).
250. C. Deenada, H. V. Keer, R. V. G. Rao and A. B. Biswas, *Brit. J. Appl. Phys.*, **17,** 1401 (1966).
251. K. Lee, A. M. Portis and G. L. Witt, *Phys. Rev.*, **132,** 144 (1963).
252. V. L. Moruzzi and D. T. Teaney, *Bull. Am. Phys. Soc.*, **8,** 383 (1963).
253. M. B. Walker and R. W. H. Stevenson, *Proc. Phys. Soc.*, **87,** 35 (1966).
254. Y. Suemune and H. Ikawa, *J. Phys. Soc. Japan*, **19,** 1686 (1964).
255. O. Beckman, I. Olovsson and K. Knox, *Acta Cryst.*, **13,** 506 (1960).
256. O. Beckman and K. Knox, *Phys. Rev.*, **121,** 376 (1961).
257. A. J. Heeger, O. Beckman and A. M. Portis, *Phys. Rev.*, **123,** 1652 (1961).
258. V. Scatturin, L. Corliss, N. Elliot and J. Hastings, *Acta Cryst.*, **14,** 19 (1961).
259. S. J. Pickart, H. A. Alperin and R. Nathans, *J. Phys. (Paris)*, **25,** 565 (1964).
260. D. T. Teaney, V. L. Moruzzi and B. E. Argyle, *J. Appl. Phys.*, **37,** 1122 (1966).
261. A. Mehra and P. Venkateswarlu, *J. Chem. Phys.*, **47,** 2334 (1967).
262. A. V. Antomov, A. I. Belyaeva and V. V. Eremenko, *Soviet Phys.—Solid State*, **8,** 2718 (1967).
263. J. Ferguson, H. J. Guggenheim and Y. Tanabe, *J. Appl. Phys.*, **36,** 1046 (1965).
264. J. Ferguson, H. J. Guggenheim and Y. Tanabe, *J. Phys. Soc. Japan*, **21,** 692 (1966).
265. K. Aoyagi, *J. Phys. Soc. Japan*, **22,** 1516 (1967).

266. B. F. Markov and R. V. Chernov, *Ukr. Khim. Zhur.*, **24,** 139 (1958).
267. S. J. Jensen, *Acta Chem. Scand.*, **18,** 2085 (1964).
268. J. W. Mellor, *A Comprehensive Treatise on Inorganic and Theoretical Chemistry*, Vol. 12, Longmans, London, 1932, p. 363.
269. H. B. Jonassen, L. J. Theriot, E. A. Bourdreaux and W. A. Ayres, *J. Inorg. Nucl. Chem.*, **26,** 595 (1964).
270. P. Paoletti and A. Vacca, *Trans. Faraday Soc.*, **60,** 50 (1964).
271. D. V. R. Rao and S. K. Naik, *Current Sci.* (*India*), **33,** 109 (1964).
272. L. Naldini and A. Sacco, *Gazz. Chim. Ital.*, **89,** 2258 (1959).
273. A. B. Blake and F. A. Cotton, *Inorg. Chem.*, **3,** 5 (1964).
274. E. A. Bourdreaux, H. B. Jonassen and L. J. Theriot, *J. Am. Chem. Soc.*, **85,** 2039 (1963).
275. N. S. Gill and R. S. Nyholm, *J. Chem. Soc.*, **1959,** 3997.
276. P. Pauling, *Inorg. Chem.*, **5,** 1498 (1966).
277. C. Furlani and A. Furlani, *J. Inorg. Nucl. Chem.*, **19,** 51 (1961).
278. H. P. de la Garanderie, *Compt. Rend.*, **255,** 2585 (1962).
279. D. M. Adams, J. Chatt, J. M. Davidson and J. Gerratt, *J. Chem. Soc.*, **1963,** 2189.
280. P. Ehrlich, F. W. Koknat and H. J. Seifert, *Z. anorg. allgem. Chem.*, **341,** 281 (1965).
281. Y. Okaya, R. Pepinsky, Y. Takeuchi, H. Kuroya, A. Shimada, P. Galletelli, N. Stemple and A. Beevers, *Acta Cryst.*, **10,** 798 (1957).
282. R. J. H. Clark and T. M. Dunn, *J. Chem. Soc.*, **1963,** 1198.
283. A. Sabatini and L. Sacconi, *J. Am. Chem. Soc.*, **86,** 17 (1964).
284. B. D. Bird and P. Day, *Chem. Commun.*, **1967,** 741.
285. F. A. Cotton, D. M. L. Goodgame and M. Goodgame, *J. Am. Chem. Soc.*, **84,** 167 (1962).
286. K. E. Lawson, *J. Chem. Phys.*, **47,** 3627 (1967).
287. H. Forstat, N. D. Love and J. N. McElearney, *Phys. Letters*, **25A,** 253 (1967).
288. W. J. Croft, M Kestigian and F. D. Leipzig, *Inorg. Chem.*, **4,** 423 (1965).
289. M. Kestigian, W. J. Croft, F. D. Leipzig and J. R. Carter, *J. Chem. Eng. Data*, **12,** 97 (1967).
290. B. Morosin and E. J. Graebner, *Acta Cryst.*, **23,** 766 (1967).
291. S. A. Shchukarev, I. V. Vasil'kova and G. M. Barvinok, *Vestn. Leningrad Univ., Ser. Fiz. i Khim.*, **20,** 145 (1965).
292. R. W. Kedzie, J. R. Shane, M. Kestigian and W. J. Croft, *J. Appl. Phys.*, **36,** 1195 (1965).
293. S. J. Jensen, P. Andersen and S. E. Rasmussen, *Acta Chem. Scand.*, **16,** 1890 (1962).
294. S. J. Jensen, *Acta Chem. Scand.*, **21,** 889 (1967).
295. J. Suss, *Z. Krist.*, **51,** 248 (1912).
296. P. Paoletti, *Trans. Faraday Soc.*, **61,** 219 (1965).
297. T. S. Briggs, *J. Inorg. Nucl. Chem.*, **30,** 2866 (1968).
298. G. Parette and K. U. Deniz, *J. Appl. Phys.*, **39,** 1232 (1968).
299. Y. Shapira and J. Zak, *Phys. Rev.*, **170,** 503 (1968).
300. N. F. Kharchenko and V. V. Eremenko, *Soviet Phys.–Solid State*, **10,** 1112 (1968).
301. Y. Tanabe and K. I. Gondaira, *J. Phys. Soc. Japan*, **22,** 573 (1967).

302. R. L. Greene, D. D. Sell, R. S. Feigelson, G. F. Imbusch and H. J. Guggenheim, *Phys. Rev.*, **171**, 600 (1968).
303. R. S. Meltzer, M. Y. Chen, D. S. McClure and M. Lowe-Pariseau, *Phys. Rev. Letters*, **21**, 913 (1968).
304. R. S. Meltzer and L. L. Lohr, *J. Chem. Phys.*, **49**, 541 (1968).
305. J. J. Foster and N. S. Gill, *J. Chem. Soc.*, **1968**, A, 2625.
306. P. L. Richards, *J. Appl. Phys.*, **38**, 1500 (1967).
307. A. Mehra, *J. Chem. Phys.*, **48**, 1871 (1968).
308. S. Guru, K. C. Dash and D. V. R. Rao, *Indian J. Chem.*, **6**, 161 (1968).
309. C. C. Houk and K. Emerson, *J. Inorg. Nucl. Chem.*, **30**, 1493 (1968).
310. H. S. Preston, C. H. L. Kennard and R. A. Plowman, *J. Inorg. Nucl. Chem.*, **30**, 1463 (1968).
311. J. S. Judge, W. M. Reiff, G. M. Intille, P. Ballway and W. A. Baker, *J. Inorg. Nucl. Chem.*, **29**, 1711 (1967).
312. F. Lions, I. G. Dance and J. Lewis, *J. Chem. Soc.*, **1967**, A, 565.
313. D. M. Bowers and A. I. Popov, *Inorg. Chem.*, **7**, 1594 (1968).
314. D. P. Madden and S. M. Nelson, *J. Chem. Soc.*, **1968**, A, 2342.
315. M. C. Feller and R. Robson, *Australian J. Chem.*, **21**, 2919 (1968).
316. G. W. A. Fowles, D. A. Rice and R. A. Walton, *J. Chem. Soc.*, **1968**, A, 1842.
317. A. W. McLellan and G. A. Melson, *J. Chem. Soc.*, **1967**, A, 137.
318. J. G. H. du Preez, W. J. A. Steyn and A. J. Basson, *J. South African Chem. Institute*, **21**, 8 (1968).
319. K. C. Dash and D. V. R. Rao, *Z. anorg. allgem. Chem.*, **350**, 207 (1967).
320. D. G. Brewer, P. T. T. Wong and M. C. Sears, *Can. J. Chem.*, **46**, 3137 (1968).
321. N. C. Collins and W. K. Glass, *Chem. Commun.*, **1967**, 211.
322. M. F. Farona and G. R. Tompkin, *Spectrochim. Acta*, **24A**, 788 (1968).
323. A. Braibanti, F. Dallavalle, M. A. Pellinghelli and E. Leporte, *Inorg. Chem.*, **7**, 1430 (1968).
324. D. M. L. Goodgame, M. Goodgame, P. J. Hayward and G. W. Rayner-Canham, *Inorg. Chem.*, **7**, 2447 (1968).
325. R. Gonano, F. Ono and K. Hirooka, *J. Appl. Phys.*, **39**, 710 (1968).
326. P. Lumme and M. T. Raivio, *Suomen Kemiistilehti.*, **41B**, 194 (1968).
327. M. Di Vaira and P. L. Orioli, *Acta Cryst.*, **24B**, 1269 (1968).
328. V. A. Schmidt and S. A. Friedberg, *J. Appl. Phys.*, **38**, 5319 (1967).
329. P. C. Moews, *Inorg. Chem.*, **5**, 5 (1966).
330. H. D. Hardt and M. Fleischer, *Z. anorg. allgem. Chem.*, **357**, 113 (1968).
331. D. M. Adams and D. M. Morris, *J. Chem. Soc.*, **1968**, A, 694.
332. N. Elliot, *J. Chem. Phys.*, **46**, 1006 (1967).
333. B. Jezowska-Trzebiatowska, S. Wajda, M. Baluka, L. Natkaniec and W. Wojcieckowski, *Inorg. Chim. Acta*, **1**, 205 (1967).
334. T. S. Davis, J. P. Fackler and M. J. Weeks, *Inorg. Chem.*, **7**, 1994 (1968).
335. B. O. Loopstra, B. van Laar and D. J. Breed, *Phys. Letters*, **26A**, 526 (1968).
336. J. H. P. Colpa, K. H. Chang and F. P. D. Obbema, *Phys. Letters*, **27A**, 380 (1968).
337. B. Golding, *Phys. Rev. Letters*, **20**, 5 (1968).

338. R. Nathans, F. Menzinger and S. S. Pickart, *J. Appl. Phys.*, **39,** 1237 (1968).
339. G. Gorodetsky, *Solid State Commun.*, **6,** 159 (1968).
340. S. L. Roscoe and H. M. Haendler, *Inorg. Chim. Acta*, **1,** 73 (1967).
341. G. A. Smolenskii, M. P. Petrov, V. V. Moskalev, V. S. Lvov, V. S. Kasperovich and E. V. Zhunova, *Soviet Phys.–Solid State*, **10,** 1040 (1968).
342. V. V. Eremenko, Y. A. Popkov, V. P. Novikov and A. I. Belyaeva, *Soviet Physics–JETP*, **25,** 297 (1967).
343. R. J. Elliot, M. F. Thorpe, G. F. Imbusch, R. J. Loudon and J. B. Parkinson, *Phys. Rev. Letters*, **21**, 147 (1968).
344. P. A. Fleury, *Phys. Rev. Letters*, **21,** 150 (1968).
345. G. F. Imbusch and H. J. Guggenheim, *Phys. Letters*, **26A,** 625 (1968).
346. K. S. Aleksandrov, L. M. Reshchikova and B. V. Beznosikov, *Soviet Phys.–Solid State*, **8,** 2904 (1967).
347. B. V. Beznosikov and N. V. Beznosikova, *Soviet Phys.–Crystallography*, **13,** 158 (1968).
348. N. S. Gill and F. B. Taylor, *Inorg. Syn.*, **9,** 136 (1967).
349. A. Furuhashi, K. Watanuki and A. Ouchi, *Bull. Chem. Soc. Japan*, **41,** 110 (1968).
350. G. A. Heath, R. L. Martin and I. M. Stewart, *Australian J. Chem.*, **22,** 83 (1969).
351. A. Furuhashi, T. Takeuchi and A. Ouchi, *Bull. Chem. Soc. Japan*, **41,** 2049 (1968).
352. H. G. M. Edwards, M. J. Ware and L. A. Woodward, *Chem. Commun.*, **1968,** 540.
353. J. S. Avery, C. D. Burbridge and D. M. L. Goodgame, *Spectrochim. Acta*, **24A,** 1721 (1968).
354. S. J. Jensen, *Acta Chem. Scand.*, **22,** 641 (1968).
355. S. J. Jensen, *Acta Chem. Scand.*, **22,** 647 (1968).
356. C. E. Taylor and J. A. Cowen, *J. Appl. Phys.*, **39,** 498 (1968).
357. H. Forstat, J. N. McElearney and B. T. Bailey, *Phys. Letters*, **27A,** 549 (1968).
358. A. Le Paillier-Mall, *Compt. Rend.*, **265B,** 657 (1967).

Chapter 6

Iron

With the exception of the octahedral iron(IV) complexes formed with *o*-phenylenebis(dimethylarsine), the halogen compounds of iron are restricted to oxidation states III and II, in both binary halides and complex halides.

HALIDES AND OXIDE HALIDES

The known halides and oxide halides of iron are given in Table 6.1.

TABLE 6.1
Halides and Oxide Halides of Iron

Oxidation state	Fluoride	Chloride	Bromide	Iodide
III	FeF_3 $FeOF$	$FeCl_3$ $FeOCl$	$FeBr_3$	
II	FeF_2	$FeCl_2$	$FeBr_2$	FeI_2

As expected, the stability of oxidation state III decreases with increasing atomic number of the halogen; iron triiodide can only be formed in the vapour phase.

Oxidation State IV

Hazeldean, Nyholm and Parish[1] have prepared octahedral iron(IV) complexes of the type $[FeX_2D_2][BF_4]_2$, where X = Cl or Br and D is *o*-phenylenebis(dimethylarsine), by oxidation of the corresponding iron(III) compounds with nitric acid. The dark green to black solids were characterized by conductance, magnetic, infrared and electronic spectral studies.

Oxidation State III

Iron trifluoride. Iron trifluoride may be prepared by reaction of iron trichloride with fluorine, bromine trifluoride, chlorine trifluoride or anhydrous hydrogen fluoride[2–4]. It has also been prepared by the fluorination of iron difluoride at 500° in a flow system[5] and by the interaction[6] of iron(III) sulphide and sulphur tetrafluoride at 5–50 atmospheres pressure and 150–350°.

Iron trifluoride is a white crystalline solid which is sparingly soluble in both water and anhydrous hydrogen fluoride[7]. Wanklyn[8] reports that iron trifluoride is thermally unstable, decomposing at 1000° to iron difluoride. Starting with hydrated iron trifluoride, Wanklyn found that at 650° he obtained a mixture of iron trifluoride and Fe_3O_4, as shown by X-ray powder diffraction data. At 1000° iron difluoride was obtained. Whether in fact the difluoride is formed by the thermal decomposition of iron trifluoride, or via hydrolysis reactions is not clear, but it is known[9] that hydrolytic reactions account for the 'thermal decomposition' of cobalt trifluoride to the difluoride. It seems probable that hydrolytic reactions account for the formation of iron difluoride from iron trifluoride.

Zmbov and Margrave[10] have shown by mass-spectrometric techniques that iron trifluoride exists in the vapour in both monomeric and dimeric forms. The vapour pressure of iron trifluoride over the temperature range 507–657° is given by the expression[10]

$$\log p_{\text{atm}} = 8.18 - 11{,}520/T$$

The heat of sublimation and formation of iron trifluoride are 60.8 and −235 kcal mole^{-1} respectively.

There is some confusion over the structure of iron trifluoride. Hepworth and coworkers[5] report a rhombohedral unit cell with $a = 5.362$ Å and $\alpha = 57.99°$. They suggested that the FeF_6 octahedra had the vanadium trifluoride-type packing arrangement On the other hand de Pape[4] reported a rhombohedral unit cell with $a = 3.734$ and $\alpha = 88.22°$ with the rhenium trioxide-type packing. The iron-fluorine stretching frequencies appear at[3] 590 and 530 cm^{-1}.

Rather surprisingly there has been little investigation of the magnetic properties of iron trifluoride. The compound is antiferromagnetic[11] and its Neel temperature is 394°K. There is a maximum in the specific heat[12] at 367°K.

The absorption spectrum of iron trifluoride shows only one *d-d* band[13] at 11,800 cm^{-1} which has been assigned to the $^4T_{1g}(G) \leftarrow {}^6A_{1g}$ transition.

Iron trifluoride trihydrate. Iron trifluoride trihydrate exists in two modifications[14]. Both are made by dissolving freshly precipitated iron(III) oxide hydrate in 40% hydrofluoric acid, the modification formed depending on the temperature of the solution. High temperatures favour the β-form. White α-$FeF_3 \cdot 3H_2O$ is isostructural with α-$AlF_3 \cdot 3H_2O$ and the unit cell has $a = 9.49$ and $c = 9.58$ Å[15]. Pink β-$FeF_3 \cdot 3H_2O$ is tetragonal[16] with $a = 7.846$ and $c = 3.877$ Å. There are two molecules in the cell of space group $P4/n$. Each iron atom in this modification is surrounded by an almost perfect octahedron of ligands. Adjacent octahedra share corners in the direction of the c axis and fluorine atoms are assumed to form these bridges. The remaining four ligands of each octahedron are two fluorines and two water molecules which occupy the other four coordination sites in a random manner. The remaining water molecule is simply entrapped in the lattice and constitutes water of hydration.

Iron oxide fluoride. Hagenmuller and coworkers[17] have prepared iron oxide fluoride by heating a mixture of iron trifluoride and iron(III) oxide at 800–900° in an atmosphere of oxygen. Chamberland and Sleight[18] prepared their sample of this compound by heating an intimate mixture of iron trifluoride and iron(III) oxide at 995–1200° in a tetrahedral anvil at 60–65 Kbar pressure.

Iron oxide fluoride has a rutile structure and Hagenmuller and coworkers report $a = 4.647$ and $c = 3.048$ Å. However Chamberland and Sleight report that slight variations in the unit-cell parameters result from small deviations from stoichiometry.

Iron trichloride. The methods of preparation of iron trichloride are given in Table 6.2.

TABLE 6.2

Preparation of Iron Trichloride

Method	Conditions	References
Fe and Cl_2	Flow system, 300–350°	19–22
Fe_2O_3 and CCl_4	Flow system, 500°	23
FeS_2 and CCl_4	Bomb, 400°	24
Fe_2O_3 and C_6Cl_6	Reflux	25
Fe_2O_3, FeS or FeS_2 and C_5Cl_8	Reflux	26
Fe_2O_3 and HCl	Flow system, 300–1000°	27–30
$FeCl_3 \cdot 6H_2O$ and $SOCl_2$	Reflux	20, 31, 32

It is not advisable to dehydrate iron trichloride hexahydrate by heating since partial hydrolysis may occur. In all cases when a product of high purity is required it is advisable to carry out a final sublimation in a stream of chlorine. Iron trichloride must not be sublimed in a vacuum at high temperatures, since it is thermally unstable under these conditions.

Iron trichloride is a dark greenish compound which is very hygroscopic. It dissolves readily in water from which a number of hydrates may be isolated. The actual product obtained depends on concentration and temperature[33,34]. Iron trichloride is also soluble in dry organic solvents such as acetone and methanol. It has been used to chlorinate organic compounds[35]. Iron trichloride and iron dichloride form a eutectic containing 13.4 mole per cent iron dichloride[22] and melting at 297.5°.

Iron trichloride sublimes very readily, vaporizing principally as the dimer[36]. As the temperature increases, the proportion of dimer relative to monomer decreases. The sublimation pressure of iron trichloride has been measured several times and the equations expressing the experimental results are

$$\log p_{\mathrm{mm}} = 14.51 - 6902/T \quad (120\text{–}150°) \quad \text{(reference 37)}$$

$$\log p_{\mathrm{mm}} = 15.11 - 7142/T \quad (160\text{–}304°) \quad \text{(reference 38)}$$

$$\log p_{\mathrm{mm}} = 13.74 - 6499/T \quad (223\text{–}304°) \quad \text{(reference 39)}$$

Iron trichloride is thermally unstable[29,38,40–42] decomposing noticeably above about 200°, but quite rapidly above 300°. The equilibrium dissociation pressure of chlorine over solid iron trichloride in the temperature range 160–210° is given by[22]

$$\log p_{\mathrm{mm}} = 11.33 - 5670/T$$

while for the temperature range 220–400° it is given by[38]

$$\log p_{\mathrm{mm}} = 11.085 - 5565/T$$

Thermodynamic data derived from these and other measurements are given in Table 6.3.

TABLE 6.3
Thermodynamic Properties of Iron Trichloride

Property	Value	References
M.p.	303 ± 3°	19, 21, 40, 42, 43
B.p.	315°	44, 45
	319°	19
	324°	42
$\Delta H^\circ_{\mathrm{form}}$ (kcal mole^{-1})	−95.7	46
	−93.4	47
ΔH_{sub} (kcal mole^{-1})	16.35	38
ΔS_{sub} (cal deg^{-1} mole^{-1})	27.99	38

Heat capacities, entropies and heat contents have been calculated[19] at various temperatures between 50° and 577°K.

Iron trichloride has a hexagonal lattice[20,48–51]. Cable and coworkers[51] give $a = 6.065$ and $c = 17.44$ Å. The iron-chlorine distance has been reported[49] to be 2.48 Å. Blairs and Shelton[20] have studied the variation of the lattice parameters with temperature, and up to 275° there was no phase transformation. At this temperature $a = 6.097$ and $c = 17.64$ Å. The structure consists of $FeCl_6$ octahedra sharing all corners. Samples of iron trichloride prepared by dehydration of iron trichloride hexahydrate using thionyl chloride show a variety of stacking faults.

Electron-diffraction studies[52–54] have confirmed the dimeric nature of iron trichloride vapour. Two iron atoms and two chlorine atoms form a planar ring and the two other pairs of chlorine atoms are in a plane at right angles to the ring. The terminal iron-chlorine distance is 2.11 Å while the bridging iron-chlorine bond is 2.28 Å.

The infrared spectrum of iron trichloride has been recorded by Wilmshurst[55]. The assignments, made on the basis of a dimeric molecule of V_h symmetry were as follows: ν_8 504, ν_9 486, ν_{13} 268 and ν_{16} 397 cm^{-1}.

Iron trichloride obeys the Curie-Weiss law from approximately 20°K to room temperature[56], with a magnetic moment of 5.73 BM and a Weiss constant of 11.5°. Neutron-diffraction measurements[51] indicate that iron trichloride becomes antiferromagnetic at 15°K with a complex magnetic structure. The diffuse-reflectance spectrum of iron trichloride has been recorded by Clark[13] who observed one band at 11,800 cm^{-1} assigned to the $^4T_{1g}(G) \leftarrow {}^6A_{1g}$ transition.

Iron trichloride reacts with tris(dialkyldithiocarbamato)iron(III) to give complexes of the type $FeCl(dtc)_2$ where dtc is the dialkyldithiocarbamato group[57]. These complexes were first prepared by Martin and coworkers[58,59] by the interaction of tris(dialkyldithiocarbamato)iron(III) and hydrochloric acid in benzene solution. Other methods of preparation include the interaction of iron trichloride and tetraalkylthiuram disulphide $[R_2N(C{=}S)S_2(C{=}S)NR_2]$ in chloroform, and the reaction between sodium dialkyldithiocarbamate and excess iron trichloride in acetone or ethanol[57]. The complexes are dark coloured crystalline solids.

A single-crystal X-ray diffraction study of monochlorobis(diethyldithiocarbamato)iron(III) by Hoskins, Martin and White[58] has shown that the compound exhibits a square-pyramidal structure. The complex is monoclinic with $a = 16.43$, $b = 9.42$, $c = 12.85$ Å and $\beta = 120.5°$. There are four molecules in the unit cell of space group $P2_1/c$. The structure consists of discrete molecules which are separated from their neighbours by the normal Van Der Waal distances. The iron atom is covalently bound to the chlorine and four sulphur atoms at the corners of a slightly distorted square pyramid as shown in Figure 6.1. The iron-chlorine and the average iron-sulphur bond distances are 2.27 and 2.32 Å respectively. The four sulphur atoms are coplanar with the iron atom 0.63 Å above this plane.

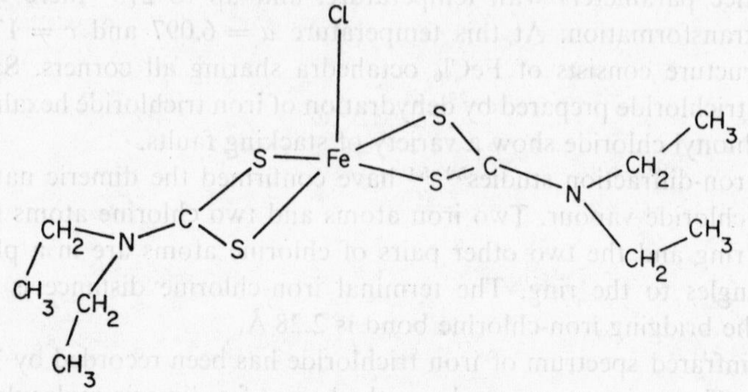

Figure 6.1 The structure of $[FeCl(S_2CN(C_2H_5)_2)_2]$. (Reproduced by permission from R. L. Martin and A. H. White, *Inorg. Chem.*, **6**, 712 (1967))

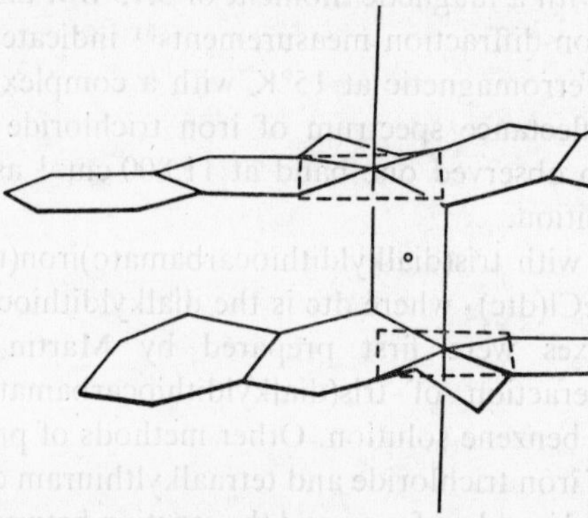

Figure 6.2 The (FeClsalen) dimer. (Reproduced by permission from M. Gerloch, J. Lewis, F. E. Mabbs and A. Richards, *Nature*, **212**, 809 (1966))

The iron-chlorine and iron-sulphur stretching frequencies occur at 309 and 353 cm^{-1} respectively. Martin and White[59] have examined the magnetic properties of a number of these dialkyldithiocarbamato complexes and have found that they all have three unpaired electrons. The spectroscopic splitting factors for the diethyl compound are $g_{\parallel} = 2.10$ and $g_{\perp} = 4.08$. The optical absorption spectra are consistent with the magnetic ground state. Wickman and coworkers[57,60] have confirmed the magnetic properties of these complexes by electron spin resonance, Mossbauer and low temperature susceptibility measurements.

Iron trichloride reacts[61] with the Schiff base *N,N'*-bis(salicylideneimine) [salen] to give dimeric $[FeCl(salen)]_2$. This compound has a monoclinic unit cell with $a = 11.37$, $b = 6.91$, $c = 19.6$ Å and $\beta = 90.5°$. The space group of the unit cell is $P2_1/c$. The structure is shown in Figure 6.2. If this

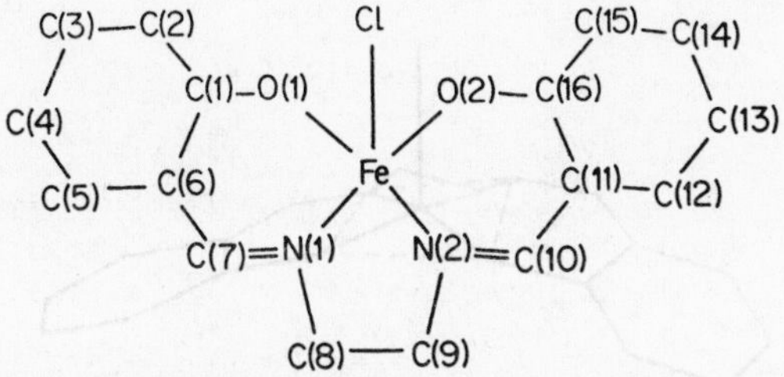

Figure 6.3 The numbering of the atoms in monomeric (FeClsalen). (Reproduced by permission from M. Gerloch and F. E. Mabbs, *J. Chem. Soc.*, **1967**, A, 1598)

TABLE 6.4

Internuclear Dimensions of [FeCl(salen)]

Distance	Value	Angle	Value
Fe—Cl	2.238 Å	Cl—Fe—O(1)	104.9°
Fe—O(1)	1.879 Å	Cl—Fe—O(2)	109.8°
Fe—O(2)	1.885 Å	Cl—Fe—N(1)	109.4°
Fe—N(1)	2.099 Å	O(2)—Fe—N(2)	97.0°
Fe—N(2)	2.064 Å		

compound is recrystallized from nitromethane another complex of formula $[FeCl(salen)] \cdot xCH_3NO_2$ may be isolated[61,62]. A single-crystal study by Gerloch and Mabbs[62] has shown this compound to be a five-coordinate square-pyramidal complex in which the nitromethane molecules pack together in the space group $Pn2_1/a$, which appears to be a superlattice resulting from a cell alternation of left and right handed [FeCl(salen)] molecules in the space group $P2_12_12_1$. The complex has an orthorhombic unit cell with $a = 13.00$, $b = 28.81$ and $c = 9.78$ Å. The important internuclear dimensions, using the notation shown in Figure 6.3, are given in Table 6.4.

The coordination about the iron atom is essentially square pyramidal with the iron atom approximately 0.46 Å above the best plane defined by the Schiff base donor atoms. The four donor atoms do not define a perfect plane but show a small tetrahedral distortion. The salen groups are bent and twisted as shown in Figure 6.4. The iron atom is not equally distant from the planes defined by the 'sal' ligand, being 0.45 Å from that of 'sal' (1) and approximately 0.12 Å from that of 'sal' (2) as shown in Figure 6.5.

Gerloch and coworkers[63] have prepared a number of complexes of

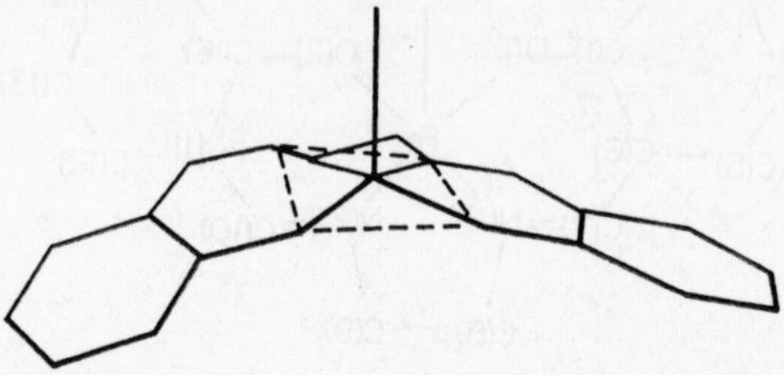

Figure 6.4 The structure of monomeric (FeClsalen). (Reproduced by permission from M. Gerloch and F. E. Mabbs, *J. Chem. Soc.*, **1967**, A, 1598)

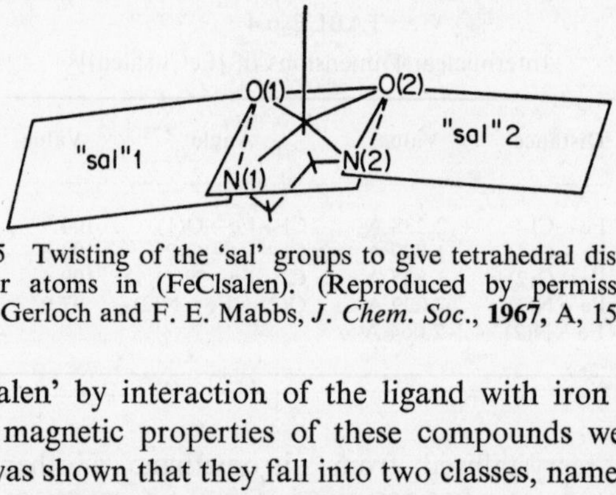

Figure 6.5 Twisting of the 'sal' groups to give tetrahedral distortion of the donor atoms in (FeClsalen). (Reproduced by permission from M. Gerloch and F. E. Mabbs, *J. Chem. Soc.*, **1967**, A, 1598)

substituted 'salen' by interaction of the ligand with iron trichloride in alcohol. The magnetic properties of these compounds were studied in detail and it was shown that they fall into two classes, namely monomeric and dimeric. Magnetic moments fall in the range 5.68–6.00 BM with small Weiss constants for the monomeric compounds and in the range 5.14–5.37 BM for the dimeric species.

Iron trichloride reacts with a large variety of other donor ligands including phosphines, alcohols, etc., and properties of some of these adducts are given in Table 6.5.

Iron trichloride hexahydrate. This hydrate is the one normally obtained by evaporation of a solution of iron trichloride in dilute hydrochloric acid[87]. It has a monoclinic lattice[87] with $a = 11.89$, $b = 7.05$, $c = 5.99$ Å and $\beta = 100.5°$. The structure contains *trans* $[FeCl_2(H_2O)_4]$ octahedral units, chloride ions and water molecules. The iron-chlorine distance is 2.30 Å and the iron-oxygen distance is 2.07 Å. The octahedra are slightly distorted, the chlorine-iron-oxygen angles being 91.2° and 88.8°. The arrangement of the complex cation, the free chloride ion and the water molecules in the lattice, indicates extensive hydrogen bonding in addition

TABLE 6.5

Adducts of the Type $FeCl_3L_n$

Compound and Properties	References
L = PCl_5, $n = 1$, green, m.p. = 332°, ionic in nitrobenzene, possibly $[PCl_4^+][FeCl_4^-]$	64–66
L = $POCl_3$, $n = 1$, brown, m.p. = 119°	67–69
L = S_2Cl_2, $n = 2$, dark solid, stable to 200°	70
L = $(CH_3)_2SO$ (DMSO), $n = 2$, yellow, infrared spectrum shows it to be $[FeCl_2(DMSO)_4^+][FeCl_4^-]$	71, 72
L = dioxan, $n = 1$, green, several hydrates also isolated	73
L = ROH, $n = 2$, R = CH_3, C_2H_5, n-C_3H_7, i-C_3H_7, etc.	74, 75
L = ketone, $n = 1$, decrease in carbonyl stretching frequency on coordination	76
L = benzanthrone, $n = 1$, carbonyl frequency lowered	77
L = $(C_6H_{11})_3P$, $n = 1$, yellow, decomposes at 240°	78
L = $(C_6H_5)_3PO$, $n = 2$, yellow, m.p. = 183°	79
L = $(C_6H_5)_3AsO$, $n = 2$, yellow, $\mu = 5.92$ BM, ionic, possibly $[FeCl_2L_4^+][FeCl_4^-]$	80
L = methylbis(3-propanedimethylarsino)arsine (TAS), $n = 1$	81
L = *N,N*-dimethylacetamide (DMA), $n = 2$, yellow	82
L = 1,2-bis(methylseleno)ethane, $n = 1$, red-brown, decomposes at 102°	83
L = 1,3-bis(methylseleno)propane, $n = \frac{1}{2}$, brown-red, decomposes at 204°	83
L = C_5H_5NO, $n = 2$, yellow, $\mu = 5.73$ BM	84
L = ethyleneurea (eu), $n = 2$, orange-brown, m.p. = 181–182°, μ = 5.91 BM, five coordinate	85
L = 1,10-phenanthroline, $n = 2$, $\mu = 5.91$ BM, ionic, $[FeCl_2(phen)_2^+]Cl^-$	86

to the normal electrostatic attraction between oppositely charged ions. The apparent hydrogen bonding system is shown in Figure 6.6.

Iron oxide chloride. Schafer[27,28,30] has made a detailed study of the action of hydrogen chloride gas on iron(III) oxide in a flow system. At relatively low temperatures of about 300°, iron oxide chloride is formed but at higher temperatures iron trichloride is produced. Other methods of preparation of iron oxide chloride include the interaction[88] of iron(III) oxide and carbon tetrachloride in a bomb at 300°, the reaction[89] between iron trichloride and a limited amount of water in a bomb at 240–275° and the thermal decomposition[22,89] of iron trichloride hexahydrate in the presence of iron trichloride at 250°. Any iron trichloride remaining in the product may be removed by washing with acetone.

Iron oxide chloride is a rust coloured crystalline solid which is thermally unstable[41], decomposing above about 400° to give iron trichloride and iron(III) oxide. The compound is isostructural with titanium oxide chloride

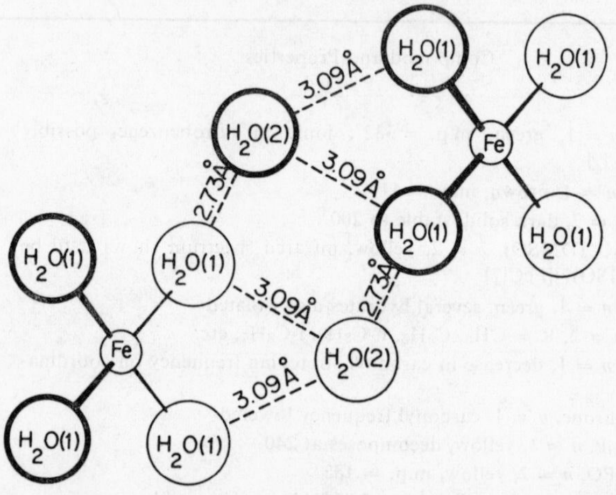

Figure 6.6 The apparent hydrogen bonds in $FeCl_3 \cdot 6H_2O$. (Reproduced by permission from M. D. Lind, *J. Chem. Phys.*, **47**, 990 (1967))

and has an orthorhombic unit cell[90–92] with $a = 3.75$, $b = 3.3$ and $c = 7.65$ Å[92]. The iron-chlorine and iron-oxygen bond distances are 2.29 Å and 1.88 Å respectively.

Iron oxide chloride reacts with ammonia and substituted amines to give both adducts and substitution compounds[93]. Some of the compounds which have been isolated include $FeO(NH_2)$ and $4FeOCl \cdot (C_2H_5)_3N$.

Iron tribromide. Although iron tribromide may be prepared[94] by direct interaction of the elements in a flow system at about 450°, the most convenient preparation is the interaction of the elements in an inverted V tube[95], with the iron at 175–200° in one arm and the bromine at 120° in the other. An alternative method is the interaction of iron trichloride and boron tribromide[96] at room temperature. The product may be purified by sublimation in an atmosphere of bromine[95].

Iron tribromide is a dark coloured hygroscopic crystalline solid which is readily soluble in water leading to the formation of a number of hydrates[95]. It is less stable thermally than the trichloride[94,95,97] and complete dissociation to iron dibromide and bromine occurs at 120–130°. The equilibrium dissociation pressure of bromine over solid iron tribromide over the temperature range 65–140° is given by the expression[95]

$$\log p_{mm} = 11.327 - 3479/T$$

The heat of dissociation is reported[95] to be 7.95 kcal mole^{-1}. Both vapour pressure[98] and absorption spectral[36] measurements indicate that iron tribromide vaporizes as a dimer. The true vapour pressure of iron tribromide over the solid is given by the expression[98]

$$\log p_{mm} = 15.24 - 7044/T$$

The heat of formation of iron tribromide[47] is −62.8 kcal mole^{-1}.

The tribromide is isomorphous with iron trichloride and the hexagonal unit cell[49] has $a = 6.42$ and $c = 18.40$ Å, with an iron-bromine bond distance of 2.63 Å. Rather suprisingly there appears to have been no study of the magnetic properties of iron tribromide[99], except that Golding and Ogel[100] report a g value of 2.052 for an acetone solution at room temperature.

Iron tribromide forms adducts with donor ligands such as bipyridyl, dimethyl sulphoxide, complex ketones, tertiary arsines etc.[71,77,81,101–103]. It also forms substitution compounds with dialkyldithiocarbamates[58,59] and Schiff bases[63].

Oxidation State II

Iron difluoride. The principal methods of preparation of iron difluoride are summarized in Table 6.6.

TABLE 6.6
Preparation of Iron Difluoride

Method	Conditions	References
Fe and anhydrous HF	Flow system, 900°	104
Fe and anhydrous HF	Bomb, 180°	105
FeF_2 hydrate and anhyd. HF	Flow system, 950–1000°	106
$FeCl_2$ and anhydrous HF[a]	Flow system, 400–600°	107, 108
$FeF_3 \cdot xH_2O$[b]	Decompose at 1000°	8
Fe_2O_3, C and anhydrous HF	Flow system, 950–1000°	106

[a] Reported in the early literature that this reaction proceeds at room temperature[22]. However higher temperatures are advisable to ensure complete reaction.

[b] Product brown to yellow suggesting oxide contamination.

Iron difluoride is a white, sparingly soluble compound which sometimes appears brown because of oxide impurity. It melts[109] at 1020° and Kent and Margrave[110] have shown by mass-spectrometric techniques that in the temperature range 692–876° iron difluoride is exclusively monomeric in the vapour phase. Over this temperature range the sublimation pressure of iron difluoride is given by the expression

$$\log p_{\mathrm{atm}} = 9.42 - 15{,}820/T$$

The heat of sublimation is 75.6 kcal mole^{-1}.

Iron difluoride has the tetragonal rutile structure[8,111–114] and Baur[112,113] gives $a = 4.696$ and $c = 3.309$ Å. The FeF_6 octahedra are tetragonally compressed and the iron-fluorine bond distances are 1.99 Å (×2) and 2.12 Å (×4).

Porto and coworkers[115] report four Raman-active modes for iron difluoride. These are (cm^{-1})

B_{1g} 73 E_g 257 A_{1g} 340 B_{2g} 496

Balkanski and coworkers[116] have observed the lattice vibrations of iron difluoride by optical transmission studies and gave the following modes (cm^{-1})

E_u 480 A_{2u} 440 E_u 320 E_u 200

The far infrared spectrum of iron difluoride below the Neel point has been investigated in some detail[117–123]. Two absorptions were observed, and close to absolute zero these occur at 52.7 and 155.4 cm^{-1}. The origin of these bands has been discussed by Halley[117].

Iron difluoride obeys the Curie-Weiss law[99] above about 100°K with a magnetic moment of 5.56 BM and a Weiss constant of 117°. The Neel temperature has been reported[115] as 78.5°K and as[124] 90°K. A specific heat maximum occurs at 78.35°K[107,109,125].

Single crystals of iron difluoride show quite marked magnetic anisotropy and at 1.5° the spectroscopic splitting factors are $g_{\parallel} = 2.25$ and $g_{\perp} = 2.05$[119]. Niira and Oguchi[126] have discussed the anisotropy from a theoretical point of view. Neutron-diffraction studies[124] have shown that iron difluoride has a similar magnetic structure to that of manganese difluoride (Figure 5.5, page 219).

There has been little investigation of the electronic absorption spectrum of iron difluoride. Clark[13] reported a band at 10,000 cm^{-1} which he assigned to the $^5E_g \leftarrow {}^5T_{2g}$ transition. On the other hand Hatfield and Piper[127] observed two bands at 10,660 and 6990 cm^{-1} which they assigned on the basis of a tetragonally distorted octahedral configuration.

Iron difluoride tetrahydrate. White iron difluoride tetrahydrate may be prepared by dissolving iron in warm hydrofluoric acid and precipitating with ethanol[128]. It is readily recrystallized from water containing a small amount of hydrofluoric acid. In air it becomes brown because of oxidation. The unit cell is trigonal[128] with $a = 9.50$ and $c = 4.82$ Å and there are three molecules in the cell of space group $R\bar{3}m$. The structure consists of discrete octahedral $FeF_2(H_2O)_4$ units with the water molecules and fluorine atoms randomly distributed over the possible sites. Considerable hydrogen bonding stabilizes the structure.

When a saturated solution of iron difluoride tetrahydrate in water is allowed to stand for several days in air, a yellow precipitate is formed[129]. This compound has the composition $Fe_2F_5 \cdot 7H_2O$ and it has a tetragonal cell with $a = 12.82$ and $c = 6.92$ Å. Four molecules of water are lost on heating to 100° in nitrogen to give a deep red trihydrate which is orthorhombic with $a = 10.94$, $b = 6.70$ and $c = 3.48$ Å. Both the hepta- and trihydrates are completely dehydrated at 180° to give hygroscopic, grey Fe_2F_5. This mixed oxidation state compound has a tetragonal cell with $a = 8.04$ and $c = 9.54$ Å.

Iron dichloride. In Table 6.7 are listed the principal methods of preparing iron dichloride.

TABLE 6.7
Preparation of Iron Dichloride

Method	Conditions	References
Fe and HCl	Flow system, red heat	130
Fe, HCl and Cl_2	Flow system, 700°	131
$FeCl_3$ and H_2	Flow system, 300–350°	22
$FeCl_3$ and H_2S	Flow system, 100°	132
$FeCl_3$ and C_6H_5Cl	Reflux	133
$FeCl_3$	Decompose at 300°	29, 42
$FeCl_2 \cdot 4H_2O$ and HCl	Flow system, 400–500°	134
$FeCl_2 \cdot 4H_2O$	Decompose at 100–200°	47, 56, 135–137

Any iron trichloride formed in the high temperature reactions may be decomposed during the subsequent purification by sublimation, either in vacuum or in a stream of hydrogen chloride. Dehydration of the tetrahydrate may give a product which is contaminated with oxide or oxide chloride, so it is advisable to sublime the sample in hydrogen chloride.

Iron dichloride is a white hygroscopic crystalline solid which is readily soluble in water. Using mass-spectrometric techniques Schoonmaker and

Porter[130] have shown iron dichloride vapour to be monomeric at low temperatures. However as the temperature is increased the proportion of dimer species increases and at the melting point approximately 25% of the vapour is dimeric.

The sublimation pressure of iron dichloride over the temperature range 397–467° is given by the expression[137]

$$\log p_{mm} = 11.10 - 9890/T$$

and for the temperature range 708–834° the expression is[138]

$$\log p_{mm} = 9.794 - 7455/T$$

Thermodynamic data for iron dichloride are given in Table 6.8.

TABLE 6.8

Thermodynamic Data for Iron Dichloride

Property	Value	References
M.p.	676 ± 1°	41, 134, 138, 139
B.p.	1012°	138
	1023°	45
ΔH°_{form} (kcal mole^{-1})	−81.5	47
	−81.86	46
ΔH_{vap} (kcal mole^{-1})	29.99	138
	32.11	45
ΔH_{sub} (kcal mole^{-1})	45.3	137
ΔS_{vap} (cal deg^{-1} mole^{-1})	23.35	138
ΔS_{sub} (cal deg^{-1} mole^{-1})	37.6	137

Iron dichloride has the cadium dichloride structure[140], the hexagonal cell having $a = 3.593$ and $c = 17.58$ Å. The $FeCl_6$ octahedra are arranged in the cubic close-packing array in samples which are sublimed and subsequently cooled very slowly[136,141]. Rapid cooling of the sublimate results in hexagonal close packing, while dehydration of the tetrahydrate at about 150° gives a randomly packed array. The last form can be converted to the cubic close-packed modification by annealing. Iron dichloride is also reported to undergo a phase transition from the cadium dichloride lattice to the cadmium diiodide-type at pressures of about 2 Kbar[142].

The asymmetrical stretching frequency[143] of monomeric iron dichloride occurs at 492 cm^{-1} while the dimer shows an absorption at 430 cm^{-1}.

Iron dichloride obeys the Curie-Weiss law[56,144] above about 100°K with a magnetic moment of 5.38 BM and a Weiss constant of 48°. At low temperatures it becomes magnetically ordered, the Neel temperature being approximately 23.5°K[144–148]. Most workers agree that iron dichloride becomes antiferromagnetic below this temperature, although Leech and Manuel[144,149] consider it to be ferromagnetic. There is a maximum in the specific heat of iron dichloride at 23.5°K indicative of a magnetic transition[56]. The magnetic structure has been studied by neutron-diffraction techniques[146,150–152] and it consists of ferromagnetic sheets of metal atoms with adjacent sheets having antiparallel spins, the spin direction being parallel to the *c* axis. The magnetic properties of iron dichloride have been discussed from a theoretical point of view by Kanamori[153].

The electronic absorption spectrum of iron dichloride vapour at approximately 1000° has been recorded by DeKock and Gruen[154]. The spectrum consists of two strong absorptions which are due to the dimeric species. The band positions and their assignments are

4600 cm^{-1}	$^5\Pi \leftarrow {}^5\Delta$
7140 cm^{-1}	$^5\Sigma \leftarrow {}^5\Delta$

Iron dichloride reacts readily with pyridine and substituted pyridines, directly or in ethanol, to give complexes with two, four or six molecules of the ligand[155–161]. Beech and coworkers[158] have studied the thermal decomposition of several of these adducts. The equations representing the decomposition reactions are given below.

$$FeCl_2L_4\,(s) \rightarrow FeCl_2L_2\,(s) + 2L\,(g) \quad (1)$$

$$FeCl_2L_2\,(s) \rightarrow FeCl_2L\,(s) + L\,(g) \quad (2)$$

$$FeCl_2L\,(s) \rightarrow FeCl_2L_{2/3}\,(s) + 1/3L\,(g) \quad (3)$$

$$FeCl_2L_{2/3}\,(s) \rightarrow FeCl_2\,(s) + 2/3L\,(g) \quad (4)$$

Thermodynamic data derived from the calorimetric studies[158] are summarized in Tables 6.9, 6.10 and 6.11. In each Table, T_i, T_p and T_f are the

TABLE 6.9

Decomposition of $FeCl_2(py)_4$

Reaction	1	2	3	4
ΔH (kcal mole^{-1})	27.0	15.2	5.3	10.1
T_i (°K)	310	400	500	545
T_p (°K)	370	450	530	585
T_f (°K)	385	465	540	600

initial temperature, peak temperature (where the rate of change of ΔH is greatest) and the final temperature, in °K, of each dissociation reaction.

The adduct $FeCl_2(py)_2$ has been studied in the greatest detail and all workers agree that this yellow compound is an octahedral polymer[155,156,159]. It has a room-temperature magnetic moment of 5.75 BM[156] and its electronic absorption spectrum has been interpreted on the basis of a tetragonally distorted configuration about the central iron atom[159].

TABLE 6.10
Decomposition of $FeCl_2(3\text{-}CH_3py)_4$

Reaction	1	2	3	4
ΔH (kcal mole^{-1})	29.2	14.4	4.1	18.3
T_i (°K)	340	410	470	525
T_p (°K)	390	450	500	590
T_f (°K)	405	460	510	600

TABLE 6.11
Decomposition of $FeCl_2(4\text{-}CH_3py)_4$

Reaction	1 + 2	3	4
ΔH (kcal mole^{-1})	47.3	5.3	15.3
T_i (°K)	360	510	550
T_p (°K)	425, 440[a]	535	600
T_f (°K)	460	545	610

[a] Two peaks observed.

Clark and Williams[162] have made a detailed examination of the infrared spectrum of grey $FeCl_2(NH_3)_2$ and reported the assignments given in Table 6.12.

Broomhead and Dwyer[163] have reported that iron dichloride reacts with terpyridyl (terpy) under suitable conditions, to give the complex $FeCl_2 \cdot$terpy. This complex has a magnetic moment of 4.60 BM. Robinson and Kennard[164] have shown by an X-ray diffraction study that the complex is isomorphous, and probably isostructural with the corresponding zinc compound which is known to have a distorted trigonal-bipyramidal configuration. The iron compound is monoclinic with $a = 16.21$, $b = 8.25$, $c = 10.97$Å and $\beta = 93.5°$.

Adams and Cornell[165] have reported a detailed infrared analysis of $FeCl_2(tu)_4$, where tu is thiourea. The complex is known[166] to be isomorphous with the corresponding nickel dichloride complex which has a *trans* octahedral configuration (Figure 8.3, page 418). The infrared study is consistent with this stereochemistry and the assignments of the various bands are given in Table 6.13.

TABLE 6.12
Infrared Assignments of $FeCl_2(NH_3)_2$

Mode		Frequency (cm^{-1})	Intensity
NH_3 deformation	degen	1600	Medium
	sym	1273	Weak
	sym	1243	Shoulder
	sym	1233	Strong
NH_3 rock		624	Medium
		554	Medium
Fe—N stretch		415	Weak
Fe—Cl stretch		235	Strong

TABLE 6.13
Infrared Assignments for $FeCl_2(tu)_4$

Mode	Frequency (cm^{-1})	Intensity
$\nu_{Fe\text{-}Cl}$	149	Strong, broad
	112	Medium
$\nu_{Fe\text{-}S}$	215	Strong
δ_{FeSC}	182	Shoulder
	163	Shoulder
δ_{SFeS}	83	Weak

The interaction of ethylenethiourea (2-imidazolidenethione) and iron dichloride in alcohol[167] results in the formation of tetrahedral $FeCl_2L_2$. The complex has a monoclinic unit cell with $a = 8.15$, $b = 2.98$, $c = 13.44$ Å and $\beta = 116.7°$.

Iron dichloride forms adducts with a wide range of other donor ligands and some properties of a number of these are given in Table 6.14.

Adducts of iron dichloride with other donor ligands such as dioxan, tetrahydrofuran and hydrazine have also been reported[179–184].

Iron dichloride tetrahydrate. This hydrate is prepared by dissolving iron metal in hydrochloric acid and crystallizing the product at room temperature[185]. It may also be isolated from the iron dichloride-hydrochloric acid-water system when there is less than about 30% hydrogen chloride.

The thermal decomposition of iron dichloride tetrahydrate has been studied by Schafer[186] and by Shen and Chang[135]. Quite good agreement

TABLE 6.14

Adducts of the Type $FeCl_2L_n$

Compound and Properties	References
L = $(C_2H_5)_2PH$, n = 2, red, m.p. = 75–80°, μ = 3.61 BM, planar	168
L = $(C_6H_{11})_2PH$, n = 2, colourless, m.p. = 95–98°, μ = 5.12 BM, tetrahedral	168
L = $(C_2H_5)_3P$, n = 2, colourless, μ = 5.00 BM, tetrahedral	168
L = $(C_6H_5)_3P$, n = 2, colourless, m.p. = 172°, μ = 4.88 BM, tetrahedral	168, 169
L = *o*-phenylenebis(dimethylarsine) (D or diars), n = 2, $\nu_{Fe\text{-}Cl}$ = 349 cm^{-1}	170
L = $(C_6H_5)_3PO$, n = 2, brown, tetrahedral	79, 171
L = $(C_6H_5)_3AsO$, n = 2, pale yellow, m.p. = 182°, $\nu_{As\text{-}O}$ = 874 cm^{-1}, $\nu_{Fe\text{-}O}$ = 384 cm^{-1}, $\nu_{Fe\text{-}Cl}$ = 315, 288 cm^{-1}	171, 172
L = bipyridyl, n = 2, μ = 5.72 BM	163, 173
L = 1,10-phenanthroline, n = 2, μ = 5.79 BM	163, 173
L = 2,9-dimethyl-1,10-phenanthroline (dmp), n = 1, pale yellow, μ = 5.18 BM, tetrahedral monomer	174
L = quinoline, n = 2, yellow, m.p. = 142°, tetrahedral	171
L = 8-aminoquinoline, n = 1, isolated as solvate with two molecules of methanol, yellow, m.p. = 77–80°, μ = 5.34 BM	175
L = di-2-pyridylamine, n = 2, μ = 5.40 BM, $\nu_{Fe\text{-}Cl}$ = 253 cm^{-1}, octahedral	176
L = di-2-pyridylamine, n = 1, μ = 5.34 BM, $\nu_{Fe\text{-}Cl}$ = 342, 313 cm^{-1}, tetrahedral	176
L = bis(2-dimethylaminoethyl)methylamine (dienMe), n = 1, mauve, μ = 5.20 BM, five coordinate	177
L = pyrazine (pyz), n = 2, decomposes at 200° and 450° to give $FeCl_2(pyz)$ and $FeCl_2$ respectively	178

was obtained, and according to Schafer the following reactions occur at the temperatures indicated.

$$FeCl_2\cdot 4H_2O \xrightarrow{105\text{–}115°} FeCl_2\cdot 2H_2O \xrightarrow{150\text{–}160°} FeCl_2\cdot H_2O \xrightarrow{220°} FeCl_2$$

The structure of iron dichloride tetrahydrate has been investigated by Penfold and Grigor[187] by single-crystal X-ray diffraction techniques. It is monoclinic with $a = 5.91$, $b = 7.17$, $c = 8.44$ Å and $\beta = 112.17°$.

There are two molecular units in the cell of space group $P2_1/c$. The structure consists of discrete distorted $FeCl_2(H_2O)_4$ units with *trans* chlorine atoms, the units being held together by multiple hydrogen bonds. The iron-chlorine distance is 2.38 Å and the iron-oxygen distances are 2.09 and 2.5 Å.

Gamo[188] has recorded the infrared spectrum of iron dichloride tetrahydrate and has discussed the effects of hydrogen bonding on the various modes of vibration of the water molecules.

Iron dichloride tetrahydrate obeys the Curie-Weiss law down to very low temperatures[99] with a magnetic moment of 5.20 BM and a Weiss constant of 11°. A low-temperature magnetic susceptibility study[189] gave the Neel temperature as 1.6°K. Nuclear magnetic resonance measurements also indicate a paramagnetic to antiferromagnetic transition at about this temperature[190–192].

Iron dichloride dihydrate. This hydrate is obtained[193,194] by evaporating a saturated aqueous solution of iron dichloride at 70°. It is also obtained by controlled dehydration of the tetrahydrate[135,186], and by dissolving iron in dilute hydrochloric acid and crystallizing the solution at about 75°[185].

The dihydrate is isostructural[193] with manganese dichloride dihydrate and cobalt dichloride dihydrate. The monoclinic lattice has $a = 7.355$, $b = 8.548$, $c = 3.637$ Å and $\beta = 98.18°$. There are two molecules in the unit cell of space group $C2/m$. The structure consists of chains of $FeCl_4(H_2O)_2$ with *trans* aquo groups, adjacent octahedra sharing edges by two bridging chlorine atoms. The adjacent chains are held together by hydrogen bonding. The iron-chlorine bond lengths are 2.488 and 2.542 Å and the iron-oxygen distance is 2.075 Å. The chlorine-iron-chlorine angle is 87.4°.

Low-temperature magnetic susceptibility measurements and ^1H NMR studies[195] indicate that at low temperatures iron dichloride dihydrate becomes magnetically ordered, the Neel temperature being approximately 23°K. The magnetic structure consists of ferromagnetic chains parallel to the c axis, coupled antiferromagnetically with adjacent chains[195].

Lawson[194] has examined the absorption spectrum of iron dichloride dihydrate in considerable detail. At room temperature there is a broad band centred at about 10,000 cm^{-1} and this has been assigned to the $^5E_g \leftarrow {^5T_{2g}}$ transition. At lower temperatures this band splits into two peaks, due no doubt to the distorted nature of the octahedral symmetry of the iron atom.

Iron dibromide. Direct interaction of the elements in a flow system at about 200° leads to a mixture of iron tribromide and iron dibromide, but the former is readily decomposed to the dibromide on resublimation in a nitrogen stream or in vacuum[137]. Thermal decomposition of iron tribromide occurs at about 100° in vacuum[47,94,95,97]. Other methods of preparation include the hydrobromination of iron(III) oxide at 200–325° in a flow system[196], passage of a mixture of hydrogen bromide and bromine over iron[131] at 690° and the dehydration of iron dibromide tetrahydrate followed by vacuum sublimation of the product[197].

Iron dibromide is a light yellow to brown hygroscopic solid which dissolves readily in water yielding various hydrates[22,197,198].

McLaren and Gregory[94] have shown by vapour pressure measurements that iron dibromide is thermally stable to at least 900°. These workers measured the vapour pressure of iron dibromide over several temperature ranges using different techniques, and their methods and results are listed below.

$\log p_{\text{mm}} = 12.045 - 10{,}294/T$	(400–689°)	Transpiration
$\log p_{\text{mm}} = 11.88 - 10{,}300/T$	(350–445°)	Effusion
$\log p_{\text{mm}} = 13.90 - 12{,}100/T$	(620–689°)	Diaphragm

The agreement between the sets of data using the different techniques is somewhat worse than expected from experimental error considerations, and no completely satisfactory explanation for the differences could be put forward. Another study[137] gave the vapour pressure of solid iron dibromide over the temperature range 397–467° as

$$\log p_{\text{mm}} = 11.95 - 10{,}220/T$$

The vapour pressure of liquid iron dibromide over the temperature range 689–909° is given by[94]

$$\log p_{\text{mm}} = 8.450 - 6912/T$$

Thermodynamic data for iron dibromide are given in Table 6.15.

Iron dibromide has a hexagonal lattice[199] with $a = 3.740$ and $c = 6.171$ Å. The structure consists of $FeBr_6$ octahedra which are arranged in a hexagonal close-packed array[197]. Iron dibromide which has been prepared by dehydration of the tetrahydrate, but has not been sublimed, has random packing of the $FeBr_6$ octahedra. Heat capacity measurements[200] indicate a structural transformation at approximately 400°. This transformation has been confirmed by Gregory and Wydeven[201] who reported that above 360° iron dibromide has bromine atoms arranged in the cubic close-packed array.

Iron dibromide obeys the Curie-Weiss law[99,201] with a magnetic moment of 5.62 BM and a Weiss constant of 6°. At approximately 11°K it becomes magnetically ordered[145,203,204] and the magnetic structure[152] is the same as that found for iron dichloride.

Iron dibromide forms adducts with a large range of donor molecules including pyridine, ammonia and triphenylphosphine, as shown in Table 6.16.

TABLE 6.15

Thermodynamic Properties of Iron Dibromide

Property	Value	Reference
M.p.	689°	94
B.p. (extrapolated)	967°	94
$\Delta H^\circ_{\text{form}}$ (kcal mole^{-1})	−58.7	47
	−59.6	196
$\Delta S^\circ_{\text{form}}$ (cal deg^{-1} mole^{-1})	33.0	196
ΔH_{fus} (kcal mole^{-1})	15.5	94
ΔH_{vap} (kcal mole^{-1})	31.6	94
ΔH_{sub} (kcal mole^{-1})	46.7	137
	47.1	94
ΔS_{sub} (cal deg^{-1} mole^{-1})	41.5	137

Adducts with other phosphines and arsines[184], phosphorus oxide tribromide[206] and dioxan[180] have also been reported.

Iron diiodide. Iron diiodide may be prepared by interaction of the elements in either a sealed tube[137,210–212] at 450–550° or under a thermal gradient[214] of 530–180°. Alternatively, the reaction between iron(III) oxide and aluminium triiodide in a sealed tube at 300° may be used[215].

Iron diiodide is a dark red-black crystalline hygroscopic solid which dissolves in water with the formation of several hydrates[216]. It is thermally unstable, and the equilibrium vapour pressure of iodine over iron diiodide is given by the expression[217]

$$\log p_{\text{mm}} = 4.14 - 6791/T$$

for the temperature range 759–858°.

Schafer and Hones[214] measured the vapour pressure of iron diiodide itself using the transpiration method and reported the following results.

$$\log p_{\text{mm}} = 13.183 - 10{,}778/T \quad (517\text{–}577^\circ)$$

$$\log p_{\text{mm}} = 9.674 - 7716/T \quad (601\text{–}686^\circ)$$

Sime and Gregory[137] using the torsion-effusion technique reported the expression

$$\log p_{\mathrm{mm}} = 11.82 - 9600/T$$

for the temperature range 397–467°. Zaugg and Gregory[218] have questioned both sets of data on the grounds that dimerization, which occurs at the temperatures used, and the presence of iron triiodide and free iodine in the gas phase, were not taken into account. Published thermodynamic data for iron diiodide are given in Table 6.17.

TABLE 6.16

Adducts of the Type $FeBr_2L_n$

Compound and Properties	References
L = pyridine or substituted pyridine, $n = 2,4,6$	157, 159, 160
L = di-2-pyridylamine, $n = 2$, $\mu = 5.22$ BM, $\nu_{\text{Fe-Br}} = 210\ \text{cm}^{-1}$, octahedral	176
L = di-2-pyridylamine, $n = 1$, $\mu = 5.40$ BM, $\nu_{\text{Fe-Br}} = 260,\ 244\ \text{cm}^{-1}$, tetrahedral	176
L = quinoline, $n = 2$, yellow, m.p. = 151°, tetrahedral	171
L = NH_3, $n = 2$, grey, $\nu_{\text{Fe-N}} = 424\ \text{cm}^{-1}$	162
L = thiourea (tu), $n = 4$, $\nu_{\text{Fe-Br}} = 149,\ 112\ \text{cm}^{-1}$, $\nu_{\text{Fe-S}} = 215\ \text{cm}^{-1}$	165
L = $(C_6H_5)_3P$, $n = 2$, m.p. = 185°, $\mu = 4.77$ BM	169
L = $(C_6H_{11})_2P\text{-}(CH_2)_3\text{-}P(C_6H_{11})_2$, $n = 1$, grey, decomposes below 200° and in solution	205
L = $(C_6H_{11})_2P\text{-}(CH_2)_5\text{-}P(C_6H_{11})_2$, $n = 1$, light grey, very air sensitive, decomposes below 165° and in solution	205
L = *o*-phenylenebis(dimethylarsine) (D or diars), $n = 2$, bright yellow, diamagnetic	102
L = $(C_6H_5)_3PO$, $n = 2$, brown, decomposes at 210°	171, 206
L = $(C_6H_5)_3AsO$, $n = 2$, m.p. = 185°, $\nu_{\text{As-O}} = 896,\ 851\ \text{cm}^{-1}$, $\nu_{\text{Fe-O}} = 403,\ 395\ \text{cm}^{-1}$, $\nu_{\text{Fe-Br}} = 225\ \text{cm}^{-1}$, tetrahedral	171, 172
L = $CH_3(C_6H_5)_2AsO$, $n = 2$, $\nu_{\text{As-O}} = 849\ \text{cm}^{-1}$, $\nu_{\text{Fe-O}} = 401\ \text{cm}^{-1}$, $\nu_{\text{Fe-Br}} = 235\ \text{cm}^{-1}$, tetrahedral	172
L = *N,N,N′,N′*-tetramethylethylenediamine (Me_4en), $n = 1$, $\mu =$ 5.38 BM, tetrahedral	207
L = *N,N,N′,N′*-tetramethyl-1,2-propylenediamine (Me_4pn), $n = 1$, $\mu = 5.41$ BM, tetrahedral	207
L = bis(2-dimethylaminoethyl)methylamine (dienMe), $n = 1$, $\Delta H^\circ_{\text{form}}$ (from constituents) = −31.78 kcal mole^{-1}, five coordinate	208
L = bis(dimethylaminoethyl)oxide (Me_4daeo), $n = 1$, pink $\mu =$ 5.26 BM, five coordinate	209

Iron diiodide has the cadmium diiodide-type hexagonal lattice[221] with $a = 4.04$ and $c = 6.75$ Å. There is an anomaly in the heat capacity between

360° and 385° suggesting that a phase change may occur, although X-ray powder diffraction studies did not confirm this suggestion.

Iron diiodide obeys the Curie-Weiss law[99] with a magnetic moment of 5.88 BM and a Weiss constant of 23°. At low temperatures it becomes antiferromagnetic[145], the Neel temperature being 3.4°K. The magnetic

TABLE 6.17

Thermodynamic Data for Iron Diiodide

Property	Value	References
M.p.	587°	219
	593°	220
	594°	137, 214
B.p. (extrapolated)	935°	137, 214
ΔH°_{form} (kcal mole^{-1})	−23.7	212
ΔS°_{form} (cal deg^{-1} mole^{-1})	41.8	212
ΔH_{sub} (kcal mole^{-1})	53.0	214
ΔH_{vap} (kcal mole^{-1})	36.76	214
ΔS_{sub} (cal deg^{-1} mole^{-1})	54.2	214
ΔS_{vap} (cal deg^{-1} mole^{-1})	37.54	214

properties have been discussed from a theoretical point of view of Kanamori[153].

Iron diiodide forms adducts with amines, pyridine, dimethyl sulphoxide, triphenylphosphine, *o*-phenylenebis(dimethylarsine), etc.[102,157,159,160,169,180,184,222,223]. In general, the properties of these adducts closely resemble those of iron dibromide.

COMPLEX HALIDES

The striking feature of the halide chemistry of iron is the frequent occurrence of tetrahedral stereochemistry. There has been surprisingly little detailed investigation of the complex halides in general, although some particular anions have been studied in considerable detail.

Oxidation State III

Hexafluoroferrates(III). The principal methods of preparation of hexafluoroferrates(III) are given in Table 6.18.

Cox and Sharpe[232] have studied the preparation of fluoroferrates(III) in hydrogen fluoride media and showed that a variety of phases, both stoichiometric and non-stoichiometric, can be isolated by this method.

The hexafluoroferrates(III) are white crystalline solids which are relatively unreactive. They are reasonably stable in air and are only slowly hydrolysed in water. Thermogravimetric studies[3] of the decomposition of the ammonium salt indicate the following mechanism:

TABLE 6.18
Preparation of Hexafluoroferrates(III)

Method	Conditions	References
FeF_3, MCl and aqueous HF		224
FeF_3, $NaHF_2$ and KHF_2	Fuse, 1:2 ratio of Na:K	2
FeF_3 and MF_2	Fuse, M = Sr, Ba	225
FeF_3, MHF_2 and $M'F_2$	Fuse, M = Rb, Cs; M′ = Mn, Fe, Co, Ni, Cu	226
$K_2FeF_5 \cdot H_2O$ and KHF_2	Fuse	227
FeF_3, KCl and F_2	Flow system, 100–375°	228
$FeCl_3 \cdot 6H_2O$ and Li_2CO_3	Interaction in aqueous HF	229
$FeBr_3$ and MF	Interaction in methanol, M = K, Rb and NH_4^+	3, 230, 231

$$(NH_4)_3FeF_6 \xrightarrow{227^\circ} NH_4FeF_4 \xrightarrow{320^\circ} FeF_3 \xrightarrow{430^\circ} Fe_2O_3$$

Heating the same salt in an ammonia atmosphere[233] gives iron difluoride at 400° and Fe_2N at 600°.

The structure of a number of hexafluoroferrates(III) have been investigated and the available data are summarized in Table 6.19.

TABLE 6.19
Symmetry and Lattice Parameters of Hexafluoroferrates(III)

Compound	Symmetry	Parameters	References
$(NH_4)_3FeF_6$	Cubic	$a = 9.10$ Å	234–237
	Tetragonal	$a = 6.39$, $c = 9.30$ Å	234
NaK_2FeF_6	Cubic	$a = 8.323$ Å	2
K_3FeF_6	Cubic	$a = 9.58$ Å	234
	Tetragonal	$a = 8.59$, $c = 8.66$ Å	227
Rb_3FeF_6	Cubic	$a = 8.88$ Å	234
$Sr_3(FeF_6)_2$	Tetragonal	$a = 14.31$, $c = 7.28$ Å	225
$Ba_3(FeF_6)_2$	Tetragonal	$a = 14.58$, $c = 7.68$ Å	225
$CsNiFeF_6$	Cubic	$a = 10.35$ Å	226
$CsCuFeF_6$	Cubic	$a = 10.38$ Å	226
$CsCoFeF_6$	Cubic	$a = 10.40$ Å	226
$CsMnFeF_6$	Cubic	$a = 10.54$ Å	226

NaK_2FeF_6 is isostructural with the corresponding chromium compound[2] (Figure 4.13, page 195) and the iron-fluorine bond distance is 1.910 Å.

The hexafluoroferrates(III) have magnetic moments in the range 5.80–5.95 BM[224]. The spectroscopic splitting factor of the hexafluoroferrate(III) anion in the cadmium telluride host lattice is 2.0029[238].

The hexafluoroferrates(III) have two infrared-active modes[3,239], ν_3 occurring at 580 cm^{-1} and ν_4 at about 465–470 cm^{-1}.

Hexafluoroferrate(III) hexahydrates. Mitra[240] has prepared a number of complexes of the general formula $MM'FeF_6 \cdot 6H_2O$ where M = K, Rb, Cu, Tl, NH_4 and M′ = Ni, Cu, Zn and Cd. He dissolved iron(III) hydroxide in 20% hydrofluoric acid, then provided a slight excess of the divalent metal as the hydroxide, carbonate or basic carbonate, and finally added the univalent metal fluoride. The mixed cation compounds crystallize on cooling. They may be dehydrated at about 100° to give a mixture of the divalent metal fluoride and an alkali-metal fluoroferrate.

Pentafluoroferrates(III). Potassium pentafluoroferrate(III) may be prepared by the interaction of iron trichloride and potassium fluoride in hydrofluoric acid[224]. The caesium salt has been prepared by the interaction of the appropriate amounts of caesium fluoride and iron tribromide in methanol[231]. De Pape and coworkers[225] have prepared the calcium, strontium and barium salts by the interaction of the alkaline-earth metal fluoride and iron trifluoride in the melt. The unit-cell parameters of these compounds are given in Table 6.20.

TABLE 6.20

Symmetry and Lattice Parameters of Some Pentafluoroferrates(III)

Compound	$CaFeF_5$	$SrFeF_5$	$BaFeF_5$
Symmetry	Orthorhombic	Monoclinic	Tetragonal
a (Å)	20.57	14.65	14.92
b (Å)	10.08	7.23	
c (Å)	7.58	7.03	7.61
β		95.5°	
Z	16	8	16

Potassium pentafluoroferrate(III) has a magnetic moment of 4.87 BM at room temperature[99].

Potassium pentafluoroaquoferrate(III), $K_2FeF_5 \cdot H_2O$, is a colourless crystalline solid which is readily prepared by the interaction of ferric alum and potassium hydrogen difluoride in water[241]. Other salts have been prepared by the interaction of an iron(III) salt and an alkali-metal or alkaline-earth metal fluoride in hydrofluoric acid[232,242–244]. The ammonium salt has a room-temperature magnetic moment of 5.91 BM[224]. Peacock[244] reports that the compounds may be dehydrated to give the anhydrous salts, which presumably contain condensed octahedra.

$Na_5Fe_3F_{14}$. This interesting compound has been isolated from the interaction of iron trifluoride and sodium fluoride in the melt[245]. Details for growing single crystals have been reported by Linares[246]. The compound exists in two modifications; the low-temperature form is monoclinic with $a = 13.23$, $b = 7.46$, $c = 12.72$ Å and $\beta = 90°$. The high-temperature form is tetragonal with $a = 7.43$ and $c = 10.38$ Å. The structure of the compound is very closely related to that of chiolite, $Na_5Al_3F_{14}$. The compound is ferromagnetic with a Curie temperature of approximately 80°K.

The electronic absorption spectrum of the complex has been recorded by Spencer and coworkers[247] and they reported the following bands and assignments.

15,000 cm^{-1}	$^4T_{1g}(G) \leftarrow {}^6A_{1g}$
20,000 cm^{-1}	$^4T_{2g}(G) \leftarrow {}^6A_{1g}$
26,000 cm^{-1}	$^4A_{1g}, {}^4E_g(G) \leftarrow {}^6A_{1g}$
29,500 cm^{-1}	$^4T_{2g}(E) \leftarrow {}^6A_{1g}$

Tetrafluoroferrates(III). Ammonium tetrafluoroferrate(III) is prepared[3] by heating a 3:1 mixture of ammonium fluoride and iron trifluoride in a stream of nitrogen at 180°, or by decomposition of the hexafluoroferrate(III) salt at 230°. Nyholm and Sharpe[224] prepared the caesium salt by the interaction of iron trichloride and caesium chloride in hydrofluoric acid, and this compound has a magnetic moment of 4.79 BM at room temperature. The ammonium compound has a tetragonal unit cell with $a = 3.785$ and $c = 6.363$ Å^3. However, it was pointed out that it is possible to index the X-ray powder diffraction data on the basis of a cell with a double the above value. Iron-fluorine stretches were observed at 575 and 500 cm^{-1} in ammonium tetrafluoroferrate(III).

Hexachloroferrates(III). It would appear that the hexachloroferrate(III) anion can be isolated only when it is stabilized by the presence of very large cations such as $[Co(NH_3)_6]^{3+}$. Hatfield and coworkers[248] have prepared several hexachloroferrates(III) by the interaction of iron trichloride hexahydrate and the appropriate complex chloride in concentrated hydrochloric acid. Mossbauer studies by Bancroft and coworkers[249] have confirmed the presence of octahedral $FeCl_6^{3-}$ units in several of these compounds.

Pentachloroaquoferrates(III). Salts of this anion are obtained by evaporation of aqueous solutions of iron trichloride hexahydrate and the appropriate chloride[248,250,251]. The caesium salt is also formed when caesium μ-trichlorohexachlorodiferrate(III) is allowed to stand in air[252].

Ammonium pentachloroaquoferrate(III) has an orthorhombic lattice with $a = 13.68$, $b = 9.88$ and $c = 7.02$ Å[253].

μ-Trichlorohexachlorodiferrate(III). The caesium salt of this anion is prepared by direct interaction of the appropriate amounts of caesium chloride and iron trichloride in concentrated hydrochloric acid[252]. The interaction of caesium chloride and iron trichloride in a sealed tube at 450° gives two compounds of empirical formula $Cs_3Fe_2Cl_9$[254]. The α-form is orange and isostructural with the corresponding chromium salt, while the β-form is yellow and has, as will be described later, a different structure.

α-$Cs_3Fe_2Cl_9$ is hexagonal[252] with $a = 7.28$ Å and $c = 8.90$ Å. It obeys the Curie-Weiss law[254] with a magnetic moment of 6.03 BM and a Weiss constant of 22°. The iron-chlorine stretching frequency occurs[255] at 316 cm^{-1}.

The compound reported[256,257] as $(pyH)_3Fe_2Cl_9$, where pyH is the pyridinium cation, has been examined independently by Ginsberg and Robin[254] and by Clark and Taylor[255]. Both groups showed conclusively that the compound does not contain the dimeric μ-trichlorohexachlorodiferrate(III) anion (see below).

Tetrachloroferrates(III). Although Johnstone and coworkers[40] reported the sodium tetrachloroferrate(III) could not be made by direct interaction of the two chlorides in the melt, more recent work has shown that in fact it does form under these conditions[258–260]. The potassium, caesium and ammonium salts have also been prepared by this method[258,261–263]. Substituted ammonium, phosphonium and arsonium salts are readily prepared by the interaction of iron trichloride and the appropriate chloride in alcohol or nitromethane solution[254,264–267]. The tetraethylammonium salt has been prepared by refluxing a mixture of tetraethylammonium chloride and iron trichloride in thionyl chloride[268].

A large range of other tetrachloroferrates(III) with complex cations has been prepared, and some properties of a number of these compounds are given in Table 6.21.

Blatt and coworkers[276–278] have prepared a number of tetrachloroferrates(III) with heterocyclic cations. They used these compounds as a method of identification of the heterocycle. The tetrachloroferrate(III) anion exists in concentrated hydrochloric acid solutions and is one of the major species involved in the extraction of iron from these solutions by ether[279]. Substituted diazonium and boronium complexes have been prepared[280–290]. Tetrachloroferrates(III) may also be formed in solvents such as triethyl phosphate and phosphorus oxide trichloride[265,291–293].

The alkali-metal tetrachloroferrates(III) are pale yellow crystalline solids. Cook and Dunn[258] have shown that the sodium and potassium salts are

TABLE 6.21
Properties of Some Tetrachloroferrates(III)

Complex	Properties	References
$[Fe(CH_3CN)_6^{2+}][FeCl_4^-]_2$	Yellow, μ = 5.92 BM, octahedral cation, tetrahedral anion	269
$[FeCl_2L^+][FeCl_4^-]$	L = bis(diisopropylphosphinyl)-methane, yellow-orange, m.p. = 112.5–113°, μ = 6.06 BM	270
$[FeL_3^{3+}][FeCl_4^-]_3$	L = bis(di-n-butylphosphinyl)-methane, yellow, m.p. = 238°, μ = 6.02 BM	270
$[Ti(acac)_3^+][FeCl_4^-]$	Orange-red, m.p. = 170–172°	249, 266, 271, 272
$[Si(acac)_3^+][FeCl_4^-]$	Cream, m.p. = 189–190°, ν_{Fe-Cl} = 385 cm^{-1}	249, 271, 273
$[Ge(acac)_3^+][FeCl_4^-]$	Yellow, m.p. = 184–185°, ν_{Fe-Cl} = 385 cm^{-1}	249, 271, 273
$[FeCl_2L_4^+][FeCl_4^-]$	L = ε-caprolactam, yellow-green, m.p. = 167°, ligand bound through oxygen atom	274
$[Fe(OMPA)_3^{3+}][FeCl_4^-]_3$	OMPA = octamethylpyrophosphoramide, yellow, m.p. = 234°	275
$[FeCl_2D_2^+][FeCl_4^-]$	D = *o*-phenylenebis(dimethylarsine), crimson, m.p. = 219°, μ = 4.55 BM per iron atom	102

quite thermally stable and vaporize as the molecular unit. The vapour pressure of sodium tetrachloroferrate(III) is given by the expression

$$\log p_{mm} = 7.496 - 5904/T$$

for the temperature range 477–827°. In the case of the potassium salt, the expression is

$$\log p_{mm} = 5.675 - 4517/T$$

for the temperature range 577–827°.

Thermodynamic data for a number of alkali-metal tetrachloroferrates(III) are given in Table 6.22.

Sodium tetrachloroferrate(III) has an orthorhombic unit cell[260] with $a = 10.34$, $b = 9.880$ and $c = 6.235$ Å. There are four molecular units in the cell of space group $P2_12_12_1$. The structure consists of sodium ions and discrete, slightly distorted, tetrachloroferrate(III) tetrahedra, as shown in Figure 6.7.

The tetraphenylarsonium salt has a tetragonal unit cell[294] with $a = 13.16$ and $c = 7.15$ Å and there are two molecular units in the cell of space

TABLE 6.22

Thermodynamic Data for Some Tetrachloroferrates(III)

Property	Compound	Value	Reference
M.p.	$NaFeCl_4$	163°	258
	$KFeCl_4$	243°	263
		249°	258
		250°	262
	$CsFeCl_4$	382°	262
	NH_4FeCl_4	295°	261
ΔH°_{form} (kcal mole^{-1})[a]	$NaFeCl_4$	−0.8	258
	$KFeCl_4$	−7.2	258
ΔH°_{form} (kcal mole^{-1})[b]	$NaFeCl_4$	−392	259
ΔH_{fus} (kcal mole^{-1})	$NaFeCl_4$	4.30	258
	$KFeCl_4$	3.83	258
ΔS_{fus} (cal deg^{-1} mole^{-1})	$NaFeCl_4$	9.86	258
	$KFeCl_4$	7.36	258

[a] From constituent chlorides.
[b] From the elements.

group $I\bar{4}$. The structure consists of the cation and discrete, but flattened $FeCl_4^-$ tetrahedra. The iron-chlorine distance is 2.19 Å and the chlorine-iron-chlorine angles are 107° and 114.5°. The cation is considerably distorted compared with that in tetraphenylarsonium iodide.

One of the most interesting complexes containing the tetrachloroferrate(III) anion is that of empirical formula $FeCl_3 \cdot 2(CH_3)_2SO$. Bennett, Cotton and Weaver[72] have shown by single-crystal X-ray diffraction studies that this adduct is in fact ionic and contains the *trans* $FeCl_2[(CH_3)_2SO]_4$ cation and the tetrachloroferrate(III) anion. The cation, whose structure is shown in Figure 6.8, is a tetragonally elongated octahedron. The iron-chlorine distance is 2.366 Å while the iron-oxygen bond length is 2.006 Å. The anion is slightly distorted and the iron-chlorine distance is 2.162 Å. The anions and cations are arranged in the NaTl type array as shown in Figure 6.9.

o-Methoxybenzenediazonium tetrachloroferrate(III) has an orthorhombic unit cell[287] with $a = 11.07$, $b = 16.28$ and $c = 7.10$ Å. There are four molecules in the unit cell of space group *Pbnm*. The structure consists of

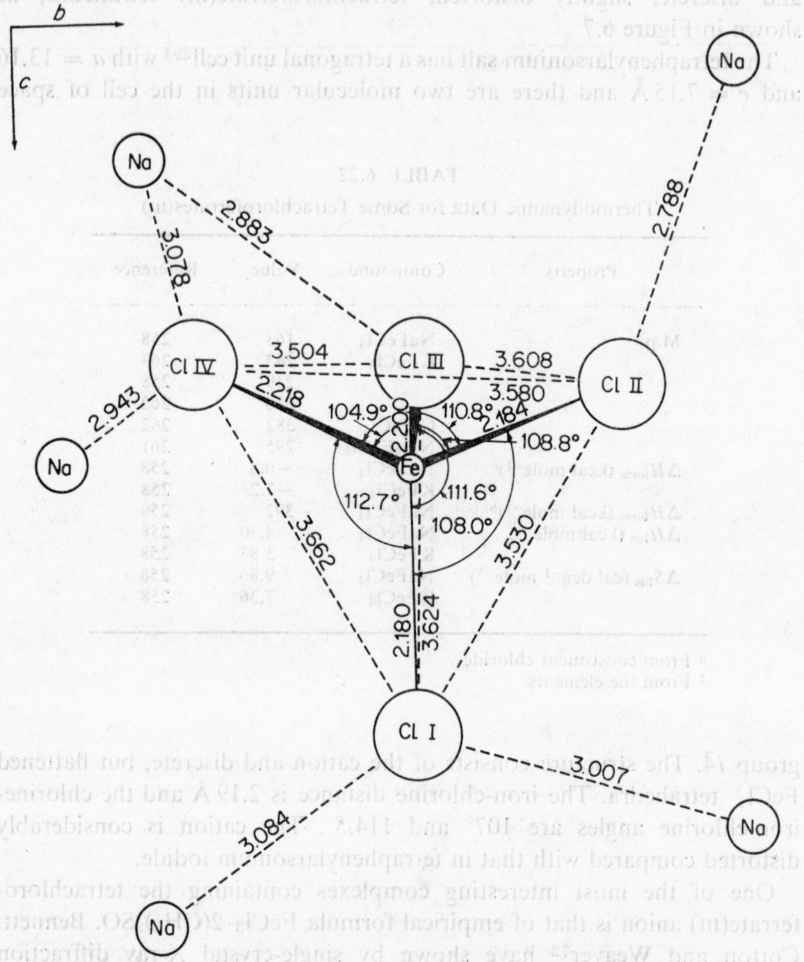

Figure 6.7 Internuclear dimensions of $NaFeCl_4$. (Reproduced by permission from R. R. Richards and N. W. Gregory, *J. Phys. Chem.*, **69**, 239 (1964))

tetrahedral tetrachloroferrate(III) anions and planar *o*-$CH_3OC_6H_4N_2$ cations.

Woodward and Taylor[295] have examined the Raman spectrum of the tetrachloroferrate(III) anion extracted from hydrochloric acid solutions of iron trichloride with ether, and assigned the spectrum on the basis of a regular tetrahedron. The following assignments were reported:

ν_1 330 ν_2 106 ν_3 385 ν_4 133 cm^{-1}

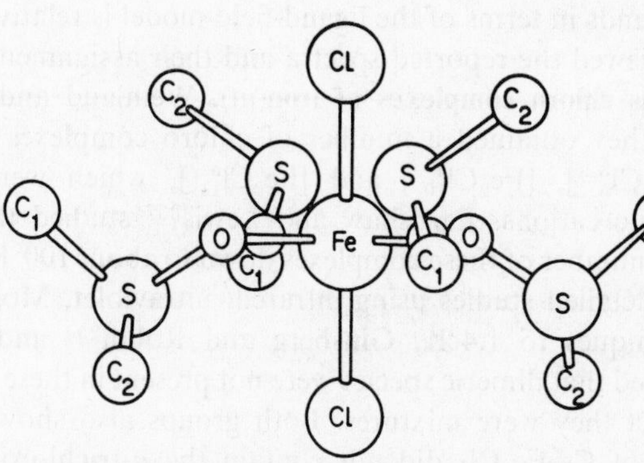

Figure 6.8 The structure of *trans* $[FeCl_2(DMSO)_4]^+$. (Reproduced by permission from M. J. Bennett, F. A. Cotton and D. L. Weaver, *Acta Cryst.*, **23**, 581 (1967))

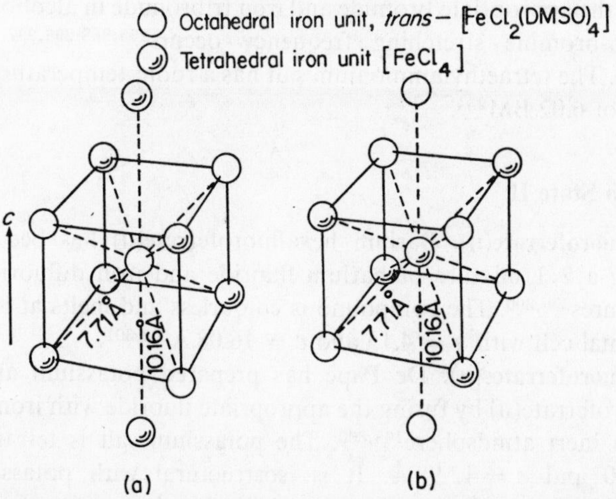

Figure 6.9 The NaTl-type packing of the Fe atoms in *trans* $[FeCl_2(DMSO)_4][FeCl_4]$. (Reproduced by permission from M. J. Bennett, F. A. Cotton and D. L. Weaver, *Acta Cryst.*, **23**, 581 (1967))

The infrared-active iron-chlorine stretching frequency occurs[255,268,273,296,297] in the region 369–385 cm^{-1}. The caesium salt has a room-temperature magnetic moment of 5.95 BM[289] and the tetraethylammonium compound obeys the Curie-Weiss law[254] with a magnetic moment of 6.00 BM and a Weiss constant of 2°.

Although the electronic absorption spectrum of the tetrachloroferrate(III) anion has been recorded in various media[254,262,267], the interpretation of the observed bands in terms of the ligand-field model is relatively difficult. Balt[298] has reviewed the reported spectra and their assignments.

Miscellaneous chloro complexes of iron(III). Weinland and Kissling[256] reported that they obtained a number of chloro complexes of iron(III), including $[Fe_2Cl_9^{3-}]$, $[Fe_2Cl_{10}^{4-}]$ and $[Fe_2Cl_{11}^{5-}]$, which were stabilized by large organic cations. Earnshaw and Lewis[257] studied the magnetic properties of a number of these complexes down to about 100°K. However, in later more detailed studies using infrared, ultraviolet, Mossbauer and magnetic techniques to 1.4°K, Ginsberg and Robin[254] and Clark and Taylor[255] showed that dimeric species were not present in these compounds and that in fact they were mixtures. Both groups also showed that the yellow β-form of $Cs_3Fe_2Cl_9$ did not contain the μ-trichlorohexachlorodiferrate(III) anion.

Tetrabromoferrates(III). Substituted ammonium, phosphonium and arsonium salts of the tetrabromoferrate(III) anion are prepared by the interaction of the appropriate bromide and iron tribromide in alcohol[254,264,267]. The iron-bromine stretching frequency occurs[255,268,296,297] at about 290 cm^{-1}. The tetraethylammonium salt has a room-temperature magnetic moment of 6.02 BM[254].

Oxidation State II

Hexafluoroferrate(II). Barium hexafluoroferrate(II) has been prepared by fusing a 2:1 mixture of barium fluoride and iron difluoride at high temperatures[299,300]. The compound is colourless and melts at 830°. It has a tetragonal cell with $a = 4.15$ and $c = 16.08$ Å[300,301].

Tetrafluoroferrates(II). De Pape has prepared potassium and barium tetrafluoroferrate(II) by fusing the appropriate fluoride with iron difluoride under an inert atmosphere[108,300]. The potassium salt is tetragonal with $a = 4.140$ and $c = 12.98$ Å. It is isostructural with potassium tetrafluoronickelate(II) (Figure 8.13, page 446). The barium compound melts at 760° and is orthorhombic with $a = 14.77$, $b = 4.25$ and $c = 5.78$ Å.

Trifluoroferrates(II). Trifluoroferrates(II) may be prepared by the interaction of equimolar amounts of alkali-metal fluoride and iron difluoride in the melt[4,104,106,108,302]. Alternatively, interaction of the same reagents in a boiling, slightly acidic solution may be used[303,304]. The ammonium salt has been prepared by the interaction of ammonium fluoride and iron dibromide in methanol[230]. Details for preparing single crystals have been reported[106,305].

The trifluoroferrates(II) are white to pale brown solids. The rubidium and caesium salts melt at 880° and 708° respectively[106].

The structures of a number of trifluoroferrates(II) have been investigated by X-ray powder diffraction techniques. The potassium and rubidium compounds adopt the perovskite structure (Figure 5.7, page 241) at room temperature, but the caesium salt is distorted. The lattice parameters, at room temperature, are given in Table 6.23.

TABLE 6.23
Symmetry and Parameters of Trifluoroferrates(II)

Compound	Symmetry	Parameters	References
$KFeF_3$	Cubic	$a = 4.12$ Å	4, 305–308
$RbFeF_3$	Cubic	$a = 4.174$ Å	106, 302
$CsFeF_3$	Hexagonal	$a = 6.158$, $c = 14.885$ Å	106

According to Okazaki and coworkers[306], potassium trifluoroferrate(II) has a rhombohedral structure at 78°K, with $a = 4.108$ Å and $\alpha = 89.85°$.

Above about 120°K the trifluoroferrates(II) obey the Curie-Weiss law, but below this temperature they become antiferromagnetic. Some magnetic properties of the trifluoroferrates(II) are summarized in Table 6.24.

TABLE 6.24
Magnetic Properties of Trifluoroferrates(II)

Compound	μ (BM)	$\theta°$	T_N (°K)	References
$NaFeF_3$	4.53	100		304
$KFeF_3$	4.27[a]		112	303
			113	309
			121	306
$RbFeF_3$	5.28	200	75	104, 302

[a] Effective magnetic moment at 300°K.

The magnetic structure of potassium trifluoroferrate(II) has been investigated by Scatturin and coworkers[310] by neutron-diffraction techniques. At 4.2°K, the neutron-diffraction pattern could be indexed on the basis of a cubic cell with $a = 8.176$ Å, that is, approximately twice the size of the room-temperature crystallographic cell.

Testardi and coworkers[311] showed that at very low temperatures rubidium trifluoroferrate(II) becomes ferromagnetic. These workers reported that there are three crystallographic transformations in this complex, each associated with a magnetic structure. Tetragonal distortion of the room-temperature cubic cell begins at 97°K where $c/a = 1.0012$. This ratio increases to 1.0034 at 86°K; at this temperature new diffraction lines appear. At 45°K a different set of diffraction lines appear. The diffraction lines at 86°K and 45°K both indicate lowering of the symmetry of the unit cell. The magnetic and structural properties, together with the thermodynamic order of the phase changes, are shown diagrammatically in Figure 6.10.

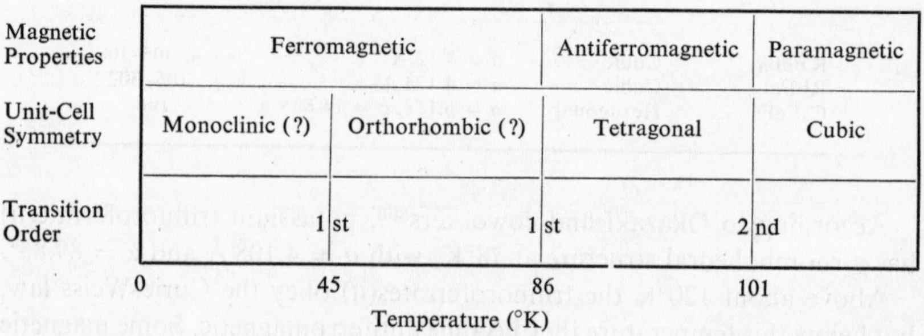

Magnetic Properties
Ferromagnetic
Antiferromagnetic
Paramagnetic
Unit-Cell Symmetry
Monoclinic (?)
Orthorhombic (?)
Tetragonal
Cubic
Transition Order
1 st
1 st
2 nd
0
45
86
101
Temperature (°K)

Figure 6.10. Structural and magnetic properties of $RbFeF_3$ as a function of temperature.

The iron-fluorine stretching frequency in potassium trifluoroferrate(II) has been observed[239] at 431 cm^{-1}.

Pentachloroferrates(II). Rubidium and caesium pentachloroferrate(II) are prepared by the interactions of the appropriate amounts of alkali-metal chloride and iron dichloride in the melt[312]. The rubidium salt melts at 467° and the caesium salt at 524°. Both compounds have tetragonal unit cells: for the rubidium compound $a = 8.794$ and $c = 14.14$ Å while for the caesium salt, $a = 9.224$ and $c = 14.741$ Å. Both salts are isostructural with caesium pentachlorocobaltate(II) (Figure 7.16, page 371).

Tetrachloroferrates(II). Alkali-metal tetrachloroferrates(II) are made by the interaction of the exact proportions of the appropriate alkali-metal chloride and iron dichloride in the melt[134,312–314]. Substituted ammonium, phosphonium and arsonium tetrachloroferrates(II) are best prepared by reaction of the chloride with iron dichloride in alcohol[267,315–318]. Other tetrachloroferrates(II) have been prepared by various techniques[268,319–321], for example, by interaction of quaternary ammonium chloride and iron dichloride in thionyl chloride.

Thermodynamic data for some tetrachloroferrates(II) are given in Table 6.25.

TABLE 6.25
Thermodynamic Properties of Tetrachloroferrates(II)

Property	Compound	Value	References
M.p.	Na_2FeCl_4	400°	314
	K_2FeCl_4	367°	313
		374°	314
		399°	134
	Rb_2FeCl_4	489°	312
	Cs_2FeCl_4	557 ± 1°	312, 313
ΔH°_{form} (kcal mole^{-1})[a]	K_2FeCl_4	−3.9	322
	$[(CH_3)_4N]_2FeCl_4$	−8.26	323
	$[(C_2H_5)_4N]_2FeCl_4$	−14.41	323

[a] From constituent chlorides.

Caesium tetrachloroferrate(II) has an orthorhombic lattice[312] with $a = 9.75$, $b = 13.02$ and $c = 7.48$ Å. The triphenylmethylarsonium salt is cubic with $a = 15.65$ Å and is isomorphous with the corresponding nickel compound[318].

The infrared spectra of some tetrachloroferrates(II) have been recorded by several groups[268,297,324], the most detailed study being that of Sabatini and Sacconi[324]. The reported[324] fundamental frequencies are given in Table 6.26.

TABLE 6.26
Infrared Spectra (cm^{-1}) of Some Tetrachloroferrates(II)

Mode	$[(CH_3)_4N]_2FeCl_4$	$[(C_2H_5)_4N]_2FeCl_4$
ν_3 (Fe—Cl stretch)	284*s*	286*s*
ν_4 (Cl—Fe—Cl bend)	126*m*	119*m*
ν_2 (Cl—Fe—Cl bend)	81*w*	77*w*

s = strong *m* = medium *w* = weak.

The tetrachloroferrates(II) have magnetic moments in the range 5.3–5.4 BM at room temperature[171,267]. Clark and coworkers[325] have made a detailed study of the magnetic properties of these compounds and suggested that the high magnetic moments can be explained if contribution from the $3d^54s^1$ configuration is assumed.

Little work has been reported on the electronic absorption spectrum of the tetrachloroferrate(II) anion. Furlani and coworkers[317] observed a single intense band at approximately 4100 cm^{-1} which they assigned to the $^5T_{2g} \leftarrow {}^5E_g$ transition on the basis of T_d symmetry.

Trichloroferrates(II). Alkali-metal trichloroferrates(II) are prepared by interaction of the stoichiometric quantities of the two chlorides in the melt[134,312–314]. The potassium salt is reported to undergo a phase transformation[313] at 305°. Both the rubidium and caesium salts have the hexagonal caesium trichloronickelate(II) structure[312] and contain condensed $FeCl_6$ octahedra. The rubidium salt has a = 7.060 and c = 6.020 Å while for caesium trichloroferrate(II) a = 7.235 and c = 6.050 Å. For the caesium salt the iron-chlorine distance is 2.49 Å and the chlorine-iron-chlorine angle is 87.06°.

Potassium trichloroferrate(II) has a room-temperature magnetic moment of 4.88 BM[326].

Tetrabromoferrates(II). These salts are prepared by methods analogous to those described for the tetrachloroferrate(II) salts[267,268,324,325]. The heat of formation of the tetraethylammonium salt from its constituent bromides[327] is −10.07 kcal mole^{-1}. The infrared spectra of several tetrabromoferrates(II) have been recorded[268,297,324] and the iron-bromine stretching frequency occurs at about 220 cm^{-1}.

The salts resemble the tetrachloroferrates(II) in having room-temperature magnetic moments of about 5.4 BM[267,325] and these values have been discussed in detail by Clark and coworkers[325]. Hamer[328] has recorded the absorption spectrum of the tetraethylammonium salt in anhydrous acetonitrile and observed a single broad band at about 3000 cm^{-1}.

Tetraiodoferrates(II). Little work has been reported on the characterization of these salts. However, the preparation and properties of the tetraiodoferrates(II) that have been reported are analogous to those of the tetrachloroferrates(II)[169,267,318,324,325].

ADDENDA

Halides

Iron trifluoride may be prepared[329] by passing anhydrous hydrogen fluoride over hydrated iron(III) nitrate, the temperature being slowly increased to 900°. The Neel temperature of the trifluoride, as determined by Mossbauer measurements, has been given as 362°K[330], 362.4°K[331] and 363.11°K[329], the agreement obviously being extremely good. Shane and Kestigian[332] report that the trifluoride has a phase transformation at 405°, where there is a change from rhombohedral to cubic symmetry.

Balt and coworkers[333] recorded the electronic spectra of iron trifluoride, ammonium hexafluoroferrate(III) and a number of aquofluoroferrates(III). Their results are summarized in Table 6.27.

TABLE 6.27
Electronic Spectra (cm^{-1}) of Iron(III) Fluoro Compounds

Compound	$^4T_{1g} \leftarrow {}^6A_{1g}$	$^4T_{2g} \leftarrow {}^6A_{1g}$	$^4E_g \leftarrow {}^6A_{1g}$
FeF_3	16,350	21,750	25,000
$[FeF_6]^{3-}$	14,600	20,000	25,500
$[FeF_3(H_2O)_3]$	13,700	19,800	24,600
$[FeF_2(H_2O)_4]^+$	13,150	18,500	24,600
$[FeF(H_2O)_5]^{2+}$	12,800	18,500	24,600

A number of halo alkoxides of iron(III) have been reported in the literature[334,335], the most interesting being those characterized by Kakos and Winter[335]. These workers prepared three different types of complex: (a) $Fe_4X_6(OCH_3)_6 \cdot 4CH_3OH$ (X = Cl, Br), (b) $Fe_4X_3(OCH_3)_9$ (X = Cl, Br) and (c) $FeCl(OCH_3)_2$. On the basis of very detailed magnetic studies it was concluded that in (a) the iron atoms are at the corners of a rhombus, that in (b) they are at the corners of a tetrahedron, while in (c) the iron atoms are situated at the corners of an isosceles triangle.

West and coworkers[336–339] have prepared a number of *N*-substituted salicylaldimine derivatives of iron trichloride. Complexes of the types $FeCl_3(SalH\text{-}NR)_2$, $FeCl(Sal\text{-}NR)_2$ and FeClL, where SalH-NR = *N*-substituted salicylaldimine, Sal-NR = the *N*-substituted salicylaldiminato group and LH_2 = *N*-(2-hydroxyphenyl)salcylaldimine, were characterized by magnetic, spectral, conductimetric and mass spectrometric techniques. The first two types of complex are monomeric, the ligand acting as a bidentate and they are probably five coordinate. For the third type of complex, the ligand acts as a tridentate and the magnetic properties suggest a dinuclear five-coordinate structure. It was also found[339] that when $[FeClL]_2$ was refluxed in pyridine (py), black crystalline $FeClL(py)_3$ was isolated. It was suggested that in this case the iron atom was octahedral with one molecule of lattice pyridine. Some properties of a number of the complexes described above are summarized in Table 6.28. Baker and coworkers[340] have reported detailed magnetic, Mossbauer and spectroscopic studies on $[Fe(salen)Cl]_2$.

Singh and Rivest[341] have prepared a number of adducts of the type $FeCl_3L$, where L is a nitrogen, phosphorus or arsenic donor molecule, by reacting triiron decacarbonyl with the appropriate ligand in chloroform solution. Some properties of these adducts, which appear to be monomeric and tetrahedral, are given in Table 6.29.

TABLE 6.28

Iron(III) Derivatives of *N*-Substituted Salicylaldimines

Type	R	M.p.	μ(BM)	$\theta°$	ν_{Fe-Cl} (cm^{-1})
$FeCl_3(SalH\text{-}NR)_2$	Phenyl	230°	5.95	25	
	p-Tolyl	222°	6.0		344, 325
	Ethyl	192°	5.90	20	
$FeCl(Sal\text{-}NR)_2$	n-Propyl	151°	5.91	6	354
	n-Butyl	109°	6.05		
	Phenyl	168°	6.0		
	Benzyl	181°			
$[FeClL]_2$			5.04		315
$FeClL(py)_3$			5.95		

TABLE 6.29

Adducts of the Type $FeCl_3L$

L	Colour	M.p.	ν_{Fe-Cl} (cm^{-1})
Aniline	Yellow		375, 290
Benzamide	Brown	158°	375, 288
Triphenylphosphine	Dark yellow	118°	370, 320
Triphenylarsine	Dark yellow	130°	360, 330

The green 1:1 adduct between iron trichloride and 1,4-dioxan has been assigned[342] a trigonal-bipyramidal structure. It has a room-temperature magnetic moment of 6.06 BM and an iron-chlorine stretching frequency of 380 cm^{-1}.

Fitzsimmons and coworkers[343] have used Mossbauer spectra in the elucidation of the structures of $FeCl_3(phen)$ and $Fe_2Cl_6(phen)_3$. Two isomers of the former complex were prepared and the experimental evidence indicated the presence of dimeric chloro-bridged species, although the nature of the differences between the two isomers was not understood. For the latter complex $[FeCl_2(phen)_2]^+[FeCl_4(phen)^-]$ was proposed as the structure. The Mossbauer, electronic and infrared spectra and the magnetic properties of the dinuclear oxygen-bridged species $[(phen)_2FeClOClFe(phen)_2]Cl_2 \cdot xH_2O$ were also investigated[343,344].

Other complexes of iron trichloride to be investigated include the phosphorus oxide trichloride[345], 1,2-bis(methylsulphonyl)ethane[346], 2,2′,2″-terpyridyl[344] and pyrazole[347] adducts and the *N,N*-diisopropyl-dithiocarbamato substitution derivative[348].

Lumme and Junkkarinen[349] have studied the thermal decomposition of iron trichloride hexahydrate in both static and dynamic systems. The decomposition temperatures are given in Table 6.30.

TABLE 6.30

Thermal Decomposition of Iron Trichloride Hexahydrate

Reaction	Temperature Range	
	Static Air	Dynamic Nitrogen
$FeCl_3 \cdot 6H_2O \rightarrow FeCl_3 \cdot 3H_2O + 3H_2O$	62–150°	
$FeCl_3 \cdot 6H_2O \rightarrow FeCl_3 \cdot 2.5H_2O + 3.5H_2O$		40–155°
$FeCl_3 \cdot 3H_2O \rightarrow FeOCl + HCl + 2H_2O$	150–250°	
$FeCl_3 \cdot 2.5H_2O \rightarrow FeOCl + HCl + 1.5H_2O$		155–300°
$FeOCl + 0.25O_2 \rightarrow 0.5Fe_2O_3 + 0.5Cl_2$	250–487°	
$FeOCl \rightarrow Fe + 0.5O_2 + 0.5Cl_2$		300–635°

Iron tribromide may be prepared by the action of liquid bromine on iron dibromide at room temperature[335]. Substituted salcylaldimine derivatives of iron tribromide, similar to those described for the chloride, have been reported[336–339].

Iron difluoride has been prepared[350] by heating iron powder in anhydrous hydrogen fluoride at 900°. There have been further studies on the electronic[351] and far infrared[352] spectra and magnetic properties[353–357] of iron difluoride.

Anhydrous iron dichloride has been prepared[358] by dissolving iron wire in hydrochloric acid and dehydrating the product at 200° in vacuo. The dibromide and diiodide have also been prepared in the same way using the appropriate hydrohalic acid[358]. Johnson and Dash[359] have made a detailed Mossbauer study of iron dichloride as a function of temperature, while the Jahn-Teller splitting of the $^5E_g \leftarrow {}^5T_{2g}$ transition has been investigated by Winter[358] and by Jones[351].

Bancroft and coworkers[360] have prepared a number of new complexes of low-spin iron(II) and have shown that the Mossbauer quadrupole splittings are useful in determining stereochemical arrangements.

The preparation and characterization of a number of adducts of iron dichloride have been reported and some properties of a number of these compounds are summarized in Table 6.31.

TABLE 6.31

Adducts of the Type $FeCl_2L_n$

Compound and Properties	References
L = acetamide, $n = 2$, white, m.p. = 116°, $\nu_{Fe-Cl} = 350$ cm^{-1}, $\nu_{Fe-O} = 420$ cm^{-1}, tetrahedral	341
L = formamide, $n = 2$, yellow, m.p. = 200°, $\nu_{Fe-Cl} = 373, 330$ cm^{-1}, $\nu_{Fe-O} = 400$ cm^{-1}, tetrahedral	341
L = *N*-methylformamide, $n = 2$, white, m.p. = 140°, $\nu_{Fe-Cl} = 380, 360$ cm^{-1}, $\nu_{Fe-O} = 475$ cm^{-1}, tetrahedral	341
L = quinoline, $n = 2$, yellow, $\mu = 5.20$ BM, $\nu_{Fe-Cl} = 330, 298$ cm^{-1}, tetrahedral	361
L = 3-methylisoquinoline, $n = 2$, $\mu = 5.26$ BM, $\nu_{Fe-Cl} = 334, 299$ cm^{-1}, tetrahedral	361
L = 2,2′2″-terpyridyl (terpy), $n = 1$, $\mu = 3.85$ BM, five coordinate	362
L = bis[2-(pyridyl)ethyl]amine (dpea), $n = 1$, mustard, $\mu = 5.28$ BM, five coordinate	363
L = pentamethylenetetrazole (PMT), $n = 1$, yellow, m.p. = 140°, $\mu = 5.32$ BM, octahedral	364
L = tris(*o*-diphenylphosphinophenyl)phosphine (QP), $n = 1$, violet, m.p. = 270–274°, $\mu = 3.2$ BM, ionic with trigonal-bipyramidal cation, [FeCl(QP)]Cl, perchlorate also isolated	365
L = di-2-pyridyl ketone (DPK), $n = 1$, $\nu_{Fe-Cl} = 324, 302$ cm^{-1}, tetrahedral	366
L = 1,4-dioxan, $n = 1$, white, $\mu = 5.56$ BM, $\nu_{Fe-Cl} = 235, 210$ cm^{-1}, octahedral	342
L = thioacetamide, $n = 1$, $\nu_{Fe-Cl} = 324, 302$ cm^{-1}, tetrahedral	367

Other ligands that have been reacted with iron dichloride include methanol, pyridine and iosquinoline[358], hydrazine[368], 2,9-dimethyl-1,10-phenanthroline[369], pyrazole[347], acrylonitrile[370], triphenylphosphine[371], triphenylphosphine oxide[361], *o*-phenylenebis(dimethylarsine)[372] and benzothiazole[373].

The thermal decomposition of iron dichloride tetrahydrate has been reinvestigated[349] while Johnson and Ridout[374] have examined the Mossbauer spectra of this hydrate in the antiferromagnetic state. There has been a further study[375] of the low temperature magnetic properties of iron dichloride dihydrate, while Winter[358] has investigated the $^5E_g \leftarrow {}^5T_{2g}$ transition as a function of temperature.

Both iron dibromide and diiodide react with a number of donor molecules to form adducts which resemble those of the dichloride discussed above[358,361,362,365,367,371–373]. For example, Venanzi and coworkers[365] report the isolation of adducts of the types [FeX(QP)]X and [FeX(QP)]BPh_4

where X = Br or I and QP = tris(*o*-diphenylphosphinophenyl)phosphine. Magnetic and spectral studies indicate that the cation has a trigonal-bipyramidal configuration.

The first iron(II) compound to have a trigonal-bipyramidal configuration confirmed by X-ray analysis is $FeBr_2(Me_6tren)$[376], where Me_6tren is tris(2-dimethylaminoethyl)amine. The complex is isomorphous with the corresponding cobalt complex (Figure 7.14, page 360) having a cubic unit cell with a = 12.185 Å. The internuclear dimensions are given in Table 6.32.

TABLE 6.32

Internuclear Dimensions in $[FeBr(Me_6tren)]^+$

Distance	Value (Å)	Angle	Value
Fe—Br(1)	2.482	N(1)—Fe—N(2)	81.3°
Fe—N(1)	2.21	N(2)—Fe—Br(1)	98.7°
Fe—N(2)	2.15	N(2)—Fe—N(2′)	117.8°

Iron diiodide forms a green 1:1 adduct with the Schiff base formed between 2,6-disulphenamidopyridine and 2-pyridinealdehyde which has a room-temperature magnetic moment of 4.9 BM[377]. The magnetic moment and electronic spectrum are consistent with octahedral symmetry.

Burbridge and Goodgame[378] have reported the electronic and Mossbauer spectra of a number of iron dihalide hydrates. Their results show quite clearly that $FeBr_2 \cdot 6H_2O$ is in fact $[FeBr_2(H_2O)_4] \cdot 2H_2O$.

Complex Halides

Babel[379] has prepared several penta- and tetrafluoroferrates(III) by the interaction of the appropriate fluorides in the melt. The unit-cell parameters and magnetic properties of these complexes are given in Table 6.33.

TABLE 6.33

Properties of Fluoroferrates(III)

Complex	Symmetry	a (Å)	b (Å)	c (Å)	μ (BM)[a]	T_N (°K)
K_2FeF_5					4.82	100
$KFeF_4$	Orthorhombic	7.596	7.768	12.27	3.86	230
$RbFeF_4$	Tetragonal	7.633		6.273	3.96	200

[a] Room temperature value.

Adams and Morris[380] report iron–chlorine vibrational frequencies of 272, 248, 227, 181 and 166 cm^{-1} for the complex $[Co(NH_3)_6][FeCl_6]$. The first three bands were considered to be associated with the ν_3 mode while the remaining two were assigned to the ν_4 mode. The Mossbauer spectrum of this complex has been recorded[343]. Yamatera and Kato[381] have recorded the electronic spectra of $[NH_4]_4[FeCl_6][SbCl_6]$ and $Cs_3Fe_2Cl_9$. Bands were observed at 8,900 and 12,800 cm^{-1} for the ammonium salt and at 9,800 and 13,600 cm^{-1} for the caesium salt of the dinuclear anion. The lower energy band was assigned to the ${}^4T_{1g}(G) \leftarrow {}^6A_{1g}$ transition while the higher energy band was assigned to the ${}^4T_{2g}(G) \leftarrow {}^6A_{1g}$ transition.

Iron trichloride reacts with *cis*-1,2-bis(dimethylarsino)ethylene (*cis*-edas) to give a crimson complex which has been shown[382] to be $[FeCl_2(cis\text{-edas})_2][FeCl_4]$. The complex with the *trans* form of the ligand does not form, or it is too soluble to be precipitated[382]. The electronic spectrum of the complex clearly shows that the cation contains octahedrally co-ordinated iron(III).

The crystal structure of tetrachlorophosphonium tetrachloroferrate(III), $PCl_4^+FeCl_4^-$, has been investigated by Kistenmacher and Stucky[383]. The yellow complex has an orthorhombic unit cell of space group *Pbcm*, containing four formula units, with $a = 6.231$, $b = 13.479$ and $c = 14.078$ Å. The structure consists of independent, approximately tetrahedral PCl_4^+ and $FeCl_4^-$ ions. The important internuclear dimensions are given in Table 6.34.

TABLE 6.34

Internuclear Dimensions in $[PCl_4^+][FeCl_4^-]$

Distance	Value (Å)	Angle	Value
Fe—Cl(1)	2.182	Cl(1)—Fe—Cl(2)	108.98°
Fe—Cl(2)	2.187	Cl(1)—Fe—Cl(3)	108.20°
Fe—Cl(3)	2.187	Cl(1)—Fe—Cl(1′)	113.56°
		Cl(2)—Fe—Cl(3)	108.83°

The complex has an iron-chlorine stretching frequency[383] of 376 cm^{-1}.

The infrared and Raman spectra of tetraethylammonium tetrachloro- and tetrabromoferrate(III) have been recorded by Avery and coworkers[384]. Their results are tabulated in Table 6.35.

Reedijk and Groeneveld[385] have prepared a number of complex tetrachloroferrates(III) by reacting a suitable chloride with iron trichloride in acetonitrile. Typical complexes isolated were $[Mg(CH_3CN)_6^{2+}][FeCl_4^-]_2$, $[Ba(CH_3CN)_8^{2+}][FeCl_4^-]_2$ and $[Zn(CH_3CN)_6^{2+}][FeCl_4^-]_2$. These complexes were characterized by X-ray diffraction, electronic and infrared spectra, and in some cases by Mossbauer spectroscopy. The iron-chlorine stretching frequencies of these complexes were found to lie in the range 376–370 cm^{-1}. The Mossbauer spectra of several other tetrahaloferrates(III) have been

TABLE 6.35
Infrared (IR) and Raman (R) Spectra (cm^{-1}) of Tetrahaloferrates(III)

Complex	Method	State of Sample	ν_1	ν_2	ν_3	ν_4
$(C_2H_5)_4NFeCl_4$	R	Solid	330	114	378	136
	R	CH_3NO_2 solution	330	106		136
	IR	Solid	330		378	138
$(C_2H_5)_4NFeBr_4$	R	Solid	200		290	
	R	CH_3NO_2 solution	201		285	95
	IR	Solid			284	118

recorded[343,386]. Both tetraethylammonium tetrachloro- and tetrabromoferrate(III) become antiferromagnetic at low temperatures, the Neel temperatures being 3.0°K and 3.9°K respectively[386]. Balt[387,388] has made further studies on the electronic spectra of the tetrachloro- and tetrabromoferrate(III) anions.

A number of tetrafluoro- and trifluoroferrates(II), prepared by direct interaction in the melt, have been the subject of crystallographic studies, the results of which are summarized in Table 6.36.

TABLE 6.36
Crystallographic Data for Some Fluoroferrates(II)

Compound	Symmetry	*a* (Å)	*b* (Å)	*c* (Å)	Reference
Rb_2FeF_4	Tetragonal	4.20		13.38	389
		4.176		13.6	350
$BaFeF_4$	Orthorhombic	14.238	14.837	5.829	390
$NaFeF_3$	Orthorhombic	5.495	5.672	7.890	389
$TlFeF_3$	Cubic	4.188			389

Sodium trifluoroferrate(II) decomposes[389] at 788° while rubidium tetrafluoroferrate(II) melts[389] at 730°. In the potassium fluoride-iron difluoride system, $K_3Fe_2F_7$, as well as potassium tetra- and trifluoroferrate(II), has been isolated[389].

Rubidium tetrafluoroferrate(II) has been shown[350] to have a K_2NiF_4 type magnetic structure. The compound has a broad maximum in its magnetic susceptibility at 90°K and has a large magnetic anisotropy.

There have been further magnetic studies on potassium[391] and rubidium[392,393] trifluoroferrate(II). Jones[351] has investigated the Jahn-Teller splitting of the $^5E_g(D) \leftarrow {}^5T_{2g}(G)$ transition in potassium trifluoroferrate(II).

Gill and Taylor[394] have reported the preparation of tetraethylammonium tetrachloro- and tetrabromoferrate(II) by interaction of the appropriate halides in alcohol. The room temperature magnetic moments of these complexes are 5.39 BM and 5.44 BM respectively[361]. The Mossbauer spectra of several tetrahaloferrates(II) have been reported[361,395].

Heath, Martin and Stewart[396,397] have made a very thorough investigation of the nature of the products obtained by passing hydrogen sulphide through iron trichloride-acetylacetone-ethanol-hydrochloric acid and iron dibromide-acetylacetone-ethanol-hydrobromic acid solutions. The dark violet coloured solids, of empirical formula $FeX_4(C_5H_7S_2)_2$, where $C_5H_7S_2$ is analytically equivalent to dithioacetylacetone, were first reported by Knauer and coworkers[398] and by Ouchi and his group[349,400]. These workers suggested that the iron atoms were in an octahedral environment, the ligand acting either as a mono- or bidentate. Heath, Martin and Stewart[396,397] showed quite conclusively that such a structure is incorrect and that the compounds were in fact tetrahaloferrate(II) salts of the 3,5-dimethyldithiolium cation, that is, they should be formulated as $[(C_5H_7S_2)^+]_2[FeX_4^{2-}]$. In addition, 4-phenyl-1,2-dithiolium tetrachloroferrate(II), $[C_9H_7S_2^+]_2[FeCl_4^{2-}]$ has been prepared[397]. Ouchi and coworkers[399] have prepared a derivative from dithiobenzoylacetone, $C_{10}H_9S_2$, and it would appear that this complex is also of the dithiolium type. Some properties of the above complexes are summarized in Table 6.37.

TABLE 6.37

Dithiolium Tetrahaloferrates(II)

Compound	μ (BM)	θ°	ν_{Fe-Cl} (cm^{-1})	References
$(C_5H_7S_2)_2FeCl_4$	5.40	20	280	397
$(C_9H_7S_2)_2FeCl_4$	5.40	11	283	397
$(C_{10}H_9S_2)_2FeCl_4$	5.10			399
$(C_5H_7S_2)_2FeBr_4$	5.46	12	217	397
	5.51			399

Mossbauer and electronic spectra have also been reported for these complexes.

The formulation of the compounds as dithiolium complexes[396,397] has been confirmed by a single-crystal study of 3,5-dimethyldithiolium tetrachloroferrate(II) by Freeman and coworkers[401]. The complex is monoclinic with $a = 17.71$, $b = 7.76$, $c = 15.86$ Å and $\beta = 122.2°$. There are four molecules in the unit cell of space group $C2/c$. The structure consists of planar dithiolium cations and approximately tetrahedral tetrachloroferrate(II) anions. The iron-chlorine bond distances in the anion are 2.34 Å and 2.31 Å while the chlorine-iron-chlorine angles are 116° and 112°. The anions and cations are arranged in such a way that the tetrachloroferrate(II) anion makes six chlorine-sulphur contacts which are significantly shorter than the sum of the covalent radii. This feature is the reason for the dark colour of the complex.

Burbridge and Goodgame[361] have prepared several complexes of the type $M^+FeX_3L^-$, where M^+ is a suitable cation such as tetraethylammonium, X is Cl, Br or I, and L is a neutral ligand such as quinoline. These complexes were prepared by adding an ethanolic solution of iron dihalide to an ethanol-2,2-dimethoxypropane solution of the neutral ligand and quaternary ammonium halide. The magnetic moments and iron-halogen stretching frequencies of these pale cream to dark red tetrahedral complexes are given in Table 6.38.

TABLE 6.38

Substituted Tetraethylammonium Tetrahaloferrates(II)

Halogen	Ligand	μ (BM)	ν_{Fe-X} (cm^{-1})
Cl	Quinoline	5.30	306, 289, 272
Cl	3-methylisoquinoline	5.36	300, 285, 272
Br	Quinoline	5.35	241, 226, 208
Br	3-methylisoquinoline	5.36	234

REFERENCES

1. G. S. F. Hazeldean, R. S. Hyholm and R. V. Parish, *J. Chem. Soc.*, **1966,** A, 162.
2. K. Knox and D. W. Mitchell, *J. Inorg. Nucl. Chem.* **21,** 253 (1961).
3. D. B. Shinn, D. S. Crocket and H. M. Haendler, *Inorg. Chem.*, **5,** 1927 (1966).
4. R. de Pape, *Compt. Rend.*, **260,** 4527 (1965).
5. M. A. Hepworth, K. H. Jack, R. D. Peacock and G. J. Westland, *Acta Cryst.*, **10,** 63 (1957).

6. *U.S. Patent*, 2,952,514 (1960).
7. A. W. Jache and G. H. Cady, *J. Phys. Chem.*, **56,** 1106 (1952).
8. B. M. Wanklyn, *J. Inorg. Nucl. Chem.*, **27,** 481 (1965).
9. D. F. Stewart, unpublished observations.
10. Z. F. Zmbov and J. L. Margrave, *J. Inorg. Nucl. Chem.*, **29,** 673 (1967).
11. E. O. Wollan, H. R. Child, W. C. Koehler and M. K. Wilkinson, *Phys. Rev.*, **112,** 1132 (1958).
12. H. Bizette, R. Mainard and J. Picard, *Compt. Rend.*, **260,** 5508 (1965).
13. R. J. H. Clark, *J. Chem. Soc.*, **1964,** 417.
14. A. H. Nielsen, *Z. anorg. allgem. Chem.*, **244,** 85 (1940).
15. I. Maak, P. Eckerlin and A. Rabenau, *Naturwissenschaften*, **48,** 218 (1964).
16. G. Teufer, *Acta Cryst.*, **17,** 1480 (1964).
17. P. Hagenmuller, J. Portier, J. Cadiou and R. de Pape, *Compt. Rend.*, **260,** 4768 (1965).
18. B. L. Chamberland and A. W. Sleight, *Solid State Commun.*, **5,** 765 (1967).
19. S. S. Todd and J. P. Coughlin, *J. Am. Chem. Soc.*, **73,** 4184 (1951).
20. S. Blairs and R. A. J. Shelton, *J. Inorg. Nucl. Chem.*, **28,** 1855 (1966).
21. B. P. Tarr, *Inorg. Syn.*, **3,** 181 (1950).
22. G. Braur, *Handbook of Preparative Inorganic Chemistry*, 2nd ed. Academic Press, New York, 1963.
23. Y. I. Ivashentsev, *Dokl. 7–i Nauch. Konf. Posvyashchen.* 40–*letiya Velikoi Oktyatr. Sots. Revolyutsii, Tomsk. Univ.*, **1957,** 157.
24. A. B. Bardawil, F. N. Collier and S. Y. Tyree, *Inorg. Chem.*, **3,** 149 (1964).
25 T. E. Austin and S. Y. Tyree, *J. Inorg. Nucl. Chem.*, **14,** 141 (1960).
26. A. B. Bardawil, F. N. Collier and S. Y. Tyree, *J. Less-Common Metals*, **9,** 20 (1965).
27. H. Schafer, *Z. anorg. allgem. Chem.*, **259,** 53 (1949).
28. H. Schafer, *Z. anorg. allgem. Chem.*, **259,** 75 (1949).
29. H. Schafer, *Z. anorg. allgem. Chem.*, **266,** 269 (1951).
30. H. Schafer, *Z. anorg. allgem. Chem.*, **259,** 265 (1949).
31. A. R. Pray, *Inorg. Syn.*, **5,** 153 (1957).
32. J. H. Freeman and J. L. Smith, *J. Inorg. Nucl. Chem.*, **7,** 224 (1958).
33. J. W. Mellor, *A Comprehensive Treatise on Inorganic and Theoretical Chemistry*, Vol. 14, Longmans, London, p. 43.
34. W. F. Linke, *J. Phys. Chem.*, **60,** 91 (1956).
35. P. Kovacic and N. O. Brace, *J. Am. Chem. Soc.*, **76,** 5491 (1954).
36. J. D. Christian and N. W. Gregory, *J. Phys. Chem.*, **71,** 1579 (1967).
37. R. R. Hammer and N. W. Gregory, *J. Phys. Chem.*, **66,** 1705 (1962).
38. L. E. Wilson and N. W. Gregory, *J. Phys. Chem.*, **62,** 433 (1958).
39. K. Sano, *J. Chem. Soc. Japan*, **59,** 1069 (1938).
40. H. F. Johnstone, H. C. Weingartner and W. E. Winsche, *J. Am. Chem. Soc.*, **64,** 241 (1942).
41. V. V. Pechkovskii and N. N. Vorob'ev, *Russ. J. Inorg. Chem.*, **9,** 6 (1964).
42. W. Kangro and E. Petersen, *Z. anorg. allgem. Chem.*, **261,** 157 (1950).
43. H. Schafer and L. Bayer, *Z. anorg. allgem. Chem.*, **271,** 338 (1953).
44. C. M. Cook, *J. Phys. Chem.*, **66,** 219 (1962).
45. C. G. Maier, *Bur. of Mines, Tech. Paper*, **1925,** 360.
46. M. F. Koehler and J. P. Coughlin, *J. Phys. Chem.*, **63,** 605 (1959).
47. J. C. M. Li and N. W. Gregory, *J. Am. Chem. Soc.*, **74,** 4670 (1952).

48. N. Wooster, *Z. Krist.*, **83,** 35 (1932).
49. N. W. Gregory, *J. Am. Chem. Soc.*, **73,** 472 (1951).
50. K. Geissberger, *Z. anorg. allgem. Chem.*, **258,** 361 (1949).
51. J. W. Cable, M. K. Wilkinson, E. O. Wollan and W. C. Koehler, *Phys. Rev.*, **127,** 714 (1962).
52. O. Hassel and H. Viervoll, *Tids. Kjemi, Bergvesen Met.*, **3,** 97 (1943).
53. E. Z. Zasorin, N. G. Rambidi and P. A. Akishin, *Zh. Strukt. Khim.*, **4,** 910 (1963).
54. O. Hassel and H. Viervoll, *Acta Chem. Scand.*, **1,** 149 (1947).
55. J. K. Wilmshurst, *J. Mol. Spectroscopy*, **5,** 343 (1960).
56. C. Starr, F. Bitter and A. R. Kaufmann, *Phys. Rev.*, **58,** 977 (1940).
57. H. H. Wickman and A. M. Trozzolo, *Inorg. Chem.*, **7,** 63 (1968).
58. B. F. Hoskins, R. L. Martin and A. H. White, *Nature*, **211,** 627 (1966).
59. R. L. Martin and A. H. White, *Inorg. Chem.*, **6,** 712 (1967).
60. H. H. Wickman, A. M. Trozzolo, H. J. Williams, G. W. Hull and F. R. Merritt, *Phys. Rev.*, **155,** 563 (1967).
61. M. Gerloch, J. Lewis, F. E. Mabbs and A. Richards, *Nature*, **212,** 809 (1966).
62. M. Gerloch and F. E. Mabbs, *J. Chem. Soc.*, **1967,** A, 1598.
63. M. Gerloch, J. Lewis, F. E. Mabbs and A. Richards, *J. Chem. Soc.*, **1968,** A, 112.
64. W. L. Groeneveld, *Rec. Trav. Chim.*, **71,** 1152 (1952).
65. Y. A. Fialkov and Y. S. Bur'yanov, *Dokl. Akad. Nauk SSSR.*, **92,** 585 (1953).
66. Y. A. Fialkov and Y. B. Bur'yanov, *Zh. Obshch. Khim.*, **25,** 2391 (1955).
67. B. A. Voitovich, *Titan i ego Splavy, Akad. Nauk SSSR, Inst. Met.*, **1961,** 188.
68. M. Baaz, V. Gutmann and L. Huebner, *Monatsh. Chem.*, **92,** 707 (1961).
69. V. V. Dadape and M. R. A. Rao, *J. Am. Chem. Soc.*, **77,** 6192 (1955).
70. R. Chand, G. S. Handard and K. Lal, *J. Indian Chem. Soc.*, **35,** 28 (1958).
71. F. A. Cotton and R. Francis, *J. Am. Chem. Soc.*, **82,** 2986 (1960).
72. M. J. Bennett, F. A. Cotton and D. L. Weaver, *Acta Cryst.*, **23,** 581 (1967).
73. P. A. McCusker, T. J. Lane and S. Kennard, *J. Am. Chem. Soc.*, **81,** 2974 (1959).
74. R. K. Multani, *Indian J. Chem.*, **2,** 506 (1964).
75. E. Loyd, C. B. Brown, D. Glynwyn, R. Bonnell and W. J. Jones, *J. Chem. Soc.*, **1928,** 658.
76. B. P. Susz and P. Chalandon, *Helv. Chim. Acta*, **41,** 1332 (1958).
77 R. C. Paul, R. Parkash and S. S. Sandhu, *Z. anorg. allgem. Chem.*, **352,** 322 (1967).
78. K. Issleib and A. Brack, *Z. anorg. allgem. Chem.*, **277,** 258 (1954).
79. M. J. Frazer, W. Gerrard and R. Twaits, *J. Inorg. Nucl. Chem.*, **25,** 637 (1963).
80. D. J. Phillips and S. Y. Tyree, *J. Am. Chem. Soc.*, **83,** 1806 (1961).
81. G. A. Barclay and R. S. Nyholm, *Chem. Ind.* (*London*), **1953,** 378.
82. W. E. Bull, S. K. Madan and J. E. Willis, *Inorg. Chem.*, **2,** 303 (1963).
83. E. E. Aynsley, N. N. Greenwood and J. B. Leach, *Chem. Ind.* (*London*), **1966,** 379.
84. K. Issleib and A. Krebich, *Z. anorg. allgem. Chem.*, **313,** 338 (1961).

85. R. J. Berni, R. R. Benerito, W. M. Ayres and H. B. Jonassen, *J. Inorg. Nucl. Chem.*, **25,** 807 (1963).
86. C. M. Harris and T. N. Lockyer, *Chem. Ind.* (*London*), **1958,** 1231.
87. M. D. Lind, *J. Chem. Phys.*, **47,** 990 (1967).
88. H. Schafer, *Z. anorg. allgem. Chem.*, **264,** 249 (1951).
89. H. Schafer, *Z. anorg. allgem. Chem.*, **260,** 279 (1949).
90. S. Goldsztaub, *Compt. Rend.*, **198,** 667 (1934).
91. S. Goldsztaub, *Bull. Soc. France Mineral.*, **58,** 6 (1935).
92. R. W. G. Wyckoff, *Crystal Structures*, Vol. 1, 2nd ed. Interscience, New York, 1963.
93. P. Hagenmuller, J. Portier, B. Barbe and P. Bouchier, *Z. anorg. allgem. Chem.*, **355,** 209 (1967).
94. R. O. MacLaren and N. W. Gregory, *J. Phys. Chem.*, **59,** 184 (1955).
95. N. W. Gregory and B. A. Thackrey, *J. Am. Chem. Soc.*, **72,** 3176 (1950).
96. P. M. Druce, M. F. Lappert and P. N. K. Riley, *Chem. Commun.*, **1967,** 486.
97. R. R. Hammer and N. W. Gregory, *J. Phys. Chem.*, **68,** 963 (1964).
98. N. W. Gregory and R. O. MacLaren, *J. Phys. Chem.*, **59,** 110 (1955).
99. E. Konig, Landolt-Bornstein, *Magnetic Properties of Coordination and Organometallic Transition Metal Compounds*, Vol. 2, Springer-Verlag, New York, 1966.
100. R. M. Golding and L. E. Orgel, *J. Chem. Soc.*, **1962,** 363.
101. P. J. Crowley and H. M. Haendler, *Inorg. Chem.*, **1,** 904 (1962).
102. R. S. Nyholm, *J. Chem. Soc.*, **1950,** 851.
103. S. K. Dhar and F. Basolo, *J. Inorg. Nucl. Chem.*, **25,** 27 (1963).
104. G. K. Wertheim, H. J. Guggenheim, H. J. Williams and D. N. E. Buchanan, *Phys. Rev.*, **158,** 446 (1967).
105. E. L. Muetterties and J. E. Castle, *J. Inorg. Nucl. Chem.*, **18,** 148 (1961).
106. M. Kestigian, F. D. Leipzig, W. J. Croft and R. Guidoboni, *Inorg. Chem.*, **5,** 1462 (1966).
107. E. Catalano and J. W. Stout, *J. Chem. Phys.*, **23,** 1803 (1955).
108. R. de Pape, *Bull. Soc. Chim. France*, **1965,** 3489.
109. J. W. Stout and E. Catalano, *J. Chem. Phys.*, **23,** 2013 (1955).
110. R. A. Kent and J. L. Margrave, *J. Am. Chem. Soc.*, **87,** 4754 (1965).
111. J. W. Stout and S. A. Reed, *J. Am. Chem. Soc.*, **76,** 5279 (1954).
112. W. H. Baur, *Acta Cryst.*, **11,** 488 (1958).
113. W. H. Baur, *Naturwissenschaften*, **44,** 349 (1957).
114. T. Moriya, K. Motizuki, J. Kanamori and T. Nagamiya, *J. Phys. Soc. Japan*, **11,** 211 (1956).
115. S. P. S. Porto, P. A. Fleury, T. C. Damen, *Phys. Rev.*, **154,** 522 (1967).
116. M. Balkanski, P. Moch and G. Parisot, *J. Chem. Phys.*, **44,** 940 (1966).
117. J. W. Halley, *Phys. Rev.*, **149,** 423 (1966).
118. J. W. Halley and I. Silvera, *J. Appl. Phys.*, **37,** 1226 (1966).
119. R. C. Ohlmann and M. Tinkham, *Phys. Rev.*, **123,** 425 (1961).
120. M. Tinkham, *J. Appl. Phys.*, **33,** 1248 (1962).
121. P. A. Fleury, S. P. S. Porto, L. E. Cheesman and H. J. Guggenheim, *Phys. Rev. Letters*, **17,** 87 (1966).
122. I. Silvera and J. W. Halley, *Phys. Rev.*, **149,** 415 (1966).
123. J. W. Halley and I. Silvera, *Phys. Rev. Letters*, **15,** 654 (1965).
124. R. A. Erickson, *Phys. Rev.*, **90,** 779 (1953).

125. J. W. Stout and E. Catalano, *Phys. Rev.*, **92,** 1575 (1953).
126. K. Niira and T. Oguchi, *Prog. Theoret. Phys.*, **11,** 425 (1954).
127. W. E. Hatfield and T. S. Piper, *Inorg. Chem.*, **3,** 1295 (1964).
128. B. R. Penfold and M. R. Taylor, *Acta Cryst.*, **13,** 953 (1960).
129. G. Brauer and M. Eichner, *Z. anorg. allgem. Chem.*, **296,** 13 (1958).
130. R. C. Schoonmaker and R. F. Porter, *J. Chem. Phys.*, **29,** 116 (1958).
131. H. Kueknl and W. Ernst, *Z. anorg. allgem. Chem.*, **317,** 84 (1962).
132. N. H. Bhuiyan, *Pakistan J. Res. Sci.*, **13,** 16 (1961).
133. P. Kovacic and N. O. Brace, *Inorg. Syn.*, **6,** 172 (1960).
134. H. L. Pinch and J. M. Hirshon, *J. Am. Chem. Soc.*, **79,** 6149 (1957).
135. C. S. Shen and M. H. Chang, *Hua Hsueh Hsueh Pao*, **26,** 124 (1960).
136. F. L. Oetting and N. W. Gregory, *J. Phys. Chem.*, **65,** 138 (1961).
137. R. J. Sime and N. W. Gregory, *J. Phys. Chem.*, **64,** 86 (1960).
138. H. Schafer, L. Bayer, G. Breil, K. Etzel and K. Krehl, *Z. anorg. allgem. Chem.*, **278,** 300 (1959).
139. N. G. Korzhukov, M. I. Ozerova, K. G. Khomyakov and L. D. Onikienko, *Russ. J. Inorg. Chem.*, **11,** 110 (1966).
140. A. Ferrari, A. Braibanti and G. Bigliardi, *Acta Cryst.*, **16,** 846 (1963).
141. R. O. MacLaren and N. W. Gregory, *J. Am. Chem. Soc.*, **76,** 5874 (1954).
142. A. Narath and J. E. Schirber, *J. Appl. Phys.*, **37,** 1124 (1966).
143. G. E. Leroi, T. C. James, J. T. Hougen, and W. Klemperer, *J. Chem. Phys.*, **36,** 2879 (1962).
144. J. W. Leech and A. J. Manuel, *Proc. Phys. Soc.*, **69B,** 210 (1956).
145. L. G. van Uitert, H. J. Williams, R. C. Sherwood and J. J. Rubin, *J. Appl. Phys.*, **36,** 1029 (1965).
146. H. Bizette, C. Terrier and B. Tsai, *Compt. Rend.*, **243,** 895 (1956).
147. H. Bizette, C. Terrier and B. Tsai, *Compt. Rend.*, **261,** 653 (1965).
148. I. S. Jacobs and P. E. Lawrence, *Phys. Rev.*, **164,** 866 (1967).
149. J. W. Leech and A. J. Manuel, *Proc. Phys. Soc.*, **69B,** 220 (1956).
150. W. C. Koehler, M. K. Wilkinson, J. W. Cable and E. O. Wollan, *J. Phys. Radium*, **20,** 180 (1959).
151. A. Herpin and P. Meriel, *Compt. Rend.*, **245,** 650 (1957).
152. M. K. Wilkinson, J. W. Cable, E. O. Wollan and W. C. Koehler, *Phys. Rev.*, **113,** 497 (1959).
153. J. Kanamori, *Prog. Theoret. Phys.*, **20,** 890 (1958).
154. C. W. DeKock and D. M. Gruen, *J. Chem. Phys.*, **44,** 4387 (1966).
155. R. J. H. Clark and C. S. Williams, *Inorg. Chem.*, **4,** 350 (1965).
156. N. S. Gill, R. S. Nyholm, G. A. Barclay, T. I. Christie and P. J. Pauling, *J. Inorg. Nucl. Chem.*, **18,** 88 (1961).
157. C. D. Burbridge, D. M. L. Goodgame and M. Goodgame, *J. Chem. Soc.*, **1967,** A, 349.
158. G. Beech, C. T. Mortimer and E. G. Tyler, *J. Chem. Soc.*, **1967,** A, 1111.
159. D. M. L. Goodgame, M. Goodgame, M. A. Hitchman and M. J. Weeks, *Inorg. Chem.*, **5,** 635 (1966).
160. J. R. Allan, D. H. Brown, R. H. Nuttall and D. W. A. Sharp, *J. Chem. Soc.*, **1966,** A, 1031.
161. O. Baudisch and W. H. Hartung, *Inorg. Syn.*, **1,** 184 (1939).
162. R. J. H. Clark and C. S. Williams, *J. Chem. Soc.*, **1966,** A, 1425.
163. J. A. Broomhead and F. P. Dwyer, *Australian J. Chem.*, **14,** 250 (1961).

164. D. J. Robinson and C. H. L. Kennard, *Australian J. Chem.*, **19,** 1285 (1966).
165. D. M. Adams and J. B. Cornell, *J. Chem. Soc.*, **1967,** A, 884.
166. M. Nardelli, L. Cavalca and A. Braibanti, *Gazz. Chim. Ital.*, **86,** 867 (1956).
167. M. Nardelli, I. Chierci and A. Braibanti, *Gazz. Chim. Ital.*, **88,** 37 (1958).
168. K. Issleib and G. Doll, *Z. anorg. allgem. Chem.*, **305,** 1 (1960).
169. L. Naldini, *Gazz. Chim. Ital.*, **90,** 391 (1960).
170. J. Lewis, R. S. Nyholm and G. A. Rodley, *J. Chem. Soc.*, **1965,** 1483.
171. D. Forster and D. M. L. Goodgame, *J. Chem. Soc.*, **1965,** 454.
172. G. A. Rodley, D. M. L. Goodgame and F. A. Cotton, *J. Chem. Soc.*, **1965,** 1499.
173. F. Basolo and F. P. Dwyer, *J. Am. Chem. Soc.*, **76,** 1454 (1954).
174. D. B. Fox, J. R. Hall and R. A. Plowman, *Australian J. Chem.*, **15,** 235 (1962).
175. J. C. Fanning and L. T. Taylor, *J. Inorg. Nucl. Chem.*, **27,** 2217 (1965).
176. C. D. Burbridge and D. M. L. Goodgame, *J. Chem. Soc.*, **1967,** A, 694.
177. M. Ciampolini and G. P. Speroni, *Inorg. Chem.*, **4,** 45 (1966).
178. G. Beech and C. T. Mortimer, *J. Chem. Soc.*, **1967,** A, 1115.
179. R. J. Kern, *J. Inorg. Nucl. Chem.*, **25,** 5 (1963).
180. H. Rheinboldt, A. Luyken and H. Schmittmann, *J. Prakt. Chem.*, **149,** 30 (1937).
181. S. Herzog, K. Gustav, E. Krueger, H. Oberenda and R. Schuster, *Z. Chem.*, **3,** 428 (1963).
182. L. Sacconi and A. Sabatini, *J. Inorg. Nucl. Chem.*, **25,** 1389 (1965).
183. J. D. Curry, M. A. Robinson and D. H. Busch, *Inorg. Chem.*, **6,** 1570 (1967).
184. L. M. Venanzi, *Angew. Chem., Intern. Ed. Engl.*, **3,** 453 (1964).
185. K. H. Gayer and L. Woontner, *Inorg. Syn.*, **5,** 179 (1957).
186. H. Schafer, *Z. anorg. allgem. Chem.*, **258,** 69 (1949).
187. B. R. Penfold and J. A. Grigor, *Acta Cryst.*, **12,** 850 (1959).
188. I. Gamo, *Bull. Chem. Soc. Japan*, **34,** 1433 (1961).
189. R. D. Pierce and S. A. Friedberg, *J. Appl. Phys.*, **32,** 66S (1961).
190. R. D. Spence, R. Au and P. A. van Dalen, *Physica*, **30,** 1612 (1964).
191. S. A. Friedberg and J. T. Schriempf, *J. Appl. Phys.*, **35,** 100 (1964).
192. J. T. Schriempf and S. A. Friedberg, *Phys. Rev.*, **136,** 518 (1964).
193. B. Morosin and E. J. Graebner, *J. Chem. Phys.*, **42,** 898 (1965).
194. K. E. Lawson, *J. Chem. Phys.*, **44,** 4159 (1966).
195. A. Narath, *Phys. Rev.*, **139A,** 1221 (1965).
196. J. D. Christian and N. W. Gregory, *J. Phys. Chem.*, **71,** 1583 (1967).
197. N. W. Gregory, *J. Phys. Chem.*, **61,** 369 (1967).
198. V. P. Il'inskii and A. T. Uverskaya, *Sbornik Trudov Gos. Inst. Priklad. Khim.*, **1958,** 112.
199. A. Ferrari and F. Giorgi, *Atti Acad. Lincei*, **9,** 1134 (1929).
200. N. W. Gregory and H. E. O'Neal, *J. Am. Chem. Soc.*, **81,** 2649 (1959).
201. N. W. Gregory and T. Wydeven, *J. Phys. Chem.*, **67,** 927 (1963).
202. W. Klemm and W. Schuth, *Z. anorg. allgem. Chem.*, **210,** 33 (1933).
203. I. S. Jacobs and P. E. Lawrence, *J. Appl. Phys.*, **35,** 996 (1964).
204. H. Bizette, C. Terrier and B. Tsai, *Compt. Rend.*, **245,** 507 (1957).
205. K. Issleib and G. Hohlfeld, *Z. anorg. allgem. Chem.*, **312,** 169 (1961).
206. J. C. Sheldon and S. Y. Tyree, *J. Am. Chem. Soc.*, **80,** 4775 (1958).
207. I. Bertini and F. Mani, *Inorg. Chem.*, **6,** 2032 (1967).
208. P. Paoletti and M. Ciampolini, *Inorg. Chem.*, **6,** 64 (1967).

209. M. Ciampolini and N. Nardi, *Inorg. Chem.*, **6,** 445 (1967).
210. H. W. Foote and M. Fleischer, *J. Phys. Chem.*, **44,** 647 (1940).
211. F. L. Oetting and N. W. Gregory, *J. Phys. Chem.*, **65,** 173 (1961).
212. W. E. Zaugg and N. W. Gregory, *J. Phys. Chem.*, **70,** 486 (1966).
213. N. V. Sidgwick, *The Chemical Elements and Their Compounds,* O.U.P., London, 1950, p. 1332.
214. H. Schafer and W. J. Hones, *Z. anorg. allgem. Chem.*, **288,** 62 (1956).
215. M. Chaigneau, *Bull. Soc. Chim. France*, **1957,** 886.
216. J. W. Mellor, *A Comprehensive Treatise on Inorganic and Theoretical Chemistry*, Longmans, London, Vol. 14, p. 128.
217. S. A. Shchukarev, M. A. Oranskaya and T. S. Bartnitskaya, *Vestn. Leningrad. Univ.*, **11,** *Ser. Fiz. i Khim.*, 104 (1956).
218. W. E. Zaugg and N. W. Gregory, *J. Phys. Chem.*, **70,** 490 (1966).
219. W. Fischer and R. Gewher, *Z. anorg. allgem. Chem.*, **222,** 303 (1935).
220. V. V. Pechkovskii and A. V. Sofronova, *Izv. Vyssh. Uchebn. Zaved., Khim. i Khim. Tekhnol.*, **10,** 603 (1967).
221. A. Ferrari and F. Giorgi, *Atti Acad. Lincei*, **10,** 522 (1929).
222. S. Prasad and D. R. Krishnamurty, *J. Indian Chem. Soc.*, **34,** 563 (1957).
223. F. A. Cotton, R. Francis and W. D. Horrocks, *J. Phys. Chem.*, **64,** 1534 (1960).
224. R. S. Nyholm and A. G. Sharpe, *J. Chem. Soc.*, **1952,** 3579.
225. J. Ravez, J. Viollet, R. de Pape and P. Hagenmuller, *Bull. Soc. Chim. France*, **1967,** 1325.
226. D. Babel, G. Pausemwang and W. Viebahn, *Z. Naturforsch.*, **22b,** 1219 (1967).
227. R. D. Peacock, *J. Chem. Soc.*, **1957,** 4684.
228. W. Klemm and E. Huss, *Z. anorg. allgem. Chem.*, **258,** 221 (1949).
229. A. H. Nielsen, *Z. anorg. allgem. Chem.*, **224,** 84 (1935).
230. H. M. Haendler, F. A. Johnson and D. S. Crocket, *J. Am. Chem. Soc.*, **80,** 2662 (1958).
231. D. S. Crocket and H. M. Haendler, *J. Am. Chem. Soc.*, **82,** 4158 (1960).
232. B. Cox and A. G. Sharpe, *J. Chem. Soc.*, **1954,** 1798.
233. H. Funk and H. Bohland, *Z. anorg. allgem. Chem.*, **334,** 155 (1964).
234. H. Bode and E. Voss, *Z. anorg. allgem. Chem.*, **290,** 1 (1957).
235. E. G. Steward and H. P. Rooksby, *Acta Cryst.*, **6,** 49 (1953).
236. H. M. Rice, R. C. Turner and J. E. Brydon, *Nature*, **169,** 749 (1952).
237. L. Pauling, *J. Am. Chem. Soc.*, **46,** 2738 (1924).
238. L. C. Kravitz and W. W. Piper, *Phys. Rev.*, **146A,** 322 (1966).
239. R. D. Peacock and D. W. A. Sharp, *J. Chem. Soc.*, **1959,** 2762.
240. G. Mitra, *J. Indian Chem. Soc.*, **35,** 257 (1958).
241. W. G. Palmer, *Experimental Inorganic Chemistry*, Cambridge, London, 1962.
242. A. H. Nielsen, *Z. anorg. allgem. Chem.*, **226,** 222 (1936).
243. A. H. Nielsen, *Z. anorg. allgem. Chem.*, **227,** 423 (1936).
244. R. D. Peacock, *Progress in Inorganic Chemistry* (Ed. F. A. Cotton), Vol. 2, Interscience, New York, 1960.
245. K. Knox and S. Geller, *Phys. Rev.*, **110,** 771 (1958).
246. R. C. Linares, *J. Appl. Phys.*, **37,** 2195 (1966).

247. E. G. Spencer, S. B. Berger, R. C. Linares and P. V. Lenzo, *Phys. Rev. Letters*, **10**, 236 (1963).
248. W. E. Hatfield, R. C. Fay, C. E. Pfluger and T. S. Piper, *J. Am. Chem. Soc.*, **85**, 265 (1963).
249. G. M. Bancroft, A. G. Maddock, W. K. Ong and R. H. Prince, *J. Chem. Soc.*, **1966**, A, 723.
250. I. Lindqvist, *Arkiv Kemi Min. Geol.*, **1946**, 24A.
251. J. W. Mellor, *A Comprehensive Treatise on Inorganic and Theoretica Chemistry*, Vol. 14, Longmans, London, p. 98.
252. H. Yamatera and K. Nakatsu, *Bull. Chem. Soc., Japan*, **27**, 244 (1954).
253. I. Linqvist, *Acta Chem. Scand.*, **2**, 530 (1948).
254. A. P. Ginsberg and M. B. Robin, *Inorg. Chem.*, **2**, 817 (1963).
255. R. J. H. Clark and F. B. Taylor, *J. Chem. Soc.*, **1967**, A, 693.
256. R. F. Weinland and A. Kissling, *Z. anorg. allgem. Chem.*, **120**, 209 (1922).
257. A. Earnshaw and J. Lewis, *J. Chem. Soc.*, **1961**, 396.
258. C. M. Cook and W. E. Dunn, *J. Phys. Chem.*, **65**, 1505 (1961).
259. R. R. Richards and N. W. Gregory, *J. Phys. Chem.*, **68**, 3089 (1964).
260. R. R. Richards and N. W. Gregory, *J. Phys. Chem.*, **69**, 239 (1965).
261. H. L. Friedman and H. Taube, *J. Am. Chem. Soc.*, **72**, 2236 (1950).
262. H. L. Friedman, *J. Am. Chem. Soc.*, **74**, 5 (1952).
263. I. S. Morozov and L. Tsegledi, *Zh. Neorg. Khim.*, **6**, 2766 (1961).
264. J. C. Sheldon and S. Y. Tyree, *J. Am. Chem. Soc.*, **80**, 2117 (1958).
265. D. W. Meek and R. S. Drago, *J. Am. Chem. Soc.*, **83**, 4322 (1961).
266. R. J. Woodruff, J. L. Marini and J. P. Fackler, *Inorg. Chem.*, **3**, 687 (1964).
267. N. S. Gill, *J. Chem. Soc.*, **1961**, 3512.
268. D. M. Adams, J. Chatt, J. M. Davidson and J. Gerratt, *J. Chem. Soc.*, **1963**, 2189.
269. B. J. Hathaway and D. G. Holah, *J. Chem. Soc.*, **1964**, 2408.
270. J. A. Walmsley and S. Y. Tyree, *Inorg. Chem.*, **2**, 312 (1963).
271. M. Cox, J. Lewis and R. S. Nyholm, *J. Chem. Soc.*, **1964**, 6113.
272. D. M. Puri and R. C. Mehrotra, *J. Less-Common Metals*, **5**, 2 (1963).
273. W. K. Ong and R. H. Prince, *J. Inorg. Nucl. Chem.*, **27**, 1037 (1965).
274. S. K. Madan and H. H. Denk, *J. Inorg. Nucl. Chem.*, **27**, 1049 (1965).
275. M. J. Joesten and K. M. Nykerk, *Inorg. Chem.*, **3**, 548 (1964).
276. A. H. Blatt and N. Gross, *J. Am. Chem. Soc.*, **77**, 5424 (1955).
277. A. H. Blatt, *J. Am. Chem. Soc.*, **70**, 1861 (1949).
278. E. P. Kohler and A. H. Blatt, *J. Am. Chem. Soc.*, **50**, 1217 (1928).
279. G. H. Morrison and H. Freiser, *Solvent Extraction in Analytical Chemistry*, Wiley, New York, 1957.
280. A. N. Nesmeyanov, K. A. Kocheshkov, V. A. Klemova and N. K. Gipp, *Ber.*, **68B**, 1877 (1935).
281. K. A. Kocheshkov and A. N. Nesmeyanov, *J. Gen. Chem. USSR.*, **6**, 144 (1936).
282. H. B. Jonassen, L. J. Theriot, E. A. Bourdreaux and W. A. Ayres, *J. Inorg. Nucl. Chem.*, **26**, 595 (1964).
283. H. Noeth, S. Lukas and P. Schweizer, *Chem. Ber.*, **98**, 962 (1965).
284. H. Noeth and S. Lukas, *Chem. Ber.*, **95**, 1505 (1962).
285. L. A. Kazitsyna, O. A. Reutov and Z. F. Buchkovskii, *Zh. Fiz. Khim.*, **34**, 850 (1960).

286. B. M. Mikhailov, N. S. Fedotov, T. A. Shchegoleva and V. D. Shchudyakov, *Dokl. Akad. Nauk SSSR.*, **145,** 340 (1962).
287. N. G. Bokii, T. N. Polynova, M. A. Porai-Koshits, B. S. Kikot and L. A. Kazitsyna, *Zh. Strukt. Khim.*, **4,** 453 (1963).
288. H. Noeth and P. Fritz, *Z. anorg. allgem. Chem.*, **322,** 297 (1961).
289. E. A. Bourdreaux, H. B. Jonassen and L. J. Theriot, *J. Am. Chem. Soc.*, **85,** 2039 (1963).
290. A. F. Gremillion, H. B. Jonassen and R. J. O'Conner, *J. Am. Chem. Soc.*, **81,** 6134 (1959).
291. M. Baaz, V. Gutmann and J. R. Masaguer, *Monatsh. Chem.*, **92,** 582 (1961).
292. M. Baaz, V. Gutmann and J. M. Masaguer, *Monatsh. Chem.*, **92,** 590 (1961).
293. V. Gutmann, *Rec. Trav. Chim.*, **75,** 605 (1956).
294. B. Zaslow and R. E. Rundle, *J. Phys. Chem.*, **61,** 490 (1957).
295. J. A. Woodward and M. J. Taylor, *J. Chem. Soc.*, **1960,** 4473.
296. R. J. H. Clark, *Spectrochim. Acta*, **21,** 955 (1965).
297. R. J. H. Clark and T. M. Dunn, *J. Chem. Soc.*, **1963,** 1198.
298. S. Balt, *Rec. Trav. Chim.*, **86,** 1025 (1967).
299. H. G. Schnering, *Z. anorg. allgem. Chem.*, **353,** 1 (1967).
300. R. de Pape and J. Ravez, *Bull. Soc. Chim. France*, **1966,** 3283.
301. H. G. Schnering, *Z. anorg. allgem. Chem.*, **353,** 13 (1967).
302. F. F. Y. Wang and M. Kestigian, *J. Appl. Phys.*, **37,** 975 (1966).
303. D. J. Machin, R. L. Martin and R. S. Nyholm, *J. Chem. Soc.*, **1963,** 1490.
304. D. J. Machin and R. S. Nyholm, *J. Chem. Soc.*, **1963,** 1500.
305. K. Knox, *Acta Cryst.*, **14,** 583 (1961).
306. A. Okazaki, Y. Suemune and T. Fuchikami, *J. Phys. Soc. Japan.*, **14,** 1823 (1959).
307. R. L. Martin, R. S. Nyholm and N. C. Stephenson, *Chem. Ind. (London)*, **1956,** 83.
308. A. Okazaki and Y. Suemune, *J. Phys. Soc. Japan*, **16,** 671 (1961).
309. K. Hirakawa, K. Hirakawa and T. Hashimoto, *J. Phys. Soc. Japan*, **15,** 2063 (1960).
310. V. Scatturin, L. Corliss, N. Elliot and J. Hastings, *Acta Cryst.*, **14,** 19 (1961).
311. L. R. Testardi, H. J. Levinstein and H. J. Guggenheim, *Phys. Rev. Letters*, **19,** 503 (1967).
312. H. J. Seifert and K. Klatyk, *Z. anorg. allgem. Chem.*, **342,** 1 (1966).
313. V. I. Ionov, I. S. Morozov and B. G. Korshunov, *Russ. J. Inorg. Chem.*, **5,** 602 (1960).
314. N. V. Galitsky, V. I. Borodin and A. I. Lystsov, *Ukr. Khim. Zh.*, **32,** 695 (1966).
315. K. C. Dash and D. V. R. Rao, *Current Sci. (India)*, **34,** 245 (1965).
316. A. B. Blake and F. A. Cotton, *Inorg. Chem.*, **3,** 5 (1964).
317. C. Furlani, E. Cervone and V. Valenti, *J. Inorg. Nucl. Chem.*, **25,** 159 (1963).
318. P. Pauling, *Inorg. Chem.*, **5,** 1498 (1966).
319. L. A. Kazitsyna, O. A. Reutov and B. S. Kikot, *J. Gen. Chem. USSR*, **31,** 2751 (1961).
320. D. M. Gruen and C. A. Angell, *Inorg. Nucl. Chem. Letters*, **2,** 75 (1966).
321. D. H. Busch and J. C. Bailar, *J. Am. Chem. Soc.*, **78,** 1137 (1956).

322. S. A. Shchukarev, I. V. Vasil'kova and G. M. Barvinok, *Vestn. Leningrad. Univ.*, **20,** *Ser. Fiz. i Khim.*, 145 (1965).
323. P. Paoletti and A. Vacca, *Trans. Faraday Soc.*, **60,** 50 (1964).
324. A. Sabatini and L. Sacconi, *J. Am. Chem. Soc.*, **86,** 17 (1964).
325. R. J. H. Clark, R. S. Nyholm and F. B. Taylor, *J. Chem. Soc.*, **1967,** A, 1802.
326. G. M. Barvinok and V. Vintruff, *Vestn. Leningrad. Univ.*, **21,** *Ser. Fiz. i Khim.*, 111 (1966).
327. P. Paoletti, *Trans. Faraday Soc.*, **61,** 219 (1965).
328. N. K. Hamer, *Mol. Phys.*, **6,** 257 (1963).
329. G. K. Wertheim, H. J. Guggenheim and D. N. E. Buchanan, *Phys. Rev.*, **169,** 465 (1968).
330. L. M. Levinson, *J. Phys. Chem. Solids*, **29,** 1331 (1968).
331. U. Bertelsen, J. M. Knudsen and H. Krogh, *Phys. Status Solidi*, **22,** 59 (1967).
332. J. R. Shane and M. Kestigian, *J. Appl. Phys.*, **39,** 1027 (1968).
333. I. W. Bremen, A. M. A. Verwy and S. Balt, *Spectrochim. Acta.*, **24A,** 1623 (1968).
334. P. P. Sharma and R. C. Mehrotra, *J. Indian Chem. Soc.*, **44,** 74 (1967).
335. G. A. Kakos and G. Winter, *Australian J. Chem.*, **22,** 97 (1969).
336. A. van den Bergen, K. S. Murray, M. J. O'Connor, N. Pekak and B. O. West, *Australian J. Chem.*, **21,** 1505 (1968).
337. A. van den Bergen, K. S. Murray and B. O. West, *Australian J. Chem.*, **21,** 1517 (1968).
338. K. S. Murray, A. van den Bergen, M. J. O'Connor, N. Pekak and B. O. West, *Inorg. Nucl. Chem. Letters*, **4,** 87 (1968).
339. K. S. Murray and B. O. West, *Inorg. Nucl. Chem. Letters*, **4,** 439 (1968).
340. W. M. Reiff, G. J. Long and W. A. Baker, *J. Am. Chem. Soc.*, **90,** 5347 (1968).
341. P. P. Singh and R. Rivest, *Can. J. Chem.*, **46,** 1773 (1968).
342. G. W. A. Fowles, D. A. Rice and R. A. Walton, *J. Chem. Soc.*, **1968,** A, 1842.
343. R. R. Berrett, B. W. Fitzsimmons and A. A. Owusu, *J. Chem. Soc.*, **1968,** A, 1575.
344. W. M. Reiff, W. A. Baker and N. E. Erickson, *J. Am. Chem. Soc.*, **90,** 4794 (1968).
345. M. T. Rogers and J. A. Ryan, *J. Phys. Chem.*, **72,** 1340 (1968).
346. J. G. H. du Preez, W. A. Steyn and A. J. Basso, *J. South African Chem. Institute*, **21,** 8 (1968).
347. N. A. Daugherty and J. H. Swisher, *Inorg. Chem.*, **7,** 1651 (1968).
348. H. H. Wickman and F. R. Merritt, *Chem. Phys. Letters*, **1,** 117 (1967).
349. P. Lumme and H. Junkkarinen, *Suomen Kemistilehti*, **41B,** 220 (1968).
350. G. K. Wertheim, H. J. Guggenheim, H. J. Levinstein, D. N. E. Buchanan and R. C. Sherwood, *Phys. Rev.*, **173,** 614 (1968).
351. G. D. Jones, *Phys. Rev.*, **155,** 259 (1967).
352. P. L. Richards, *J. Appl. Phys.*, **38,** 1500 (1967).
353. J. W. Stout, M. I. Steinfeld and M. Yuzuri, *J. Appl. Phys.*, **39,** 1141 (1968).
354. H. J. Guggenheim, M. T. Hutchings and B. D. Raiford, *J. Appl. Phys.*, **39,** 1120 (1968).

355. G. W. Wertheim, *J. Appl. Phys.*, **38,** 971 (1967).
356. T. Tanaka, L. Libelo and R. Kligman, *Phys. Rev.*, **171,** 531 (1968).
357. J. H. Barry, *Phys. Rev.*, **174,** 531 (1968).
358. G. Winter, *Australian J. Chem.*, **21,** 2859 (1968).
359. D. P. Johnson and J. G. Dash, *Phys. Rev.*, **172,** 983 (1968).
360. G. M. Bancroft, M. J. Mays and B. E. Prater, *Chem. Commun.*, **1968,** 1374.
361. C. D. Burbridge and D. M. L. Goodgame, *J. Chem. Soc.*, **1968,** A, 1074.
362. J. S. Judge, W. M. Reiff, G. M. Intille, P. Ballway and W. A. Baker, *J. Inorg. Nucl. Chem.*, **29,** 1711 (1967).
363. D. P. Madden and S. M. Nelson, *J. Chem. Soc.*, **1968,** A, 2342.
364. D. M. Bowers and A. I. Popov, *Inorg. Chem.*, **7,** 1594 (1968).
365. M. T. Halfpenny, J. G. Hartley and L. M. Venanzi, *J. Chem. Soc.*, **1967,** A, 627.
366. M. C. Feller and R. Robson, *Australian J. Chem.*, **21,** 2919 (1968).
367. C. D. Flint and M. Goodgame, *J. Chem. Soc.*, **1968,** A, 750.
368. A. Braibanti, F. Dallavalle, M. A. Pellinghelli and E. Leporti, *Inorg. Chem.*, **7,** 1430 (1968).
369. H. S. Preston, C. H. L. Kennard and R. A. Plowman, *J. Inorg. Nucl. Chem.*, **30,** 1463 (1968).
370. M. F. Farona and G. R. Tompkin, *Spectrochim. Acta*, **24A,** 788 (1968).
371. L. H. Pignolet, D. Forster and W. deW. Horrocks, *Inorg. Chem.*, **7,** 828 (1968).
372. R. D. Feltham and W. Silverthorn, *Inorg. Chem.*, **7,** 1154 (1968).
373. N. N. Y. Chan, M. Goodgame and M. J. Weeks, *J. Chem. Soc.*, **1968,** A, 2499.
374. C. E. Johnson and M. S. Ridout, *J. Appl. Phys.*, **38,** 1272 (1967).
375. A. C. Daniel, A. W. Bevan and R. J. Mahler, *J. Appl. Phys.*, **39,** 496 (1968).
376. M. Di Vaira and P. L. Orioli, *Acta Cryst.*, **24B,** 1269 (1968).
377. C. W. Kaufman and M. A. Robinson, *J. Inorg. Nucl. Chem.*, **30,** 2475 (1968).
378. C. D. Burbridge and D. M. L. Goodgame, *J. Chem. Soc.*, **1968,** A, 1410.
379. D. Babel, *Z. Naturforsch.*, **23a,** 1417 (1968).
380. D. M. Adams and D. M. Morris, *J. Chem. Soc.*, **1968,** A, 694.
381. H. Yamatera and A. Kato, *Bull. Chem. Soc. Japan*, **41,** 2220 (1968).
382. R. D. Feltham, H. G. Metzger and W. Silverthorn, *Inorg. Chem.*, **7,** 2003 (1968).
383. T. J. Kistenmacher and G. D. Stucky, *Inorg. Chem.*, **7,** 2150 (1968).
384. J. S. Avery, C. D. Burbridge and D. M. L. Goodgame, *Spectrochim. Acta*, **24A,** 1721 (1968).
385. J. Reedijk and W. L. Groeneveld, *Rec. Trav. Chim.*, **87**, 513 (1968).
386. P. R. Edwards and C. E. Johnson, *J. Chem. Phys.*, **49,** 211 (1968).
387. S. Balt, *Mol. Phys.*, **14,** 233 (1968).
388. I. W. Bremen and S. Balt, *Rec. Trav. Chim.*, **87,** 349 (1968).
389. A. Tressaud, R. de Pape, J. Portier and P. Hagenmuller, *Compt. Rend.*, **266C,** 984 (1968).
390. H. G. Schnering and P. Bleckmann, *Naturwissenschaften*, **55,** 342 (1968).
391. R. Fatehally, N. P. Sastry and R. Nagarajan, *Phys. Status Solidi*, **26,** 91 (1968).
392. U. Ganiel, S. Shtrikman and M. Kestigian, *J. Appl. Phys.*, **39,** 1254 (1968).

393. R. B. Flippen, *Phys. Rev. Letters*, **21**, 1079 (1968).
394. N. S. Gill and F. B. Taylor, *Inorg. Syn.*, **9**, 136 (1967).
395. P. R. Edwards, C. E. Johnson and R. J. P. Williams, *J. Chem. Phys.*, **47**, 2074 (1967).
396. G. A. Heath, R. L. Martin and I. M. Stewart, *Chem. Commun.*, **1969**, 54.
397. G. A. Heath, R. L. Martin and I. M. Stewart, *Australian J. Chem.*, **22**, 83 (1969).
398. K. Knauer, P. Hemmerich and J. D. W. van Voorst, *Angew. Chem., Intern. Ed. Engl.*, **6**, 262 (1967).
399. A. Ouchi, M. Nakatani and Y. Takahashi, *Bull. Chem. Soc., Japan*, **41**, 2044 (1968).
400. A. Furuhashi, T. Takeuchi and A. Ouchi, *Bull. Chem. Soc. Japan*, **41**, 2049 (1968).
401. H. C. Freeman, G. H. W. Milburn, C. E. Nockolds, P. Hemmerich and K. H. Knauer, *Chem. Commun.*, **1969**, 54.

Chapter 7

Cobalt

The outstanding feature of the halide chemistry of cobalt, compared with that of the previous elements, is the almost complete domination of a single oxidation state in the binary halides, which is reflected also in the complex halides. Cobalt trifluoride is the only binary halide of cobalt(III), but all of the binary halides of cobalt(II) are known. Although cobalt trichloride itself has not been prepared, a number of complexes which may be considered to be derived from it are known, and this is also the case for the bromide and iodide. The complex halide chemistry of cobalt is restricted entirely to oxidation state(II) except for fluoro complexes.

HALIDES

Oxidation State III

Cobalt trifluoride. Cobalt trifluoride is most conveniently prepared by fluorination of either cobalt difluoride or dichloride at 200–300° in a flow system[1–6]. Alternative methods include the fluorination of cobalt oxides at 150–300° in a flow system[3,7], the reaction of chlorine trifluoride with cobalt dichloride at 250° in a flow system[8,9] and finally refluxing cobalt dichloride with bromine trifluoride[10]. This latter method is not strongly recommended because of the difficulty of removing the last traces of the fluorinating agent.

Cobalt trifluoride is a brown, very hygroscopic and very reactive compound. Although Ruff and Ascher[5] claimed that cobalt trifluoride is thermally unstable, Stewart[11] has shown conclusively that cobalt difluoride is formed by hydrolytic reactions and not by thermal decomposition. Cobalt trifluoride has been widely utilized for fluorination of organic substrates. It may be used on the laboratory scale but its most important applications are in industrial processes involving the synthesis of fluorocarbons[2–4,12–20].

Hepworth and coworkers[1] have shown that it has a rhombohedral lattice with $a = 5.279$ Å and $\alpha = 56.97°$. The structure consists of CoF_6

octahedra which are regular within experimental error and are packed in the same manner as in vanadium trifluoride. The cobalt-fluorine bond distance is 1.89 Å. The magnetic moment of cobalt trifluoride is 2.46 BM at 293°K and falls to 1.60 BM at 90°K[21].

Cobalt trifluoride hydrates. Electrolytic oxidation of cobalt difluoride in 40% hydrofluoric acid results in the formation of blue $CoF_3 \cdot 3 \cdot 5H_2 \cdot O$[22,23]. Clark and coworkers[23] have measured the magnetic susceptibility of this hydrate over a wide temperature range, and it was found to obey the Curie-Weiss law with a magnetic moment of 4.47 BM and a Weiss constant of 60°. On the basis of these measurements it was concluded that the most probable formulation of the compound is $[CoF_3(H_2O)_3] \cdot 0 \cdot 5H_2O$ and not $[Co(H_2O)_6][CoF_6] \cdot H_2O$.

Maak and coworkers[24] report that cobalt trifluoride trihydrate also exists and that it is isomorphous with the α-form of aluminium trifluoride trihydrate.

Other cobalt(III) halo compounds. Although it is not possible to isolate cobalt trichloride and the corresponding bromide and iodide, it is well known that there is a very large number of cobalt(III) complexes which formally are derivatives of these halides, but which do not fall into the category of complex halide anions. It is convenient therefore, to deal with these compounds in a separate section at this juncture. However, we will only consider those compounds in which at least one halogen atom is directly coordinated to the cobalt atom.

Most of these derivatives are chloro compounds, usually with amine or phosphine ligands attached to the cobalt atom, and usually with one or more of the halogen atoms not directly coordinated to the metal. Examples are $[Co(en)_2Cl_2]Cl$ and $CoCl_3[(C_2H_5)_3P]_2$[25–48]. The majority of the investigations on these compounds have been concerned with spectroscopic and kinetic studies. Much of this work has been reviewed[49–52] and will not be considered here.

In the majority of cases, and particularly with the amine series, the complexes appear, on the basis of their absorption spectra, to contain octahedrally coordinated cobalt, and in several cases this has been confirmed by structural determinations.

Shiegeta and coworkers[53] have shown that $[Co(NH_3)_5Cl]Cl_2$ has an orthorhombic unit cell with $a = 13.34$, $b = 10.33$ and $c = 6.73$ Å. The space group is *Pnma* and contains four molecules. The cobalt-chlorine distance is 2.27 Å while three cobalt-nitrogen bonds are identical at 1.97 Å, with the other two at 1.91 and 1.98 Å.

Becker and coworkers[54] report that preliminary single-crystal data on *trans* $[Co(en)_2Cl_2]Cl$ indicate that the four nitrogen atoms and the cobalt atom are coplanar and that the carbon atoms of the ethylenediamine rings lie outside this plane. However, full details of the structure are not yet available.

Nakahara, Saito and Kuroya[55] have reported that *trans* dichlorobis(ethylenendiamine)cobalt(III) chloride hydrochloride dihydrate, $[CoCl_2(en)_2]Cl \cdot HCl \cdot 2H_2O$, has a monoclinic unit cell with $a = 10.68$, $b = 7.89$, $c = 9.09$ Å and $\beta = 110.43°$. There are two molecules in the unit cell of space group $P2_1/c$. The structure of the cation is shown in Figure 7.1. The cobalt-chlorine bond distance is 2.33 Å and the cobalt-nitrogen distances are 1.98 and 2.00 Å. A noteworthy feature of the structure is that both ethylenediamine rings are in the *gauche* form and that there are two kinds of *gauche* molecules in the crystal, one being the mirror image of the other. These two kinds of molecule are coordinated to the cobalt atom is such a way that one is related to the other by a centre of symmetry. The complex cations are arranged in layers which are parallel to the (100) plane, the chlorine-cobalt-chlorine bond direction being inclined at an angle of 14.5° to the (010) plane. There is strong hydrogen bonding between the remaining structural entities which are arranged as follows.

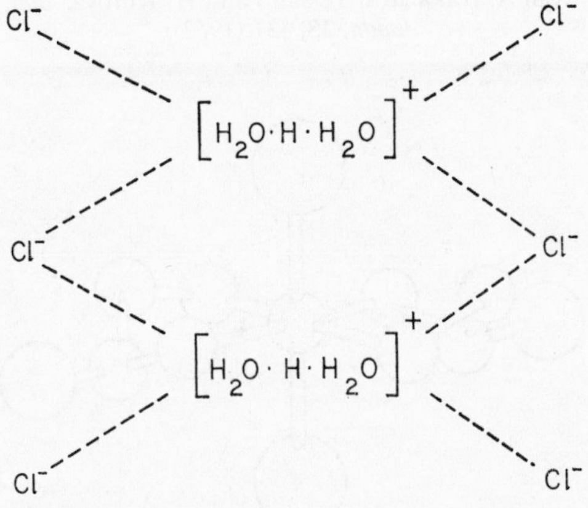

The corresponding bromine compound has a similar structure with $a = 10.98$, $b = 8.18$, $c = 9.46$ Å and $\beta = 113.2°$. The cobalt-bromine and cobalt-nitrogen bond distances are 2.44 and 2.00 Å respectively[33].

Saito and Iwasaki[56] have examined the crystal structure of *trans* dichlorobis(*l*-propylenediamine)cobalt(III) chloride hydrochloride dihydrate, $[CoCl_2(l\text{-}pn)_2]Cl \cdot HCl \cdot 2H_2O$, by single-crystal X-ray diffraction techniques. The monoclinic unit cell has $a = 22.092$, $b = 8.406$, $c = 9.373$ Å and $\beta = 99.65°$. There are four formula units in the cell of space group

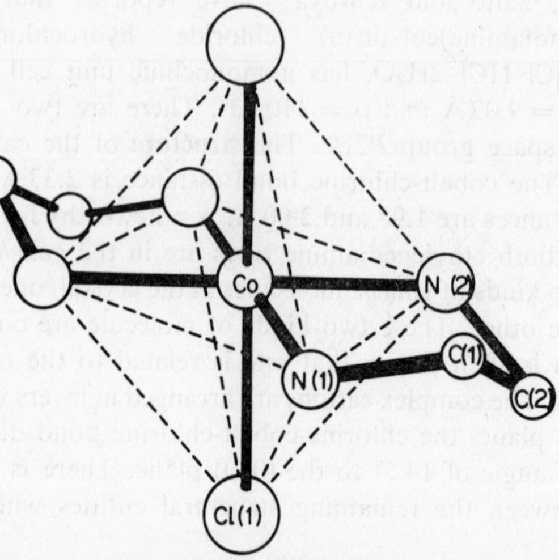

Figure 7.1 Perspective drawing of the $[CoCl_2en_2]^+$ ion. (Reproduced by permission from A. Nakahara, Y. Saito and H. Kuroya, *Bull. Chem. Soc. Japan*, **25**, 331 (1952))

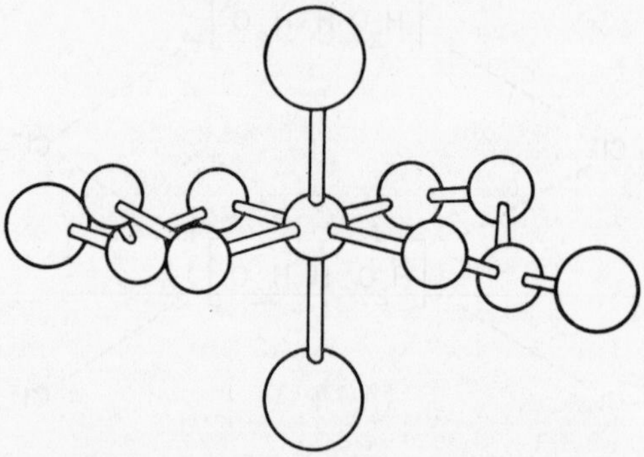

Figure 7.2 Perspective drawing of the $[CoCl_2(l\text{-pn})_2]^+$ ion. (Reproduced by permission from Y. Saito and H. Iwasaki, *Bull. Chem. Soc. Japan*, **35**, 1131 (1962))

*C*2. The structure consists of approximately octahedral cations, chloride ions and $H_5O_2^+$ groups. The configuration of the cation is shown in Figure 7.2 and the cobalt-chlorine bond distances are 2.29 Å, while the cobalt-nitrogen bond lengths are all different and range from 1.94 to 2.02 Å. The amine ligand is in the *gauche* conformation, the C—CH_3 groups being equatorial with respect to the chlorine-cobalt-chlorine direction. The arrangement of the other species is as described for the ethylenediamine compound above.

TABLE 7.1
Internuclear Dimensions (Å) of $[Co(phen)_2Cl_2]Cl \cdot 3H_2O$

Distance	Value	Distance	Value
Co—Cl_1	2.23	Co—N_2	1.98
Co—Cl_2	2.26	Co—N_3	2.03
Co—N_1	1.96	Co—N_4	1.94

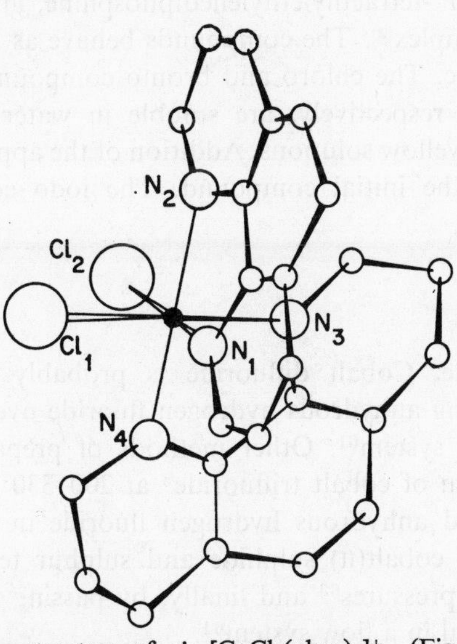

Figure 7.3 The structure of *cis* $[CoCl_2(phen)_2]^+$. (Fig. 2 from A. V. Ablov, Aykon and T. I. Malinovskii *Dokl. Akad. Nauk. SSSR*, **167**, 1051 (1966))

The structure of *cis* dichlorobis(1,10-phenathroline)cobalt(III) chloride trihydrate has been examined by Ablov and coworkers[57]. The monoclinic unit cell has $a = 15.46$, $b = 13.50$, $c = 12.28$ Å and $\beta = 90.17°$. There are four formula units in the cell of space group $C2$. The structure of the cation is shown in Figure 7.3 and the internuclear distances are given in Table 7.1.

Jensen, Nygaard and Pedersen[58] report that complexes of the type $CoX_2(R_3P)_2$ (X = Cl, Br) may be oxidized with the corresponding nitrosyl halide in benzene at −80° to give complexes of the type $CoX_3(R_3P)_2$. The compounds give very intensly coloured solutions in organic solvents, and molecular weight measurements show them to be monomeric, so they are five coordinate. $CoCl_3[(C_2H_5)_3P]_2$ is a violet-black solid which melts at 102–104°. Its dipole moment is zero, indicating that the phosphine groups are equivalent and presumably in the axial positions of a trigonal bipyramid[58]. Similar properties of analogous complexes have suggested that they are also trigonal bipyramids.

Issleib and Mitscherling[59] have prepared green $[CoI_2(R_3PO)_2]I$, where $R = C_6H_{11}$, by oxidizing $CoI_2(R_3PO)_2$ with iodine in boiling ethanol. The compound behaves as a 1:1 electrolyte and the cation contains square-planar cobalt(III).

Oxidation by atmospheric oxygen of CoX_2(tep), where X = Cl, Br and I and tep = *P,P,P′,P′*-tetraethylethylenediphosphine, gives the corresponding CoX_3(tep) complex[60]. The compounds behave as 1:1 electrolytes and are all diamagnetic. The chloro and bromo compounds, which are dark brown and green respectively, are soluble in water but undergo slow hydrolysis to give yellow solutions. Addition of the appropriate hydrohalic acid regenerates the initial compound. The iodo complex however is insoluble in water.

Oxidation State II

Cobalt difluoride. Cobalt difluoride is probably most conveniently prepared by passing anhydrous hydrogen fluoride over cobalt dichloride at 300° in a flow system[61]. Other methods of preparation include the hydrogen reduction of cobalt trifluoride[5] at 200–330°, the interaction of cobalt powder and anhydrous hydrogen fluoride in a bomb[62] at 180°, the interaction of cobalt(II) sulphide and sulphur tetrafluoride at high temperatures and pressures[63] and finally, by passing nitryl fluoride over heated cobalt metal in a flow system[64].

Cobalt difluoride is a pink crystalline solid. Mikhailov and Petrov[65] have given a detailed description of the preparation of single crystals of cobalt difluoride suitable for magnetic and spectral studies. Cobalt difluoride is appreciably soluble in water and from its solutions several hydrates may be isolated. These hydrates may also be prepared by dissolving cobalt carbonate in hydrofluoric acid[66–68]. The melting point of cobalt difluoride has been reported as[69] 1127° and as[70] 1250°. The heat of fusion has been given[69] as 10.72 kcal $mole^{-1}$. Cobalt difluoride vaporizes as the monomer and Buchler and coworkers[71] have shown by mass-spectrometric techniques that the molecule is linear. Margrave and coworkers[72] have measured the sublimation pressure of cobalt difluoride by the Knudsden and the Langmuir methods over the temperature range 699–968°. The vapour pressure equations deduced were:

$$\log p_{\text{atm}} = 8.54 - 15{,}320/T$$

$$\log p_{\text{atm}} = 8.23 - 15{,}110/T$$

for the Knudsden and Langmuir methods respectively. The heat of sublimation was given as 78 kcal mole^{-1}.

Cobalt difluoride has the tetragonal rutile-type structure[73–75]. Baur[73] gives $a = 4.695$ and $c = 3.179$ Å. The structure consists of slightly tetragonally compressed CoF_6 units.

Balkanski, Moch and Parisot[76] have recorded the infrared spectrum of a single crystal of cobalt difluoride by reflection, optical transmission and Kramers-Kronig inversion techniques The results are given in Table 7.2. The oscillation parameters for cobalt difluoride were also calculated[76].

TABLE 7.2

Infrared Spectrum (cm^{-1}) of Cobalt Difluoride

Lattice Mode	Transmission	Reflection	K-K
E_u	420	410	412
A_{2u}	360	340	
E_u	270	265	268
E_u	205	190	196

The magnetic properties of cobalt difluoride have been investigated by a number of workers[21,77–85]. Above about 50°K the Curie-Weiss law is obeyed[80,86] with a magnetic moment of 5.15 BM and a Weiss constant of 50°. At approximately 38°K cobalt difluoride becomes antiferromagnetic[79–81] and at the same temperature there is an anomaly in the specific heat[70,79,87,88] and in the thermal conductivity[89]. The pressure dependence of the temperature of the magnetic transition has also been investigated[77].

Neutron-diffraction studies[82] indicate that cobalt difluoride has the same magnetic structure as that of manganese difluoride (Figure 5.5, page 219). Cobalt difluoride is magnetically anisotropic both in the paramagnetic and antiferromagnetic states. Studies by Stout and Matarrese[85] and by Astrov and coworkers[80] show that at the Neel temperature

$\chi_\perp = 4.8\chi_\parallel$. Below the Neel temperature $\chi_\perp$ remains practically constant while $\chi_\parallel$ decreases rapidly, approaching zero as the temperature approaches zero, as shown in Figure 7.4. Lines[84] has discussed the magnetic properties of cobalt difluoride from a theoretical point of view.

In the antiferromagnetic state, two far-infrared absorptions occurring

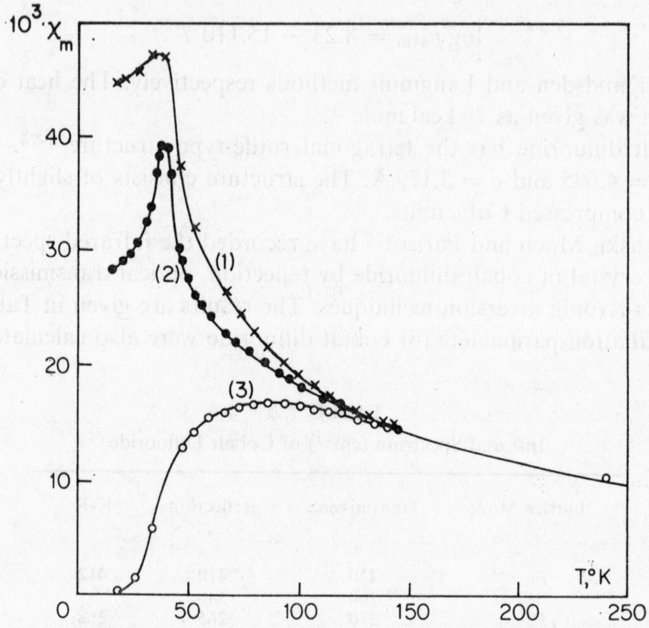

Figure 7.4 The magnetic susceptibility of cobalt difluoride. (1) Perpendicular. (2) Powder. (3) Parallel. (Reproduced by permission from P. N. Astrov, A. S. Borovik-Romanov and M. P. Oilova, *Soviet Physics–JETP*, **6**, 626 (1958))

at 36 and 28.5 cm^{-1} have been observed by Richards[81,90] at temperatures close to absolute zero. The theory of far-infrared antiferromagnetic resonance has been discussed by Kamimura[91].

Ludi and Feitnecht[92] have investigated the electronic absorption spectrum of cobalt difluoride and the observed bands were assigned on the basis of octahedral symmetry as follows:

$$6{,}800\ cm^{-1} \quad {}^4T_{2g}(F) \leftarrow {}^4T_{1g}(F)$$

$$14{,}100\ cm^{-1} \quad {}^4A_{2g}(F) \leftarrow {}^4T_{1g}(F)$$

$$20{,}400\ cm^{-1} \quad {}^4T_{1g}(P) \leftarrow {}^4T_{1g}((F)$$

Cobalt difluoride tetrahydrate. This pink compound is readily prepared by dissolving cobalt carbonate in hydrofluoric acid at room temperature[66,67]. It obeys the Curie-Weiss law over a wide temperature range[67] with a magnetic moment of 5.2 BM and a Weiss constant of 28°.

Cobalt dichloride. The principal methods of preparation of cobalt dichloride are given in Table 7.3.

TABLE 7.3
Preparation of Cobalt Dichloride

Method	Conditions	Reference
Co and Cl_2	Suspended in ether or alcohol	93
Dehydration of $CoCl_2 \cdot 6H_2O$	HCl gas, flow system, 400°	94
$CoCl_2 \cdot 6H_2O$ and $SOCl_2$	Reflux	95
Co(II) salt and $SOCl_2$[a]	Reflux	96
$Co(OOCH_3)_2$ and CH_3COCl	Interaction in benzene	97
Co_3O_4 and CCl_4	Flow system, 300–900°	98

[a] Suitable salts are the acetate, carbonate, formate and nitrate.

The dehydration of cobalt dichloride hexahydrate in vacuum or in a nitrogen stream often produces a product contaminated with oxides and is not recommended. Purification of anhydrous cobalt dichloride is readily achieved by sublimation in anhydrous hydrogen chloride at about 500°.

Cobalt dichloride is a pale blue crystalline solid which is extremely hygroscopic; on standing in air it is readily converted to the hexahydrate. It is very soluble in water, from which a number of hydrates have been isolated, and it is also soluble in a range of organic solvents such as acetone, alcohols and acetonitrile.

Mass-spectrometric studies at very high temperatures[99] indicate that cobalt dichloride vaporizes both as the monomer and the dimer. Schafer and coworkers[100] give the vapour pressure equation for solid cobalt dichloride as

$$\log p_{\text{mm}} = 12.91 - 12{,}040/T \qquad (632\text{–}737°)$$

and for liquid cobalt dichloride

$$\log p_{\text{mm}} = 9.67 - 8770/T \qquad (746\text{–}800°)$$

Thermodynamic data for cobalt dichloride derived from the vapour pressure data and from other sources are given in Table 7.4.

Cobalt dichloride has the cadmium dichloride-type hexagonal lattice with $a = 3.553$ and $c = 17.39$ Å[102]. Solid cobalt dichloride shows cobalt-chlorine stretching frequencies[103] at 272 and 220 cm^{-1}. Leroi and co-workers[104] have studied the gas-phase infrared spectrum of cobalt dichloride at high temperatures and report cobalt-chlorine stretching frequencies at 493 and 422 cm^{-1} for the monomer and dimer respectively.

Magnetic susceptibility measurements show that cobalt dichloride obeys the Curie-Weiss law above about 30°K with a magnetic moment[105,106] of 5.29 BM and a Weiss constant[83,105–107] in the range 19–38°. Below 30°K

TABLE 7.4
Thermodynamic Data for Cobalt Dichloride

Property	Value	References
M.p	724°	95
	735°	61
	740°	100
B.p.	1049°	61, 95, 101
	1053°	100
ΔH°_{form} (kcal mole^{-1})	−77.8	61
ΔH_{vap} (kcal mole^{-1})	30.61	101
	34.62	100
ΔS_{vap} (cal deg^{-1} mole^{-1})	26.11	100

cobalt dichloride becomes magnetically ordered[105,107–109]. There is also an anomaly in the specific heat in this temperature region[105,110]. Cobalt dichloride has quite a marked magnetic anisotropy[109,111] and Bizette and coworkers[109] reported that between 200 and 300°K $\chi_\perp$ is greater than $\chi_\parallel$, both values showing a maximum at 25°K.

There is some controversy as to the low-temperature magnetic structure of cobalt dichloride. Neutron-diffraction measurements[112] suggest anti-ferromagnetic behaviour, but Leech and Manuel[105,113] consider cobalt dichloride to be a ferromagnet. A theoretical treatment of the magnetic properties of cobalt dichloride has been given by Kanamori[114].

The absorption spectrum of a single crystal of cobalt dichloride at several temperatures down to 4.2°K has been recorded by Ferguson, Wood and Knox[115]. Both spin-allowed and spin-forbidden transitions were assigned on the basis of octahedral symmetry. Similar results have been obtained from a less detailed study by Ludi and Feitnechet[92]. Trutia and Musa[116] have recorded the absorption spectrum of solid cobalt dichloride and report that on melting, the change in the absorption spectrum is dramatic and consistent with a change from octahedral to tetrahedral coordination of the metal. The gas-phase absorption spectrum of cobalt dichloride has been recorded by Hougen, Leroi and James[117] and by DeKock and Gruen[118]. Very recently, Trutia and Musa[119] have questioned the results of Hougen and coworkers and suggested that these workers had the melt present as well as the vapour, since their spectrum showed the characteristics of the spectra of both the melt and the pure vapour.

Cobalt dichloride reacts with pyridine in solvents such as alcohol, chloroform and benzene to give adducts of the type $CoCl_2(py)_n$ where n is 2, 4 or 6[120–123]. Very little is known about dark red $CoCl_2(py)_6$ except that it decomposes under vacuum at room temperature to give $CoCl_2(py)_4$[120].

The adduct $CoCl_2(py)_4$ is a pink to red crystalline compound which is thermally unstable[123,124], decomposing at 110° to give the blue form of $CoCl_2(py)_2$. The crystal structure of $CoCl_2(py)_4$ has been determined by Porai-Koshits and Antsishkina[125–127] using single-crystal X-ray diffraction techniques. The tetragonal unit cell, of space group *I*4/*acd* containing eight molecular units, has a = 15.9 and c = 17.0 Å. The structure consists of *trans* octahedral $CoCl_2(py)_4$ units. All four pyridine rings are rotated relative to the plane of coordination to the same extent of approximately 45°. Half of the molecules in the unit cell have the pyridine rings rotated in a clockwise direction, the other half anti-clockwise. The cobalt-chlorine and cobalt-nitrogen bond distances are 2.32 and 1.99 Å respectively.

The adduct $CoCl_2(py)_4$ has a magnetic moment of 4.72 BM at room temperature and Clark and Williams[128] report cobalt-chlorine and cobalt-nitrogen stretching frequencies at 230 and 217 cm^{-1} respectively. The electronic absorption spectrum[123] of $CoCl_2(py)_4$ is consistent with the octahedral structure and the three observed bands were assigned as follows:

$$11{,}110\ \mathrm{cm^{-1}} \quad {}^2E_g \leftarrow {}^4T_{1g}(F)$$

$$16{,}130\ \mathrm{cm^{-1}} \quad {}^4A_{2g}(F) \leftarrow {}^4T_{1g}(F)$$

$$19{,}230\ \mathrm{cm^{-1}} \quad {}^4T_{1g}(P) \leftarrow {}^4T_{1g}(F)$$

The most extensively studied adduct of cobalt(II) is $CoCl_2(py)_2$. It was quickly established that this adduct exists in two forms, one blue and the other violet. The violet α-form is the one normally prepared and is readily obtained at room temperature[121]. On heating to 120° in a sealed tube it changes to the blue β-form[121,129]. Alternatively, the β-form may be prepared by dissolving α-$CoCl_2(py)_2$ in chloroform and precipitating with ligroin. The original supposition[121] that the violet α-form had a *trans* square-planar structure was proved incorrect by magnetic measurements[130].

Dunitz[131] showed by single-crystal X-ray diffraction studies that α-$CoCl_2(py)_2$ is in fact polymeric and contains octahedrally coordinated cobalt atoms. Although the crystal structure of β-$CoCl_2(py)_2$ has not been reported there is little doubt that it contains tetrahedrally coordinated cobalt. Furthermore, it has been shown[132] that blue $CoCl_2(py)_2$ is isostructural with $CoBr_2(py)_2$ which is known to be tetrahedral.

α-$CoCl_2(py)_2$ has a monoclinic (pseudo-orthorhombic) unit cell[131,133,134] with $a = 34.42$, $b = 17.38$, $c = 3.66$ Å and $\gamma = 90°$. The primitive space group is $P2/b$ and it contains eight molecules. The structure consists of

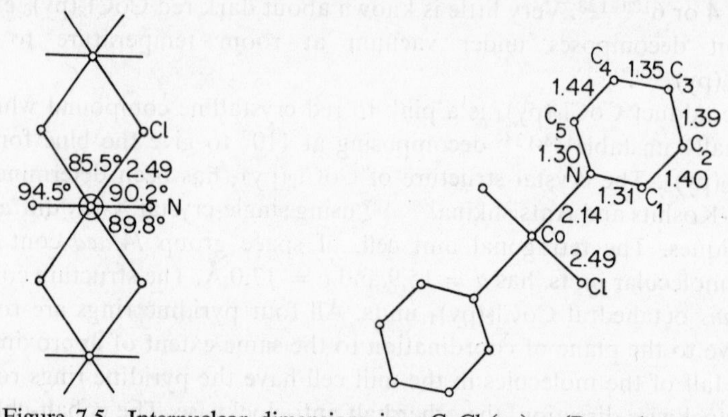

Figure 7.5 Internuclear dimensions in α-$CoCl_2(py)_2$. (Reproduced by permission from J. D. Dunitz, *Acta Cryst.*, **10**, 307 (1957))

chains, running parallel to the c axis, of octahedral $CoCl_4(py)_2$ units sharing two chlorine atoms with each of their two neighbouring units. The structure is shown in Figure 7.5 which gives the internuclear dimensions. An important feature to note is that the chlorine atoms are equidistant from each of the cobalt atoms in adjacent pairs.

The infrared spectral and magnetic properties of both forms of $CoCl_2(py)_2$ are summarized in Table 7.5.

Gill and her associates[138] have investigated the infrared spectrum of the coordinated pyridine molecule in α-$CoCl_2(py)_2$ and report that there is very little alteration of the spectrum of pyridine on complex formation. Thus the electron density on the pyridine ring remains almost unchanged and it was suggested that back-bonding is of considerable importance.

The electronic spectra of both forms of $CoCl_2(py)_2$ have been recorded[139–142]. The results have been interpreted on the basis of octahedral (α) and distorted tetrahedral (β) configurations. In solution, both forms give the spectrum of tetrahedrally coordinated cobalt, indicating breakdown of the octahedral polymer.

The equilibrium

$$CoCl_2(py)_2 + 2py \rightleftharpoons CoCl_2(py)_4$$

in non-aqueous solvents such as chloroform and benzene has been investigated and various thermodynamic data have been calculated[143–145].

As noted previously, $CoCl_2(py)_2$ is thermally unstable and decomposition studies[123,124,146] indicate that the mode of decomposition is

$$CoCl_2(py)_2\,(s) \rightarrow CoCl_2(py)\,(s) + py\,(g)$$
$$CoCl_2(py)\,(s) \rightarrow CoCl_2(py)_{2/3}\,(s) + 1/3py\,(g)$$
$$CoCl_2(py)_{2/3}\,(s) \rightarrow CoCl_2\,(s) + 2/3py\,(g)$$

TABLE 7.5
Properties of $CoCl_2(py)_2$

Property	α-$CoCl_2(py)_2$	Reference	β-$CoCl_2(py)_2$	References
$\nu_{Co\text{-}Cl}$	234 cm^{-1}	135	344, 304 cm^{-1}	128, 136
			349, 306 cm^{-1}	137
$\nu_{Co\text{-}N}$	227 cm^{-1}	135	252 cm^{-1}	128
			254, 190 cm^{-1}	137
δ_{NCoN}			142 cm^{-1}	137
δ_{ClCoCl}			101 cm^{-1}	137
μ_{eff} (RT)	5.15 BM	122	4.42 BM	122

The temperature of these reactions have been reported as[123] 210°, 250° and 350° respectively and as[146] 223°, 284° and 370° respectively. Beech and coworkers[147] give the heats of dissociation for the reaction

$$CoCl_2(py)_2\,(s) \rightleftharpoons CoCl_2\,(s) + 2py\,(g)$$

at 303°K as 45.1 kcal mole^{-1} for the α-form and as 42.3 kcal mole^{-1} for the β-form. The heat of transition for the α-β change is reported[148] to be 3.02 kcal mole^{-1} at 400°K. Absorption spectra[123] show quite conclusively that $CoCl_2(py)$ and $CoCl_2(py)_{2/3}$ both contain octahedrally coordinated cobalt.

A vast amount of work has been reported on the preparation and properties of cobalt dichloride adducts with substituted pyridines, aniline and closely related nitrogen bases. The adducts have been prepared in two general ways, namely, direct interaction, often in refluxing ligand, and secondly by mixing ethanolic solutions of the reactants.

Two types of complexes are formed, namely $CoCl_2L_4$ and $CoCl_2L_2$, the latter being obtained from $CoCl_2L_4$ in certain cases by controlled thermal decomposition. With some ligands it is not possible to obtain adducts of the type $CoCl_2L_4$ and the reasons for this have been discussed in terms of steric hindrance. King and coworkers[143,144,149,150] have studied the equilibrium

$$CoCl_2L_4 \rightleftharpoons CoCl_2L_2 + 2L$$

and have shown that the position of the equilibrium depends quite markedly on the nature of the ligand L.

For adducts of the type $CoCl_2L_2$, two different stereochemistries may be obtained, namely tetrahedral and octahedral, corresponding to the stereochemistries of the two bis(pyridine) adducts described above. However, in some cases it is the nitrogen base which acts as the bridging ligand in the octahedral polymer form. The two stereochemistries can be readily distinguished by their absorption spectra, their infrared spectra or their magnetic properties. There has been much discussion on the factors which influence the stereochemistry with a particular ligand. As with pyridine itself, it is possible to obtain both stereochemistries with the one ligand. For example, Malinovskii[151,152] has shown that there are two forms of $CoCl_2$(*p*-toluidine)$_2$, the blue form being tetrahedral and the violet form octahedral. The tetrahedral form has a monoclinic unit cell with $a = 12.30$, $b = 4.59$, $c = 26.10$ Å and $\beta = 93.75°$. There are four molecules in the unit cell of space group *I*2/*a*. The cobalt-chlorine and cobalt-nitrogen bond distances are 2.24 and 1.92 Å respectively.

There have been several studies of the thermal decomposition of a number of substituted pyridine and similar adducts of cobalt dichloride[123,141,147,148,153]. The adducts are reported to decompose according to the following general reactions.

$$CoCl_2L_4\,(s) \xrightarrow{T_1} CoCl_2L_2\,(s) + 2L\,(g)$$

$$CoCl_2L_2\,(s) \xrightarrow{T_2} CoCl_2L\,(s) + L\,(g)$$

$$CoCl_2L\,(s) \xrightarrow{T_3} CoCl_2L_{2/3}\,(s) + 1/3L\,(g)$$

$$CoCl_2L_{2/3}\,(s) \xrightarrow{T_4} CoCl_2\,(s) + 2/3L\,(g)$$

$$CoCl_2L_2\,(s) \xrightarrow{T_5} CoCl_2\,(s) + 2L\,(g)$$

$$CoCl_2L_2\,(s) \xrightarrow{T_6} CoCl_2L_{2/3}\,(s) + 4/3L\,(g)$$

$$CoCl_2L\,(s) \xrightarrow{T_7} CoCl_2\,(s) + L\,(g)$$

The approximate temperatures for the decomposition reactions of a number of cobalt dichloride adducts are given in Table 7.6.

The relative stabilities of the various adducts and their modes of decomposition have been discussed in terms of the basicity of the ligand, steric effects and σ- and π-bonding between the metal and the ligand.

The general properties and stereochemistries of a number of adducts of the types $CoCl_2L_4$ and $CoCl_2L_2$ are given in Table 7.7 and 7.8. Except in the case of $CoCl_2$(*p*-toluidine)$_2$ the structures have been deduced from

TABLE 7.6

Decomposition Temperatures of Some Cobalt Dichloride Adducts

Compound	Temperatures
$CoCl_2$(2-methylpyridine)$_2$	$T_5 = 220°$
$CoCl_2$(3-methylpyridine)$_4$	$T_1 = 110°$, $T_6 = 230°$, $T_4 = 333°$
$CoCl_2$(4-methylpyridine)$_4$	$T_1 = 140°$, $T_2 = 215°$, $T_3 = 260°$, $T_4 = 330°$
$CoCl_2$(2,6-lutidine)	$T_7 = 240°$
$CoCl_2$(2,4,6-collidine)$_2$	$T_2 = 160°$, $T_7 = 290°$
$CoCl_2$(aniline)$_2$	$T_2 = 190°$, $T_3 = 250°$
$CoCl_2$(acridine)$_2$	$T_2 = 280°$, $T_7 = 335°$

TABLE 7.7

Properties of $CoCl_2L_4$ Adducts

Compound and Properties	References
L = 3-methylpyridine, pink, μ = 4.94 BM, octahedral	123, 149
L = 4-methylpyridine, pink, μ = 4.94 BM, octahedral	123, 154
L = 3-ethylpyridine, pink, μ = 5.13 BM, octahedral	149
L = 4-ethylpyridine, pink, μ = 5.19 BM, octahedral	149
L = 4-propylpyridine, pink, μ = 5.21 BM, octahedral	149
L = 3,4-lutidine, pink, μ = 5.20 BM, octahedral	149
L = 3,5-lutidine, pink, μ = 5.04 BM, octahedral	149
L = 4-vinylpyridine, pink, μ = 4.66 BM, octahedral	155, 156
L = isoquinoline, pink, μ = 4.95 BM, octahedral	150

infrared, magnetic and spectral studies and their similarity with compounds of known structure.

Lee, Griswold and Kleinberg[168] have studied the thermal decomposition using differential thermal analysis techniques, of $CoCl_2(bipy)_3 \cdot H_2O$ and isolated a number of decomposition products. The decomposition scheme is shown diagrammatically in Figure 7.6. The decomposition products

TABLE 7.8
Properties of $CoCl_2L_2$ Adducts

Compound and Properties	References
L = 2-methylpyridine, blue, m.p. = 162°, μ = 4.48 BM, $\nu_{Co\text{-}Cl}$ = 334, 306 cm^{-1}, $\nu_{Co\text{-}N}$ = 240, 224 cm^{-1}, tetrahedral	123, 140, 143, 154, 157–159
L = 3-methylpyridine, blue, m.p. = 144°, μ = 4.49 BM, $\nu_{Co\text{-}Cl}$ = 342, 303 cm^{-1}, $\nu_{Co\text{-}N}$ = 239 cm^{-1}, tetrahedral	123, 149, 157, 158
L = 4-methylpyridine, blue, m.p. = 148°, μ = 4.48 BM, $\nu_{Co\text{-}Cl}$ = 343, 305 cm^{-1}, $\nu_{Co\text{-}N}$ = 247, 234 cm^{-1}, tetrahedral	123, 154, 157, 158
L = 3-ethylpyridine, blue, μ = 4.48 BM, tetrahedral	140, 149
L = 4-ethylpyridine, blue, μ = 4.47 BM, tetrahedral	149
L = 4-propylpyridine, blue, μ = 4.53 BM, tetrahedral	149
L = 2-pentylpyridine, blue, μ = 4.48 BM, tetrahedral	140
L = 2-vinylpyridine, blue, μ = 4.55 BM, tetrahedral	155
L = 4-vinylpyridine, blue, μ = 4.50 BM, tetrahedral	155, 156
L = 2-chloropyridine, blue, μ = 4.53 BM, $\nu_{Co\text{-}Cl}$ = 332, 312 cm^{-1}, $\nu_{Co\text{-}N}$ = 227 cm^{-1}, tetrahedral	157, 160, 161
L = 4-chloropyridine, lilac, m.p. = 148°, μ = 5.12 BM, octahedral	157, 161, 162
L = 2-bromopyridine, blue, μ = 4.52 BM, $\nu_{Co\text{-}Cl}$ = 327, 319 cm^{-1}, $\nu_{Co\text{-}N}$ = 222 cm^{-1}, tetrahedral	157, 160, 161
L = 3-bromopyridine, pale violet, μ = 5.25 BM, octahedral	161
L = 4-bromopyridine, lilac, m.p. = 155°, μ = 5.33 BM, octahedral	157, 162
L = 3,4-lutidine, blue, μ = 4.60 BM, tetrahedral	149
L = 3,5-lutidine, blue, μ = 4.59 BM, tetrahedral	149
L = 2,4,6-collidine, blue, m.p. = 160°, $\nu_{Co\text{-}Cl}$ = 318, 274 cm^{-1}	123, 158
L = aniline, $\nu_{Co\text{-}Cl}$ = 318, 299 cm^{-1}, tetrahedral	153
L = *o*-toluidine, $\nu_{Co\text{-}Cl}$ = 325, 292 cm^{-1}, tetrahedral	163
L = *m*-toluidine, $\nu_{Co\text{-}Cl}$ = 320, 298 cm^{-1}, tetrahedral	163
L = *p*-toluidine, α-form, blue, μ = 4.72 BM, $\nu_{Co\text{-}Cl}$ = 318, 292 cm^{-1}, tetrahedral	128, 151, 152, 163, 164
L = acridine, blue, m.p. = 264°, octahedral	123
L = quinoline, blue, μ = 4.42 BM, tetrahedral	140, 141, 150
L = isoquinoline, μ = 4.54 BM, tetrahedral	140, 150
L = morpholine, blue, $\nu_{Co\text{-}Cl}$ = 337, 304 cm^{-1}, tetrahedral	165
L = quinoxaline, blue, μ = 4.55 BM, tetrahedral	166
L = 2-methylthiomethylpyridine, lilac, μ = 4.78 BM, octahedral	167

were investigated by study of absorption spectra, magnetic properties, conductance and X-ray powder diffraction patterns. It was shown that α-$CoCl_2$(bipy) contains octahedrally coordinated cobalt while the β-form is tetrahedral. Clark and Williams[169] recorded the infrared spectra of the various decomposition products and assigned, where possible, the cobalt-chlorine stretching frequencies. They also showed that $CoCl_2(bipy)_{1.33}$ should be formulated with a dimeric octahedral cation and a tetrahedral anion, that is as $[(bipy)_2Co_2Cl_2(bipy)_2^{2+}][CoCl_4^{2-}]$.

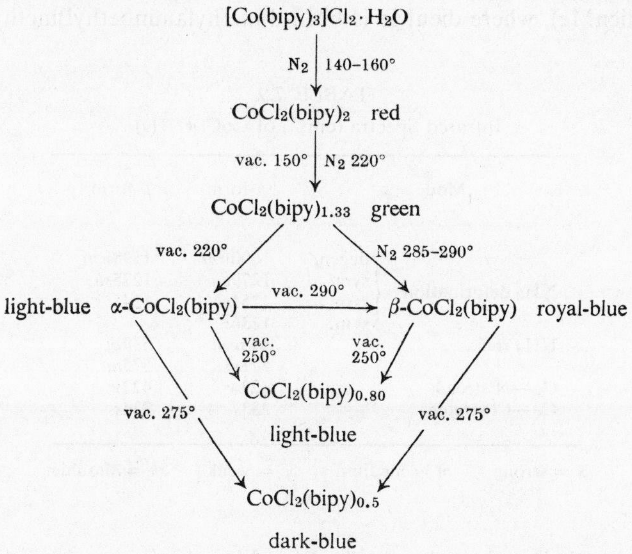

Figure 7.6. Decomposition of cobalt dichloride-bipyridyl adducts.

When anhydrous ammonia is passed into a boiling solution of cobalt dichloride in amyl alcohol, or passed over $CoCl_2 \cdot 6NH_3$ at approximately 140°, rose-red $CoCl_2(NH_3)_2$ is formed[170]. The less stable blue-violet β-form may be prepared by heating the hexamine to 66° in a vacuum in the presence of concentrated sulphuric acid[170]. The β-form is converted to the α-form on heating to 210° in an atmosphere of ammonia[170]. The magnetic moment for α-$CoCl_2(NH_3)_2$ is 5.50 BM and for the β-form is 5.41 BM[171], indicating octahedral coordination of the cobalt. X-ray powder diffraction patterns[172] suggest that the α- and β-forms of $CoCl_2(NH_3)_2$ have similar but not identical structures. Clark and Williams[172] have recorded the infrared spectra of both forms of $CoCl_2(NH_3)_2$ and their assignments are given in Table 7.9.

Sacco and Freni[173] report that the adducts $CoCl_2(CH_3CN)_4$ and $CoCl_2(CH_3CN)_2$ may be isolated from the direct interaction in alcohol of the two constituents under suitable conditions. Violet $CoCl_2(CH_3CN)_4$ has a magnetic moment of 1.89 BM[174] and is thought to contain octahedrally coordinated cobalt. $CoCl_2(CH_3CN)_2$ is blue[175] and obeys the Curie-Weiss law[174] with a magnetic moment of 4.94 BM and a Weiss constant of 20°. Consideration of magnetic and spectral data[174,176] indicates that the compound is $[Co(CH_3CN)_4^{2+}][CoCl_4^{2-}]$.

Ciampolini and Speroni[177] isolated a violet crystalline complex, $CoCl_2$(dienMe), where dienMe is bis(2-dimethylaminoethyl)methylamine,

TABLE 7.9

Infrared Spectra (cm^{-1}) of $CoCl_2(NH_3)_2$

Mode		α-form	β-form
NH_3 deformation	degen.	1600*wm*	1598*wm*
	sym.	1279*vw*	1275*sh*
	sym.	1251*sh*	1245*vs*
	sym.	1236*s*	
NH_3 rock		645*wm*	639*m*
		571*m*	575*m*
Co—N stretch		423*w*	422*w*
Co—Cl stretch		233*s*	234*s*

s = strong *m* = medium *w* = weak *sh* = shoulder

by mixing cobalt dichloride and the ligand in a hot butanol solution. The compound had a room-temperature magnetic moment of 4.60 BM and was considered to be five coordinate because of its diffuse-reflectance spectrum and its non-conductance in chloroform and nitrobenzene. A single-crystal X-ray diffraction study by Di Vaira and Orioli[178] has confirmed this stereochemistry. The complex has a monoclinic unit cell, of space group $P2_1/c$ containing eight molecular units, with $a = 8.38$, $b = 29.19$, $c = 12.01$ Å and $\beta = 102.1°$. The structure, shown in Figure 7.7, may be described as a distorted square-pyramid or as a distorted trigonal bipyramid. The bond angles in $CoCl_2$(dienMe) are given in Table 7.10.

2,2′-Diaminoethylamine (dien) reacts[179] with cobalt dichloride to give blue $CoCl_2$(dien). This compound has a magnetic moment of 4.74 BM and Barclay and Barnard[179] suggested that it should be formulated either as [Co(dien)Cl]Cl or as $[Co(dien)_2][CoCl_4]$. The latter formulation, containing both octahedral and tetrahedral cobalt, has been confirmed by Ciampolini and Speroni[177] by diffuse-reflectance spectral studies.

By interacting 1,1,7,7-tetraethyldiethylenetriamine, Et_4dien, with cobalt

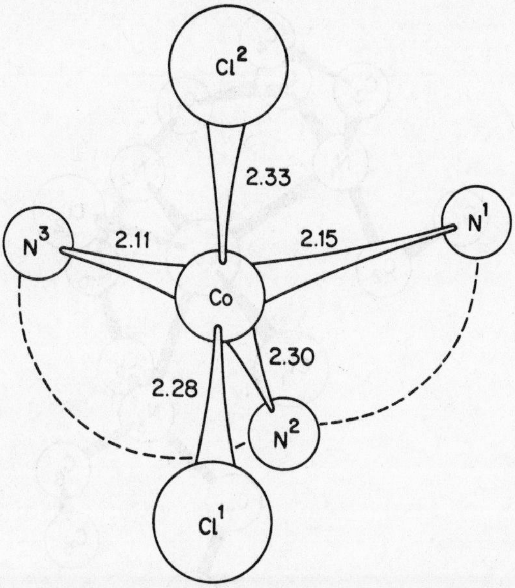

Figure 7.7 The structure of $CoCl_2$(dienMe). The dotted lines represent the CH_2—CH_2 chains. (Reproduced by permission from M. Di Vaira and P. Orioli, *Chem. Commun.*, **1965**, 590)

TABLE 7.10
Bond Angles in $CoCl_2$(dienMe)

Angle	Value	Angle	Value
Cl^1—Co—Cl^2	104°	Cl^2—Co—N^2	162°
Cl^1—Co—N^1	107°	Cl^2—Co—N^3	94°
Cl^1—Co—N^2	93°	N^1—Co—N^2	78°
Cl^1—Co—N^3	113°	N^2—Co—N^3	80°
Cl^2—Co—N^1	95°	N^1—Co—N^3	135°

dichloride in alcohol, Dori and Gray[180] isolated red-violet $CoCl_2(Et_4$dien). This compound has a magnetic moment of 4.71 BM, and together with its absorption spectrum it was suggested that the adduct was five coordinate both in the solid state and in solution. This stereochemistry has now been confirmed for the solid state by single-crystal X-ray diffraction studies[181].

The compound has a triclinic unit cell with $a = 7.04$, $b = 12.99$, $c = 9.90$ Å, $\alpha = 68.1°$, $\beta = 80.9°$ and $\gamma = 79.7°$. The space group is $P\bar{1}$ and it contains two molecules. The structure is shown in Figure 7.8. The molecular

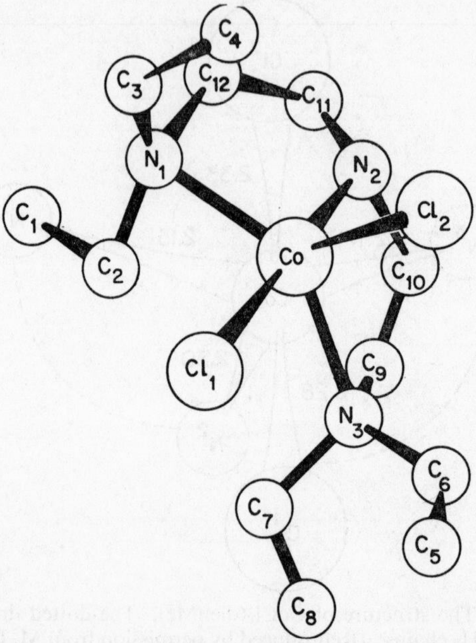

Figure 7.8 The structure of $CoCl_2(Et_4dien)$. (Reproduced by permission from Z. Dori, R. Eisenberg and H. B. Gray, *Inorg. Chem.*, **6,** 483 (1967))

TABLE 7.11
Internuclear Dimensions of $CoCl_2(Et_4dien)$

Distance	Value	Angle	Value	Angle	Value
Co—Cl$_1$	2.319 Å	N$_2$—Co—N$_3$	80.7°	Cl$_2$—Co—N$_2$	83.3°
Co—Cl$_2$	2.357 Å	Cl$_1$—Co—N$_1$	93.1°	Cl$_2$—Co—N$_3$	105.9°
Co—N$_1$	2.21 Å	Cl$_1$—Co—N$_2$	173.4°	Cl$_1$—Co—Cl$_2$	101.8°
Co—N$_2$	2.16 Å	Cl$_1$—Co—N$_3$	101.9°	N$_1$—Co—N$_2$	80.1°
Co—N$_3$	2.19 Å	Cl$_2$—Co—N$_1$	126.8°	N$_1$—Co—N$_3$	120.4°

geometry is quite irregular and can be viewed either as a distorted square pyramid or as a distorted trigonal bipyramid. The important bond distances and angles are given in Table 7.11.

A number of products have been isolated from the reaction between cobalt dichloride and 1,3-bis(2′-pyridyl)-2,3-diaza-1-propene (paphy) in ethanol[182,183]. Two of these have the empirical formula $CoCl_2$(paphy). The α-form is obtained as red plates and the β-form as dark green needles. The β-form is converted to the α-form by heating in ethanol. The α-form has a magnetic moment of 5.04 BM and its diffuse-reflectance spectrum suggests that it contains cobalt in an octahedral environment[182,183], probably with halogen bridging atoms.

The magnetic moment[182,183] of β-$CoCl_2$(paphy) is 4.84 BM at room temperature and falls to 4.56 BM at 80°K. A single-crystal X-ray diffraction study[182,183] has revealed square-pyramidal stereochemistry for the

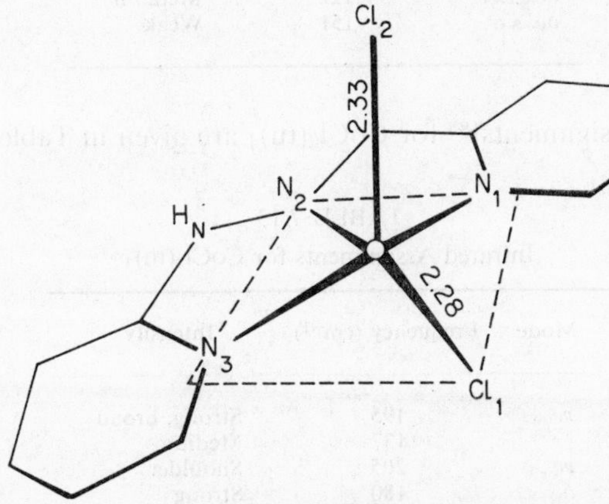

Figure 7.9 The structure of β-$CoCl_2$(paphy). (Reproduced by permission from I. G. Dance, M. Gerloch, J. Lewis, F. S. Stephens and F. Lions, *Nature*, **210**, 298 (1966))

cobalt atom in this complex. The compound is monoclinic with $a = 8.72$, $b = 11.93$, $c = 12.64$ Å and $\beta = 98°$. There are four molecules in the unit cell of space group $P2_1/n$. The structure of the complex is shown in Figure 7.9. The cobalt atom is 0.39 Å above the plane containing the atoms N_1, N_2, N_3 and Cl_1. The average cobalt-nitrogen bond distance is 2.12 Å and the angles about the cobalt atom are: N_1—Co—N_2 = N_3—Co—N_2 = 74°, N_2—Co—Cl_1 = 157° and N_2—Co—Cl_2 = 94°.

Thiourea (tu) reacts with cobalt dichloride in alcohol to give complexes of the types $CoCl_2(tu)_2$ and $CoCl_2(tu)_4$[184–186]. Blue $CoCl_2(tu)_2$ melts at 148° and obeys the Curie-Weiss law[184] with a magnetic moment of 4.45 BM and a Weiss constant of 18°. The compound is non-conducting in solution and its absorption spectrum indicates tetrahedral geometry. Infrared studies[184–186] indicate that the sulphur atom of the ligand acts as the donor site. Adams and Cornell[185] report the assignments shown in Table 7.12.

TABLE 7.12
Infrared Assignments of $CoCl_2(tu)_2$

Mode	Frequency (cm^{-1})	Intensity
$\nu_{Co\text{-}Cl}$	316	Strong
	295	Strong
$\nu_{Co\text{-}S}$	290	Shoulder
	252	Strong
$\delta_{Cl\text{-}Co\text{-}Cl}$	122	Medium
$\delta_{Co\text{-}S\text{-}C}$	151	Weak

The infrared assignments[185] for $CoCl_2(tu)_4$ are given in Table 7.13.

TABLE 7.13
Infrared Assignments for $CoCl_2(tu)_4$

Mode	Frequency (cm^{-1})	Intensity
$\nu_{Co\text{-}Cl}$	195	Strong, broad
	137	Medium
$\nu_{Co\text{-}S}$	205	Shoulder
$\delta_{Co\text{-}S\text{-}C}$	180	Strong
	166	Shoulder

Nardelli and coworkers[187] and Carlin and Holt[188] report that the blue adduct $CoCl_2(etu)_2$, where etu is ethylenethiourea, may be isolated when the ligand and cobalt dichloride are reacted together in alcohol. The compound has an orthorhombic unit cell[187] with $a = 13.59$, $b = 7.73$ and $c = 11.98$ Å. The compound is a non-electrolyte and obeys the Curie-Weiss law[188] with a magnetic moment of 4.51 BM and a Weiss constant of 9°. The electronic absorption spectrum has been interpreted on the basis of tetrahedral symmetry[188]. Adams and Cornell[185] have shown that the sulphur atom of the ligand acts as the donor site. The infrared assignments reported for this compound are given in Table 7.14.

Cobalt dichloride reacts[189,190] with a large range of phosphines, arsines and several phosphine and arsine oxides to give, in general, distorted tetrahedral complexes of the type $CoCl_2L_2$. These complexes have been studied by various techniques including infrared and absorption spectra and magnetic studies. The triphenylarsine oxide adduct exists in two forms, having quite different infrared spectra, although magnetic and absorption spectral studies indicate that both are tetrahedral. Some properties of a number of these complexes are given in Table 7.15.

TABLE 7.14
Infrared Assignments for $CoCl_2(etu)_2$

Mode	Frequency (cm^{-1})	Intensity
$\nu_{Co\text{-}Cl}$	313	Strong, broad
	291	Strong
$\nu_{Co\text{-}S}$	281	Strong
	235	Strong
$\delta_{Cl\text{-}Co\text{-}Cl}$	151	Medium
$\delta_{Co\text{-}C\text{-}S}$	139	Medium

TABLE 7.15
$CoCl_2L_n$ Adducts with Phosphines, Arsines and the Corresponding Oxides

Compound and Properties	References
L = $(CH_3)_3P$, $n = 2$, green, m.p. = 130–132°	191
L = $C_2H_5PH_2$, $n = 2$, μ = 4.30 BM, tetrahedral	192
L = $(C_2H_5)_2PH$, $n = 2$, green-black, tetrahedral	192
L = $(C_2H_5)_3P$, $n = 2$, blue, m.p. = 101–102°, μ = 4.39 BM, tetrahedral	191, 193
L = $(C_6H_5)_3P$, $n = 2$, blue, m.p. = 231–232°, μ = 4.41 BM, θ = 2°, $\nu_{Co\text{-}Cl}$ = 342, 318 cm^{-1}, $\nu_{Co\text{-}P}$ = 187, 151 cm^{-1}, tetrahedral	151, 158, 191, 194–196
L = $(C_6H_{11})_2PH$, $n = 2$, light blue, m.p. = 112–114°, μ = 4.85 BM, tetrahedral	197
L = $(C_6H_{11})_2P(CH_2)_5P(C_6H_{11})_2$, $n = 1$, blue, air stable, tetrahedral	198
L = $(C_6H_{11})_2PP(C_6H_{11})_2$, $n = 1$, blue, m.p. = 128–130°, μ = 4.48 BM, tetrahedral	199
L = 2-phenylisophosphindoline, $n = 3$, isolated as a monohydrate, deep green-black, m.p. = 183–184°, monomeric, non-electrolyte, probably square-pyramidal	200
L = $(CH_3)_3As$, $n = 2$, μ = 4.75 BM, tetrahedral, looses one molecule of arsine to give halogen-bridged polymer	193
L = $(C_6H_5)_3PO$, $n = 2$, deep blue, m.p. = 238–240°, μ = 4.79 BM, $\nu_{Co\text{-}Cl}$ = 342, 317 cm^{-1}, $\nu_{P\text{-}O}$ = 1155 cm^{-1}, tetrahedral	59, 158, 164, 196, 201, 202
L = bis(diphenylphosphinoethyl)oxide, $n = 1$, blue, stable in air, μ = 4.50 BM, tetrahedral	203
L = $(C_6H_5)_2CH_3AsO$, $n = 2$, $\nu_{As\text{-}O}$ = 851 cm^{-1}, $\nu_{Co\text{-}O}$ = 418 cm^{-1}, $\nu_{Co\text{-}Cl}$ = 315, 287 cm^{-1}, tetrahedral	204
L = $(C_6H_5)_3AsO$, $n = 2$, Form A: blue, μ = 4.60 BM, θ = 6°, $\nu_{As\text{-}O}$ = 889 cm^{-1}, $\nu_{Co\text{-}O}$ = 382, $\nu_{Co\text{-}Cl}$ = 325, 302 cm^{-1}, tetrahedral: Form B: blue, μ = 4.77 BM, $\nu_{As\text{-}O}$ = 886, 867 cm^{-1}, $\nu_{Co\text{-}O}$ = 404, 392 cm^{-1}, $\nu_{Co\text{-}Cl}$ = 323, 288 cm^{-1}, tetrahedral	196, 204–206
L = $(C_6H_{11})_3AsO$, $n = 2$, $\nu_{As\text{-}O}$ = 852 cm^{-1}	207

TABLE 7.16
Properties of $CoCl_2L_n$ Adducts

Compound and Properties	References
L = quinoxaline (Q), $n = 1$, $\mu = 5.31$ BM, octahedral	166
L = 2-methylquinoxaline (Mq), $n = 1$, blue-grey, $\mu = 5.07$ BM, octahedral	166
L = 2,3-dimethylquinoxaline (Dmq), $n = 1$, $\mu = 4.90$ BM, possibly dimeric tetrahedral	166
L = pyrazine (pyz), $n = 1$, pink, $\mu = 4.92$ BM, octahedral	212
L = methylpyrazine (mpyz), $n = 1$, violet, $\mu = 5.36$ BM, octahedral	212
L = 2,5-dimethylpyrazine (2,5-dmp), $n = 1$, violet, $\mu = 5.4$ BM, octahedral	212, 213
L = 2,6-dimethylpyrazine (2,6-dmp), $n = 1$, violet, $\mu = 5.46$ BM, octahedral	212, 214
L = 2-methyl-8-methylthioquinoline (mmtq), $n = 1$, blue, $\mu = 4.56$ BM, tetrahedral	215
L = 2,9-dimethyl-1,10-phenanthroline (dmp), $n = 1$, blue, $\mu = 4.53$ BM, tetrahedral	216
L = $(C_2H_5)_3N$, $n = 2$, $\mu = 4.66$ BM, tetrahedral	94
L = $(C_2H_5)_2NH$, $n = 2$, $\mu = 4.52$ BM, tetrahedral	94
L = 2,2′-bipyridylamine (bipyam), $n = 1$, $\mu = 4.51$ BM, $\nu_{Co-Cl} = 377$, 318 cm^{-1}, tetrahedral	217, 218
L = *N,N,N′,N′*-tetramethylethylenediamine (Me_4en), $n = 1$, blue, m.p. = 176–180°, $\mu = 4.65$ BM, tetrahedral	219
L = *N,N,N′,N′*-tetramethyl-1,2-propylenediamine(Me_4pn), $n = 1$, blue, m.p. = 208–214°, $\mu = 4.70$ BM, tetrahedral	219
L = *N,N,N′,N′*-tetramethyltrimethylenediamine (Me_4tn), $n = 1$, blue, m.p. = 232–241°, $\mu = 4.60$ BM, tetrahedral	219
L = tris(2-dimethylaminoethyl)amine (Me_6tren), $n = 1$, grey-violet, $\mu =$ 4.45 BM, shown to be $[CoCl(Me_6tren)]^+Cl^-$ with five-coordinate cation	220
L = bis(2-dimethylaminoethyl)oxide (Me_4daeo), $n = 1$, blue, $\mu = 4.70$ BM, monomeric, five-coordinate	221
L = bis(2-dimethylaminoethyl)sulphide (Me_4daes), $n = 1$, $\mu = 4.55$ BM, five-coordinate	222
L = Schiff base formed between *N*-methyl-*o*-aminobenzaldehyde and *N,N*-diethylethylenediamine (MABen-NEt_2), $n = 1$, dark green, $\mu = 4.82$ BM, five-coordinate	223
L = Schiff base formed between *o*-methylthiobenzaldehyde and *N,N*-diethylethylenediamine (MSBen-NEt_2), $n = 1$, blue, m.p. = 173–177°, $\mu = 4.66$, tetrahedral	224
L = *N,N*-ethylenebis(2-thenylidineimine) (SNNS), $n = 1$, blue, $\mu = 4.58$ BM, tetrahedral	225
L = CH_3NHNH_2, $n = 2$, pale violet, $\mu = 4.89$ BM, octahedral	226
L = $(CH_3)_2NNH_2$, $n = 2$, dark blue, $\mu = 4.38$ BM, tetrahedral	226
L = 1,2-di-(2-pyridyl)ethylene, $n = 1$, blue, $\mu = 4.77$ BM, tetrahedral	227
L = 1,2-di-(4-pyridyl)ethylene, $n = 1$, blue-grey, $\mu = 5.30$ BM, octahedral	227
L = imidazole (imz), $n = 2$, $\mu = 4.51$ BM, tetrahedral	228
L = 2-methylimidazole (Mlz), $n = 1$, blue, $\mu = 4.58$ BM, tetrahedral	229
L = 1,2-dithiocyanatoethane, $n = 1$, pink, m.p. = 169–173°, $\mu = 4.97$ BM, octahedral	230
L = ethylene urea (2-imidazolidinone, eu), $n = 3$, dark blue, m.p. = 160–161°, $\mu = 4.35$ BM, five-coordinate	231
L = *N,N*-dimethylacetamide (DMA), $n = 2$, deep blue, $\mu = 4.73$ BM	232
L = β(bisphenylphosphino)ethyl-pyridine, $n = 1$, blue, m.p. = 224°, $\mu =$ 4.69 BM, tetrahedral	233
L = hexamethylphosphoramide, $n = 2$, $\mu = 4.72$ BM, tetrahedral	234

Magnetic and spectral studies indicate that cobalt dichloride reacts with pyridine *N*-oxide and several other *N*-oxides to give, in general, tetrahedral complexes[208–211].

A large range of other ligands has been reacted with cobalt dichloride to give adducts containing four-, five- and six-coordinate cobalt. Most of the adducts are prepared either by direct interaction, or by mixing alcohol solutions of the two constituents. Some properties of a number of these adducts are summarized in Table 7.16.

Adducts with tetrahydrofuran, urea, dioxan, alcohols, etc., have also been reported[235–247].

Cobalt dichloride hexahydrate. The dehydration of this blue compound, which is readily isolated from saturated solutions of cobalt dichloride in dilute hydrochloric acid, has been studied by several groups[242,248,249]. Four molecules of water are lost very readily, either at about 50° or even on standing over concentrated sulphuric acid. Borchardt and Daniels[248] give the following decomposition mechanism and temperatures.

$$CoCl_2 \cdot 6H_2O\ (s) \xrightarrow{49^\circ} CoCl_2 \cdot 2H_2O\ (s) + 4H_2O\ (l)$$

$$CoCl_2 \cdot 2H_2O\ (s) \xrightarrow{137^\circ} CoCl_2 \cdot H_2O\ (s) + H_2O\ (g)$$

$$CoCl_2 \cdot H_2O\ (s) \xrightarrow{175^\circ} CoCl_2\ (s) + H_2O\ (g)$$

The structure of cobalt dichloride hexahydrate has been studied by Mizuno and coworkers[250,251] and by Stroganov and his group[252]. The unit-cell parameters reported by the two groups are given in Table 7.17.

TABLE 7.17
Unit-cell Parameters of Cobalt Dichloride Hexahydrate

	Mizuno[250]	Stroganov *et al.*[252]
Symmetry	Monoclinic	Monoclinic
a (Å)	10.34	6.58
b (Å)	7.06	7.03
c (Å)	6.67	11.50
β	122.33°	49.5°
Space group	*C*2/*m*	*P*2$_1$/*c*

Although both groups agreed that the structure consists of *trans* $[CoCl_2(H_2O)_4]$ groups with two molecules of water of hydration, some discrepancies exist as to the stereochemistry about the cobalt atom. Mizuno's results[250] are to be preferred since his structure is more compatible with ^{1}H NMR studies[253,254]. The structure of cobalt dichloride hexahydrate is shown in Figure 7.10. The four water molecules bound to the cobalt atom form a square, the cobalt-oxygen bond length being 2.12 Å. The cobalt-chlorine bonds, which are at right angles to this plane, are

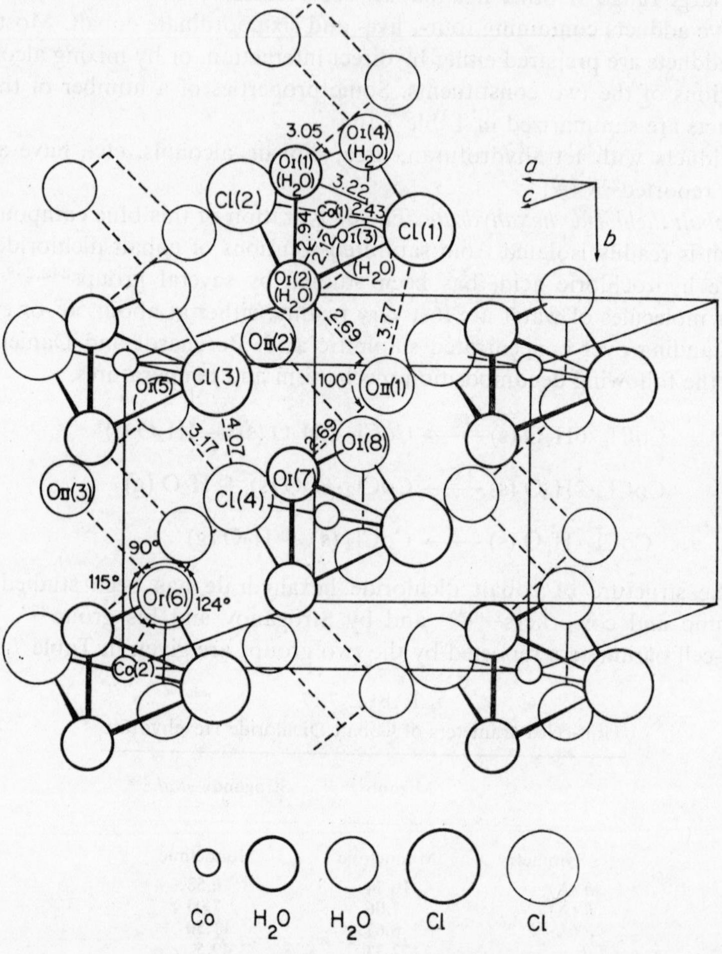

Figure 7.10 The structure of cobalt dichloride hexahydrate. (Reproduced by permission from J. Mizuno, *J. Phys. Soc. Japan*, **15**, 1412 (1960))

2.43 Å in length. The $[CoCl_2(H_2O)_4]$ octahedra are held together by hydrogen bonds to give chains parallel to the *b* axis.

The infrared spectrum of cobalt dichloride hexahydrate has been studied by Ferraro and Walker[103] and Gamo[255], particular emphasis being placed on the modes of vibration of the water molecules.

The magnetic properties of cobalt dichloride hexahydrate have been examined in great detail. It obeys the Curie-Weiss law down to very low temperatures[86], with a magnetic moment of 4.61 BM and a Weiss constant of 20°. At low temperature it becomes antiferromagnetic and the Neel temperature is approximately 2°K[253,256–262]. A maximum in the specific heat occurs at 2.29°K[263,264]. Investigation of the magnetic structure[265] and the spectroscopic splitting factors[266] have also been reported.

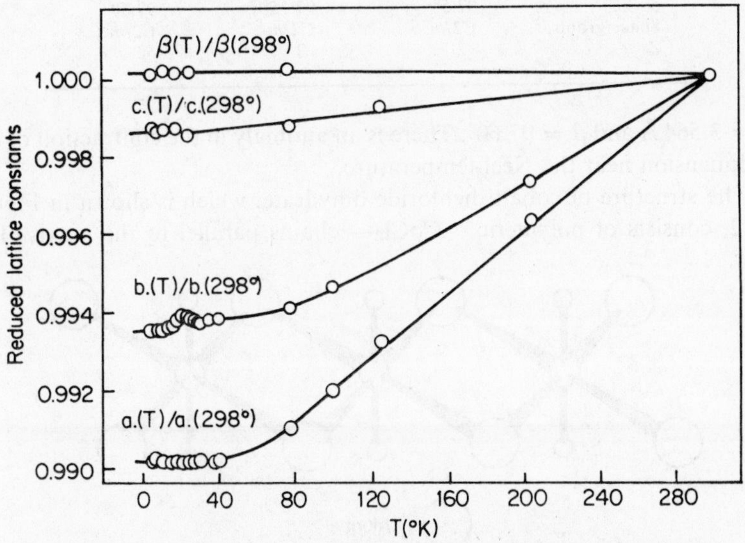

Figure 7.11 The lattice parameters of $CoCl_2 \cdot 2H_2O$ as a function of temperature. (Reproduced by permission from B. Morosin, *J. Chem. Phys.*, **44**, 252 (1966))

Cobalt dichloride dihydrate. This hydrate may be prepared by evaporating an aqueous solution of cobalt dichloride[267–269] at about 70°. It is also formed during the thermal decomposition of the corresponding hexahydrate (see above).

The structure of cobalt dichloride dihydrate has been investigated by electron-diffraction and single-crystal X-ray diffraction techniques. The reported unit-cell data are given in Table 7.18.

The very small differences in the unit-cell parameters quoted in Table 7.18 may have arisen because of different temperatures at which the measurements were made. Morosin[269] has shown that the unit cell of cobalt dichloride dihydrate contracts anisotropically on cooling as shown in Figure 7.11. At 5°K the unit-cell dimensions are $a = 7.206$, $b = 8.498$,

TABLE 7.18
Unit-cell Parameters of Cobalt Dichloride Dihydrate

	Vainshtein[270]	Morosin and Graebner[267]	Morosin[269]
Symmetry	Monoclinic	Monoclinic	Monoclinic
a (Å)	7.315	7.256	7.279
b (Å)	8.544	8.575	8.553
c (Å)	3.581	3.554	3.569
β	97.5°	97.55°	97.58°
Space group	$C2/m$	$C2/m$	$C2/m$
Z	2	2	2

$c = 3.564$ Å and $\beta = 97.60°$. There is an anomaly in the contraction of the b dimension near the Neel temperature.

The structure of cobalt dichloride dihydrate, which is shown in Figure 7.12, consists of polymeric —$CoCl_2$— chains parallel to the c axis. The

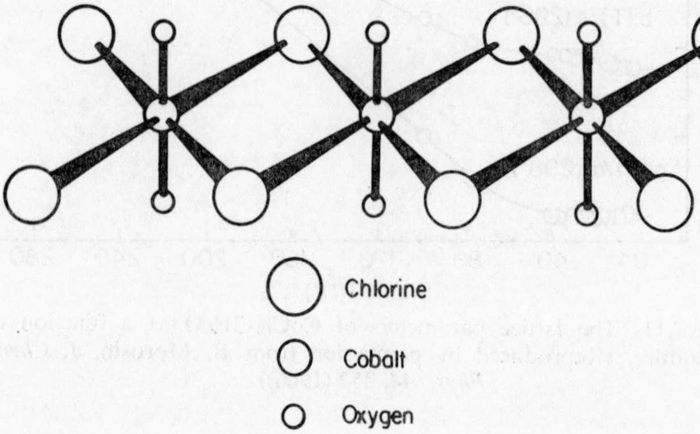

Figure 7.12 The chain structure of cobalt dichloride dihydrate

water molecules complete the octahedral coordination about each cobalt atom. The chains are held together by hydrogen bonds between an oxygen atom of one chain and a chlorine atom of the next chain. The important internuclear dimensions at 298°K and 5°K are given in Table 7.19.

The proton positions in cobalt dichloride dihydrate have been deduced from ^1H NMR studies by van Tiggelen and coworkers[271], as shown in Figure 7.13.

The infrared spectrum of cobalt dichloride dihydrate has been recorded by Gamo[255] who investigated the effect of coordination and hydrogen bonding on the various modes of vibration of the water molecules.

Cobalt dichloride dihydrate becomes antiferromagnetic at low temperatures, the Neel temperature being approximately 17.5°K[272,273]. There is a maximum in the specific heat at approximately the same temperature[274,275]. The magnetic structure[276–278] of cobalt dichloride dihydrate

TABLE 7.19

Internuclear Dimensions of Cobalt Dichloride Dihydrate

Dimension	298°K	5°K
Co—Cl(1)	2.459 Å	2.460 Å
Co—Cl(2)	2.487 Å	2.471 Å
Co—O	2.034 Å	2.036 Å
Cl(1)—Co—Cl(2)	87.64°	87.42°

consists of ferromagnetically ordered chains which are coupled antiferromagnetically with adjacent chains.

Lawson[268] has examined the electronic absorption spectrum of a single crystal of cobalt dichloride dihydrate at several temperatures. Assignments were made on the basis of octahedral symmetry.

Cobalt dibromide. Cobalt dibromide may be prepared by the interaction of cobalt metal and hydrogen bromide at red heat in a flow system[99]. However, it is probably more conveniently prepared[279] by dissolving the carbonate in hydrobromic acid, dehydrating the product at 130° and subliming it in vacuum at 550°, or by passing hydrogen bromide over cobalt dibromide hexahydrate at 500° in a flow system[94]. It has also been reported that the anhydrous bromide results from the interaction of cobalt acetate and acetyl bromide[97] or the dehydration of the hexahydrate over concentrated sulphuric acid at room temperature[61].

Cobalt dibromide is a green hygroscopic solid which dissolves readily in water. It melts[280] at 678° and its heat of formation has been reported[61] as -63.8 kcal mole^{-1}.

Mass-spectrometric studies[99] show that cobalt dibromide vaporizes to give monomeric and dimeric species. The infrared spectrum of cobalt dibromide vapour at approximately 900° shows two strong absorptions at 396 and 325 cm^{-1} which have been assigned to the cobalt-bromine stretching frequencies of the monomer and dimer respectively[104].

At room temperature cobalt dibromide has a hexagonal lattice with $a = 3.865$ and $c = 6.120$ Å[279,281]. The $CoBr_6$ octahedra are packed in the hexagonal close-packed arrangement[279]. Heat capacity measurements[279] indicate a phase change at about 370° to a structure with cubic close-packing of the $CoBr_6$ octahedra. The heat of transition is 40 cal mole^{-1}. Dehydration of the hexahydrate of cobalt dibromide at 130° gives a hexagonal lattice in which the $CoBr_6$ octahedra are randomly arranged[279].

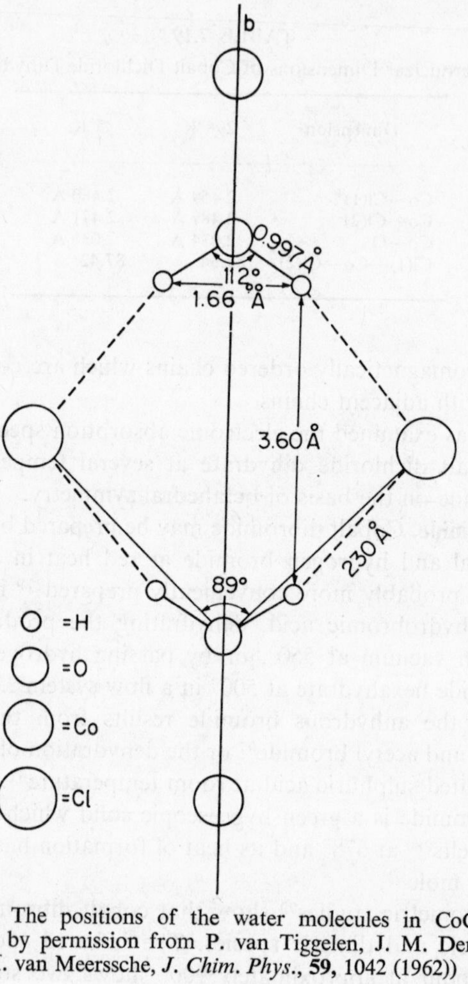

Figure 7.13 The positions of the water molecules in $CoCl_2{\cdot}2H_2O$. (Reproduced by permission from P. van Tiggelen, J. M. Dereppe and M. van Meerssche, *J. Chim. Phys.*, **59**, 1042 (1962))

The structure does not become ordered even on heating to 500°. This suggests that the ordered form is produced only by condensation from the vapour[279].

Cobalt dibromide obeys the Curie-Weiss law[83] above about 60°K with a magnetic moment of 5.3 BM and a Weiss constant of 6°. A Neel temperature of 19°K has also been reported[282,283].

The electronic absorption spectrum of a single crystal of cobalt dibromide has been investigated in detail by Ferguson and coworkers[115]. The spectrum shows the same general features as that of cobalt dichloride. Ludi and Feitneicht[92] have also investigated the electronic spectrum of cobalt dibromide and the following spin-allowed transitions were reported.

$$\sim 6{,}000\ \text{cm}^{-1} \quad {}^4T_{2g}(F) \leftarrow {}^4T_{1g}(F)$$

$$\sim 12{,}000\ \text{cm}^{-1} \quad {}^4A_{2g}(F) \leftarrow {}^4T_{1g}(F)$$

$$\sim 18{,}000\ \text{cm}^{-1} \quad {}^4T_{1g}(P) \leftarrow {}^4T_{1g}(F)$$

Trutia and Musa[116] have investigated the absorption spectrum of cobalt dibromide in the solid and liquid phases. As with cobalt dichloride there appears to be a change from octahedral to tetrahedral coordination on melting.

As one would expect, cobalt dibromide reacts with pyridine, aniline and their substitution products in an analogous manner to cobalt dichloride. For the cobalt dibromide adducts however, there is less information available. In general, most of the complexes are of the type $CoBr_2L_2$, although compounds of the type $CoBr_2L_4$ are also known. In one case, a compound of the general formula $CoBr_2L_5$ has been reported[162]. The equilibrium between octahedral and tetrahedral forms in solution, the effects of the size and basicity of the ligand and the influence of π-bonding on the stability and structure of the various compounds have been discussed in some detail. As with cobalt dichloride, most of the structures of the cobalt dibromide adducts have been deduced from magnetic and spectral studies and comparison with compounds of known structure. Only the structures of the two pyridine adducts have been examined in detail.

$CoBr_2(py)_4$ is orthorhombic with $a = 15.9$, $b = 9.5$ and $c = 14.2$ Å[284]. The structure consists of *trans* $CoBr_2(py)_4$ groups in which the plane of the pyridine rings is at an angle of approximately 50° to the plane containing the cobalt atom and the four nitrogen atoms. The dipyridinate is isomorphous[132] with β-$CoCl_2(py)_2$. Its monoclinic unit cell, of space group $P2_1/c$, has $a = 8.40$, $b = 18.0$, $c = 8.52$ Å and $\beta = 101.25°$, and contains four molecules.

The thermal decomposition of several of these adducts has been investigated and the reactions describing the various decomposition steps can be written as follows.

$$CoBr_2L_4\ (s) \xrightarrow{T_1} CoBr_2L_2\ (s) + 2L\ (g)$$

$$CoBr_2L_2\ (s) \xrightarrow{T_2} CoBr_2L_{4/3}\ (s) + 2/3L\ (g)$$

$$CoBr_2L_{4/3}\ (s) \xrightarrow{T_3} CoBr_2L\ (s) + 1/3L\ (g)$$

$$CoBr_2L_2\ (s) \xrightarrow{T_4} CoBr_2L\ (s) + L\ (g)$$

$$CoBr_2L\ (s) \xrightarrow{T_5} CoBr_2\ (s) + L\ (g)$$

$$CoBr_2L_2\ (s) \xrightarrow{T_6} CoBr_2\ (s) + 2L\ (g)$$

The reaction temperatures for a number of adducts are given in Table 7.20.

TABLE 7.20
Decomposition Temperatures of Some Cobalt Dibromide Adducts

Compound	Temperatures	References
$CoBr_2$(pyridine)$_4$	$T_1 = 120°$, $T_6 = 250°$	123, 148
$CoBr_2$(2-methylpyridine)$_2$	$T_6 = 250°$	123, 148
$CoBr_2$(3-methylpyridine)$_4$	$T_1 = 120°$, $T_6 = 250°$	123, 148
$CoBr_2$(4-methylpyridine)$_4$	$T_1 = 120°$, $T_4 = 185°$, $T_5 = 290°$	123, 148
$CoBr_2$(2,6-lutidine)	$T_5 = 160°$	123
$CoBr_2$(2,4,6-collidine)$_4$	$T_1 = 100°$, $T_6 = 150°$	123
$CoBr_2$(aniline)$_2$	$T_2 = 210°$, $T_3 = 320°$	153
$CoBr_2$(quinoline)$_2$	$T_4 = 280°$, $T_5 = 335°$	141

The properties of the cobalt dibromide adducts of pyridine, aniline and their substitution derivatives are summarized in Table 7.21.

As with cobalt dichloride, two forms of $CoBr_2(NH_3)_2$ are known[288]. The α-form is rose coloured while the β-form is blue-violet. Clark and Williams[172] have investigated the infrared and electronic spectra of the two forms of this compound. In both cases there is a very close resemblance to the corresponding chloro compound[172].

Hot ethanolic solutions of cobalt dibromide and tris(2-dimethylaminoethyl)amine, trenMe$_6$, react[220] to give violet $CoBr_2$(trenMe$_6$). This compound has a room-temperature magnetic moment of 4.47 BM and behaves as a 1:1 electrolyte in nitrobenzene and nitromethane[220]. Thus the compound should be formulated as [CoBr(trenMe$_6$)]Br. Absorption spectra suggested that the central cobalt atom was in a trigonal-bipyramidal configuration[220].

TABLE 7.21
Adducts of the Type $CoBr_2L_n$

Compound and Properties	References
L = pyridine, $n = 4$, pink, $\mu = 4.56$ BM, octahedral	242, 284
L = 3-methylpyridine, $n = 4$, pink, $\mu = 5.07$ BM, octahedral	149
L = 4-methylpyridine, $n = 4$, pink, $\mu = 5.04$ BM, octahedral	154
L = 3-ethylpyridine, $n = 4$, pink, $\mu = 5.02$ BM, octahedral	149
L = 4-ethylpyridine, $n = 4$, pink, $\mu = 5.07$ BM, octahedral	149
L = 4-vinylpyridine, $n = 4$, pink, $\mu = 5.2$ BM, octahedral	156
L = 4-cyanopyridine, $n = 4$, pink, m.p. = 108°, $\mu = 4.94$ BM, octahedral	162
L = isoquinoline, $n = 4$, pink, $\mu = 5.20$ BM, octahedral	150
L = pyridine, $n = 2$, blue, m.p. = 208°, $\mu = 4.50$ BM, $\nu_{Co\text{-}Br} =$ 274, 242 cm^{-1}, $\nu_{Co\text{-}N} = 250$ cm^{-1}, tetrahedral	122, 123, 128, 136, 140, 159, 242
L = 2-methylpyridine, $n = 2$, blue, m.p. = 165°, $\mu = 4.56$ BM, $\nu_{Co\text{-}Br} = 264$, 244 cm^{-1}, $\nu_{Co\text{-}N} = 237$, 222 cm^{-1}, tetrahedral	123, 140, 143, 154, 157–159, 285
L = 3-methylpyridine, $n = 2$, blue, m.p. = 122°, $\nu_{Co\text{-}Br} = 273$ cm^{-1}, $\nu_{Co\text{-}N} = 242$ cm^{-1}, tetrahedral	123, 157, 158
L = 4-methylpyridine, $n = 2$, blue, m.p. = 145°, $\mu = 4.48$ BM, $\nu_{Co\text{-}Br} = 276$, 231 cm^{-1}, $\nu_{Co\text{-}N} = 255$, 248 cm^{-1}, tetrahedral	123, 154, 157, 158, 285, 286
L = 2-ethylpyridine, $n = 2$, blue, $\mu = 4.61$ BM, tetrahedral	140, 149
L = 3-ethylpyridine, $n = 2$, blue, $\mu = 4.72$ BM, tetrahedral	140, 149
L = 4-ethylpyridine, $n = 2$, blue, $\mu = 4.45$ BM, tetrahedral	140, 149
L = 4-vinylpyridine, $n = 2$, blue, $\mu = 4.4$ BM, tetrahedral	156
L = 2-chloropyridine, $n = 2$, blue, $\mu = 4.45$ BM, $\nu_{Co\text{-}Br} =$ 260, 242 cm^{-1}, $\nu_{Co\text{-}N} = 224$ cm^{-1}, tetrahedral	157, 160, 161
L = 4-chloropyridine, $n = 2$, lilac, m.p. = 130°, $\mu = 5.29$ BM, octahedral	157, 162
L = 2-bromopyridine, $n = 2$, blue, $\mu = 4.61$ BM, $\nu_{Co\text{-}Br} =$ 251, 236 cm^{-1}, $\nu_{Co\text{-}N} = 215$ cm^{-1}, tetrahedral	157, 160, 161
L = 3-bromopyridine, $n = 2$, pale violet, $\mu = 5.19$ BM, octahedral	157, 161
L = 4-bromopyridine, $n = 2$, lilac, m.p. = 132°, $\mu = 5.36$ BM, octahedral	157, 162
L = 2,6-lutidine, $n = 1$, blue, octahedral	123, 157
L = 2,4,6-collidine, $n = 2$, blue, m.p. = 168°, $\nu_{Co\text{-}Br} = 278$ cm^{-1}, tetrahedral	123, 157
L = aniline, $n = 2$, $\nu_{Co\text{-}Br} = 245$ cm^{-1}, tetrahedral	153, 287
L = *o*-toluidine, $n = 2$, $\nu_{Co\text{-}Br} = 245$ cm^{-1}, tetrahedral	163, 287
L = *m*-toluidine, $n = 2$, $\nu_{Co\text{-}Br} = 246$ cm^{-1}, tetrahedral	163, 287
L = *p*-toluidine, $n = 2$, blue, $\mu = 4.84$ BM, $\nu_{Co\text{-}Br} = 248$ cm^{-1}, tetrahedral	128, 163, 164, 287
L = quinoline, $n = 2$, blue, $\mu = 4.43$ BM, $\nu_{Co\text{-}Br} = 260$ cm^{-1}, tetrahedral	140, 141, 150, 158, 196

This stereochemistry has been confirmed by Di Vaira and Orioli[289]. The compound has a cubic unit cell with $a = 12.088$ Å. There are four molecules in the cell of space group $P2_13$. The structure consists of $[CoBr(trenMe_6)]^+$ and Br^- ions arranged in a distorted sodium chloride-type lattice. The cation, shown in Figure 7.14, has a trigonal-bipyramidal

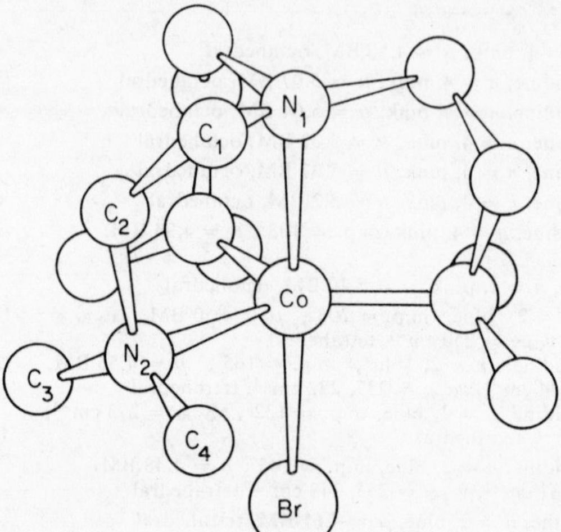

Figure 7.14 The structure of the $[CoBr(trenMe_6)]^+$ cation. (Reproduced by permission from M. Di Vaira and P. L. Orioli, *Inorg. Chem.*, **6**, 955 (1967))

TABLE 7.22
Internuclear Dimensions of $[CoBr(trenMe_6)]^+$

Distance	Value	Angle	Value
Co—N(1)	2.151 Å	N(1)—Co—N(2)	81.1°
Co—N(2)	2.080 Å	N(2)—Co—Br	98.9°
Co—Br	2.431 Å	N(2)—Co—N(2′)	117.6°

structure of C_3 symmetry. The important internuclear dimensions are given in Table 7.22.

Cobalt dibromide reacts[184,186] readily with thiourea (tu) to give green-blue $CoBr_2(tu)_2$. This adduct melts at 163°, obeys the Curie-Weiss law with a magnetic moment of 4.5 BM and a Weiss constant of 6°, and has been shown to contain cobalt in a distorted tetrahedral configuration. Infrared studies[184,185] indicate that the thiourea is bound to the cobalt atom through its sulphur atom. Adams and Cornell[185] reported the infrared assignments shown in Table 7.23.

TABLE 7.23
Infrared Assignments of $CoBr_2(tu)_2$

Mode	Frequency (cm^{-1})	Intensity
$\nu_{Co\text{-}Br}$	235	Strong
	190	Medium
$\nu_{Co\text{-}S}$	276	Strong
	251	Strong

Green-blue $CoBr_2(etu)_2$, where etu is ethylenethiourea, obeys[188] the Curie-Weiss law with a magnetic moment of 4.54 BM and a Weiss constant of 7°. It is a non-electrolyte in acetone and its absorption spectrum is consistent with a tetrahedral configuration. The infrared spectrum of this compound has been investigated[185]; cobalt-bromine stretches at 212 and 186 cm^{-1} and cobalt-nitrogen stretches at 255 and 246 cm^{-1} were reported.

Issleib and Wenschuh[290] report that three products are obtained from the reaction between cobalt dibromide and diphenylphosphine in methanol. Some properties of these three compounds are given in Table 7.24.

TABLE 7.24
Diphenylphosphine Adducts of Cobalt Dibromide

Compound[a]	Colour	M.p.	μ (BM)	Structure
$[CoBr_2(Ph_2PH)_3]$	Brown	163–165°	2.01	Trigonal bipyramidal
$[CoBr(Ph_2PH)_3]Br$	Green	141–143°	3.37	Tetrahedral
$[CoBr(Ph_2PH)_4]_2Br_2$	Yellow	135–136°	0	Octahedral with bromine bridges

[a] Ph = phenyl = C_6H_5.

A single-crystal X-ray diffraction study by Bertrand and Plymale[291] of $CoBr_2[(C_6H_5)_2PH]_3$ has confirmed the five-coordinate structure of this complex but has revealed that the stereochemistry about the cobalt atom is not trigonal bipyramidal, but is intermediate between this configuration and square pyramidal. The compound is triclinic with $a = 11.05, b = 11.47,$

$c = 15.41$ Å, $\alpha = 98.0°$, $\beta = 82.7°$ and $\gamma = 118.5°$. The structure is shown in Figure 7.15. An interesting feature to note is that the two bromine atoms and one phosphorus atom (P_3) are essentially coplanar with atoms (P_1) and (P_2) in the axial positions. Since there are two bromines and one phosphorus atom in the equatorial positions, the structure

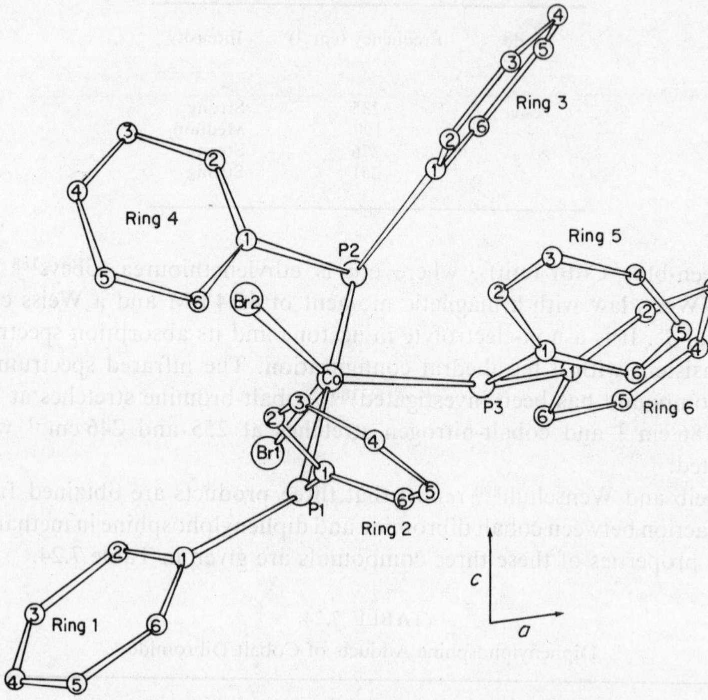

Figure 7.15 The structure of $CoBr_2[(C_6H_5)_2PH]_3$. (Reproduced by permission from J. A. Bertrand and D. L. Plymale, *Inorg. Chem.*, **5**, 879 (1966))

TABLE 7.25

Internuclear Dimensions of $CoBr_2[(C_6H_5)_2PH]_3$

Distance	Value	Angle	Value
Co—Br$_1$	2.54 Å	Br$_1$—Co—Br$_2$	125.6°
Co—Br$_2$	2.33 Å	Br$_1$—Co—P$_2$	89.2°
Co—P$_1$	2.23 Å	Br$_2$—Co—P$_3$	136.3°
Co—P$_2$	2.20 Å	P$_1$—Co—P$_2$	175.9°
Co—P$_3$	2.18 Å	P$_2$—Co—P$_3$	91.3°

cannot be a trigonal bipyramid. The arrangement of atoms about the cobalt atom is also unusual in that the more electronegative atoms are not both in the axial positions. Some of the more important internuclear dimensions are given in Table 7.25.

A number of other phosphines[189,190], as well as phosphine and arsine oxides, have been found to react with cobalt dibromide and properties of some of the adducts formed are given in Table 7.26.

TABLE 7.26

$CoBr_2L_n$ Adducts with Phosphines, Phosphine Oxides and Arsine Oxides

Compound and Properties	References
L = $(CH_3)_3P$, $n = 2$, green, m.p. = 89–90°	191
L = $(C_2H_5)_3P$, $n = 4$, green, m.p. = 108–112°, $\mu = 1.97$ BM	197
L = $(C_2H_5)_3P$, $n = 2$, blue-green, m.p. = 118–120°, $\mu = 4.76$ BM, tetrahedral	191, 193
L = $(C_6H_5)_3P$, $n = 2$, green, m.p. = 218–219°, $\mu = 4.52$ BM, $\theta = 1°$, $\nu_{Co\text{-}Br} = 276, 239\ cm^{-1}$, $\nu_{Co\text{-}P} = 187, 142\ cm^{-1}$, tetrahedral	137, 158, 191, 194–196, 292, 293
L = $(C_6H_{11})_3P$, $n = 2$, blue, m.p. = 205°, $\mu = 4.61$ BM, tetrahedral	194, 294
L = $(C_6H_{11})_2P(CH_2)_3P(C_6H_{11})_2$, $n = 2$, green, m.p. = 207–212°, $\mu = 4.4$ BM, tetrahedral, air and moisture sensitive as solid	198
L = $(C_6H_{11})_2PP(C_6H_{11})_2$, $n = 2$, green isomer; m.p. = 212–216°, $\mu = 3.52$ BM, square planar: blue isomer; m.p. = 157–160°, $\mu = 3.80$ BM, tetrahedral	199
L = 2-phenylisophosphindoline, $n = 3$, deep purple, m.p. = 181-183°, monomeric, non-electrolyte, probably square pyramidal	200
L = 1,4-bis(diphenylphosphino)butane (PC_4P), $n = 1$, blue, $\mu = 4.50$ BM, tetrahedral	203
L = 1,5-bis(diphenylphosphino)pentane (PC_5P), $n = 1$, blue, $\mu = 4.59$ BM, tetrahedral	203
L = $(C_6H_5)_3PO$, $n = 2$, blue, m.p. = 223°, $\mu = 4.84$ BM, $\nu_{Co\text{-}Br} = 249, 233\ cm^{-1}$, $\nu_{P\text{-}O} = 1153\ cm^{-1}$, tetrahedral	59, 123, 164, 196, 202
L = bis(2-diphenylphosphinoethyl)oxide, $n = 1$, blue, $\mu = 4.55$ BM, tetrahedral	203
L = $(C_6H_5)_2CH_3AsO$, $n = 2$, $\nu_{As\text{-}O} = 855, 840\ cm^{-1}$, $\nu_{Co\text{-}O} = 425, 416\ cm^{-1}$, $\nu_{Co\text{-}Bo} = 240, 222\ cm^{-1}$, tetrahedral	204
L = $(C_6H_5)_3AsO$, $n = 2$, blue, $\mu = 4.63$ BM, $\theta = 6°$, $\nu_{As\text{-}O} = 870\ cm^{-1}$, $\nu_{Co\text{-}O} = 405\ cm^{-1}$, $\nu_{Co\text{-}Br} = 243, 220\ cm^{-1}$ tetrahedral	204–206

Cobalt dibromide reacts with a large range of other ligands and some properties of a number of these adducts are given in Table 7.27.

A number of other adducts of cobalt dibromide with donor molecules such as alcohols, sulphur ligands, amines and substituted amines have been reported[94,179,198,209,224,229,230,233,234,238,240,242,296–299].

Cobalt dibromide hexahydrate. This red crystalline solid which melts at 47–48° may be prepared by dissolving the carbonate in concentrated

TABLE 7.27

Adducts of the Type $CoBr_2L_n$

Compound and Properties	References
L = CH_3CN, n = 2, blue, m.p. = 243°, μ = 4.84 BM	173, 175, 176
L = quinoxaline, n = 2, blue-green, μ = 4.61 BM, tetrahedral	166
L = 2-methylquinoxaline (Mq), n = 1, μ = 4.90 BM, tetrahedral	166
L = 2,3-dimethylquinoxaline (Dmq), n = 1, μ = 4.83 BM, tetrahedral	166
L = pyrazine (pyz), n = 1, pink, μ = 4.92 BM, octahedral	212
L = methylpyrazine (mpz), n = 1, blue, μ = 4.68 BM, tetrahedral dimer	212, 214
L = 2,5-dimethylpyrazine (2,5-dmp), n = 1, blue, μ = 4.65 BM, tetrahedral dimer	212, 214
L = 2,6-dimethylpyrazine (2,6-dmp), n = 2, violet, μ = 5.34 BM, octahedral	212, 214
L = 2-methyl-8-methylthioquinoline (mmtq), n = 1, blue-green, μ = 4.75 BM, tetrahedral	215
L = 2-methylthiomethylpyridine (mmp), n = 2, lilac, μ = 4.92 BM, octahedral	167
L = 2,9-dimethyl-1,10-phenanthroline (dmp), n = 1, green-blue, μ = 4.68 BM, tetrahedral	216
L = di-2-pyridylamine, n = 1, blue-green, μ = 4.48 BM, θ = 2°, tetrahedral	217, 218
L = *N,N,N′,N′*-tetramethylethylenediamine (Me_4en), n = 1, blue, m.p. = 186–189°, μ = 4.76 BM, tetrahedral	219
L = bis(2-dimethylaminoethyl)methylamine (Me_5dien), n = 1, violet, μ = 4.80 BM, five-coordinate	177, 295
L = bis(2-dimethylaminoethyl)oxide (Me_4daeo), n = 1, blue, μ = 4.73 BM, five-coordinate	221
L = bis(2-dimethylaminoethyl)sulphide (Me_4daes), n = 1, μ = 4.50 BM, five-coordinate	222
L = Schiff base formed between *N*-methyl-*o*-aminobenzaldehyde and *N,N*-diethylethylenediamine (MABen-NEt_2), n = 1, dark green, μ = 4.70 BM, non-electrolyte, monomeric, five-coordinate	223
L = Schiff base formed between *o*-methylthiobenzaldehyde and *N,N*-diethylethylenediamine (MSBen-NEt_2), n = 1, green-blue, m.p. = 192–194°, μ = 4.71 BM, tetrahedral	224
L = *N,N*-ethylenebis(2-thenylideneimine), n = 1, green, μ = 4.58 BM, tetrahedral	225
L = 1,2-di-(4-pyridyl)ethylene, μ = 5.13 BM, octahedral	227

hydrobromic acid and crystallizing the product on a steam bath[61,68,279]. The compound is antiferromagnetic at low temperatures, the Neel temperature being 3.2°K[300]. There is a maximum in the specific heat at

3.07°K[301]. The absorption spectrum[116] of cobalt dibromide hexahydrate is very similar to that of the corresponding chloride, indicating that four water molecules are coordinated to the cobalt atom.

Cobalt dibromide dihydrate. Evaporation of a saturated aqueous solution of cobalt dibromide at about 100° yields this hydrate, the structure of which has been determined by single-crystal X-ray diffraction techniques[302]. It is isostructural with manganese dibromide dihydrate having an orthorhombic unit cell with $a = 7.630$, $b = 8.770$ and $c = 3.765$ Å. There are two molecules in the cell of space group $C2/m$. On cooling, there is a transition to a different packing arrangement and at 77°K the unit cell is monoclinic with $a = 14.2$, $b = 7.36$, $c = 9.04$ Å and $\beta = 91°$. Cobalt dibromide dihydrate becomes magnetically ordered at low temperatures[273,276].

Cobalt diiodide. The more convenient methods of preparing cobalt diiodide are given in Table 7.28.

TABLE 7.28
Preparation of Cobalt Diiodide

Method	Conditions	References
$CoCO_3$ and aqueous HI	Heat product in HI/N_2 stream at 200°	94
Co and I_2	Flow system, red heat	68, 303
Co and HI	Sealed tube, 500°	61, 303
Co_3O_4 and AlI_3	Sealed tube, 230°	304, 305
CoS_2 and AlI_3	Sealed tube, 230°	306

All of the above methods of preparation give the black α-form of cobalt diiodide. A less common yellow β-form has been reported[61] to result when the α-form is sublimed at 570° in high vacuum.

The melting point of cobalt diiodide has been reported[280] as 495° and as[307] 515°. It is thermally unstable and the equilibrium dissociation pressure of iodine over the temperature range 631–704° is given by the expression

$$\log p = 3.785 - 5203/T$$

The heat of formation has been given as[61] -39.13 kcal mole^{-1}.

Cobalt diiodide has the cadmium diiodide structure[308] with $a = 3.96$ and $c = 6.65$ Å. It obeys the Curie-Weiss law[86] over a wide temperature range with a magnetic moment of 5.18 BM and a Weiss constant of 40°. A Neel temperature of 12° has been reported[282] although earlier measurements[283] gave a maximum in $\chi_\perp$ at 3°K.

Cobalt diiodide reacts with a large range of donor molecules such as pyridine, phosphines, Schiff bases, etc., to give in general, adducts which resemble very closely those of cobalt dichloride and cobalt dibromide. As with the chloro and bromo compounds, the cobalt diiodide adducts have been studied using a variety of techniques including infrared, visible and ultraviolet spectra, magnetochemistry, conductivity and X-ray crystallography. The literature for many of these compounds is summarized in Table 7.29.

TABLE 7.29
Adducts of Cobalt Diiodide

Donor Molecule	References
Pyridine	122, 123, 128, 138, 140, 148, 158, 242, 309
Substituted pyridines	123, 140, 143, 148, 154, 156–158, 161, 162, 285, 286, 309
Aniline, substituted anilines and related amines	128, 140, 150, 153, 163–165, 196, 287 310
Pyrazine and substitution derivatives	140, 212, 214
Acetonitrile	173, 175, 176
Phosphines and arsines	26, 60, 189–191, 193–196, 198–200, 203, 205, 291, 293, 294, 311
Phosphine and arsine oxides	59, 164, 202, 205
Substituted urea and thiourea	185, 186, 188, 298
Miscellaneous	43, 94, 166, 172, 177, 179, 215, 216, 219, 220, 225, 227, 229, 230, 234, 238, 297, 299, 312–315

Cobalt diiodide hexahydrate. This dark red, very hygroscopic crystalline compound may be prepared by dissolving cobalt carbonate in hydriodic acid and evaporating the solution at low temperatures[61,316]. An X-ray diffraction study by Shchukarev and coworkers[316] has shown that all six water molecules are coordinated to the metal.

COMPLEX HALIDES

Apart from the fluoro complexes of cobalt(IV) and cobalt(III), the chemistry of the complex halides of cobalt is restricted to that of cobalt(II).

Oxidation State IV

Hexafluorocobaltate(IV). Direct fluorination of caesium tetrachlorocobaltate(II) at 300° in a flow system yields yellow caesium hexafluorocobaltate(IV)[317]. The compound has a cubic lattice with $a = 8.91$ Å and a room-temperature magnetic moment of 2.97 BM.

Oxidation State III

Hexafluorocobaltates(III). Potassium hexafluorocobaltate(III) has been prepared by the interaction of cobalt trifluoride and potassium fluoride in hydrofluoric acid[318] and also by the fluorination of a 3:1 mixture of potassium chloride and cobalt dichloride in a flow system at elevated temperatures[319]. Hoppe[317] and also Meyers and Cotton[320] have prepared a number of hexafluorocobaltates(III) by fluorinating the corresponding alkali-metal hexacyanocobaltate(III) in a flow system at 300–350°. The barium salt and the mixed alkali-metal salt, K_2NaCoF_6, have also been prepared by this method.

The X-ray powder diffraction data for several hexafluorocobaltates(III) are given in Table 7.30.

TABLE 7.30
Lattice Parameters of Some Hexafluorocobaltates(III)

Compound	*a* (Å)	References
Na_3CoF_6	8.13	317
K_2NaCoF_6	8.22	320
K_3CoF_6	8.55	317, 320
Rb_3CoF_6	8.86	317
Cs_3CoF_6	9.22	317

Although all of the compounds in the table were originally reported to be cubic, Meyers and Cotton[320] report that only K_2NaCoF_6 is truly cubic, and that the other salts show varying amounts of distortion of the lattice and of the CoF_6^{3-} octahedra.

The cobalt-fluorine stretching frequency[320,321] in several hexafluoro-cobaltates(III) occurs in the range 480–510 cm^{-1}. Both the potassium and barium salts obey the Curie-Weiss law[322] with magnetic moments of 5.63 and 5.27 BM and Weiss constants of 10° and 39° respectively.

The electronic absorption spectra of a number of hexafluorocobaltates(III) have been recorded by Cotton and Meyers[322]. Exceptionally large splittings were observed which were claimed to be due to a dynamic Jahn-Teller distortion.

Oxidation State II

Hexafluorocobaltate(II). Lilac barium hexafluorocobaltate(II), Ba_2CoF_6, has been prepared by Schnering[323] by fusing a 2:1 mixture of barium fluoride and cobalt difluoride at 1250°. The compound has a tetragonal unit cell with $a = 4.101$ and $c = 16.284$ Å[324]. The CoF_6^{4-} octahedra are tetragonally elongated and the cobalt-fluorine bond distances are 2.05 and 2.13 Å. The compound is isostructural with Bi_2NbO_5F and may be described by the formula $(BaF)_2(CoF_4)$. It consists of complex CoF_4^{2-} ions which form infinite layers of the type $[CoF_{4/2}F_2]$ so that the central cobalt atom is octahedrally coordinated. The compound obeys the Curie-Weiss law[323] with a magnetic moment of 5.22 BM and a Weiss constant of 115°. It becomes antiferromagnetic at low temperatures.

Tetrafluorocobaltates(II). Fusion of cobalt difluoride and the appropriate fluoride or hydrogen difluoride in a 2:1 proportion in an inert atmosphere, or in a stream of anhydrous hydrogen fluoride, yields the corresponding tetrafluorocobaltate(II)[67,325]. The ammonium salt has been prepared by heating and compressing a 2:1 mixture of ammonium fluoride and cobalt difluoride[326].

The tetrafluorocobaltates(II) are pale coloured compounds and the potassium, rubidium and thallium salts have tetragonal lattices with the unit-cell parameters given in Table 7.31.

TABLE 7.31
Lattice Parameters of Some Tetrafluorocobaltates(II)

Compound	*a* (Å)	*b* (Å)	References
K_2CoF_4	4.074	13.085	67, 325
Rb_2CoF_4	4.135	13.67	67, 325
Tl_2CoF_4	4.114	14.05	67, 325

These compounds become antiferromagnetic at low temperatures[67] and the effective magnetic moments at room temperature and the Neel temperatures are given in Table 7.32.

TABLE 7.32
Magnetic Properties of Some Tetrafluorocobaltates(II)

Compound	μ_{eff} (BM)	T_N (°K)
K_2CoF_4	4.20	145
Rb_2CoF_4	4.41	130
Tl_2CoF_4	4.56	130

Srivastava[327] has made a detailed study of the magnetic properties of a single crystal of potassium tetrafluorocobaltate(II) and reported a Neel temperature of 125°. Even above this temperature the compound is magnetically anisotropic.

Trifluorocobaltates(II). These salts have been prepared by both wet and dry methods. The interaction of cobalt dibromide and the fluorides of ammonium, potassium and rubidium in methanol[328,329]; the evaporation at 90° of an aqueous solution of cobalt dichloride and potassium hydrogen difluoride[330], or the mixing of saturated solutions of cobalt difluoride and alkali-metal fluoride and heating the resulting solution to the boiling point[331,332], all produce salts of the trifluorocobaltate(II) anion. Fusion of 1:1 mixtures of the appropriate fluoride and cobalt difluoride also produces trifluorocobaltates(II)[67,325]. Russian workers[333,334] have made single crystals of trifluorocobaltates(II) by fusing mixtures of either alkali-metal chloride and cobalt difluoride, or alkali-metal fluoride and cobalt dichloride in an inert atmosphere. The preparation of single crystals from products obtained by the wet method has also been described[335].

With the exception of the sodium salt[67,325,332], the trifluorocobaltates(II) have the ideal perovskite structure (Figure 5.7, page 241) at room temperature. The symmetry and lattice parameters of the trifluorocobaltates(II) at room temperature are given in Table 7.33.

TABLE 7.33

Symmetry and Lattice Parameters of Trifluorocobaltates(II)

Compound	Symmetry	Parameters	References
$NaCoF_3$	Orthorhombic	$a = 5.542$, $b = 5.603$, $c = 7.793$ Å	67, 325
$KCoF_3$	Cubic	$a = 4.060$–4.078 Å	67, 325, 329, 335–339
$RbCoF_3$	Cubic	$a = 4.138$ Å	67, 325
		$a = 4.141$ Å	329
$TlCoF_3$	Cubic	$a = 4.116$ Å	67, 325
NH_4CoF_3	Cubic	$a = 4.127$ Å	67, 325
		$a = 4.129$ Å	329

For the potassium salt at room temperature, the cobalt-fluorine bond distance is 2.04 Å. Okazaki and coworkers[336,338] have studied the crystal structure of potassium trifluorocobaltate(II) as a function of temperature and report that at 78°K it has a tetragonal unit cell with $a = 4.057$ and $c = 4.049$ Å. The potassium salt has a cobalt-fluorine stretching frequency[321] of 349 cm^{-1}.

The magnetic properties of the trifluorocobaltates(II) have been investigated by various workers. The compounds become quite strongly antiferromagnetic at low temperatures. There is, in some cases, considerable disagreement over the value of the Weiss constant and this probably arises because of the presence of small amounts of impurities. The magnetic properties of the trifluorocobaltates(II) are given in Table 7.34. For a number of compounds only the room temperature effective magnetic moment is available.

TABLE 7.34
Magnetic Properties of Trifluorocobaltates(II)

Compound	μ (BM)	$\theta°$	T_N (°K)	References
$NaCoF_3$	4.83			67
	4.41	207		332
			35	333
$KCoF_3$	4.01			67
	4.03			337
	3.92	502	135	331, 332
	4.95	125	114	330
$RbCoF_3$	4.18			67
$CsCoF_3$	4.87	60		67
$TlCoF_3$	4.49			67
NH_4CoF_3	4.17			67

There is an anomaly in the thermal conductivity of the potassium salt at 114°K[340], while there is a maximum in the heat capacity[339] of the same compound at 109.5°K. Neutron-diffraction measurements of potassium trifluorocobaltate(II)[341] show that the magnetic super lattice is cubic with $a = 8.084$ Å at liquid helium temperatures, that is, twice the edge of the crystallographic unit cell at room temperature. The measurements also indicate that the compound has the *G*-type magnetic structure in which each magnetic ion is coupled antiferromagnetically with its six nearest neighbours.

The electronic absorption spectra of several trifluorocobaltates(II) have been studied at various temperatures by single-crystal techniques[115,333,334]. Three major spin-allowed transitions were observed and assigned as follows:

$$\sim 7{,}200 \text{ cm}^{-1} \quad {}^4T_{2g}(F) \leftarrow {}^4T_{1g}(F)$$

$$\sim 15{,}000 \text{ cm}^{-1} \quad {}^4A_{2g}(F) \leftarrow {}^4T_{1g}(F)$$

$$\sim 19{,}300 \text{ cm}^{-1} \quad {}^4A_{1g}(P) \leftarrow {}^4T_{1g}(F)$$

Pentachlorocobaltates(II). Rubidium and caesium pentachlorocobaltates(II) may be prepared by fusing 3:1 mixtures of alkali-metal chloride and cobalt dichloride[342]. The rubidium salt has also been prepared by passing anhydrous hydrogen chloride over the appropriate mixture of rubidium chloride and cobalt dichloride at 600° in a flow system[343], while the caesium salt is more commonly prepared from aqueous solutions[142,344–346]. Both the rubidium and caesium salts are blue crystalline solids, their melting points being 506° and 549° respectively[342].

The structure of caesium pentachlorocobaltate(II) was first determined by Powell and Wells[345] and was later refined by Figgis and coworkers[347]. The tetragonal unit cell has $a = 9.219$ and $c = 14.554$ Å and

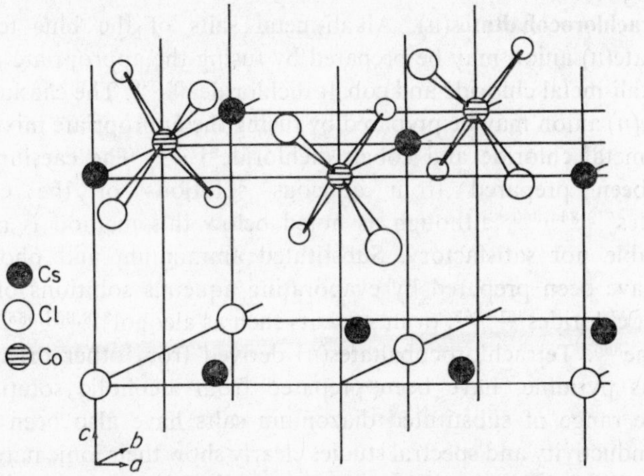

Figure 7.16 The structure of Cs_3CoCl_5. (Reproduced by permission from B. N. Figgis, M. Gerloch and R. Mason, *Acta Cryst.*, **17**, 506 (1964))

there are four formula units in the cell of space group *I*4/*mcm*. The structure does *not* contain discrete $CoCl_5^{3-}$ units, but contains an equal number of tetrachlorocobaltate(II) ions and free chloride ions. The arrangement of the ions in the lattice is shown in Figure 7.16. The $CoCl_4^{2-}$ tetrahedra are not regular; the bond lengths are equal at 2.252 Å but distortion arises because of crystal packing effects and the chlorine-cobalt-chlorine angles are 106.0° and 111.1°. The rubidium[343] salt is isostructural with the caesium compound, the tetragonal unit cell having $a = 8.799$ and $c = 14.239$ Å.

The magnetic properties of caesium pentachlorocobaltate(II) have been studied in detail[348]. It obeys the Curie-Weiss law with a magnetic moment of 4.56 BM and a Weiss constant of 4°. Room temperature effective magnetic moments of 4.49 BM[346] and 4.66 BM[349] have also been reported. The Neel temperature is 0.52°K[350]. The magnetic anisotropy has been examined in detail[349,351,352] and was found to be relatively small. At 90°K $g_{\parallel} = 2.32$ and $g_{\perp} = 2.27$[352] while at 77°K $g_{\parallel} = 2.40$ and $g_{\perp} = 2.30$[346].

The absorption spectrum of caesium pentachlorocobaltate(II) has been examined in some detail, both as a function of temperature and of applied magnetic field strength[346,353–358]. The ground state is $^4A_{2g}(F)$ and some of the observed bands are apparently split, which is consistent with the observed crystallographic distortion of the tetrahedra. A further lowering of the symmetry of the already slightly distorted tetrahedral environment of the cobalt atom occurs in the range 20–80°K, as shown by Zeeman splitting studies.

Tetrachlorocobaltates(II). Alkali-metal salts of the blue tetrachloro-cobaltate(II) anion may be prepared by fusing the appropriate mixture of the alkali-metal chloride and cobalt dichloride[342,359]. The caesium salt has baltate(II) anion may be prepared by fusing the appropriate mixture of the alkali-metal chloride and cobalt dichloride[342,359]. The caesium salt has also been prepared from aqueous solutions of the constituent chlorides[142,344,360], although as noted below this method is not widely applicable nor satisfactory. Substituted ammonium and phosphonium salts have been prepared by evaporating aqueous solutions of the constituent chlorides[361–363], or in solvents such as alcohol[348,364,365] and nitromethane[366]. Tetrachlorocobaltates(II) derived from other organic bases, such as pyridine, have been prepared from alcoholic solutions[120,367]. A wide range of substituted diazonium salts have also been prepared, and conductivity and spectral studies clearly show their ionic nature[368–372]. Cobalt dichloride reacts with a range of other ligands such as dimethyl-sulphoxide, and physico-chemical studies show that the products are ionic and normally contain an octahedral cobalt cation and the tetrachloro-cobaltate(II) anion[25,313,314,373–378].

Cotton and coworkers[348] report that the use of non-aqueous solvents is highly desirable for the preparation of salts of the tetrachlorocobaltate(II) anion, since they showed spectrophotometrically that aqueous solutions of cobalt dichloride, even when saturated with hydrogen chloride gas or lithium chloride, contain no appreciable quantities of the tetrachlorocobaltate(II) anion. However, as noted above, evaporation of these solutions in the presence of caesium chloride yields the caesium salt.

Thermodynamic data for a number of tetrachlorocobaltates(II) are given in Table 7.35.

The structure of caesium tetrachlorocobaltate(II) has been studied by Porai-Koshits[127] and also by Tischenko and Pinsker[381]. Both groups found that there are four molecules in the orthorhombic unit cell of space group *Pnma*, although they obtained slightly different unit-cell parameters on differently orientated axes, as shown below.

$a = 9.737 \quad b = 12.972 \quad c = 7.302$ Å (reference 127)

$a = 9.72 \quad b = 7.38 \quad c = 12.94$ Å (reference 381)

TABLE 7.35

Thermodynamic Data for Some Tetrachlorocobaltates(II)

Property	Compound	Value	Reference
M.p.	K_2CoCl_4	436°	342
	Rb_2CoCl_4	538°	342
	Cs_2CoCl_4	597°	342
	$[(C_4H_9)_4N]_2CoCl_4$	182.5°	348
ΔH°_{form} (kcal mole^{-1})[a]	K_2CoCl_4	−1.5	379
	Rb_2CoCl_4	−5.79	380
	Cs_2CoCl_4	−10.40	380
	$[(CH_3)_4N]_2CoCl_4$	−10.21	380
	$[(C_2H_5)_4N]_2CoCl_4$	−16.34	380

[a] From constituent chlorides.

The structure, which is shown in Figure 7.17, consists of slightly distorted $CoCl_4^{2-}$ tetrahedra with an average cobalt-chlorine bond distance of 2.23 Å, together with caesium ions[127].

The tetramethylammonium salt has an orthorhombic unit cell with $a = 12.276$, $b = 9.001$ and $c = 15.539$ Å[361]. There are four formula units in the cell of space group *Pnma*. The structure consists of tetramethylammonium cations and distorted $CoCl_4^{2-}$ anions. The distortion is apparently due to crystal packing effects and not to the Jahn-Teller effect since the corresponding zinc compound shows a similar distortion. The cobalt-chlorine bond distances, corrected for thermal motion, fall in the range 2.248–2.306 Å, while the chlorine-cobalt-chlorine angles lie between 108.3° and 112.8°.

Tetraethylammonium tetrachlorocobaltate(II) is tetragonal[366] with $a = 9.00$ and $c = 14.97$ Å and the triphenylmethylarsonium salt, which is isomorphous with the corresponding nickel compound, has a cubic cell with $a = 15.53$ Å[365].

The infrared spectra of several tetrachlorocobaltates(II) have been reported by various groups[382–385]. The most detailed study is that by Sabatini and Sacconi[384] and their results are given in Table 7.36.

TABLE 7.36
Infrared Spectra (cm^{-1}) of Some Tetrachlorocobaltates(II)

Mode	$[(CH_3)_4N]_2CoCl_4$	$[(C_2H_5)_4N]_2CoCl_4$
ν_3 (Co—Cl stretch)	296*s*	297*s*
ν_4 (Cl—Co—Cl bend)	131*m*	130*m*
ν_2 (Cl—Co—Cl bend)	85*w*	82*w*

s = strong *m* = medium *w* = weak.

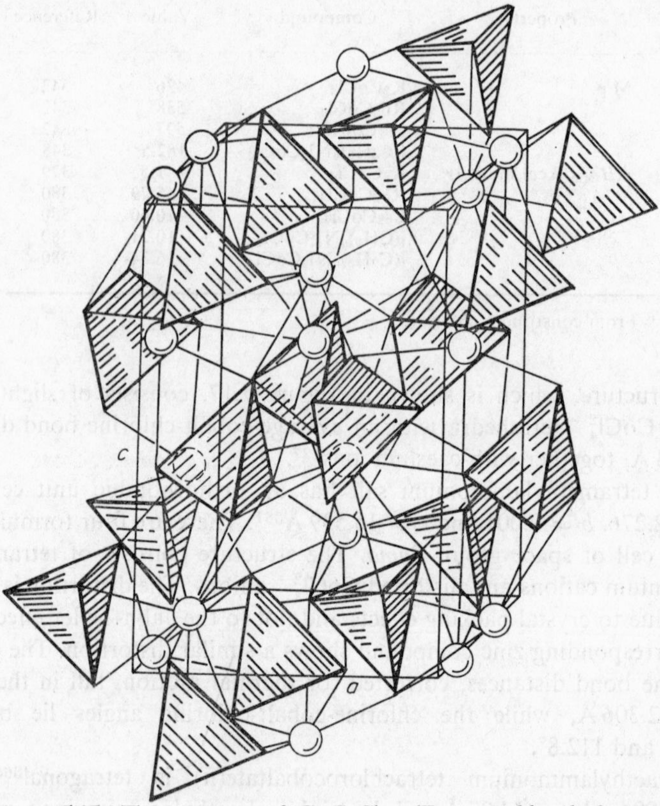

Figure 7.17 The structure of Cs_2CoCl_4. (From *Dokl. Akad. Nauk SSSR*, **100**, 913 (1955))

The magnetic properties of the tetrachlorocobaltates(II) have been discussed in detail by Cotton and his group[348,386]. The compounds obey the Curie-Weiss law with small Weiss constants, as shown in Table 7.37.

The electronic absorption spectrum of the tetrachlorocobaltate(II) anion has been investigated by various groups[142,360,364,378,388,389], the most detailed study being that of Cotton, Goodgame and Goodgame[348]. These workers recorded the spectra of a number of salts in several media and found that two absorptions were invariably present. The first of these occurs in the range 5,200–5,700 cm^{-1} and was attributed to the $^4T_{1g}(F) \leftarrow {}^4A_{2g}(F)$ transition, while the second, falling in the range 14,600–14,800 cm^{-1} was assigned to the $^4T_{1g}(P) \leftarrow {}^4A_{2g}(F)$ transition.

TABLE 7.37

Magnetic Properties of Some Tetrachlorocobaltates(II)

Compound	μ (BM)	$\theta°$	References
K_2CoCl_4	4.83[a]		387
Cs_2CoCl_4	4.71[a]		349
$(NH_4)_2CoCl_4$	4.77	22	362
$[(C_4H_9)_4N]_2CoCl_4$	4.72	12	348
$[(C_6H_5)_3CH_3As]_2CoCl_4$	4.57	1	348, 386
$(C_9H_8N)_2CoCl_4$	4.60	5	348, 386

[a] Room temperature measurement only.

Sundheim and Kukk[359] have shown by means of absorption spectrometry that the tetrachlorocobaltate(II) anion is the predominant species in alkali-metal chloride-cobalt dichloride melts.

Tetrachlorocobaltate(II) hydrates. Several of these compounds have been prepared from aqueous solutions of cobalt dichloride and the appropriate chloride, but with the exception of the ammonium salt, very little is known about them[344,362,390,391].

On the basis of the absorption spectrum, which is characteristic of octahedrally coordinated cobalt, Fogel and coworkers[362] have shown that ammonium tetrachlorocobaltate(II) dihydrate must be considered as $(NH_4)_2[CoCl_4(H_2O)_2]$, even though the water is readily removed. For the temperature range 15–60° the equilibrium vapour pressure of water over the hydrate is given by the expression

$$\log p_{mm} = 13.380 - 4140/T$$

The heat and entropy of dissociation are 38.1 kcal mole^{-1} and 100.9 cal deg^{-1} mole^{-1} respectively. The compound obeys the Curie-Weiss law with a magnetic moment of 5.18 BM and a Weiss constant of 21°.

Substituted tetrachlorocobaltates(II). A number of substituted tetrachlorocobaltates(II) of the general type $[CoCl_3L]^-$, where L is a neutral ligand such as triphenylphosphine, are known[137]. The infrared spectral assignments[137] of two of these compounds are given in Table 7.38.

TABLE 7.38

Infrared Spectra (cm^{-1}) of $[(C_2H_5)_4N][CoCl_3L]$

Mode	$L = C_5H_5N$	$L = (C_6H_5)_3P$
$\nu_{Co\text{-}Cl}$	326*s*	320*s*
	289*m*	282*s*
$\nu_{Co\text{-}N}$	226*m*	
	186*w*	
	162*w*	
$\nu_{Co\text{-}P}$		167*m*
$\nu_{Co\text{-}Cl_3}$	128*s*	112*s*

s = strong *m* = medium *w* = weak.

Trichlorocobaltates(II). Potassium, rubidium and caesium trichlorocobaltates(II) may be prepared in the melt and the reaction is best carried out in an atmosphere of hydrogen chloride[342,343]. The melting points of these salts are 343°, 485° and 547° respectively. Gruen and DeKock[392] have shown by spectral studies that caesium trichlorocobaltate(II) exists as the molecular species in the vapour phase at approximately 1100°. Engberg and Soling[343] have examined the structure of rubidium trichlorocobaltate(II) by single-crystal X-ray diffraction techniques and reported that the compound is hexagonal with $a = 6.999$ and $c = 5.996$ Å. The space group is $P6_3/mmc$ and contains two formula units. The structure, which is shown in Figure 7.18, consists of infinite chains of distorted $CoCl_6$ octahedra sharing faces in the *c* axis direction. Thus the compound is isostructural with caesium trichloronickelate(II). The cobalt-chlorine and cobalt-cobalt distances are 2.46 and 2.998 Å respectively, and the chlorine-cobalt-chlorine angles are 86.7° and 93.3°. The caesium salt is also isostructural with the corresponding nickel compound and the unit-cell parameters are $a = 7.19$ and $c = 6.03$ Å[342].

A number of hydrated trichlorocobaltates(II) have been isolated from aqueous solutions of cobalt dichloride and alkali-metal chlorides[344,390,391,393].

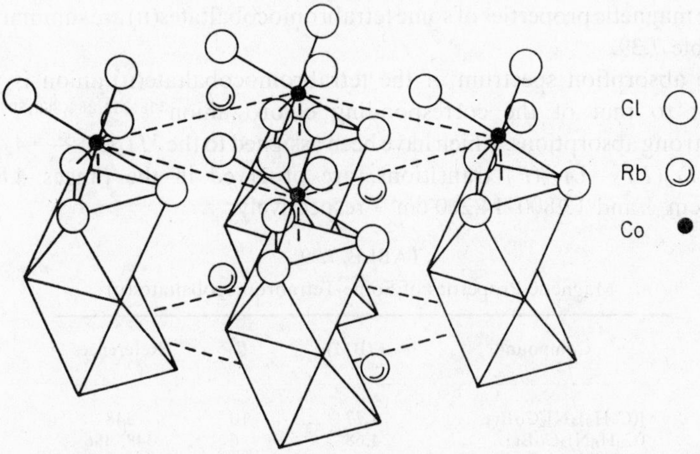

Figure 7.18 The crystal structure of $RbCoCl_3$. (Reproduced by permission from A. Engberg and H. Soling, *Acta Chem. Scand.*, **21**, 168 (1967))

Pentabromocobaltate(II). Van Stapele and coworkers[346] have prepared caesium pentabromocobaltate(II) by slow evaporation of an aqueous solution saturated with cobalt dibromide containing a stoichiometric excess of caesium bromide. The complex is isomorphous with the corresponding chloro compound and the tetragonal unit cell has $a = 9.619$ and $c = 15.163$ Å. It has a room-temperature magnetic moment of 4.58 BM and at 4°K $g_{\parallel} = 2.42$ and $g_{\perp} = 2.32$.

Tetrabromocobaltates(II). Substituted ammonium and arsonium tetrabromocobaltates(II) are most conveniently prepared in ethanol by reaction of stoichiometric quantities of the constituent bromides[348,364,365,384,386,394]. The caesium salt has been prepared in the melt[359]. As with the corresponding chloro complex, the complex bromide anion does not appear to be present to any appreciable extent in aqueous solutions containing hydrogen bromide or lithium bromide[348]. However, evaporation of a solution containing cobalt dibromide and caesium bromide does lead to the isolation of the caesium complex[360].

The tetrabromocobaltates(II) are pale blue solids and the tetrabutylammonium salt melts[348] at 104.5°. The heat of formation of the tetraethylammonium compound from its constituent bromides is reported[395] to be -11.14 kcal mole^{-1}.

Pauling[365] has shown that triphenylmethylarsonium tetrabromocobaltate(II) is cubic and isomorphous with the corresponding tetrachloronickelate(II). The cobalt-bromine stretching frequency[382–384] in the tetrabromocobaltate(II) salts occurs at about 230 cm^{-1}.

The magnetic properties of some tetrabromocobaltates(II) are summarized in Table 7.39.

The absorption spectrum of the tetrabromocobaltate(II) anion is very similar to that of the corresponding chloro anion[348,359,364,389,394,396]. Two strong absorptions, which have been assigned to the $^4T_{1g}(F) \leftarrow {}^4A_{2g}(F)$ and $^4T_{1g}(P) \leftarrow {}^4A_{2g}(F)$ transitions, are observed in the ranges 4,800–5,080 cm^{-1} and 13,800–14,200 cm^{-1} respectively.

TABLE 7.39
Magnetic Properties of Some Tetrabromocobaltates(II)

Compound	μ (BM)	$\theta°$	References
$[(C_4H_9)_4N]_2CoBr_4$	4.77	10	348
$(C_9H_8N)_2CoBr_4$	4.68	6	348, 386
$(C_5H_6N)_2CoBr_4$	4.63	0	348, 386
$[(C_6H_5)_3CH_3As]_2CoBr_4$	4.69[a]		364

[a] Room temperature measurement only.

Substituted tetrabromocobaltates(II). A number of complexes of the type $[CoBr_3L]^-$ are known where L is a neutral ligand such as pyridine, triphenylphosphine and benzimidizole. The magnetic properties of these compounds have been examined by Cotton and coworkers[194,397] and they obey the Curie-Weiss law with very small Weiss constants, and magnetic moments of about 4.5 BM. The absorption spectra have been discussed by Goodgame and Goodgame[196]. Bradbury and coworkers[137] report cobalt-bromine stretching frequencies for these compounds of about 260 cm^{-1}.

Tetraiodocobaltates(II). The tetraiodocobaltates(II) are usually prepared by interaction of a suitable iodide with cobalt diiodide in a non-aqueous solvent, in the manner described for the chloro and bromo analogues[175,314,348,364,365,374,384]. In general the magnetic, spectral and chemical properties of these complexes are very similar to those of the chloro and bromo salts[348,364]. Substituted complexes of the type $[CoI_3L]^-$ are also known[194,196,397].

ADDENDA

Halides

As one would expect there has been a considerable amount of work reported in the literature over the last year on the preparation, characterization and reactions of cobalt(III) complexes in which at least one halogen atom is bound directly to the central cobalt atom[398–429].

The structure of $[Co(NH_3)_5Cl]Cl_2$ has been redetermined by Messmer and Amma[430]. The orthorhombic unit cell has $a = 13.26$, $b = 10.34$ and $c = 6.72$ Å. The cobalt-chlorine distance was found to be 2.286 Å and the cobalt-nitrogen distances are 1.962 Å (×2), 1.964 Å, 1.978 Å and 1.998 Å. It was suggested that the nitrogen atom of the longest cobalt-nitrogen bond was involved in nitrogen-chlorine intermolecular hydrogen bonding.

Cowley and coworkers[431–433] have reported further studies on the magnetic properties and magnetic structure of cobalt difluoride, particularly near the Neel temperature. Richards[434] has discussed the far-infrared antiferromagnetic resonance of the same compound, while van der Ziel and Guggenheim[435] have reported a very detailed investigation of its electronic spectrum. These workers used a single crystal of the difluoride and measured the temperature dependence of the spectrum.

The melting point of cobalt dichloride has been reported[436] as 720°. Jacobs and coworkers[437] reported that this compound shows a far-infrared antiferromagnetic resonance band of 19 cm^{-1} at 4.2°K. At this temperature $g_{\parallel} = 6.0$ and $g_{\perp} = 3.1$.

The preparation and characterization of a large range of adducts of cobalt dichloride has appeared in the literature since the completion of the main discussion on the dichloride (page 337). The work reported may be roughly divided into two sections. First, there are the papers on the further investigation of previously known adducts. For example, there have been detailed investigations[438,439] of the ^{1}H NMR spectra of $CoCl_2(py)_4$ and $CoCl_2(py)_2$, and the electronic spectra of both of these adducts and the analogous complexes with substituted pyridines[440,441]. Secondly, there is the preparation and characterization of adducts with previously unreported donor ligands. Particular interest has been shown in the development of ligands which contain two or more different donor atoms. Ligands of this type include *o*-dimethylaminophenyldiphenylphosphine, the Schiff base formed between *o*-methoxybenzaldehyde and 1,2-propylenediamine, tri(2-cyanoethyl)phosphine, and complex amines.

There has also been a continuing interest in the preparation of five-coordinate complexes. Sacconi and coworkers[442,443] have prepared a number of cobalt dihalide adducts with ligands based on the substituted amine

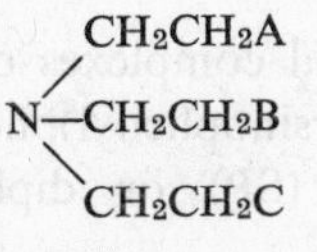

where A, B and C stands for any of the groups $N(C_2H_5)_2$, $P(C_6H_5)_2$, $As(C_6H_5)_2$ and SCH_3. Eight different ligands, having the following sets of donor atoms N_3P, N_3As, N_3S, N_2P_2, N_2As_2, N_2S_2, N_2SP and NP_3, were prepared. Adducts of the type [CoXL]Y, where Y is halogen, perchlorate or tetraphenylborate, were prepared for all of the above ligands except N_2As_2. Magnetic, conductimetric and spectroscopic studies showed quite conclusively that the cation contains cobalt in a trigonal-bipyramidal environment. The properties[443] of the tris(2-diphenylphosphinoethyl)amine (NP_3) complexes are given in Table 7.40.

TABLE 7.40
Properties of Some Five-Coordinate Cobalt(II) Complexes

Compound	Colour	μ (BM)
$[CoCl(NP_3)]BPh_4$	Magenta	4.26
$[CoBr(NP_3)]BPh_4$	Crimson	4.35
$[CoBr(NP_3)]Br$	Magenta	4.30
$[CoI(NP_3)]BPh_4$	Dark brown	4.36
$[CoI(NP_3)]I$	Brown	2.02

An interesting point to note is the change over from high-spin to low-spin in the iodo complexes. In nitromethane solution $[CoI(NP_3)]I$ exists in both the high- and low-spin forms, as shown by the electronic spectrum[443].

Another group of five-coordinate cobalt(II) compounds have been prepared by Venanzi and his group[413] using the tetradentate phosphine tris(*o*-diphenylphosphinophenyl)phosphine (QP). Some properties of these complexes are given in Table 7.41.

The structure of $[CoCl(QP)]BPh_4$ has been determined[441] using three dimensional single-crystal X-ray diffraction techniques. The compound has an orthorhombic unit cell, of space group $P2_1nb$ and containing four formula units, with $a = 18.85$, $b = 18.25$ and $c = 18.60$ Å. The cation is considerably distorted, the cobalt-phosphorus bond distances being 2.02, 2.26, 2.27 and 2.32 Å, while the equatorial phosphorus-cobalt-phosphorus angles are 108.7°, 112.7° and 137.6°. Since the ligand has trigonal symmetry it was considered that the distortion is due to a Jahn-Teller effect. A similar effect was noted for $CoBr_2[P(C_6H_5)_3]_3$ (Figure 7.15, page 362).

Deep red, square-pyramidal complexes of the type $[CoXL_2]Y$, where L is diphenyl(*o*-diphenylarsinophenyl)phosphine (AP), diphenyl(*o*-methylthiophenyl)phosphine (SP) or diphenyl(*o*-methyl-selenophenyl)-

phosphine (SeP) and Y = ClO_4^-, $SnCl_3^-$ or $SnBr_3^-$, have been prepared by Dyer and Meek[445]. The magnetic moments of these complexes are in the range 2.14–2.6 BM.

TABLE 7.41
Tris(*o*-diphenylphosphinophenyl)phosphine (QP) Complexes

Compound	Colour	Decomposition Temperature	μ (BM)
$[CoCl(QP)]_2[CoCl_4]$	Blue-black	323–325°	55.5[a]
$[CoCl(QP)]ClO_4\cdot 3C_2H_5OH$	Blue-black	314–316°	2.05
$[CoCl(QP)]BPh_4$	Red-black	262–263°	1.99
$[CoBr(QP)]_2[CoBr_4]$	Blue-black	335–338°	5.54[b]
$[CoBr(QP)]ClO_4\cdot 4C_2H_5OH$	Blue-black	319–321°	2.00
$[CoBr(QP)]BPh_4$	Blue-black	258–259°	1.98
$[CoI(QP)]I$	Blue-black	350–352°	2.09
$[CoI(QP)]ClO_4\cdot 4C_2H_5OH$	Blue-black	334–335°	2.09
$[CoI(QP)]BPh_4$	Blue-black	265–267°	1.94

[a] μ per cation = 1.94 BM   [b] μ per cation = 1.89 BM.

Houk and Emerson[446] report that cobalt dichloride forms 1:1 and 1:3 adducts with 2-pyridone (TP), the stable tautomer of 2-hydroxypyridine. The properties of the complexes are given below.

$CoCl_2(TP)$ pale blue, μ = 5.34 BM, octahedral
$CoCl_2(TP)_3$ blue, μ = 4.85 BM, octahedral

The 1:1 adduct is isomorphous with the corresponding adducts of nickel, manganese and copper dichlorides[446], the unit cell being monoclinic with a = 16.78, b = 8.65, c = 24.80 Å and β = 94.2°.

Although the complexes $CoX_2(thiourea)_3$ (X = Cl, Br, I) were originally reported[447] to be monomeric five-coordinate complexes, a more recent study[448] has shown that they should be formulated as $[CoX(thiourea)_3]X$, the cobalt atom being in a pseudo-tetrahedral environment.

O'Connor and Amma[449] have reported the results of a three dimensional single-crystal X-ray diffraction study of $CoCl_2(thiourea)_4$. The adduct has a tetragonal unit cell of space group $P4_2/n$ containing four formula units, with a = 13.508 and c = 9.106 Å. The central cobalt atom has a *trans* octahedral configuration with the following dimensions.

Co—Cl = 2.469 Å   S—Co—S = 88.0° and 92.0°
Co—S = 2.502 and 2.533 Å   Cl—Co—S = 84.6°, 87.1° and 95.4°

Two structurally different forms of the complex $CoCl_2(4\text{-vinylpyridine})_2$

have been identified, one having octahedral[450] and the other tetrahedrally[451] coordinated cobalt. The crystallographic details are given in Table 7.42.

TABLE 7.42
Crystallographic Details for $CoCl_2$(4-vinylpyridine)$_2$

Parameter	Octahedral Form	Tetrahedral Form
Symmetry	Monoclinic	Monoclinic
a (Å)	17.72	20.20
b (Å)	3.65	7.74
c (Å)	23.49	14.61
β	111.5°	126°
Space group	$A2/a$	$C2/c$
Z	4	4
Co—Cl (Å)		2.22
Co—N (Å)		2.01

The properties of a number of other adducts of cobalt dichloride are given in Table 7.43.

TABLE 7.43
Adducts of the Type $CoCl_2L_n$

Compound and Properties	References
L = 2-methoxyaniline, $n = 1$, blue, $\mu = 4.54$ BM, $\nu_{Co-O} = 393$ cm^{-1}, $\nu_{Co-Cl} = 318, 306$ cm^{-1}, tetrahedral	452
L = *p*-nitroso-*N*,*N*-dimethylaniline (NODMA), $n = 2$, m.p. = 166°, $\mu = 4.25$ BM, tetrahedral	453
L = trimethylamine *N*-oxide, $n = 2$, purple, $\mu = 4.71$ BM, tetrahedral	454
L = imidazole, $n = 2$, $\mu = 4.66$ BM, $\nu_{Co-Cl} = 321, 308$ cm^{-1}, tetrahedral	455
L = 2,2′,2″-terpyridyl (terpy), $n = 1$, green, $\mu = 5.03$ BM, five coordinate	456–458
L = bis[2-(2-pyridyl)ethyl]amine (dpea), $n = 1$, violet, five coordinate	459
L = isopropylaminomethylpyridine-(2), $n = 2$, yellow, m.p. = 210°, $\mu = 4.91$ BM, octahedral	460
L = isopropylaminomethylpyridine-(2), $n = 1$, dark blue, m.p. = 218°, $\mu = 4.64$ BM, tetrahedral	460
L = dimethylaminomethylpyridine-(2), $n = 1$, violet-blue, m.p. = 183°, $\mu = 4.78$ BM, tetrahedral	460
L = diethylaminomethylpyridine-(2), $n = 1$, blue, tetrahedral	460
L = pentamethylenetetrazole (PMT), $n = 1$, pink, $\mu = 5.12$ BM, octahedral	461
L = 1-methylbenzimadazole, $n = 2$, blue, $\mu = 4.73$ BM, tetrahedral	462
L = 4(5)-bromoimidazole, $n = 2$, blue, $\mu = 4.76$ BM, $\nu_{Co-N} = 250$ cm^{-1}, $\nu_{Co-Cl} = 327, 302, 290$ cm^{-1}, tetrahedral	463
L = benzothiazole (BT), $n = 2$, blue, $\mu = 4.53$, $\nu_{Co-Cl} = 345, 323$ cm^{-1}, tetrahedral	464, 465
L = 2-(2-pyridyl)benzothiazole (pbt), $n = 2$, brown, $\mu = 5.10$ BM, octahedral	466
L = 2-(2-pyridyl)benzothiazole (pbt), $n = 1$, dark blue, $\mu = 4.76$ BM, tetrahedral	466

TABLE 7.43 (*contd.*)

Adducts of the Type $CoCl_2L_n$

L = benzoxazole, $n = 2$, pink, $\mu = 5.31$ BM, $\nu_{Co-N} = 232\ cm^{-1}$, $\nu_{Co-Cl} = 254\ cm^{-1}$, octahedral polymer with bridging halogens	467
L = tri(2-cyanoethyl)phosphine (CEP), $n = 2/3$, royal blue, $\mu = 4.68$ BM, tetrahedral	468
L = benzene-1,2-diamine, $n = 1$, green, $\mu = 4.60$ BM, tetrahedral	469
L = benzene-1,3-diamine, $n = 1$, blue, $\mu = 4.43$ BM, $\nu_{Co-Cl} = 319, 299\ cm^{-1}$, tetrahedral	470
L = benzene-1,4-diamine, $n = 1$, blue, $\mu = 4.53$ BM, $\nu_{Co-Cl} = 308, 300\ cm^{-1}$, tetrahedral	471
L = Schiff based formed between furan-2-carboxaldehyde and *N,N*-dimethylethylenediamine ($FuAenNMe_2$), $n = 1$, m.p. = 206–209°, $\mu = 4.74$ BM, tetrahedral	472
L = Schiff base formed between *o*-methoxybenzaldehyde and *N,N*-diethylethylenediamine ($MOBenNEt_2$), $n = 1$, blue, m.p. = 204–206°, $\mu = 4.50$ BM, tetrahedral	473
L = Schiff base formed between 2,6-diacetylpyridine and *s*-butylamine [$py(Bu^s)_2$], $n = 1$, green, m.p. = 266–271°, $\mu = 4.85$ BM, five coordinate	474
L = Schiff base formed between 2,6-diacetylpyridine and cyclohexylamine [$py(cy)_2$], $n = 1$, grey-green, m.p. = 320°, $\mu = 4.85$ BM, five coordinate	474
L = 2,2′-bis(dimethylamino)azobenzene, $n = 1$, blue-black, $\mu = 4.64$ BM, ionic, [CoClL]L, tetrahedral cation	462
L = tris(2-methylthioethylamine) (TSN), $n = 2/3$, amethyst, $\mu = 4.45$ BM, ionic with five-coordinate cation and tetrahedral anion, $[CoCl(TSN)]_2CoCl_4$	475
L = 1-(2-pyridyl)-2-(3-pyridyl)ethylene, $n = 1$, royal blue, $\mu = 4.25$ BM, tetrahedral polymer	476
L = 1-(2-pyridyl)-2-(4-pyridyl)ethylene, $n = 1$, pale blue, $\mu = 5.02$ BM, octahedral	476
L = pyridine-2-aldehyde 2′-pyridylhydrazone (paphy), $n = 1$; α-form, red-fawn, $\mu = 5.04$ BM, octahedral; β-form, green, $\mu = 4.84$ BM, five coordinate	456
L = 1,2-bis(diphenylphosphino)ethane (DPE), $n = 2$, green, $\mu = 2.12$ BM, ionic, $[CoCl(DPE)_2]Cl$, five-coordinate cation	477
L = 1,2-bis(diphenylphosphino)ethane (DPE), $n = 1$, blue, $\mu = 4.41$ BM, tetrahedral	477
L = 1,3-bis(diphenylphosphino)propane (DPP), $n = 1$, light blue, $\mu = 4.43$ BM, tetrahedral	477
L = 1,4-bis(diphenylphosphino)butane (DPB), $n = 1$, light blue, $\mu = 4.31$ BM, tetrahedral	478
L = bis-phenylphosphinomethylpyridine(2), $n = 2$, violet, m.p. = 4.67 BM, octahedral	479
L = *o*-dimethylaminophenyldiphenylphosphine (PN), $n = 1$, royal blue, m.p. = 254–2550°, $\mu = 4.46$ BM, $\nu_{Co-Cl} = 350, 309\ cm^{-1}$, tetrahedral	480
L = bis(*o*-dimethylaminophenyl)phenylphosphine (PDN), $n = 1$, royal blue, m.p. = 210–211°, $\mu = 4.67$ BM, $\nu_{Co-Cl} = 354, 305\ cm^{-1}$, tetrahedral	480
L = tris(*o*-dimethylaminophenyl)phosphine (PTN), $n = 1$, royal blue, m.p. = 196–197°, $\mu = 4.71$ BM, $\nu_{Co-Cl} = 337, 330, 288\ cm^{-1}$, tetrahedral	480
L = 2,6-pyridinedisulphenamide, $n = 2$, brown, $\mu = 4.6$ BM, octahedral	481
L = biuret, $n = 2$, violet, $\mu = 4.96$ BM, octahedral	482
L = trimethylamine oxide, $n = 2$, $\nu_{Co-O} = 569, 559\ cm^{-1}$, $\nu_{Co-Cl} = 310, 280\ cm^{-1}$, tetrahedral	483
L = triethylamine oxide, $n = 2$, $\nu_{Co-Cl} = 312, 279\ cm^{-1}$, tetrahedral	483
L = trimethylphosphine oxide, $n = 2$, $\mu = 4.34$ BM, $\nu_{P-O} = 1125, 1094\ cm^{-1}$, $\nu_{Co-O} = 444\ cm^{-1}$, $\nu_{Co-Cl} = 319, 290\ cm^{-1}$, tetrahedral	483, 484

TABLE 7.43 (*contd.*)

Adducts of the Type $CoCl_2L_n$

L = triphenylphosphine oxide, $n = 2$, blue, m.p. = 240°, $\nu_{Co-Cl} = 340, 315$ cm^{-1}, tetrahedral	485
L = trimethylarsine oxide, $n = 2$, $\mu = 4.42$ BM, $\nu_{As-O} = 874, 952, 831$ cm^{-1}, $\nu_{Co-O} = 429$ cm^{-1}, $\nu_{Co-Cl} = 296, 280$ cm^{-1}, tetrahedral	486
L = di-*t*-butyl nitrogen oxide, $n = 2$, m.p. = 154°, $\mu = 2.96$ BM, $\theta = 20°$, $\nu_{NO} = 1326$ cm^{-1}, tetrahedral	487
L = diphenylcyclopropenone, $n = 2$, pale blue, m.p. = 193–195°, $\mu = 5.30$ BM, $\nu_{Co-Cl} = 313, 297$ cm^{-1}, octahedral	488
L = 1,4-dioxan, $n = 1$, lavender, $\mu = 5.32$ BM, $\nu_{Co-Cl} = 240$ cm^{-1}, octahedral	489
L = 2,5-dithiahexane, $n = 2$, $\nu_{Co-Cl} = 266$ cm^{-1}, octahedral	489
L = 2,5-dithiahexane, $n = 1$, $\nu_{Co-Cl} = 296, 281, 267$ cm^{-1}, polymeric with tetrahedral and octahedral cobalt	489
L = 1,2-bis(methylsulphonyl)ethane (BMSE), $n = 1$, blue, m.p. = 148°, tetrahedral	490
L = *N,N'*-dibutylthiourea, $n = 2$, deep blue, $\mu = 4.55$ BM, $\theta = 12°$, tetrahedral	491
L = 2-thiazolidinethione, $n = 2$, dark blue, m.p. = 142°, $\mu = 4.88$ BM, tetrahedral	492
L = trimethylphosphine sulphide, $n = 2$, $\mu = 4.32$ BM, $\nu_{P-S} = 538, 533$ cm^{-1}, tetrahedral	484
L = trimethylarsine sulphide, $n = 2$, $\mu = 4.31$ BM, $\nu_{As-S} = 442$ cm^{-1}	486
L = selenourea, $n = 2$, blue-green, tetrahedral	491
L = 1,2-diselenocyanatoethane, $n = 1$, pink, $\mu = 5.01$ BM, octahedral	493

Other adducts of cobalt dichloride include those with 2-methylpyridine[494], 4-methylpyridine[495], hydrazine[496], aniline[497], *p*-toluidine[498], ethylenethiourea[498], pyrazole[499], hexamethylphosphoramide[500], 2,9-dimethyl-1,10-phenanthroline[501], dipyridyl ketone[502], 1,1,7,7-tetraethyldiethylenetriamine[503], 1-ethyl-2-methylbenzimadazole[462], benzil mono(2'-pyridyl)-hydrazone[504], thioacetamide[505], *N,N*-disubstituted thioureas[491] and dissymetric phosphines[506] as donor ligands.

Kakos and Winter[507] have prepared cobalt(II) halo alkoxides of the type $CoX(OCH_3)$ (X = Cl, Br, I). The magnetic and spectral properties of these complexes indicate that the chloro derivative has an octahedral configuration while the other two are tetrahedral.

Cohen-Adad and coworkers[436,508,509] have studied the cobalt dichloride-water system in considerable detail. Mono-, di- and hexahydrates were isolated. The thermal decomposition of cobalt dichloride hexahydrate has been investigated in static air and dynamic nitrogen atmospheres[510] and the results are given in Table 7.44. There have been further magnetic studies on cobalt dichloride hexahydrate[511–513] and dihydrate[514,515]. The hexahydrate has been reported[513] to have a room-temperature magnetic moment of 4.92 BM.

The reactions of cobalt dibromide resemble closely, as one would expect, those of cobalt dichloride, although in certain cases the stereochemistry adopted by the cobalt atom changes. One interesting reaction of cobalt dibromide that has been reported is that with acetylacetone[516]. The direct interaction of the constituents, or the action of bromine on tris-(acetylacetonato)cobalt(III) in dichloromethane at 0°, leads to the formation of pale blue $CoBr_2(C_5H_8O_2)$. This compound has a room temperature magnetic moment of 4.25 BM suggesting a tetrahedral configuration, and its infrared spectrum is indicative of the presence of the ketonic form

TABLE 7.44

Thermal Decomposition of Cobalt Dichloride Hexahydrate

Reaction	Temperature Range	
	Static Air	Dynamic Nitrogen
$CoCl_2 \cdot 6H_2O \rightarrow CoCl_2 \cdot 2 \cdot 5H_2O + 3 \cdot 5H_2O$	<50–120°	<50–123°
$CoCl_2 \cdot 2 \cdot 5H_2O \rightarrow CoCl_2 \cdot 2H_2O + 0.5H_2O$	120–136°	
$CoCl_2 \cdot 2 \cdot 5H_2O \rightarrow CoCl_2 \cdot H_2O + 1.5H_2O$		123–162°
$CoCl_2 \cdot 2H_2O \rightarrow CoCl_2 \cdot H_2O + H_2O$	136–166°	
$CoCl_2 \cdot H_2O \rightarrow CoCl_2 + H_2O$	166–208°	162–200°
$CoCl_2 \rightarrow Co + Cl_2$		550–921°

of acetylacetone. If this adduct is reacted with an equimolar amount of sodium acetylacetonate at 0° then a blue compound of formula CoBr(HA), where H_2A is acetylacetone, may be isolated. The infrared spectrum of this complex indicates that the ligand is present as the enol form, while the electronic spectrum suggests tetrahedral cobalt. On this basis a bromine bridged structure was proposed.

Some properties of a number of other adducts of cobalt dibromide are given in Table 7.45.

TABLE 7.45

Adducts of the Type $CoBr_2L_n$

Compound and Properties	References
L = 2-methoxyaniline, $n = 1$, blue, $\mu = 4.48$ BM, $\nu_{Co-O} = 448$ cm^{-1}, $\nu_{Co-Br} = 248$ cm^{-1}, tetrahedral	452
L = trimethylamine *N*-oxide, $n = 2$, purple, $\mu = 4.77$ BM, tetrahedral	454
L = imidazole, $n = 2$, $\mu = 4.63$ BM, tetrahedral	455
L = 2,2′,2″-terpyridyl (terpy), $n = 1$, dark green, $\mu = 4.99$ BM, five coordinate	456–458
L = bis[2-(2-pyridyl)ethyl]amine, $n = 1$, violet, $\mu = 4.57$ BM, five coordinate	459

TABLE 7.45 (*contd.*)
Adducts of the Type $CoBr_2L_n$

L = isopropylaminomethylpyridine-(2), n = 1, dark blue, m.p. = 209°, μ = 4.67 BM, tetrahedral	460
L = dimethylaminomethylpyridine-(2), n = 1, violet-blue, m.p. = 186°, μ = 4.74 BM, tetrahedral	460
L = diethylaminomethylpyridine-(2), n = 1, blue, tetrahedral	460
L = pentamethylenetetrazole (PMT), n = 2, royal blue, m.p. = 159–161°, μ = 4.48 BM, tetrahedral	461
L = 4(5)-bromoimidazole, n = 2, blue, μ = 4.56 BM, ν_{Co-N} = 250 cm^{-1}, ν_{Co-Br} = 250 cm^{-1}, tetrahedral	463
L = benzothiazole (BT), n = 2, blue, μ = 4.60 BM, ν_{Co-Br} = 260, 255 cm^{-1}, tetrahedral	464, 465
L = 2-(2-pyridyl)benzothiazole (pbt), n = 2, brown, μ = 4.83 BM, octahedral	466
L = 2-(2-pyridyl)benzothiazole (pbt), n = 1, green-blue, μ = 4.76 BM, tetrahedral	466
L = benzoxazole, n = 2, blue, μ = 4.59 BM, ν_{Co-N} = 238 cm^{-1}, ν_{Co-Br} = 256 cm^{-1}, tetrahedral	467
L = benzene-1,2-diamine, n = 1, green, μ = 4.57 BM, tetrahedral	469
L = benzene-1,3-diamine, n = 1, green-blue, μ = 4.42 BM, tetrahedral	470
L = benzene-1,4-diamine, n = 1, blue, μ = 4.53 BM, tetrahedral	471
L = Schiff base formed between pyridine-2-carboxaldehyde and *N,N*-diethylethylenediamine ($PyAenNEt_2$), n = 1, m.p. = 194–196°, μ = 4.92 BM, five coordinate	472
L = Schiff base formed between furan-2-carboxaldehyde and *N,N*-dimethylethylenediamine ($FuAenNMe_2$), n = 1, m.p. = 226–233°, μ = 4.62 BM, tetrahedral	472
L = Schiff base formed between *o*-methoxybenzaldehyde and *N,N*-dimethylethylenediamine ($MOBenNMe_2$), n = 1, blue, m.p. = 203–206°, μ = 4.59 BM, tetrahedral	473
L = Schiff base formed between *o*-methoxybenzaldehyde and *N,N*-diethylethylenediamine ($MOBenNEt_2$), n = 1, blue, m.p. = 204–207°, μ = 4.61 BM, tetrahedral	473
L = Schiff base formed between 2,6-diacetylpyridine and *s*-butylamine [py(Bu^s)$_2$], n = 1, green, m.p. = 261–264°, μ = 4.84 BM, five coordinate	474
L = Schiff base formed between 2,6-diacetylpyridine and cyclohexylamine [py(cy)$_2$], n = 1, grey-green, m.p. = 327°, μ = 4.94 BM, five coordinate	474
L = tris(2-methylthioethyl)amine (TSN), n = 2/3, purple-red, μ = 4.55 BM, ionic with five-coordinate cation and tetrahedral anion, [CoBr(TSN)]$_2$[$CoBr_4$]	475
L = 1-(2-pyridyl)-2-(3-pyridyl)ethylene, n = 1, blue, μ = 4.33 BM, tetrahedral	476
L = 1-(2-pyridyl)-2-(4-pyridyl)ethylene, n = 1, royal blue, μ = 4.55 BM, tetrahedral	476
L = 1,2-bis(diphenylphosphino)ethane (DPE), n = 2, green, μ = 1.86 BM, ionic, [CoBr(DPE)$_2$]Br, five-coordinate cation	477
L = 1,2-bis(diphenylphosphino)ethane (DPE), n = 1, light blue, μ = 4.46 BM, tetrahedral	477
L = 1,3-bis(diphenylphosphino)propane (DPP), n = 1, blue-green, μ = 4.47 BM, tetrahedral	477
L = 1,4-bis(diphenylphosphino)butane (DPB), n = 1, green-blue, m.p. = 293–296°, μ = 4.56 BM, tetrahedral	478
L = bis-phenylphosphinomethylpyridine-(2), n = 2, violet, m.p. = 237°, μ = 4.84 BM, octahedral	479
L = *o*-dimethylaminophenyldiphenylphosphine (PN), n = 1, royal blue, m.p. = 219–220°, μ = 4.46 BM, ν_{Co-Br} = 280, 256 cm^{-1}, tetrahedral	480

TABLE 7.45 (*contd.*)
Adducts of the Type $CoBr_2L_n$

L = bis(*o*-dimethylaminophenyl)phenylphosphine (PDN), $n = 1$, blue-green, m.p. = 192–193°, $\mu = 4.60$ BM, $\nu_{Co-Br} = 281, 242$ cm^{-1}, tetrahedral	480
L = tris(*o*-dimethylaminophenyl)phosphine (PTN), $n = 1$, blue, m.p. = 182–183°, $\mu = 4.74$ BM, $\nu_{Co-Br} = 270, 241$ cm^{-1}, tetrahedral	480
L = trimethylamine oxide, $n = 2$, $\nu_{Co-O} = 573, 555$ cm^{-1}, $\nu_{Co-Br} = 232$ cm^{-1}, tetrahedral	483
L = triethylamine oxide, $n = 2$, $\nu_{Co-Br} = 237$ cm^{-1}, tetrahedral	483
L = trimethylphosphine oxide, $n = 2$, $\mu = 4.46$ BM, $\nu_{P-O} = 1125, 1094$ cm^{-1}, $\nu_{Co-O} = 436$ cm^{-1}, $\nu_{Co-Br} = 246$ cm^{-1}, tetrahedral	483, 484
L = trimethylarsine oxide, $n = 2$, $\nu_{As-O} = 872, 851, 825$ cm^{-1}, $\nu_{Co-O} = 427$ cm^{-1}, $\nu_{Co-Br} = 234, 222$ cm^{-1}, tetrahedral	486
L = di-*t*-butyl nitrogen oxide, $n = 2$, m.p. = 135°, $\mu = 2.73$ BM, $\theta = 32°$, $\nu_{NO} = 1321$ cm^{-1}, tetrahedral	487
L = diphenylcyclopropenone, $n = 2$, deep blue, m.p. = 210°, $\mu = 4.78$ BM, $\nu_{Co-Br} = 262, 233$ cm^{-1}, tetrahedral	488
L = 1,4-dioxan, $n = 2$, lavender, $\mu = 5.19$ BM, $\nu_{Co-Br} = 208$ cm^{-1}, octahedral	517
L = 1,4-dioxan, $n = 1$, light blue, $\mu = 5.17$ BM, octahedral	517
L = 1,2-di(isopropylthio)ethane, $n = 2$, $\nu_{Co-Br} = 220$ cm^{-1}, octahedral	489
L = *N,N'*-diethylthiourea, $n = 2$, blue, $\mu = 4.68$ BM, $\theta = 23°$, tetrahedral	491
L = 2-thiazolidinethione, $n = 2$, blue-green, m.p. = 175°, $\mu = 4.97$ BM, tetrahedral	492
L = trimethylphosphine sulphide, $n = 2$, $\mu = 4.41$ BM, $\nu_{P-S} = 536$ cm^{-1}, $\nu_{Co-S} = 316, 310$ cm^{-1}, $\nu_{Co-Br} = 246, 240$ cm^{-1}, tetrahedral	484
L = triphenylphosphine sulphide, $n = 2$, tetrahedral	484
L = trimethylarsine sulphide, $n = 2$, $\mu = 4.34$ BM, tetrahedral	486
L = 2-ethylthiocyclohexylamine, $n = 2$, pink, $\mu = 4.79$ BM, octahedral	518

Other adducts of cobalt dibromide that have been studied include those with pyridine and its derivatives[438–441], substituted amines with nitrogen, phosphorus, arsenic and sulphur donor atoms[442], tri(2-cyanoethyl)-phosphine[468], 1,1,7,7-tetraethyldiethylenetriamine[503], benzil mono(2'-pyridyl)hydrazone[504], disymmetric phosphines[506], thioacetamide[505], hexa-methylphosphoramide[500] and 2,5-dithiahexane[489].

The magnetic properties of cobalt dibromide hexahydrate have been studied in some detail[519,520]. The Neel temperature is reported[519] to be 3.2°K and there is a large magnetic anisotropy[520]. The dihydrate[510], prepared by dissolving cobalt carbonate in the calculated amount of concentrated hydrobromic acid, looses both molecules of water between 80 and 190°.

Some properties of a number of cobalt diiodide adducts are summarized in Table 7.46.

TABLE 7.46
Adducts of the Type CoI_2L_n

Compound and Properties	References
L = trimethylamine *N*-oxide, $n = 2$, purple-lilac, $\mu = 4.88$ BM, tetrahedral	454
L = imidazole, $n = 2$, $\mu = 4.65$ BM, tetrahedral	455
L = 2,2′,2″-terpyridyl (terpy), $n = 1$, $\mu = 4.79$ BM, five coordinate	457, 458
L = isopropylaminomethylpyridine-(2), $n = 2$, grey, $\mu = 4.65$ BM, ionic, five-coordinate cation	460
L = isopropylaminomethylpyridine-(2), $n = 1$, green, m.p. = 203°, $\mu = 4.64$ BM, tetrahedral	460
L = dimethylaminomethylpyridine-(2), $n = 2$, grey, $\mu = 4{\cdot}62$ BM, ionic, five-coordinate cation	460
L = dimethylaminomethylpyridine-(2), $n = 1$, green, m.p. = 215°, $\mu = 4{\cdot}68$ BM, tetrahedral	460
L = diethylaminomethylpyridine-(2), $n = 2$, red-violet, m.p. = 120°, $\mu = 4.62$ BM, ionic, five-coordinate cation	460
L = diethylaminomethylpyridine-(2), $n = 1$, green, $\mu = 4.67$ BM, tetrahedral	460
L = 4(5)-bromoimidazole, $n = 2$, royal blue, $\mu = 4.79$ BM, $\nu_{Co-N} = 246$ cm^{-1}, $\nu_{Co-I} = 210$ cm^{-1}, tetrahedral	463
L = benzothiazole (BT), $n = 2$, blue, $\mu = 4.58$ BM, $\nu_{Co-I} = 235$, 226 cm^{-1}, tetrahedral	464, 465
L = 2-(2-pyridyl)benzothiazole (pbt), $n = 2$, brown, $\mu = 4.79$ BM, octahedral	466
L = benzoxazole, $n = 2$, $\mu = 4.56$ BM, $\nu_{Co-N} = 241$ cm^{-1}, $\nu_{Co-I} = 233$ cm^{-1}, tetrahedral	467
L = benzene-1,3-diamine, $n = 1$, green, $\mu = 4.52$ BM, tetrahedral	470
L = benzene-1,4-diamine, $n = 1$, green, $\mu = 4.59$ BM, tetrahedral	471
L = Schiff base formed between pyridine-2-carboxaldehyde and *N,N*-dimethylethylenediamine ($PyAenNMe_2$), $n = 1$, m.p. = 222–225°, $\mu = 5.05$ BM, five coordinate	472
L = Schiff base formed between pyridine-2-carboxaldehyde and *N,N*-diethylethylenediamine ($PyAenNEt_2$), $n = 1$, m.p. = 199–204°, $\mu = 4.77$ BM, five coordinate	472
L = Schiff base formed between furan-2-carboxaldehyde and *N,N*-dimethylethylenediamine ($FuAenNMe_2$), $n = 1$, m.p. = 220–225°, $\mu = 4.78$ BM, tetrahedral	472
L = Schiff base formed between *o*-methoxybenzaldehyde and *N,N*-dimethylethylenediamine ($MOBenNMe_2$), $n = 1$, blue, m.p. = 186–189°, $\mu = 4.67$ BM, tetrahedral	473
L = Schiff base formed between 2-pyridinealdehyde and 2,6-disulphenamidopyridine (PDSPA), $n = 1$, brown, $\mu = 5.2$ BM, octahedral	481
L = tris(2-methylthioethyl)amine (TSN), $n = 2/3$, brown, $\mu = 4.45$ BM, ionic with five-coordinate cation and tetrahedral anion, $[CoI(TSN)]_2[CoBr_4]$	475
L = 1-(2-pyridyl)-2-(3-pyridyl)ethylene, $n = 1$, bright green, $\mu = 4.52$ BM, tetrahedral	476
L = 1-(2-pyridyl)-2-(4-pyridyl)ethylene, $n = 1$, green, $\mu = 4.61$ BM, tetrahedral	476
L = 1,2-bis(diphenylphosphino)ethane (DPE), $n = 2$, dark brown, $\mu = 1.92$ BM, ionic, $[CoI(DPE)_2]I$, five-coordinate cation	477
L = 1,2-bis(diphenylphosphino)ethane (DPE), $n = 1$, light blue, $\mu = 4.63$ BM, tetrahedral	477

TABLE 7.46 (*contd.*)
Adducts of the Type CoI_2L_n

L = 1,3-bis(diphenylphosphino)propane (DPP), $n = 1$, light brown, $\mu = 4.56$ BM, tetrahedral	477
L = *o*-dimethylaminophenyldiphenylphosphine (PN), $n = 1$, green, m.p. = 226–227°, $\mu = 4.64$ BM, $\nu_{Co-I} = 240, 223$ cm^{-1}, tetrahedral	480
L = bis(*o*-dimethylaminophenyl)phenylphosphine (PDN), $n = 1$, green, m.p. = 233–234°, $\mu = 4.93$ BM, $\nu_{Co-I} = 235, 210$ cm^{-1}, tetrahedral	480
L = tris(*o*-dimethylaminophenyl)phosphine (PTN), $n = 1$, dark brown, m.p. = 179–180°, $\mu = 4.87$ BM, $\nu_{Co-I} = 235, 205$ cm^{-1}, tetrahedral	480
L = trimethylphosphine oxide, $n = 2$, $\mu = 4.46$ BM, $\nu_{P-O} = 1134, 1107$ cm^{-1}, $\nu_{Co-O} = 423$ cm^{-1}, tetrahedral	483, 484
L = di-*t*-butyl nitrogen oxide, $n = 2$, m.p. = 94°, $\mu = 2.70$ BM, $\theta = 23°$, $\nu_{NO} = 1316$ cm^{-1}, tetrahedral	487
L = trimethylphosphine sulphide, $n = 2$, $\mu = 4.42$ BM, $\nu_{P-S} = 531$ cm^{-1}, $\nu_{Co-S} = 317, 311$ cm^{-1}, tetrahedral	484
L = *N,N'*-diethylthiourea, $n = 2$, green, $\mu = 4.79$ BM, $\theta = 26°$, tetrahedral	491
L = 2-thiazolidinethione, $n = 2$, light green, m.p. = 164°, $\mu = 5.01$ BM, tetrahedral	492

Other donor ligands that have been reacted with cobalt diiodide include pyridine and its derivatives[438–441], substituted amines with nitrogen, phosphorus, arsenic and sulphur donor atoms[442], tri(2-cyanoethyl)-phosphine[468], trimethylarsine oxide and sulphide[486], 1,1,7,7-tetra-methyldiethylenetriamine[503], disymmetric phosphines[506], thioaceta-mide[505], 2,5-dithiahexane[489] and 1,2-di(isopropylthio)ethane[489].

Cobalt diiodide dihydrate[510], prepared by dissolving the carbonate in the appropriate amount of 57% hydriodic acid, looses both molecules of water between 40 and 150°.

Complex Halides

The magnetic and spectral properties of several trifluorocobaltates(II) have been investigated[521,522].

Huiskamp and coworkers[523,524] have studied the magnetic properties of caesium pentachloro- and pentabromocobaltate(II) in considerable detail. Anomalies in the specific heats of the salts give Neel temperatures of 0.523°K and 0.282°K for the chloro and bromo complexes respectively. Magnetic susceptibility measurements on single crystals of these two compounds reveal maxima at temperatures slightly above the respective Neel temperatures,

$$Cs_3CoCl_5 \qquad \chi_{max} \text{ at } T = 1.05T_N$$
$$Cs_3CoBr_5 \qquad \chi_{max} \text{ at } T = 1.5T_N$$

The magnetic structures of both complexes were discussed in detail.

The preparation and characterization of a number of tetrahalocobaltates(II) with both simple and complex cations have been reported[413,456,462,475,525–532]. Some of the compounds with complex cations, for example, those with tris(*o*-diphenylphosphinophenyl)-phosphine[413] and tris(2-methylthioethyl)amine[475], have already been discussed. Heath, Martin and Stewart[526,527] have reported that the complexes derived from the cobalt dihalides and dithioacetylacetone are similar to those of iron(II), (page 314), and should be formulated as $[C_5H_7S_2]_2CoX_4$. The tetrachlorocobaltate(II) salt of the hitherto unknown 3,5-dimethyl-1,2-diselenolium ion was also prepared and characterized[526,532]. Some properties of the dithiolium and diselenolium tetrahalocobaltates(II) are given in Table 7.47.

TABLE 7.47

Properties of Some Complex Tetrahalocobaltates(II)

Compound	μ (BM)	$\theta°$	ν_{Co-X} (cm^{-1})
$[C_5H_7S_2]_2CoCl_4$	4.69	8	295
$[C_9H_7S_2]_2CoCl_4$	4.65		293
$[C_5H_7S_2]_2CoBr_4$	4.85	12	222
$[C_5H_7Se_2]_2CoCl_4$	4.6		

Ouchi and coworkers[529,530] have also investigated the complexes derived from dithioacetylacetone and dithiobenzoylacetone, although as noted

TABLE 7.48

Properties of Substituted Tetrahalocobaltates(II)[a]

Compound	μ (BM)	ν_{Co-X} (cm^{-1})	δ_{Co-X} (cm^{-1})	ν_{Co-N} (cm^{-1})
$[Et_4N][pyCoCl_3]$	4.50	326*s*, 289*s*	128*s*	226*m*, 186*w*, 162*w*
$[py\cdot al][pyCoCl_3]$	4.54	320*s*, 286*s*	118*s*	226*m*, 162*w*
$[pic\cdot al][picCoCl_3]$	4.92	309*s*, 287*s*	120*s*	230*m*, 215*m*
$[Et_4N][pyCoBr_3]$	4.68	257*s*, 238*sh*	96*m*	174*s*, 150*w*
$[py\cdot pr][pyCoBr_3]$	4.70	250*s*, 228*sh*	94*s*	175*m*, 153*w*
$[pic\cdot pr][picCoBr_3]$	4.99	244*s*	97*m*	204*sh*, 170*m*, 155*w*
$[py\cdot pr][pyCoI_3]$		244*sh*, 238*s*	68*m*	161*s*, 134*m*

[a] py = pyridine, al = allyl, pic = α-picoline = 2-methylpyridine, pr = n-propyl, *s* = strong, *m* = medium, *w* = weak, *sh* = shoulder.

previously (page 314), these complexes were incorrectly formulated by these workers.

Mitra[531] has reinvestigated the magnetic anisotropy of caesium tetrachlorocobaltate(II) originally studied by Figgis and coworkers[349]. Mitra shows that the previous work is in error and he reports that the molecular anisotropy ($K_{\parallel} - K_{\perp}$) ranges from 641 to 676 $\times 10^{-6}$ cgs depending upon the orientation of the crystal with respect to the magnetic field.

Brown, Forrest, Nuttall and Sharp[533] have prepared and characterized a large range of substituted tetrahalocobaltates(II). Some properties of these tetrahedral complexes are summarized in Table 7.48.

REFERENCES

1. M. A. Hepworth, K. H. Jack, R. D. Peacock and G. J. Westland, *Acta Cryst.*, **10,** 63 (1957).
2. H. J. Emeleus and G. L. Hurst, *J. Chem. Soc.*, **1964,** 396.
3. E. T. McBee, B. W. Hotten, L. R. Evans, A. A. Alberts, Z. D. Welch, W. B. Ligett, R. C. Schreyer and K. W. Krantz, *Ind. Eng. Chem.*, **39,** 310 (1947).
4. W. B. Burford, R. D. Fowler, J. M. Hamilton, H. C. Anderson, C. E. Weber and R. G. Sweet, *Ind. Eng. Chem.*, **39,** 321 (1947).
5. O. Ruff and E. Ascher, *Z. anorg. allgem. Chem.*, **183,** 193 (1929).
6. H. F. Priest, *Inorg. Syn.*, **3,** 173 (1950).
7. N. Watanabe, Y. Ohara, T. Tanigawa and S. Yoshizawa, *Kogyo Kagaku Zasshi*, **65,** 1168 (1962).
8. E. G. Rochow and I. Kukin, *J. Am. Chem. Soc.*, **74,** 1615 (1952).
9. *British Patent*, 660,522 (1951).
10. A. G. Sharpe and H. J. Emeleus, *J. Chem. Soc.*, **1948,** 2135.
11. D. F. Stewart, unpublished observation.
12. G. Fuller, M. Stacey, J. C. Tatlow and C. R. Thomas, *Tetrahedron*, **18,** 123 (1962).
13. M. Stacey and J. C. Tatlow, *Advances in Fluorine Chemistry*, Vol. 1, Butterworths, London, 1960.
14. V. S. Plashkin, G. P. Tataurov and S. V. Sokolov, *J. Gen. Chem. USSR*, **36,** 1705 (1966).
15. D. A. Rausch, R. A. Davis and D. W. Osborne, *J. Org. Chem.*, **28,** 494 (1963).
16. P. L. Coe, B. T. Croll and C. R. Patrick, *Tetrahedron*, **20,** 2097 (1964).
17. R. D. Fowler, H. C. Anderson, J. M. Hamilton, W. B. Burford, A. Spadetti, S. B. Bitterlich and I. Litant, *Ind. Eng. Chem.*, **39,** 343 (1947).
18. R. G. Benner, A. F. Benning, F. B. Downing, C. F. Irwin, K. C. Johnson, A. L. Linch, H. M. Parmelee and W. V. Wirth, *Ind. Eng. Chem.*, **39,** 329 (1947).
19. W. T. Miller, A. L. Dittman, R. L. Ehrenfeld and A. M. Prober, *Ind. Eng. Chem.*, **39,** 333 (1947).
20. E. A. Belmore, W. M. Ewalt and B. H. Wojcik, *Ind. Eng. Chem.*, **39,** 338 (1947).
21. P. Henkel and W. Klemm, *Z. anorg. allgem. Chem.*, **222,** 74 (1935).
22. G. A. Barbieri and F. Calzolari, *Z. anorg. allgem. Chem.*, **170,** 109 (1928).

23. H. C. Clark, B. Cox and A. G. Sharpe, *J. Chem. Soc.*, **1957,** 4132.
24. I. Maak, P. Eckerlin and A. Rabenau, *Naturwissenschaften*, **48,** 218 (1964).
25. R. G. Cunningham, R. S. Nyholm and M. L. Tobe, *J. Chem. Soc.*, **1964,** 5800.
26. G. A. Barclay and R. S. Nyholm, *Chem. Ind. (London)*, **1953,** 378.
27. A. A. Vlcek, *Inorg. Chem.*, **6,** 1425 (1967).
28. B. Bosnich, C. K. Poon and M. L. Tobe, *Inorg. Chem.*, **4,** 1102 (1965).
29. B. Bosnich, R. D. Gillard, E. D. McKenzie and G. A. Webb, *J. Chem. Soc.*, **1966,** A, 1331.
30. F. Basolo, *J. Am. Chem. Soc.*, **70,** 2634 (1948).
31. J. C. Bailar and J. B. Work, *J. Am. Chem. Soc.*, **68,** 232 (1946).
32. R. G. Pearson, R. E. Meeker and F. Basolo, *J. Inorg. Nucl. Chem.*, **1,** 341 (1955).
33. S. Ooi, Y. Komiyama, Y. Saito and H. Kuroya, *Bull. Chem. Soc. Japan*, **32,** 263 (1959).
34. T. B. Jackson and J. O. Edwards, *Inorg. Chem.*, **1,** 398 (1962).
35. S. C. Chan, *J. Chem. Soc.*, **1966,** A, 142.
36. R. G. Pearson, R. E. Meeker and F. Basolo, *J. Am. Chem. Soc.*, **78,** 709 (1956).
37. F. Basolo, *J. Am. Chem. Soc.*, **75,** 227 (1953).
38. S. C. Chan and M. L. Tobe, *J. Chem. Soc.*, **1962,** 4531.
39. G. Morgan and F. H. Burstall, *J. Chem. Soc.*, **1937,** 1649.
40. M. L. Morris and D. H. Busch, *J. Am. Chem. Soc.*, **82,** 1521 (1960).
41. J. B. Work, *Inorg. Syn.*, **2,** 221 (1946).
42. J. C. Bailar, *Inorg. Syn.*, **2,** 223 (1946).
43. G. J. Sutton, *Australian J. Chem.*, **13,** 473 (1960).
44. W. R. Fitzgerald and D. W. Watts, *Australian J. Chem.*, **21,** 595 (1968).
45. W. R. Fitzgerald and D. W. Watts, *Australian J. Chem.*, **19,** 935 (1966).
46. R. D. Gillard and G. Wilkinson, *J. Chem. Soc.*, **1963,** 3193.
47. E. W. Gillow and G. M. Harris, *Inorg. Chem.*, **7,** 394 (1968).
48. W. A. Hynes, L. K. Yanowski and M. Shiller, *J. Am. Chem. Soc.*, **60,** 3053 (1938).
49. F. Basolo and R. G. Pearson, *Mechanisms of Inorganic Reactions*, John Wiley, New York, 1967.
50. R. D. Gillard, *Progress In Inorganic Chemistry* (Ed. F. A. Cotton), Vol. 7, John Wiley, New York, 1966.
51. J. H. Dunlop and R. D. Gillard, *Advances in Inorganic Chemistry and Radiochemistry* (Ed. H. J. Emeleus and A. G. Sharpe), Vol. 9, Academic Press, London, 1966.
52. J. Lewis and R. G. Wilkins, *Modern Coordination Chemistry*, Interscience, New York, 1960.
53. Y. Shiegeta, Y. Komiyama and H. Kuroya, *Bull. Chem. Soc. Japan*, **36,** 1159 (1963).
54. K. A. Becker, G. Grosse and K. Plieth, *Z. Krist.*, **112,** 375 (1959).
55. A. Nakahara, Y. Saito and H. Kuroya, *Bull. Chem. Soc. Japan*, **25,** 331 (1952).
56. Y. Saito and H. Iwasaki, *Bull. Chem. Soc. Japan*, **35,** 1131 (1962).
57. A. V. Ablov, A. Y. Kon and T. I. Malinovskii, *Dokl. Akad. Nauk SSSR*, **167,** 1051 (1966).

58. K. A. Jensen, B. Nygaard and C. T. Pedersen, *Acta Chem. Scand.*, **17,** 1126 (1963).
59. K. Issleib and B. Mitscherling, *Z. anorg. allgem. Chem.*, **304,** 73 (1960).
60. C. E. Whymore and J. C. Bailar, *J. Inorg. Nucl. Chem.*, **14,** 42 (1960).
61. G. Braur, *Handbook of Preparative Inorganic Chemistry*, 2nd ed., Academic Press, New York, 1963.
62. E. L. Muetterties and J. E. Castle, *J. Inorg. Nucl. Chem.*, **18,** 148 (1961).
63. *U.S. Patent*, 2,952,514 (1960).
64. E. E. Aynsley, G. Hetherington and P. L. Robinson, *J. Chem. Soc.*, **1954,** 1119.
65. N. N. Mikhailov and S. V. Petrov, *Kristallografiya*, **11,** 443 (1966).
66. A. Kurtenacker, W. Finger and F. Hey, *Z. anorg. allgem. Chem.*, **211,** 83 (1933).
67. W. Rudorff, G. Lincke and D. Babel, *Z. anorg. allgem. Chem.*, **320,** 150 (1963).
68. G. L. Clark and K. K. Bukner, *J. Am. Chem. Soc.*, **44,** 230 (1922).
69. J. S. Binford, J. M. Strohmenger and T. H. Hebert, *J. Phys. Chem.*, **71,** 2404 (1967).
70. J. W. Stout and E. Catalano, *J. Chem. Phys.*, **23,** 2103 (1955).
71. A. Buchler, J. L. Stauffer and W. Klemperer, *J. Chem. Phys.*, **40,** 3471 (1964).
72. J. L. Margrave, A. S. Kana'an and G. Besenbruch, *J. Inorg. Nucl. Chem.*, **28,** 1035 (1966).
73. W. H. Baur, *Acta Cryst.*, **11,** 488 (1958).
74. W. H. Baur, *Naturwissenschaften*, **44,** 349 (1957).
75. T. Moriya, K. Motizuki, J. Kanamori and T. Nagamiya, *J. Phys. Soc. Japan*, **11,** 211 (1956).
76. M. Balkanski, P. Moch and G. Parisot, *J. Chem. Phys.*, **44,** 940 (1966).
77. D. N. Astrov, S. I. Novikova and M. P. Orlova, *Zh. Eksptl. i Theoret. Fiz.*, **37,** 1197 (1959).
78. S. Foner, *J. Phys. Radium*, **20,** 336 (1959).
79. H. Bizette and B. Tsai, *Bull. Inst. Intern. Froid Annexe*, **1955,** 149.
80. D. N. Astrov, A. S. Borovik-Romanov and M. P. Orlova, *Soviet Phys.-JETP*, **6,** 626 (1958).
81. P. L. Richards, *J. Appl. Phys.*, **35,** 850 (1964).
82. R. A. Erickson, *Phys. Rev.*, **90,** 779 (1953).
83. W. J. de Haas and B. H. Schultz, *Physica*, **6,** 481 (1939).
84. M. E. Lines, *Phys. Rev.*, **137,** 982 (1965).
85. J. W. Stout and L. M. Matarrese, *Rev. Modern Phys.*, **25,** 338 (1953).
86. E. Konig, *Landolt-Bornstein, Magnetic Properties of Coordination and Organometallic Transition Metal Compounds*, Vol. 2, Springer-Verlag, New York, 1966.
87. E. Catalano and J. W. Stout, *J. Chem. Phys.*, **23,** 1803 (1955).
88. J. W. Stout and E. Catalano, *Phys. Rev.*, **92,** 1575 (1953).
89. G. A. Slack, *Phys. Rev.*, **122,** 1451 (1961).
90. P. L. Richards, *J. Appl. Phys.*, **34,** 1237 (1963).
91. H. Kamimura, *J. Appl. Phys.*, **35,** 844 (1964).
92. A. Ludi and W. Feitnecht, *Helv. Chim. Acta*, **46,** 2226 (1963).
93. R. C. Osthoff and R. C. West, *J. Am. Chem. Soc.*, **76,** 4732 (1954).

94. W. E. Hatfield and J. T. Yoke, *Inorg. Chem.*, **1,** 463 (1962).
95. A. R. Pray, *Inorg. Syn.*, **5,** 153 (1957).
96. D. Khristov, S. Karaivanov and V. Koluschki, *Godishnik Sofiskyia Univ., Khim. Fak.*, **55,** 49 (1960–61).
97. G. W. Watt, P. S. Gentile and E. P. Helvenston, *J. Am. Chem. Soc.*, **77,** 2752 (1955).
98. Y. I. Ivashentsev, *Tr. Tomskogo Gos. Univ., Ser. Khim.*, **157,** 77 (1963).
99. R. C. Schoonmaker, A. H. Friedman and R. F. Porter, *J. Chem. Phys.*, **31,** 1586 (1959).
100. H. Schafer, L. Bayer, G. Briel, K. Etzel and K. Krehl, *Z. anorg. allgem. Chem.*, **278,** 300 (1959).
101. C. G. Maier, *Bur. of Mines Tech. Paper*, **1925,** 360.
102. A. Ferrari, A. Braibanti and G. Bigliardi, *Acta Cryst.*, **16,** 846 (1963).
103. J. R. Ferraro and A. Walker, *J. Chem. Phys.*, **42,** 1278 (1965).
104. G. E. Leroi, T. C. James, J. T. Hougen and W. Klemperer, *J. Chem. Phys.*, **36,** 2879 (1962).
105. J. W. Leech and A. J. Manuel, *Proc. Phys. Soc.*, **69B,** 210 (1956).
106. C. Starr, F. Bitter and A. R. Kaufmann, *Phys. Rev.*, **58,** 977 (1940).
107. R. Stevenson, *Can. J. Phys.*, **40,** 1385 (1962).
108. R. T. Mina, *Soviet Phys.-JETP*, **13,** 911 (1961).
109. H. Bizette, C. Terrier and B. Tsai, *Compt. Rend.*, **243,** 1295 (1956).
110. R. C. Chisholm and J. W. Stout, *J. Chem. Phys.*, **36,** 972 (1962).
111. M. E. Lines, *Phys. Rev.*, **131,** 546 (1963).
112. M. K. Wilkinson, J. W. Cable, E. O. Wollan and W. C. Koehler, *Phys. Rev.*, **113,** 497 (1959).
113. J. W. Leech and A. J. Manuel, *Proc. Phys. Soc.*, **69B,** 220 (1956).
114. J. Kanamori, *Prog. Theoret. Phys.*, **20,** 890 (1958).
115. J. Ferguson, D. L. Wood and K. Knox, *J. Chem. Phys.*, **39,** 881 (1963).
116. A. Trutia and M. Musa, *Phys. Status Solidi*, **8,** 663 (1965).
117. J. T. Hougen, G. E. Leroi and T. C. James, *J. Chem. Phys.*, **34,** 1670 (1961).
118. C. W. DeKock and D. M. Gruen, *J. Chem. Phys.*, **44,** 4387 (1966).
119. A. Trutia and M. Musa, *Spectrochim. Acta*, **23A,** 1165 (1967).
120. L. I. Katzin, J. R. Ferraro and E. Gebert, *J. Am. Chem. Soc.*, **72,** 5471 (1950).
121. E. G. Cox, A. J. Shorter, W. Wardlaw and W. R. J. Way, *J. Chem. Soc.*, **1937,** 1556.
122. N. S. Gill, R. S. Nyholm, G. A. Barclay, T. I. Christie and P. J. Pauling, *J. Inorg. Nucl. Chem.*, **18,** 88 (1961).
123. J. R. Allan, D. H. Brown, R. H. Nuttall and D. W. A. Sharp, *J. Inorg. Nucl. Chem.*, **26,** 1895 (1964).
124. I. G. Murgulescu, E. Segal and D. Fatu, *J. Inorg. Nucl. Chem.*, **27,** 2677 (1965).
125. M. A. Porai-Koshits and A. S. Antsishkina, *Dokl. Akad. Nauk SSSR*, **92,** 333 (1953).
126. M. A. Porai-Koshits and A. S. Antsishkina, *Izv. Sektora Platiny i Drug. Blagrod. Metal., Inst. Obshchei i Neorg. Khim., Akad. Nauk SSSR*, **1955,** 19.
127. M. A. Porai-Koshits, *Trudy Inst. Krist., Akad. Nauk SSSR*, **10,** 117 (1954).
128. R. J. H. Clark and C. S. Williams, *Inorg. Chem.*, **4,** 350 (1965).

129. E. D. P. Barkworth and S. Sugden, *Nature*, **139**, 374 (1937).
130. D. P. Mellor and C. R. Coryell, *J. Am. Chem. Soc.*, **60**, 1786 (1938).
131. J. D. Dunitz, *Acta Cryst.*, **10**, 307 (1957).
132. M. A. Porai-Koshits, L. O. Atovmyan and G. N. Tishchenko, *Zh. Strukt. Khim.*, **1**, 337 (1960).
133. E. Ferroni and E. Bondi, *J. Inorg. Nucl. Chem.*, **8**, 458 (1958).
134. R. Zannetti and R. Serra, *Gazz. Chim. Ital.*, **90**, 328 (1960).
135. C. W. Frank and L. B. Rogers, *Inorg. Chem.*, **5**, 615 (1966).
136. R. J. H. Clark and C. S. Williams, *Chem. Ind.* (*London*), **1964**, 1317.
137. J. Bradbury, K. P. Forrest, R. H. Nuttall and D. W. A. Sharp, *Spectrochim. Acta*, **23A**, 2701 (1967).
138. N. S. Gill, R. H. Nuttall, D. E. Scaife and D. W. A. Sharp, *J. Inorg. Nucl. Chem.*, **18**, 79 (1961).
139. E. Konig and H. L. Schlaffer, *Z. Physik. Chem.* (*Frankfurt*), **26**, 371 (1960).
140. A. B. P. Lever and S. M. Nelson, *J. Chem. Soc.*, **1966**, A, 859.
141. D. H. Brown, R. H. Nuttall and D. W. A. Sharp, *J. Inorg. Nucl. Chem.*, **26**, 1151 (1964).
142. L. I. Katzin and E. Gebert, *J. Am. Chem. Soc.*, **75**, 2830 (1953).
143. H. C. A. King, E. Koros and S. M. Nelson, *J. Chem. Soc.*, **1963**, **5449**.
144. H. C. A. King, E. Koros and S. M. Nelson, *Nature*, **196**, 572 (1962).
145. W. Libus and I. Uruska, *Inorg. Chem.*, **5**, 256 (1966).
146. L. R. Ocone, J. R. Soulen and B. P. Block, *J. Inorg. Nucl. Chem.*, **15**, 76, (1960).
147. G. Beech, S. J. Ashcroft and C. T. Mortimer, *J. Chem. Soc.*, **1967**, A, 929.
148. G. Beech, C. T. Mortimer and E. G. Tyler, *J. Chem. Soc.*, **1967**, A, 925.
149. J. deO. Cabral, H. C. A. King, S. M. Nelson, T. M. Shepard and E. Koros, *J. Chem. Soc.*, **1966**, A, 1348.
150. H. C. A. King, E. Koros and S. M. Nelson, *J. Chem. Soc.*, **1964**, 4832.
151. T. I. Malinovskii, *Kristallografiya*, **2**, 734 (1957).
152. G. B. Boki, T. I. Malinovskii and A. V. Ablov, *Kristallografiya*, **1**, 49 (1956).
153. I. S. Ahuja, D. H. Brown, R. H. Nuttall and D. W. A. Sharp, *J. Inorg. Nucl. Chem.*, **27**, 1105 (1965).
154. D. P. Graddon and E. C. Watton, *Australian J. Chem.*, **18**, 507 (1965).
155. N. H. Agnew and L. F. Larkworthy, *J. Chem. Soc.*, **1965**, 4669.
156. K. C. Dash and D. V. R. Rao, *Indian J. Chem.*, **5**, 569 (1967).
157. N. S. Gill and H. J. Kingdon, *Australian J. Chem.*, **19**, 2197 (1966).
158. J. R. Allan, D. H. Brown, R. H. Nuttall and D. W. A. Sharp, *J. Inorg. Nucl. Chem.*, **27**, 1305 (1965).
159. C. Postmus, K. Nakamoto and J. R. Ferraro, *Inorg. Chem.*, **6**, 2194 (1967).
160. W. R. McWhinnie, *J. Inorg. Nucl. Chem.*, **27**, 2573 (1965).
161. D. E. Billing and A. E. Underhill, *J. Chem. Soc.*, **1968**, A, 29.
162. D. P. Graddon, K. B. Heng and E. C. Watton, *Australian J. Chem.*, **21**, 121 (1968).
163. I. S. Ahuja, D. H. Brown, R. H. Nuttall and D. W. A. Sharp, *J. Inorg. Nucl. Chem.*, **27**, 1625 (1965).
164. R. H. Holm and F. A. Cotton, *J. Chem. Phys.*, **32**, 1168 (1960).
165. I. S. Ahuja, *J. Inorg. Nucl. Chem.*, **29**, 2091 (1967).
166. D. E. Billing and A. E. Underhill, *J. Chem. Soc.*, **1968**, A, 5.

167. P. S. K. Chia, S. E. Livingstone and T. N. Lockyer, *Australian J. Chem.*, **20,** 239 (1967).
168. R. H. Lee, E. Griswold and J. Kleinberg, *Inorg. Chem.*, **3,** 1278 (1964).
169. R. J. H. Clark and C. S. Williams, *Spectrochim. Acta*, **23A,** 1055 (1967).
170. J. W. Mellor, *A Comprehensive Treatise on Inorganic and Theoretical Chemistry*, Vol. 14, Longmans, London, p. 631.
171. W. Klemm and W. Schuth, *Z. anorg. allgem. Chem.*, **210,** 33 (1933).
172. R. J. H. Clark and C. S. Williams, *J. Chem. Soc.*, **1966,** A, 1425.
173. A. Sacco and M. Freni, *Gazz. Chim. Ital.*, **89,** 1800 (1959).
174. A. Sacco and F. A. Cotton, *J. Am. Chem. Soc.*, **84,** 2043 (1962).
175. B. J. Hathaway and D. G. Holah, *J. Chem. Soc.*, **1964,** 2400.
176. F. A. Cotton and R. H. Holm, *J. Am. Chem. Soc.*, **82,** 2983 (1960).
177. M. Ciampolini and G. P. Speroni, *Inorg. Chem.*, **5,** 45 (1966).
178. M. Di Vaira and P. L. Orioli, *Chem. Commun.*, **1965,** 590.
179. G. A. Barclay and A. K. Barnard, *J. Chem. Soc.*, **1958,** 2540.
180. Z. Dori and H. B. Gray, *J. Am. Chem. Soc.*, **88,** 1394 (1966).
181. Z. Dori, R. Eisenberg and H. B. Gray, *Inorg. Chem.*, **6,** 483 (1967).
182. I. G. Dance, M. Gerloch, J. Lewis, F. S. Stephens and F. Lions, *Nature*, **210,** 298 (1966).
183. M. Gerloch, *J. Chem. Soc* , **1966,** A, 1317.
184. F. A. Cotton, O. D. Faut and J. T. Mague, *Inorg. Chem.*, **3,** 17 (1964).
185. D. M. Adams and J. B. Cornell, *J. Chem. Soc.*, **1967,** A, 884.
186. C. D. Flint and M. Goodgame, *J. Chem. Soc.*, **1966,** A, 744.
187. M. Nardelli, I. Chierci and A. Braibanti, *Gazz. Chim. Ital.*, **88,** 37 (1958).
188. R. L. Carlin and S. L. Holt, *Inorg. Chem.*, **2,** 849 (1963).
189. G. Booth, *Advances in Inorganic Chemistry and Radiochemistry* (Ed. H. J. Emeleus and A. G. Sharpe), Vol. 6, Academic Press, London, 1964.
190. L. M. Venanzi, *Angew. Chem., Intern. Ed. Engl.*, **3,** 453 (1964).
191. K. A. Jensen, P. H. Nielsen and C. T. Pedersen, *Acta Chem. Scand.*, **17,** 1115 (1963).
192. W. E. Hatfield and J. T. Yoke, *Inorg. Chem.*, **1,** 470 (1962).
193. W. E. Hatfield and J. T. Yoke, *Inorg. Chem.*, **1,** 475 (1962).
194. F. A. Cotton, O. D. Faut, D. M. L. Goodgame and R. H. Holm, *J. Am. Chem. Soc.*, **83,** 1780 (1961).
195. M. C. Browning, R. F. B. Davies, D. J. Morgan, L. E. Sutton and L. M. Venanzi, *J. Chem. Soc.*, **1961,** 4816.
196. D. M. L. Goodgame and M. Goodgame, *Inorg. Chem.*, **4,** 139 (1965).
197. K. Issleib and G. Doll, *Z. anorg. allgem. Chem.*, **305,** 1 (1960).
198. K. Issleib and G. Hohlfeld, *Z. anorg. allgem. Chem.*, **312,** 169 (1961).
199. K. Issleib and G. Schwager, *Z. anorg. allgem. Chem.*, **311,** 83 (1961).
200. J. W. Collier and F. G. Mann, *J. Chem. Soc.*, **1964,** 1815.
201. R. H. Richard and J. Kenyon, *J. Chem. Soc.*, **89,** 262 (1906).
202. F. A. Cotton, R. D. Barnes and E. Bannister, *J. Chem. Soc.*, **1960,** 2199.
203. L. Sacconi and J. Gelsomini, *Inorg. Chem.*, **7,** 291 (1968).
204. G. A. Rodley, D. M. L. Goodgame and F. A. Cotton, *J. Chem. Soc.*, **1965,** 1499.
205. D. M. L. Goodgame, M. Goodgame and F. A. Cotton, *Inorg. Chem.*, **1,** 239 (1962).
206. D. J. Phillips and S. Y. Tyree, *J. Am. Chem. Soc.*, **83,** 1806 (1961).

207. A. Merijanian and R. A. Zingaro, *Inorg. Chem.*, **5,** 187 (1966).
208. J. V. Quagliano, J. Fujita, G. Franz, D. J. Phillips, J. A. Walmsley and S. Y. Tyree, *J. Am. Chem. Soc.*, **83,** 3770 (1961).
209. H. N. Ramaswamy and H. B. Jonassen, *Inorg. Chem.*, **4,** 1595 (1965).
210. H. N. Ramaswamy and H. B. Jonassen, *J. Inorg. Nucl. Chem.*, **27,** 740 (1965).
211. J. T. Summers and J. V. Quagliano, *Inorg. Chem.*, **3,** 1767 (1964).
212. A. B. P. Lever, J. Lewis and R. S. Nyholm, *J. Chem. Soc.*, **1962,** 1235.
213. G. Beech and C. T. Mortimer, *J. Chem. Soc.*, **1967,** A, 1115.
214. A. B. P. Lever, J. Lewis and R. S. Nyholm, *Nature*, **189,** 58 (1961).
215. P. S. K. Chia and S. E. Livingstone, *Australian J. Chem.*, **21,** 339 (1968).
216. D. B. Fox, J. R. Hall and R. A. Plowman, *Australian J. Chem.*, **15,** 235 (1962).
217. W. R. McWhinnie, *J. Inorg. Nucl. Chem.*, **27,** 1619 (1965).
218. M. Goodgame, *J. Chem. Soc.*, **1966,** A, 63.
219. L. Sacconi, I. Bertini and F. Mani, *Inorg. Chem.*, **6,** 262 (1967).
220. M. Ciampolini and N. Nardi, *Inorg. Chem.*, **5,** 41 (1966).
221. M. Ciampolini and N. Nardi, *Inorg. Chem.*, **6,** 445 (1967).
222. M. Ciampolini and J. Gelsomini, *Inorg. Chem.*, **6,** 1821 (1967).
223. L. Sacconi, I. Bertini and R. Morassi, *Inorg. Chem.*, **6,** 1548 (1967).
224. L. Sacconi and G. P. Speroni, *Inorg. Chem.*, **7,** 295 (1968).
225. R. K. Y. Ho and S. E. Livingstone, *Australian J. Chem.*, **18,** 659 (1965).
226. D. Nicholls, M. Rowley and R. Swindells, *J. Chem. Soc.*, **1966,** A, 950.
227. M. Brierley and W. J. Geary, *J. Chem. Soc.*, **1967,** A, 963.
228. W. J. Eilbeck, F. Holmes and A. E. Underhill, *J. Chem. Soc.*, **1967,** A, 757.
229. W. J. Eilbeck, F. Holmes, C. E. Taylor and A. E. Underhill, *J. Chem. Soc.*, **1968,** A, 128.
230. D. C. Goodall, *J. Chem. Soc.*, **1967,** A, 203.
231. R. J. Berni, R. R. Benerito, W. M. Ayres and H. B. Jonassen, *J. Inorg. Nucl. Chem.*, **25,** 807 (1963).
232. W. E. Bull, S. K. Madan and J. E. Willis, *Inorg. Chem.*, **2,** 303 (1963).
233. E. Uhlig and M. Maaser, *Z. anorg. allgem. Chem.*, **344,** 205 (1964).
234. J. T. Donoghue and R. S. Drago, *Inorg. Chem.*, **2,** 572 (1963).
235. R. Chand, G. S. Handard and K. Lal, *J. Indian Chem. Soc.*, **35,** 28 (1958).
236. S. Herzog, K. Gustav, E. Krueger, H. Oberenda and R. Schuster, *Z. Chem.*, **3,** 428 (1963).
237. A. Y. Deich and V. S. Nasonov, *Zh. Neorg. Khim.*, **4,** 1198 (1959).
238. R. Juhasz and L. F. Yntema, *J. Am. Chem. Soc.*, **62,** 3522 (1940).
239. M. Nardelli, A. Braibanti and I. Chierici, *Gazz. Chim. Ital.*, **87,** 1226 (1957).
240. E. Loyd, C. B. Brown, D. Glynwyn, R. Bonnell and W. J. Jones, *J. Chem. Soc.*, **1928,** 658.
241. L. Sacconi and A. Sabatini, *J. Inorg. Nucl. Chem.*, **25,** 1389 (1965).
242. A. Hantzsch, *Z. anorg. allgem. Chem.*, **159,** 273 (1927).
243. J. D. Curry, M. A. Robinson and D. H. Busch, *Inorg. Chem.*, **6,** 1570 (1967).
244. C. H. Langford and P. O. Langford, *Inorg. Chem.*, **1,** 184 (1962).
245. R. Gomer and C. N. Tyson, *J. Am. Chem. Soc.*, **66,** 2685 (1944).
246. E. Uhlig, P. Schueler and D. Diehlman, *Z. anorg. allgem. Chem.*, **335,** 156 (1965).

247. R. J. Kern, *J. Inorg. Nucl. Chem.*, **25,** 5 (1963).
248. H. J. Borchardt and F. Daniels, *J. Phys. Chem.*, **61,** 917 (1957).
249. C. S. Shen and M. H. Chang, *Hua Hsueh Husch Pao*, **26,** 124 (1960).
250. J. Mizuno, *J. Phys. Soc. Japan*, **15,** 1412 (1960).
251. J. Mizuno, K. Ukei and T. Sugawara, *J. Phys. Soc., Japan*, **14,** 383 (1959).
252. E. V. Strogonov, I. I. Kozhina and S. N. Andreev, *Vestn. Leningrad. Univ.*, **13,** *Ser. Fiz. i Khim.*, 109 (1958).
253. T. Sugawara, *J. Phys. Soc. Japan*, **14,** 1248 (1959).
254. Z. M. El Saffar, *J. Phys. Soc. Japan*, **17,** 1334 (1962).
255. I. Gamo, *Bull. Chem. Soc. Japan*, **34,** 1430 (1961).
256. W. van der Lugt and N. J. Poulis, *Physica*, **26,** 917 (1960).
257. E. Sawatzky and M. Bloom, *Phys. Letters*, **2,** 28 (1962).
258. R. B. Flippen and S. A. Friedberg, *J. Appl. Phys.*, **31,** 338S (1960).
259. D. S. Sahri, *Phys. Letters*, **19,** 625 (1965).
260. J. Skalyo, A. F. Cohen, S. A. Friedberg and R. B. Griffiths, *Phys. Rev.*, **164,** 705 (1967).
261. E. Sawatzky and M. Bloom, *J. Phys. Soc. Japan*, **17,** 507S (1962).
262. T. Haseda and E. Kanda, *J. Phys. Soc. Japan*, **12,** 1051 (1957).
263. J. Skalyo and S. A. Friedberg, *Phys. Rev. Letters*, **13,** 133 (1964).
264. W. K. Robinson and S. A. Friedberg, *Phys. Rev.*, **117,** 402 (1960).
265. R. D. Spence, P. Middents, Z. El Saffar and R. Kleinberg, *J. Appl. Phys.*, **35,** 854 (1964).
266. M. Date, *J. Phys. Soc. Japan*, **14,** 1244 (1959).
267. B. Morosin and E. J. Graebner, *Acta Cryst.*, **16,** 1176 (1963).
268. K. E. Lawson, *J. Chem. Phys.*, **44,** 4159 (1966).
269. B. Morosin, *J. Chem. Phys.*, **44,** 252 (1966).
270. B. K. Vainshtein, *Dokl. Akad. Nauk SSSR*, **68,** 301 (1949).
271. P. van Tiggelen, J. M. Dereppe and M. van Meerssche, *J. Chim. Phys.*, **59,** 1042 (1962).
272. H. Kobayashi and T. Haseda, *J. Phys. Soc. Japan*, **19,** 765 (1964).
273. A. Narath, *Phys. Letters*, **13,** 12 (1964).
274. T. Shinoda, H. Chihara and S. Seki, *J. Phys. Soc. Japan*, **19,** 1637 (1964).
275. T. Shinoda, H. Chihara and S. Seki, *J. Phys. Soc. Japan*, **18,** 1088 (1964).
276. A. Narath, *J. Phys. Soc. Japan*, **19,** 2244 (1964).
277. A. Narath, *Phys. Rev.*, **136,** 766 (1964).
278. A. Narath, *Phys. Rev.*, **140,** 552 (1965).
279. T. J. Wydeven and N. W. Gregory, *J. Phys. Chem.*, **68,** 3249 (1964).
280. G. Devoto and A. Guzzi, *Gazz. Chim. Ital.*, **59,** 591 (1929).
281. A. Ferrari and F. Giorgi, *Atti Acad. Lincei*, **9,** 1134 (1929).
282. L. G. van Uitert, H. J. Williams, R. C. Sherwood and J. J. Rubin, *J. Appl. Phys.*, **36,** 1029 (1965).
283. H. Bizette, C. Terrier and T. Belling, *Compt. Rend.*, **246,** 250 (1958).
284. A. S. Antsishkina and M. A. Porai-Koshits, *Kristallografiya*, **3,** 676 (1958).
285. D. P. Graddon, R. Schulz, E. C. Watton and D. G. Weeden, *Nature*, **198,** 1299 (1963).
286. M. Goodgame and P. J. Hayward, *J. Chem. Soc.*, **1966,** A, 632.
287. A. V. Ablov, Z. P. Burnasheva and E. G. Levitskaya, *Zh. Neorg. Khim.*, **1,** 2465 (1956).

288. J. W. Mellor, *A Comprehensive Treatise on Inorganic and Theoretical Chemistry*, Vol. 14, Longmans, London, p. 715.
289. M. Di Vaira and P. L. Orioli, *Inorg. Chem.*, **6,** 955 (1967).
290. K. Issleib and E. Wenschuh, *Z. anorg. allgem. Chem.*, **305,** 15 (1960).
291. J. A. Bertrand and D. L. Plymale, *Inorg. Chem.*, **5,** 879 (1966).
292. W. D. Horrocks and G. N. La Mar, *J. Am. Chem. Soc.*, **85,** 3512 (1963).
293. G. N. La Mar, W. D. Horrocks and L. C. Allen, *J. Chem. Phys.*, **41,** 2126 (1964).
294. K. Issleib and A. Brack, *Z. anorg. allgem. Chem.*, **277,** 258 (1954).
295. P. Paoletti and M. Ciampolini, *Inorg. Chem.*, **6,** 64 (1967).
296. P. J. Crowley and H. M. Haendler, *Inorg. Chem.*, **1,** 904 (1962).
297. H. Rheinboldt, A. Luyken and H. Schmittmann, *J. Prakt. Chem.*, **149,** 30 (1937).
298. M. Schafer and C. Curran, *Inorg. Chem.*, **5,** 265 (1966).
299. K. Issleib and A. Krebich, *Z. anorg. allgem. Chem.*, **313,** 338 (1961).
300. M. Garber, *J. Phys. Soc. Japan*, **15,** 734 (1960).
301. H. Forstat, G. Taylor and R. D. Spence, *Phys. Rev.*, **116,** 897 (1959).
302. B. Morosin, *J. Chem. Phys.*, **47,** 417 (1967).
303. J. W. Mellor, *A Comprehensive Treatise on Inorganic and Theoretical Chemistry*, Vol. 14, Longmans, London, p. 737.
304. M. Chaigneau, *Bull. Soc. Chim. France*, **1957,** 886.
305. M. Chaigneau, *Compt. Rend.*, **242,** 263 (1956).
306. M. Chaigneau and M. Chastagnier, *Bull. Soc. Chim. France*, **1958,** 1192.
307. V. V. Pechkovskii and A. V. Sofronova, *Russ. J. Inorg. Chem.*, **11,** 828 (1966).
308. A. Ferrari and F. Giorgi, *Atti Acad. Lincei*, **10,** 522 (1929).
309. S. M. Nelson and T. M. Shepard, *J. Chem. Soc.*, **1965,** 3284.
310. T. I. Malinovskii, *Kristallografiya*, **3,** 364 (1958).
311. G. N. La Mar, *J. Phys. Chem.*, **69,** 3212 (1965).
312. F. A. Cotton and R. Francis, *J. Am. Chem. Soc.*, **82,** 2986 (1960).
313. F. A. Cotton and R. Francis, *J. Inorg. Nucl. Chem.*, **17,** 62 (1961).
314. F. A. Cotton, R. Francis and W. D. Horrocks, *J. Phys. Chem.*, **64,** 1534 (1960).
315. M. Ciampolini and P. Paoletti, *Inorg. Chem.*, **6,** 1261 (1967).
316. S. A. Shchukarev, E. V. Stroganov, S. N. Andreev and O. F. Purvinskii, *Zh. Strukt. Khim.*, **4,** 63 (1963).
317. R. Hoppe, *Rec. Trav. Chim.*, **75,** 569 (1956).
318. J. T. Grey, *J. Am. Chem. Soc.*, **68,** 605 (1946).
319. W. Klemm and E. Huss, *Z. anorg. allgem. Chem.*, **258,** 221 (1949).
320. M. D. Meyers and F. A. Cotton, *J. Am. Chem. Soc.*, **82,** 5027 (1960).
321. R. D. Peacock and D. W. A. Sharp, *J. Chem. Soc.*, **1959,** 2762.
322. F. A. Cotton and M. D. Meyers, *J. Am. Chem. Soc.*, **82,** 5033 (1960).
323. H. G. Schnering, *Z. anorg. allgem. Chem.*, **353,** 1 (1967).
324. H. G. Schnering, *Z. anorg. allgem. Chem.*, **353,** 13 (1967).
325. W. Rudorff, J. Kandler, G. Lincke and D. Babel, *Angew. Chem.*, **71,** 672 (1959)
326. D. S. Crocket and R. A. Grossman, *Inorg. Chem.*, **3,** 644 (1964).
327. K. G. Srivastava, *Phys. Letters*, **4,** 55 (1963).

328. H. M. Haendler, F. A. Johnson and D. S. Crocket, *J. Am. Chem. Soc.*, **80,** 2662 (1958).
329. D. S. Crocket and H. M. Haendler, *J. Am. Chem. Soc.*, **82,** 4158 (1960).
330. K. Hirakawa, K. Hirakawa and T. Hashimoto, *J. Phys. Soc. Japan*, **15,** 2063 (1960).
331. D. J. Machin, R. L. Martin and R. S. Nyholm, *J. Chem. Soc.*, **1963,** 1490.
332. D. J. Machin and R. S. Nyholm, *J. Chem. Soc.*, **1963,** 1500.
333. R. V. Pisarev, A. I. Belyaeva and P. P. Syrnikov, *Soviet Phys.-Solid State*, **8,** 627 (1966).
334. N. N. Nesterova, I. G. Sinii, R. V. Pisarev and P. P. Syrnikov, *Soviet Phys.-Solid State*, **9,** 15 (1967).
335. K. Knox, *Acta Cryst.*, **14,** 583 (1961).
336. A. Okazaki, Y. Suemune and T. Fuchikami, *J. Phys. Soc. Japan*, **14,** 1823 (1959).
337. R. L. Martin, R. S. Nyholm and N. C. Stephenson, *Chem. Ind. (London)*, **1956,** 83.
338. A. Okazaki and Y. Suemune, *J. Phys. Soc. Japan*, **16,** 671 (1961).
339. C. Deenadas, H. V. Kerr, R. V. Gopala Rao and A. B. Biswas, *British J. Appl. Phys.*, **18,** 1833 (1967).
340. Y. Suemune and H. Ikawa, *J. Phys. Soc. Japan*, **19,** 1686 (1964).
341. V. Scatturin, L. Corliss, N. Elliot and J. Hastings, *Acta Cryst.*, **14,** 19 (1961).
342. H. J. Seifert, *Z. anorg. allgem. Chem.*, **307,** 137 (1960).
343. A. Engberg and H. Soling, *Acta Chem. Scand.*, **21,** 168 (1967).
344. A. Benrath, *Z. anorg. allgem. Chem.*, **163,** 396 (1927).
345. H. M. Powell and A. F. Wells, *J. Chem. Soc.*, **1935,** 359.
346. R. P. van Stapele, H. J. Beljers, P. F. Bongers and H. Zijlstra, *J. Chem. Phys.*, **44,** 3719 (1966).
347. B. N. Figgis, M. Gerloch and R. Mason, *Acta Cryst.*, **17,** 506 (1964).
348. F. A. Cotton, D. M. L. Goodgame and M. Goodgame, *J. Am. Chem. Soc.*, **83,** 4690 (1961).
349. B. N. Figgis, M. Gerloch and R. Mason, *Proc. Roy. Soc. (London)*, **279A,** 210 (1964).
350. K. S. Krishnan and A. Mookherji, *Phys. Rev.*, **51,** 528 (1937).
351. K. S. Krishnan and A. Mookherji, *Phys. Rev.*, **51,** 774 (1937).
352. A. Bose, R. Rai, S. Kumar and S. Mitra, *Physica*, **32,** 1437 (1966).
353. B. R. Judd, *Proc. Phys. Soc.*, **84,** 1036 (1964).
354. N. Pelletier-Allard, *Compt. Rend.*, **252,** 3970 (1961).
355. N. Pelletier-Allard, *Compt. Rend.*, **260,** 2170 (1965).
356. N. Pelletier-Allard, *Compt. Rend.*, **258,** 1215 (1964).
357. N. Pelletier-Allard, *Compt. Rend.*, **256,** 115 (1963).
358. J. Ferguson, *J. Chem. Phys.*, **32,** 528 (1960).
359. B. R. Sundheim and M. Kukk, *Discussions Faraday Soc.*, **32,** 49 (1961).
360. D. R. Stephens and H. G. Drickamer, *J. Chem. Phys.*, **35,** 429 (1961).
361. J. R. Wiesner, R. C. Srivastava, C. H. L. Kennard, M. Di Vaira and E. C. Lingafelter, *Acta Cryst.*, **23,** 565 (1967).
362. N. Fogel, C. C. Lin, C. Ford and W. Grindstaff, *Inorg. Chem.*, **3,** 720 (1964).
363. B. Morosin and E. C. Lingafelter, *Acta Cryst.*, **12,** 611 (1959).
364. N. S. Gill and R. S. Nyholm, *J. Chem. Soc.*, **1959,** 3997.
365. P. Pauling, *Inorg. Chem.*, **5,** 1498 (1966).

366. G. D. Stucky, J. B. Folkers and T. J. Kistenmacher, *Acta Cryst.*, **23,** 1064 (1967).
367. E. G. V. Percival and W. Wardlaw, *J. Chem. Soc.*, **1929,** 1505.
368. H. B. Jonassen, L. J. Theriot, E. A. Bourdreaux and W. M. Ayres, *J. Inorg. Nucl. Chem.*, **26,** 595 (1964).
369. L. A. Kazitsyna, N. B. Kupletskaya, V. A. Ptitsyna and O. A. Reutov, *J. Gen. Chem. USSR*, **33,** 3170 (1963).
370. E. A. Bourdreaux, H. B. Jonassen and L. J. Theriot, *J. Am. Chem. Soc.*, **85,** 2039 (1963).
371. L. A. Kazitsyna, O. A. Reutov and B. S. Kikot, *J. Gen. Chem. USSR*, **31,** 2751 (1961).
372. L. A. Kazitsyna, O. A. Reutov and Z. F. Buchkovskii, *J. Gen. Chem. USSR*, **31,** 2745 (1961).
373. R. S. Drago and D. Meek, *J. Phys. Chem.*, **65,** 1446 (1961).
374. D. W. Meek, D. K. Staub and R. S. Drago, *J. Am. Chem. Soc.*, **82,** 6013 (1960).
375. S. K. Madan and H. H. Denk, *J. Inorg. Nucl. Chem.*, **27,** 1049 (1965).
376. M. J. Joesten and K. M. Nykerk, *Inorg. Chem.*, **3,** 548 (1964).
377. J. A. Costamagna and R. Levitus, *J. Inorg. Nucl. Chem.*, **28,** 2685 (1966).
378. D. M. Gruen and C. A. Angell, *Inorg. Nucl. Chem. Letters*, **2,** 75, (1966).
379. S. A. Shchukarev, I. V. Vasil'kova and G. M. Barvinok, *Vestn. Leningrad. Univ.*, **20,** *Ser. Fiz. i Khim.*, 145 (1965).
380. P. Paoletti and A. Vacca, *Trans. Faraday Soc.*, **60,** 50 (1964).
381. G. N. Tishchenko and Z. G. Pinsker, *Dokl. Akad. Nauk. SSSR*, **100,** 913 (1955).
382. D. M. Adams, J. Chatt, J. M. Davidson and J. Gerratt, *J. Chem. Soc.*, **1963,** 2189.
383. R. J. H. Clark and T. M. Dunn, *J. Chem. Soc.*, **1963,** 1198.
384. A. Sabatini and L. Sacconi, *J. Am. Chem. Soc.*, **86,** 17 (1964).
385. D. Forster, *Chem. Commun.*, **1967,** 113.
386. R. H. Holm and F. A. Cotton, *J. Chem. Phys.*, **31,** 788 (1959).
387. G. M. Barvinok and V. Vintruff, *Vestn. Lenigrad. Univ.*, **21,** *Ser. Fiz. i Khim.*, 111 (1966).
388. J. Ferguson, *J. Chem. Phys.*, **39,** 116 (1963).
389. B. D. Bird and P. Day, *Chem. Commun.*, **1967,** 741.
390. A. Benrath and E. Neumann, *Z. anorg. allgem. Chem.*, **243,** 174 (1939).
391. A. Benrath, *Z. anorg. allgem. Chem.*, **240,** 87 (1938).
392. D. M. Gruen and C. W. DeKock, *J. Chem. Phys.*, **43,** 3395 (1965).
393. A. Benrath and G. Ritter, *J. Prakt. Chem.*, **154,** 2 (1939).
394. P. Ros, *Rec. Trav. Chim.*, **82,** 823 (1963).
395. P. Paoletti, *Trans. Faraday Soc.*, **61,** 219 (1965).
396. N. K. Hamer, *Mol. Phys.*, **6,** 257 (1963).
397. M. Goodgame and F. A. Cotton, *J. Am. Chem. Soc.*, **84,** 1543 (1962).
398. S. K. Madan and J. Peone, *Inorg. Chem.*, **7,** 824 (1968).
399. S. C. Chan and F. Leh, *J. Chem. Soc.*, **1968,** A, 1079.
400. K. Garbett and R. D. Gillard, *J. Chem. Soc.*, **1968,** A, 979.
401. S. Asperger and M. Flogel, *J. Chem. Soc.*, **1968,** A, 769.
402. J. A. Friend and D. J. Stabb, *Australian J. Chem.*, **21,** 1905 (1968).
403. R. D. Feltham and W. Silverthorn, *Inorg. Chem.*, **7,** 1154 (1968).

404. I. Watanabe and Y. Yamagata, *J. Chem. Phys.*, **46,** 407 (1967).
405. D. A. Buckingham, L. Durham and A. M. Sargeson, *Australian J. Chem.*, **20,** 257 (1967).
406. J. W. Vaughn and R. D. Lindholm, *Inorg. Syn.*, **9,** 163 (1967).
407. G. G. Schlessinger, *Inorg. Syn.*, **9,** 160 (1967).
408. H. B. Mark and F. C. Anson, *Inorg. Syn.*, **9,** 176 (1967).
409. D. A. Buckingham, P. A. Marzilli and A. M. Sargeson, *Inorg. Chem.*, **6,** 1032 (1967).
410. N. Sadasivan and J. H. Kernohan, *Inorg. Chem.*, **6,** 770 (1967).
411. C. K. Poon and M. L. Tobe, *J. Chem. Soc.*, **1968,** A, 1549.
412. S. Asperger and M. Pribanic, *J. Chem. Soc.*, **1968,** A, 1503.
413. J. G. Hartley, D. G. E. Kerfoot and L. M. Venanzi, *Inorg. Chim. Acta*, **1,** 145 (1967).
414. C. Y. Hsu and C. S. Garner, *Inorg. Chim. Acta*, **1,** 17 (1967).
415. A. R. Norris and M. L. Tobe, *Inorg. Chim. Acta*, **1,** 41 (1967).
416. S. C. Chan and K. Y. Hui, *J. Chem. Soc.*, **1968,** A, 1741.
417. A. T. Phillip, *Australian J. Chem.*, **21,** 2797 (1968).
418. W. R. Fitzgerald, A. J. Parker and D. W. Watts, *J. Am. Chem. Soc.*, **90,** 5744 (1968).
419. P. W. Schneider and J. P. Collman, *Inorg. Chem.*, **7,** 2010 (1968).
420. A. R. Gainsford and D. A. House, *Inorg. Nucl. Chem. Letters*, **4,** 621 (1968).
421. V. I. Muscat and J. B. Walker, *J. Inorg. Nucl. Chem.*, **30,** 2765 (1968).
422. B. Bosnich and D. W. Watts, *J. Am. Chem. Soc.*, **90,** 6228 (1968).
423. D. A. Palmer and D. W. Watts, *Australian J. Chem.*, **21,** 2895 (1968).
424. D. M. Palade, *Russ. J. Inorg. Chem.*, **12,** 1695 (1967).
425. R. D. Feltham, H. G. Metzger and W. Silverthorn, *Inorg. Chem.*, **7,** 2003 (1968).
426. R. G. Link, *Inorg. Chem.*, **7,** 2394 (1968).
427. D. A. Buckingham, I. I. Olsen and A. M. Sargeson, *J. Am. Chem. Soc.*, **90,** 6539 (1968).
428. D. A. House and C. S. Garner, *Inorg. Chem.*, **6,** 272 (1967).
429. R. D. Cannon, B. Chiswell and L. M. Venanzi, *J. Chem. Soc.*, **1967**, A, 1277.
430. G. G. Messmer and E. L. Amma, *Acta Cryst.*, **24B,** 417 (1968).
431. R. A. Cowley and P. Martel, *Phys. Rev. Letters*, **18,** 162 (1967).
432. P. Martel, R. A. Cowley and R. W. H. Stevenson, *J. Appl. Phys.*, **39,** 1116 (1968).
433. P. Martel, R. A. Cowley and R. W. H. Stevenson, *Can. J. Phys.*, **46,** 1355 (1968).
434. P. L. Richards, *J. Appl. Phys.*, **38,** 1500 (1967).
435. J. P. van der Ziel and H. J. Guggenheim, *Phys. Rev.*, **166,** 479 (1968).
436. R. Cohen-Adad, A. Tranquard and A. Mairchand, *Bull. Soc. Chim. France*, **1968,** 65.
437. I. S. Jacobs, S. Roberts and S. D. Silverstein, *J. Appl. Phys.*, **39,** 816 (1968).
438. R. S. Drago and B. B. Wayland, *Inorg. Chem.*, **7,** 628 (1968).
439. D. Forster, *Inorg. Chim. Acta*, **2,** 116 (1968).
440. Y. Bando and S. Nagakura, *Inorg. Chem.*, **7,** 893 (1968).
441. D. P. Graddon and G. M. Mockler, *Australian J. Chem.*, **21,** 1775 (1968).

442. L. Sacconi and R. Morassi, *Inorg. Nucl. Chem. Letters*, **4,** 449 (1968).
443. L. Sacconi and I. Bertini, *J. Am. Chem. Soc.*, **90,** 5443 (1968).
444. T. L. Blundell, H. M. Powell and L.M. Venanzi, *Chem. Commun.*, **1967,** 763.
445. G. Dyer and D. W. Meek, *J. Am. Chem. Soc.*, **89,** 3983 (1967).
446. C. C. Houk and K. Emerson, *J. Inorg. Nucl. Chem.*, **30,** 1493 (1968).
447. K. C. Dash and D. V. R. Rao, *Z. anorg. allgem. Chem.*, **345,** 217 (1966).
448. N. Mences, C. P. MacColl and R. Levitus, *Inorg. Nucl. Chem. Letters*, **4,** 597 (1968).
449. J. E. O'Conner and E. L. Amma, *Chem. Commun.*, **1968,** 892.
450. M. Laing and E. Horsfield, paper presented at South African Chemical Institute meeting, Cape Town, July 1968.
451. L. J. Admiral and G. Gafner, *Chem. Commun.*, **1968,** 1221.
452. E. J. Duff, *J. Chem. Soc.*, **1968,** A, 1812.
453. C. J. Popp and R. O. Ragsdale, *Inorg. Chem.*, **7,** 1845 (1968).
454. D. W. Herlockler and R. S. Drago, *Inorg. Chem.*, **7,** 1479 (1968).
455. D. M. L. Goodgame, M. Goodgame, P. J. Hayward and G. W. Rayner-Canham, *Inorg. Chem.*, **7,** 2447 (1968).
456. F. Lions, I. G. Dance and J. Lewis, *J. Chem. Soc.*, **1967,** A, 565.
457. J. S. Judge, W. M. Reiff, G. M. Intille, P. Ballway and W. A. Baker, *J. Inorg. Nucl. Chem.*, **29,** 1711 (1967).
458. J. S. Judge and W. A. Baker, *Inorg. Chim. Acta*, **1,** 239 (1967).
459. D. P. Madden and S. M. Nelson, *J. Chem. Soc.*, **1968,** A, 2342.
460. E. Uhlig, R. Krahmer and H. Wolf, *Z. anorg. allgem. Chem.*, **361,** 157 (1968).
461. D. M. Dowers and A. I. Popov, *Inorg. Chem.*, **7,** 1594 (1968).
462. R. Price, *J. Chem. Soc.*, **1967,** A, 521.
463. W. J. Eilbeck, F. Holmes, C. E. Taylor and A. E. Underhill, *J. Chem. Soc.*, **1968,** A, 1189.
464. E. J. Duff, M. N. Hughes and K. J. Rutt, *J. Chem. Soc.*, **1968,** A, 2354.
465. N. N. Y. Chan, M. Goodgame and M. J. Weeks, *J. Chem. Soc.*, **1968,** A, 2499.
466. L. F. Lindoy and S. E. Livingstone, *Inorg. Chim. Acta*, **2,** 119 (1968).
467. E. J. Duff and M. N. Hughes, *J. Chem. Soc.*, **1968,** A, 2144.
468. R. A. Walton and R. Whyman, *J. Chem. Soc.*, **1968,** A, 1394.
469. E. J. Duff, *J. Chem. Soc.*, **1968,** A, 434.
470. E. J. Duff, *J. Inorg. Nucl. Chem.*, **30,** 1257 (1968).
471. E. J. Duff, *J. Inorg. Nucl. Chem.*, **30,** 861 (1968).
472. G. Zakrzewski and L. Sacconi, *Inorg. Chem.*, **7,** 1034 (1968).
473. L. Sacconi and I. Bertini, *Inorg. Chem.*, **7,** 1178 (1968).
474. L. Sacconi, R. Morassi and S. Midollini, *J. Chem. Soc.*, **1968,** A, 1510.
475. M. Ciampolini, J. Gelsomini and N. Nardi, *Inorg. Chim. Acta*, **2,** 343 (1968).
476. M. Brierley and W. J. Gearey, *J. Chem. Soc.*, **1968,** A, 1641.
477. W. DeW. Horrocks, G. R. van Hecke and D. W. Hall, *Inorg. Chem.*, **6,** 694 (1967).
478. S. S. Sandhu and M. Gupta, *Chem. Ind.* (*London*), **1967,** 1876.
479. E. Uhlig and M. Schafer, *Z. anorg. allgem. Chem.*, **359,** 67 (1968).
480. R. E. Cristopher, I. R. Gordon and L. M. Venanzi, *J. Chem. Soc.*, **1968,** A, 205.

481. C. W. Kauffman and M. A. Robinson, *J. Inorg. Nucl. Chem.*, **30**, 2475 (1968).
482. A. W. McLellan and G. A. Melson, *J. Chem. Soc.*, **1967**, A, 137.
483. S. H. Hunter, V. M. Langford, G. A. Rodley and C. J. Wilkins, *J. Chem. Soc.*, **1968**, A, 305.
484. A. M. Brodie, S. H. Hunter, G. A. Rodley and C. J. Wilkins, *J. Chem. Soc.*, **1968**, A, 2039.
485. E. Lindner, R. Lehner and H. Scheer, *Chem. Ber.*, **100**, 1331 (1967).
486. A. M. Brodie, S. H. Hunter, G. A. Rodley and C. J. Wilkins, *J. Chem. Soc.*, **1968**, A, 987.
487. W. Beck, K. Schmidtner and H. J. Keller, *Chem. Ber.*, **100**, 503 (1967).
488. C. W. Bird and E. M. Briggs, *J. Chem. Soc.*, **1967**, A, 1004.
489. C. D. Flint and M. Goodgame, *J. Chem. Soc.*, **1968**, A, 2178.
490. J. G. H. du Preez, W. A. Steyn and A. J. Basson, *J. South African Chem. Institute*, **21**, 8 (1968).
491. O. Piovesana and C. Furlani, *J. Inorg. Nucl. Chem.*, **30**, 1249 (1968).
492. P. P. Singh and R. Rivest, *Can. J. Chem.*, **46**, 2361 (1968).
493. D. C. Goodall, *J. Inorg. Nucl. Chem.*, **30**, 2483 (1968).
494. S. S. Zumdahl and R. S. Drago, *J. Am. Chem. Soc.*, **89**, 4319 (1967).
495. D. G. Brewer, P. T. T. Wong and M. C. Sears, *Can. J. Chem.*, **46**, 3137 (1968).
496. A. Braibanti, F. Dallavalle, M. A. Pellinghelli and E. Leporti, *Inorg. Chem.*, **7**, 1430 (1968).
497. N. C. Collins and W. K. Glass, *Chem. Commun.*, **1967**, 211.
498. A. M. Elabrash and M. E. Sobier, *J. Chem. U.A.R.*, **10**, 388 (1967).
499. N. A. Daugherty and J. H. Swisher, *Inorg. Chem.*, **7**, 1651 (1968).
500. S. S. Zumdahl and R. S. Drago, *Inorg. Chem.*, **7**, 2162 (1968).
501. H. S. Preston, C. H. L. Kennard and R. A. Plowman, *J. Inorg. Nucl. Chem.*, **30**, 1463 (1968).
502. M. C. Feller and R. Robson, *Australian J. Chem.*, **21**, 2919 (1968).
503. Z. Dori and H. B. Gray, *Inorg. Chem.*, **7**, 889 (1968).
504. B. Chiswell, *Australian J. Chem.*, **21**, 2561 (1968).
505. C. D. Flint and M. Goodgame, *J. Chem. Soc.*, **1968**, A, 750.
506. L. H. Pignolet and W. DeW. Horrocks, *Chem. Commun.*, **1968**, 1012.
507. G. A. Kakos and G. Winter, *Australian J. Chem.*, **20**, 2343 (1967).
508. R. Cohen-Adad and J. Said, *Bull. Soc. Chim. France*, **1967**, 564.
509. R. Cohen-Adad, J. Said, M. T. Saugier and A. Sebaoum, *Compt. Rend.*, **267C**, 962 (1968).
510. P. Lumme and K. Junkkarinen, *Suomen Kemistilehti*, **41B**, 114 (1968).
511. J. P. Legrand and J. P. Renard, *Compt. Rend.*, **266B**, 1165 (1968).
512. K. Kasumata and M. Date, *J. Phys. Soc. Japan*, **24**, 751 (1968).
513. B. N. Figgis, M. Gerloch, J. Lewis, F. E. Mabbs and G. A. Webb, *J. Chem. Soc.*, **1968**, A, 2086.
514. J. B. Torrance and H. Tinkham, *J. Appl. Phys.*, **39**, 822 (1968).
515. A. C. Daniel, A. W. Bevan and R. J. Mahler, *J. Appl. Phys.*, **39**, 496 (1968).
516. Y. Nakamura and S. Kawaguchi, *Chem. Commun.*, **1968**, 716.
517. G. W. A. Fowles, D. A. Price and R. A. Walton, *J. Chem. Soc.*, **1968**, A, 1842.

518. E. Wenschuh, B. Fritzsche and E. Mach, *Z. anorg. allgem. Chem.*, **358,** 233 (1968).
519. H. Forstatt, J. N. McElearney and P. T. Bailey, *Phys. Letters*, **27A,** 70 (1968).
520. T. E. Murray and G. K. Wessel, *J. Phys. Soc. Japan*, **24,** 738 (1968).
521. M. P. Petrov and G. M. Nedlin, *J. Appl. Phys.*, **39,** 1012 (1968).
522. G. A. Smolenskii, M. P. Petrov and R. V. Pisarev, *J. Appl. Phys.*, **38,** 1269 (1967).
523. K. W. Mess, E. Lagenijk, D. A. Curtis and W. J. Huiskamp, *Physica*, **34,** 126 (1967).
524. R. F. Wielinga, H. W. J. Blote, J. A. Roest and W. J. Huiskamp, *Physica*, **34,** 223 (1967).
525. P. Bamfield, *J. Chem. Soc.*, **1967,** A, 804.
526. G. A. Heath, R. L. Martin and I. M. Stewart, *Australian J. Chem.*, **22,** 83 (1969).
527. G. A. Heath, R. L. Martin and I. M. Stewart, *Chem. Commun.*, **1969,** 54.
528. N. S. Gill and F. B. Taylor, *Inorg. Syn.*, **9,** 136 (1967).
529. A. Ouchi, M. Nakatani and Y. Takahashi, *Bull. Chem. Soc. Japan*, **41,** 2044 (1968).
530. A. Furuhashi, T. Takeuchi and A. Ouchi, *Bull. Chem. Soc. Japan*, **41,** 2049 (1968).
531. S. Mitra, *J. Chem. Phys.*, **49,** 4724 (1968).
532. G. A. Heath, I. M. Stewart and R. L. Martin, *Inorg. Nucl. Chem. Letters*, in the press.
533. D. H. Brown, K. P. Forrest, R. H. Nuttall and D. W. A. Sharp, *J. Chem. Soc.*, **1968,** A, 2146.

Chapter 8

Nickel

Nickel is the only element in the first-row transition series whose binary halides are restricted to a single oxidation state, although some adducts of the unknown trihalides have been prepared indirectly using ligands such as *o*-phenylenebis(dimethylarsine). In fluoro complexes oxidation states III and IV are obtained quite easily.

HALIDES

Oxidation State III

Although binary halides of nickel(III) have not been prepared, a number of halo derivatives of nickel(III) with halogen atoms directly bound to the metal atom, are known. It is convenient in this section to deal with chloro, bromo and iodo complexes together.

One of the first derivatives of this type prepared was $NiCl_3D_2$, where D = *o*-phenylenebis(dimethylarsine). Nyholm[1–3] prepared this compound by refluxing a mixture of nickel dichloride and the ligand in alcoholic 10 M hydrochloric acid, with oxygen passing through the solution. The yellow-brown complex has a room temperature magnetic moment of 1.89 BM and in non-aqueous solvents it is a 1:1 electrolyte containing a six-coordinate cation.

The crystal structure of this complex has been determined by Kreisman and coworkers[4] using single-crystal X-ray diffraction techniques. It has a monoclinic unit cell with $a = 9.33$, $b = 9.69$, $c = 14.87$ Å and $\beta = 98.6°$. The unit cell of space group $P2_1/c$ contains two molecular units. The structure of the cation is shown in Figure 8.1 and has the chlorine atoms in a *trans* configuration. The important bond dimensions are given in Table 8.1.

Kreisman and coworkers[4] have examined the electron spin resonance spectrum of this complex and also of the uncomplexed diarsine radical cation. The spectrum of the nickel complex shows hyperfine structure which can only be explained on the basis of some delocalization of the unpaired electron onto the arsenic atoms, and the rhombic *g* tensor is similar to that observed for the radical cation. These preliminary results seem to favour description of the molecule as containing the metal stabilized radical, rather than a simple nickel(III) formulation.

The nickel-chlorine stretching frequency in $NiCl_3D_2$ has been reported[5] as occurring at 282 cm^{-1}.

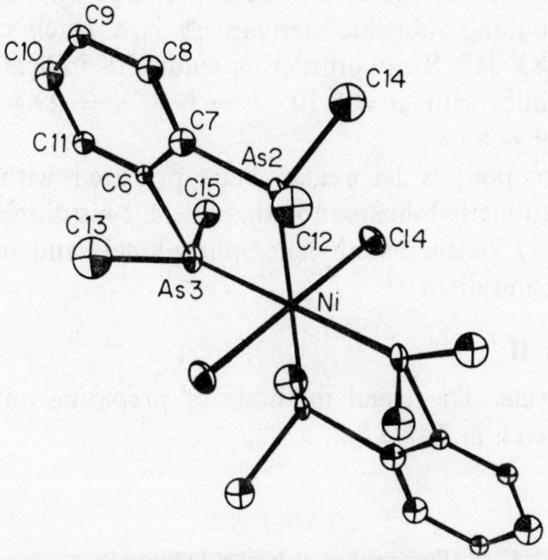

Figure 8.1 The structure of $NiCl_2D_2^+$. (Reproduced by permission from P. Kreisman, R. Marsh, J. R. Preer and H. B. Gray, *J. Am. Chem. Soc.*, **90**, 1067 (1968))

TABLE 8.1
Dimensions of the $[NiCl_2D_2]^+$ Cation

Distance	Value (Å)	Angle	Value
$Ni—As_2$	2.37	$As_2—Ni—As_3$	86.2°
$Ni—As_3$	2.34	$As_2—Ni—Cl_4$	92.9°
$Ni—Cl_4$	2.43	$As_3—Ni—Cl_4$	92.0°

The corresponding nickel bromide adduct of *o*-phenylenebis(dimethylarsine) is also known[1–3,6].

Both the nickel(III) chloro and bromo derivatives of *o*-phenylenebis(dimethylarsine) may be oxidized with concentrated nitric acid to nickel(IV) complexes which were obtained as the perchlorates[2]. The deep blue-green compounds, $[NiX_2D_2][ClO_4]_2$, are diamagnetic[2].

Jensen, Nygaard and Pedersen[7] prepared a series of compounds of the general type $NiX_3(R_3P)_2$, where X is Cl or Br, by oxidizing the corresponding dihalo compound with nitrosyl halide at −80°. The compounds isolated were dark coloured solids and molecular weight and dipole measurements indicated that they were monomeric with either *trans* trigonal-bipyramidal or *trans* square-pyramidal stereochemistry.

$NiBr_3[(C_2H_5)_3P]_2$, which has also been prepared by bromine oxidation of the corresponding dibromo derivative[8], is a black crystalline solid which melts at 83–84°. X-ray diffraction studies[9,10] indicate that the compound is triclinic with $a = 8.10$, $b = 17.8$, $c = 29.4$ Å, $\alpha = 91.25°$, $\beta = 90.0°$ and $\gamma = 99.83°$.

Nickel(III) compounds have also been prepared with other ligands such as 1,2-bis(dimethylphosphino)ethane[11], 1,2-bis(diphenylphosphino)-ethane[12], *P,P,P′,P′*-tetraethylethylenediphosphine[13] and methylbis(3-propanedimethylarsino)arsine[14].

Oxidation State II

Nickel difluoride. The usual methods of preparing anhydrous nickel difluoride are given in Table 8.2.

TABLE 8.2
Preparation of Nickel Difluoride

Method	Conditions	References
Ni and anhydrous HF	Bomb, 225°	15
$NiCl_2$ and F_2	Flow system, 350–400°	16–18
$NiCl_2$ and anhydrous HF	Flow system, >500°	19, 20
$NiCl_2$ and ClF_3	Flow system 250°	21
NiO and F_2	Flow system, 375°	22
NiO and ClF_3	Flow system, 200°	23
NiS and F_2	Flow system, 375°	22
NiS and SF_4	Bomb, 150–300°, 5–50 atm	24
$NiF_2 \cdot 4H_2O$ and HF	Flow system, 350–400°	25–27
$NiF_2 \cdot 4H_2O$ and SiF_4	Flow system, 400°	28

The reaction of fluorine on metallic nickel, either in the form of sheet or powder, is slow[29–31] because the difluoride is involatile and has good cohesive stability. Nickel difluoride can be prepared by interaction of fluorine and finely powdered nickel, but the preparation is laborious and is complete only after several separate fluorinations at 550° with grinding of the sample between each fluorination[22]. As a consequence of the slow rate of fluorination of nickel, the metal itself and some of its alloys, such as Monel, have been used widely to handle fluorine and other corrosive fluorides[32].

Nickel difluoride is a yellow solid which is only very slightly soluble in anhydrous hydrogen fluoride[33]. It melts at[34] 1450° and sublimes to give a linear monomeric molecule in the gas phase[35,36]. Margrave and co-workers[35] used mass-spectrometric techniques to study the vapour pressure of nickel difluoride as a function of temperature, and found the vapour pressure, over the temperature range 781–833°, to be given by the expression

$$\log p_{\text{atm}} = 10.17 - 16{,}900/T$$

The heat of sublimation[35] and the heat of formation[37] are 79.4 and −157.2 kcal mole^{-1} respectively.

Nickel difluoride has the tetragonal rutile structure[22,38–40] adopted by manganese difluoride (Figure 1.1, page 18) and several other difluorides. The unit-cell parameters are $a = 4.650$ and $c = 3.083$ Å. The NiF_6 octahedra are only slightly tetragonally compressed, the nickel-fluorine bond distances being 2.01 and 1.98 Å. Haefner and coworkers[40] report that below the temperature of the magnetic anomaly (see below) nickel difluoride becomes orthorhombic.

The room temperature magnetic moment of nickel difluoride has been reported as 2.85 BM[41], but it becomes magnetically ordered at low temperatures, the Neel temperature being reported[42–44] as 73.3°K. Maxima in the heat capacity and specific heat of nickel difluoride also occur at about this temperature[25,34,45].

Detailed magnetic studies[40,46–49] show that nickel difluoride has quite different magnetic properties from the isomorphous difluorides of manganese, iron and cobalt. For nickel difluoride $\chi_{\perp}$ is greater than $\chi_{\parallel}$ and at 301.5°K the anisotropy is 1.102×10^{-4} while at 90.07°K the anisotropy is 1.890×10^{-4}. The magnetic structure is quite complex and recent results[40,48,49] show an earlier simpler description to be incorrect[47].

Balkanski, Moch and Parisot[50,51] have reported the infrared spectrum of a single crystal of nickel difluoride using optical transmission, reflectance and Kramers-Kronig inversion techniques. The fundamental modes are given in Table 8.3.

Tsuchida[52] has also investigated the far-infrared spectrum of a single crystal of nickel difluoride and reported a nickel-fluorine stretch at 446 cm^{-1}. Milligan, Jacox and McKinley[28] have recorded the infrared spectrum of nickel difluoride vapour in argon at 800–900°. The spectrum observed consisted of the nickel-fluorine stretching frequency, but the band was resolved into absorptions for the individual isotopes of nickel as shown in Table 8.4.

No other nickel-fluorine stretching frequency was observed and this is consistent with the linear nature of the molcule in the gas phase.

The far-infrared antiferromagnetic resonance of nickel difluoride has been observed by Richards[44,53]. At 0°K the extrapolated values are 3.33 and 31.14 cm^{-1}.

TABLE 8.3
Infrared Spectrum (cm^{-1}) of Nickel Difluoride

Mode	Transmission	Reflectance	Kramers-Kronig
E_u	445	445	440
A_{2u}	375	370	
E_u	285	285	286
E_u	225	225	228

TABLE 8.4
Infrared Spectrum of Nickel Difluoride Vapour

Molecule	Frequency (cm^{-1})
$^{64}Ni^{19}F_2$	765.5
$^{62}Ni^{19}F_2$	770.0
$^{60}Ni^{19}F_2$	774.7
$^{58}Ni^{19}F_2$	779.7

The absorption spectrum of nickel difluoride has been investigated in detail by several workers[54–56]. The observed bands have been assigned on the basis of octahedral symmetry with a $^3A_{2g}$ ground state. Both spin-allowed and spin-forbidden transitions have been observed and assigned. Ferguson and coworkers[55] have studied the absorption spectrum of nickel difluoride in magnesium difluoride and zinc difluoride host lattices. Various concentrations of nickel were used and the origin of the fine-structure was discussed in some detail.

Nickel difluoride tetrahydrate. This pale green hydrate may be prepared by dissolving the carbonate in hydrofluoric acid[25,57]. It has also been isolated from the nickel difluoride-hydrogen fluoride-water system[58]. The compound may be completely dehydrated by heating alone, but better in an atmosphere of hydrogen fluoride[25,58]. Several intermediate hydrates have been reported to be formed in this process.

Nickel difluoride tetrahydrate obeys the Curie-Weiss law over a wide temperature range with a Weiss constant of 19° and a magnetic moment of 3.26 BM[59].

Nickel dichloride. The methods of preparing nickel dichloride are given in Table 8.5.

TABLE 8.5
Preparation of Nickel Dichloride

Method	Conditions	References
Ni and Cl_2	Flow system, elevated temp.	60, 61
Ni and Cl_2	Interaction in ethanol at 20°	62
NiO, C and Cl_2	Flow system, >300°	63
$NiCl_2 \cdot 6H_2O$ and HCl	Flow system, 400–550°	28, 64–66
$NiCl_2 \cdot 6H_2O$ and $SOCl_2$	Reflux	20
Ni salt and $SOCl_2$[a]	Reflux	67
$Ni(OOCH_3)_2$ and CH_3COCl	Reaction in benzene	68

[a] Suitable salts are the acetate, carbonate, formate and nitrate.

In all preparations it is advisable to resublime the crude product at high temperatures in either hydrogen chloride gas or high vacuum.

Nickel dichloride is a yellow-brown crystalline solid which is extremely hygroscopic. Normally the hexahydrate is obtained from aqueous solutions at room temperature, although other hydrates are known.

Although an early mass-spectrometric study[69] of nickel dichloride vapour over the temperature range 440–700° indicated that it vaporizes only as the monomer, later infrared studies[28,70] clearly show the presence of dimeric species in the vapour. Over the temperature range 700–783° the vapour pressure of nickel dichloride is given by the expression[71]

$$\log p_{\text{mm}} = 12.051 - 11{,}499/T$$

The available thermodynamic data for nickel dichloride are given in Table 8.6.

Heat capacities have been calculated[73] from room temperature to 1050°.

Nickel dichloride is isostructural[75] with the dichlorides of manganese, iron and cobalt, the hexagonal unit cell having $a = 3.478$ and $c = 17.41$ Å. The infrared spectrum of gaseous nickel dichloride has been studied by Leroi and coworkers[70] and also by Milligan and his group[28]. The latter workers used a nickel dichloride-argon matrix at 425° and deduced isotopic metal-chlorine stretching frequencies for the linear monomer as shown in Table 8.7.

TABLE 8.6

Thermodynamic Data for Nickel Dichloride

Property	Value	Reference
M.p.	1001°	72
	1009°	65
	1030°	73
ΔH°_{form} (kcal mole^{-1})	−74.51	74
ΔH_{fus} (kcal mole^{-1})	8.47	73
ΔH_{vap} (kcal mole^{-1})	53.81	71
ΔS_{fus} (cal deg^{-1} mole^{-1})	14.18	73
ΔS_{vap} (cal deg^{-1} mole^{-1})	43.28	71

TABLE 8.7

Infrared Spectrum of Linear Nickel Dichloride Species

Molecule	Frequency (cm^{-1})
$^{60}Ni^{37}Cl_2$	509.0
$^{60}Ni^{37}Cl^{35}Cl$	512.8
$^{60}Ni^{35}Cl_2$	516.6
$^{58}Ni^{37}Cl^{35}Cl$	517.4
$^{58}Ni^{35}Cl_2$	520.6

A broad band centred at 439 cm^{-1} was assigned to a stretching frequency of dimeric nickel dichloride[28].

Nickel dichloride obeys the Curie-Weiss law at temperatures down to about 60°K with a Weiss constant of 67° and a magnetic moment of 3.32 BM[76,77]. It becomes magnetically ordered at about 50°K[76–79] and there is also an anomaly in the heat capacity at this temperature[80]. Magnetic susceptibility studies on a single crystal show that nickel dichloride is magnetically isotropic[79] with a *g* value of 2.21[81].

Theoretical implications of the magnetic properties of nickel dichloride have been discussed by several groups[77,82,83].

The diffuse-reflectance spectrum of nickel dichloride has been examined[84,85] and the observed bands have been assigned on the basis of octahedral symmetry. The major absorptions occur at 11,600, 12,900, 19,400, 22,100 and 26,300 cm^{-1}. The absorption spectrum of nickel dichloride in the gaseous phase has also been reported[86–88].

As with the dihalides of iron and cobalt, nickel dichloride reacts readily with pyridine to give $NiCl_2(py)_4$. The blue compound has a tetragonal unit cell[89] with $a = 15.8$ and $c = 16.9$ Å. There are eight molecules in the cell of space group $I4/acd$. The complex is isomorphous with $CoCl_2(py)_4$, that is, it has a *trans* octahedral structure. The nickel-chlorine and nickel-nitrogen bond lengths are 2.39 Å and 2.00 Å respectively[89,90]. The infrared and electronic absorption spectra of $NiCl_2(py)_4$ have been reported[85,91–96] and have been assigned on the basis of the above structure. Clark and Williams[91] report nickel-chlorine and nickel-nitrogen stretching frequencies of 246 and 236 cm^{-1} respectively, while Frank and Rogers[92] give a value of 242 cm^{-1} for the nickel-nitrogen mode. The distortion about the nickel atom in this adduct causes splitting of the $^3T_{2g} \leftarrow {}^3A_{2g}$ and $^3T_{1g}(F) \leftarrow {}^3A_{2g}$ transitions for octahedral nickel(II) and Rowley and Drago[93] report the following assignments at liquid nitrogen temperature.

9,042 cm^{-1}	$^3E_g(F) \leftarrow {}^3B_{1g}$
11,730 cm^{-1}	$^3B_{2g}(F) \leftarrow {}^3B_{1g}$
12,620 cm^{-1}	$^1A_{1g}(D) \leftarrow {}^3B_{1g}$
12,804 cm^{-1}	$^1B_{1g}(D) \leftarrow {}^3B_{1g}$
14,930 cm^{-1}	$^3A_{2g}(F) \leftarrow {}^3B_{1g}$
16,818 cm^{-1}	$^3E_g(F) \leftarrow {}^3B_{1g}$
26,759 cm^{-1}	$^3E_g(P) \leftarrow {}^3B_{1g}$

The adduct has a room temperature magnetic moment of 3.11 BM[94]. $NiCl_2(py)_4$ is thermally unstable, decomposing initially to $NiCl_2(py)_2$ at about 110–150° (see below)[97–101].

Nickel dichloride reacts with a large range of substituted pyridines, aniline, and similar donor ligands to form adducts of the types $NiCl_2L_4$ and $NiCl_2L_2$. The thermal stability, stereochemistry and empirical formulae of these complexes have been discussed in terms of steric hindrance, π-acceptor properties and ligand polarizability. The proposed structures for many of the adducts have been deduced from magnetic and spectral studies since virtually no crystallographic data have been reported. Complexes of the type $NiCl_2L_4$ are relatively few in number[94–97,100,102–104]. A number of these adducts are thermally unstable, decomposing to $NiCl_2L_2$. This can be a useful way of preparing these complexes, although a large number of $NiCl_2L_2$ adducts may be prepared by direct interaction of the constituents. Some of these 1:2 adducts are also thermally unstable and much thermal stability work has been reported. The reported modes of decomposition of the $NiCl_2L_4$ and $NiCl_2L_2$ adducts may be summarized by the following equations.

$$NiCl_2L_4\,(s) \xrightarrow{T_1} NiCl_2L_2\,(s) + 2L\,(g)$$

$$NiCl_2L_2\,(s) \xrightarrow{T_2} NiCl_2L\,(s) + L\,(g)$$

$$NiCl_2L\,(s) \xrightarrow{T_3} NiCl_2\,(s) + L\,(g)$$

$$NiCl_2L\,(s) \xrightarrow{T_4} NiCl_2L_{2/3}\,(s) + 1/3L\,(g)$$

$$NiCl_2L_{2/3}\,(s) \xrightarrow{T_5} NiCl_2\,(s) + 2/3L\,(g)$$

$$NiCl_2L_2\,(s) \xrightarrow{T_6} NiCl_2\,(s) + 2L\,(g)$$

$$NiCl_2L_4\,(s) \xrightarrow{T_7} NiCl_2L\,(s) + 3L\,(g)$$

The approximate temperatures at which these reactions take place are given in Table 8.8.

TABLE 8.8

Decomposition Temperatures of Nickel Dichloride Adducts

Compound	Temperatures[a]	Reference
$NiCl_2$(pyridine)$_4$	$T_1 = 145°$, $T_2 = 200°$, $T_3 = 340°$	101
	$T_1 = 180°$, $T_2 = 220°$, $T_3 = 350°$	97
	$T_1 = 110°$, $T_2 = 190°$, $T_3 = 310°$	99
	$T_1 = 110°$, $T_2 = 190°$, $T_3 = 270°$	100
$NiCl_2$(pyridine)$_2$	$T_2 = 170°$, $T_3 = 290°$	100
$NiCl_2$(2-methylpyridine)$_2$	$T_2 = 130°$, $T_3 = 300°$	101
	$T_2 = 200°$, $T_3 = 320°$	97
	$T_6 = 60°$	100
$NiCl_2$(3-methylpyridine)$_4$	$T_1 = 160°$, $T_2 = 220°$, $T_4 = 280°$, $T_5 = 330°$	101
	$T_1 = 160°$, $T_2 = 200°$, $T_4 = 250°$, $T_5 = 320°$	97
	$T_1 = 115°$, $T_2 = 175°$, $T_3 = 260°$	100
$NiCl_2$(3-methylpyridine)$_2$	$T_2 = 110°$, $T_3 = 260°$	100
$NiCl_2$(4-methylpyridine)$_4$	$T_7 = 200°$, $T_3 = 335°$	101
	$T_7 = 180°$, $T_3 = 320°$	97
	$T_7 = 145°$, $T_3 = 280°$	100
$NiCl_2$(4-methylpyridine)$_2$	$T_2 = 150°$, $T_3 = 280°$	100
$NiCl_2$(2,6-lutidine)$_2$	$T_2 = 230°$, $T_3 = 300°$	97
$NiCl_2$(2,4,6-collidine)$_4$	$T_7 = 260°$, $T_3 = 340°$	97
$NiCl_2$(aniline)$_4$	$T_1 = 75°$, $T_6 = 232°$	100
	$T_6 = 230°$	101
$NiCl_2$(quinoline)$_2$	$T_2 = 190°$, $T_4 = 245°$, $T_5 = 320°$	105
	$T_2 = 175°$, $T_3 = 250°$	100

[a] The temperature quoted for reference 101 is that at which the rate of change of the heat of decomposition is greatest.

Murgulescu and coworkers[98] reported that they obtained $NiCl_2(py)_{2/3}$ as an intermediate product of the thermal decomposition of $NiCl_2(py)_4$. However this product was not identified in a more recent study by Beech and coworkers[101]. In general, there is good agreement between the results of Beech, Mortimer and Tyler[101] and Allan and coworkers[97], but there are some discrepancies between these workers and Majumdar and coworkers[100] as to the temperature of decomposition, and in some cases even the mode of decomposition. Magnetic and spectral studies of the intermediate products of decomposition show clearly that in each case the nickel is in an octahedral environment[94,97].

The properties of a number of pyridine, substituted pyridine, aniline and substituted aniline adducts of nickel dichloride are given in Table 8.9.

TABLE 8.9

Pyridine Type Adducts of the Series $NiCl_2L_n$

Compound and Properties	References
L = pyridine, $n = 2$, yellow-green, $\mu = 3.37$ BM, $\nu_{Ni-Cl} = 246$ cm^{-1}, octahedral	91, 92, 97–100, 106–111
L = 2-methylpyridine, $n = 2$, dark blue, $\mu = 3.48$ BM, $\nu_{Ni-Cl} = 327, 297$ cm^{-1}, $\nu_{Ni-N} = 239, 227$ cm^{-1}, tetrahedral	91, 95, 97, 100, 103, 112
L = 3-methylpyridine, $n = 2$, yellow, $\mu = 3.45$ BM, octahedral	94–97, 100, 102, 110, 111
L = 4-methylpyridine, $n = 2$, $\mu = 3.35$ BM, octahedral	94, 100, 110, 111
L = 4-chloropyridine, $n = 2$, $\mu = 3.35$ BM, octahedral	113
L = 3-bromopyridine, $n = 2$, $\mu = 3.24$ BM, octahedral	113
L = 2,3-lutidine, $n = 2$, blue, m.p. = 99°, $\mu = 3.61$ BM, tetrahedral	114
L = 2,4-lutidine, $n = 2$, blue, m.p. = 98°, $\mu = 3.60$ BM, tetrahedral	114
L = 2,6-lutidine, $n = 2$, blue, $\nu_{Ni-Cl} = 343$ cm^{-1}, tetrahedral	95, 97
L = aniline, $n = 2$, pale green, $\mu = 3.36$ BM, octahedral	100, 110, 111, 115
L = *p*-toluidine, $n = 2$, blue-green, $\mu = 3.43$ BM, $\nu_{Ni-Cl} = 238$ cm^{-1}, octahedral	91, 100, 111, 116
L = quinoline, $n = 2$, yellow isomer; $\mu = 3.41$ BM, $\nu_{Ni-Cl} = 363$ cm^{-1}, $\nu_{Ni-N} = 223$ cm^{-1}, octahedral: blue isomer; $\mu = 3.54$ BM, tetrahedral	91, 92, 95, 100, 110, 117–119

Complexes of the types $NiCl_2L_2$ and $NiCl_2L$ are known for both bipyridyl and *o*-phenanthroline[120–122]. Pfeiffer and Tappermann[121] prepared the $NiCl_2L_2$ adducts by heating the corresponding $[NiL_3]Cl_2 \cdot xH_2O$ complexes. Both compounds are green and have room-temperature magnetic moments of 3.17 BM, consistent with octahedral coordination about the metal atom[120]. Broomhead and Dwyer[122] prepared the mono ligand derivatives by direct interaction of nickel dichloride and the ligand in boiling dimethylformamide. Both of their complexes have magnetic moments of about 3.3 BM at room temperature, indicating octahedral stereochemistry for the central nickel atom[122].

Lee and coworkers[123] have made a detailed study of the thermal decomposition of $[Ni(bipy)_3]Cl_2 \cdot 7H_2O$ and their results may be summarized by the following scheme:

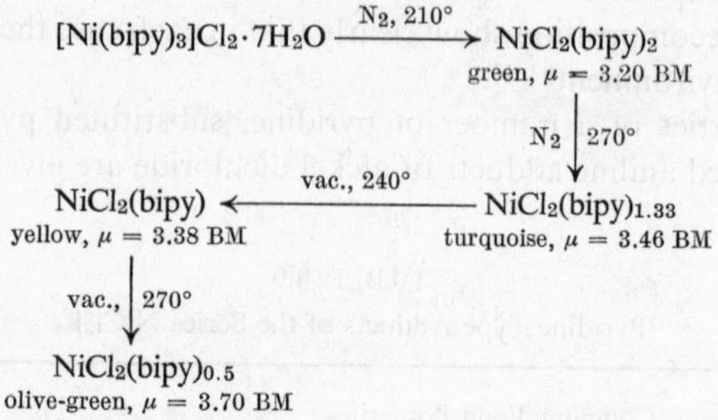

The decomposition products were characterized by means of their electronic spectra, magnetic susceptibility and electrolytic conductance. The octahedral stereochemistry of $NiCl_2(bipy)_2$ and $NiCl_2(bipy)$ was confirmed by these studies[123]. An X-ray diffraction pattern of $NiCl_2(bipy)_{1.33}$ showed conclusively that it was a single compound, not a mixture. Its absorption spectrum showed characteristics of both tetrahedrally and octahedrally coordinated nickel(II)[123] but it was not possible to determine unambiguously the structure from the data available. However, Clark and Williams[124] in a later infrared study showed that the compound contains a dimeric cation and should be formulated as $[(bipy)_2NiCl_2Ni(bipy)_2]^{2+}$ $[NiCl_4]^{2-}$.

Uhlig and Staiger[125,126] have studied the preparation and properties of complexes of nickel dichloride with a number of aliphatic amines such as CH_3NH_2, $C_2H_5NH_2$, $C_3H_7NH_2$ and $(C_2H_5)_2NH$. Some were prepared by direct interaction of the amine and nickel dichloride, others by thermal decomposition. With the exception of $NiCl_2[(C_2H_5)_2NH]_2$, which is four-coordinate, all of the amine complexes are octahedral as deduced from magnetic, conductance and spectral studies.

Pale blue bis(ethylenediamine)nickel(II) dichloride, $NiCl_2(en)_2$, may be prepared[127] by reacting $Ni(en)_3Cl_2 \cdot 2H_2O$ with nickel dichloride dihydrate in methanol and precipitating the product with acetone. The compound is soluble in water and begins to decompose without melting at about 290°. Antsishkina and Porai-Koshits[128] have shown by X-ray diffraction studies that the compound is ionic with a dimeric cation and should be formulated as $[Ni_2(en)_4Cl_2]Cl_2$. It has a monoclinic unit cell, of space group $P2_1/n$ containing four molecules, with $a = 13.9$, $b = 11.3$, $c = 6.3$ Å and $\beta = 92.5°$. The chlorine atoms in the cation are in a *cis* arrangement and act as bridging groups. The nickel-chlorine and nickel-nitrogen bond distances are 2.5 and 2.0 Å respectively.

The interaction of 1,4,8,11-tetraazacyclotetradecane (cyclam) with hydrated nickel dichloride in alcohol leads to the formation of mauve $NiCl_2$(cyclam)[129]. The complex has a room temperature magnetic moment

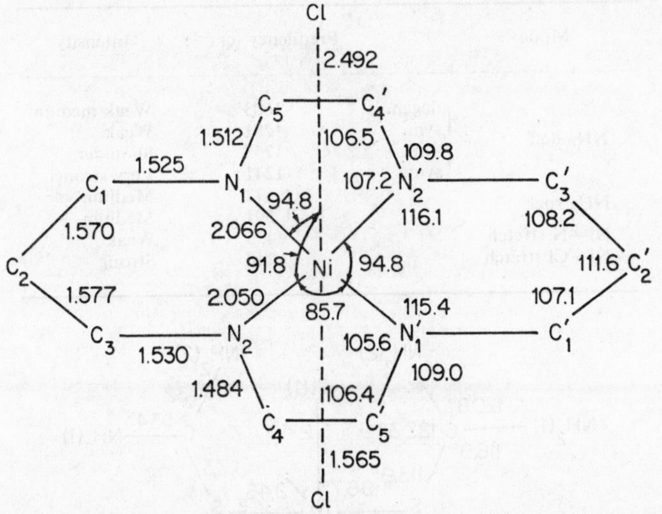

Figure 8.2 The structure of dichloro-1,4,8,11-tetraazacyclotetradecane-nickel (II). (Reproduced by permission from B. Bosnich, R. Mason, P. J. Pauling, G. B. Robertson and M. L. Tobe, *Chem. Commun.*, **1965,** 97)

of 3.09 BM and acts as a 2:1 electrolyte in water and methanol[129]. It was concluded[129] that the complex contained the nickel atom in an octahedral environment, and this geometry has been confirmed by a single-crystal X-ray diffraction study by Bosnich and coworkers[130]. The monoclinic lattice has $a = 13.644$, $b = 8.355$, $c = 6.541$ Å and $\beta = 104.58°$. There are two molecules in the unit cell of space group $P2_1/a$. The structure of the compound, which is shown in Figure 8.2, has a distorted octahedral configuration, the symmetry of the ligand field about the central nickel atom being less than D_{4h}. The 1,3-diaminopropane rings have the chair conformation while the ethylenediamine residues are in the *gauche* form, their carbon atoms being equatorial substituents on the nitrogen atoms of the 1,3-diaminopropane rings.

With 1,4,8,11-tetraazaundecane, Bosnich and coworkers[131] have isolated the adduct $NiCl_2L$. This violet-pink compound has a room temperature magnetic moment of about 3.1 BM and like the above closely related compound, is thought to have a *trans* octahedral configuration.

Solid nickel dichloride readily reacts with gaseous ammonia to give the complex $NiCl_2(NH_3)_6$[132]. On heating to 120° this complex loses four

TABLE 8.10
Infrared Spectrum of $NiCl_2(NH_3)_2$

Mode		Frequency (cm^{-1})	Intensity
NH_3 def.	degen.	1603	Weak medium
	sym.	1281	Weak
	sym.	1253	Shoulder
	sym.	1241	Very strong
NH_3 rock		672	Medium
		591	Medium
Ni—N stretch		435	Weak
Ni—Cl stretch		240	Strong

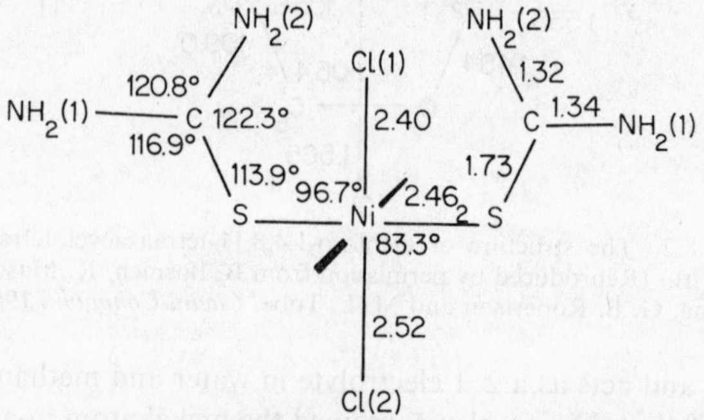

Figure 8.3 The structure and dimensions of $NiCl_2(tu)_4$. (Reproduced by permission from A. Lopez-Castro and M. R. Turner, *J. Chem. Soc.*, **1963**, 1309)

molecules of ammonia to give yellow-green $NiCl_2(NH_3)_2$[132]. Magnetic and spectral studies[94,110,111,133] on the latter compound clearly indicate that the nickel atom is in a distorted octahedral environment. Clark and Williams[133] have studied the infrared spectrum of this compound and the reported assignments are given in Table 8.10.

Thiourea (tu) reacts with nickel dichloride in alcohol to give yellow $NiCl_2(tu)_4$[134–136]. An early crystallographic determination[135] indicated a *trans* configuration for the complex. This was confirmed by a more recent single-crystal X-ray study by Lopez-Castro and Turner[137]. The compound has a tetragonal unit cell of space group *I*4, containing two molecular units, the parameters at 110°K being $a = 9.558$ and $c = 8.981$ Å. The structure of the complex is shown in Figures 8.3 and 8.4. It was suggested

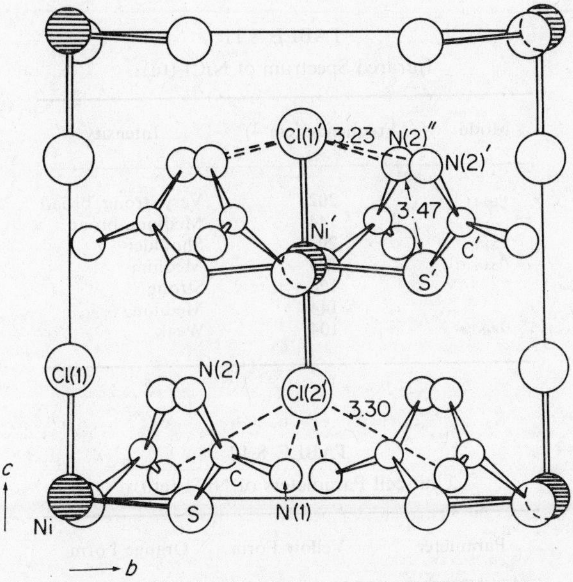

Figure 8.4 The structure of $NiCl_2(tu)_4$ projected down the *a* axis. (Reproduced by permission from A. Lopez-Castro and M. R. Turner, *J. Chem. Soc.*, **1963**, 1309)

that the discrete units are held together by hydrogen bonds as well as by van der Waals forces. Infrared studies[134,136,138,139] reveal that the thiourea ligands are bound to the nickel atom through the sulphur atoms. Adams and Cornell[138] report the infrared assignments given in Table 8.11.

Holt and Carlin[140] have reported that the interaction of nickel dichloride and ethylenethiourea (etu) yields yellow $NiCl_2(etu)_4$. Digestion of this complex in ethanol gives an orange compound of the same empirical formula[140]. Both isomers obey the Curie-Weiss law; the magnetic moments and Weiss constants are 3.29 BM and 15° for the yellow isomer and 3.33 BM and 13° for the orange isomer. X-ray powder diffraction data[140] showed that the two isomers are not isomorphous. Holt and Carlin[140] concluded that both had an octahedral configuration and that there were differences in the orientation of the ligands in the two forms. These conclusions were confirmed by single-crystal X-ray diffraction studies by Robinson and coworkers[141]. The unit-cell parameters of both forms of $NiCl_2(etu)_4$ are given in Table 8.12. The unit-cell parameters of the orange form correspond to those of the yellow form reported by Nardelli and coworkers[142].

TABLE 8.11
Infrared Spectrum of $NiCl_2(tu)_4$

Mode	Frequency (cm^{-1})	Intensity
$\nu_{Ni\text{-}Cl}$	202	Very strong, broad
	144	Medium, broad
$\nu_{Ni\text{-}S}$	202	Shoulder
$\delta_{Ni\text{-}S\text{-}C}$	174	Medium
	156	Strong
	144	Medium
$\delta_{S\text{-}Ni\text{-}S}$	104	Weak

TABLE 8.12
Unit-cell Parameters of $NiCl_2(etu)_4$

Parameter	Yellow Form	Orange Form
Symmetry	Monoclinic	Triclinic
a (Å)	17.57	8.46
b (Å)	8.35	8.41
c (Å)	16.14	8.81
α		107.0°
β	119.5°	117.5°
γ		90.5°
Z	4	1
Space group	$C2/c$	$P\bar{1}$

Both forms of $NiCl_2(etu)_4$ have the same basic NiS_4 skeleton including two unequal independent nickel-sulphur bonds, but they differ in the orientation and lengths of the nickel-chlorine bonds and also in the orientation of the ethylenethiourea molecules. The structures of the two forms of $NiCl_2(etu)_4$ are shown in Figure 8.5, while the important internuclear dimensions are given in Table 8.13. No evidence was obtained either for intermolecular or intramolecular hydrogen bonding.

Adams and Cornell[138] have recorded the infrared spectrum of an unspecified form of $NiCl_2(etu)_4$ and the assignments reported are given in Table 8.14.

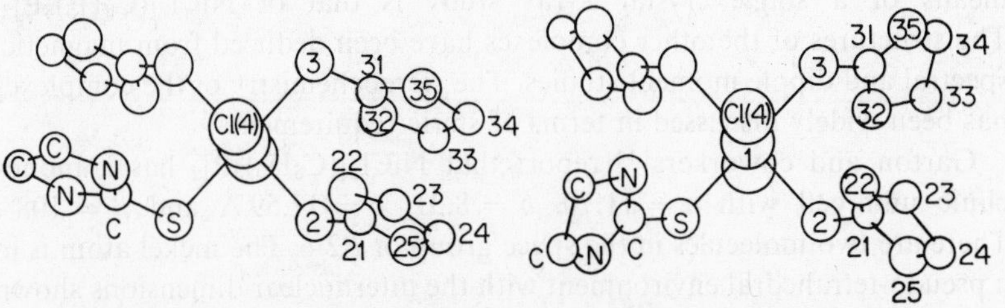

Figure 8.5 The structures of the triclinic (left) and monoclinic (right) forms of $NiCl_2(etu)_4$ projected normally onto planes defined by their sulphur atoms. (Reproduced by permission from W. T. Robinson, S. L. Holt and G. B. Carpenter, *Inorg. Chem.*, **6**, 603 (1967))

TABLE 8.13
Internuclear Dimensions of $NiCl_2(etu)_4$

Dimensions	Yellow Form	Orange Form
Ni(1)—Cl(4)	2.496 Å	2.457 Å
Ni(1)—S(3)	2.443 Å	2.490 Å
Ni(1)—S(2)	2.482 Å	2.451 Å
Mean S—C	1.697 Å	1.683 Å
Mean C(1)—N	1.340 Å	1.324 Å
Mean N—C(ethylene)	1.528 Å	1.467 Å
Mean C—C	1.583 Å	1.557 Å
S(2)—Ni(1)—S(3)	92.7°	93.6°
Cl(4)—Ni(1)—S(2)	101.1°	99.1°
Cl(4)—Ni(1)—S(3)	80.9°	87.3°
Ni(1)—S(2)—C(21)	113.6°	115.0°
Ni(1)—S(3)—C(31)	117.9°	107.1°

TABLE 8.14
Infrared Spectrum of $NiCl_2(etu)_4$

Mode	Frequency (cm^{-1})	Intensity
$\nu_{Ni\text{-}Cl}$	192	Very strong, broad
	145	Medium
$\nu_{Ni\text{-}S}$	239	Medium
$\delta_{Ni\text{-}S\text{-}C}$	205	Strong, broad
	140	Shoulder

Nickel dichloride reacts with a large range of phosphines[143] to give, in general, four-coordinate adducts. Both tetrahedral and square-planar complexes are known. The only structure that has been established by means of a single crystal X-ray study is that of $NiCl_2[(C_6H_5)_3P]_2$. The structures of the other complexes have been deduced from magnetic, spectral and dipole moment studies. The stereochemistry of the complexes has been widely discussed in terms of steric requirements.

Garton and coworkers[144] report that $NiCl_2[(C_6H_5)_3P]_2$ has a monoclinic unit cell with $a = 11.70$, $b = 8.31$, $c = 17.59$ Å and $\beta = 108°$. There are two molecules in the space group of $P2/c$. The nickel atom is in a pseudo-tetrahedral environment with the internuclear dimensions shown in Table 8.15.

TABLE 8.15
Internuclear Dimensions of $NiCl_2[(C_6H_5)_3P]_2$

Distance	Value	Angle	Value
Ni—Cl	2.27 Å	Cl—Ni—Cl	123°
Ni—P	2.28 Å	P—Ni—P	117°

Properties of some other phosphine adducts of nickel dichloride are given in Table 8.16.

Nickel dichloride also reacts with several phosphine and arsine oxides to give products with properties as shown in Table 8.17. As with the phosphine adducts, the structures of these compounds have been deduced from magnetic, spectral and other physico-chemical studies.

TABLE 8.17
Phosphine and Arsine Oxide Adducts of the Type $NiCl_2L_n$

Compound and Properties	References
L = $(C_2H_5)_3PO$, $n = 2$, blue-violet, m.p. = 93°, tetrahedral	166
L = $(C_6H_{11})_3PO$, $n = 2$, blue, m.p. = 199–201°, tetrahedral	166
L = $(C_6H_5)_3PO$, $n = 2$, blue, $\mu = 3.7$ BM, tetrahedral	119, 166–168
L = $(C_6H_5)_3AsO$, $n = 2$, blue, $\mu = 3.95$ BM, $\nu_{As\text{-}O} = 865, 845$ cm^{-1}, $\nu_{Ni\text{-}O} = 432$ cm^{-1}, $\nu_{Ni\text{-}Cl} = 310, 284$ cm^{-1}, tetrahedral	119, 169–171

As would be expected, nickel dichloride reacts with a large range of other donor molecules to give adducts of nickel(II) in a variety of stereochemical environments. Some properties of a number of these adducts are given in Table 8.18.

TABLE 8.16

Phosphine Adducts of the Type $NiCl_2L_n$

Compound and Properties	References
L = $(C_6H_5)_2PH$, $n = 3$, dark brown, m.p. = 98°, five coordinate	145
L = $(CH_3)_2C_6H_5P$, $n = 2$, red, m.p. = 153°, square planar	146
L = $(C_2H_5)_2C_6H_5P$, $n = 2$, red, m.p. = 112–113°, square planar	147
L = $(n\text{-}C_3H_7)_2C_6H_5P$, $n = 2$, blue, m.p. = 149–150°, $\mu = 3.35$ BM, tetrahedral	148
L = $CH_3(C_6H_5)_2P$, $n = 2$, red-violet, m.p. = 148–150°, diamagnetic, square planar	149
L = $C_2H_5(C_6H_5)_2P$, $n = 2$, red, m.p. = 146–151°, diamagnetic, square planar	149, 150
L = $i\text{-}C_3H_7(C_6H_5)_2P$, $n = 2$, red, m.p. = 160–165°, diamagnetic, square planar	149
L = $C_5H_{11}(C_6H_5)_2P$, $n = 2$, red, m.p. = 77–79°, diamagnetic, square planar	149
L = $n\text{-}C_5H_{11}(C_6H_5)_2P$, $n = 2$, blue-purple, m.p. = 92–100°, $\mu = 3.12$ BM, tetrahedral	149
L = $C_7H_7(C_6H_5)_2P$, $n = 2$, blue-green, m.p. = 190–215°, $\mu = 3.23$ BM, tetrahedral	151
L = $C_7H_7(C_6H_5)_2P$, $n = 2$, red, m.p. = 190–215°, diamagnetic, square planar	151
L = $(C_7H_7)_2C_6H_5P$, $n = 2$, red, m.p. = 216–219°, diamagnetic, square planar	151
L = $C_2H_5(C_6H_{11})_2P$, $n = 2$, red, m.p. = 169–170°, square planar	152
L = $(C_2H_5)_2C_6H_{11}P$, $n = 2$, red, m.p. = 97–98°, square planar	152
L = $(C_2H_5)_3P$, $n = 2$, red, m.p. = 112–113°, diamagnetic, square planar	147, 153, 154
L = $(i\text{-}C_3H_7)_3P$, $n = 2$, red, m.p. = 184–186° diamagnetic, square planar	155
L = $(s\text{-}C_4H_9)_3P$, $n = 2$, red, m.p. = 103–105°, diamagnetic, square planar	147, 154, 155
L = $(C_6H_5)_3P$, $n = 2$, blue, m.p. = 247–250°, $\mu = 3.07$ BM, $\nu_{Ni\text{-}Cl} = 338, 302\ cm^{-1}$, $\nu_{Ni\text{-}P} = 189, 166\ cm^{-1}$, tetrahedral	119, 144, 154, 156–159
L = $(C_6H_{11})_3P$, $n = 2$, red, m.p. = 227°, diamagnetic, square planar	154, 160
L = bis(diphenylphosphino)methane (DPM), $n = 2$, brown, diamagnetic, square planar	12
L = 1,2-bis(diphenylphosphino)ethane (DPE), $n = 1$, red, diamagnetic, $\nu_{Ni\text{-}Cl} = 341, 229\ cm^{-1}$, square planar	11, 12, 161
L = 1,3-bis(diphenylphosphino)propane (DPP), $n = 1$, diamagnetic, square planar	12
L = *P,P,P′,P′*-tetraethylethylenediphosphine (TEP), $n = 1$, yellow-brown, m.p. = 245–249°, diamagnetic, *cis* square planar	13
L = $(C_6H_{11})_2P(CH_2)_5P(C_6H_{11})_2$, $n = 1$, red, m.p. = 163–165°, square planar	162
L = tetracyclohexyldiphosphine, $n = 1$, red, m.p. = 201–205°, diamagnetic, *cis* square planar	163
L = 2-phenylisophosphindoline, $n = 2$, red, m.p. = 218–220°, square planar	164
L = bis(*o*-methylthiophenyl)phenylphosphine, $n = 1$, purple, diamagnetic, five coordinate	165
L = diphenyl(*o*-methylthiophenyl)phosphine, $n = 2$, pale green, $\mu = 3.15$ BM, octahedral	165

TABLE 8.18

Adducts of the Type $NiCl_2L_n$

Compound and Properties	References
L = quinoxaline (Q), $n = 1$, pale yellow, $\mu = 3.44$ BM, octahedral	110, 111, 172, 173
L = pyrazine (pyz), $n = 2$, pale green, $\mu = 3.18$ BM, octahedral with bridging chlorines, decomposes to $NiCl_2(pyz)$ and subsequently $NiCl_2$ at approximately 280° and 480° respectively	174–176
L = 2-methylpyrazine (mpz), $n = 5$, yellow-green, $\mu = 3.30$ BM, octahedral, fifth molecule of ligand acts as solvate	174, 175
L = 2-methylpyrazine (mpz), $n = 1$, pale green, $\mu = 3.42$ BM, octahedral	110, 174, 175
L = 2,5-dimethylpyrazine (2,5-dmp), $n = 1$, pale yellow, $\mu = 3.4$ BM, octahedral, decomposes to $NiCl_2$ at approximately 300°	110, 111, 174, 175, 177, 178
L = 2,6-dimethylpyrazine (2,6-dmp), $n = 2$, pale green, $\mu = 3.38$ BM, octahedral	110, 111, 174, 175
L = 2-methylthiomethylpyridine (mmp), $n = 2$, green, $\mu = 3.18$ BM, octahedral	179
L = 2-methyl-8-methylthioquinoline (mmtq), $n = 1$, yellow-green, $\mu = 3.29$ BM, tetrahedral	180
L = *N,N*-dimethylethylenediamine (Me_2en), $n = 2$, blue, $\mu = 3.21$ BM, octahedral	176
L = *N,N*-diethylethylenediamine (Et_2en), $n = 2$, green, m.p. = 186–191°, $\mu = 3.29$ BM, octahedral	181
L = *N,N,N′,N′*-tetramethylethylenediamine (Me_4en), $n = 1$, green, $\mu = 3.45$ BM, octahedral	182, 183
L = *N,N,N′,N′*-tetramethyl-1,2-propylenediamine (Me_4pn), $n = 1$, green, m.p. = 300°, $\mu = 3.45$ BM, octahedral	182, 183
L = *N,N,N′,N′*-tetramethyltrimethylenediamine (Me_4tn), $n = 1$, green, m.p. = 222–228°, $\mu = 3.37$ BM, tetrahedral	182, 183
L = 1,1,7,7-tetraethyldiethylenetriamine (Et_4dien), $n = 1$, red, diamagnetic, $[NiCl(Et_4dien)]^+Cl^-$	184
L = bis(2-dimethylaminoethyl)methylamine (dienMe), $n = 1$, ochre, $\mu = 3.38$ BM, five coordinate	185
L = tris(2-dimethylaminoethyl)amine (trenMe), $n = 1$, light olive-green, $\mu = 3.42$ BM, five-coordinate cation, $[NiCl(trenMe)]^+Cl^-$	186
L = bis(2-dimethylaminoethyl)oxide (Me_4daeo), $n = 1$, red, $\mu = 3.40$ BM, five coordinate	187
L = bis(2-dimethylaminoethyl)sulphide (Me_4daes), $n = 1$, $\mu = 3.24$ BM, five coordinate	188
L = Schiff base formed between *N*-methyl-*o*-aminobenzaldehyde and *N,N*-diethylethylenediamine (MABen-NEt_2), $n = 1$, amber-yellow, $\mu = 3.32$ BM, five coordinate	189
L = Schiff base formed between *o*-methylthiobenzaldehyde and *N,N*-diethylethylenediamine (MSBen-NEt_2), $n = 1$, light green, m.p. = 182–184°, $\mu = 3.32$ BM, five coordinate in solid state	190
L = 2-methylimidazole (mlz), $n = 2$, blue, $\mu = 3.41$ BM, $\nu_{Ni\text{-}N} = 270$ cm^{-1}, octahedral	191
L = 2-pyridinaldoxime (HPOX), $n = 2$, green, $\mu = 3.16$ BM, octahedral	192, 193
L = bis(2-pyridyl-β-ethyl)amine, $n = 1$, green, $\mu = 3.25$ BM, five coordinate	194
L = bis(2-pyridyl-β-ethyl)sulphide, $n = 1$, green, $\mu = 3.20$ BM, five coordinate, not dimeric as originally proposed	194, 195
L = 2-[β(dimethylamino)ethyl]pyridine, $n = 1$, yellow-brown, m.p. = 221°, $\mu = 3.27$ BM, octahedral	196
L = 2-[β(isopropylamino)ethyl]pyridine, $n = 1$, yellow, m.p. = 227°, $\mu = 3.15$ BM, octahedral	196
L = β(bisphenylphosphino)ethylpyridine, $n = 1$, red-violet, m.p. = 196–197°, $\mu = 3.29$ BM, octahedral	197

Other adducts of nickel dichloride with nitriles[110,111,198,199], dioxan[62,200–202], alcohols[62,203], tetrahydrofuran[204,205], dimethylsulphoxide[206–210], substituted amines[211–213], *N*-oxides[214–216], and numerous other donor molecules[118,173,217–225] have also been prepared and studied using the standard physico-chemical techniques.

Nickel dichloride hexahydrate. This hydrate has been isolated from saturated aqueous solutions of nickel dichloride[226,227]. The compound is thermally unstable[27,228,229], beginning to lose water at about 50°. It would appear[27,229] that four molecules of water are lost very readily and that the anhydrous compound is formed at about 210°.

Nickel dichloride hexahydrate is isostructural with the corresponding cobalt compound[230,231]. The unit-cell parameters are given in Table 8.19.

The description of the structure given by Mizuno[231] is probably the more acceptable. The structure of the compound, which is shown in Figure 8.6, consists of *trans* $[NiCl_2(H_2O)_4]$ octahedra and water of crystallization. The nickel-chlorine and nickel-oxygen bond lengths are 2.38 and 2.10 Å respectively.

Nakagawa and Shimanouchi[232] have investigated the infrared spectrum of nickel dichloride hexahydrate and report H_2O rocking, H_2O wagging and nickel-oxygen stretching frequencies of 755, 645 and 405 cm^{-1} respectively.

Nickel dichloride hexahydrate obeys the Curie-Weiss law down to quite low temperatures[233–235] with a magnetic moment of 3.11 BM and a Weiss constant of approximately 10°. Haseda and coworkers[233] report that the compound becomes antiferromagnetic at 6.2°K. A specific heat anomaly[236] at 5.34°K is associated with the magnetic transition. Spence and coworkers[237] deduced a magnetic structure of space group C_c2/c for nickel dichloride hexahydrate from ^1H NMR measurements, while a more recent study by Kleinberg[238] gives the space group as I_c2/c. The latter magnetic cell is shown in Figure 8.7 and consists of antiferromagnetic (001) planes with antiferromagnetic coupling between planes. The spin

TABLE 8.19

Unit-cell Parameters of $NiCl_2 \cdot 6H_2O$

	Stroganov *et al.*[230]	Mizuno[231]
Symmetry	Monoclinic	Monoclinic
a (Å)	6.68	10.23
b (Å)	7.13	7.05
c (Å)	11.76	5.67
β	49.5°	112.17°
Z	2	2
Space group	$P2_1/c$	$C2/m$

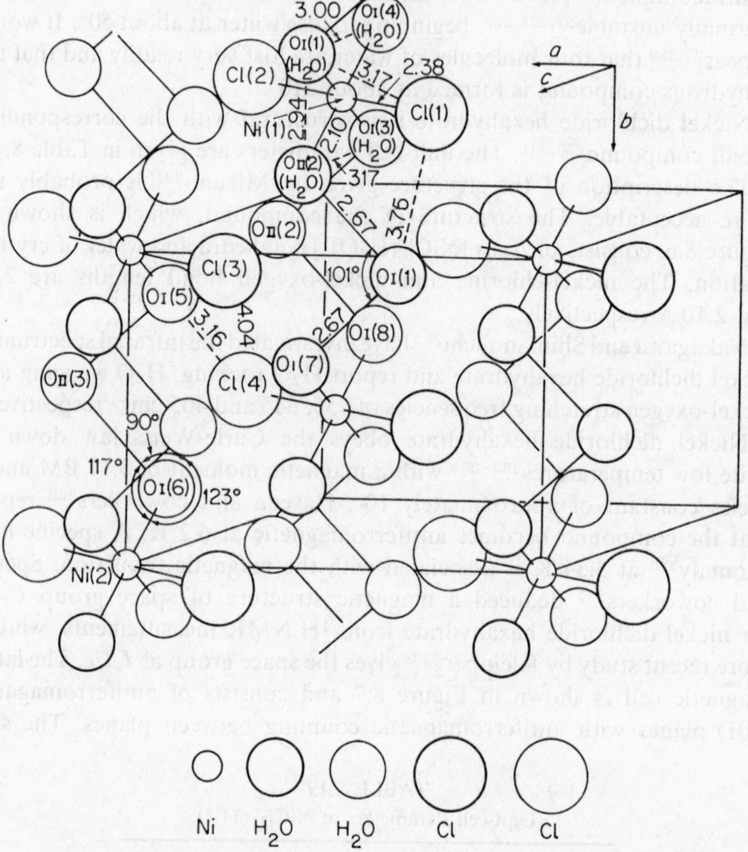

Figure 8.6 The structure of nickel dichloride hexahydrate. (Reproduced by permission from J. Mizuno, *J. Phys. Soc. Japan*, **16**, 1574 (1961))

direction is not along the a' axis (a' is perpendicular to c) but approximately 22.5° from the a' axis towards a.

Nickel dichloride tetrahydrate. Evaporation of an aqueous solution of nickel dichloride at about 40° is probably the most convenient method of preparing this compound[226]. It has also been prepared by passing hydrogen chloride gas into a saturated aqueous solution of nickel dichloride[27].

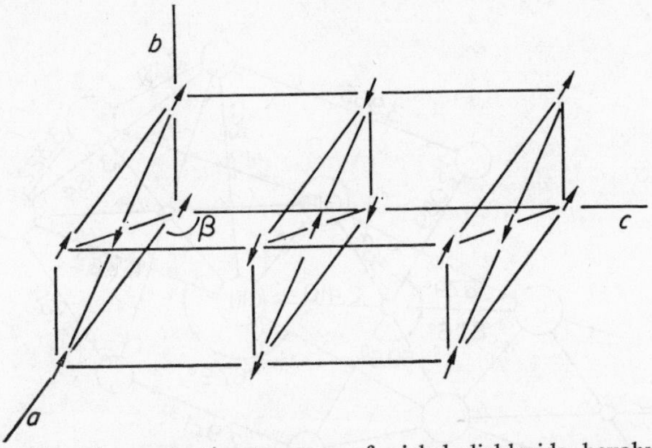

Figure 8.7 The magnetic structure of nickel dichloride hexahydrate. (Reproduced by permission from R. Kleinberg, *J. Appl. Phys.*, **38**, 1453 (1967))

Stroganov and coworkers[239] report that the compound has a tetragonal unit cell with $a = 6.62$ and $c = 13.23$ Å.

Nickel dichloride dihydrate. Nickel dichloride dihydrate may be prepared[226] by evaporation of an aqueous solution of nickel dichloride at about 75°, or by the controlled thermal decomposition of nickel dichloride hexahydrate[27,228,229], the former method being the more convenient.

The structure of nickel dichloride dihydrate was first investigated by Vainshtein[240] and a more refined analysis of the data has been recently given by Morosin[241]. The compound has a monoclinic unit cell[241] with $a = 6.909$, $b = 6.886$, $c = 8.830$ Å and $\beta = 92.25°$. There are four molecular units in the space group $I2/m$. The structure consists of polymeric chains in which each nickel atom is surrounded by four chlorine atoms and two water molecules. The four chlorine atoms form a distorted square plane about the nickel atom and each square plane is tilted with respect to the plane about an adjacent nickel atom to give an angle of 167.94° between the two planes. The chains are held together by hydrogen bonds. The structure with bond dimensions is shown in Figure 8.8. Because of the orientation of the planes, the oxygen atoms of the water molecules are not equally spaced. This results in the formation of two different types of hydrogen bond. One is located along the bisector of the chlorine-oxygen-chlorine angle, while the other is located so that the oxygen-hydrogen vector points towards the centre of a triangle, defined by an oxygen atom and a chlorine atom on an adjacent chain and the nearer oxygen atom of the same chain (2.921 Å). The lengths of the bonds to the oxygen and chlorine atoms in the adjacent chain are 2.941 and 2.542 Å respectively. The positions of the hydrogen bonds are shown in Figure 8.9.

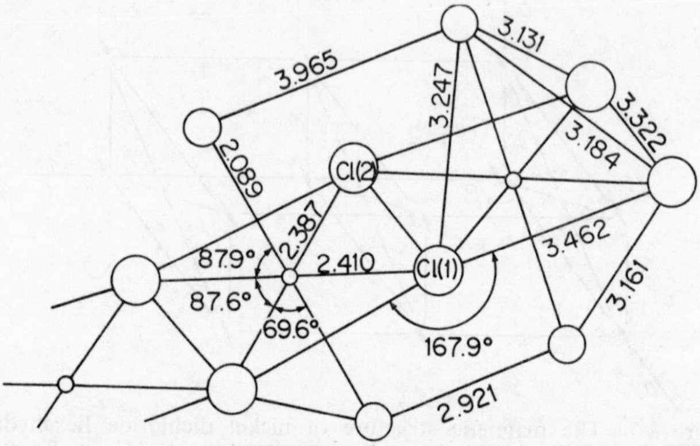

Figure 8.8 The structure and dimensions of the nickel dichloride dihydrate chain. (Reproduced by permission from B. Morosin, *Acta Cryst.*, **23**, 630 (1967))

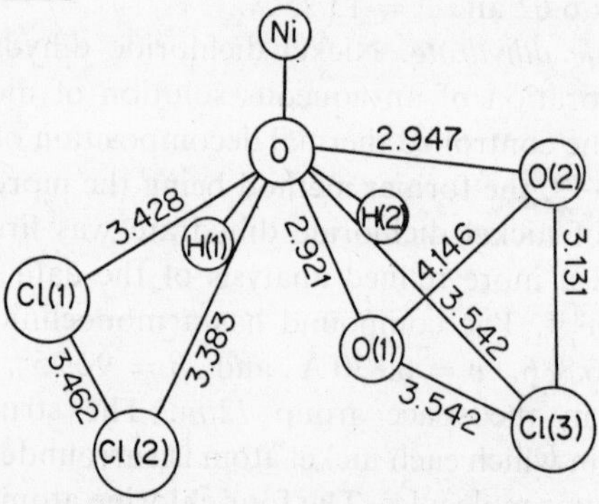

Figure 8.9 Oxygen and its near neighbours in nickel dichloride dihydrate. (Reproduced by permission from B. Morosin, *Acta Cryst.*, **23**, 630 (1967))

Nickel dibromide. Probably the most convenient methods of preparing nickel dibromide are the interaction of nickel and bromine in ether[242] and the dehydration of the hexahydrate[242] at about 140°. For both methods of preparation it is advisable to sublime the product at about 900° in a stream of hydrogen bromide. Other reported methods of preparation include the passage of hydrogen bromide over nickel at red heat[69], the bromination of nickel tetracarbonyl at 1100° in a flow system[243] and the reaction between nickel acetate and acetyl bromide in benzene[68].

Nickel dibromide is a yellow crystalline solid which melts[72] in a sealed tube at 965°. Mass-spectrometric studies[69] indicate that it vaporizes as the monomer in the temperature range 440–700°. According to Schafer and Jacob[243] the vapour pressure of solid nickel dibromide is given by the expression

$$\log p_{\text{atm}} = 16.681 - 13{,}112/T - 0.3497 \times 10^{-3}T - 1.711 \log T$$

The heat and entropy of vaporization of nickel dibromide are 54.84 kcal mole^{-1} and 52.60 cal deg^{-1} mole^{-1} respectively.

Nickel dibromide obeys the Curie-Weiss law[225] with a magnetic moment of 3.0 BM and a Weiss constant of 20°. A Neel temperature of 60°K has been reported[244–246]. It is magnetically anisotropic[244]; at 300°K the values of $g_{\parallel}$ and $g_{\perp}$ are 2.19 and 2.21 respectively, while at 77°K $g_{\parallel} = 2.16$ and $g_{\perp} = 2.23$.

The diffuse-reflectance spectrum of nickel dibromide[84,85] is very similar to that of nickel dichloride. The principal bands occur at 10,300, 12,100, 16,800 and 20,700 cm^{-1} and have been assigned on the basis of octahedral symmetry.

As would be expected, nickel dibromide reacts with pyridine, substituted pyridines, aniline and similar nitrogen base donor molecules to give adducts of the types $NiBr_2L_4$ and $NiBr_2L_2$. As with the corresponding chloro complexes, and indeed with the majority of the transition-metal halide adducts of these ligands, virtually no crystallographic data are available, and the proposed structures have been deduced from magnetic and spectral studies. The relative stabilities and stereochemistries of the nickel dibromide adducts have been discussed by some authors in the same terms as for the nickel dichloride complexes (page 413).

The only nickel dibromide adduct of this type which has been subjected to a full single crystal X-ray diffraction study is $NiBr_2(py)_4$. This green compound is isostructural with the corresponding chloro derivative[90,247] and also with $CoBr_2(py)_4$[248]. The orthorhombic unit cell has the parameters $a = 15.8$, $b = 9.3$ and $c = 14.2$ Å. There are four molecules in the cell of space group *Pna*. The structure consists of *trans* $NiBr_2(py)_4$ octahedral units in which the plane of the pyridine rings is rotated at about 50° relative to the NiN_4 plane. The nitrogen-bromine and nickel-nitrogen bond lengths are 2.58 and 2.00 Å respectively.

The thermal decomposition of a number of adducts of nickel dibromide with pyridine and substituted pyridines has been investigated[97,100,101,105].

The most detailed study is that by Allan and coworkers[97] whose results are summarized by the following equations.

$$NiBr_2(py)_4 \xrightarrow{140^\circ} NiBr_2(py)_2 \xrightarrow{200^\circ} NiBr_2(py) \xrightarrow{300^\circ} NiBr_2$$

$$NiBr_2(2\text{-}CH_3py)_2 \xrightarrow{250^\circ} NiBr_2$$

$$NiBr_2(3\text{-}CH_3py)_4 \xrightarrow{160^\circ} NiBr_2(3\text{-}CH_3py)_2 \xrightarrow{200^\circ} NiBr_2(3\text{-}CH_3py \xrightarrow{230^\circ} NiBr_2(3\text{-}CH_3py)_{2/3} \xrightarrow{300^\circ} NiBr_2$$

$$NiBr_2(4\text{-}CH_3py)_2 \xrightarrow{190^\circ} NiBr_2(4\text{-}CH_3py) \xrightarrow{240^\circ} NiBr_2(4\text{-}CH_3py)_{2/3} \xrightarrow{340^\circ} NiBr_2$$

$$NiBr_2(2,6\text{-}lut)_2 \xrightarrow{100^\circ} NiBr_2(2,6\text{-}lut) \xrightarrow{200^\circ} NiBr_2$$

$$NiBr_2(2,4,6\text{-}coll)_4 \xrightarrow{220^\circ} NiBr_2(2,4,6\text{-}coll) \xrightarrow{310^\circ} NiBr_2$$

Magnetic and spectral studies[94,97] indicate that the compounds of the types $NiBr_2L$ and $NiBr_2L_{2/3}$ have polymeric structures in which the coordination about the nickel atom is octahedral.

Some properties of a number of nickel dibromide adducts with pyridine, substituted pyridines and related ligands are given in Table 8.20.

Nickel dibromide adducts with 2,4,6-collidine[95,97], aniline[100,110,111,115], *o*-toluidine[100,116], *m*-toluidine[110,111,116] and 3,4-xylidine[116] have also been reported.

Uhlig and Staiger[125,126] have prepared a number of adducts of nickel dibromide with aliphatic amines as the donor molecule. These complexes have similar properties to the previously described chloro derivatives. Bis(ethylenediamine)nickel(II) dibromide has a monoclinic unit cell[128] with $a = 14.2$, $b = 11.3$, $c = 6.6$ Å and $\beta = 94°$. Like the corresponding chloro compound, it should be formulated as an ionic compound with a dimeric cation, that is $[Ni_2(en)_4Br_2]Br_2$. The nickel-bromine and nickel-nitrogen bond distances are 2.7 and 2.0 Å respectively.

Nickel dibromide reacts with 1,4,8,11-tetraazacyclotetradecane[129] and 1,4,8,11-tetraazaundecane[131] to form mauve 1:1 adducts. The properties of these two compounds closely resemble those of the nickel dichloride compounds, indicating that they contain octahedrally coordinated nickel(II).

The reaction between hydrated nickel dibromide and bis(2-pyridyl-β-ethyl)amine (BPEA) in anhydrous ethanol leads to the formation of green $NiBr_2(BPEA)$[194]. The adduct has a monoclinic unit cell with $a = 13.88$, $b = 14.51$, $c = 7.98$ Å and $\beta = 100.4°$ and has a room-temperature magnetic moment of 3.28 BM. Molecular weight, conductance and spectral studies show that the compound has a five-coordinate structure[194]. Reaction of nickel dibromide with bis(2-pyridyl-β-ethyl)-sulphide (BPES) under the same conditions results in the precipitation of yellow-green $NiBr_2(BPES)$[194]. This compound has a room-temperature magnetic moment of 3.29 BM and is also five coordinate[194].

Nickel dibromide reacts with a wide range of phosphines[143,251] to give

TABLE 8.20

Pyridine Type Adducts of the Series $NiBr_2L_n$

Compound and Properties	References
L = pyridine, $n = 4$, light blue-green, $\nu_{Ni-N} = 235$ cm^{-1}, octahedral	91, 93, 95–97, 100, 101, 106, 107, 111, 248
L = 3-methylpyridine, $n = 4$, light blue-green, $\mu = 3.17$ BM, octahedral	95–97, 103, 111
L = 4-methylpyridine, $n = 4$, pale blue, $\mu = 3.30$ BM, octahedral	95–97, 100, 103, 111, 249
L = pyridine, $n = 2$, yellow-green, $\mu = 3.30$ BM, octahedral	91, 95, 97, 108, 110, 111
L = 2-methylpyridine, $n = 2$, deep blue, $\mu = 3.47$ BM, $\nu_{Ni-Br} =$ 256 cm^{-1}, $\nu_{Ni-N} = 239$ cm^{-1}, tetrahedral	91, 95, 97, 103, 112
L = 3-methylpyridine, $n = 2$, yellow, $\mu = 3.43$ BM, octahedral	95–97, 103, 110, 111
L = 4-methylpyridine, $n = 2$, yellow, $\mu = 3.38$ BM, octahedral	95–97, 103, 110, 111
L = 4-chloropyridine, $n = 2$, $\mu = 3.21$ BM, octahedral	113
L = 3-bromopyridine, $n = 2$, $\mu = 3.20$ BM, octahedral	113
L = 2,3-lutidine, $n = 2$, blue, m.p. = 119°, $\mu = 3.54$ BM, tetrahedral	114
L = 2,4-lutidine, $n = 2$, blue, m.p. = 105°, $\mu = 3.53$ BM, tetrahedral	114
L = 2,5-lutidine, $n = 2$, blue, m.p. = 105°, diamagnetic, square planar	114, 250
L = 2,6-lutidine, $n = 2$, dark blue, tetrahedral	97, 250
L = 3,4-lutidine, $n = 2$, blue-violet, $\mu = 3.60$ BM, tetrahedral	104, 250
L = *p*-toluidine, $n = 2$, blue-green, $\mu = 3.42$ BM, octahedral	91, 110, 111, 116
L = quinoline, $n = 2$, dark blue, $\nu_{Ni-Br} = 263$, 252 cm^{-1}, ν_{Ni-N} = 212 cm^{-1}, tetrahedral	91, 95, 100, 105, 111, 117, 119, 250

four-, five- and six-coordinate nickel(II). Arsines also react readily with nickel dibromide but there have been fewer studies. Probably one of the most interesting features of this section of the chemistry of nickel dibromide are the reports by Venanzi and coworkers[151] and by Hayter and Humiec[149] of a square planar-tetrahedral isomerism in certain diphenylalkylphosphine complexes. The particular isomer isolated from the reaction between nickel dibromide and the appropriate phosphine in alcohol depends on experimental conditions such as temperature, and the solvent used for recrystallization. A similar type of isomer equilibrium has been reported by van Hecke and Horrocks[12] with dibromo[1,3-bis(diphenylphosphino)-propane]nickel(II) in dichloromethane solution.

Refluxing a mixture of anhydrous nickel dibromide and diphenylphosphine in a non-ionizing solvent leads to the formation of brown, diamagnetic $NiBr_2[(C_6H_5)_2PH]_3$[145]. Spectral and other studies were consistent with the formulation of this compound as a five-coordinate monomer[145]. This stereochemistry has been confirmed by Bertrand and Plymale[252] who showed that the compound is isomorphous and isostructural with the corresponding cobalt dibromide complex, for which full structural data are available (Figure 7.15, page 362).

Another interesting structure is that of the green form of dibromobis-(benzyldiphenylphosphine)nickel(II). A preliminary report by Kilbourn, Powell and Darbyshire[253] indicates that the triclinic unit cell contains three single molecules. One of these has a *trans* square-planar configuration while the other two contain tetrahedrally coordinated nickel.

The triethylphosphine, triphenylphosphine and tricyclohexylphosphine adducts of nickel dibromide have been investigated by X-ray powder diffraction techniques and the lattice parameters of the adducts are given in Table 8.21.

TABLE 8.21
Lattice Parameters of Nickel Dibromide-Phosphine Adducts of the Type $NiBr_2L_2$

	$L = (C_2H_5)_3P$[10,254]	$L = (C_6H_5)_3P$[144]	$L = (C_6H_{11})_3P$[160]
Symmetry	Monoclinic	Monoclinic	Triclinic
a (Å)	7.60	10.3	9.88
b (Å)	11.52	9.1	10.28
c (Å)	13.53	39.2	10.74
α			112.7°
β	124.7°	111°	109.7°
γ			90.7°
Z	2	4	1
Space group	$P2_1/c$		$P\bar{1}$

The triethylphosphine adduct was shown to contain nickel in a *trans* square-planar arrangement with nickel-phosphorus and nickel-bromine bond distances of 2.26 and 2.30 Å respectively. The bromine-nickel-phosphorus angle was reported to be 90.5°. The results of the study on the triphenylphosphine adduct suggested tetrahedrally coordinated nickel.

A single-crystal study[255] of the complex $NiBr_2TAS$[14], where TAS is the tridentate arsine methyl bis(3-propanedimethylarsino)arsine, has revealed a distorted square pyramidal stereochemistry for the nickel atom. The unit cell is orthorhombic with $a = 16.37$, $b = 15.35$ and $c = 15.28$ Å. There are two molecular units in the cell of space group *Pbca*. The structure

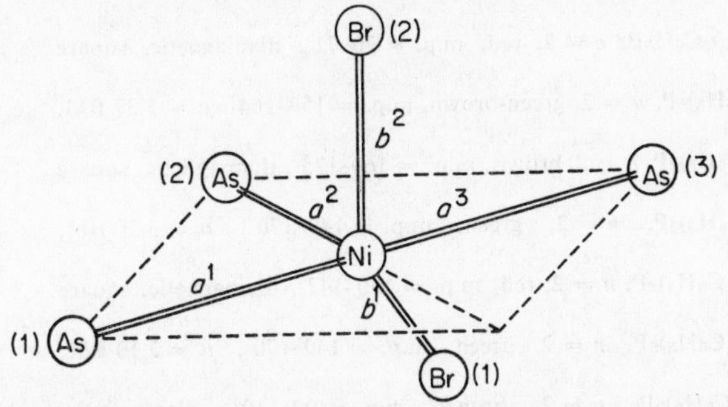

Figure 8.10 The structure of $NiBr_2 \cdot TAS$. (Reproduced by permission from G. A. Mair, H. M. Powell and D. E. Henn, *Proc. Chem. Soc.*, **1960,** 415)

of the compound, shown in Figure 8.10, consists of a distorted square pyramid in which one bromine atom is at the apex of the pyramid, while the other bromine atom is below the approximate plane of the nickel and arsenic atoms by about 20°. The important internuclear dimensions are given in Table 8.22.

TABLE 8.22

Internuclear Dimensions of $NiBr_2 \cdot TAS$

Distance	Value (Å)	Angle	Value	Angle	Value
Ni—As(1)	2.27	Br(2)—Ni—As(1)	92°	Br(1)—Ni—As(2)	154°
Ni—As(2)	2.26	Br(2)—Ni—As(2)	95°	As(3)—Ni—As(2)	90°
Ni—As(3)	2.26	Br(2)—Ni—As(3)	93°	As(3)—Ni—As(1)	175°
Ni—Br(1)	2.37	Br(2)—Ni—Br(1)	111°	As(2)—Ni—As(1)	91°
Ni—Br(2)	2.69	Br(1)—Ni—As(1)	88°		

The abnormal NiBr(2) bond length appears to be caused by steric effects of the terminal methyl groups on the arsine.

Some properties of a number of other phosphine and arsine adducts of nickel dibromide are given in Table 8.23.

TABLE 8.23

Phosphine and Arsine Adducts of the Type $NiBr_2L_n$

Compound and Properties	References
$L = (C_6H_5)_2PH$, $n = 3$, dark brown, m.p. = 98°, diamagnetic, five coordinate	145, 252
$L = (C_2H_5)_2C_6H_5P$, $n = 2$, red, m.p. = 225–227°, diamagnetic, square planar	164
$L = (n\text{-}C_4H_9)_2C_6H_5P$, $n = 2$, red, m.p. = 70–71°, diamagnetic, square planar	148
$L = CH_3(C_6H_5)_2P$, $n = 2$, green-brown, m.p. = 159–164°, $\mu = 3.37$ BM, tetrahedral	149
$L = C_2H_5(C_6H_5)_2P$, $n = 2$, brown, m.p. = 160–175°, diamagnetic, square planar	149, 150
$L = C_2H_5(C_6H_5)_2P$, $n = 2$, green, m.p. = 149–170°, $\mu = 3.20$ BM, tetrahedral	149, 150
$L = n\text{-}C_3H_7(C_6H_5)_2P$, $n = 2$, red, m.p. = 130–147°, diamagnetic, square planar	149
$L = n\text{-}C_3H_7(C_6H_5)_2P$, $n = 2$, green, m.p. = 149–170°, $\mu = 3.10$ BM, tetrahedral	149
$L = i\text{-}C_3H_7(C_6H_5)_2P$, $n = 2$, brown, m.p. = 90–110°, diamagnetic, square planar	149
$L = i\text{-}C_3H_7(C_6H_5)_2P$, $n = 2$, green, m.p. = 171–177°, $\mu = 3.05$ BM, tetrahedral	149
$L = n\text{-}C_4H_9(C_6H_5)_2P$, $n = 2$, brown, m.p. = 90–160°, diamagnetic, square planar	149
$L = n\text{-}C_4H_9(C_6H_5)_2P$, $n = 2$, green, m.p. = 161–162°, $\mu = 3.30$ BM, tetrahedral	148
$L = i\text{-}C_4H_9(C_6H_5)_2P$, $n = 2$, brown, m.p. = 70–74°, square planar	149
$L = i\text{-}C_4H_9(C_6H_5)_2P$, $n = 2$, green, m.p. = 131–150°, tetrahedral	149
$L = s\text{-}C_4H_9(C_6H_5)_2P$, $n = 2$, brown, m.p. = 85–90°, square planar	149
$L = s\text{-}C_4H_9(C_6H_5)_2P$, $n = 2$, green, m.p. = 158–169°, tetrahedral	149
$L = t\text{-}C_4H_9(C_6H_5)_2P$, $n = 2$, green, m.p. = 193–203°, tetrahedral	149
$L = C_7H_7(C_6H_5)_2P$, $n = 2$, green, m.p. = 192–193°, $\mu = 2.70$ BM	151, 253
$L = C_7H_7(C_6H_5)_2P$, $n = 2$, red, m.p. = 192–193°, diamagnetic, square planar	151
$L = (C_7H_7)_2C_6H_5P$, $n = 2$, red, m.p. = 223–225°, diamagnetic, square planar	148
$L = (C_7H_7)_3P$, $n = 2$, brown, m.p. = 134–137°, diamagnetic, square planar	151
$L = C_2H_5(C_6H_{11})_2P$, $n = 2$, red, m.p. = 150–152°, diamagnetic, square planar	152
$L = (C_2H_5)_2C_6H_{11}P$, $n = 2$, red, m.p. = 82–83°, diamagnetic, square planar	152
$L = (C_2H_5)_3P$, $n = 2$, red-violet, m.p. = 106–107°, diamagnetic, square planar	8, 10, 147, 154, 254
$L = (i\text{-}C_3H_7)_3P$, $n = 2$, red, m.p. = 151–154°, diamagnetic, square planar	155
$L = (s\text{-}C_3H_7)_3P$, $n = 2$, red, m.p. = 91–93°, diamagnetic, square planar	155

TABLE 8.23 (*contd.*)

Phosphine and Arsine Adducts of the Type $NiBr_2L_n$

L = $(C_6H_5)_3P$, $n = 2$, dark green, m.p. = 223°, $\mu = 2.97$ BM, $\nu_{Ni\text{-}Br} = 267, 258, 218\ cm^{-1}$, $\nu_{Ni\text{-}P} = 190, 148\ cm^{-1}$, tetrahedral	119, 144, 154, 156–159, 256, 257
L = $(C_6H_{11})_3P$, $n = 2$, olive-green, m.p. = 207°, diamagnetic, square planar	160, 258
L = bis(diphenylphosphino)methane (DPM), $n = 2$, red, m.p. = 115–117°, diamagnetic, square planar	11, 12
L = 1,2-bis(diphenylphosphino)ethane (DPE), $n = 1$, red, $\nu_{Ni\text{-}Br} = 288, 265\ cm^{-1}$, diamagnetic, square planar	11, 12, 161
L = 1,3-bis(diphenylphosphino)propane (DPP), $n = 1$, red, diamagnetic, square planar	12
L = 1,4-bis(diphenylphosphino)butane (PC_4P), $n = 1$, dark green, $\mu = 3.30$ BM, tetrahedral	259
L = 1,5-bis(diphenylphosphino)pentane (PC_5P), $n = 1$, dark green, $\mu = 3.28$ BM, tetrahedral	259
L = tetracyclohexyldiphosphine, $n = 1$, red, m.p. = 192–194°, diamagnetic, *cis* square planar	163
L = *P,P,P′,P′*-tetraethylethylenediphosphine, $n = 1$, brown, m.p. = 213°, diamagnetic, square planar	13
L = $(C_6H_{11})_2P(CH_2)_2P(C_6H_{11})_2$, $n = 1$, red, m.p. = 235°, *cis* square planar	162
L = 2-phenylisophosphindoline, $n = 2$, red, m.p. = 225–227°, square planar	164
L = bis(*o*-methylthiophenyl)phenylphosphine, $n = 1$, purple, diamagnetic, five coordinate	165
L = *o*-phenylenebis(dimethylarsine), $n = 1$, pale red, diamagnetic, square planar	161
L = *o*-phenylenebis(dimethylarsine), $n = 2$, brown, diamagnetic, 1:1 electrolyte, isomorphous with $NiCl_2L_2$	260

A number of other phosphines and arsines such as diphenyl(*o*-diphenylarsinophenyl)phosphine and bis(3-dimethylarsinylpropyl)methylarsine are reported to form adducts with nickel dibromide[143,146,261–267].

A number of tetrahedral adducts of nickel dibromide have been prepared with phosphine and arsine oxides. Some properties of several of these compounds are given in Table 8.24.

Lever and coworkers[174,175,177] have reported that 2,5-dimethylpyrazine reacts readily with nickel dibromide to give purple diamagnetic $NiBr_2L$. The structure of this adduct has been investigated by Ayres and coworkers[268] by single-crystal X-ray diffraction techniques. The monoclinic unit cell has $a = 9.09$, $b = 8.20$, $c = 6.52$ Å and $\beta = 108.6°$. The unit cell of space group $C2/m$ contains two molecules of the complex. The structure consists of infinite chains of nickel atoms linked together via the nitrogen atoms of the pyrazine ligand, with the bromine atoms at right angles to the plane of the pyrazine ring. The configuration about the nickel atom is thus *trans* square planar. The nickel-bromine and nickel-nitrogen bond distances are 2.31 and 1.85 Å respectively.

Other adducts of nickel dibromide using both pyrazine itself and substituted pyrazines have been prepared, as shown in Table 8.25.

TABLE 8.24
Phosphine and Arsine Oxide Adducts of the Type $NiBr_2L_n$

Compound and Properties	References
L = $(C_2H_5)_3PO$, $n = 2$, blue-violet, m.p. = 93°	166
L = $(C_6H_5)_3PO$, $n = 2$, blue, $\mu = 3.98$ BM	167, 168
L = $(C_6H_{11})_3PO$, $n = 2$, blue, m.p. = 199–201°	166
L = bis(diphenylphosphinoethyl)oxide (POP), $n = 1$, brown, $\mu = 3.23$ BM	259
L = $(C_6H_5)_3AsO$, $n = 2$, blue, $\mu = 3.96$ BM, $\nu_{As\text{-}O} = 846$ cm^{-1}, $\nu_{Ni\text{-}Br} = 234$ cm^{-1}	167, 169, 171

TABLE 8.25
Adducts of the Type $NiBr_2L_n$ with Pyrazine and Related Ligands

Compound and Properties	References
L = pyrazine, $n = 2$, pale green, $\mu = 3.13$ BM, octahedral	174, 175, 177
L = 2-methylpyrazine, $n = 5$, yellow-green, $\mu = 3.24$ BM, octahedral, fifth molecule of ligand acts as solvate	174, 175
L = 2-methylpyrazine, $n = 1$, pale blue, $\mu = 3.27$ BM, octahedral	110, 111, 174, 175, 177
L = 2,6-dimethylpyrazine, $n = 1$, pale blue, $\mu = 3.22$ BM, octahedral	110, 111, 174, 175, 177

Froust and Soderberg[269] have reported that the reaction between bis(diphenylglyoximato)nickel(II) and bromine gives black $NiBr_2(dpg)_2$. The tetragonal unit cell of space group *P4/ncc* has $a = 19.51$ and $c = 6.72$ Å and contains four molecules. Although it was originally suggested that the compound contained nickel(IV), the crystallographic evidence suggests the presence of bromine molecules surrounded by, and interacting with, the phenyl groups in a $Ni(dpg)_2$ host lattice.

Goodgame and Venanzi[176,181] have prepared nickel dibromide adducts of *N,N*-dimethylethylenediamine (Me_2en) and *N,N*-diethylethylenediamine (Et_2en) of the general type $NiBr_2L_2$. The Me_2en adduct is green and has a room-temperature magnetic moment of 3.23 BM, whereas the Et_2en complex, which is orange, is diamagnetic. This difference was attributed to variations in the tetragonal distortion of the adducts.

TABLE 8.26
Adducts of the Type $NiBr_2L_n$

Compound and Properties	References
L = 2-methylquinoxaline, $n = 1$, grey-pink, $\mu = 3.17$ BM, octahedral	173
L = 2-methylthiomethylpyridine, $n = 2$, green, $\mu = 3.28$ BM, octahedral	179
L = 2-methyl-8-methylthioquinoline, $n = 1$, green-brown, $\mu = 3.20$ BM, tetrahedral	180
L = *N,N,N′,N′*-tetramethylethylenediamine, $n = 1$, violet, m.p. = 198–200°, $\mu = 3.26$ BM, tetrahedral	182, 183
L = *N,N,N′,N′*-tetramethyl-1,2-propylenediamine, $n = 1$, violet, m.p. = 222–225°, $\mu = 3.40$ BM, tetrahedral	182, 183
L = *N,N,N′,N′*-tetramethyltrimethylenediamine, $n = 1$, violet, m.p. = 265–272°, $\mu = 3.32$ BM, tetrahedral	182, 183
L = 1,1,7,7-tetraethyldiethylenetriamine, $n = 1$, red-violet, diamagnetic, ionic with five coordinate cation	184
L = bis(2-dimethylaminoethyl)methylamine (dienMe), $n = 1$, ochre, $\mu = 3.38$ BM, five coordinate	185, 270
L = tris(2-dimethylaminoethyl)amine (trienMe), $n = 1$, mustard, $\mu = 3.42$ BM, ionic with five coordinate cation	186
L = bis(2-dimethylaminoethyl)oxide (Me_4daeo), $n = 1$, red, $\mu = 3.40$ BM, five coordinate	187
L = bis(2-dimethylaminoethyl)sulphide (Me_4daes), $n = 1$, $\mu = 3.30$ BM, five coordinate	188
L = Schiff base formed between *N*-methyl-*o*-aminobenzaldehyde and *N,N*-diethylethylenediamine, $n = 1$, yellow, $\mu = 3.26$ BM, five coordinate	189
L = Schiff base formed between *o*-methylthiobenzaldehyde and *N,N*-diethylethylenediamine, $n = 1$, yellow-green, m.p. = 182–185°, $\mu = 3.14$ BM, five coordinate	190
L = *N*-2-aminoethyl-2-thenylideneimine, $n = 2$, yellow-green, $\mu = 3.15$ BM, *trans* octahedral	271
L = 2-methylimidazole (MIz), $n = 2$, blue, $\mu = 3.59$ BM, $\nu_{Ni\text{-}Br} = 241$, 227 cm^{-1}, $\nu_{Ni\text{-}N} = 272$ cm^{-1}, tetrahedral	181
L = 2[β-(phenylamino)ethyl]pyridine, $n = 1$, red-brown, m.p. = 205°, $\mu = 3.28$ BM, octahedral	196
L = 2[β-(isopropylamino)ethyl]pyridine, $n = 1$, red-brown, m.p. = 158–160°, $\mu = 3.14$ BM, octahedral	196
L = β(bisphenylphosphino)ethyl pyridine, $n = 1$, blue, m.p. = 196°, $\mu = 3.33$ BM, octahedral	197
L = NH_3, $n = 2$, yellow-green, $\mu = 3.29$ BM, $\nu_{Ni\text{-}N} = 434$ cm^{-1}, octahedral	110, 111, 138, 272
L = ethylenethiourea, $n = 4$, yellow, $\mu = 3.30$ BM, $\theta = 15°$, octahedral, orange isomer also known	140
L = 1-(1-naphthyl)-2-thiourea, $n = 2$, green, $\mu = 3.32$ BM, tetrahedral	140

Nickel dibromide reacts readily with excess quinoxaline (Q) at 100° to give yellow $NiBr_2Q_2$[110,111,172,173], which has a room-temperature magnetic moment of 3.45 BM. Its diffuse-reflectance spectrum is consistent with nickel in a tetragonally distorted octahedral configuration. If the yellow 1:2 adduct is boiled with quinoxaline, then a dark brown isomer of the compound may be isolated[172]. This adduct has a magnetic moment of 3.51 BM and a very complex electronic absorption spectrum. It was suggested that the compound, which appeared to be polymeric, contained both tetrahedrally and octahedrally coordinated nickel(II).

Some properties of a number of other adducts of nickel dibromide are given in Table 8.26.

Other donor ligands that have been reacted with nickel dibromide include bipyridyl, dimethyl sulphoxide, dioxan etc.[118,120,124,146,194,195,199–201,209,211,212,216,223,224,273–278].

Nickel dibromide hydrates. Green nickel dibromide hexahydrate may be prepared by dissolving the carbonate or hydroxide in hydrobromic acid at room temperature, and has also been isolated from the nickel dibromide-water system[242,279]. The hexahydrate looses water very readily. Above about 29° the green trihydrate may be isolated from aqueous solutions[242,279]. A dihydrate has also been reported[280]; this was prepared by standing the hexahydrate over concentrated sulphuric acid at 5°.

The structures of the dihydrate and hexahydrate of nickel dibromide have been investigated by Weigel[280]. Both compounds have monoclinic unit cells with the parameters shown in Table 8.27.

TABLE 8.27
Unit-cell Parameters of $NiBr_2 \cdot 2H_2O$ and $NiBr_2 \cdot 6H_2O$

Parameter	$NiBr_2 \cdot 2H_2O$	$NiBr_2 \cdot 6H_2O$
a (Å)	7.21	6.83
b (Å)	7.23	7.18
c (Å)	9.17	11.63
β	92.7°	51.1°

For the hexahydrate, two bromine atoms and four water molecules are coordinated to each nickel atom, whereas in the dihydrate there are four bromine atoms and two water molecules about the central metal atom. For both hydrates the nickel-bromine and nickel-oxygen bond distances are 2.6 and 2.0 Å respectively.

Nickel diiodide. Probably the most convenient method of preparing nickel diiodide in solution is the reaction between a nickel salt, such as the chloride or nitrate, and sodium iodide in ethanol[172,181]. Alternatively, the hexahydrate, which is readily formed by dissolving nickel carbonate in hydrodic acid, may be dehydrated quite readily, the product being sublimed[242] at 500–600°. Chaigneau[281,282] has interacted aluminium triiodide and nickel oxide in a sealed tube at 230° to prepare anhydrous nickel diiodide.

Nickel diiodide is a dark crystalline solid which melts without decomposition at about 780° in a sealed tube, whereas in an open system it tends to decompose on melting[72,283]. Nickel diiodide has a hexagonal unit cell[280] with $a = 3.88$ and $c = 19.6$ Å.

Nickel diiodide obeys the Curie-Weiss law[235] with a magnetic moment of 3.25 BM and a Weiss constant of 42°. At low temperatures it becomes magnetically ordered and a Neel temperature of 75°K has been reported[245].

In general, the chemistry of nickel diiodide closely resembles that of the corresponding chloride and bromide. Thus Sharp and coworkers[95,97,105,116] have investigated the preparation and properties of a large range of adducts with pyridine, substituted pyridines, aniline and related donor molecules. Like the chloro and bromo derivatives, many of these adducts are thermally unstable[97,100,112] and the following decomposition reactions have been reported.

$$NiI_2(\text{pyridine})_6 \xrightarrow{100°} NiI_2(\text{pyridine})_4 \xrightarrow{160°} NiI_2(\text{pyridine})_2$$

$$NiI_2(\text{2-methylpyridine})_2 \xrightarrow{180°} NiI_2$$

$$NiI_2(\text{3-methylpyridine})_4 \xrightarrow{150°} NiI_2(\text{3-methylpyridine})_2$$

$$NiI_2(\text{4-methylpyridine})_6 \xrightarrow{100°} NiI_2(\text{4-methylpyridine})_4 \xrightarrow{180°} NiI_2(\text{4-methylpyridine})_2$$

$$NiI_2(\text{2,6-lutidine})_4 \xrightarrow{100°} NiI_2(\text{2,6-lutidine})_2$$

$$NiI_2(\text{2,4,6-collidine})_4 \xrightarrow{180°} NiI_2(\text{2,4,6-collidine})_2$$

The electronic absorption spectra of a large range of nickel diiodide adducts with pyridine and related ligands have been investigated by Ludwig and Wittman[250]. Diffuse-reflectance and solution spectra showed that the complexes of the type NiI_2L_2 could be divided into two groups. Those with α-substituted pyridine rings were of D_{2h} symmetry with a singlet ground state, while those in which the pyridine ring was not substituted in the α position were of C_{2v} symmetry with a triplet ground state.

Some properties of a number of nickel diiodide adducts of pyridine and related ligands are given in Table 8.28.

TABLE 8.28
Pyridine and Related Ligand Adducts of the Type NiI_2L_n

Compound and Properties	References
L = pyridine, n = 4, green-blue, μ = 3.21 BM, octahedral	94–97, 107
L = 3-methylpyridine, n = 4, yellow-green, μ = 3.21 BM, octahedral	94, 97, 100, 103
L = 4-methylpyridine, n = 4, yellow-green, μ = 3.24 BM, octahedral	94, 97, 100, 103, 249
L = 2,6-lutidine, n = 4, light green, octahedral	95, 97
L = 3,5-lutidine, n = 4, green, μ = 3.29 BM, octahedral	104
L = pyridine, n = 2, dark green, μ = 3.44 BM, tetrahedral	94–97, 106, 108, 112
L = 2-methylpyridine, n = 2, dark green, diamagnetic square planar	95, 97, 103
L = 3-methylpyridine, n = 2, dark green, μ = 3.47 BM, tetrahedral	95, 97, 103, 112
L = 4-methylpyridine, n = 2, dark green, μ = 3.38 BM, tetrahedral	94, 95, 97, 103, 249
L = 2,3-lutidine, n = 2, dark green, m.p. = 112°, diamagnetic, square planar	114
L = 2,4-lutidine, n = 2, dark green, m.p. = 110°, diamagnetic, square planar	114
L = 2,5-lutidine, n = 2, dark green, m.p. = 117°, diamagnetic, square planar	114
L = quinoline, n = 2, olive-green, diamagnetic, square planar	105, 117

Adducts with aniline and substituted anilines have also been reported[100,115,116].

The preparation and properties of a number of simple amine adducts of nickel diiodide have been investigated by Uhlig and Staiger[125,126]. Some of the decomposition reactions reported are:

$$NiI_2[(CH_3)_2NH]_4 \xrightarrow{120°} NiI_2[(CH_3)_2NH]_2$$

$$NiI_2[C_2H_5NH_2]_6 \xrightarrow{110°} NiI_2[C_2H_5NH_2]_4$$

$$NiI_2[C_3H_7NH_2]_6 \xrightarrow{80°} NiI_2[C_3H_7NH_2]_4$$

Nickel diiodide has been reacted with quite a wide variety of phosphine and arsine ligands[143,251]. As with the corresponding nickel dibromide complexes, Hayter and Humiec[149] have reported a square planar-tetrahedral isomerism with the diphenylalkylphosphines. In addition an unusual, but unidentified type of isomerism was observed in cases where the alkyl group of the phosphine ligand is branched at the α carbon atom. Another square planar-tetrahedral equilibrium[12] was shown to exist in dichloromethane solutions of bisiodobis[1,3-bis(diphenylphosphino)-propane] nickel(II).

$NiI_2[(C_6H_5)_3P]_2$, a brown solid melting at 218–220°, has a room-temperature magnetic moment of 2.92 BM and is considered to contain tetrahedrally coordinated nickel[119,154,156,157,159,257]. Preliminary X-ray crystallographic studies[144] indicate a monoclinic unit cell with $a = 19.6$, $b = 10.3$, $c = 18.2$ Å and $\beta = 112°$. The space group is consistent with a tetrahedral configuration about the nickel atom.

Nickel diiodide reacts readily with diphenylphosphine in refluxing non-ionic solvents to give a dark blue 1:3 adduct[145]. The compound melts at 115° and its spectrum and magnetism indicated that it contained five-coordinate nickel. This stereochemistry has been confirmed by Bertrand and Plymale[252] by showing that the compound is isostructural with the corresponding cobalt dibromide adduct, for which full crystallographic details are available (Figure 7.15, page 362). The nickel diiodide complex has a triclinic unit cell with $a = 11.16$, $b = 11.93$, $c = 15.53$ Å, $\alpha = 81.7°$, $\beta = 100.8°$ and $\gamma = 118.3°$. The compound has a room-temperature magnetic moment of 1.29 BM and this low moment is considered to be due to proximity to the cross-over situation between the singlet ground state and a triplet ground state.

Nickel diiodide reacts with *o*-phenylenebis(dimethylarsine) (D) to give brown NiI_2D_2[284]. The adduct has a monoclinic unit cell with $a = 9.49$, $b = 9.25$, $c = 16.94$ Å and $\beta = 114°$. There are two molecules in the unit cell of space group $P2_1/c$. The compound is monomeric and contains *trans* iodo groups as shown in Figure 8.11 which also shows the important internuclear dimensions. The nickel-arsenic bond is appreciably shorter than the sum of the covalent radii and Stephenson[284] has suggested that this fact indicates the presence of $d_\pi - d_\pi$ bonding. In addition there is quite strong interaction between the iodine atoms and the methyl groups of the arsenic ligand, resulting in distortion of the tetrahedral arrangement of the bonds about the arsenic atoms as shown in Figure 8.12.

Table 8.29 summarizes the properties of a number of complexes of nickel diiodide with other phosphines.

A number of complexes with phosphines and arsines, other than those described above, have also been reported[143,148,154,164,257,261,264,265,285,286].

Many derivatives of nickel diiodide have been prepared with

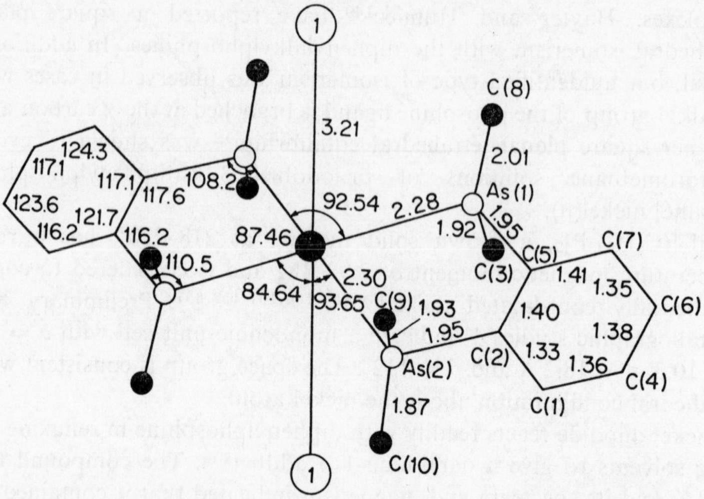

Figure 8.11 Internuclear dimensions of NiI_2D_2. (Reproduced by permission from N. C. Stephenson, *Acta Cryst.*, 17, 592 (1964))

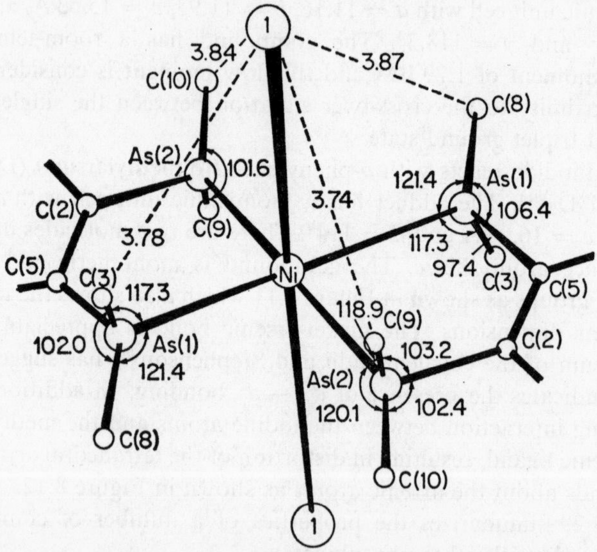

Figure 8.12 Schematic representation of the approach of the iodine atoms to the nickel–arsenic plane in NiI_2D_2 and resultant distortion of the tetrahedral configuration about the arsenic atoms. (Reproduced by permission from N. C. Stephenson, *Acta Cryst.*, 17, 592 (1964))

miscellaneous ligands and in general the compounds are similar to the corresponding chloro and bromo derivatives described earlier[118,120,129,131,133,138,140,146,165,167,168,172–177,180–183,185,186,189–191,194,196,197,199–201,209,211,212,222–224,262,269,271,276,277].

TABLE 8.29

Phosphine Adducts of the Type NiI_2L_n

Compound and Properties	References
L = $(n\text{-}C_4H_9)_2C_6H_5P$, $n = 2$, bronze, m.p. = 87–88°, diamagnetic, square planar	148
L = $CH_3(C_6H_5)_2P$, $n = 2$, red-brown, m.p. = 145–147°, μ = 3.32 BM, tetrahedral	149
L = $C_2H_5(C_6H_5)_2P$, $n = 2$, red-brown, m.p. = 127–138°, μ = 3.10 BM, tetrahedral	149, 150
L = $n\text{-}C_3H_7(C_6H_5)_2P$, $n = 2$, red-brown, m.p. = 120–129°, μ = 3.06 BM, tetrahedral	149
L = $i\text{-}C_3H_7(C_6H_5)_2P$, $n = 2$, brown, m.p. = 100–140°, μ = 2.82 BM, tetrahedral	149
L = $i\text{-}C_3H_7(C_6H_5)_2P$, $n = 2$, green, m.p. = 105–140°, diamagnetic, unknown structure	149
L = $i\text{-}C_4H_9(C_6H_5)_2P$, $n = 2$, brown, m.p. = 118–133°, tetrahedral	149
L = $s\text{-}C_4H_9(C_6H_5)_2P$, $n = 2$, brown, m.p. = 110–132°, tetrahedral	149
L = $s\text{-}C_4H_9(C_6H_5)_2P$, $n = 2$, green, m.p. = 108–115°, diamagnetic, unknown structure	149
L = $t\text{-}C_4H_9(C_6H_5)_2P$, $n = 2$, brown, m.p. = 198–208°, tetrahedral	149
L = $C_7H_7(C_6H_5)_2P$, $n = 2$, two dark red isomers isolated, one diamagnetic, the other paramagnetic	151
L = $(C_7H_7)_2C_6H_5P$, $n = 2$, brown, m.p. = 180–185°, diamagnetic, square planar	151
L = bis(diphenylphosphino)methane (DPM), $n = 2$, purple, diamagnetic, square planar with ligand acting as monodentate	12
L = 1,2-bis(diphenylphosphino)ethane (DPE), $n = 1$, purple, diamagnetic, square planar	11, 12, 161
L = 1,2-bis(diphenylphosphino)ethane (DPE), $n = 2$, yellow, diamagnetic, octahedral	12
L = 1,3-bis(diphenylphosphino)propane (DPP), $n = 1$, purple, diamagnetic, square planar	12
L = 1,5-bis(diphenylphosphino)pentane (PC_5P), $n = 1$, maroon, μ = 3.24 BM, tetrahedral	259

Nickel diiodide hexahydrate. The blue-green deliquescent hydrate is readily obtained by dissolving nickel carbonate or hydroxide in hydriodic acid, and evaporating the resulting solution at low temperatures[242,287]. The compound is thermally unstable, decomposing with loss of water at quite low temperatures. It has a hexagonal unit cell[280,287] with a = 7.67

and $c = 4.87$ Å. The lattice contains the $[Ni(H_2O)_6]^{2+}$ cation and iodide anions.

COMPLEX HALIDES

Although binary fluorides of nickel(IV) and nickel(III) are unknown, these oxidation states can be stabilized relatively easily in complex fluorides. All fluoro complexes of nickel are octahedral, but with the other halogens all of the discrete anionic complexes are tetrahedral. Octahedral coordination with the heavy halogens appears to be limited to polymeric anions.

Oxidation State IV

Hexafluoronickelates(IV). Salts of the hexafluoronickelate(IV) anion may be prepared by fluorination either of a 1:2 mixture of nickel dichloride and the appropriate alkali-metal chloride[288], or of alkali-metal tetrachloronickelate(II)[289]. The red diamagnetic compounds have cubic lattices with $a = 8.11$, 8.45 and 8.92 Å for the potassium, rubidium and caesium salts respectively[288,289]. The potassium salt has a nickel-fluorine stretching frequency[290] of 654 cm^{-1}.

Oxidation State III

Hexafluoronickelate(III). Fluorination of a 3:1 mixture of potassium chloride and nickel dichloride at 310–320° in a flow system gives potassium hexafluoronickelate(III)[291]. The compound has a cubic lattice with $a = 8.44$ Å[291] and has a room-temperature magnetic moment of 2.54 BM[292].

Oxidation State II

Hexafluoronickelate(II). Schnering[293] has prepared yellow barium hexafluoronickelate(II) by fusing a 2:1 mixture of barium fluoride and nickel difluoride at 1200°. The compound has a tetragonal cell with $a = 4.054$ and $c = 16.341$ Å[294]. The arrangement of the ions in the structure is the same as that previously described for the corresponding cobalt compound (page 367). The nickel-fluorine bond distances are 2.03 and 1.97 Å.

Barium hexafluoronickelate(II) becomes antiferromagnetic at low temperatures[293] but at higher temperatures the Curie-Weiss law is obeyed with a magnetic moment of 3.76 BM and a Weiss constant of 385°.

Tetrafluoronickelates(II). Alkali-metal tetrafluoronickelates(II) are most conveniently prepared by fusing nickel difluoride with the appropriate amount of alkali-metal or alkaline-earth metal fluoride or hydrogen difluoride either in vacuum, in an atmosphere of anhydrous hydrogen fluoride, or in fluorine[59,295–299]. The ammonium salt has been prepared by the reaction between nickel dibromide and ammonium fluoride in methanol[300] and by compressing an intimate mixture of nickel difluoride and ammonium fluoride at a very high pressure[301].

The structures of a number of the yellow tetrafluoronickelates(II) have been investigated by X-ray powder diffraction techniques and the available lattice parameters are given in Table 8.30.

TABLE 8.30

Symmetry and Lattice Parameters of Tetrafluoronickelates(II)

Compound	Symmetry	Parameters	References
Li_2NiF_4	Cubic	$a = 8.313$ Å	20, 59, 298
K_2NiF_4	Tetragonal	$a = 4.006$, $c = 13.076$ Å	59, 297
Rb_2NiF_4	Tetragonal	$a = 4.087$, $c = 13.71$ Å	59, 295
$(NH_4)_2NiF_4$	Tetragonal	$a = 4.084$, $c = 13.79$ Å	59, 295
Tl_2NiF_4	Tetragonal	$a = 4.051$, $c = 14.22$ Å	59, 295
$BaNiF_4$	Orthorhombic	$a = 14.46$, $b = 4.15$, $c = 5.80$ Å	299

Balz and Plieth[297] have examined the structure of potassium tetrafluoronickelate(II) by single-crystal X-ray techniques. The structure of this complex is shown in Figure 8.13.

The magnetic properties of the tetrafluoronickelates(II) have been the subject of considerable experimental and theoretical study. The gross features of the magnetic properties are summarized in Table 8.31.

TABLE 8.31

Magnetic Properties of Tetrafluoronickelates(II)

Compound	μ (BM)	$\theta°$	Reference
Li_2NiF_4	3.18	0	298
	3.18	5	59
K_2NiF_4[a]	2.00		59
Rb_2NiF_4[a]	2.05		59
$(NH_4)_2NiF_4$	3.30	13	59
Tl_2NiF_4[a]	2.03		59

[a] These compounds exhibit considerable antiferromagnetic interaction and the value reported is the effective magnetic moment at room temperature.

Potassium tetrafluoronickelate(II) has been subjected to the most detailed investigation[302–308]. Srivastava[302] has shown that there is a maximum in the magnetic susceptibility at about 250°K, but magnetic anisotropy does not become apparent until temperatures below 110°K. Legrand and Plumier[303] have made a neutron-diffraction study of the potassium compound as a function of temperature and the results show

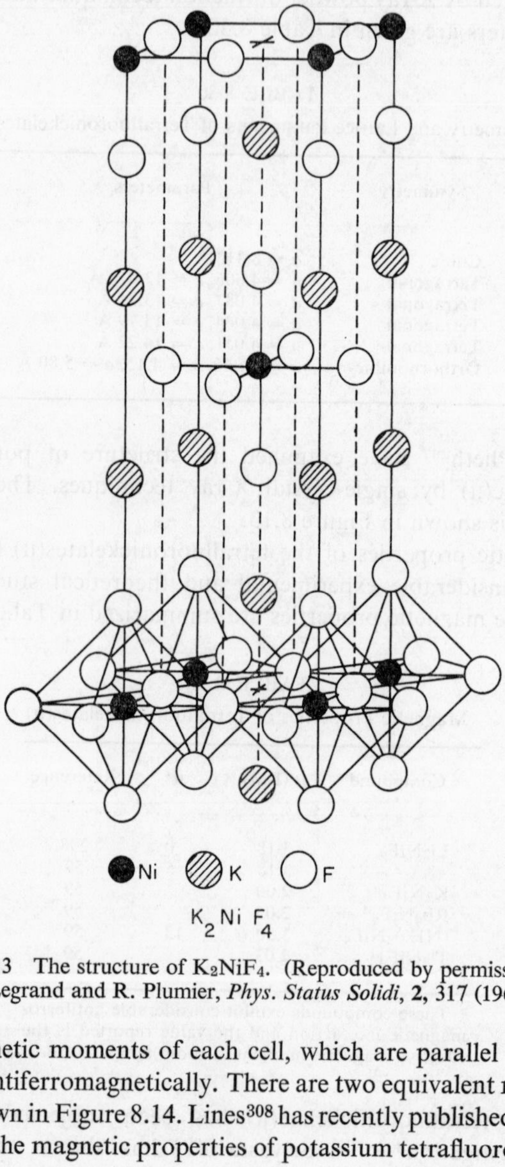

Figure 8.13 The structure of K_2NiF_4. (Reproduced by permission from E. Legrand and R. Plumier, *Phys. Status Solidi*, **2**, 317 (1962))

that the magnetic moments of each cell, which are parallel to the *c* axis, are coupled antiferromagnetically. There are two equivalent magnetic cells which are shown in Figure 8.14. Lines[308] has recently published a theoretical discussion of the magnetic properties of potassium tetrafluoronickelate(II).

The diffuse-reflectance spectra of the lithium, potassium and rubidium tetrafluoronickelates(II) have been observed by Rudorff and coworkers[59] and the reported assignments are shown in Table 8.32.

TABLE 8.32

Diffuse Reflectance Spectra (cm^{-1}) of Tetrafluoronickelates(II)

Li_2NiF_4	K_2NiF_4	Rb_2NiF_4	Transition
13,400	13,000	12,600	$^3T_{1g}(F) \leftarrow {}^3A_{2g}(F)$
15,000	15,150	15,050	$^1E_g(D) \leftarrow {}^3A_{2g}(F)$
21,300	21,200	21,000	$^1T_{2g}(D) \leftarrow {}^3A_{2g}(F)$
24,100	24,100	23,600	$^3T_{1g}(P) \leftarrow {}^3A_{2g}(F)$

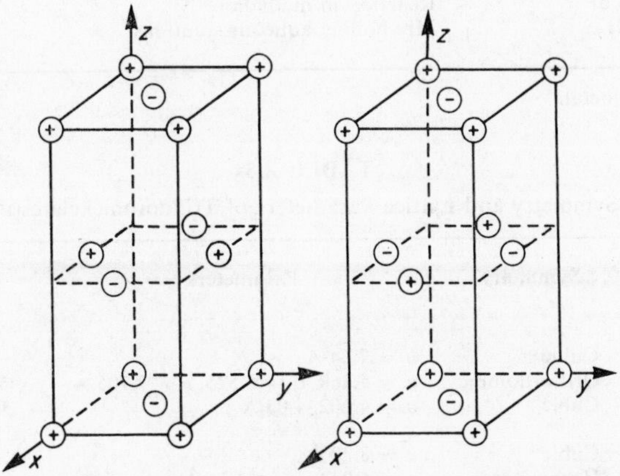

Figure 8.14 The two equivalent magnetic cells of K_2NiF_4. (Reproduced by permission from E. Legrand and R. Plumier, *Phys. Status Solidi*, **2,** 317 (1962))

Crocket and Haendler[309] have shown from an infrared study that there is considerable hydrogen bonding in ammonium tetrafluoronickelate(II).

Trifluoronickelates(II). The principal methods of preparing the pale yellow trifluoronickelates(II) are given in Table 8.33. It is important to note that in all cases the correct proportions of the constituents must be used to avoid concomitant formation of tetrafluoronickelates(II).

Of the above methods the most convenient is fusion of nickel difluoride with the appropriate fluoride, particularly if single crystals of the complex are required. Details for growing single crystals of these compounds prepared from aqueous solutions have been given by Knox[315].

The structures of most of the trifluoronickelates(II) have been examined in considerable detail by X-ray powder diffraction techniques, the major interest being in the correlation of structure and magnetic properties. The room-temperature data are summarized in Table 8.34.

TABLE 8.33

Preparation of Trifluoronickelates(II)

Method[a]	Conditions	References
NiF_2 and MF or MHF_2	Fuse in vacuum or in ahydrous hydrogen fluoride atmosphere	59, 295, 296, 310, 311
NiF_2 and MCl	Fuse	312
NiF_2 and NH_4F	Compress at high pressure	103
$NiCl_2$, MCl and HF	Flow system, 600°	2a
$NiBr_2$ and RbF	Reaction in methanol	309
NiF_2 and MF	Mix boiling aqueous solutions	313, 314

[a] M = Alkali metal.

TABLE 8.34

Symmetry and Lattice Parameters of Trifluoronickelates(II)

Compound	Symmetry	Parameters	References
$NaNiF_3$	Cubic	a = 7.64 Å	20
	Orthorhombic	a = 5.360, b = 5.525, c = 7.705 Å	59, 295, 316
$KNiF_3$	Cubic	a = 4.002–4.05 Å	59, 295, 296, 315, 317, 318
$RbNiF_3$	Cubic	a = 8.20 Å	20
	Hexagonal	a = 5.843, c = 14.31 Å	59
$CsNiF_3$	Hexagonal	a = 6.236, c = 5.225 Å	310
NH_4NiF_3	Pseudo-cubic	a = 8.15 Å	295
$TlNiF_3$	Hexagonal	a = 5.87, c = 14.37 Å	311

Potassium trifluoronickelate(II), which melts[296] at 1130°, has a perfect perovskite-type structure[315] as shown in Figure 8.15, the nickel-fluorine bond distance being 2.01 Å. Unlike the corresponding potassium complexes of manganese, iron and cobalt, the lattice symmetry of potassium trifluoronickelate(II) is not changed on cooling. At 78°K the cubic lattice has a = 4.001 Å[317,318]. The thallium salt, which has a hexagonal lattice at room temperature, becomes a cubic perovskite with a = 4.10 Å when heat treated under high pressure[311]. The reason for the existence of two modifications for the sodium and rubidium salts is not clear, but it may arise merely from the quite different experimental conditions required to prepare each modification.

The magnetic properties of some trifluoronickelates(II) are given in Table 8.35. All of the compounds are strongly antiferromagnetic with the

TABLE 8.35
Magnetic Properties of Some Trifluoronickelates(II)

Compound	μ_{eff} (BM)	T_N (°K)	Reference
$NaNiF_3$	2.40	142	314
		156	319
$KNiF_3$	2.05	300	59
	2.07	275	320
	2.19	280	313
$TlNiF_3$		150	311

Neel temperatures shown in the table. Rather surprisingly there is no anomaly in the thermal conductivity of potassium trifluoronickelate(II) at the Neel temperature, in marked contrast to the very distinct discontinuities observed for the corresponding manganese and cobalt

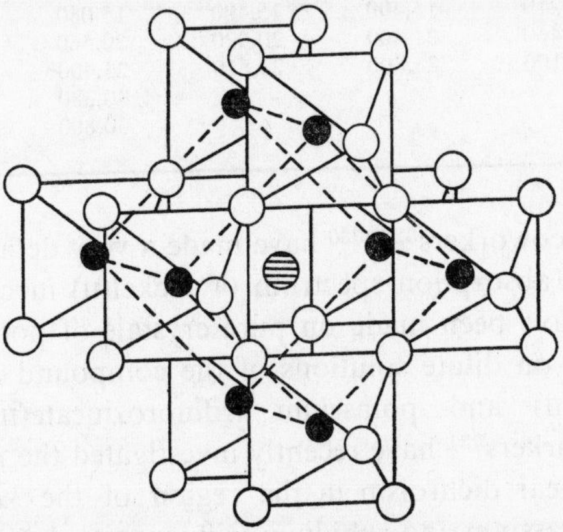

Figure 8.15 The $KNiF_3$ lattice. (Reproduced by permission from E. Legrand and R. Plumier, *Phys. Status Solidi*, **2**, 317 (1962))

complexes[321]. The magnetic properties of potassium trifluoronickelate(II) have recently been discussed by Lines[308] from a theoretical point of view. Neutron-diffraction measurements show that the potassium salt has a

'*G*-type' magnetic structure in which each magnetic ion is coupled antiferromagnetically with its six nearest neighbours[322]. At 4.2°K the magnetic superlattice lines have been indexed on the basis of a cubic cell with $a = 7.986$ Å, that is, twice the corresponding perovskite cell at room temperature[322]. Machin and coworkers[313,314] have investigated the magnetic properties of solid solutions of several trifluoronickelates(II) in the corresponding trifluorozincate(II) host lattices.

The infrared spectrum of potassium trifluoronickelate(II) has been investigated by several groups[290,323–325]. There are three infrared active modes, corresponding to stretching, bending and lattice modes, which occur at 446, 255 and 153 cm^{-1} respectively[323].

The electronic absorption spectra of a number of trifluoronickelates(II) have been recorded and the assignments, which are based on octahedral symmetry, are given in Table 8.36.

TABLE 8.36

Electronic Spectra (cm^{-1}) of Some Trifluoronickelates(II)

$NaNiF_3$[59]	$NaNiF_3$[326]	$KNiF_3$[59]	$KNiF_3$[327]	$RbNiF_3$[312]	Transition
	7,610		7,250	7.630	$^3T_{2g}(F) \leftarrow {}^3A_{2g}(F)$
12,900	12,810	12,800	12,530	12,380	$^3T_{1g}(F) \leftarrow {}^3A_{2g}(F)$
15,150	15,210	15,400	15,400	15,080	$^1E_g(D) \leftarrow {}^3A_{2g}(F)$
21,200	21,460	21,400	20,920	20,560	$^1T_{2g}(D) \leftarrow {}^3A_{2g}(F)$
23,800	24,100	23,800	23,810	23,400	$^3T_{1g}(P) \leftarrow {}^3A_{2g}(F)$
				30,380	$^1E_g(G) \leftarrow {}^3A_{2g}(F)$
				30,800	$^1T_{2g}(G) \leftarrow {}^3A_{2g}(F)$

Ferguson and coworkers[328–330] have made a very detailed investigation of the electronic absorption spectrum of nickel(II) in cubic perovskites. Measurements have been made on pure crystals of potassium trifluoronickelate(II) and on dilute solutions of the compound in potassium trifluoromagnesate(II) and potassium trifluorozincate(II) host lattices. Pisarev and coworkers[331] have recently investigated the magnetic circular and magnetic linear dichroism in the region of the $^1E_g(D) \leftarrow {}^3A_{2g}(F)$ transition for potassium and rubidium trifluoronickelate(II).

Pentachloronickelate(II). Blue caesium pentachloronickelate(II) may be isolated from the caesium chloride-nickel dichloride melt of the appropriate stoichiometry if the system is quenched rapidly[64]. Slow cooling results in decomposition of the salt giving caesium chloride and the trichloronickelate(II). The pentachloro compound is isostructural with caesium pentachlorocobaltate(II) and thus contains tetrahedrally coordinated nickel. Boston and coworkers[332] have investigated the absorption spectrum of molten caesium pentachloronickelate(II).

Tetrachloronickelates(II). The usual method of preparing salts of the tetrachloronickelate(II) anion is by the interaction of nickel dichloride and quaternary ammonium, phosphonium or arsonium chloride in alcohol or nitromethane[167,333–336]. The tetramethylammonium salt has also been prepared[337] by evaporating an aqueous solution of nickel dichloride and tetramethylammonium chloride at 80°. It is important to note that the alkali-metal tetrachloronickelates(II) cannot be prepared from the melt.

The blue tetrachloronickelates(II) are unstable towards hydrolysis. The heat of formation[338] of the tetramethyl and tetraethylammonium salts from their constituent chlorides are −0.09 and −6.33 kcal $mole^{-1}$ respectively.

The tetramethylammonium salt is isostructural with the corresponding cobalt complex[337], having an orthorhombic unit cell with $a = 12.264$, $b = 8.982$ and $c = 15.486$ Å. The space group is *Pnma* and there are four molecules in the unit cell. The structure consists of tetramethylammonium cations and discrete, but flattened tetrachloronickelate(II) anions. The distortion arises because of packing effects. The important internuclear dimensions are given in Table 8.37.

TABLE 8.37

Internuclear Dimensions in Tetramethylammonium Tetrachloronickelate(II)

Distance	Value	Angle	Value
Ni—Cl(1)	2.256 Å	Cl(1)—Ni—Cl(2)	107.8°
Ni—Cl(2)	2.279 Å	Cl(1)—Ni—Cl(3)	114.4°
Ni—Cl(3)	2.283 Å	Cl(3)—Ni—Cl(2)	107.8°
		Cl(2)—Ni—Cl(2′)	111.4°

Similarly, the tetraethylammonium salt is also isostructural with the corresponding cobalt compound[336], the cell being tetragonal with $a = 9.05$ and $c = 15.01$ Å. There are two molecules in the unit cell of space group $P4_2/nmc$. The tetrachloronickelate(II) anion is distorted so that the chlorine-nickel-chlorine bond angles are 106.83° and 110.81°. The nickel-chlorine bond distance is 2.245 Å.

Pauling[334] has shown by a single-crystal X-ray diffraction study of triphenylmethylarsonium tetrachloronickelate(II) that the lattice is cubic with $a = 15.557$ Å, having four molecules in the unit cell of space group $P2_13$. Within experimental error the tetrachloronickelate(II) anion is tetrahedral, the nickel-chlorine bond distance being 2.27 Å.

The infrared spectra of several tetrachloronickelates(II) have been investigated[339–341] and Sabatini and Sacconi[339] report a nickel-chlorine stretch at 289 cm^{-1} and chlorine-nickel-chlorine bending modes at 112 and 79 cm^{-1} for the tetraethylammonium salt. This same complex, and also the triphenylmethylarsonium compound, obey the Curie-Weiss law with the following parameters.

Compound	μ (BM)	$\theta°$	Reference
$[(C_2H_5)_4N]_2NiCl_4$	3.85	30	342
$[(C_6H_5)_3CH_3As]_2NiCl_4$	4.12	35	333

The electronic absorption spectrum of the tetrachloronickelate(II) ion has been investigated in the solid, in solution, in the melt and in several host lattices[333,335,343–354]. In general there are three spin-allowed transitions associated with the tetrahedral anion. These occur at approximately 5000, 8000 and 15,000 cm^{-1}. However, in the lithium chloride-potassium chloride eutectic melt, there is some evidence for the formation of structurally different complexes[345,350].

Trichloronickelates(II). Alkali-metal trichloronickelates(II) may be prepared from the melt[20,64], and these salts, as well as those of organic bases, may also be prepared in either concentrated hydrochloric acid solutions or in alcohol[355–358]. The trichloronickelates(II) are yellow crystalline solids, the caesium salt melting[64] at 758°. A number of hydrates of various types have also been isolated from aqueous solutions[156,359].

There is some confusion over the structure of caesium trichloronickelate(II). From an electron-diffraction study Tischenko[360] deduced a hexagonal unit cell with $a = 7.18$ and $c = 5.93$ Å, but Asmussen and Soling[356] gave a hexagonal cell with $a = 7.17$ and $c = 11.87$ Å. A recent redetermination of the structure by Stucky and coworkers[361] reproduced the parameters of Tischenko, the hexagonal cell being of space group $P6_3$ and containing two formula units. The caesium salt has been shown[361] to be isomorphous with tetramethylammonium tribromonickelate(II), for which full data from a single-crystal structural investigation are available. The anions are chain polymers of NiX_6 octahedra sharing faces (Figure 8.16, page 454). Tetramethylammonium trichloronickelate(II) is hexagonal with $a = 7.85$ and $c = 6.16$ Å[361].

The magnetic properties of a number of trichloronickelates(II) have been investigated and the reported data are summarized in Table 8.38.

The electronic absorption spectra[84,85,332,363] of a number of trichloronickelates(II) have been observed and the approximate positions of the transitions and the assignments based on octahedral symmetry are as follows.

6,500 cm^{-1}	$^3T_{2g}(F) \leftarrow {}^3A_{2g}(F)$
11,300 cm^{-1}	$^3T_{1g}(F) \leftarrow {}^3A_{2g}(F)$
21,300 cm^{-1}	$^3T_{1g}(P) \leftarrow {}^3A_{2g}(F)$

TABLE 8.38
Magnetic Properties of Some Trichloronickelates(II)

Compound	μ (BM)	$\theta°$	Reference
$KNiCl_3$	3.12[a]		362
$RbNiCl_3$	3.48	112	356
$CsNiCl_3$	3.37	76	356
$CH_3NH_3NiCl_3$	3.47	60	356
$(CH_3)_4NNiCl_3$	3.20[b]	0	363
n-$C_3H_7NH_3NiCl_3$	3.32	40	356
$C_5H_6NNiCl_3$	3.21	36	356

[a] Room temperature measurement only.
[b] This compound has a T.I.P. term of 318 × 10^{-6} esu.

Tetrabromonickelates(II). Salts of this blue tetrahedral anion are prepared by interaction of the appropriate bromides in alcohol[333–335,357]. The heat of formation of tetraethylammonium tetrabromonickelate(II) from its constituent bromides is reported[364] to be −0.98 kcal $mole^{-1}$.

Stucky and coworkers[336] have shown that the tetraethylammonium salt is isomorphous with the corresponding chloro compound; thus it contains discrete tetrahedral tetrabromonickelate(II) anions. The same compound obeys the Curie-Weiss law[333,342] with a magnetic moment of 3.80 BM and a Weiss constant of 18°.

The nickel-bromine stretching frequency in several tetrabromonickelates(II) have been recorded[339,340]. The bands tend to be split and occur at about 230 cm^{-1} with a shoulder at about 220 cm^{-1}. The electronic absorption spectrum of the tetrabromonickelate(II) anion[333,335,343,351–353,357,365] is similar to that of the corresponding chloro complex, but the absorptions occur at lower energies.

Substituted tetrabromonickelates(II). Complexes of the type $[NiBr_3L]^-$, where L is a neutral ligand such as triphenylphosphine, have been prepared by Cotton, Faut and Goodgame[357]. The distorted tetrahedral nature of these complexes has been confirmed by spectral and magnetic measurements[119,158,357]. The nickel-bromine stretching frequencies in $(C_2H_5)_4N[NiBr_3\cdot(C_6H_5)_3P]$ occur[158] at 242 and 212 cm^{-1}.

Tribromonickelates(II). These compounds are prepared by the same methods described for the trichloronickelates(II)[356,358]. The structure of tetramethylammonium tribromonickelate(II) has been determined by

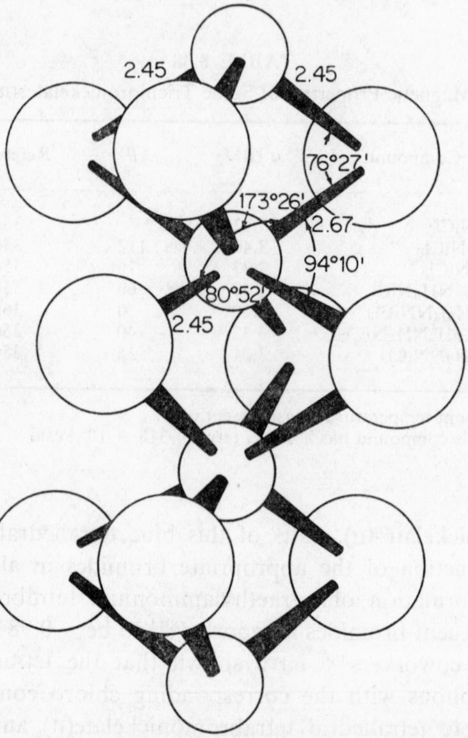

Figure 8.16 The $NiBr_6$ chain in $(CH_3)_4NNiBr_3$. (Reproduced by permission from G. Stucky, S. D'Agostino and G. McPherson, *J. Am. Chem. Soc.*, **88,** 4828 (1966))

Stucky and coworkers[361] by single-crystal X-ray diffraction techniques. The hexagonal unit cell of space group $P6_3$ contains two molecules and has $a = 9.35$ and $c = 6.35$ Å. The anion adopts a polymeric chain structure containing $NiBr_6$ octahedra sharing opposite faces as shown in Figure 8.16. The nickel-nickel distance is 3.17 Å. As with the corresponding chloro complex, there is some controversy as to the structure of the caesium salt. The reported unit-cell parameters are shown below.

Hexagonal $a = 7.50$ Å $c = 6.24$ Å reference 361

Hexagonal $a = 7.49$ Å $c = 12.48$ Å reference 356

Rubidium, caesium and pyridinium tribromonickelate(II) obey the Curie-Weiss law with the following parameters[356].

Compound	μ (BM)	$\theta°$
$RbNiBr_3$	3.51	156
$CsNiBr_3$	3.42	101
$C_5H_6NNiBr_3$	3.41	32

The electronic absorption spectrum of the tribromonickelate(II) anion has been observed and the bands assigned on the basis of octahedral symmetry[84,85,361]. As expected, the bands occur at lower energies than those for the corresponding chloro complexes.

Tetraiodonickelates(II). The preparation and physical properties of the tetraiodonickelate(II) salts are entirely analogous to those of the corresponding chloro and bromo complexes. Thus, the magnetic and spectral data clearly demonstrate the tetrahedral configuration of the anion[167,199,333,334,339,342,343,352]. In addition, substituted tetraiodonickelates(II) similar to those described above for the bromo complex, have also been prepared[119,357].

ADDENDA

Halides

Richards[366] has made a further study of the far-infrared antiferromagnetic resonance of nickel difluoride.

The sublimation pressure of nickel dichloride has been investigated in detail by McCreary and Thorn[367]. Two different types of effusion cell were used and the reported thermodynamic data are

ΔH_{sub}	52.76 and 53.59 kcal mole^{-1}
ΔS_{sub}	42.64 and 43.70 cal deg^{-1} mole^{-1}

The heat capacity of nickel dichloride has been measured[368] between 1.8 and 16°K. Gruen and coworkers[369] have observed the fluorescence spectrum of nickel dichloride vapour in an argon matrix at 500°.

As with cobalt, the recently reported chemistry of the nickel dihalides may be divided into (a) further investigation of already known complexes, and (b) the preparation and characterization of new adducts. Particular interest has been shown in the preparation of five-coordinate complexes and multidentate ligands with two or more different types of donor atom. Typical examples of (a) are the ^1H NMR study[370] of complexes of the type $NiX_2(py)_4$ (X = Cl, Br, I) and the investigation of the electronic spectra of a number of tetragonal complexes[371].

Boorman and Carty[372] have recorded the infrared spectra of a number of nickel dihalide–phosphine adducts and their results are given in Table 8.39.

TABLE 8.39
Infrared Spectra (cm^{-1}) of Nickel Dihalide–Phosphine Adducts

Compound[a]	ν_{Ni-X}	ν_{Ni-P}	Structure[b]
$NiCl_2[(CH_3)_2C_6H_5P]_2$	406	343 or 418	*Trans* S.P.
$NiCl_2[C_2H_5(C_6H_5)_2P]_2$	407	335 or 416	*Trans* S.P.
$NiCl_2[(CH_3)_3P]_2$	403	361	*Trans* S.P.
$NiCl_2[(C_2H_5)_3P]_2$	402	415	*Trans* S.P.
$NiCl_2(n\text{-}C_4H_9)_3P]_2$	403		*Trans* S.P.
$NiCl_2[(C_6H_5)_3P]_2$	339, 303	186	Tetrahedral
$NiCl_2[DPM]_2$	313		*Cis* S.P.
$NiCl_2[DPE]$	330, 320		*Cis* S.P.
$NiBr_2[(CH_3)_2C_6H_5P]_2$	314	348 or 416	*Trans* S.P.
$NiBr_2[C_2H_5(C_6H_5)_2P]_2$	260, 238	194, 168	Tetrahedral
$NiBr_2[(CH_3)_3P]_2$	340	360	*Trans* S.P.
$NiBr_2[(C_2H_5)_3P]_2$	337	412	*Trans* S.P.
$NiBr_2[(C_6H_5)_3P]_2$	265, 232	193, 184	Tetrahedral
$NiBr_2[DPM]_2$	330, 320		*Cis* S.P.
$NiBr_2[DPE]$	264, 245		*Cis* S.P.
$NiI_2[(CH_3)_2C_6H_5P]_2$	255	424	*Trans* S.P.
$NiI_2[C_2H_5(C_6H_5)_2P]_2$	220, 213	196, 168	Tetrahedral
$NiI_2[(CH_3)_3P]_2$	280	366	*Trans* S.P.
$NiI_2[(C_6H_5)_3P]_2$	215	198, 182	Tetrahedral
$NiI_2[DPM]_2$	210		*Cis* S.P.
$NiI_2[DPE]$	216		*Cis* S.P.

[a] DPM = bis(diphenylphosphino)methane.
DPE = 1,2-bis(diphenylphosphino)ethane.
[b] S.P. = square planar.

Two forms of the adduct $NiCl_2 \cdot dmp$, where dmp is 2,9-dimethyl-1,10-phenanthroline, are known[373,374]. The α-form is yellow and has a magnetic moment of 3.18 BM. The purple β-form has a magnetic moment of 3.40 BM and has been shown to be isomorphous with the corresponding zinc complex, indicating tetrahedral coordination. A single-crystal X-ray diffraction study[374] of the α-form, obtained as a tan crystalline solid containing two molecules of solvated chloroform, indicates a dimeric, square-pyramidal environment for the nickel atom. The complex has a monoclinic unit cell with $a = 11.75$, $b = 13.46$, $c = 11.27$ Å and $\beta = 91.2°$. There are two molecules in the cell of space group $P2_1/n$. The nickel-chlorine bond distances are 2.32 Å (terminal) and 2.38 and 2.39 Å (bridging), while the nickel-nitrogen distances are 2.05 and 2.06 Å. The nickel atom is displaced by approximately 0.4 Å from the square plane towards the apex of the pyramid.

Bombieri and coworkers[375] have prepared green adducts of the type $NiX_2L \cdot 2H_2O$, where X = Cl or Br and L = *N,N,N′,N′*-tetramethyl-*o*-phenylenediamine, by interaction of the hydrated halides and the ligand in ethanol. The structures of these complexes were deduced from three-dimensional single-crystal X-ray diffraction studies. The crystallographic data are given in Table 8.40.

TABLE 8.40
Crystallographic Data for $NiX_2L \cdot 2H_2O$

Parameter	X = Cl	X = Br
Symmetry	Orthorhombic	Orthorhombic
a (Å)	14.28	14.71
b (Å)	10.93	11.21
c (Å)	8.90	9.83
Z	4	4
Space group	*Cmcm*	*Cmcm*

The isomorphous compounds contain octahedrally coordinated nickel atoms and crystallographic and spectral evidence indicates that the molecules are hydrogen bonded to form chains parallel to the *c* axis. The important internuclear dimensions are listed in Table 8.41.

TABLE 8.41
Internuclear Dimensions in $NiX_2L \cdot 2H_2O$

Dimension	X = Cl	X = Br
Ni—X	2.435 Å	2.606 Å
Ni—O	2.123 Å	2.036 Å
Ni—N	2.118 Å	2.151 Å
X—Ni—X	91.8°	92.1°
X—Ni—O	89.6°	89.1°
X—Ni—N	92.2°	90.9°
O—Ni—N	90.4°	90.9°
N—Ni—N	83.9°	86.0°

Anhydrous nickel dichloride reacts[376] with trimethylenethiourea (L) to give orange-red $NiCl_2L_4$. This adduct melts at 156° and has a magnetic moment of 3.24 BM. The unit cell is triclinic[376], the parameters at 120°K being $a = 8.35$, $b = 8.69$, $c = 9.47$ Å, $\alpha = 117.9°$, $\beta = 74.3°$ and $\gamma = 93.3$. There is one molecule in the cell of space group $P\bar{1}$. The structure consists of discrete *trans* octahedral units. The internuclear dimensions are given in Table 8.42.

TABLE 8.42
Internuclear Dimensions in $NiCl_2$(trimethylenethiourea)$_4$

Distance	Value (Å)	Angle	Value	Angle	Value
Ni—Cl	2.413	Cl—Ni—S(1)	92.52°	Cl—Ni—S(1′)	87.47°
Ni—S(1)	2.466	Cl—Ni—S(2)	79.80°	Cl—Ni—S(2′)	100.30°
Ni—S(2)	2.491	S(1)—Ni—S(2)	95.30°	S(1)—Ni—S(2′)	84.70°

The deviation from 90° of the angles subtended at the nickel atom is believed to be due to an intramolecular N—H . . . Cl hydrogen bond[376].

Sacconi and coworkers[377–382] have been particularly successful in preparing adducts of nickel dihalides with tri- and tetradentate ligands containing two or more types of donor atom. These ligands include those based on substituted amines (page 379) and those utilizing Schiff bases. Of those based on the substituted amines, the complexes with tris-(2-diphenylphosphinoethyl)amine (NP_3) are the best established[377,382]. Dark blue to violet diamagnetic complexes of the types $[NiX(NP_3)]X$ and $[NiX(NP_3)]BPh_4$ were isolated and assigned a trigonal-bipyramidal structure. Complex phosphines and arsines also form five-coordinate nickel(II) complexes. For example, Meek and coworkers[383,384] have prepared diamagnetic complexes of the type $[NiXL]ClO_4$, where X = Cl, Br or I and L = tris(*o*-methylselenophenyl)phosphine or tris(3-dimethylarsinopropyl)arsine, both ligands acting as tetradentates. Another interesting study is that by Cristopher, Gordon and Venanzi[385]. These workers prepared adducts of the type NiX_2L, where X = Cl, Br or I and L = *o*-dimethylaminophenyldiphenylphosphine (PN), bis(*o*-dimethyl-aminophenyl)phenylphosphine (PDN) or tris(*o*-dimethylaminophenyl)-phosphine (PTN), and found that the coordination number adopted by the nickel atom varied according to the physical state of the complex, on the ligand and on the halogen. The coordination numbers are given in Table 8.43.

TABLE 8.43
Coordination Numbers in NiX_2L[a]

X	L	Solid	Solution[b]
Cl	PN	5	4
Cl	PDN	5	4 and 5
Cl	PTN	5	4 and 5
Br	PN	5	4
Br	PDN	5	4 and 5
Br	PTN	5	4 and 5
I	PN	4	4
I	PDN	4 and 5	4
I	PTN	4 and 5	4

[a] Four coordinate = pseudo-tetrahedral.
[b] Dichloromethane solution.

Some properties of a number of other nickel dichloride adducts are given in Table 8.44.

TABLE 8.44
Adducts of the Type $NiCl_2L_n$

Compound and Properties	References
L = hydrazine, $n = 2$, $\mu = 3.18$ BM, octahedral	386, 387
L = methylhydrazine, $n = 2$, blue, $\mu = 3.01$ BM, octahedral	387
L = *N,N*-dimethylhydrazine, $n = 4$, blue, octahedral	387
L = *N,N*-dimethylhydrazine, $n = 2$, grey, $\mu = 2.80$ BM, octahedral	387
L = *N,N*-dimethylhydrazine, $n = 1$, yellow-green, octahedral	387
L = 4-vinylpyridine, $n = 4$, light blue, $\mu = 3.05$ BM	388
L = 4-vinylpyridine, $n = 2$, yellow, $\mu = 3.2$ BM	388
L = 2-methoxyaniline, $n = 2$, yellow, $\mu = 3.59$ BM, octahedral	389
L = 2-methoxyaniline, $n = 1$, yellow, $\mu = 3.35$ BM, octahedral	389
L = *p*-nitroso-*N,N*-dimethylaniline (NODMA), $n = 1$, isolated as monohydrate, m.p. > 360°, $\mu = 3.49$ BM, octahedral	390
L = 4,4′-bipyridine, $n = 1$, yellow-green, octahedral	391
L = imidazole, $n = 4$, $\mu = 3.14$ BM, $\nu_{Ni-Cl} = 232\ cm^{-1}$, octahedral	392
L = imidazole, $n = 1$, pale yellow, $\mu = 3.44$ BM, octahedral	392
L = 2,2′,2″-terpyridyl (terpy), $n = 1$, gold, $\mu = 3.35$ BM, five coordinate	393–395
L = pentamethylenetetrazole (PMT), $n = 1$, light green, $\mu = 3.28$ BM, $\nu_{Ni-Cl} = 253\ cm^{-1}$, octahedral	396
L = 4(5)-bromoimidazole, $n = 4$, lime green, $\mu = 3.15$ BM, octahedral	397
L = 4(5)-bromoimidazole, $n = 2$, lime yellow, $\mu = 3.14$ BM, octahedral	397
L = 2-pyridone (TP), $n = 1$, salmon, $\mu = 3.28$ BM, octahedral	398
L = benzoxazole, $n = 2$, yellow, $\mu = 3.40$ BM, $\nu_{Ni-N} = 232\ cm^{-1}$, $\nu_{Ni-Cl} = 265\ cm^{-1}$, octahedral	399
L = benzothiazole (BT), $n = 2$, yellow, $\mu = 3.27$ BM, $\nu_{Ni-Cl} = 258\ cm^{-1}$, octahedral with halogen bridges	400

TABLE 8.44 (*contd.*)

Adducts of the Type $NiCl_2L_n$

L = 2-(2-pyridyl)benzothiazole (pbt), n = 2, green, μ = 3.10 BM, octahedral	401
L = 2-(2-pyridyl)benzothiazole (pbt), n = 1, yellow green, μ = 3.57 BM, octahedral	401
L = tri(2-cyanoethyl)phosphine (CEP), n = 2; α-form, red, diamagnetic, square planar; β-form, purple, μ = 3.12 BM, ν_{Ni-Cl} = 240 cm^{-1}, octahedral	402
L = isopropylaminomethylpyridine-(2), n = 2, green, μ = 3.24 BM, octahedral	403
L = isopropylaminomethylpyridine-(2), n = 1, yellow-green, m.p. = 275°, μ = 3.37 BM, octahedral	403
L = dimethylaminomethylpyridine-(2), n = 1, citron yellow, m.p. = 248°, μ = 3.38 BM, octahedral	403
L = di-2-pyridylamine, n = 2, pale blue, μ = 3.11 BM, ν_{Ni-Cl} = 268 cm^{-1}, octahedral	404
L = di-2-pyridylamine, n = 1, pale green, μ = 3.24 BM, octahedral	404
L = benzene-1,2-diamine (*o*-phenylenediamine), n = 2, blue, μ = 3.18 BM, ν_{Ni-N} = 272 cm^{-1}, ν_{Ni-Cl} = 220 cm^{-1}, δ_{Ni-Cl} = 112 cm^{-1}, *trans* octahedral	405–407
L = benzene-1,2-diamine, n = 1, green, μ = 3.02 BM, octahedral	405
L = benzene-1,3-diamine, n = 2, grey-blue, μ = 3.21 BM, octahedral	408
L = benzene-1,3-diamine, n = 1, yellow, μ = 3.01 BM, octahedral	408
L = benzene-1,4-diamine, n = 2, green-yellow, μ = 3.21 BM, octahedral	409
L = benzene-1,4-diamine, n = 1, yellow, μ = 3.10 BM, octahedral	409
L = 1,8-diaminonaphthalene (dan), n = 2, μ = 3.02 BM, ν_{Ni-N} = 300 cm^{-1}, ν_{Ni-Cl} = 144 cm^{-1}, δ_{Ni-Cl} = 92 cm^{-1}, *trans* octahedral	406
L = 2,2′-diaminobiphenyl (dabp), n = 2, μ = 3.26 BM, ν_{Ni-N} = 300 cm^{-1}, ν_{Ni-Cl} = 232 cm^{-1}, δ_{Ni-Cl} = 116 cm^{-1}, *trans* octahedral	406
L = 1-(2-pyridyl)-2-(4-pyridyl)ethylene, n = 1, pale yellow, μ = 3.40 BM, octahedral	410
L = pyridine-2-aldehyde 2′-pyridylhydrazone (paphy), n = 1, green, μ = 3.25 BM, octahedral	395
L = Schiff base formed between 2,6-diacetylpyridine and ethylamine [py(Et)$_2$], n = 1, yellow-brown, m.p. = 270–273°, μ = 3.35 BM, five coordinate	411
L = Schiff base formed between 2,6-diacetylpyridine and n-propylamine [py(Pr^n)$_2$], n = 1, olive-green, m.p. = 258–264°, μ = 3.29 BM, five coordinate	411
L = Schiff base formed between 2,6-diacetylpyridine and *i*-propylamine [py(Pr^i)$_2$], n = 1, orange-brown, m.p. = 330°, μ = 3.34 BM, five coordinate	411
L = Schiff base formed between 2,6-diacetylpyridine and *s*-butylamine [py(Bu^s)$_2$], n = 1, light brown, m.p. = 275°, μ = 3.31 BM, five coordinate	411
L = Schiff base formed between 2,6-diacetylpyridine and cyclohexylamine [py(cy)$_2$], n = 1, amber yellow, m.p. = 325°, μ = 3.22 BM, five coordinate	411
L = Schiff base formed between *o*-methoxybenzaldehyde and *N,N*-dimethylethylenediamine (MOBenNMe$_2$), n = 1, red, m.p. = 184–187°, μ = 3.44 BM, five coordinate	380
L = Schiff base formed between *o*-methoxybenzaldehyde and *N,N*-diethylethylenediamine (MOBenNEt$_2$), n = 1, purple, m.p. = 186–189°, μ = 3.48 BM, tetrahedral	380
L = Schiff base formed between *o*-methoxybenzaldehyde and *N,N*-dimethyltrimethylenediamine (MOBtnNMe$_2$), n = 1, red, m.p. = 237–240°, μ = 3.38 BM, five coordinate	380

TABLE 8.44 (*contd.*)
Adducts of the Type $NiCl_2L_n$

L = Schiff base formed between *o*-methoxybenzaldehyde and ethylenediamine [(MOB)$_2$en], n = 1, violet, m.p. = 210°, μ = 3.44 BM, tetrahedral	380
L = Schiff base formed between *o*-methoxybenzaldehyde and 1,2-propylenediamine [(MOB)$_2$pn], n = 1, violet, m.p. = 212–217°, μ = 3.37 BM, tetrahedral	380
L = 1,2-di(*o*-aminophenylthio)ethane (SN), n = 1, blue, m.p. = 257–258° μ = 3.07 BM, octahedral	412
L = tris(2-methylthioethyl)amine (TSN), n = 1, light green, μ = 3.16 BM, octahedral	413
L = 2,6-di(β-diphenylphosphinoethyl)pyridine (PNP), n = 1, brown, μ = 3.14 BM, five coordinate	414
L = bisphenylphosphinomethylpyridine(2), n = 2, yellow-green, m.p. = 292°, μ = 3.21 BM, octahedral	415
L = 2-ethylmercaptocyclohexylamine, n = 2, blue, μ = 3.00 BM, octahedral	416
L = 2,6-pyridinedisulphenamide, n = 2, yellow, μ = 3.2 BM, octahedral	417
L = trimethylphosphine, n = 2, red, m.p. = 199–200°, *trans* square planar	418
L = 1,4-bis(diphenylphosphino)butane (DPB), n = 1, light violet, m.p. = 270–272°, μ = 3.15 BM	419
L = *o*-dimethylaminophenyldiphenylphosphine (PN), n = 1, brown, m.p. = 245–246°, μ = 3.28 BM, ν_{Ni-Cl} = 311, 299, 239 cm^{-1}, five coordinate, dimeric with chlorine bridges	385
L = bis(*o*-dimethylaminophenyl)phenylphosphine (PDN), n = 1, olive green, m.p. = 238–239°, μ = 3.20 BM, ν_{Ni-Cl} = 315, 245 cm^{-1}, five coordinate, dimeric with chlorine bridges	385
L = tris(*o*-dimethylaminophenyl)phosphine (PTN), n = 1, pale green, m.p. = 253–254°, μ = 3.17 BM, ν_{Ni-Cl} = 321, 241 cm^{-1}, five coordinate, dimeric with chlorine bridges	385
L = 1,2-bis(diphenylphosphino)-*o*-carborane, $[(C_6H_5)_2P(B_{10}H_{10}C_2)P(C_6H_5)_2]$, n = 1, red, m.p. = 340°, diamagnetic, square planar	420
L = tetrakis(diphenylphosphinomethyl)methane, $C[CH_2P(C_6H_5)_2]_4$, n = 1/2, red, diamagnetic, ν_{Ni-Cl} = 340, 327 cm^{-1}, *cis* square planar, ligand bridges two nickel atoms	421
L = 1,4-dioxan, n = 1, yellow, μ = 3.46 BM, ν_{Ni-Cl} = 250 cm^{-1}, octahedral	422
L = 1,2-bis(methylsulphonyl)ethane (BMSE), n = 1, light green, m.p. = 215°, tetrahedral	423
L = 2,5-dithiahexane, n = 2, ν_{Ni-Cl} = 264 cm^{-1}, octahedral	424
L = 1,2-di(isopropylthio)ethane, n = 2, ν_{Ni-Cl} = 273 cm^{-1}, octahedral	424
L = *N*-methylthiourea, n = 2, yellow-orange, μ = 3.29 BM, octahedral	425
L = *N,N'*-dimethylthiourea, n = 4, green, octahedral	425
L = *N,N'*-diethylthiourea, n = 4, green, octahedral	425

Other donor ligands that have been reacted with nickel dichloride include pyridine[370,371,407,426], 3-methylpyridine[427], 4-methylpyridine[427,428], aniline[407,429], acetonitrile[430], aziridine[371], biuret[431,432], *N,N'*-dimethylethylenediamine[371], monoethanolamine[433], urea[434], thioacetamide[435], tris(*o*-methylselenophenyl)phosphine[383], tris(3-dimethylarsinopropyl)-arsine[384], 5,7,7,12,14,14-hexamethyl-1,4,8,11-tetracyclotetradecane[371], *N*-2-thiophenyl-2'-pyridylmethyleneimine[436] and dissymmetric phosphines[437].

The sublimation pressure of nickel dibromide, determined by two different methods[367], gave the following values of the heat and entropy of sublimation

$$\Delta H_{sub} \quad 51.97 \text{ and } 53.02 \text{ kcal mole}^{-1}$$
$$\Delta S_{sub} \quad 43.73 \text{ and } 45.11 \text{ cal deg}^{-1} \text{ mole}^{-1}$$

Direct interaction of *N,N'*-dimethylethylenediamine and hydrated nickel dibromide in a suitable non-aqueous solvent results in the formation of light blue $NiBr_2L_2$[371,438]. The complex has a monoclinic unit cell[438] with $a = 7.880$, $b = 8.664$, $c = 10.697$ Å and $\beta = 101.21°$. There are two molecules in the cell of space group $P2_1/c$. The electronic spectrum of this *trans* octahedral complex has been discussed by Rowley and Drago[371].

Di Vaira and Orioli[439] have reported the results of a three-dimensional X-ray analysis of the previously reported complex $NiBr_2(Me_6tren)$, where Me_6tren is tris(2-dimethylaminoethyl)amine. The complex has a cubic cell of space group $P2_13$ containing four formula units with $a = 12.123$ Å. The compound is isomorphous with the corresponding cobalt complex (Figure 7.14, page 360) and the important internuclear dimensions are given in Table 8.45.

TABLE 8.45

Internuclear Dimensions in $[NiBr(Me_6tren)]^+$

Distance	Value (Å)	Angle	Value
Ni—Br(1)	2.467	N(1)—Ni—N(2)	84.2°
Ni—N(1)	2.10	N(2)—Ni—Br(1)	95.8°
Ni—N(2)	2.13	N(2)—Ni—N(2')	119.0°

Another five-coordinate nickel bromide complex has also been investigated by Orioli and Di Vaira[440]. The adduct dibromo-[1-(*o*-methoxyphenyl)-2,6-diazaoctane]nickel(II), $NiBr_2(MOBH_2tnHNEt)$, where $MOBH_2tnNHEt$ is the dihydrogenated Schiff base formed between *o*-methoxybenzaldehyde and *N*-ethyltrimethylenediamine, has been prepared[379] and the above workers report that it has a monoclinic unit cell with $a = 11.130$, $b = 9.835$, $c = 16.573$ Å and $\beta = 113.53°$. There are four formula units in the cell of space group $P2_1/c$. The environment about the central nickel atom is intermediate between square pyramidal and trigonal bipyramidal. The important internuclear dimensions are given in Table 8.46.

Bis(2-diphenylphosphinoethyl)amine (PNP) reacts with nickel dibromide to give diamagnetic $NiBr_2(PNP)$[441]. This complex has an orthorhombic unit cell with $a = 15.01$, $b = 24.55$ and $c = 7.30$ Å. There are four formula units in the cell of space group *Abm*2. The nickel atom has a distorted square-pyramidal configuration. Some of the internuclear dimensions are given in Table 8.47.

TABLE 8.46

Internuclear Dimensions in $NiBr_2$(MOBH$_2$tnNHEt)

Distance	Value (Å)	Angle	Value	Angle	Value
Ni—Br(1)	2.416	Br(1)—Ni—Br(2)	149.7°	Br(2)—Ni—N(1)	107.3°
Ni—Br(2)	2.398	Br(1)—Ni—O	86.2°	Br(2)—Ni—N(2)	94.6°
Ni—O	2.318	Br(1)—Ni—N(1)	101.3°	O—Ni—N(1)	87.8°
Ni—N(1)	2.017	Br(1)—Ni—N(2)	91.8°	O—Ni—N(2)	175.3°
Ni—N(2)	2.013	Br(2)—Ni—O	85.0°	N(1)—Ni—N(2)	96.8°

TABLE 8.47

Internuclear Dimensions in $NiBr_2$(PNP)

Distance	Value (Å)	Angle	Value	Angle	Value
Ni—Br(1)	2.70	N—Ni—P	86.0°	Br(1)—Ni—N	89.1°
Ni—Br(2)	2.33	P—Ni—Br(2)	93.4°	Br(1)—Ni—P	93.1°
Ni—P	2.17	N—Ni—Br(2)	169.7°	Br(1)—Ni—Br(2)	101.1°
Ni—N	2.01				

Brooks and Glockling[442] have reported an interesting reaction at room temperature between dibromobis(triethylphosphine)nickel(II) and trimethylgermane. On cooling the reaction medium to −20° almost colourless crystals were obtained. The nickel-containing product was liquid at room temperature and smelled strongly of triethylphosphine. Its infrared spectrum showed a strong absorption at 1937 cm^{-1} and this was assigned to a nickel-hydrogen stretching mode. The compound was too unstable to allow complete characterization.

The tetrahedral configuration of dibromobis(triphenylphosphine)-nickel(II) has been confirmed by a single-crystal X-ray diffraction study by Jarvis, Mais and Owston[443]. The complex has a monoclinic unit cell with $a = 9.828$, $b = 37.178$, $c = 10.024$ Å and $\beta = 114.65°$. There are four molecules in the cell of space group $P2_1/n$. Bromine-bromine repulsion results in an enlarged bromine-nickel-bromine angle, thus distorting the tetrahedral configuration. Some of the internuclear dimensions are given in Table 8.48.

TABLE 8.48

Internuclear Dimensions in $NiBr_2[(C_6H_5)_3P]_2$

Distance	Value (Å)	Angle	Value	Angle	Value
Ni—Br(1)	2.346	Br(1)—Ni—Br(2)	126.3°	Br(2)—Ni—P(1)	101.8°
Ni—Br(2)	2.329	Br(1)—Ni—P(1)	106.5°	Br(2)—Ni—P(2)	103.1°
Ni—P(1)	2.323	Br(1)—Ni—P(2)	108.3°	P(1)—Ni—P(2)	103.1°
Ni—P(2)	2.343				

The properties of a number of other nickel dibromide adducts are summarized in Table 8.49.

TABLE 8.49

Adducts of the Type $NiBr_2L_n$

Compound and Properties	References
L = 2-methoxyaniline, n = 2, yellow, μ = 3.62 BM, octahedral	389
L = 2-methoxyaniline, n = 1, yellow, μ = 3.48 BM, octahedral	389
L = imidazole, n = 1, pale yellow, μ = 3.17 BM, octahedral	392
L = 2,2′,2″-terpyridyl (terpy), n = 1, μ = 3.19 BM, five coordinate	393, 394
L = pentamethylenetetrazole (PMT), n = 1, yellow-green, μ = 3.29 BM, octahedral	396
L = 4(5)-bromoimidazole, n = 4, lime green, μ = 3.22 BM, octahedral	397
L = 4(5)-bromoimidazole, n = 2, lime yellow, μ = 3.42 BM, ν_{Ni-N} = 249 cm^{-1}, octahedral	397
L = benzoxazole, n = 2, yellow, μ = 3.29 BM, ν_{Ni-N} = 227 cm^{-1}, octahedral	399
L = benzothiazole (BT), n = 4, pale green, μ = 2.81 BM, decomposes at 90° and 105° to give complexes with n = 2 and n = 1 respectively, octahedral	400
L = benzoxazole, n = 2, yellow, μ = 2.91 BM, octahedral with halogen bridges	400
L = benzoxazole, n = 1, pink, μ = 3.39 BM, octahedral with bridging halogen and ligand	400
L = 2-(2-pyridyl)benzothiazole (pbt), n = 2, green, μ = 3.15 BM, octahedral	401
L = tri(2-cyanoethyl)phosphine (CEP), n = 2; α-form, purple, diamagnetic, square planar; β-form, blue, μ = 3.15 BM, octahedral	402
L = tri(2-cyanoethyl)phosphine (CEP), n = 1, green, μ = 3.05 BM, ν_{Ni-Br} = 245 cm^{-1}, tetrahedral	402
L = isopropylaminomethylpyridine-(2), n = 2, m.p. = 170°, green form has μ = 3.24 BM, blue form has μ = 3.18 BM, both octahedral	403
L = dimethylaminomethylpyridine-(2), n = 2, blue-green, μ = 3.27 BM, octahedral	403
L = diethylaminomethylpyridine-(2), n = 2, green, m.p. = 115°, ionic with five-coordinate cation	403

TABLE 8.49 (*contd.*)

Adducts of the Type $NiBr_2L_n$

L = diethylaminomethylpyridine-(2), n = 1, red-violet, μ = 3.36 BM, octahedral	403
L = di-2-pyridylamine, n = 2, pale green-blue, μ = 3.13 BM, octahedral	404
L = di-2-pyridylamine, n = 1, brown, μ = 3.36 BM, ν_{Ni-Br} = 267, 235 cm^{-1}, tetrahedral	404
L = benzene-1,2-diamine (*o*-phenylenediamine), n = 2, blue, μ = 3.10 BM, ν_{Ni-N} = 266 cm^{-1}, ν_{Ni-Br} = 135 cm^{-1}, δ_{Ni-Br} = 102 cm^{-1}, octahedral	405–407
L = benzene-1,2-diamine, n = 1, green, μ = 2.98 BM, octahedral	405
L = benzene-1,3-diamine, n = 2, grey-blue, μ = 3.14 BM, octahedral	408
L = benzene-1,3-diamine, n = 1, yellow, μ = 3.00 BM, octahedral	408
L = benzene-1,4-diamine, n = 2, green-yellow, μ = 3.16 BM, octahedral	409
L = benzene-1,4-diamine, n = 1, yellow, μ = 3.16 BM, octahedral	409
L = 1,8-diaminonaphthalene (dan), n = 2, μ = 3.18 BM, ν_{Ni-N} = 300 cm^{-1}, ν_{Ni-Br} = 112 cm^{-1}, δ_{Ni-Br} = 66 cm^{-1}, *trans* octahedral	406
L = 2,2′-diaminobiphenyl (dabp), n = 2, μ = 3.17 BM, ν_{Ni-N} = 300 cm^{-1}, ν_{Ni-Br} = 144 cm^{-1}, δ_{Ni-Br} = 100 cm^{-1}, *trans* octahedral	406
L = 1-(2-pyridyl)-2-(3-pyridyl)ethylene, n = 2, pale yellow, μ = 3.34 BM, octahedral	410
L = pyridine-2-aldehyde 2′-pyridylhydrazone (paphy), n = 1, green, μ = 3.23 BM	395
L = Schiff base formed between pyridine-2-carboxaldehyde and *N,N*-dimethylethylenediamine ($PyAenNMe_2$), n = 1, m.p. = 219–223°, μ = 3.31 BM, five coordinate	444
L = Schiff base formed between pyridine-2-carboxaldehyde and *N,N*-diethylethylenediamine ($PyAenNEt_2$), n = 1, m.p. = 202–204°, μ = 3.37 BM, five coordinate	444
L = Schiff base formed between 2,6-diacetylpyridine and methylamine [$py(Me)_2$], n = 1, orange-brown, μ = 3.34 BM, five coordinate	411
L = Schiff base formed between 2,6-diacetylpyridine and ethylamine [$py(Et)_2$], n = 1, brown, m.p. = 288–293°, μ = 3.27 BM, five coordinate	411
L = Schiff base formed between 2,6-diacetylpyridine and n-propylamine [$py(Pr^n)$], n = 1, brown, m.p. = 242–244°, μ = 3.32 BM, five coordinate	411
L = Schiff base formed between 2,6-diacetylpyridine and *i*-propylamine [$py(Pr^i)$], n = 1, orange-brown, m.p. = 330°, μ = 3.31 BM, five coordinate	411
L = Schiff base formed between 2,6-diacetylpyridine and *s*-butylamine [$py(Bu^s)_2$], n = 1, grey-brown, m.p. = 277–283°, μ = 3.26 BM, five coordinate	411
L = Schiff base formed between 2,6-diacetylpyridine and cyclohexylamine ($py(cy)_2$], n = 1, brown, m.p. = 335°, μ = 3.22 BM, five coordinate	411
L = Schiff base formed between *o*-methoxybenzaldehyde and *N,N*-dimethylethylenediamine ($MOBenNMe_2$), n = 1, red, m.p. = 178–182°, μ = 3.38 BM, five coordinate	380
L = Schiff base formed between *o*-methoxybenzaldehyde and *N,N*-diethylethylenediamine ($MOBenNEt_2$), n = 1, purple, m.p. = 220–223°, μ = 3.34 BM, tetrahedral	380
L = Schiff base formed between *o*-methoxybenzaldehyde and *N,N*-dimethyltrimethylenediamine ($MOBtnNMe_2$), n = 1, red, m.p. = 220–223°, μ = 3.37 BM, five coordinate	380

TABLE 8.49 (*contd.*)
Adducts of the Type $NiBr_2L_n$

L = Schiff base formed between *o*-methoxybenzaldehyde and ethylenediamine [$(MOB)_2en$], $n = 1$, violet, m.p. = 200°, $\mu = 3.40$ BM, tetrahedral	380
L = Schiff base formed between *o*-methoxybenzaldehyde and 1,2-propylenediamine [$(MOB)_2pn$], $n = 1$, violet, m.p. = 200–207°, $\mu = 3.35$ BM, tetrahedral	380
L = bis(2-phenylarsinoethyl)amine (AsNAs), $n = 1$, crimson, m.p. = 192–194°, $\mu = 2.25$ BM, probably dimeric with both five- and six-coordinate Ni	378
L = bis(2-phenylarsinoethyl)sulphide (AsSAs), $n = 2$, light green, m.p. = 130–131°, $\mu = 3.12$ BM, octahedral	378
L = 1,2-bis(diphenylarsino)ethane (AsC_2As), $n = 1$, purple, m.p. = 215–220°, diamagnetic, square planar	378
L = Schiff base formed between *o*-methylaminobenzaldehyde and 3-aminopropyldiphenylphosphine (NNP), $n = 1$, mustard, m.p. = 212–218°, $\mu = 3.21$ BM, five coordinate	379
L = Schiff base formed between *o*-methoxybenzaldehyde and 3-aminopropyldiphenylphosphine (ONP), $n = 1$, brown, m.p. = 185–188°, $\mu = 3.28$ BM, five coordinate	379
L = Schiff base formed between *o*-methylthiobenzaldehyde and 3-aminopropyldiphenylphosphine (SNP), $n = 1$, mustard, m.p. = 214–218°, $\mu = 3.36$ BM, five coordinate	379
L = Schiff base formed between *o*-methylaminobenzaldehyde and 3-aminopropyldiphenylarsine (NNAs), $n = 1$, olive green, m.p. = 217°, $\mu = 3.29$ BM, five coordinate	379
L = Schiff base formed between *o*-methoxybenzaldehyde and 3-aminopropyldiphenylarsine (ONAs), $n = 1$, dark purple, m.p. = 184–189°, $\mu = 3.34$ BM, five coordinate	379
L = Schiff base formed between *o*-methylthiobenzaldehyde and 3-aminopropyldiphenylarsine (SNAs), $n = 1$, mustard brown, m.p. = 202–205°, $\mu = 3.14$ BM, five coordinate	379
L = 1,2-di(*o*-aminophenylthio)ethane (SN), $n = 1$, blue, m.p. = 259–260°, $\mu = 3.07$ BM, octahedral	412
L = tris(2-methylthioethyl)amine (TSN), $n = 1$, light green, $\mu = 3.12$ BM, octahedral	413
L = 2,6-di(β-diphenylphosphinoethyl)pyridine (PNP), $n = 1$, brown, temperature-independent paramagnetism of 250×10^{-6} cgs, five coordinate	414
L = bisphenylphosphinomethylpyridine-(2), $n = 2$, red-brown, diamagnetic, ionic, five-coordinate cation	415
L = 2-ethylthiocyclohexylamine, $n = 2$, yellow-blue, $\mu = 3.21$ BM, octahedral	445
L = S,S′-*o*-xylyl-2,3-butanedione bis(2-mercaptoanil), $n = 1$, green-brown, $\mu = 3.27$ BM, *trans* octahedral	446
L = S,S′-*o*-xylyl-1,2-cyclohexanedione bis(2-mercaptoanil), $n = 1$, green-brown, $\mu = 3.17$ BM, *trans* octahedral	446
L = S,S′-dibenzyl-2,3-butanedione bis(2-mercaptoanil), $n = 1$, green-brown, $\mu = 3.13$ BM, *trans* octahedral	446
L = trimethylphosphine, $n = 3$, blue-black, m.p. = 140–150°, has small temperature-independent paramagnetism, trigonal bipyramidal	418, 447
L = trimethylphosphine, $n = 2$, dark red, m.p. = 178–181°, *trans* square planar	418
L = 1,4-bis(diphenylphosphino)butane (DPB), $n = 1$, green, m.p. = 254–256°, $\mu = 3.18$ BM	419
L = *o*-dimethylaminophenyldiphenylphosphine (PN) $n = 1$, dark brown, m.p. = 238–239°, $\mu = 3.24$ BM, $\nu_{Ni-Br} = 227, 179\ cm^{-1}$, tetrahedral	385

TABLE 8.49 (*contd.*)
Adducts of the Type $NiBr_2L_n$

L = bis(*o*-dimethylaminophenyl)phenylphosphine (PDN), $n = 1$, brown, m.p. = 255–256°, $\mu = 3.25$ BM $\nu_{Ni-Br} = 254, 177$ cm^{-1}, five coordinate, dimeric with bromide bridges	385
L = tris(*o*-dimethylaminophenyl)phosphine (PTN), $n = 1$, ochre, m.p. = 238–239°, $\mu = 3.23$ BM, $\nu_{Ni-Br} = 249, 196$ cm^{-1}, five coordinate	385
L = tetrakis(diphenylphosphinomethyl)methane, $n = 1/2$, red, m.p. = 296°, diamagnetic, $\nu_{Ni-Br} = 290, 274$ cm^{-1}, *cis* square planar, ligand bridges two nickel atoms	421
L = diphenylcyclopropenone (dpcp), $n = 2$, pale green, m.p. = 240°, $\mu = 3.56$ BM, octahedral	448
L = 1,4-dioxan, $n = 2$, yellow, $\mu = 3.35$ BM, $\nu_{Ni-Br} = 205$ cm^{-1}, octahedral	422
L = 1,4-dioxan, $n = 1$, orange, $\mu = 3.46$ BM, octahedral	422
L = 1,2-di(isopropylthio)ethane, $n = 2$, $\nu_{Ni-Br} = 216$ cm^{-1}, octahedral	424
L = *N,N'*-dimethylthiourea, $n = 4$, blue-green, octahedral	425
L = *N,N'*-diethylthiourea, $n = 4$, blue-green, octahedral	425

Adducts of nickel dibromide with the following donor molecules have also been reported; pyridine[370,371,426], aniline[407], biuret[431], monoethanolamine[433], thioacetamide[435], tris(*o*-methylselenophenyl)phosphine[383], tris-(3-dimethylarsinopropyl)arsine[384], the adduct formed between boron trifluoride and diphenylurea[449], the adduct formed between boron trifluoride and diphenylthiourea[449], dissymmetric phosphines[437] and 2,5-dithiahexane[424].

McCreary and Thorn[367] have reported that attempts to determine the sublimation pressure of nickel diiodide were unsuccessful because of the decomposition of the diiodide to the monoiodide and free iodine.

The properties of a number of adducts of nickel diiodide are summarized in Table 8.50.

TABLE 8.50
Adducts of the Type NiI_2L_n

Compound and Properties	References
L = 4(5)-bromoimidazole, $n = 4$, bottle green, $\mu = 3.26$ BM, octahedral	397
L = 4(5)-bromoimidazole, $n = 2$, green, $\mu = 3.23$ BM, $\nu_{Ni-N} = 247$ cm^{-1}, $\nu_{Ni-I} = 208, 188$ cm^{-1}, tetrahedral	397
L = benzoxazole, $n = 2$, green-yellow, $\mu = 3.40$ BM, $\nu_{Ni-N} = 226$ cm^{-1}, octahedral	399
L = benzothiazole (BT), $n = 2$, brown, $\mu = 3.38$ BM, $\nu_{Ni-I} = 238, 233$ cm^{-1}, tetrahedral	400

TABLE 8.50 (*contd.*)
Adducts of the Type NiI_2L_n

L = isopropylaminomethylpyridine-(2), n = 2, green, m.p. = 200°, μ = 3.29 BM, octahedral	403
L = dimethylaminomethylpyridine-(2), n = 2, green, μ = 3.32 BM, ionic, five coordinate cation	403
L = diethylaminomethylpyridine-(2), n = 2, dark green, m.p. = 175°, μ = 3.25 BM, ionic, five-coordinate cation	403
L = diethylaminomethylpyridine-(2), n = 1, brown, μ = 3.39 BM, octahedral	403
L = di-2-pyridylamine, n = 1, black, μ = 3.35 BM, ν_{Ni-I} = 202 cm^{-1}, tetrahedral	404
L = benzene-1,2-diamine (*o*-phenylenediamine), n = 2, green, μ = 3.05 BM, ν_{Ni-N} = 262 cm^{-1}, ν_{Ni-I} = 120 cm^{-1}, δ_{Ni-I} = 80 cm^{-1}, *trans* octahedral	406, 407
L = benzene-1,3-diamine, n = 2, green, μ = 3.11 BM, octahedral	408
L = benzene-1,3-diamine, n = 1, green, μ = 3.02 BM, octahedral	408
L = benzene-1,4-diamine, n = 2, green, μ = 3.11 BM, octahedral	409
L = benzene-1,4-diamine, n = 1, green, μ = 3.14 BM, octahedral	409
L = 1,8-diaminonaphthalene (dan), n = 2, μ = 3.02 BM, ν_{Ni-N} = 290 cm^{-1}, ν_{Ni-I} = 80 cm^{-1}, δ_{Ni-I} = 52 cm^{-1}, *trans* octahedral	406
L = 2,2′-diaminobiphenyl (dabp), n = 2, μ = 3.02 BM, ν_{Ni-N} = 292 cm^{-1}, ν_{Ni-I} = 124 cm^{-1}, δ_{Ni-I} = 86 cm^{-1}, *trans* octahedral	406
L = pyridine-2-aldehyde 2′-pyridylhydrazone (paphy), n = 1, red-brown, μ = 3.23 BM	395
L = Schiff base formed between pyridine-2-carboxaldehyde and *N,N*-dimethylethylenediamine ($PyAenNMe_2$), n = 1, m.p. = 242–243°, μ = 3.28 BM, five coordinate	444
L = Schiff base formed between pyridine-2-carboxaldehyde and *N,N*-diethylethylenediamine ($PyAenNEt_2$), n = 1, m.p. = 187–191°, μ = 3.27 BM, five coordinate	444
L = Schiff base formed between *o*-methoxybenzaldehyde and *N,N*-diethylethylenediamine ($MOBenNEt_2$), n = 1, purple, m.p. = 170–176°, μ = 3.33 BM, tetrahedral	380
L = bis(2-diphenylarsinoethyl)amine (AsNAs), n = 1, dark grey, m.p. = 224–226°, diamagnetic, five coordinate	378
L = bis(2-diphenylarsinoethyl)oxide (AsOAs), n = 1, dark brown, m.p. = 218–223°, μ = 3.19 BM, tetrahedral	378
L = bis(2-diphenylarsinoethyl)sulphide (AsSAs), n = 1, green, m.p. = 216–217°, diamagnetic, five coordinate	378
L = 1,2-bis(diphenylarsino)ethane (AsC_2As), n = 1, amethyst, m.p. = 200°, diamagnetic, square planar	378
L = 1,2-bis(diphenylarsino)butane (AsC_4As), n = 1, brown, m.p. = 199–200°, μ = 3.17 BM, tetrahedral	378
L = Schiff base formed between *o*-methylaminobenzaldehyde and 3-aminopropyldiphenylphosphine (NNP), n = 1, dark brown, m.p. = 210°, diamagnetic, five coordinate	379
L = Schiff base formed between *o*-methoxybenzaldehyde and 3-aminopropyldiphenylphosphine (ONP), n = 1, brown, m.p. = 180° μ = 3.20 BM, five coordinate	379
L = Schiff base formed between *o*-methylthiobenzaldehyde and 3-aminopropyldiphenylphosphine (SNP), n = 1, dark brown, m.p. = 180°, diamagnetic five coordinate	379
L = Schiff base formed between *o*-methylaminobenzaldehyde and 3-aminopropyldiphenylarsine (NNAs), n = 1, brown, m.p. = 220°, μ = 3.38 BM, five coordinate	379

TABLE 8.50 (*contd.*)

Adducts of the Type NiI_2L_n

L = Schiff base formed between *o*-methoxybenzaldehyde and 3-aminopropyldiphenylarsine (ONAs), *n* = 1, brown, m.p. = 174–178° μ = 3.20 BM, five coordinate	379
L = Schiff base formed between *o*-methylthiobenzaldehyde and 3-aminopropyldiphenylarsine (SNAs), *n* = 1, brown, m.p. = 190–193°, μ = 3.17 BM, five coordinate	379
L = 1,2-di(*o*-aminophenylthio)ethane (SN), *n* = 1, green, m.p. = 229–230°, μ = 3.15 BM, octahedral	412
L = tris(2-methylthioethyl)amine (TSN), *n* = 1, golden brown, μ = 3.10 BM, octahedral	413
L = 2,6-di(β-diphenylphosphinoethyl)pyridine (PNP), *n* = 1, brown, temperature-independent paramagnetism of 300 × 10^{-6} cgs, five coordinate	414
L = 2-methylthioaniline, *n* = 2, olive green, μ = 3.20 BM, octahedral	436
L = bisphenylphosphinomethylpyridine(2), *n* = 2, dark green, diamagnetic, ionic, five-coordinate cation	415
L = 2-ethylmercaptocyclohexylamine, *n* = 2, yellow-green, μ = 3.00 BM, octahedral	416
L = *S,S′*-*o*-xylyl-2,3-butanedione bis(2-mercaptoanil), *n* = 1, μ = 2.92 BM, *trans* octahedral	446
L = trimethylphosphine, *n* = 3, violet-black, m.p. = 160–170°	418
L = *o*-dimethylaminophenyldiphenylphosphine (PN), *n* = 1, black, m.p. = 208–209°, μ = 3.22 BM, ν_{Ni-I} = 239, 228 cm^{-1}, tetrahedral	385
L = bis(*o*-dimethylaminophenyl)phenylphosphine (PDN), *n* = 1, black, m.p. = 209–210°, μ = 3.25 BM, mixture of four- and five-coordinate species	385
L = tris(*o*-dimethylaminophenyl)phosphine (PTN), *n* = 1, maroon, m.p. = 211–212°, μ = 3.25 BM, mixture of four- and five-coordinate species	385
L = tetrakis(diphenylphosphinomethyl)methane, *n* = 1/2, black, diamagnetic, ν_{Ni-N} = 293–274 cm^{-1}, *cis* square planar, ligand bridges two nickel atoms	421
L = 1,4-dioxan, *n* = 2, black, μ = 3.35 BM, octahedral	422
L = *N,N′*-dimethylthiourea, *n* = 4, green, octahedral	425
L = *N,N′*-diethylthiourea, *n* = 4, green, octahedral	425

The following ligands have also been reacted with nickel diiodide; pyridine[370,426], monoethanolamine[433], tris(*o*-methylselenophenyl)phosphine[383], tris(3-dimethylarsinopropyl)arsine[384], dissymmetric phosphines[437], 2,5-dithiahexane and 1,2-di(isopropylthio)ethane[424].

Porri, Gallazzi and Vitulli[450] have prepared yellow to orange complexes of the type $NiX[(C_6H_5)_3P]_3$ (X = Cl, Br, I) by reacting π-allylNiX with triphenylphosphine in the presence of norbornadiene at room temperature. The magnetic moments of the compounds fall in the range 1.9–2.0 BM at room temperature. A preliminary X-ray analysis of the bromo complex indicates tetrahedral symmetry.

Complex Halides

A number of hexafluoronickelates(IV) have been synthesised by fluorination of a suitable mixture of nickel difluoride and appropriate halide in a bomb at 350° and 350 atmospheres[451]. The infrared and electronic spectra of the complexes were recorded and the infrared spectra are given in Table 8.51.

TABLE 8.51
Infrared Spectra (cm^{-1}) of Hexafluoronickelates(IV)

Complex	ν_3	ν_4
Na_2NiF_6	658.4, 668.9	350
K_2NiF_6	662	350
Rb_2NiF_6	653.7	343
Cs_2NiF_6	654.4	337
$(NO)_2NiF_6$	647	352

The structures of yellow-green barium and strontium tetrafluoronickelate(II) have been determined[452]. Both salts are orthorhombic with $a = 4.153$, $b = 14.458$ and $c = 5.799$ Å for the barium salt and $a = 3.935$, $b = 14.435$ and $c = 5.653$ Å for the strontium compound. The structures consist of alkaline-earth cations and zig-zag chains of NiF_6 octahedra sharing *cis* edges.

Single crystals of sodium trifluoronickelate(II) have been prepared[453] by fusing a 2:3 mixture of sodium chloride and nickel difluoride at 900°. The orthorhombic symmetry of the compound was confirmed, the cell having $a = 5.361$, $b = 5.524$ and $c = 7.788$ Å.

There have been further studies on the magnetic and spectral properties of the trifluoronickelates(II)[454–461]. For example, the Neel temperature of sodium trifluoronickelate(II) is reported[454] to be 149° while neutron-diffraction measurements[455] suggest that in the rubidium salt, which is isostructural with caesium trifluoromanganate (page 240), there is different amounts of canting for the two sub-lattices of the two different cation sites. Sintani and coworkers[456] have recorded the far-infrared spectrum of potassium trifluoronickelate(II) as a function of temperature. At room temperature, the three observed lattice vibrations occur at 448.9 (ν_1), 255.5 (ν_2) and 153.0 (ν_3) cm^{-1}. The first mode is particularly temperature sensitive; above T_N the slope is -0.07 cm^{-1} deg^{-1} while below T_N the slope is -0.08 cm^{-1} deg^{-1}. The temperature dependence of ν_2 and ν_3 are relatively small.

The preparation of tetraethylammonium tetrachloro- and tetrabromonickelate(II) has been reported by Gill and Taylor[462].

Tetramethylammonium trichloronickelate(II), prepared by interaction of the constituent chlorides in nitromethane, is isostructural with the corresponding bromo derivative[463] (Figure 8.16, page 454). The hexagonal unit cell has $a = 9.019$ and $c = 6.109$ Å. The nickel-chlorine distance is 2.408 Å and the chlorine-nickel-chlorine and nickel-chlorine-nickel angles are 84.05° and 78.74° respectively.

3,5-Dimethyldithiolium tetrabromonickelate(II) is a pale blue-green solid[464,465] with a room-temperature magnetic moment of 3.89 BM and it has a nickel-bromine stretching frequency of 225 cm^{-1}.

Sharp and coworkers[466] have prepared and characterized several substituted tetrahalonickelates(II). Their study clearly shows the pseudo-tetrahedral environment of the nickel atom. Some properties of two complexes are given in Table 8.52.

TABLE 8.52

Properties of Substituted Tetrahalonickelates(II)[a]

	$[picNiBr_3]^-$	$[pyNiI_3]^-$
μ (BM)	4.01	3.72
ν_{Ni-N} (cm^{-1})	176	164, 136
ν_{Ni-X}	250	250
δ_{Ni-X}	100, 92	76

[a] pic = 2-methylpyridine, py = pyridine.

TABLE 8.53

Internuclear Dimensions in $[(n\text{-}C_4H_9)_4N][NiBr_3(C_9H_7N)]$

Distance	Value (Å)	Angle	Value	Angle	Value
Ni—Br(1)	2.379	Br(1)—Ni—N	105.1°	Br(1)—Ni—Br(2)	116.8°
Ni—Br(2)	2.373	Br(2)—Ni—N	99.7°	Br(1)—Ni—Br(3)	104.5°
Ni—Br(3)	2.372	Br(3)—Ni—N	111.5°	Br(2)—Ni—Br(3)	118.5°
Ni—N	2.029				

Horrocks, Templeton and Zalkin[467] have determined the crystal structure of tetra-*n*-butylammonium tribromo(quinoline)nickelate(II), $[(n\text{-}C_4H_9)_4N][NiBr_3(C_9H_7N)]$, and confirmed the pseudo-tetrahedral environment of the nickel atom. The above complex has a triclinic unit cell with $a = 12.282$, $b = 10.291$, $c = 12.726$ Å, $\alpha = 101.02°$, $\beta = 99.51°$ and $\gamma = 106.55°$. There are two formula units in the cell of space group $P\bar{1}$. Some of the important internuclear dimensions are given in Table 8.53. Three of the butyl chains of the cation adopt the *trans* conformation while the other is in the *gauche* conformation.

REFERENCES

1. R. S. Nyholm, *J. Chem. Soc.*, **1950,** 2061.
2. R. S. Nyholm, *J. Chem. Soc.*, **1951,** 2602.
3. R. S. Nyholm, *J. Chem. Soc.*, **1950**, 2071.
4. P. Kreisman, R. Marsh, J. R. Preer and H. B. Gray, *J. Am. Chem. Soc.*, **90,** 1067 (1968).
5. J. Lewis, R. S. Nyholm and G. A. Rodley, *J. Chem. Soc.*, **1965,** 1483.
6. R. S. Nyholm, *J. Chem. Soc.*, **1952,** 2906.
7. K. A. Jensen, B. Nygaard and C. T. Pedersen, *Acta Chem. Scand.*, **17** 1126 (1963).
8. K. A. Jensen and B. Nygaard, *Acta Chem. Scand.*, **3,** 474 (1949).
9. G. Giacometti, V. Scatturin and A. Turco, *Ric. Sci.*, **27,** 2449 (1955).
10. V. Scatturin and A. Turco, *J. Inorg. Nucl. Chem.*, **8,** 447 (1958).
11. G. Booth and J. Chatt, *J. Chem. Soc.*, **1965,** 3238.
12. G. R. van Hecke and W. D. Horrocks, *Inorg. Chem.*, **5,** 1968 (1966).
13. C. E. Wymore and J. C. Bailar, *J. Inorg. Nucl. Chem.*, **14,** 42 (1960).
14. G. A. Barclay and R. S. Nyholm, *Chem. Ind. (London)*, **1953,** 378.
15. E. L. Muetterties and J. E. Castle, *J. Inorg. Nucl. Chem.*, **18,** 148 (1961).
16. M. Faber, R. T. Meyer and J. L. Margrave, *J. Phys. Chem.*, **62,** 883 (1958).
17. O. Ruff and E. Ascher, *Z. anorg. allgem. Chem.*, **183,** 193 (1929).
18. H. F. Priest, *Inorg. Syn.*, **3,** 173 (1950).
19. G. C. Hood and M. M. Woyski, *J. Am. Chem. Soc.*, **73,** 2738 (1951).
20. P. Allamagny, *Bull. Soc. Chim. France*, **1960,** 1099.
21. E. G. Rochow and I. Kukin, *J. Am. Chem. Soc.*, **74,** 1615 (1952).
22. H. M. Haendler, W. L. Patterson and W. J. Bernard, *J. Am. Chem. Soc.*, **74,** 3167 (1952).
23. R. L. Farrar and H. A. Smith, *J. Phys. Chem.*, **59,** 763 (1955).
24. *U.S. Patent*, 2,952,514 (1960).
25. E. Catalano and J. W. Stout, *J. Chem. Phys.*, **23,** 1284 (1955).
26. A. I. Belyaeva, V. V. Eremenko, N. N. Mikhailov, V. N. Pavlov and S. V. Petrov, *Soviet Phys.-JETP*, **23,** 979 (1966).
27. L. V. My, G. Perinet and P. Bianco, *Bull. Soc. Chim. France*, **1966,** 365.
28. D. E. Milligan, M. E. Jacox and J. D. McKinley, *J. Chem. Phys.*, **42,** 902 (1965).
29. J. D. McKinley, *J. Chem. Phys.*, **45,** 1690 (1966).
30. R. L. Jarry, J. Fischer and W. H. Gunther, *J. Electrochem. Soc.*, **110,** 346 (1963).
31. Y. A. Luk'yanychev, I. I. Astakhov and N. S. Nikolaev, *Izv. Akad. Nauk SSSR, Ser. Khim.*, **1965,** 588.

32. J. H. Canterford and T. A. O'Donnell, *Technique of Inorganic Chemistry* (Ed. H. B. Jonassen and A. Weissberger), Vol. 7, Interscience, New York, 1968.
33. A. W. Jache and G. H. Cady, *J. Phys. Chem.*, **56,** 1106 (1952).
34. J. W. Stout and E. Catalano, *J. Chem. Phys.*, **23,** 2013 (1955).
35. T. C. Ehlert, R. A. Kent and J. L. Margrave, *J. Am. Chem. Soc.*, **86,** 5093 (1964).
36. A. Buchler, J. L. Stauffer and W. Klemperer, *J. Chem. Phys.*, **40,** 3471 (1964).
37. E. Rudzitis, E. H. van Deventer and W. N. Hubbard, *J. Chem. Eng. Data*, **12,** 133 (1967).
38. W. H. Baur, *Acta Cryst.*, **11,** 488 (1958).
39. W. H. Baur, *Naturwissenschaften*, **44,** 349 (1957).
40. K. Haefner, J. W. Stout and C. S. Barrett, *J. Appl. Phys.*, **37,** 449B (1966).
41. P. Henkel and W. Klemm, *Z. anorg. allgem. Chem.*, **222,** 74 (1935).
42. P. L. Richards, *J. Appl. Phys.*, **35,** 850 (1964).
43. A. H. Cooke, K. A. Gehring and R. Lazenky, *Proc. Phys., Soc.*, **85,** 967 (1965).
44. P. L. Richards, *Phys. Rev.*, **138,** 1769 (1965).
45. J. W. Stout and E. Catalano, *Phys. Rev.*, **92,** 1575 (1953).
46. L. M. Matarrese and J. W. Stout, *Phys. Rev.*, **94,** 1792 (1954).
47. R. A. Erickson, *Phys. Rev.*, **90,** 779 (1953).
48. T. Moriya, *Phys. Rev.*, **117,** 635 (1960).
49. R. G. Shulman, *Phys. Rev.*, **121,** 125 (1961).
50. M. Balkanski, P. Moch and G. Parisot, *J. Chem. Phys.*, **44,** 940 (1966).
51. M. Balkanski, P. Moch and G. Parisot, *Compt. Rend.*, **258,** 2785 (1964).
52. A. Tsuchida, *J. Phys. Soc. Japan*, **21,** 2497 (1966).
53. P. L. Richards, *J Appl. Phys.*, **34,** 1237 (1963).
54 M Balkanski, P. Moch and R. G. Shulman, *J. Chem. Phys.*, **40,** 1897 (1964).
55. J. Ferguson, H. J. Guggenheim and D. L. Wood, *J. Chem. Phys.*, **39,** 3149 (1963).
56. J. Ferguson, H. J. Guggenheim, L. F. Johnson and H. Kamimura, *J. Chem. Phys.*, **38,** 2579 (1963).
57. A. Kurtenacker, W. Finger and F. Hey, *Z. anorg. allgem. Chem.*, **211,** 83 (1933).
58. A. C. Tulemello, *Diss. Abs.*, **23,** 432 (1962).
59. W. Rudorff, J. Kandler and D. Babel, *Z. anorg. allgem. Chem.*, **317,** 261 (1962).
60. J. D. McKinley and K. E. Shuler, *J. Chem. Phys.*, **28,** 1207 (1958).
61. J. D. McKinley, *J. Chem. Phys.*, **40,** 120 (1964).
62. R. C. Osthoff and R. C. West, *J. Am. Chem. Soc.*, **76,** 4732 (1954).
63. Y. I. Ivashentsev and G. G. Bodunova, *Tr. Tomskogo Gos. Univ., Ser. Khim.*, **154,** 63 (1962).
64. E. Iberson, R. Gut and D. M. Gruen, *J. Phys. Chem.*, **66,** 65 (1962).
65. J. W. Johnson, D. Cubicciotti and C. M. Kelly, *J. Phys. Chem.*, **62,** 1107 (1958).
66. M. Nehme and S. J. Teichner, *Bull. Soc. Chim. France*, **1960,** 389.
67. D. Khristov, S. Karaivanov and V. Kolushki, *Godishnik. Sofiskyia Univ., Khim. Fak.*, **55,** 49 (1960–61).

68. G. W. Watt, P. S. Gentile and E. P. Helvenston, *J. Am. Chem. Soc.*, **77,** 2752 (1955).
69. R. C. Schoonmaker, A. H. Friedman and R. F. Porter, *J. Chem. Phys.*, **31,** 1586 (1959).
70. G. E. Leroi, T. C. James, J. T. Hougen and W. Klemperer, *J. Chem. Phys.*, **36,** 2879 (1962).
71. H. Schafer, L. Bayer, G. Briel, K. Etzel and K. Krehl, *Z. anorg. allgem. Chem.*, **278,** 300 (1959).
72. W. Fischer and R. Gewher, *Z. anorg. allgem. Chem.*, **222,** 303 (1935).
73. J. P. Coughlin, *J. Am. Chem. Soc.*, **73,** 5314 (1951).
74. K. Sano, *J. Chem. Soc. Japan*, **58,** 376 (1937).
75. A. Ferrari, A. Braibanti and G. Bigliardi, *Acta Cryst.*, **16,** 846 (1963).
76. C. Starr, F. Bitter and A. R. Kaufmann, *Phys. Rev.*, **58,** 977 (1940).
77. J. W. Leech and A. J. Manuel, *Proc. Phys. Soc.*, **69B,** 210 (1956).
78. R. Stevenson, *Can. J. Phys.*, **40,** 1385 (1962).
79. H. Bizette, C. Terrier and B. Tsai, *Compt. Rend.*, **243,** 1295 (1956).
80. R. H. Busey and W. F. Giauque, *J. Am. Chem. Soc.*, **74,** 4443 (1952).
81. Y. Ting and D. Williams, *Phys. Rev.*, **82,** 507 (1951).
82. J. W. Leech and A. J. Manuel, *Proc. Phys. Soc.*, **69B,** 220 (1956).
83. J. Kanamori, *Prog. Theoret. Phys.*, **20,** 890 (1958).
84. R. W. Asmussen and O. Bostrup, *Acta Chem. Scand.*, **11,** 745 (1957).
85. O. Bostrup and C. K. Jorgensen, *Acta Chem. Scand*, **11,** 1223 (1957).
86. J. T. Hougen, G. E. Leroi and T. C. James, *J. Chem. Phys.*, **34,** 1670 (1961).
87. C. W. DeKock and D. M. Gruen, *J. Chem. Phys.*, **44,** 4387 (1966).
88. C. W. DeKock and D. M. Gruen, *J. Chem. Phys.*, **46,** 1096 (1967).
89. M. A. Porai-Koshits and A. S. Antsishkina, *Dokl. Akad. Nauk SSSR*, **92,** 333 (1953).
90. M. A. Porai-Koshits, E. K. Yukhno, A. S. Antsishkina and L. M. Dikareva, *Kristallografiya*, **2,** 371 (1957).
91. R. J. H. Clark and C. S. Williams, *Inorg. Chem.*, **4,** 350 (1965).
92. C. W. Frank and L. B. Rogers, *Inorg. Chem.*, **5,** 615 (1966).
93. D. A. Rowley and R. S. Drago, *Inorg. Chem.*, **6,** 1092 (1967).
94. S. M. Nelson and T. M. Shepard, *J. Chem. Soc.*, **1965,** 3276.
95. J. R. Allan, D. H. Brown, R. H. Nuttall and D. W. A. Sharp, *J. Inorg. Nucl. Chem.*, **27,** 1305 (1965).
96. D. M. L. Goodgame, M. Goodgame, M. A. Hitchman and M. J. Weeks, *J. Chem. Soc.*, **1966,** A, 1769.
97. J. R. Allan, D. H. Brown, R. H. Nuttall and D. W. A. Sharp, *J. Inorg. Nucl. Chem.*, **27,** 1529 (1965).
98. I. G. Murgulescu, E. Segal and D. Fatu, *J. Inorg. Nucl. Chem.*, **27,** 2677 (1965).
99. D. H. Brown, R. H. Nuttall and D. W. A. Sharp, *J. Inorg. Nucl. Chem.*, **25,** 1067 (1963).
100. A. K. Majumda, A. K. Mukherjee and A. K. Mukherjee, *J. Inorg. Nucl. Chem.*, **26,** 2177 (1964).
101. G. Beech, C. T. Mortimer and E. G. Tyler, *J. Chem. Soc.*, **1967,** A, 1111.
102. N. S. Gill and H. J. Kingdon, *Australian J. Chem.*, **19,** 2197 (1966).
103. L. M. Vallarino, W. E. Hill and J. V. Quagliano, *Inorg. Chem.*, **4,** 1598 (1965).

104. S. Buffagni, L. M. Vallarino and J. V. Quagliano, *Inorg. Chem.*, **3,** 671 (1964).
105. D. H. Brown, R. H. Nuttall and D. W. A. Sharp, *J. Inorg. Nucl. Chem.*, **26,** 1151 (1964).
106. N. S. Gill, R. H. Nuttall, D. E. Scaife and D. W. A. Sharp, *J. Inorg. Nucl. Chem.*, **18,** 79 (1961).
107. N. S. Gill, R. S. Nyholm, G. A. Barclay, T. I. Christie and P. J. Pauling, *J. Inorg. Nucl. Chem.*, **18,** 88 (1961).
108. H. C. A. King, E. Koros and S. M. Nelson, *J. Chem. Soc.*, **1963,** 5449.
109. E. Konig and H. L. Schlaffer, *Z. Physik. Chem. (Frankfurt)*, **26,** 371 (1960).
110. A. B. P. Lever, *Inorg. Chem.*, **4,** 763 (1965).
111. A. B. P. Lever, S. M. Nelson and T. M. Shepard, *Inorg. Chem.*, **4,** 810 (1965).
112. M. D. Glonek, C. Curran and J. V. Quagliano, *Inorg. Chem.*, **84,** 2014 (1962).
113. D. E. Billing and A. E. Underhill, *J. Chem. Soc.*, **1968,** A, 29.
114. S. Buffagni, L. M. Vallarino and J. V. Quagliano, *Inorg. Chem.*, **3,** 480 (1964).
115. I. S. Ahuja, D. H. Brown, R. H. Nuttall and D. W. A. Sharp, *J. Inorg. Nucl. Chem.*, **27,** 1105 (1965).
116. I. S. Ahuja, D. H. Brown, R. H. Nuttall and D. W. A. Sharp, *J. Inorg. Nucl. Chem.*, **27,** 1625 (1965).
117. D. M. L. Goodgame and M. Goodgame, *J. Chem. Soc.*, **1963,** 207.
118. M. Goodgame and M. J. Weeks, *J. Chem. Soc.*, **1966,** A, 1156.
119. D. M. L. Goodgame and M. Goodgame, *Inorg. Chem.*, **4,** 139 (1965).
120. C. M. Harris and E. D. McKenzie, *J. Inorg. Nucl. Chem.*, **19,** 372 (1961).
121. P. Pfeiffer and F. Tappermann, *Z. anorg. allgem. Chem.*, **215,** 273 (1933).
122. J. A. Broomhead and F. P. Dwyer, *Australian J. Chem.*, **14,** 250 (1961).
123. R. H. Lee, E. Griswold and J. Kleinberg, *Inorg. Chem.*, **3,** 1278 (1964).
124. R. J. H. Clark and C. S. Williams, *Spectrochim. Acta*, **23A,** 1055 (1967).
125. E. Uhlig and K. Staiger, *Z. anorg. allgem. Chem.*, **336,** 42 (1965).
126. E. Uhlig and K. Staiger, *Z. anorg. allgem. Chem.*, **336,** 179 (1965).
127. H. M. State, *Inorg. Syn.*, **6,** 198 (1960).
128. A. S. Antsishkina and M. A. Porai-Koshits, *Dokl. Akad. Nauk SSSR*, **143,** 105 (1962).
129. B. Bosnich, M. L. Tobe and G. A. Webb, *Inorg. Chem.*, **4,** 1109 (1965).
130. B. Bosnich, R. Mason, P. J. Pauling, G. B. Robertson and M. L. Tobe, *Chem. Commun.*, **1965,** 97.
131. B. Bosnich, R. D. Gillard, E. D. McKenzie and G. A. Webb, *J. Chem. Soc.*, **1966,** A, 1331.
132. J. W. Mellor, *A Comprehensive Treatise on Inorganic and Theoretical Chemistry*, Vol. 15, Longmans, London, p. 414.
133. R. J. H. Clark and C. S. Williams, *J. Chem. Soc.*, **1966,** A, 1425.
134. C. D. Flint and M. Goodgame, *J. Chem. Soc.*, **1966,** A, 744.
135. L. Cavalca, M. Nardelli and A. Braibanti, *Gazz. Chim. Ital.*, **86,** 942 (1956).
136. K. Swaminathan and H. M. N. H. Irving, *J. Inorg. Nucl. Chem.*, **26,** 1291 (1964).
137. A. Lopez-Castro and M. R. Turner, *J. Chem. Soc.*, **1963,** 1309.
138. D. M. Adams and J. B. Cornell, *J. Chem. Soc.*, **1967,** A, 884.
139. R. W. Olliff, *J. Chem. Soc.*, **1965,** 2036.

140. S. L. Holt and R. L. Carlin, *J. Am. Chem. Soc.*, **86**, 3017 (1964).
141. W. T. Robinson, S. L. Holt and G. E. Carpenter, *Inorg. Chem.*, **6**, 605 (1967).
142. M. Nardelli, I. Chierchi and A. Braibanti, *Gazz. Chim. Ital.*, **88**, 37 (1958).
143. G. Booth, *Advances in Inorganic Chemistry and Radiochemistry* (Ed. H. J. Emeleus and A. G. Sharpe), Vol. 6, Academic Press, London, 1964.
144. G. Garton, D. H. Henn, H. M. Powell and L. M. Venanzi, *J. Chem. Soc.*, **1963**, 3625.
145. R. G. Hayter, *Inorg. Chem.*, **2**, 932 (1963).
146. R. C. Cass, G. E. Coates and R. G. Hayter, *J. Chem. Soc.*, **1955**, 4007.
147. K. A. Jensen, *Z. anorg. allgem. Chem.*, **229**, 265 (1936).
148. C. R. C. Coussmaker, M. H. Hutchinson, J. R. Mellor, L. E. Sutton and L. M. Venanzi, *J. Chem. Soc.*, **1961**, 2705.
149. R. G. Hayter and F. S. Humiec, *Inorg. Chem.*, **4**, 1701 (1965).
150. R. G. Hayter and F. S. Humiec, *J. Am. Chem. Soc.*, **84**, 2004 (1962).
151. M. C. Browning, J. R. Mellor, D. J. Morgan, S. A. J. Pratt, L. E. Sutton and L. M. Venanzi, *J. Chem. Soc.*, **1962**, 693.
152. P. Rigo, C. Pecile and A. Turco, *Inorg. Chem.*, **6**, 1636 (1967).
153. R. W. Asmussen, K. A. Jensen and H. Soling, *Acta Chem. Scand.*, **9**, 1391 (1955).
154. K. A. Jensen, P. H. Nielsen and C. T. Pedersen, *Acta Chem. Scand.*, **17**, 1115 (1963).
155. G. Giacometti and A. Turco, *J. Inorg. Nucl. Chem.*, **15**, 242 (1960).
156. L. M. Venanzi, *J. Chem. Soc.*, **1958**, 719.
157. L. M. Venanzi, *J. Inorg. Nucl. Chem.*, **8**, 137 (1958).
158. J. Bradbury, K. P. Forrest, R. H. Nuttall and D. W. A. Sharp, *Spectrochim. Acta*, **23A**, 2701 (1967).
159. K. Yamamoto, *Bull. Chem. Soc. Japan*, **27**, 501 (1954).
160. A. Turco, V. Scatturin and G. Giacometti, *Nature*, **183**, 601 (1959).
161. M. J. Hudson, R. S. Nyholm and M. H. B. Stiddard, *J. Chem. Soc.*, **1968**, A, 40.
162. K. Issleib and G. Hohfeld, *Z. anorg. allgem. Chem.*, **312**, 169 (1961).
163. K. Issleib and G. Schwager, *Z. anorg. allgem. Chem.*, **311**, 83 (1961).
164. J. W. Collier, F. G. Mann, D. G. Watson and H. R. Watson, *J. Chem. Soc.*, **1964**, 1803.
165. M. O. Workman, G. Dyer and D. W. Meek, *Inorg. Chem.*, **6**, 1543 (1967).
166. K. Issleib and B. Mitscherling, *Z. anorg. allgem. Chem.*, **304**, 73 (1960).
167. D. M. L. Goodgame, M. Goodgame and F. A. Cotton, *J. Am. Chem. Soc.*, **83**, 4161 (1961)
168. F. A. Cotton and D. M. L. Goodgame, *J. Am. Chem. Soc.*, **82**, 5771 (1960).
169. G. A. Rodley, D. M. L. Goodgame and F. A. Cotton, *J. Chem. Soc.*, **1965**, 1499.
170. D. J. Phillips and S. Y. Tyree, *J. Am. Chem. Soc.*, **83**, 1806 (1961).
171. D. M. L. Goodgame and F. A. Cotton, *J. Am. Chem. Soc.*, **82**, 5774 (1960).
172. A. B. P. Lever, *J. Inorg. Nucl. Chem.*, **27**, 149 (1965).
173. D. E. Billing, A. E. Underhill and G. M. Smart, *J. Chem. Soc.*, **1968**, A, 8.
174. A. B. P. Lever, J. Lewis and R. S. Nyholm, *J. Chem. Soc.*, **1963**, 5042.
175. A. B. P. Lever, J. Lewis and R. S. Nyholm, *J. Chem. Soc.*, **1964**, 4761.
176. D. M. L. Goodgame and L. M. Venanzi, *J. Chem. Soc.*, **1963**, 5909.

177. A. B. P. Lever, J. Lewis and R. S. Nyholm, *Nature*, **189**, 58 (1961).
178. G. Beech and C. T. Mortimer, *J. Chem. Soc.*, **1967**, A, 1115.
179. P. S. K. Chia, S. E. Livingstone and T. N. Lockyer, *Australian J. Chem.*, **20**, 239 (1967).
180. P. S. K. Chia and S. E. Livingstone, *Australian J. Chem.*, **21**, 339 (1968).
181. D. M. L. Goodgame and L. M. Venanzi, *J. Chem. Soc.*, **1963**, 616.
182. L. Sacconi and I. Bertini, *Inorg. Nucl. Chem. Letters*, **2**, 29, (1966).
183. L. Sacconi, I. Bertini and F. Mani, *Inorg. Chem.*, **6**, 262 (1967).
184. Z. Dori and H. B. Gray, *J. Am. Chem. Soc.*, **88**, 1394 (1966).
185. M. Ciampolini and G. P. Speroni, *Inorg. Chem.*, **5**, 45 (1966).
186. M. Ciampolini and N. Nardi, *Inorg. Chem.*, **5**, 41 (1966).
187. M. Ciampolini and N. Nardi, *Inorg. Chem.*, **6**, 445 (1967).
188. M. Ciampolini and J. Gelsomini, *Inorg. Chem.*, **6**, 1821 (1967).
189. L. Sacconi, I. Bertini and R. Morassi, *Inorg. Chem.*, **6**, 1548 (1967).
190. L. Sacconi and G. P. Speroni, *Inorg. Chem.*, **7**, 295 (1968).
191. W. J. Eilbeck, F. Holmes, C. E. Taylor and A. E. Underhill, *J. Chem. Soc.*, **1968**, A, 128.
192. R. A. Krause and D. H. Busch, *J. Am. Chem. Soc.*, **82**, 4830 (1960).
193. R. A. Krause, N. B. Colthup and D. H. Busch, *J. Phys. Chem.*, **65**, 2216 (1961).
194. S. M. Nelson and J. Rodgers, *Inorg. Chem.*, **6**, 1390 (1967).
195. E. Uhlig and G. Heinrich, *Z. anorg. allgem. Chem.*, **330**, 40 (1964).
196. E. Uhlig, J. Csaszar and M. Maaser, *Z. anorg. allgem. Chem.*, **331**, 324 (1964).
197. E. Uhlig and M. Maaser, *Z. anorg. allgem. Chem.*, **344**, 205 (1966).
198. R. J. Kern, *J. Inorg. Nucl. Chem.*, **25**, 5 (1963).
199. B. J. Hathaway and D. G. Holah, *J. Chem. Soc.*, **1964**, 2400.
200. H. Rheinboldt, A. Lukyen and H. Schmittmann, *J. Prakt. Chem.*, **149**, 30 (1937).
201. R. Juhasz and L. F. Yntema, *J. Am. Chem. Soc.*, **62**, 3522 (1940).
202. P. J. Hendra and D. B. Powell, *J. Chem. Soc.*, **1960**, 5105.
203. D. R. Chesterman and A. S. Nickelson, *J. Chem. Soc.*, **1936**, 1300.
204. S. Herzog, K. Gustav, E. Krueger, H. Oberenda and R. Schuster, *Z. Chem.*, **3**, 428 (1963).
205. R. J. Kern, *J. Inorg. Nucl. Chem.*, **24**, 1105 (1962).
206. E. E. Weaver and W. Keim, *Proc. Indian Acad. Sci.*, **70**, 123 (1960).
207. F. A. Cotton and R. Francis, *J. Am. Chem. Soc.*, **82**, 2986 (1960).
208. R. S. Drago and D. Meek, *J. Phys. Chem.*, **65**, 1446 (1961).
209. F. A. Cotton, R. Francis and W. D. Horrocks, *J. Phys. Chem.*, **64**, 1534 (1960).
210. D. W. Meek, D. K. Staub and R. S. Drago, *J. Am. Chem. Soc.*, **82**, 6013 (1960).
211. E. Uhlig, P. Schueler and D. Diehlman, *Z. anorg. allgem. Chem.*, **335**, 156 (1965).
212. W. Hieber and E. Levy, *Z. anorg. allgem. Chem.*, **219**, 225 (1934).
213. M. E. Farago, J. M. James and V. C. G. Trew, *J. Chem. Soc.*, **1968**, A, 48.
214. J. V Quagliano, J. Fujita, G. Franz, D. J. Phillips, J. A. Walmsley and S. Y. Tyree, *J. Am. Chem. Soc.*, **83**, 3770 (1961).
215. J. T. Summers and J. V. Quagliano, *Inorg. Chem.*, **3**, 1767 (1964).

216. S. Kida, J. V. Quagliano, J. A. Walmsley and S. Y. Tyree, *Spectrochim. Acta*, **19,** 189 (1963).
217. S. K. Madan and H. H. Denk, *J. Inorg. Nucl. Chem.*, **27,** 1049 (1965).
218. W. E. Bull, S. K. Madan and J. E. Willis, *Inorg. Chem.*, **2,** 303 (1963).
219. L. Sacconi and A. Sabatini, *J. Inorg. Nucl. Chem.*, **25,** 1389 (1963).
220. J. A. Scruggs and M. A. Robinson, *Inorg. Chem.*, **6,** 1007 (1967).
221. J. D. Curry, M. A. Robinson and D. H. Busch, *Inorg. Chem.*, **6,** 1570 (1967).
222. J. T. Donoghue and R. S. Drago, *Inorg. Chem.*, **2,** 572 (1963).
223. G. J. Sutton, *Australian J. Chem.*, **13,** 74 (1960).
224. E. Uhlig and M. Maaser, *Z. anorg. allgem. Chem.*, **322,** 25 (1963).
225. G. J. Sutton, *Australian J. Chem.*, **19,** 733 (1966).
226. J. W. Mellor, *A Comprehensive Treatise on Inorganic and Theoretical Chemistry*, Vol. 15, Longmans, London, p. 413.
227. N. P. Fedot'ev and Z. I. Dmitreshova, *Zh. Priklad. Khim.*, **30,** 221 (1957).
228. C. S. Shen and M. H. Chang, *Hua Hsueh Hsueh Pao*, **26,** 124 (1960).
229. G. A. Evlanov, *Izv. Yysshikh Uchebn. Zavadenii, Khim. i Khim. Tekhnol.*, **6,** 3 (1963).
230. E. V. Stroganov, I. I. Kozhina and S. N. Andreev, *Vestn. Leningrad Univ.*, **15,** *Ser. Fiz. i Khim.*, 109 (1960).
231. J. Mizuno, *J. Phys. Soc. Japan*, **16,** 1574 (1961).
232 I. Nakagawa and T. Shimanouchi, *Spectrochim. Acta*, **20,** 429 (1964).
233. T. Haseda, H. Kobayashi and M. Date, *J. Phys. Soc. Japan*, **14,** 1724 (1959).
234. T. Haseda and M. Date, *J. Phys. Soc. Japan*, **13,** 175 (1958).
235. E. Konig, Landolt-Bornstein, *Magnetic Properties of Coordination and Organometallic Transition Metal Compounds*, Vol. 2, Springer-Verlag, New York, 1966.
236. W. K. Robinson and S. A. Friedberg, *Phys. Rev.*, **117,** 402 (1960).
237. R. D. Spence, P. Middents, Z. El Saffar and R. Kleinberg, *J. Appl. Phys.*, **35,** 854 (1964).
238. R. Kleinberg, *J. Appl. Phys.*, **38,** 1453 (1967).
239. E. V. Stroganov, I. I. Kozhina, S. N. Andreev and A. B. Kolyadin, *Vestn. Leningrad Univ.*, **15,** *Ser. Fiz. i Khim.*, 130 (1960).
240. B. K. Vainshtein, *Zh. Fiz. Khim.*, **26,** 1774 (1952).
241. B. Morosin, *Acta Cryst.*, **23,** 630 (1967).
242. G. Braur, *Handbook of Preparative Inorganic Chemistry*, Academic Press, New York,, 1963.
243. H. Schafer and H. Jacob, *Z. anorg. allgem. Chem.*, **286,** 56 (1956).
244. J. S. Wells and D. R. Winder, *J. Chem. Phys.*, **45,** 410 (1966).
245. L. G. Van Uitert, H. J. Williams, R. C. Sherwood and J. J. Rubin, *J. Appl. Phys.*, **36,** 1029 (1965).
246. I. Tsubokawa, *J. Phys. Soc. Japan*, **15,** 2109 (1960).
247. M. A. Paraj-Kojic, A. S. Antsishkina, L. M. Dickareva and E. K. Jakhov, *Acta Cryst.*, **10,** 784 (1957).
248. A. S. Antsishkina and M. A. Porai-Koshits, *Kristallografiya*, **3,** 676 (1958).
249. M. Goodgame and P. J. Hayward, *J. Chem. Soc.*, **1966,** A, 632.
250. W. Ludwig and G. Wittman, *Helv. Chim. Acta*, **47,** 1265 (1964).
251. L. M. Venanzi, *Angew. Chem., Intern. Ed. Engl.*, **3,** 453 (1964).
252. J. A. Bertrand and D. L. Plymale, *Inorg. Chem.*, **5,** 879 (1966).

253. B. T. Kilbourn, H. M. Powell and J. A. C. Darbyshire, *Proc. Chem. Soc.*, **1963,** 207.
254. G. Giacometti, V. Scatturin and A. Turco, *Gazz. Chim. Ital.*, **88,** 434 (1958).
255. G. A. Mair, H. M. Powell and D. E. Henn, *Proc. Chem. Soc.*, **1960,** 415.
256. W. Reppe and W. J. Schweckendiek, *Ann. Chim. (Paris)*, **560,** 104 (1948).
257. G. N. La Mar, W. D. Horrocks and L. C. Allen, *J. Chem. Phys.*, **41,** 2126 (1964).
258. K. Issleib and A. Brack, *Z. anorg. allgem. Chem.*, **277,** 258 (1954).
259. L. Sacconi and J. Gelsomini, *Inorg. Chem.*, **7,** 291 (1968).
260. C. M. Harris, R. S. Nyholm and D. J. Phillips, *J. Chem. Soc.*, **1960,** 4379.
261. M. C. Browning, R. F. B. Davies, D. J. Morgan, L. E. Sutton and L. M. Venanzi, *J. Chem. Soc.*, **1961,** 4816.
262. T. D. DuBois and D. W. Meek, *Inorg. Chem.*, **6,** 1395 (1967).
263. K. Issleib and G. Doll, *Z. anorg. allgem. Chem.*, **305,** 1 (1960).
264. G. A. Barclay, R. S. Nyholm and R. V. Parish, *J. Chem. Soc.*, **1961,** 4433.
265. G. Dyer, J. G. Hartley and L. M. Venanzi, *J. Chem. Soc.*, **1965,** 1293.
266. K. Issleib and D. W. Muller, *Chem. Ber.*, **92,** 3175 (1959).
267. K. Issleib and E. Priebe, *Chem. Ber.*, **92,** 3183 (1959).
268. F. D. Ayres, P. Pauling and G. B. Robertson, *Inorg. Chem.*, **3,** 1303 (1964).
269. A. S. Foust and R. H. Soderberg, *J. Am. Chem. Soc.*, **89,** 5501 (1967).
270. P. Paoletti and M. Ciampolini, *Inorg. Chem.*, **6,** 64 (1967).
271. R. K. Y. Ho and S. E. Livingstone, *Australian J. Chem.*, **18,** 659 (1965).
272. W. Klemm and W. Schuth, *Z. anorg. allgem. Chem.*, **210,** 33 (1933).
273. P. J. Crowley and H. M. Haendler, *Inorg. Chem.*, **1,** 904 (1962).
274. E. Loyd, C. B. Brown, D. Glynwyn, R. Bonnell and W. J. Jones, *J. Chem. Soc.*, **1928,** 658.
275. F. A. Cotton and R. Francis, *J. Inorg. Nucl. Chem.*, **17,** 62 (1961).
276. M. Brierley and W. J. Geary, *J. Chem. Soc.*, **1967,** A, 963.
277. L. F. Lindoy, S. E. Livingstone and T. N. Lockyer, *Australian J. Chem.*, **19,** 1391 (1966).
278. S. K. Dhar and F. Basolo, *J. Inorg. Nucl. Chem.*, **25,** 37 (1963).
279. J. W. Mellor, *A Comprehensive Treatise on Inorganic and Theoretical Chemistry*, Vol. 15, Longmans, London, p. 426.
280. B. Weigel, *Bull. Soc. Chim. France*, **1963,** 2087.
281. M. Chaigneau, *Bull. Soc. Chim. France*, **1957,** 886.
282. M. Chaigneau, *Compt. Rend.*, **242,** 263 (1956).
283. V. V. Pechkovskii and A. V. Sofronova, *Russ. J. Inorg. Chem.*, **11,** 828 (1966).
284. N. C. Stephenson, *Acta Cryst.*, **17,** 592 (1964).
285. G. S. Benner, W. E. Hatfield and D. W. Meek, *Inorg. Chem.*, **3,** 1544 (1964).
286. G. N. La Mar, *J. Phys. Chem.*, **69,** 3212 (1965).
287. M. Gewdin-Louer and D. Weigel, *Compt. Rend.*, **264B,** 895 (1967).
288. W. Klemm and E. Huss, *Z. anorg. allgem. Chem.*, **258,** 221 (1949).
289. H. Bode and E. Voss, *Z. anorg. allgem. Chem.*, **286,** 136 (1956).
290. R. D. Peacock and D. W. A. Sharp, *J. Chem. Soc.*, **1959,** 2762.
291. H. Bode and E. Voss, *Z. anorg. allgem. Chem.*, **290,** 1 (1957).
292. R. Hoppe, *Rec. Trav. Chim.*, **75,** 569 (1956).
293. H. G. Schnering, *Z. anorg. allgem. Chem.*, **353,** 1 (1967).
294. H. G. Schnering, *Z. anorg. allgem. Chem.*, **353,** 13 (1967).

295. W. Rudorff, J. Kandler, G. Lincke and D. Babel, *Angew. Chem.*, **71,** 672 (1959).
296. G. Wagner and D. Balz, *Z. Elektrochem.*, **56,** 574 (1952).
297. D. Balz and K. Plieth, *Z. Elektrochem.*, **59,** 545 (1955).
298. W. Rudorff and J. Kandler, *Naturwissenschaften*, **44,** 418 (1957).
299. J. C. Cousseins and M. Samouel, *Compt. Rend.*, **265C,** 1121 (1967).
300. H. M. Haendler, F. A. Johnson and D. S. Crocket, *J. Am. Chem. Soc.*, **80,** 2662 (1958).
301. D. S. Crockett and R. A. Grossman, *Inorg. Chem.*, **3,** 644 (1964).
302. K. G. Srivastava, *Phys. Letters*, **4,** 55 (1963).
303. E. Legrand and R. Plumier, *Phys. Status Solidi*, **2,** 317 (1962).
304. R. Plumier, *J. Appl. Phys.*, **35,** 950 (1964).
305. R. Plumier, *J. Phys. (Paris)*, **24,** 741 (1963).
306. R. Plumier, *J. Phys. (Paris)*, **25,** 859 (1964).
307. R. Plumier and E. Legrand, *J. Phys. Radium*, **23,** 474 (1962).
308. M. E. Lines, *Phys. Rev.*, **164,** 736 (1967).
309. D. S. Crocket and H. M. Haendler, *J. Am. Chem. Soc.*, **82,** 4158 (1960).
310. D. Babel, *Z. Naturforsch.*, **20a,** 165 (1965).
311. K. Kohn, R. Fukada and S. Iida, *J. Phys. Soc. Japan*, **22,** 333 (1967).
312. R. V. Pisarev, *Soviet Phys.-Solid State*, **8,** 1836 (1966).
313. D. J. Machin, R. L. Martin and R. S. Nyholm, *J. Chem. Soc.*, **1963,** 1490.
314. D. J. Machin and R. S. Nyholm, *J. Chem. Soc.*, **1963,** 1500.
315. K. Knox, *Acta Cryst.*, **14,** 583 (1961).
316. S. Ogawa, *J. Phys. Soc. Japan*, **15,** 2361 (1960).
317. A. Okazaki and Y. Suemune, *J. Phys. Soc. Japan*, **16,** 671 (1961).
318. A. Okazaki, Y. Suemune and T. Fuchikami, *J. Phys. Soc. Japan*, **14,** 1823 (1959).
319. V. M. Yudin and A. B. Sherman, *Phys. Status Solidi*, **20,** 759 (1967).
320. R. L. Martin, R. S. Nyholm and N. C. Stephenson, *Chem. Ind. (London)*, **1956,** 83.
321. Y. Suemune and H. Ikawa, *J. Phys. Soc. Japan*, **19,** 1686 (1964).
322. V. Scatturin, L. Corlis, N. Elliot and J. Hastings, *Acta Cryst.*, **14,** 19 (1961).
323. I. Nakagawa, A. Tsuchida and T. Shimanouchi, *J. Chem. Phys.*, **47,** 982 (1967).
324. A. Tsuchida and I. Nakagawa, *J. Phys. Soc. Japan*, **20,** 1726 (1965).
325. M. Balkanski, R. Le Toullec, P. Moch and M. Teng, *Compt. Rend.*, **261,** 1492 (1965).
326. R. V. Pisarev, *Soviet Phys.-Solid State*, **7,** 1114 (1965).
327. K. Knox, R. G. Shulman and S. Sugano, *Phys. Rev.*, **130,** 512 (1963).
328. J. Ferguson, H. J. Guggenheim and D. L. Wood, *J. Chem. Phys.*, **40,** 822 (1964).
329. J. Ferguson and H. J. Guggenheim, *J. Chem. Phys.*, **44,** 1095 (1966).
330. J. Ferguson, *Australian J. Chem.*, **21,** 323 (1968).
331. R. V. Pisarev, I. G. Sing and G. A. Smolensky, *Solid State Commun.*, **5,** 959 (1967).
332. C. R. Boston, J. Brynestead and G. P. Smith, *J. Chem. Phys.*, **47,** 3193 (1967).
333. N. S. Gill and R. S. Nyholm, *J. Chem. Soc.*, **1959,** 3997.
334. P. Pauling, *Inorg. Chem.*, **5,** 1498 (1966).
335. C. Furlani and G. Morpurgo, *Z. Physik. Chem. (Frankfurt)*, **29,** 93 (1961).

336. G. D. Stucky, J. B. Folkers and T. J. Kistenmacher, *Acta Cryst.*, **23,** 1064 (1967).
337. J. R. Wiesner, R. C. Srivastava, C. H. L. Kennard, M. Di Vaira and E. C. Lingafelter, *Acta Cryst.*, **23,** 565 (1967).
338. P. Paoletti and A. Vacca, *Trans. Faraday Soc.*, **60,** 50 (1964).
339. A. Sabatini and L. Sacconi, *J. Am. Chem. Soc.*, **86,** 17 (1964).
340. R. J. H. Clark and T. M. Dunn, *J. Chem. Soc.*, **1963,** 1198.
341. D. M. Adams, J. Chatt, J. M. Davidson and J. Gerratt, *J. Chem. Soc.*, **1963,** 2189.
342. L. Sacconi, M. Ciampolini and U. Campigli, *Inorg. Chem.*, **4,** 407 (1965).
343. G. P. Smith, C. H. Liu and T. R. Griffiths, *J. Am. Chem. Soc.*, **86,** 4796 (1964).
344. J. Brynestead, C. R. Boston and G. P. Smith, *J. Chem. Phys.*, **47,** 3179 (1967).
345. C. R. Boston and G. P. Smith, *J. Phys. Chem.*, **62,** 409 (1958).
346. D. M. Gruen and R. L. McBeth, *J. Phys. Chem.*, **63,** 393 (1959).
347. H. A. Weaklien, *J. Chem. Phys.*, **36,** 2117 (1962).
348. C. R. Boston and G. P. Smith, *J. Am. Chem. Soc.*, **85,** 1006 (1963).
349. C. K. Jorgenson, *Mol. Phys.*, **1,** 410 (1958).
350. B. F. Sundheim and G. Harrington, *J. Chem. Phys.*, **31,** 700 (1959).
351. G. P. Smith and C. R. Boston, *J. Chem. Phys.*, **43,** 4051 (1965).
352. B. D. Bird and P. Day, *Chem. Commun.*, **1967,** 741.
353. N. K. Hamer, *Mol. Phys.*, **6,** 257 (1963).
354. D. R. Stephens and H. G. Drickamer, *J. Chem. Phys.*, **35,** 429 (1961).
355. H. Remy and F. Meyer, *Ber.*, **77B,** 679 (1944).
356. R. W. Asmussen and H. Soling, *Z. anorg. allgem. Chem.*, **283,** 3 (1956).
357. F. A. Cotton, O. D. Faut and D. M. L. Goodgame, *J. Am. Chem. Soc.*, **83,** 344 (1961).
358. N. Blondet and D. Colaitis, *Bull. Soc. Chim. France*, **1963,** 83.
359. J. W. Mellor, *A Comprehensive Treatise on Inorganic and Theoretical Chemistry*, Vol. 15, Longmans, London, p. 419.
360. G. N. Tischenko, *Tr. Inst. Krist. Akad. Nauk SSSR*, **11,** 93 (1955).
361. G. Stucky, S. D'Agostino and G. McPherson, *J. Am. Chem. Soc.*, **88,** 4828 (1966).
362. G. M. Barvinok and V. Vintruff, *Vestn. Leningrad Univ.*, **21,** *Ser. Fiz. i Khim.*, 111 (1966).
363. D. M. L. Goodgame, M. Goodgame and M. J. Weeks, *J. Chem. Soc.*, **1964,** 5194.
364. P. Paoletti, *Trans. Faraday Soc.*, **61,** 219 (1965).
365. P. Ros, *Rec. Trav. Chim.*, **82,** 823 (1963).
366. P. L. Richards, *J. Appl. Phys.*, **38,** 1500 (1967).
367. J. R. McCreary and R. J. Thorn, *J. Chem. Phys.*, **48,** 3290 (1968).
368. M. O. Kistryakova and O. A. Zarubina, *Zh. Eksp. Thoer. Fiz., Pis'ma Redaktsiyu*, **7,** 16 (1968).
369. D. M. Gruen, J. R. Clifton and C. W. DeKock, *J. Chem. Phys.*, **48,** 1394 (1968).
370. D. Forster, *Inorg. Chim. Acta*, **2,** 116 (1968).
371. D. A. Rowley and R. S. Drago, *Inorg. Chem.*, **7,** 795 (1968).
372. P. M. Boorman and A. J. Carty, *Inorg. Nucl. Chem. Letters*, **4,** 105 (1968).

373. H. S. Preston, C. H. L. Kennard and R. A. Plowman, *J. Inorg. Nucl. Chem.*, **30,** 1463 (1968).
374. H. S. Preston and C. H. L. Kennard, *Chem. Commun.*, **1968,** 819.
375. G. Bombieri, Forsellini, G. Bandoli, L. Sindellari, R. Graziani and C. Panattoni, *Inorg. Chim. Acta*, **2,** 27 (1968).
376. H. Luth and M. R. Truter, *J. Chem. Soc.*, **1968,** A, 1879.
377. L. Sacconi and I. Bertini, *J. Am. Chem. Soc.*, **89,** 2235 (1967).
378. L. Sacconi, I. Bertini and F. Mani, *Inorg. Chem.*, **7,** 1417 (1968).
379. L. Sacconi, G. P. Speroni and R. Morassi, *Inorg. Chem.*, **7,** 1521 (1968).
380. L. Sacconi and I. Bertini, *Inorg. Chem.*, **7,** 1178 (1968).
381. L. Sacconi and R. Morassi, *Inorg. Nucl. Chem. Letters*, **4,** 449 (1968).
382. L. Sacconi and I. Bertini, *J. Am. Chem. Soc.*, **90,** 5443 (1968).
383. G. Dyer and D. W. Meek, *Inorg. Chem.*, **6,** 149 (1967).
384. G. S. Benner and D. W. Meek, *Inorg. Chem.*, **6,** 1399 (1967).
385. R. E. Cristopher, I. R. Gordon and L. M. Venanzi, *J. Chem. Soc.*, **1968,** A, 205.
386. A. Braibanti, F. Dallavalle, M. A. Pellinghelli and E. Leporti, *Inorg. Chem.*, **7,** 1430 (1968).
387. D. Nicholls and R. Swindells, *J. Inorg. Nucl. Chem.*, **30,** 2211 (1968).
388. R. N. Patel and D. V. R. Rao, *Z. anorg. allgem. Chem.*, **351,** 68 (1967).
389. E. J. Duff, *J. Chem. Soc.*, **1968,** A, 1812.
390. C. J. Popp and R. O. Ragsdale, *Inorg. Chem.*, **7,** 1845 (1968).
391. T. R. Musgrave and C. E. Mattson, *Inorg. Chem.*, **7,** 1433 (1968).
392. D. M. L. Goodgame, M. Goodgame, P. J. Hayward and G. W. Rayner-Canham, *Inorg. Chem.*, **7,** 2447 (1968).
393. J. S. Judge, W. M. Reiff, G. M. Intille, P. Ballway and W. A. Baker, *J. Inorg. Nucl. Chem.*, **29,** 1711 (1967).
394. J. S. Judge and W. A. Baker, *Inorg. Chim. Acta*, **1,** 239 (1967).
395. F. Lions, I. G. Dance and J. Lewis, *J. Chem. Soc.*, **1967,** A, 565.
396. D. M. Bowers and A. I. Popov, *Inorg. Chem.*, **7,** 1594 (1968).
397. W. J. Eilbeck, F. Holmes, C. E. Taylor and A. E. Underhill, *J. Chem. Soc.*, **1968,** A, 1189.
398. C. C. Houk and K. Emerson, *J. Inorg. Nucl. Chem.*, **30,** 1493 (1968).
399. E. J. Duff and M. N. Hughes, *J. Chem. Soc.*, **1968,** A, 2144.
400. E. J. Duff, M. N. Hughes and K. J. Rutt, *J. Chem. Soc.*, **1968,** A, 2354.
401. L. F. Lindoy and S. E. Livingstone, *Inorg. Chim. Acta*, **2,** 119 (1968).
402. R. A. Walton and R. Whyman, *J. Chem. Soc.*, **1968,** A, 1394.
403. E. Uhlig, R. Krahmer and H. Wolf, *Z. anorg. allgem. Chem.*, **361,** 157 (1968).
404. C. D. Burbridge and D. M. L. Goodgame, *J. Chem. Soc.*, **1968,** A, 237.
405. E. J. Duff, *J. Chem. Soc.*, **1968,** A, 434.
406. B. J. A. Kakazia and G. A. Melson, *Inorg. Chim. Acta*, **2,** 186 (1968).
407. D. R. Marks, D. J. Phillips and J. P. Redfern, *J. Chem. Soc.*, **1967,** A, 1464.
408. E. J. Duff, *J. Inorg. Nucl. Chem.*, **30,** 1257 (1968).
409. E. J. Duff, *J. Inorg. Nucl. Chem.*, **30,** 861 (1968).
410. M. Brierley and W. J. Geary, *J. Chem. Soc.*, **1968,** A, 1641.
411. L. Sacconi, R. Morassi and S. Midollini, *J. Chem. Soc.*, **1968,** A, 1968.
412. R. D. Cannon, B. Chiswell and L. M. Venanzi, *J. Chem. Soc.*, **1967,** A, 1277.
413. M. Ciampolini, J. Gelsomini and N. Nardi, *Inorg. Chim. Acta*, **2,** 343 (1968).

414. S. M. Nelson and W. S. J. Kelly, *Chem. Commun.*, **1968,** 436.
415. E. Uhlig and M. Schafer, *Z. anorg. allgem. Chem.*, **359,** 67 (1968).
416. E. Wenschuh, *Z. Naturforsch.*, **23b,** 595 (1968).
417. C. W. Kaufman and M. A. Robinson, *J. Inorg. Nucl. Chem.*, **30,** 2475 (1968).
418. K. A. Jensen and O. Dahl, *Acta Chem. Scand.*, **22,** 1044 (1968).
419. S. S. Sandhu and M. Gupta, *Chem. Ind. (London)*, **1967,** 1876.
420. H. D. Smith, M. A. Robinson and S. Papetti, *Inorg. Chem.*, **6,** 1014 (1967).
421. J. Ellermann and W. H. Gruber, *Chem. Ber.*, **101,** 3234 (1968).
422. G. W. A. Fowles, D. A. Rice and R. A. Walton, *J. Chem. Soc.*, **1968,** A, 1842.
423. J. G. H. du Preez, W. J. A. Steyn and A. J. Basson, *J. South African Chem. Institute*, **21,** 8 (1968).
424. C. D. Flint and M. Goodgame, *J. Chem. Soc.*, **1968,** A, 2178.
425. R. A. Bailey and T. R. Peterson, *Can. J. Chem.*, **46,** 3119 (1968).
426. A. A. Knyazeva, I. I. Kalinichenko and T. A. Degtyareva, *Russ. J. Inorg. Chem.*, **12,** 642 (1967).
427. N. Hurduc, L. Odochian, E. Segal and I. A. Schneider, *Z. Physik. Chem. (Leipzig)*, **237,** 90 (1968).
428. D. G. Brewer, P. T. T. Wong and M. C. Sears, *Can. J. Chem.*, **46,** 3137 (1968).
429. N. C. Collins and W. K. Glass, *Chem. Commun.*, **1967,** 211.
430. J. Reedijk and W. L. Groeneveld, *Rec. Trav. Chim.*, **87,** 552 (1968).
431. A. W. McLellan and G. A. Melson, *J. Chem. Soc.*, **1967,** A, 137.
432. B. B. Kedzia, P. X. Armendarez and K. Nakamoto, *J. Inorg. Nucl. Chem.*, **30,** 849 (1968).
433. V. V. Udovenko and U. S. Duchinskii, *Russ. J. Inorg. Chem.*, **12,** 510 (1967).
434. I. G. Druzhinen, B. Murzubraimov and K. Rysmendeev, *Russ. J. Inorg. Chem.*, **12,** 1037 (1967).
435. C. D. Flint and M. Goodgame, *J. Chem. Soc.*, **1968,** A, 750.
436. L. F. Lindoy and S. E. Livingstone, *Inorg. Chem.*, **7,** 1149 (1968).
437. L. H. Pignolet and W. DeW. Horrocks, *Chem. Commun.*, **1968,** 1012.
438. R. Nasanen, L. Lemmetti, U. Lamminsivu and H. Abdalla, *Suomen Kemistilehti*, **41B,** 1 (1968).
439. M. Di Vaira and P. L. Orioli, *Acta Cryst.*, **24B,** 595 (1968).
440. P. L. Orioli and M. Di Vaira, *J. Chem. Soc.*, **1968,** A, 2078.
441. P. L. Oriola and L. Sacconi, *Chem. Commun.*, **1968,** 1310.
442. E. H. Brooks and F. Glockling, *J. Chem. Soc.*, **1967,** A, 1030.
443. J. A. J. Jarvis, R. H. B. Mais and P. G. Owston, *J. Chem. Soc.*, **1968,** A, 1473.
444. G. Zakrzewski and L. Sacconi, *Inorg. Chem.*, **7,** 1034 (1968).
445. E. Wenschuh, B. Fritzsche and H. Mach, *Z. anorg. allgem. Chem.*, **358,** 233 (1968).
446. M. S. Elder, G. M. Prinz, P. Thornton and D. H. Busch, *Inorg. Chem.*, **7,** 2426 (1968).
447. B. B. Chastain, D. W. Meek, E. Billig, J. E. Hix and H. B. Gray, *Inorg. Chem.*, **7,** 2412 (1968).
448. C. W. Bird and E. M. Briggs, *J. Chem. Soc.*, **1967,** A, 1004.
449. N. N. Greenwood and B. H. Robinson, *J. Chem. Soc.*, **1967,** A, 511.
450. L. Porri, M. C. Gallazzi and G. Vitulli, *Chem. Commun.*, **1967,** 228.

451. R. Bougon, *Compt. Rend.*, **267C,** 681 (1968).
452. H. G. Schnering and P. Bleckmann, *Naturwissenschaften*, **55,** 342 (1968).
453. A. G. Tutov and P. P. Syrnikov, *Soviet Phys.-Crystallography*, **12,** 619 (1968).
454. A. Epstein, J. Makovsky, M. Melamud and H. Skakid, *Phys. Rev.*, **174,** 560 (1968).
455. S. J. Pickart and H. A. Alperin, *J. Appl. Phys.*, **39,** 1332 (1968).
456. K. Sintani, Y. Tomono, A. Tsuchida and K. Siratori, *J. Phys. Soc. Japan*, **25,** 99 (1968).
457. T. R. McGuire and M. W. Schafer, *J. Appl. Phys.*, **39,** 1130 (1968).
458. R. V. Pisarev and S. D. Prochorova, *Phys. Letters*, **26A,** 356 (1968).
459. G. A. Smolenskii, R. V. Pisarev, M. P. Petrov, V. V. Moskulev, I. G. Siny and V. M. Judin, *J. Appl. Phys.*, **39,** 568 (1968).
460. G. A. Smolenskii, M. P. Petrov and R. V. Pisarev, *J. Appl. Phys.*, **38,** 1269 (1967).
461. A. T. Starovortov, V. I. Ozhogin and V. A. Boko, *Soviet Phys.-Solid State*, **10,** 1228 (1968).
462. N. S. Gill and F. B. Taylor, *Inorg. Syn.*, **9,** 136 (1967).
463. G. D. Stucky, *Acta Cryst.*, **24B,** 330 (1968).
464. G. A. Heath, R. L. Martin and I. M. Stewart, *Australian J. Chem.*, **22,** 83 (1969).
465. G. A. Heath, R. L. Martin and I. M. Stewart, *Chem. Commun.*, **1969,** 54.
466. D. H. Brown, K. P. Forrest, R. H. Nuttall and D. W. A. Sharp, *J. Chem. Soc.*, **1968,** A, 2146.
477. W. DeW. Horrocks, D. H. Templeton and A. Zalkin, *Inorg. Chem.*, **7,** 2303 (1968).

Chapter 9

Copper

With the exception of potassium hexafluorocuprate(III), K_3CuF_6, all of the halo compounds of copper are restricted to oxidation states II and I. By far the greatest amount of information available is on copper(II) compounds.

One of the most notable features of the halogen chemistry of copper(II) is that the element displays practically every conceivable type of stereochemistry in coordination numbers four, five and six. The interest in this phenomenon is reflected in the large number of full structural determinations which have been carried out on copper halo compounds.

HALIDES

The known halides of copper are given in Table 9.1.

TABLE 9.1
Halides of Copper

Oxidation State	Fluoride	Chloride	Bromide	Iodide
II	CuF_2	$CuCl_2$	$CuBr_2$	
I	(CuF)	$CuCl$	$CuBr$	CuI

Copper diiodide cannot be prepared because iodide reduces copper(II) to copper (I). There is some evidence that copper monofluoride may exist in the vapour phase, but it has never been isolated as a solid.

Oxidation State II

Copper difluoride. Copper difluoride may be prepared by reaction of copper metal, or the oxides, sulphides and other halides with either fluorine or anhydrous hydrogen fluoride in a flow system at elevated temperatures[1–5]. Copper difluoride is a white hygroscopic solid which melts at 785° and which vaporizes as a monomeric species[6]. Over the temperature range 624–753° the vapour pressure is given by the expression[6]

$$\log p_{\text{atm}} = 8.58 - 13{,}000/T$$

The heat of sublimation is 63 kcal mole^{-1}.

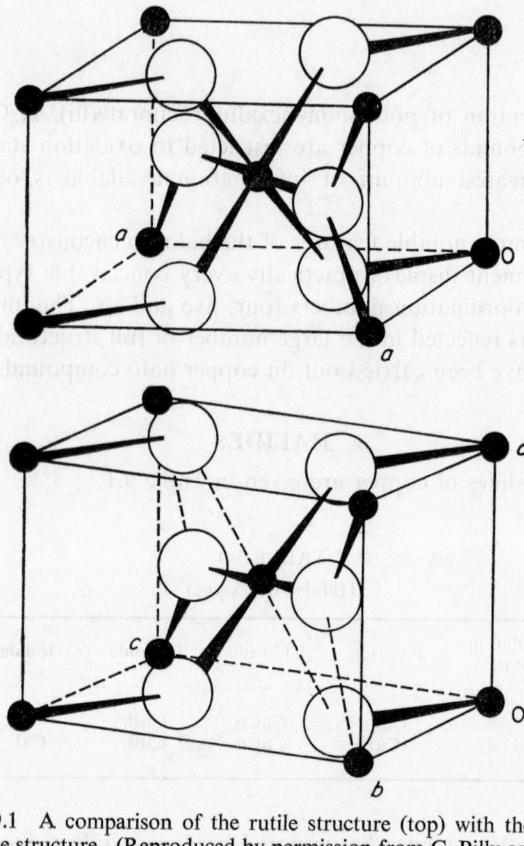

Figure 9.1 A comparison of the rutile structure (top) with the copper difluoride structure. (Reproduced by permission from C. Billy and H. M. Haendler, *J. Am. Chem. Soc.*, **79**, 1049 (1957))

Billy and Haendler[7] have shown by single-crystal X-ray diffraction studies that copper difluoride has a distorted rutile structure. The monoclinic unit cell has $a = 3.32$, $b = 4.54$, $c = 4.59$ Å and $\beta = 83.6°$. The space group is $P2_1/n$ and there are two formula units in the cell. There is a very pronounced tetragonal distortion in the CuF_6 octahedra, as shown in Figure 9.1. The copper-fluorine bond distances are 1.93 and 2.27 Å.

The copper-fluorine asymmetric stretching frequency is reported[8] to occur at 765 cm^{-1} in a neon matrix at 4°K.

The magnetic properties of copper difluoride have been studied over a wide temperature range by a number of investigators[8–14]. The compound becomes quite strongly antiferromagnetic below about 75°K. The spectroscopic splitting factor has been reported by several workers[8,11,12] and the most recent measurements by Kasai and coworkers[8] give a value for $g_{\parallel}$ of 1.913 and for $g_{\perp}$ of 2.601 for copper difluoride in an argon matrix at 4°K.

The diffuse-reflectance spectrum of copper difluoride has been observed by Oelkrug[15] and by Schmitz-DuMont and Grimm[16]. The following assignments were reported by Oelkrug[15], a tetragonally distorted octahedral configuration being assumed.

7,500 cm^{-1}	$^2A_{1g} \leftarrow {}^2B_{1g}$
8,800 cm^{-1}	$^2B_{2g} \leftarrow {}^2B_{1g}$
11,350 cm^{-1}	$^2E_g \leftarrow {}^2B_{1g}$

Copper difluoride dihydrate. This blue hydrate is most conveniently prepared[3,17–21] by dissolving copper carbonate, basic carbonate or oxide in 40 % hydrofluoric acid and evaporating the resultant solution at about 70°. It may be recrystallized readily from dilute hydrofluoric acid. The compound is thermally unstable[18,22] decomposing firstly at about 130° to Cu(OH)F and subsequently at about 400° to copper(II) oxide and copper difluoride.

TABLE 9.2

Lattice Parameters of Copper Difluoride Dihydrate

Parameter	Room Temperature	4.2°K
a (Å)	6.417	6.370
b (Å)	7.396	7.425
c (Å)	3.300	3.244
β	99.65°	101.25°

The structure of copper difluoride dihydrate has been investigated both by X-ray[23] and neutron-diffraction[20] techniques on single crystals. Extremely good agreement was obtained for the monoclinic unit-cell parameters at room temperature by the two methods. In addition, Abrahams[24] has reported a neutron-diffraction study at 4.2°K and the unit-cell dimensions deduced from the neutron-diffraction studies at the two temperatures are given in Table 9.2. There are two molecules in the unit cell of space group *I2/m*.

The structure of copper difluoride dihydrate consists of puckered sheets of square-planar *trans* $CuF_2(H_2O)_2$ groups which are linked together by quite strong hydrogen bonding in such a way as to give a tetragonally elongated octahedral coordination of four fluorine atoms and two water molecules about the copper atom, as shown in Figure 9.2. The important internuclear dimensions deduced from the neutron-diffraction studies are given in Table 9.3.

TABLE 9.3

Internuclear Dimensions in Copper Difluoride Dihydrate

Dimension	Room Temperature	4.2°K
Cu—O	1.941 Å	1.945 Å
Cu—F	1.898 Å	1.899 Å
Cu—F′	2.465 Å	2.391 Å
O—H	0.980 Å	0.959 Å
O . . . F	2.715 Å	2.718 Å
Cu—O—H	122.4°	124.8°
H—O—H	115.5°	110.1°

Steckhanov and coworkers[25] have investigated the various infrared-active modes of the water molecules in copper difluoride dihydrate as a function of temperature, and from the splittings have deduced that there are two different types of hydrogen bond present.

Copper difluoride dihydrate obeys the Curie-Weiss law[21,26] down to quite low temperatures with a magnetic moment of 1.9 BM and a Weiss constant of 37°. Susceptibility measurements[21,27] indicate that the molar susceptibility both parallel and perpendicular to the *c* axis reaches a maximum at about 25°K, although ^1H and ^{19}F NMR measurements[19], as well as specific heat studies[19], indicate a Neel temperature of 11°K. The rather large difference between the magnetic susceptibility maximum and the Neel temperature has been discussed by Abrahams[24]. The magnetic structure, shown in Figure 9.3, deduced from low temperature neutron-diffraction studies[24], consists of (101) planes containing individual $CuF_2(H_2O)_2$ formula units that form antiferromagnetic sheets which

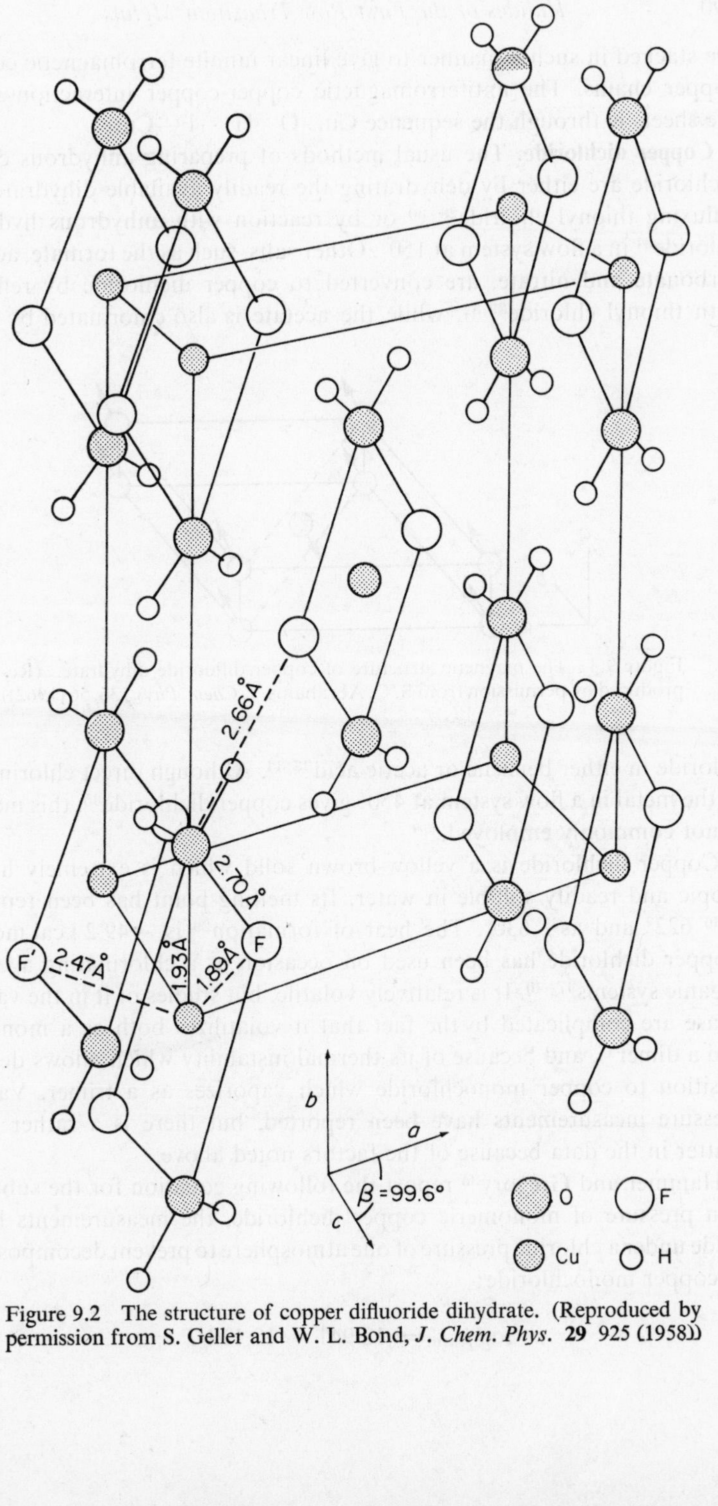

Figure 9.2 The structure of copper difluoride dihydrate. (Reproduced by permission from S. Geller and W. L. Bond, *J. Chem. Phys.* **29** 925 (1958))

are stacked in such a manner to give linear infinite ferromagnetic copper-copper chains. The antiferromagnetic copper-copper interaction within the sheets is through the sequence Cu—O—H···F—Cu.

Copper dichloride. The usual methods of preparing anhydrous copper dichloride are either by dehydrating the readily available dihydrate with refluxing thionyl cholride[28–30] or by reaction with anhydrous hydrogen chloride[31] in a flow system at 150°. Other salts, such as the formate, acetate, carbonate and nitrate, are converted to copper dichloride by refluxing with thionyl chloride[29,32], while the acetate is also chlorinated by acetyl

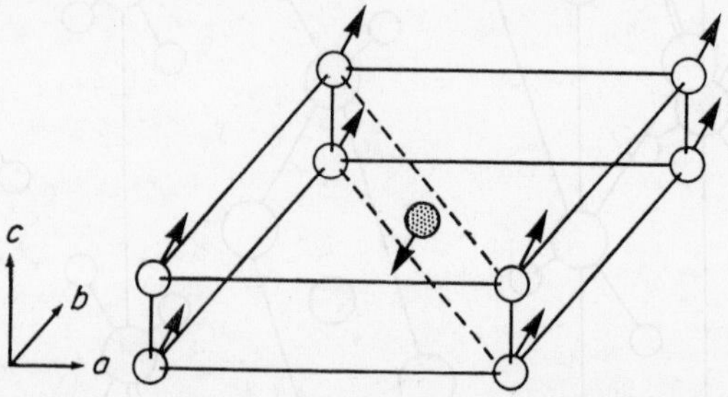

Figure 9.3 The magnetic structure of copper difluoride dihydrate. (Reproduced by permission from S. C. Abrahams, *J. Chem. Phys.*, **36**, 56 (1962))

chloride in either benzene or acetic acid[33,34]. Although direct chlorination of the metal in a flow system at 450° gives copper dichloride[35], this method is not commonly employed.

Copper dichloride is a yellow-brown solid which is extremely hygroscopic and readily soluble in water. Its melting point has been reported as[36] 622° and as[29] 630°. The heat of formation[29] is −49.2 kcal mole^{-1}. Copper dichloride has been used on occasion as a chlorinating agent in organic systems[37–40]. It is relatively volatile, but studies of it in the vapour phase are complicated by the fact that it volatilizes both as a monomer and a dimer[41], and because of its thermal instability which allows decomposition to copper monochloride which vaporizes as a trimer. Vapour pressure measurements have been reported, but there is a rather large scatter in the data because of the factors noted above.

Hammer and Gregory[35] report the following equation for the sublimation pressure of monomeric copper dichloride, the measurements being made under a chlorine pressure of one atmosphere to prevent decomposition to copper monochloride:

$$\log p_{\mathrm{mm}} = 12.941 - 10{,}119/T$$

The heat and entropy of sublimation at 440° were calculated to be 46.3 kcal $mole^{-1}$ and 46.0 cal deg^{-1} $mole^{-1}$ respectively. The same workers[35] reported that the dissociation pressure of chlorine over heated copper dichloride is given by the expression

$$\log p_{\text{mm}} = 12.741 - 8672/T$$

The heat and entropy of dissociation are 37.9 kcal $mole^{-1}$ and 45.1 cal deg^{-1} $mole^{-1}$ respectively. Shchukarev and Oranskaya[42] have reported the following dissociation pressure equations:

$$\log p_{\text{mm}} = 3.061 - 3114/T \quad (219\text{–}246°)$$

$$\log p_{\text{mm}} = 8.508 - 6979/T \quad (>436°)$$

Wells[43] has reported that copper dichloride has a monoclinic unit cell with $a = 6.85$, $b = 3.30$, $c = 6.70$ Å and $\beta = 121.0°$. The structure

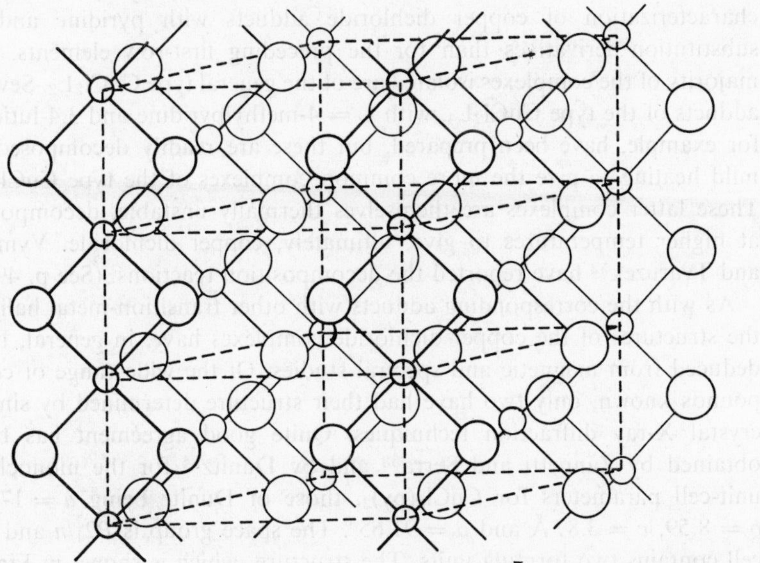

Figure 9.4 The crystal structure of copper dichloride. (Reproduced by permission from A. F. Wells, *J. Chem. Soc.*, **1947**, 1670)

consists of infinite chains of planar $CuCl_4$ groups in which the copper-chlorine distance is 2.3 Å and the chlorine-copper-chlorine angle is 87°. The chains are arranged as shown in Figure 9.4 in such a way that the chains are parallel to the *b* axis and each copper has two additional chlorine contacts at 2.95 Å so as to form a tetragonally elongated octahedral environment.

The infrared spectrum of solid copper dichloride has been recorded[44,45] and strong absorptions were observed at 328 and 275 cm^{-1}. In the gas phase[41] the copper-chlorine stretching frequency of the monomeric species occurs at 496 cm^{-1} while that of the dimeric species is at 417 cm^{-1}.

The magnetic properties of copper dichloride have been investigated by several workers[26,46,47]. At low temperatures it becomes antiferromagnetic. A magnetic moment of 1.97 BM and a Weiss constant of 78° have been reported[14].

The electronic absorption spectrum in the solid and gaseous phases, and also in various solvents, have been studied in some detail[48–51]. In the solid state a single *d-d* transition at about 12,200 cm^{-1} is observed[48,49].

Considerably less work has been reported on the preparation and characterization of copper dichloride adducts with pyridine and its substitution derivatives than for the preceding first-row elements. The majority of the complexes isolated are of the general type $CuCl_2L_2$. Several adducts of the type $CuCl_2L_4$, with L = 4-methylpyridine and 3,4-lutidine for example, have been prepared, but these are readily decomposed by mild heating to give the more common complexes of the type $CuCl_2L_2$. These latter complexes are themselves thermally unstable, decomposing at higher temperatures to give, ultimately, copper dichloride. Vymetal and Tvaruzek[52] have reported the decomposition reactions. (See p. 493.)

As with the corresponding adducts with other transition-metal halides, the structures of the copper dichloride complexes have, in general, been deduced from magnetic and spectral studies. Of the wide range of compounds known, only two have had their structure determined by single-crystal X-ray diffraction techniques. Quite good agreement has been obtained by Zannetti and Serra[53] and by Dunitz[54] for the monoclinic unit-cell parameters for $CuCl_2(py)_2$, those of Dunitz being $a = 17.00$, $b = 8.59$, $c = 3.87$ Å and $\beta = 91.65°$. The space group is $P2_1/n$ and the cell contains two formula units. The structure, which is shown in Figure 9.5, consists of *trans* square-planar $CuCl_2(py)_2$ groups which are linked together to form chains, parallel to the *c* axis, by weak copper-chlorine interactions so as to give distorted octahedral symmetry to the copper atoms.

$CuCl_2$(4-methylpyridine)$_2$ has recently been reported by Duckworth and coworkers[55] to have a triclinic unit cell with $a = 8.58$, $b = 9.20$, $c = 10.99$ Å, $\alpha = 86.8°$, $\beta = 110.1°$ and $\gamma = 123°$. There are two molecules in the unit cell of space group $P\bar{1}$. The two molecules, which are

$$CuCl_2(pyridine)_2 \xrightarrow{235^\circ} CuCl_2(pyridine) \xrightarrow{255^\circ} CuCl_2(pyridine)_{1/2} \xrightarrow{290^\circ} CuCl_2$$

$$CuCl_2(2\text{-methylpyridine})_2 \xrightarrow{175^\circ} CuCl_2(2\text{-methylpyridine})_{3/2} \xrightarrow{195^\circ} CuCl_2(2\text{-methylpyridine}) \xrightarrow{230^\circ} CuCl_2(2\text{-methylpyridine})_{1/2} \xrightarrow{290^\circ} CuCl_2$$

$$CuCl_2(3\text{-methylpyridine})_2 \xrightarrow{220^\circ} CuCl_2(3\text{-methylpyridine}) \xrightarrow{260^\circ} CuCl_2(3\text{-methylpyridine})_{3/4} \xrightarrow{280^\circ} CuCl_2(3\text{-methylpyridine})_{1/2} \xrightarrow{300^\circ} CuCl_2$$

$$CuCl_2(4\text{-methylpyridine})_2 \xrightarrow{200^\circ} CuCl_2(4\text{-methylpyridine})_{5/4} \xrightarrow{250^\circ} CuCl_2(4\text{-methylpyridine}) \xrightarrow{270^\circ} CuCl_2(4\text{-methylpyridine}) \xrightarrow{350^\circ} CuCl_2$$

$$CuCl_2(2\text{-ethylpyridine})_2 \xrightarrow{170^\circ} CuCl_2(2\text{-ethylpyridine})_{3/2} \xrightarrow{230^\circ} CuCl_2(2\text{-ethylpyridine}) \xrightarrow{280^\circ} CuCl_2$$

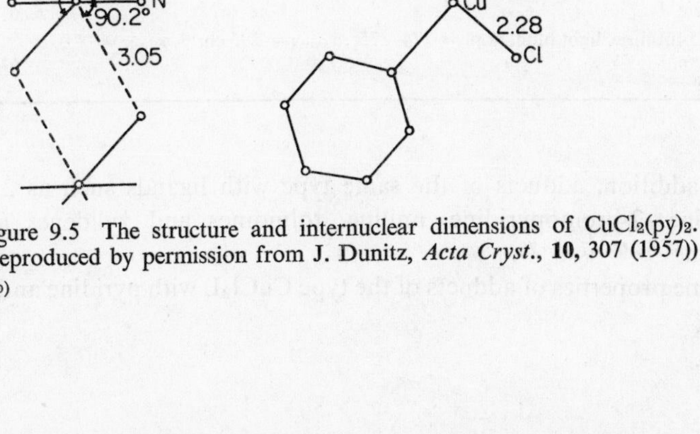

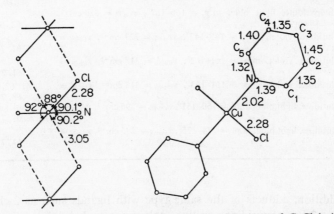

Figure 9.5 The structure and internuclear dimensions of $CuCl_2(py)_2$. (Reproduced by permission from J. Dunitz, *Acta Cryst.*, **10**, 307 (1957))

17 (24 pp)

related by a centre of symmetry, form a dimeric unit in which each copper atom is surrounded by two chlorine and two nitrogen atoms in the *trans* positions of an approximately square-planar arrangement. The two molecules are associated through two weak intermolecular bonds between the copper atoms and a bridging chlorine atom, so that each copper atom has a distorted tetragonal-pyramidal environment. The copper-chlorine bond distances are 2.21 and 2.26 Å while the copper-nitrogen lengths are 1.98 and 2.02 Å. The methyl groups of the pyridine rings assume *cis* positions with respect to the square-plane about the copper atom and effectively hinder coordination in one of the axial positions.

Some properties of a number of copper dichloride adducts with pyridine and related ligands are summarized in Table 9.4.

TABLE 9.4
Adducts of the Type $CuCl_2L_2$ with Pyridine and Related Ligands

Compound and Properties	References
L = pyridine, pale blue, m.p. = 258–259°, μ = 1.85 BM, $\nu_{Cu\text{-}Cl}$ = 294, 235 cm^{-1}, $\nu_{Cu\text{-}N}$ = 268 cm^{-1}, octahedral	52, 54, 56–64
L = 2-methylpyridine, dark blue, m.p. = 148–150°, μ = 1.87 BM, $\nu_{Cu\text{-}Cl}$ = 309 cm^{-1}, $\nu_{Cu\text{-}N}$ = 261 cm^{-1}	52, 57, 61, 62, 64–68
L = 3-methylpyridine, light blue, m.p. = 215–216°, $\nu_{Cu\text{-}Cl}$ = 294 cm^{-1}, $\nu_{Cu\text{-}N}$ = 267 cm^{-1}	52, 61, 64, 69
L = 4-methylpyridine, light blue, m.p. = 178–179°, $\nu_{Cu\text{-}Cl}$ = 299 cm^{-1}, $\nu_{Cu\text{-}N}$ = 266 cm^{-1}	52, 55, 61, 64, 66, 69, 70
L = 2-ethylpyridine, light blue, m.p. = 166–167°, $\nu_{Cu\text{-}Cl}$ = 320 cm^{-1}, $\nu_{Cu\text{-}N}$ = 246 cm^{-1}	52, 62
L = 2,3-lutidine, violet, m.p. = 145–147°, $\nu_{Cu\text{-}Cl}$ = 323 cm^{-1}, $\nu_{Cu\text{-}N}$ = 270 cm^{-1}	61, 71
L = 2,5-lutidine, violet, m.p. = 161–162°, $\nu_{Cu\text{-}Cl}$ = 318 cm^{-1}, $\nu_{Cu\text{-}N}$ = 257 cm^{-1}	61, 71
L = 2,6-lutidine, violet, m.p. = 161–163°, $\nu_{Cu\text{-}Cl}$ = 315 cm^{-1}, $\nu_{Cu\text{-}N}$ = 242 cm^{-1}	61, 62, 64, 68, 71
L = 3,4-lutidine, light blue, m.p. = 150–151°, $\nu_{Cu\text{-}Cl}$ = 295 cm^{-1}, $\nu_{Cu\text{-}N}$ = 259 cm^{-1}	61, 71
L = 3,5-lutidine, light blue, m.p. = 274–275°, $\nu_{Cu\text{-}Cl}$ = 295 cm^{-1}, $\nu_{Cu\text{-}N}$ = 250 cm^{-1}	61, 71

In addition, adducts of the same type with ligands such as 2-chloropyridine, 2-bromopyridine, aniline, toluidines and xylidenes are also known[61,62,64,67,72–76].

Some properties of adducts of the type $CuCl_2L$ with pyridine and related ligands which are derived from the bis-ligand derivatives discussed above are given in Table 9.5[52,71].

TABLE 9.5
Adducts of the Type $CuCl_2L$ with Pyridine and Related Ligands

Compound and Properties
L = pyridine, grey-green, m.p. = 249–250°
L = 2-methylpyridine, dark blue, m.p. = 181–182°
L = 3-methylpyridine, grey-green, m.p. = 234–236°
L = 4-methylpyridine, grey-white, m.p. = 178–180°
L = 2-ethylpyridine, brown-green, m.p. = 158–160°
L = 2,3-lutidine, dark green, m.p. = 152–154°
L = 2,4-lutidine, dark blue, m.p. = 143–144°
L = 2,5-lutidine, light green, m.p. = 163–164°

Emerson[77] has reported that 1,8-naphthyridine reacts with copper dichloride in water to give bright green $CuCl_2L_2$. The compound has a monoclinic unit cell with $a = 13.674$, $b = 8.198$, $c = 16.280$ Å and $\beta = 121.66°$. There are four molecules in the unit cell of space group $C2/c$. Only one nitrogen atom of the ligand is coordinated to the copper atom, and rather surprisingly, the compound has a *cis* planar configuration. The copper-chlorine and copper-nitrogen distances are 2.24 and 2.0 Å respectively.

2,2′,2″-tripyridyl (terpy) reacts[78] with copper dichloride to give $CuCl_2(terpy)\cdot 2H_2O$. Corbridge and Cox[79] have shown by a crystallographic study that the compound is isomorphous with $ZnCl_2(terpy)$ which is known to be five-coordinate. The monoclinic unit cell is of space group $P2_1/a$, contains four formula units, and has the parameters $a = 16.20$, $b = 8.26$, $c = 10.66$ Å and $\beta = 95°$.

Copper dichloride reacts very readily with pyridine *N*-oxide and closely related ligands to give complexes which exhibit somewhat unusual magnetic behaviour. Two kinds of complexes, $CuCl_2L_2$ and $CuCl_2L$ are formed, and for the former compounds there are two types of magnetic behaviour that may be correlated with the structure of the compounds.

The complexes of the type $CuCl_2L$ have sub-normal magnetic moments, they have terminal chlorine groups and appear to be dinuclear N-oxide-bridged species[80]. For the 1:2 adducts, there are in general two isomers. The green isomers have *trans* square-planar structures and thus show normal copper-chlorine stretching frequencies, whereas the yellow isomers have copper-chlorine stretching frequencies at much lower energies.

The structures of these latter complexes are unknown, although it is possible that they have a distorted tetrahedral structure[81]. The properties

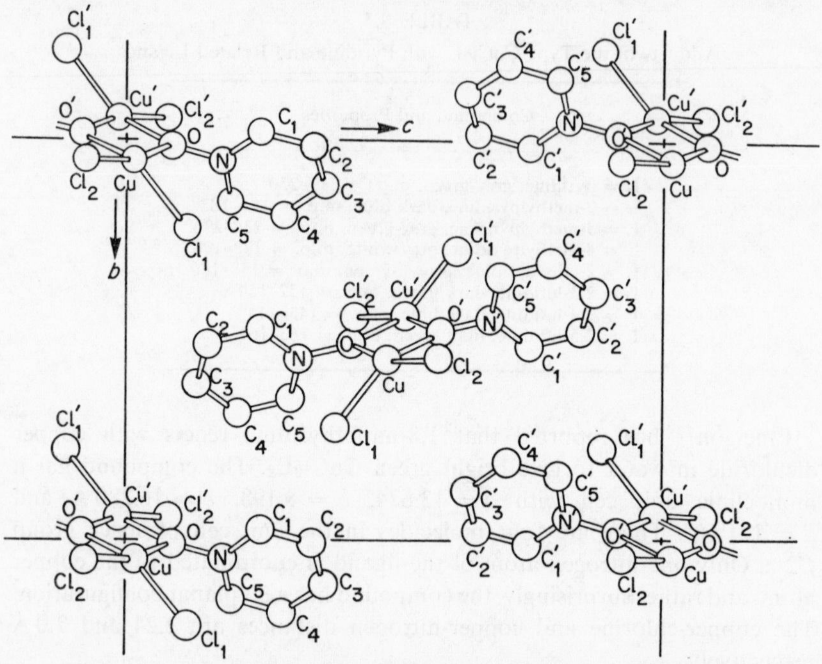

Figure 9.6 The structure of $[(C_5H_5NO)CuCl_2]_2$ projected along the *a* axis. (Reproduced by permission from R. S. Sager, R. J. Williams and W. H. Watson, *Inorg. Chem.*, **6**, 951 (1967))

TABLE 9.6
Internuclear Dimensions in $[CuCl_2 \cdot C_5H_5NO]_2$

Distance	Value (Å)	Angle	Value
Cu—Cl(1)	2.206	Cl(1)—Cu—Cl(2)	99.4°
Cu—Cl(2)	2.217	Cu—O—Cu′	107.8°
Cu—O	1.979	O—Cu—O′	72.9°
Cu—O′	2.036	Cu—O—N	123.5°
Cu—Cu	3.25	Cl(1)—Cu—O	97.6°

of the pyridine *N*-oxide type adducts have been correlated with steric effects, and in particular with the value of the pK_a for the ligand[80,82].

Schafer, Morrow and Smith[83] showed that $CuCl_2 \cdot C_5H_5NO$ has a monoclinic unit cell with $a = 5.844$, $b = 10.049$, $c = 13.643$ Å and $\gamma = 104.87°$. There are four formula units in the cell of space group $P2_1/b$. The structure which is in Figure 9.6 shows that the complex is dimeric with bridging oxygen atoms, the copper atoms having a distorted

TABLE 9.7

Crystallographic Data for Copper Dichloride-Pyridine *N*-Oxide Type Adducts

Compound	Symmetry	Parameters
$CuCl_2$·3-methylpyridine *N*-oxide	Monoclinic	$a = 5.97$, $b = 20.90$, $c = 8.13$ Å, $\beta = 98.2°$, $Z = 4$, $P2_1/c$
$CuCl_2$·4-methylpyridine *N*-oxide	Triclinic	$a = 9.94$, $b = 9.35$ $c = 5.93$ Å, $\alpha = 98.8°$ $\beta = 106.5°$, $\gamma = 116.85°$, $Z = 2$, $P1$ or $P\bar{1}$
$CuCl_2$·2,6-lutidine *N*-oxide	Monoclinic	$a = 10.64$, $b = 8.41$, $c = 17.98$ Å, $\beta = 143.63°$, $Z = 2$, $P2_1/c$
$CuCl_2$[2-methylpyridine *N*-oxide]$_2$	Monoclinic	$a = 6.73$, $b = 10.39$, $c = 10.73$ Å, $\beta = 101.37°$, $Z = 4$, Cc or $C2/c$
$CuCl_2$[3-methylpyridine *N*-oxide]$_2$		
yellow isomer	Monoclinic	$a = 7.75$, $b = 17.23$, $c = 12.04$ Å, $\beta = 111.1°$, $Z = 4$, $P2_1/c$
green isomer	Monoclinic	$a = 10.67$, $b = 11.01$, $c = 14.77$ Å, $\beta = 120.43°$, $Z = 4$, $P2_1/c$
$CuCl_2$[4-methylpyridine *N*-oxide]$_2$		
yellow isomer	Monoclinic	$a = 10.93$, $b = 20.07$, $c = 7.47$ Å, $\beta = 111.43°$, $Z = 4$, $P2_1/c$
green isomer	Monoclinic	$a = 6.29$, $b = 10.67$, $c = 11.51$ Å, $\beta = 101.75°$, $Z = 2$, $P2_1/c$
$CuCl_2$[2,6-lutidine *N*-oxide]$_2$		
yellow isomer	Orthorhombic	$a = 13.89$, $b = 7.68$, $c = 16.29$ Å, $Z = 4$, $Pna\,2_1$
green isomer	Monoclinic	$a = 15.59$, $b = 7.61$, $c = 14.53$ Å, $\beta = 96.27°$, $Z = 4$, Cc or $C2/c$

square-planar environment. The pyridine rings are approximately coplanar and make an angle of approximately 79° with the plane of the copper and oxygen atoms. The structure has been refined by Sager and coworkers[84] who gave the internuclear dimensions shown in Table 9.6.

Kidd, Sager and Watson[81] have made a very extensive crystallographic study of a wide variety of copper dichloride adducts with pyridine N-oxide and its substitution derivatives, and have shown that in a number of cases isomers are formed. The crystallographic data are summarized in Table 9.7.

Kidd and coworkers[81] have also reported that the yellow-green complex Cu_3Cl_6[2-methylpyridine*N*-oxide]$_2 \cdot 2H_2O$ may be isolated from the reaction between copper dichloride dihydrate and 2-methylpyridine *N*-oxide. It decomposes at 232–235° and has a triclinic unit cell with $a = 9.55$, $b = 7.23$, $c = 9.42$ Å, $\alpha = 92.84°$, $\beta = 115.87°$ and $\gamma = 105.6°$. The space group is $P1$ or $P\bar{1}$ and there is one molecule in the unit cell. The following structure, which is very similar to that of $Cu_3Cl_6[CH_3CN]_2$, was suggested for this compound.

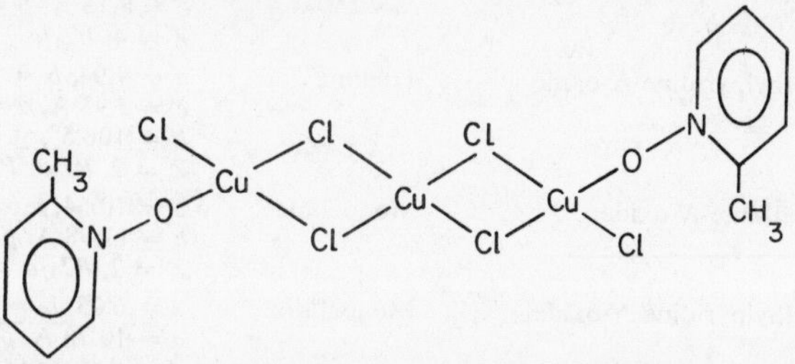

Some properties of the adducts of copper dichloride with pyridine *N*-oxide and related ligands are summarized in Table 9.8.

Other pyridine *N*-oxides, particularly those substituted in the 4-position, have been reacted with copper dichloride to give adducts similar to those described above[80,81,88–90].

Hatfield and coworkers[82,91] have prepared a number of quinoline *N*-oxide adducts of copper dichloride. As well as the expected complexes of the types $CuCl_2L_2$ and $CuCl_2L$, several adducts having the compositions $(CuCl_2)_3L_2$ and $(CuCl_2)_4L_3$ were isolated. The latter complexes are probably halogen-bridged polymers. The general properties of the 1:2 and 1:1 compounds are very similar to those of the pyridine *N*-oxide adducts described above. In the $CuCl_2L$ series dinuclear species are produced and there has been good correlation between the structure of the adducts and the pK_a of the quinoline *N*-oxide.

A number of substitution compounds of the type CuCl(OR), where ROH is an aminoalcohol which acts as a chelating group, have been prepared[92–94]. A detailed magnetic and spectral study of a range of these complexes by Uhlig and Staiger[94] has revealed that there are three types of complex. These are:

(a) Compounds with sub-normal magnetic moments. These appear to be dinuclear oxygen-bridged species in which the copper atom is in a distorted tetrahedral environment.

(b) Magnetically normal compounds in which the copper atom has a higher coordination number than four.
(c) Intermediate compounds.

Copper dichloride reacts very readily with acetonitrile at room temperature[95] to give the yellow adduct $Cu_2Cl_4(CH_3CN)_2$. Near the boiling point

TABLE 9.8

$CuCl_2L_n$ Adducts with Pyridine *N*-Oxide and Related Ligands

Compound and Properties[a]	References
L = pyridine *N*-oxide, $n = 1$, yellow-green, m.p. = 241°, $\mu = 0.77$ BM, $\nu_{Cu\text{-}Cl} = 330, 311\ cm^{-1}$	80–83, 85–88
L = pyridine *N*-oxide, $n = 2$, yellow, m.p. = 172–174°, $\mu = 0.63$ BM, $\nu_{Cu\text{-}Cl} = 310, 286\ cm^{-1}$	80–82, 86, 87
L = 2-methylpyridine *N*-oxide, $n = 2$, green, m.p. = 149–150°, $\mu =$ 1.87 BM, $\nu_{Cu\text{-}Cl} = 328\ cm^{-1}$, $\nu_{Cu\text{-}O} = 380\ cm^{-1}$	80, 81
L = 3-methylpyridine *N*-oxide, $n = 1$, yellow, m.p. = 238°, $\mu =$ 0.55 BM, $\nu_{Cu\text{-}Cl} = 316\ cm^{-1}$	80, 81, 89
L = 3-methylpyridine *N*-oxide, $n = 2$; yellow isomer, m.p. = 145–146°: green isomer, m.p. = 145–146°, $\nu_{Cu\text{-}Cl} = 322\ cm^{-1}$, $\nu_{Cu\text{-}O} = 373\ cm^{-1}$	80, 81
L = 4-methylpyridine *N*-oxide, $n = 1$, m.p. = 214°, $\mu = 0.52$ BM, $\nu_{Cu\text{-}Cl} = 327\ cm^{-1}$	80, 81, 88, 90
L = 4-methylpyridine *N*-oxide, $n = 2$, green, m.p. = 176°, $\mu =$ 1.99 BM, $\nu_{Cu\text{-}Cl} = 339, 316, 296\ cm^{-1}$, $\nu_{Cu\text{-}O} = 426\ cm^{-1}$	80, 81
L = 2,6-lutidine *N*-oxide, $n = 1$, dark green, m.p. = 215°, $\nu_{Cu\text{-}Cl} =$ 342, 326 cm^{-1}	80, 81, 89
L = 2,6-lutidine *N*-oxide, $n = 2$, yellow, m.p. = 198°, $\mu = 1.74$ BM, $\nu_{Cu\text{-}Cl} = 319, 297\ cm^{-1}$, $\nu_{Cu\text{-}O} = 408, 370\ cm^{-1}$	80, 81, 89

[a] Magnetic moments are per dimer.

TABLE 9.9

Lattice Parameters of Copper Dichloride-Acetonitrile Adducts

Parameter	$Cu_2Cl_4(CH_3CN)_2$	$Cu_3Cl_6(CH_3CN)_2$
a (Å)	3.86	6.78
b (Å)	7.91	6.13
c (Å)	18.37	16.51
β	91.1°	105.7°
Z	2	2
Space group	$P2_1/c$	$P2_1/c$

of acetonitrile a second compound, $Cu_3Cl_6(CH_3CN)_2$, is formed[95]. The structures of these two interesting compounds have been studied by Willett and Rundle[95] by single-crystal X-ray diffraction techniques Both compounds are of monoclinic symmetry and the lattice parameters are given in Table 9.9.

Both structures contain discrete planar molecules of approximately C_{2h} symmetry as shown in Figures 9.7 and 9.8. In each compound the nitrile groups are *trans* with respect to the copper-chlorine bridging

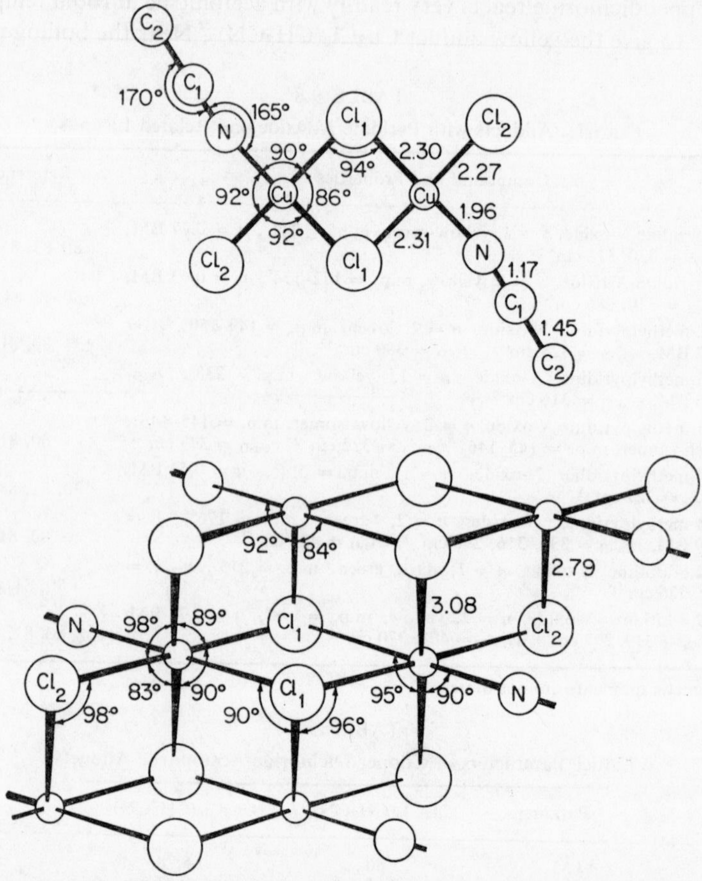

Figure 9.7 The structure of $Cu_2Cl_4(CH_3CN)_2$. (Reproduced by permission from R. D. Willett and R. E. Rundle, *J. Chem. Phys.*, **40**, 838 (1964))

systems and each copper atom is octahedrally coordinated by bonding to adjacent molecules.

Complexes with propionitrile and acrylonitrile have also been reported[96] but the structures of these complexes are unknown.

Ammonia reacts readily with a solution of copper nitrate and lithium chloride to give two isomers of formula $CuCl_2(NH_3)_2$[97,98]. The green isomer has also been prepared by pumping $CuCl_2 \cdot 6NH_3$ at room temperature[99].

The light blue α-isomer has a cubic lattice[97] with $a = 3.907$ Å and its structure is similar to that of the corresponding copper dibromide adduct for which full structural information is available, as described later (page 513). Both isomers have a polymeric octahedral structure and Clark and Williams[98] have made a detailed examination of the infrared and

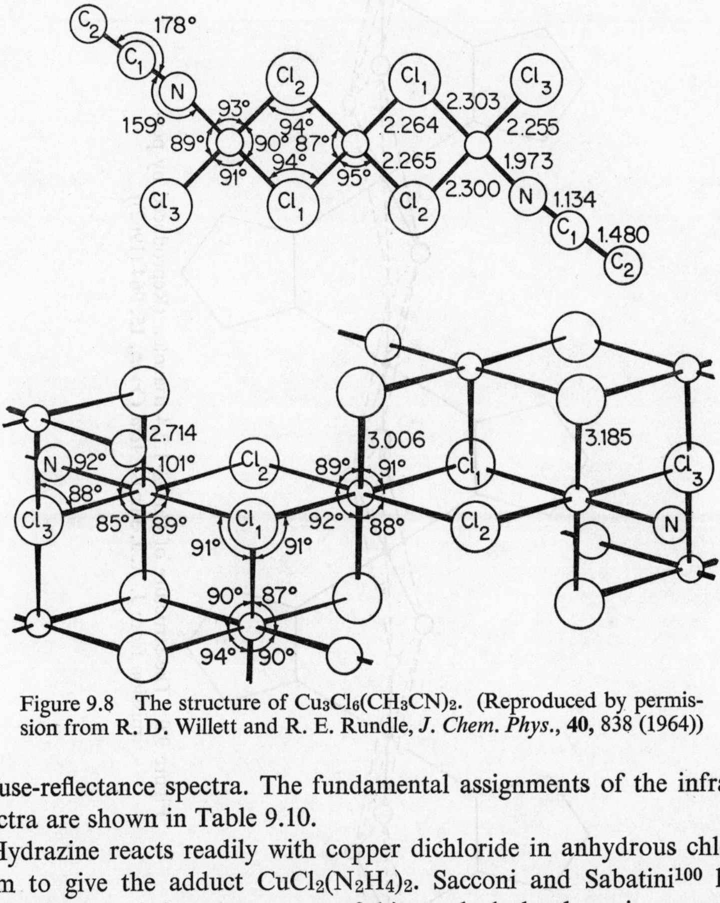

Figure 9.8 The structure of $Cu_3Cl_6(CH_3CN)_2$. (Reproduced by permission from R. D. Willett and R. E. Rundle, *J. Chem. Phys.*, **40**, 838 (1964))

diffuse-reflectance spectra. The fundamental assignments of the infrared spectra are shown in Table 9.10.

Hydrazine reacts readily with copper dichloride in anhydrous chloroform to give the adduct $CuCl_2(N_2H_4)_2$. Sacconi and Sabatini[100] have investigated the infrared spectrum of this octahedral polymeric compound and report a copper-nitrogen stretching frequency of 440 cm^{-1}.

A 1:1 complex of copper dichloride and 1,2,4-triazole is readily precipitated by dissolving the corresponding copper sulphate adduct in 20% hydrochloric acid solution[101]. The compound, which has a room-temperature magnetic moment of 1.81 BM, may also be made by direct interaction in ethanol[102]. Sanero[103] showed that the compound had a

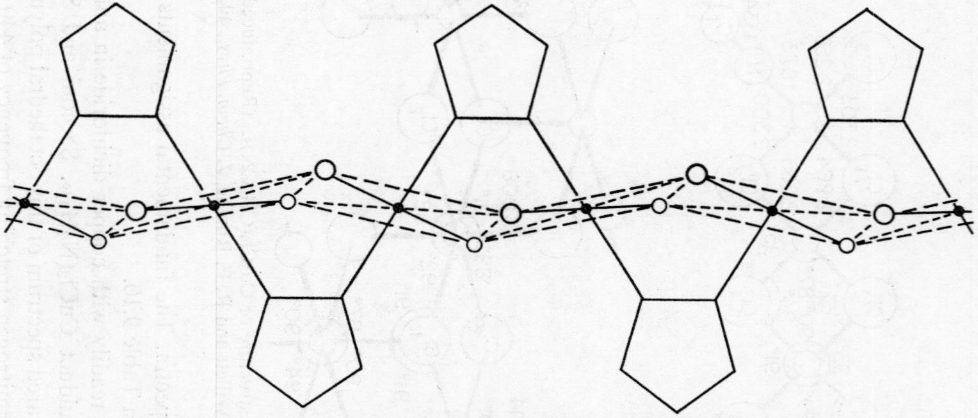

Figure 9.9 The structure of $CuCl_2 \cdot$1,2,4-triazole. (Reproduced by permission from J. A. J. Jarvis, *Acta Cryst.*, **15**, 964 (1962))

monoclinic cell. The structure has since been refined by Jarvis[104] who gave the following unit-cell parameters: $a = 6.81$, $b = 11.39$, $c = 7.13$ Å and $\beta = 96.97°$. There are four molecules in the unit cell of space group

TABLE 9.10
Infrared Spectra (cm^{-1}) of $CuCl_2(NH_3)_2$ Isomers

Mode		α-isomer	β-isomer
NH_3 def.	degen.	1603*m*	1603*m*
	sym.		1277*sh*
	sym.	1250*vs*	1253*s*
NH_3 rock		716*m*	718*w*
		666*m*	666*m*
Cu—N stretch		480*w*	482*w*
Cu—Cl stretch		267*s, br*	257*s, br*

vs = very strong *s* = strong *m* = medium
w = weak *sh* = shoulder *br* = broad

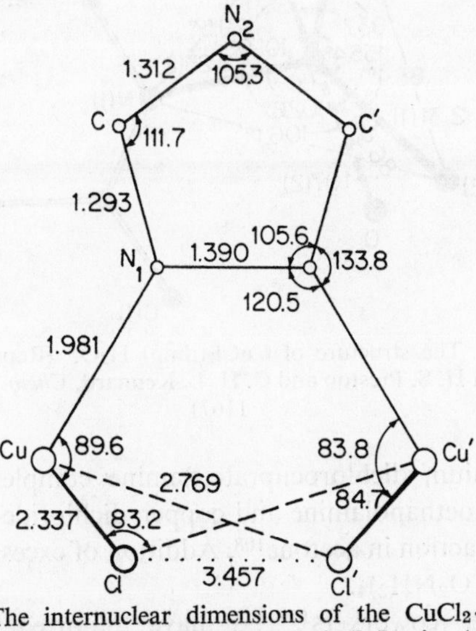

Figure 9.10 The internuclear dimensions of the $CuCl_2 \cdot 1,2,4$-triazole adduct. (Reproduced by permission from J. A. J. Jarvis, *Acta Cryst.*, **15**, 964 (1962))

*I*2/*c*. The structure, which is shown in Figure 9.9, contains chains of planar $CuCl_4$ groups which are linked together by the triazole molecules to give a distorted octahedral environment to the copper atoms. The internuclear dimensions are shown in Figure 9.10.

Prasad and Sharma[105,106] and Yoke and coworkers[99,107] have studied the reactions between copper dichloride and primary, secondary and tertiary amines. The adduct $CuCl_2(C_2H_5NH_2)_2$ obeys the Curie-Weiss law with a magnetic moment of 1.76 BM and a Weiss constant of 6°, and the behaviour of the corresponding diethylamine adduct is similar[99]. Both of these bis(amine) adducts decompose on heating to give unstable 1:1 adducts. Decomposition occurs through an oxidation-reduction reaction, the ethyl groups being dehydrogenated. The resulting solid products

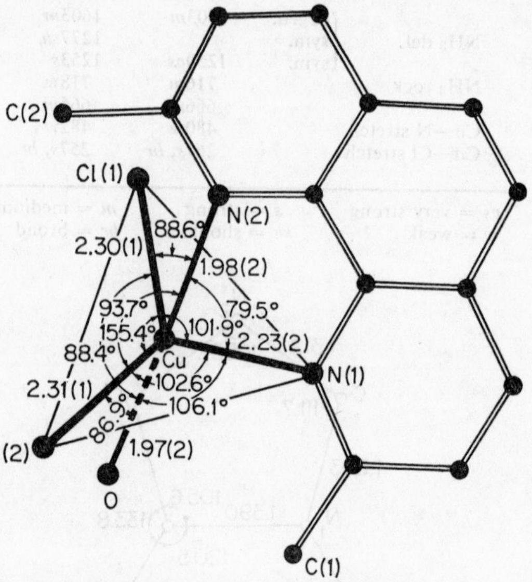

Figure 9.11 The structure of $CuCl_2(dmp)\cdot H_2O$. (Reproduced by permission from H. S. Preston and C. H. L. Kennard, *Chem. Commun.*, **1967**, 1167)

are alkylammonium dichlorocuprate(I)amine complexes. Stoichiometric amounts of monoethanolamine and copper dichloride form a red-brown 1:1 adduct on reaction in acetone[108]. Addition of excess of the ligand gives $CuCl_2(CH_2OHCO_2NH_2)_4$.

Plowman and coworkers[109–111] have reported that yellow-green $CuCl_2(dmp)\cdot H_2O$, where dmp is 2,9-dimethyl-1,10-phenanthroline, may be isolated by reaction of the constituents in hydrochloric acid solution or by boiling $CuCl_2(dmp)_2$ in water. Preston and Kennard[112] have studied the crystal structure of this compound; it is monoclinic with $a = 9.88$, $b = 8.17$, $c = 9.53$ Å and $\beta = 105.5°$. There are two molecules in the unit cell of space group $P2_1$. The structure, which is shown in Figure 9.11, contains discrete five-coordinate molecular units. The rather large difference between the equatorial and axial copper-nitrogen bonds is probably due to the interaction of the methyl group [C(1)] on the ligand with the water molecule.

Sutton[113] has reported the isolation of $CuCl_2(Mepic)_2$, where Mepic is 6-methyl-2-picolyamine. The blue compound has a room-temperature magnetic moment of 2.02 BM and behaves as a 1:1 electrolyte in non-aqueous solvents, so that the compound should be formulated as

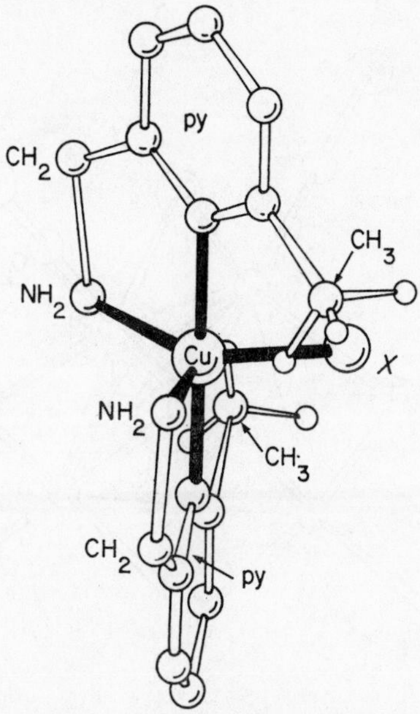

Figure 9.12 The probable structure of $[Cu(mepic)_2X]^+$. (Reproduced by permission from S. Utsuno and K. Sone, *J. Inorg. Nucl. Chem.*, **28,** 2647 (1966))

$[CuCl(Mepic)_2]Cl$. The work of Sutton on the copper dichloride-Mepic system has been supported by Utsuno and Sone[114], although these workers isolated their compound as a trihydrate. The ionic nature of the complex was confirmed and the probable structure, deduced from the similarity of the compound to $[CuI(bipy)_2]I$, is shown in Figure 9.12.

Copper dichloride reacts readily with a range of phosphine and arsine oxides in ethanol to give complexes of the type $CuCl_2L_2$[115–118]. Magnetic and spectral data indicate that the phosphine oxide adducts contain tetrahedrally coordinated copper. With triphenylarsine oxide however, spectral studies[117] suggest that the compound should be formulated as

$[Cu(Ph_3AsO)_4][CuCl_4]$. Bertrand[119] attempted to prepare single crystals of $CuCl_2[(C_6H_5)_3PO]_2$ suitable for crystallographic analysis by recrystallization of the compound from methyl isobutyl ketone. However, he found that oxygen was abstracted from the solvent and the compound isolated from solution had the formula $Cu_4OCl_6[(C_6H_5)_3PO]_4$. Bertrand also found that the same compound could be prepared by refluxing the stoichiometric amounts of copper dichloride, copper(II) oxide and triphenylphosphine oxide in nitromethane. The orange compound has a cubic lattice and there is one molecule in the unit cell of space group $P\bar{4}3m$. The structure, which

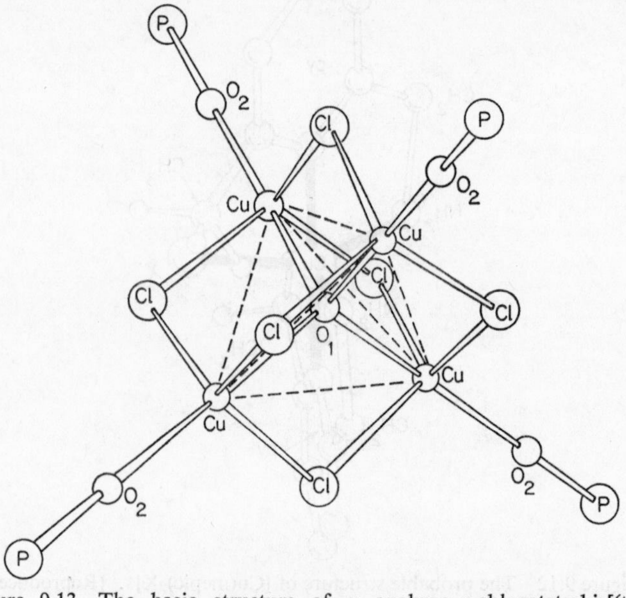

Figure 9.13 The basic structure of μ_4-oxohexa-μ-chlorotetrakis[(triphenylphosphine oxide) copper (II)]. (Reproduced by permission from J. A. Bertrand, *Inorg. Chem.*, **6**, 495 (1967))

is shown in Figure 9.13, consists of a tetrahedral arrangement of copper atoms about a central oxygen atom with each pair of copper atoms being bridged by a chlorine atom. The fifth coordination position in the slightly distorted trigonal-bipyramidal arrangement about each copper atom is occupied by the oxygen of the phosphine oxide molecule. Thus the compound should be formulated as μ_4-oxohexa-μ-chlorotetrakis[(triphenylphosphine oxide)copper(II)]. The linear nature of the phosphorus-oxygen-copper system indicates π-bonding and this suggestion is supported by a high phosphorus-oxygen stretching frequency of 1194 cm^{-1}. This is to be compared with a value of 1195 cm^{-1} for triphenylphosphine oxide itself[120], and Cotton and coworkers[120] and Sheldon and Tyree[121] report phosphorus-oxygen stretching frequencies for a wide range of triphenylphosphine oxide complexes of about 1150 cm^{-1}.

Copper dichloride reacts readily with a number of alcohols to give adducts of various stoichiometries[95,122–124]. Probably the most interesting adduct of this type is that formed with n-propanol of empirical formula $Cu_5Cl_{10}(C_3H_7OH)_2$[95]. The compound is monoclinic with $a = 10.17$, $b = 6.04$, $c = 18.30$ Å and $\beta = 94.6°$. There are two molecules in the unit cell of space group $P2_1/n$. The structure, which is shown in Figure 9.14, consists of an extended chain of square-planar coordinated copper atoms linked by shared edges, with the propanol groups arranged in a *trans*

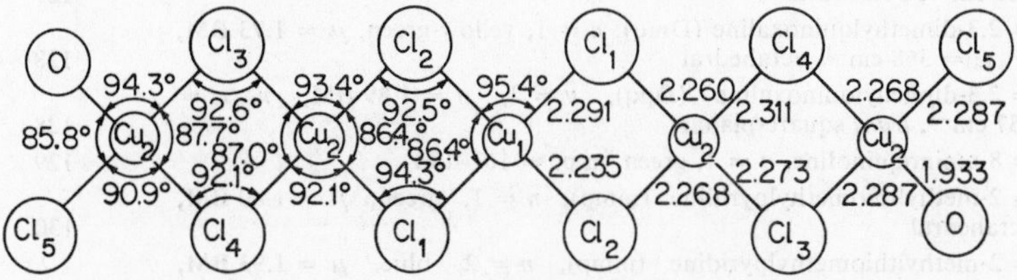

Figure 9.14 The structure of $Cu_5Cl_{10}(C_3H_7OH)_2$. (Reproduced by permission from R. D. Willett and R. E. Rundle, *J. Chem. Phys.*, **40**, 838 (1964))

configuration on the two extreme copper atoms. Thus the compound has a structure similar to those of the acetonitrile adducts described earlier (Figures 9.7 and 9.8, pages 500, 501). Again like these adducts, the copper atoms in the propanol derivative attain octahedral coordination by means of long-range interaction with chlorine atoms in a neighbouring molecule.

Copper dichloride forms a large number of other adducts with nitrogen, oxygen and sulphur ligands and the properties of some of these adducts are given in Table 9.11. It may be noted at this point that tertiary phosphines and arsines react with copper dichloride with concomitant reduction to copper(I)[125] and therefore these compounds will be described later.

A number of other adducts of copper dichloride with donor molecules such as 2-pyridinealdoxime, imidazole, thioxan, dimethylsulphoxide, have been reported[31,142–180].

Copper dichloride dihydrate. This hydrate is readily isolated from saturated aqueous solutions of copper dichloride. The compound looses water very readily on heating[22,34,181] to give the anhydrous chloride. Perret[181] reports the dissociation pressure of water over heated copper dichloride dihydrate for the temperature range 30–140° to be given by the expression

$$\log p_{atm} = 7.545 - 3050/T$$

The heat and entropy of dissociation are 13.95 kcal $mole^{-1}$ and 34.5 cal

TABLE 9.11
Adducts of the Type $CuCl_2L_n$

Compound and Properties	References
L = biuret, n = 2, pale green, μ = 1.95 BM, *trans* octahedral	126
L = quinoxaline, n = 1, green, μ = 1.77 BM, $\nu_{Cu\text{-}Cl}$ = 324 cm^{-1}, octahedral	127, 128
L = 2-methylquinoxaline (Mq), n = 2, green, μ = 1.84 BM, $\nu_{Cu\text{-}Cl}$ = 320 cm^{-1}, octahedral	128
L = 2,3-dimethylquinoxaline (Dmq), n = 1, yellow-green, μ = 1.73 BM, $\nu_{Cu\text{-}Cl}$ = 368 cm^{-1}, octahedral	128
L = 2,3-diphenylquinoxaline (Dpq), n = 2, μ = 1.89 BM, $\nu_{Cu\text{-}Cl}$ = 337 cm^{-1}, *trans* square-planar	128
L = 8-aminoquinoline, n = 1, green, m.p. = 167–180°	129
L = 2-methylthiomethylpyridine (mmp), n = 1, green, μ = 1.80 BM, octahedral	130
L = 2-methylthiomethylpyridine (mmp), n = 2, blue, μ = 1.93 BM, octahedral	130
L = *o*-methylthioaniline (mta), n = 1, green, μ = 1.81 BM, octahedral	131
L = 8-methylthioquinoline (mtq), n = 1, green, μ = 1.78 BM	132
L = 2-methyl-8-methylthioquinoline (mmtq), n = 1, brown, μ = 1.82 BM, tetrahedral	133
L = 2-(2′-methylthioethyl)pyridine (mtp), n = 1, green, μ = 1.84 BM, octahedral	134
L = *N,N,N′,N′*-tetramethylethylenediamine (Me_4en), n = 1, m.p. = 145–148°, μ = 1.91 BM, square planar	135
L = *N,N,N′,N′*-tetramethyl-1,2-propylenediamine (Me_4pn), n = 1, m.p. = 127–130°, μ = 1.88 BM, square planar	135
L = bis(2-dimethylaminoethyl)methylamine (dienMe), n = 1, light blue, μ = 1.84 BM, five coordinate	136
L = bis(2-dimethylaminoethyl)oxide (Me_4daeo), n = 1, green, μ = 1.75 BM, five coordinate	137
L = bipyridyl, n = 2, ionic with five-coordinate cation, $g_\parallel$ = 2.02, $g_\perp$ = 2.19	138, 139
L = ethyleneurea (eu), n = 2, light green, m.p. = 144–145°, μ = 1.89 BM	140
L = 1,3-di(2′-pyridyl)-1,2-diaza-2-propenato (pdpo), n = 1, m.p. = 208–209°, μ = 2.3 BM, square planar	141

deg^{-1} $mole^{-1}$ respectively. Perret[181] reports a heat of formation of copper dichloride dihydrate of −77·1 kcal $mole^{-1}$.

The crystal structure of copper dichloride dihydrate was first investigated by Harker[182] who reported an orthorhombic unit cell with

$a = 7.38$, $b = 8.04$ and $c = 3.72$ Å, of space group *Pbmn* containing two molecules. The structure consists of infinite chains of square-planar $CuCl_4$ units sharing all halogens, with *trans* water molecules completing the octahedral arrangement about the copper atoms.

Adams and Lock[44] report a copper-chlorine stretching frequency of 299 cm^{-1} for copper dichloride dihydrate.

The magnetic properties of copper dichloride dihydrate have been studied by many groups using a wide variety of techniques. The compound obeys the Curie-Weiss law and van den Handel and coworkers[183] have shown that it is anisotropic. The magnetic moment of a single crystal orientated parallel to the *a* axis is 1.89 BM and parallel to the *b* axis is 1.76 BM. The Weiss constant was given as 5°, although a value of 24° has also been reported[26].

There is a maximum[184] in the susceptibility for powdered copper dichloride dihydrate at 5.2°K, but for a single crystal orientated parallel to the *a* axis the maximum occurs at 4.8°K. Neutron-diffraction and NMR measurements[185–191] indicate a paramagnetic to antiferromagnetic transition at 4.3°K. In addition, Poulis and Hardeman[189] have shown that the Neel temperature is dependent on both field strength and crystal alignment. There is a maximum in the specific heat of copper dichloride dihydrate in the same temperature region[184,192].

The electron spin resonance spectra of both powdered and single-crystal samples of copper dichloride dihydrate have been investigated by several groups[11,193–196]. The single-crystal data are given in Table 9.12.

TABLE 9.12

ESR Data for Single Crystals of Copper Dichloride Dihydrate

g_a	g_b	g_c	Reference
2.190	2.050	2.248	194
2.195	2.075	2.26	195
2.197	2.037	2.252	196

The magnetic structure of copper dichloride dihydrate was first investigated by Poulis and Hardeman[188] using ^1H NMR techniques. The conclusions of this work were questioned by Rundle and coworkers[197] and by Peterson and Levy[198]. The latter workers carried out a neutron-diffraction study of the compound at room temperature and the dimensions shown in Table 9.13 were deduced.

These results show that the bisector of the hydrogen-oxygen-hydrogen angle is collinear with the copper-oxygen linkage, thus indicating that the

TABLE 9.13
Dimensions in Copper Dichloride Dihydrate at Room Temperature by Neutron-Diffraction Techniques

Distance	Value (Å)	Angle	Value
Cu—Cl	2.275	Cu—O—H	125.99°
Cu—O	1.925	H—O—H	108.05°
O—H	0.948	O—H—Cl	164.4°
H—Cl	2.258		

interpretation of the hydrogen-oxygen-hydrogen bending mode put forward by Rundle and coworkers[197] was incorrect. Rundle[199] then used the data of Petersen and Levy to give an alternative interpretation of the

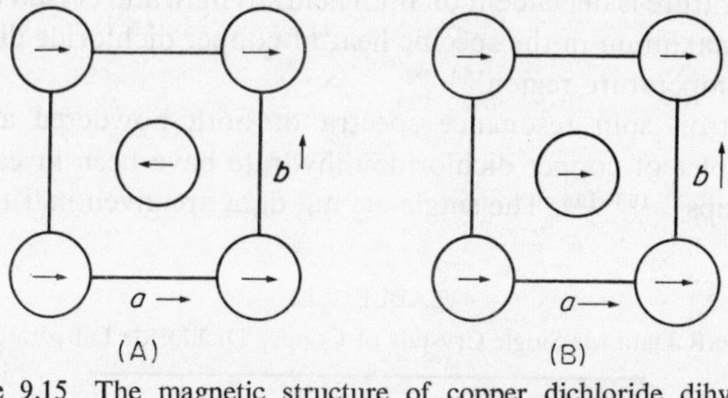

Figure 9.15 The magnetic structure of copper dichloride dihydrate: (A) proposed by Rundle, (B) proposed by Poulis and Hardeman. (Reproduced by permission from R. E. Rundle, *J. Am. Chem. Soc.*, **79,** 3372 (1957))

magnetic structure to that given by Poulis and Hardeman[188]. The two alternative structures are shown in Figure 9.15 The most recent work[190] has added to the confusion over the magnetic structure of copper dichloride dihydrate. Neutron-diffraction studies at 1.5 and 4.5°K confirm the original interpretation of the magnetic structure by Poulis and Hardeman. It would now appear that Poulis and Hardeman deduced the correct magnetic structure but their estimate of the proton positions was erroneous. Theoretical implications of the magnetic properties of copper dichloride dihydrate have been discussed by several workers[200–203].

Copper dibromide. This black compound may be prepared by the interaction of the elements at 300° in a sealed tube[204], by halogen exchange with copper dichloride and boron tribromide[205], from the reaction between copper acetate and acetyl bromide in benzene or acetic acid[33,34], or by dissolving the carbonate in hydrobromic acid and evaporating the solution to dryness[31].

Copper dibromide is very soluble in water from which the di- and tetrahydrates may be isolated under controlled conditions. Copper dibromide has been widely used as a brominating agent in organic systems[37–39,206–208]. It is thermally unstable, decomposing very readily to copper monobromide and bromine[34,42,204,209–212]. Hammer and Gregory[204] have studied this decomposition over the temperature range 50–110° by effusion techniques, and give the following expression for the equilibrium dissociation pressure of bromine over copper dibromide

$$\log p_{\text{mm}} = 10.48 - 5470/T$$

The heat and entropy of dissociation were given as 23.2 kcal mole^{-1} and 41.8 cal deg^{-1} mole^{-1} respectively[204]. Shchukarev and Oranskaya[42] give the expression

$$\log p_{\text{mm}} = 6.172 - 3598/T$$

for the equilibrium dissociation pressure over the temperature range 130–316°. Finally, Barret and coworkers[212] give for the temperature range 150–250°

$$\log p_{\text{atm}} = 7.52 - 4270/T$$

Helmoltz[213] has shown that copper dibromide has a monoclinic unit cell with $a = 7.14$, $b = 3.46$, $c = 7.18$ Å and $\beta = 121°$. The structure consists of square-planar $CuBr_4$ units joined by sharing edges, the chains being parallel to the *b* axis. The chains are so arranged in the lattice that each copper atom has a grossly distorted octahedral configuration. The copper-bromine bond lengths are 2.40 and 3.18 Å. Copper-bromine stretching frequencies of 250 and 219 cm^{-1} have been reported[44,45].

Copper dibromide obeys the Curie-Weiss law above about 250°K[14,26,46] with a magnetic moment of 1.91 BM and a Weiss constant of 246°. Neel temperatures of 170°K[26] and 226°K[47] have been reported. A room temperature *g* value of 2.33 has also been obtained[11].

The electronic absorption spectrum of copper dibromide has been reported by Howard and Keeton[49] and *d-d* transitions were observed at 15,600 and 12,100 cm^{-1}.

In general terms, the variety and number of adducts reported for copper dibromide is similar to those already described for copper dichloride.

Thus copper dibromide reacts readily with pyridine[56–62,64,69,214] to give green $CuBr_2(py)_2$ which melts at 235–236°. It has a magnetic moment of 1.79 BM and the infrared spectrum shows copper-bromine stretching frequencies at 255 and 202 cm^{-1} and a copper-nitrogen mode at 269 cm^{-1}.

Kupcik and Durovic[214] have examined the structure of $CuBr_2(py)_2$ by single-crystal X-ray diffraction techniques. The compound is monoclinic with $a = 8.30$, $b = 17.72$, $c = 4.04$ Å and $\beta = 96°$. The space group is $P2_1/n$ and the cell contains two molecules. The structure is very similar to that of the corresponding copper dichloride adduct, that is, the copper atom is in a distorted octahedral environment being surrounded by four bromines and two pyridine molecules. The copper-nitrogen bond distance is 1.99 Å and the copper-bromine bond lengths are 2.46 and 3.19 Å.

Adducts of copper dibromide with aniline, substituted anilines and substituted pyridines have also been reported[62,66,72–74]. Goldstein and coworkers[62] have investigated the infrared spectra of some of these adducts and the reported assignments are given in Table 9.14.

TABLE 9.14
Infrared Spectra (cm^{-1}) of Some $CuBr_2L_2$ Adducts

Mode	L = 2-methylpyridine	L = 2,6-lutidine	L = 2-ethylpyridine	L = quinoline
$\nu_{Cu\text{-}Br}$[a]	231	230	251	266
$\delta_{Cu\text{-}Br}$	116	108	107	110
$\nu_{Cu\text{-}N}$[a]	(268, 259)	244	251	256
$\delta_{Cu\text{-}N}$	194	140	180	140
$\gamma_{Cu\text{-}Br}$	58	59	46	52

[a] asymmetric stretching mode.

Torgonskaya and Zonov[215] have investigated the reaction between copper dibromide and ammonia as a function of both temperature and pressure. Their results indicated the formation of the adducts $CuBr_2(NH_3)_n$ where n is 3, 2, 1.5 and 1. Hanic and Cakajdova[97] have also investigated the reaction between copper dibromide and ammonia and report that as with copper dichloride, two bis(amine) complexes may be isolated. These are light green α-$CuBr_2(NH_3)_2$ and dark green, unstable β-$CuBr_2(NH_3)_2$. The α-form has a monoclinic unit cell[216] with $a = 8.18$, $b = 8.15$, $c = 4.05$ Å and $\beta = 94.8°$. There are two molecules in the unit cell of space group $C2/m$. The β-form is cubic with $a = 4.068$ Å[97]. The structures of both forms of $CuBr_2(NH_3)_2$ are shown in Figure 9.16. For the β-form, the copper-bromine and copper-nitrogen bond distances are 2.876 and 2.034 Å respectively.

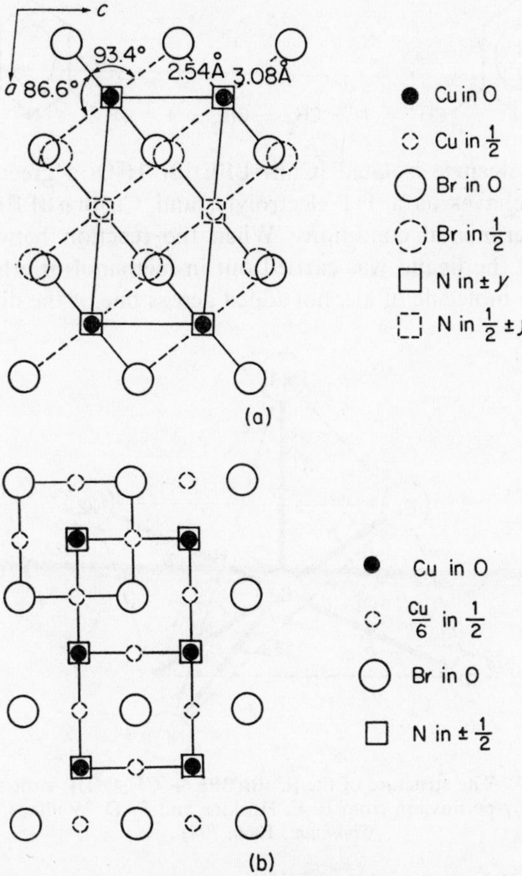

Figure 9.16 (a) The structure of α-$CuBr_2(NH_3)_2$. (b) The structure of β-$CuBr_2(NH_3)_2$. (Reproduced by permission from F. Hanic, *Acta Cryst.*, **12,** 739 (1959))

The infrared and diffuse-reflectance spectra of both forms of $CuBr_2(NH_3)_2$ have been recorded by Clark and Williams[98] and they strongly resemble those of the corresponding chloro complexes. The copper-bromine stretching frequency occurs at 218 cm^{-1} in α-$CuBr_2(NH_3)_2$ while for the β-form, a split band at 222 and 205 cm^{-1} was observed[98].

Harris and McKenzie[217] have investigated the reaction between copper dibromide and the Schiff base 1,2-bis(2′-pyridylmethyleneamino)ethane (BPE) in acetonitrile or nitromethane solution. The Schiff base is represented diagrammatically below.

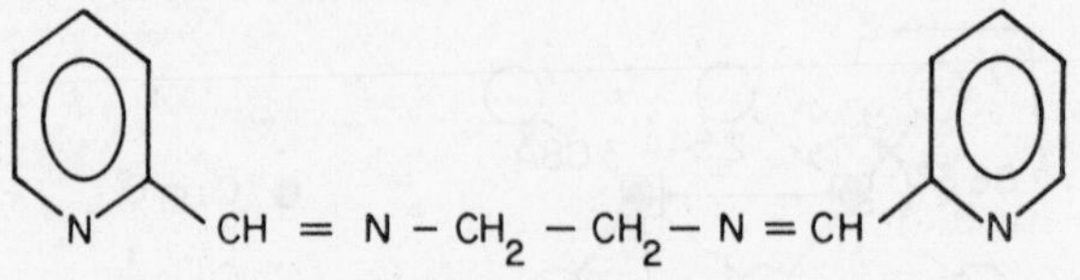

Harris and McKenzie isolated [CuBr(BPE)]Br·H_2O, a green crystalline solid which behaves as a 1:1 electrolyte, and Cu_2Br_4(BPE) depending upon the experimental conditions. When the reaction between copper dibromide and the ligand was carried out in methanol or ethanol it was found that one molecule of alcohol added across one of the double bonds

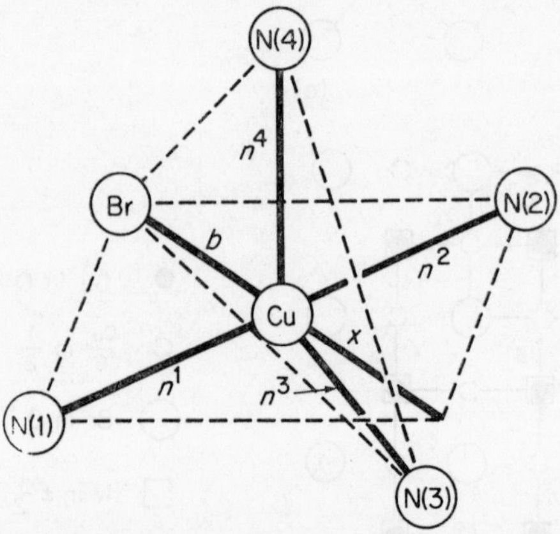

Figure 9.17 The structure of the [CuBr(BPE + CH_3OH)]$^+$ cation. (Reproduced by permission from B. F. Hoskins and F. D. Whillans, *Chem. Commun.*, **1966,** 798)

TABLE 9.15
Internuclear Dimensions of the [CuBr(BPE+CH_3OH)]$^+$ ion

Distance	Value (Å)	Angle	Value	Angle	Value
Cu—Br	2.40	Br—Cu—N(1)	99°	N(2)—Cu—N(4)	100°
Cu—N(1)	2.04	Br—Cu—N(2)	100°	N(3)—Cu—N(4)	112°
Cu—N(2)	2.01	Br—Cu—N(3)	145°	N(1)—Cu—N(4)	80°
Cu—N(3)	1.94	Br—Cu—N(4)	102°	N(1)—Cu—N(3)	83°
Cu—N(4)	2.12	N(2)—Cu—N(3)	79°	N(1)—Cu—N(2)	160°

of the ligand, and a new complex was isolated as the perchlorate salt. The structure of $[CuBr(BPE + CH_3OH)]ClO_4$ has been determined by Hoskins and Williams[218] by single-crystal X-ray diffraction techniques. The compound is monoclinic with $a = 9.30$, $b = 13.14$, $c = 16.03$ Å and $\beta = 98.3°$. The space group is $P2_1/c$ and the cell contains four formula units. The structure of the cation, which is shown in Figure 9.17, shows that the stereochemistry about the copper atom is intermediate between square pyramidal and trigonal bipyramidal. The important internuclear dimensions are given in Table 9.15.

Copper dibromide reacts with pyridine *N*-oxides and quinoline *N*-oxides to give adducts which closely resemble those of copper dichloride[80,82,85,88–91,219]. The magnetic properties of these complexes have been discussed in detail, particularly by Hatfield and his associates, and together with other physico-chemical data have been used to establish possible structures.

Some properties of a number of other adducts of copper dibromide are summarized in Table 9.16.

Other donor ligands that have been reacted with copper dibromide include dioxan, dimethyl sulphoxide, triphenylphosphine oxide and substituted ethylenes[31,110,111,116,142,146,150,152–155,157,159,161–163,168,169,173–176,178,179,223].

Copper dibromide dihydrate. Little is known about this compound which may be isolated from aqueous solutions of copper dibromide. Date[224] reports that it becomes antiferromagnetic below about 6°K.

Copper(II) iodide complexes. Although copper diiodide itself cannot be isolated[225,226], a number of adducts of it may be prepared using suitable techniques[113,114,138,139,141,173,174,227,228].

The best established complex of this type is that with bipyridyl, $[CuI(bipy)_2]I$, which is prepared by mixing bis(bipyridyl)copper(II) perchlorate and sodium iodide in a suitable solvent[138,227]. The brown compound, which is a 1:1 electrolyte and has a normal magnetic moment, is of triclinic symmetry[227,228] with $a = 10.66$, $b = 14.37$, $c = 7.44$ Å, $\alpha = 93.3°$, $\beta = 101.1°$ and $\gamma = 107.6°$. There are two formula units in the cell of space group $P\bar{1}$. The cation has a trigonal-bipyramidal configuration, as shown in Figure 9.18. The electron spin resonance and absorption spectra of this complex have been investigated by Elliot and coworkers[189] who report that $g_{\parallel} = 2.025$ and $g_{\perp} = 2.17$.

Oxidation State I

Copper monofluoride. Although there are reports in the early literature[229,230] on the preparation and properties of copper monofluoride,

TABLE 9.16
Miscellaneous Adducts of the Type $CuBr_2L_n$

Compound and Properties	References
L = quinoxaline, $n = 1$, dark brown, $\mu = 1.74$ BM, $\nu_{Cu\text{-}Br} = 225\ cm^{-1}$, octahedral	127, 128
L = 2-methylquinoxaline (Mq), $n = 2$, dark brown, $\mu = 1.75$ BM, $\nu_{Cu\text{-}Br} = 252\ cm^{-1}$, octahedral	128
L = 2,3-dimethylquinoxaline (Dmq), $n = 2$, brown, $\mu = 1.55$ BM, $\nu_{Cu\text{-}Br} = 278\ cm^{-1}$, octahedral	128
L = 2,3-diphenylquinoxaline (Dpq), $n = 2$, yellow-brown, $\mu =$ 1.84 BM, $\nu_{Cu\text{-}Br} = 269\ cm^{-1}$, *trans* square planar	128
L = 8-aminoquinoline, $n = 1$, green, m.p. = 170–180°	129
L = 2-methylthiomethylpyridine (mmp), $n = 1$, brown, $\mu = 1.75$ BM, octahedral	130
L = *o*-methylthioaniline (mta), $n = 1$, dark green, $\mu = 1.73$ BM, octahedral	131
L = 8-methylthioquinoline (mtq), $n = 1$, red-brown, $\mu = 1.76$ BM	132
L = 2-methyl-8-methylthioquinoline (mmtq), $n = 1$, dark brown, $\mu =$ 1.79 BM, tetrahedral	133
L = 2-(2-methylthioethyl)pyridine (mtp), $n = 1$, brown, $\mu = 1.68$ BM, octahedral	134
L = *N,N,N′,N′*-tetramethylethylenediamine (Me_4en), $n = 1$, m.p. = 112–113°, $\mu = 1.82$ BM, square planar	135
L = *N,N,N′,N′*-tetramethyl-1,2-propylenediamine (Me_4pn), $n = 1$, m.p. = 100–105°, $\mu = 1.96$ BM, square planar	135
L = *N,N,N′,N′*-tetramethyltrimethylenediamine (Me_4tn), $n = 1$, m.p. = 95–99°, $\mu = 2.06$ BM, tetrahedral	135
L = bis(2-dimethylaminoethyl)methylamine (dienMe), $n = 1$, bright green, ΔH°_{form} (from constituents) = -36.13 kcal mole^{-1}, five coordinate	136, 220
L = tris(2-dimethylaminoethyl)amine (trenMe), $n = 1$, light green, $\mu =$ 1.86 BM, five coordinate	221
L = bis(2-dimethylaminoethyl)oxide (Me_4daeo), $n = 1$, light green, $\mu =$ 1.71 BM, five coordinate	137
L = bipyridyl, $n = 2$, ionic, five-coordinate cation, $g_\parallel = 2.02$, $g_\perp = 2.18$	138, 139, 222
L = 2-methyl-6-aminomethylpyridine (Mepic), $n = 2$, blue, $\mu =$ 2.02 BM, five coordinate cation	113, 114
L = 1,3-di(2-pyridyl)-1,2-diaza-2-propene (pdp), $n = 1$, m.p. = 232–235°, $\mu = 1.8$ BM, five coordinate	141
L = 1,3-di(2-pyridyl)-1,2-diaza-2-propenato (pdpo), $n = 1$, m.p. = 200°, square planar	141
L = 3,5,6-tri(2-pyridyl)-1,2,4-triazine (tpt), $n = 1$, m.p. = 300°, $\mu =$ 1.9 BM, trigonal bipyramidal	178
L = thiosemicarbazide (tsc), $n = 1$, brown, $\mu = 1.55$ BM, octahedral	160
L = thiosemicarbazide (tsc), $n = 2$, brown, $\mu = 1.78$ BM, octahedral	160
L = benzimidazole, $n = 4$, blue, m.p. = 175°, $\mu = 1.99$ BM, *trans* octahedral	170

more recent work[3,231] has shown that the compound does not exist as a stable entity at room temperature. However, Kent, McDonald and Margrave[6] have reported the presence of ionized copper monofluoride in their mass-spectrometric study of the sublimation pressure of copper difluoride.

Copper monochloride. White copper monochloride is probably most conveniently prepared by reducing copper dichloride or copper sulphate in hydrochloric acid. Suitable reductants include very pure copper metal[232],

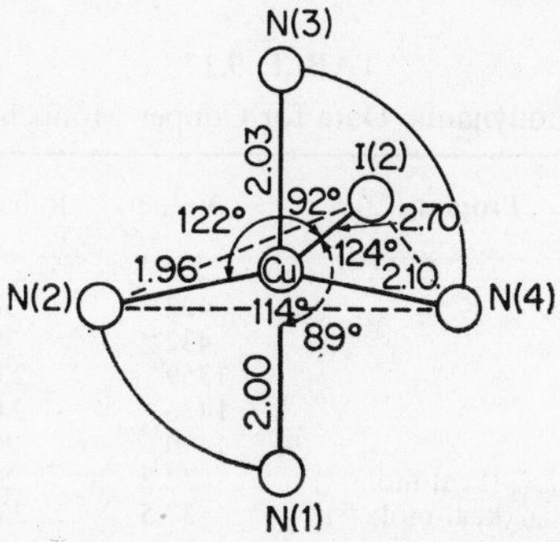

Figure 9.18 The structure of the $[CuI(bipy)_2]^+$ cation. (Reproduced by permission from G. A. Barclay, B. F. Hoskins and C. H. L. Kennard, *J. Chem. Soc.*, **1963**, 5691)

sulphur dioxide[29,233–235] and ascorbic acid[236]. Other methods of preparation include the passage of hydrogen chloride gas over copper metal at 800–1000° in a flow system[237,238] and the thermal decomposition of copper dichloride[34,35,42].

Mass-spectrometric[239], electron-diffraction[240] and molecular weight studies[237] indicate that copper monochloride vaporizes principally as a trimer. Wong and Schomaker[240] interpreted the electron-diffraction patterns of copper monochloride vapour on the basis of a six-membered puckered ring in which the copper and chlorine atoms alternate. The copper-chlorine bond distance is 2.160 Å while the copper-chlorine-copper angle is approximately 90°. The sublimation pressure of the low temperature form of copper monochloride (see below) is given by the expression[234]

$$\log p_{\text{mm}} = 11.235 - 8156/T$$

Thermodynamic data derived from the pressure measurements and from other sources are given in Table 9.17.

At room temperature, copper monochloride has a cubic structure[242,243] with $a = 5.4057$ Å; this is normally referred to as the γ-form of copper monochloride. At 407° there is a phase transition to β-copper monochloride which has a hexagonal unit cell with $a = 3.91$ and $c = 6.42$ Å[242]. Badachhape and Goswami[244] report that slow evaporation of γ-copper monochloride in high vacuum results in a new form of copper monochloride which has a cubic spinel-type structure with $a = 8.265$ Å.

Klemperer and his associates[245] have studied the infrared spectrum of gaseous copper monochloride in equilibrium with liquid copper monochloride. A strong absorption, attributed to a stretching frequency of the cyclic trimeric species, was observed at 350 cm^{-1}.

TABLE 9.17
Thermodynamic Data for Copper Monochloride

Property	Value	Reference
M.p.	422°	36
	432°	29
B.p.	1359°	237
	1436°	241
	1490°	29
ΔH°_{form} (kcal mole^{-1})	−32.2	29
ΔH_{sub} (kcal mole^{-1})	37.5	234

The electronic absorption spectrum of copper monochloride has been recorded and discussed by Nikitine and coworkers[246,247].

Copper monochloride reacts with a wide range of nitrogen, phosphorus, arsenic, oxygen and sulphur donor ligands to give adducts which, in general, are polymeric[102,110,111,142,233,248–271].

Brown and Dunitz[252] have determined the crystal structure of the 1:1 copper monochloride-azomethane adduct prepared initially by Diels and Koll[253]. The lattice is triclinic[252] with $a = 6.87$, $b = 7.03$, $c = 3.82$ Å, $\alpha = 97.1°$, $\beta = 95.25°$ and $\gamma = 111.7°$. The unit cell contains two formula units and the space group is $P\bar{1}$. The structure consists of copper-chlorine chains, with a copper-chlorine distance of 2.35 Å, running parallel to the c axis, these chains being joined by weaker copper-chlorine bonds of length 2.55 Å, as shown in Figure 9.19. The chains are further linked together through *trans* azomethane molecules, so that each copper atom has a distorted tetrahedral coordination. The copper-nitrogen bond distance is 1.99 Å.

Copper dichloride reacts with substituted pyrazines to give orange adducts of the type 2CuCl·L[254,255]. Lever and coworkers investigated the magnetic and spectral properties of these complexes and deduced from their data that the ligand bridges the two copper monochloride molecules.

Billing and Underhill[266] have prepared similar complexes with quinoxaline and substituted quinoxalines.

Cohn and Parry[261] have investigated the reaction between copper monochloride and dimethylaminodifluorophosphine, $(CH_3)_2NPF_2$. Two colourless, diamagnetic solids, in which the ratio of copper monochloride to ligand were 1:1 and 1:2, were isolated. Both have small dissociation vapour pressures at room temperature and are soluble in non-polar solvents. Molecular weight measurements of $CuCl[(CH_3)_2NPF_2]$ in bromoform

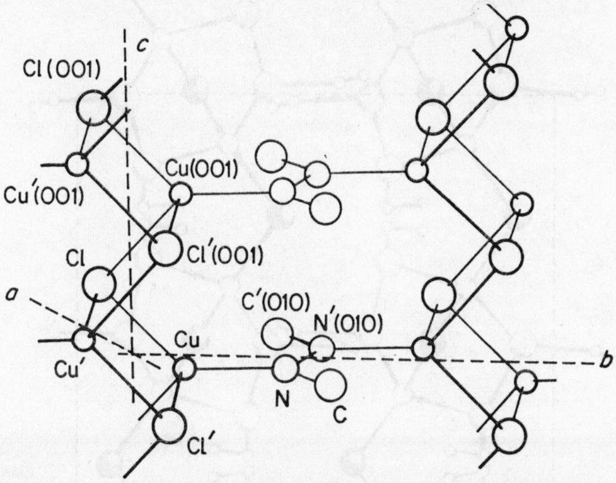

Figure 9.19 The structure of the copper dichloride–azomethane adduct. (Reproduced by permission from I. D. Brown and J. D. Dunitz, *Acta Cryst.*, **13**, 28 (1960))

indicated a tetramer. NMR and infrared spectra indicated that the phosphorus atom is coordinated to the copper atom, and Cohn and Parry[261] have suggested that this adduct has the same type of structure as $[CuI \cdot (CH_3)_3As]_4$ (page 524).

Arbuzov and Zorvastrova[267] have studied the reaction between copper monochloride and several phosphate esters. Molecular weight determinations of the 1:1 adducts with isopropylphosphate ester and phenylphosphate ester, which melt at 112–114° and 95–96° respectively, indicate the formation of trimeric species.

The reaction between allyl alcohol, $CH_2{=}CHCH_2OH$, and copper monochloride results in the formation of a compound of 1:1 stoichiometry[233]. There is a drop of 95 cm^{-1} in the carbon-carbon double bond stretching frequency and a drop of 15 cm^{-1} in the carbon-oxygen stretching frequency of the ligand, suggesting that the alcohol coordinates to the copper atom mainly through the carbon-carbon double bond, although the hydroxy group also participates in the coordination. A similar bonding mechanism occurs in the 1:1 adduct of copper monochloride with 2-allylpyridine[257]. This grey-green compound, which decomposes over the temperature range 113–125°, has a carbon-carbon double bond stretching frequency of the olefin which is 87 cm^{-1} lower than that of the free ligand.

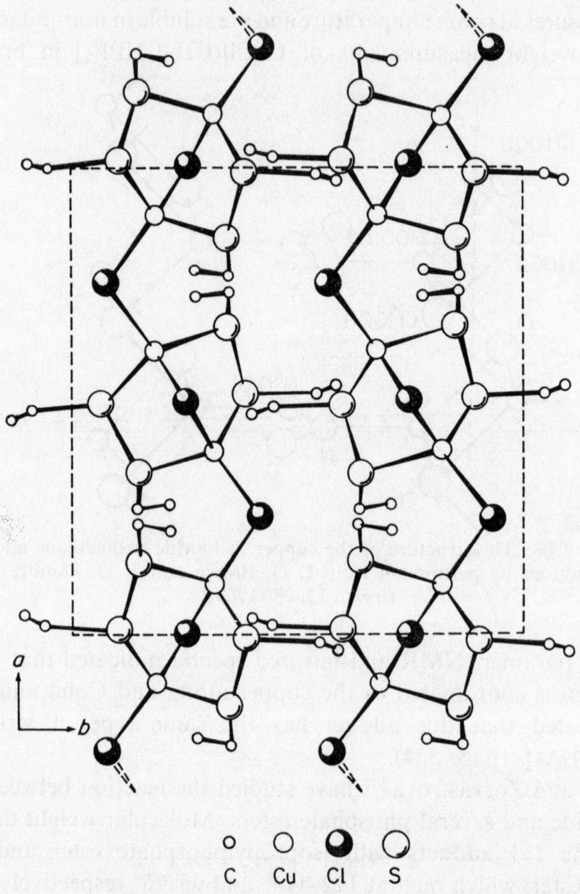

Figure 9.20 The structure of the copper dichloride–diethyl sulphide adduct. (Reproduced by permission from C. I. Brańden, *Acta Chem. Scand.*, **21**, 1000 (1967))

Direct interaction of copper monochloride and diethylsulphide results in the formation of a 1:2 adduct[251]. The compound has an orthorhombic lattice with $a = 11.04$, $b = 12.64$ and $c = 12.54$ Å. The unit cell is of space group $C222_1$ and contains eight molecules. The structure, which is shown in Figure 9.20, consists of copper-chlorine chains which are parallel to the *a* axis, the copper atoms of the chains being grouped in pairs. Each pair of copper atoms binds two diethylsulphide molecules through covalent copper-sulphur bonds. The copper atoms are tetrahedrally coordinated. The copper-chlorine bond distances are 2.29 and 2.34 Å while the copper-sulphur bonds are 2.34 and 2.40 Å.

Dalziel and coworkers[272], investigating the reaction between triphenylphosphine sulphide and copper monochloride, isolated triclinic $CuCl[(C_6H_5)_3PS]$ and reported that it has a unit cell with $a = 9.72$, $b = 11.45$, $c = 13.31$ Å, $\alpha = 126.75°$, $\beta = 94.67°$ and $\gamma = 117.97°$. There are two molecules in the unit cell of space group $P1$. A strong band at 604 cm^{-1} and a shoulder at 616 cm^{-1} were assigned to the phosphorus-sulphur stretching mode.

Copper monobromide. This compound is normally prepared either by the thermal decomposition of copper dibromide in a flow system[34,204,210,234] or by reduction of copper sulphate in an aqueous bromide solution with sulphur dioxide[29,273]. Other methods of preparation include the electrolytic reduction at a controlled potential, of copper dibromide in aqueous sodium bromide solutions[274], the action of ethyl bromide on a heated aqueous solution of copper sulphate[275], and addition of water to the yellow oil which results from the reaction between copper metal and hydrogen bromide in ether[276]. As some of the above methods of preparation indicate, copper monobromide is insoluble in water. It is readily purified by vacuum sublimation.

Copper monobromide undergoes the following phase transformations below its melting point[210,234,277–279].

$$\gamma\text{-CuBr} \xrightarrow{385°} \beta\text{-CuBr} \xrightarrow{470°} \alpha\text{-CuBr}$$

Shelton[234] has made a detailed study of the variation of the sublimation pressures of the three forms of copper monobromide, which vaporize principally as trimeric species[239], and his results are expressed by the following equations.

$$\gamma\text{-CuBr} \quad \log p_{\text{mm}} = 9.693 - 7674/T \quad (<385°)$$

$$\beta\text{-CuBr} \quad \log p_{\text{mm}} = 8.299 - 6756/T \quad (385\text{–}470°)$$

$$\alpha\text{-CuBr} \quad \log p_{\text{mm}} = 7.787 - 6309/T \quad (>470°)$$

Thermodynamic data for copper monobromide derived from the sublimation pressure measurements and from other sources are summarized in Table 9.18.

There is quite good agreement in the literature as to the structures of the three modifications of copper monobromide[277–280]. α-Copper monobromide has a body-centred cubic lattice, β-CuBr is hexagonal with the wurtzite-type packing, while γ-CuBr has the zinc blende-type cubic lattice. The unit-cell parameters of the three modifications of copper monobromide reported by Hoshino[280] are given in Table 9.19.

TABLE 9.18
Thermodynamic Properties of Copper Monobromide

Property		Value	References
M.p.		480°	42
		488°	279
		496°	210
		498°	29
B.p.		1342°	29
ΔH°_{form} (kcal mole^{-1})		−24.85	273
ΔS°_{form} (cal deg^{-1} mole^{-1})		3.5	273
ΔH_{sub} (kcal mole^{-1})	γ	35.1	234
	β	30.9	234
	α	28.8	234
ΔH_{trans} (kcal mole^{-1})	γ → β	1.4	234, 280
	β → α	0.7	243, 280
ΔS_{trans} (cal deg^{-1} mole^{-1})	γ → β	2.1	280
	β → α	0.9	280

TABLE 9.19
Unit-cell Parameters of Copper Monobromide

Form	Temperature	*a* (Å)	*c* (Å)
α	480°	4.53	
β	430°	4.04	6.58
μ	20°	5.679	

Kurdyumova and Semiletov[281] report that a metastable tetragonal form of copper monobromide with $a = 3.02$ and $c = 4.25$ Å is formed when copper monobromide is annealed at 120–150°.

Copper monobromide reacts with donor molecules to give adducts which closely resemble those of copper monochloride. Thus with amines, 2:1, 1:1 and 1:2 adducts are readily formed[250,263,264,282]. Molecular weight measurements[263] of some of the white 1:1 adducts indicate them to be tetrameric.

A tetrameric structure has also been proposed[283] for the 1:1 adduct with diphenylphosphine. The compound is colourless, diamagnetic and decomposes at 160–162°. With phosphate esters[267], trimeric species are indicated by molecular weight measurements. The phosphorus-sulphur stretching frequency in $CuBr[(C_6H_5)_3PS]$ is split, strong absorptions being observed[272] at 602 and 592 cm^{-1}, with a shoulder at 620 cm^{-1}. This adduct has monoclinic cell, of space group $P2_1/c$ and containing eight formula units, with $a = 9.78$, $b = 20.48$, $c = 19.58$ Å and $\beta = 126.75°$.

The yellow-orange 2:1 adducts of copper monobromide with pyrazine, quinoxaline and their substitution derivatives[254,255,266] have very similar properties to the corresponding copper monochloride derivatives. Thus it would appear that the ligand bridges the two copper monobromide groups.

Other ligands that have been reacted with copper monobromide include pyridine *N*-oxide[219], 2,9-dimethyl-1,10-phenanthroline[110,111], 4,6,4′,6′-tetramethyl-2,2′-bipyridine[265], phosphines[269–271,284,285], arsines[258], thiourea[260] and dithian[142].

Copper monoiodide. Copper monoiodide is most conveniently prepared by reducing copper sulphate in an aqueous solution of potassium iodide with sulphur dioxide[29,286]. Direct interaction of the elements[238] in a thermal gradient of 450–100°, and the reaction between copper sulphide and aluminium triiodide at 230° in a sealed tube[287] also lead to the formation of copper monoiodide. Alternatively, copper acetate may be reacted with acetyl iodide in acetic acid[34], or with the stoichiometric amounts of iodine and acetone in acetic acid[288] at 70–80°.

Copper monoiodide is a white solid which is insoluble in water. It may be purified by vacuum sublimation. As with copper monobromide, there are three modifications of crystalline copper monoiodide and the structures of these forms are comparable to those of the bromide (see below). Rosenstock and coworkers[239] have shown from mass-spectrometric studies that copper monoiodide vaporizes principally as a trimer. Shelton[234] has reported the sublimation pressures of the three modifications of copper monoiodide, assuming the species present in the vapour are trimeric. His results are expressed by the following equations.

$$\gamma\text{-CuI} \quad \log p_{mm} = 11.141 - 9463/T \quad (<369°)$$

$$\beta\text{-CuI} \quad \log p_{mm} = 9.410 - 8351/T \quad (369\text{–}407°)$$

$$\alpha\text{-CuI} \quad \log p_{mm} = 8.678 - 7853/T \quad (>407°)$$

Thermodynamic data available for copper monoiodide are given in Table 9.20.

The reported[210,234,289–291] temperatures for the $\gamma \rightarrow \beta$ and $\beta \rightarrow \alpha$ phase transformations are in the ranges 367–375° and 405–412° respectively. The unit-cell parameters of the three modifications are:

γ-CuI	$a = 6.10$ Å	Reference 292
β-CuI	$a = 4.31$ and $c = 7.09$ Å	Reference 279
α-CuI	$a = 6.15$ Å	Reference 279

Kurdyumova and Baranova[293] report the formation of a metastable hexagonal form of copper monoiodide with $a = 4.25$ and $c = 20.86$ Å, which results from careful evaporation of copper monoiodide.

TABLE 9.20
Thermodynamic Data for Copper Monoiodide

Property		Value	Reference
M.p.		588°	286
		600°	210
		602°	279
		605°	29
B.p.		1293°	286
		1336°	29
ΔH°_{form} (kcal mole^{-1})		−16.2	29
ΔH_{sub} (kcal mole^{-1})	γ	43.3	234
	β	38.2	234
	α	35.9	234
ΔH_{trans} (kcal mole^{-1})	$\gamma \rightarrow \beta$	1.4	234
	$\beta \rightarrow \alpha$	0.77	234

Copper monoiodide reacts with a wide range of phosphines and arsines[125,258,269,284,294–296]. The adducts are readily prepared by adding the ligand to a solution of copper monoiodide saturated with potassium iodide, or from ethanolic solutions of the constituents. Molecular weight measurements[296] of the 1:1 triphenylphosphine and triphenylarsine adducts indicate the formation of tetrameric species, and this has been confirmed for $CuI[(CH_3)_3As]$ by a single-crystal X-ray diffraction study[296]. The compound is cubic with $a = 13.08$ Å and there are eight formula units in the cell of space group $I\bar{4}3m$. The structure consists of a tetrahedral array of copper atoms which are linked together by bridging iodine atoms. The copper-copper, copper-arsenic and copper-iodine bond distances are 2.60, 2.50 and 2.66 Å respectively. The arsenic atoms lie on the elongation of the axis joining the centre of the tetrahedron to the copper atoms, while the methyl groups are joined to the arsenic atoms so that the latter attain a tetrahedral configuration.

Some properties of a number of phosphine and arsine adducts of copper monoiodide are given in Table 9.21.

TABLE 9.21
Phosphine and Arsine Adducts of the Type $CuIL_n$

Compound and Properties	Reference
L = $(C_2H_5)_3P$, $n = 1$, m.p. = 236–240°, cubic with a = 13.05 Å	296
L = $(n\text{-}C_3H_7)_3P$, $n = 1$, m.p. = 207°, monoclinic with a = 18.9, b = 15.4, c = 19.9 Å and β = 78°	296
L = $(n\text{-}C_4H_9)_3P$, $n = 1$, m.p. = 75°	296
L = $(C_2H_5)_3P$, $n = 2$, m.p. = 37–39°, dimeric with iodine bridges	296
L = $(CH_3)_2C_6H_5P$, $n = 1$, m.p. = 98.5–99°	295
L = $(C_2H_5)_2C_6H_5P$, $n = 1$, m.p. = 153.5–155°	295
L = $C_6H_4(CH_2)_2PC_6H_5$ = 2-phenylisophosphindoline, m.p. = 195.5–196°, tetrameric	269
L = $(n\text{-}C_3H_7)_3As$, $n = 1$, m.p. = 202–212°, hexagonal with a = 22.4 and c = 23.7 Å	296
L = $(n\text{-}C_4H_9)_3As$, $n = 1$, m.p. = 61.5°, orthorhombic with a = 26.3, b = 22.3 and c = 12.3 Å	296
L = $C_{11}H_{27}As_3$ = methylbis(3-propanedimethylarsine)arsine, $n = 1$, diamagnetic, tetrahedral monomer	294

Malik[297] has prepared a number of copper monoiodide adducts with pyridine and related ligands by adding excess of the ligand to an aqueous solution of copper monoiodide containing excess potassium iodide. All of the adducts prepared were diamagnetic as expected for copper(I). Some properties of several of these adducts are listed in Table 9.22.

TABLE 9.22
Pyridine and Related Ligand Adducts of the Type CuIL

Compound and Properties
L = pyridine, amethyst, m.p. = 190°
L = 2-methylpyridine, light brown, m.p. = 155°
L = 3-methylpyridine, light brown, m.p. = 150°
L = 4-methylpyridine, dark brown, m.p. = 240°
L = 2,4-lutidine, light buff, m.p. = 145°
L = 2,6-lutidine, brown-black, m.p. = 145–150°
L = quinoline, light yellow, m.p. = 155°

Other ligands that have been reacted with copper monoiodide include amines[250,263,264], 8-methylthioquinoline[130], 2,9-dimethyl-1,10-phenanthroline[110,111], pyrazines[254,255], quinoxalines[266], di(2-pyridyl)sulphide[223], thiourea[260], dithian[142], tetrahydrothiapyran[249] and triphenylstibine sulphide[298].

COMPLEX HALIDES

The only complex halide containing copper in an oxidation state other than II or I is the hexafluorocuprate(III) anion. An interesting feature which emerges is the fact that certain anions, particularly the tetrahalocuprates(II), display two distinct types of anion stereochemistry depending on the nature of the cation.

Oxidation State III

Hexafluorocuprate(III). Klemm and Huss[299] have prepared pale green potassium hexafluorocuprate(III) by fluorinating a 3:1 mixture of potassium chloride and copper dichloride at 250° in a flow system. It reacts violently with water and has a room-temperature magnetic moment of 2.8 BM[300].

Oxidation State II

Hexafluorocuprate(II). Colourless barium hexafluorocuprate(II), Ba_2CuF_6, has been prepared[301] by fusing a 2:1 mixture of barium fluoride and copper difluoride at 1100°. The compound has a monoclinic unit cell[302] with $a = 4.163$, $b = 4.163$, $c = 15.850$ Å and $\beta = 90.98°$. The space group of the cell is $I2/m$. The structure is analogous to that described for the corresponding cobalt complex (page 367) except that in this case, the CuF_6 octahedra are tetragonally compressed, the copper-fluorine bond lengths being 1.85 and 2.08 Å[302]. The compound obeys the Curie-Weiss law over a wide temperature range[301] with a magnetic moment of 1.85 BM and a Weiss constant of 15°. ESR measurements give a g value of 2.20[301].

Tetrafluorocuprates(II). White alkali-metal tetrafluorocuprates(II) may be prepared quite readily by fusing the stoichiometric amounts of copper difluoride and an appropriate alkali-metal fluoride or hydrogen difluoride[16,303]. Ammonium tetrafluorocuprate(II) has been prepared[304] by subjecting an intimate mixture of ammonium fluoride and copper difluoride to a pressure of 100,000 psi at 100°, and also by adding the calculated quantity of ammonium fluoride to a solution of copper nitrate in methanol[305].

Sodium tetrafluorocuprate(II) has a monoclinic unit cell[303,306,307] with $a = 3.26$, $b = 9.35$, $c = 5.60$ Å and $\beta = 87.54°$. There are two formula units in the cell of space group $P2_1/c$. The structure, which is shown in

Figure 9.21, consists of CuF_6 octahedra which share *trans* edges to form chains parallel to the *a* axis, the chains being separated by the sodium ions. Some important internuclear dimensions are given in Table 9.23[306].

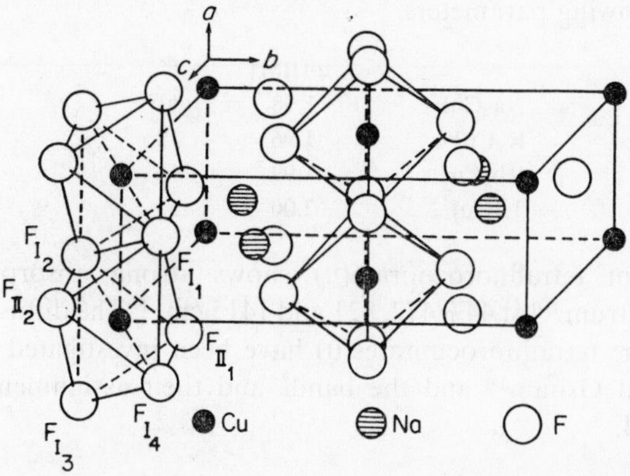

Figure 9.21 The structure of sodium tetrafluorocuprate (II). (Reproduced by permission from D. Babel, *Z. anorg. allgem. Chem.*, **336**, 200 (1965))

TABLE 9.23
Internuclear Dimensions in Sodium Tetrafluorocuprate(II)

Distance	Value (Å)	Angle	Value
Cu—F_{I_1}	1.917	F_{I_1}—Cu—F_{I_2}	81.4°
Cu—F_{I_2}	2.367	F_{I_1}—CuF_{I}—$_{I_1}$	91.0°
Cu—F_{II}	1.901	F_{I_2}—Cu—F_{II_1}	88.1°

Knox[308] has shown by a single-crystal X-ray diffraction study that potassium tetrafluorocuprate(II) is isostructural with the corresponding nickel complex (Figure 8.13, page 446). The tetragonal unit cell of space group *I4/mmm* contains two formula units and has $a = 4.155$ and $c = 12.74$ Å. The copper-fluorine bond distances of the tetragonally compressed CuF_6 octahedra are 1.95 and 2.08 Å. This was the first example of tetragonal flattening of octahedral copper to be proved by X-ray structural analysis, although it is to be noted that Liehr and Ballhausen[309] had predicted that this phenomenon would occur in this compound prior to the structural determination.

Both rubidium and thallium tetrafluorocuprate(II) have tetragonal unit cells[303,307] with $a = 4.24$ and $c = 13.28$ Å for the rubidium complex and $a = 4.20$ and $c = 13.66$ Å for the latter compound.

The alkali-metal tetrafluorocuprates(II) obey the Curie-Weiss law[303,307] with the following parameters.

	μ (BM)	$\theta°$
Na_2CuF_4	1.98	11
K_2CuF_4	1.96	12
Rb_2CuF_4	1.99	13
Tl_2CuF_4	2.00	15

Ammonium tetrafluorocuprate(II) shows strong absorptions in its infrared spectrum[310] at 447, 477, 821 and 1415 cm^{-1}. The diffuse-reflectance spectra of the tetrafluorocuprates(II) have been investigated by Schmitz-DuMont and Grimm[16] and the bands and their assignments are given in Table 9.24.

TABLE 9.24
Diffuse-reflectance Spectra (cm^{-1}) of Tetrafluorocuprates(II)

Complex	${}^2B_{2g} \leftarrow {}^2B_{1g}$	${}^2E_g \leftarrow {}^2B_{1g}$
Na_2CuF_4	9750	12150
K_2CuF_4	9250	11700
Rb_2CuF_4	9650	12200
Cs_2CuF_4		11850

Oelkrug[15] gives values of 9,550 and 12,200 cm^{-1} for the sodium salt.

Trifluorocuprates(II). The first reported method of preparing trifluorocuprates(II) was by dissolving copper carbonate or dichloride in hydrofluoric acid and adding excess of the alkali-metal fluoride[311–314]. Machin, Martin and Nyholm[312] investigated the optimum conditions for formation of the salts and reported that it is essential that the free acid concentration be kept to a minimum, and also the separate solutions be heated close to the boiling point before mixing, and then maintained at the boiling point for about ten seconds after mixing. A second convenient route to the alkali-metal trifluorocuprates(II) is by fusion of the stoichiometric quantities of copper difluoride and alkali-metal fluoride[16,303]. Ammonium trifluorocuprate(II) may be prepared by compressing[304] an intimate mixture of copper difluoride and ammonium fluoride to 100,000 psi at about 100°, by adding ammonium fluoride to a solution of copper nitrate in methanol[305], or by reacting ammonium fluoride and copper dibromide in methanol[315]. The latter method may also be used for the alkali-metal salts[316].

The structures of the alkali-metal trifluorocuprates(II) have been investigated by X-ray powder diffraction techniques. All workers agree that the trifluorocuprates(II) exhibit a markedly distorted perovskite structure in which the CuF_6 octahedra are tetragonally compressed. However, there are discrepancies in the literature over the room temperature unit-cell parameters of several of these complexes, as shown in Table 9.25.

TABLE 9.25

Room-temperature Unit-cell Parameters of Some Trifluorocuprates(II)

Compound	Symmetry	Parameters	References
$NaCuF_3$	Monoclinic	$a = 10.01, b = 11.37, c = 7.52$ Å $\beta = 86.9°$	303, 307
$KCuF_3$	Tetragonal	$a = 4.14, c = 3.92$ Å	311, 317–320
	Tetragonal	$a = 5.855, c = 7.85$ Å	303, 321
$RbCuF_3$	Tetragonal	$a = 4.26, c = 3.95$ Å	319
	Tetragonal	$a = 6.001, c = 7.894$ Å	303
$CsCuF_3$	Hexagonal	$a = 12.56, c = 11.56$ Å	322
$TlCuF_3$	Tetragonal	$a = 4.30, c = 3.93$ Å	307
	Tetragonal	$a = 6.083, c = 7.866$ Å	303

Two copper-fluorine distances for potassium trifluorocuprate(II) are reported by Edwards and Peacock[311] as 1.96 Å while the other four are 2.07 Å. Okazaki and coworkers[318,320] report that at 78°K the potassium salt has a tetragonal unit cell with $a = 4.121$ and $c = 3.911$ Å.

The magnetic properties of several trifluorocuprates(II) are summarized in Table 9.26.

TABLE 9.26

Magnetic Properties of Some Trifluorocuprates(II)

Compound	μ (BM)	T_N (°K)	Reference
$NaCuF_3$	1.76		303
$KCuF_3$	1.17	220	303
	1.38	215	312
	2.16[a]	243	313
$TlCuF_3$	1.37	235	303

[a] Weiss constant of 355° included in this result.

Although Scatturin and coworkers[323] found no evidence for antiferromagnetic behaviour from neutron-diffraction studies at 4.2°K, detailed magnetic susceptibility, anisotropy, specific heat and ^{19}F NMR spectroscopy measurements by Hirakawa and coworkers[314,324] show quite conclusively that potassium trifluorocuprate(II) is a one dimensional antiferromagnet.

Peacock and Sharp[325] report a copper-fluorine stretching frequency in potassium trifluorocuprate(II) of 489 cm^{-1}. The diffuse-reflectance spectra[15,16] of the trifluorocuprates(II) resemble those of the tetrafluorocuprates(II). Thus two bands are observed at about 9,000 and 11,500 cm^{-1} which have been assigned to the $^2B_{2g} \leftarrow {}^2B_{1g}$ and $^2E_g \leftarrow {}^2B_{1g}$ transitions respectively.

Pentachlorocuprates(II). Mori and coworkers[326–330], as well as several other workers[331,332], have reported the preparation of a number of compounds containing the pentachlorocuprate(II) anion, with either complex cations of transition metals such as $[Cr(NH_3)_6]^{3+}$ or with large organic cations. These complexes have been characterized by magnetic, spectral, X-ray diffraction and ESR studies[48,49,327,328,331,333–335]. Adams and Lock[44] point out however, that the infrared spectra show only *one* copper-chlorine stretching frequency and they suggested that these compounds may be similar to salts such as caesium pentachlorocobaltate(II); that is, the compounds may contain tetrachlorocuprate(II) groups with free chloride ions. This suggestion has been confirmed in one case by Zaslo and Ferguson[336] who studied the compound originally formulated as $(dienH)_3CuCl_5$ by single-crystal X-ray diffraction techniques. They showed that in fact the compound is bis(2-aminoethyl)ammonium monochloride tetrachlorocuprate(II), $[(NH_3CH_2CH_2)_2\,NH_2]Cl[CuCl_4]$. The compound is orthorhombic with $a = 7.117$, $b = 23.78$ and $c = 7.342$ Å. There are four molecules in the unit cell of space group *Pnma*. The structure, which is shown in Figure 9.22, consists of discrete square-planar tetrachlorocuprate(II) units with an equal number of free chloride ions and cations. Obviously it is necessary that a complete reinvestigation of all of the supposed pentachlorocuprates(II) be undertaken in view of this structure determination.

Tetrachlorocuprates(II). The usual method of preparation of tetrachlorocuprates(II) is to react an amine hydrochloride and copper dichloride in either water or alcohol[108,332,337–348]. The arsonium salts may be prepared in a similar manner[344,349]. Caesium tetrachlorocuprate(II) has been prepared from the melt[350] and from aqueous solution[351,352]. Gruen and Angell[353] have identified the tetrachlorocuprate(II) anion in a dilute solution of copper dichloride in molten magnesium dichloride hexahydrate.

The tetrachlorocuprates(II) are yellow crystalline solids. The potassium salt melts[36] at 330° and its heat of formation[354] from the constituent chlorides is −2.6 kcal mole^{-1}. The heats of formation from the constituent chlorides of the tetramethylammonium and tetraethylammonium salts are −6.33 and −12.08 kcal mole^{-1} respectively[355]. The ammonium,

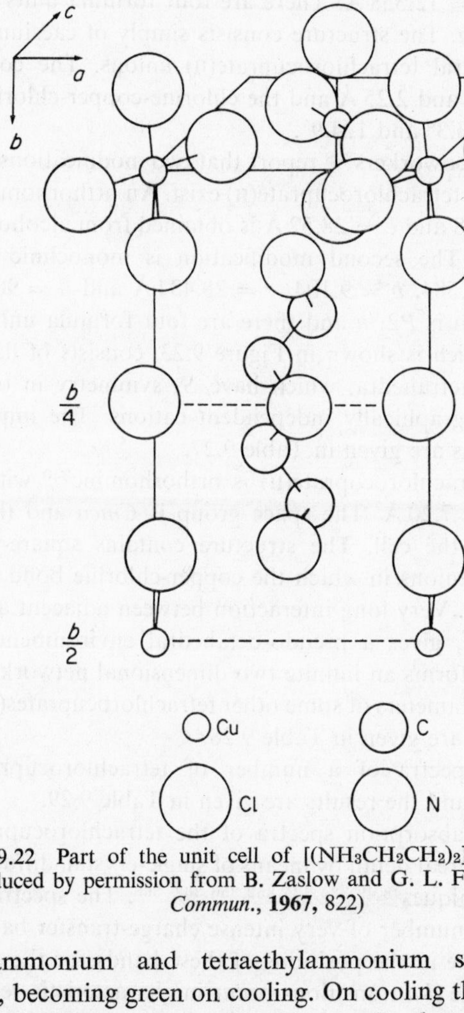

Figure 9.22 Part of the unit cell of $[(NH_3CH_2CH_2)_2]NH_2Cl[CuCl_4]$. (Reproduced by permission from B. Zaslow and G. L. Ferguson, *Chem. Commun.*, **1967**, 822)

tetramethylammonium and tetraethylammonium salts are thermochroic[340,356], becoming green on cooling. On cooling the symmetry about the central copper atom changes from D_{4h} to C_{4v}, but only slight changes are observed in the electronic spectra.

The structures of the tetrachlorocuprates(II) fall into two distinct classes, those containing square-planar anions and those with a tetrahedral anion.

These two structural arrangements have been indentified by single-crystal X-ray diffraction studies in several cases. Infrared and electronic absorption spectra have proved to be useful as diagnostic tests for identifying the nature of the anion.

Caesium tetrachlorocuprate(II) is orthorhombic[341,351,357] with $a = 9.719$, $b = 7.658$ and $c = 12.358$ Å. There are four formula units in the cell of space group *Pnma*. The structure consists simply of caesium cations and flattened tetrahedral tetrachlorocuprate(II) anions. The copper-chlorine distances are 2.18 and 2.25 Å and the chlorine-copper-chlorine angles are 102.5°, 102.9°, 123.3° and 124.9°.

Bonamico and coworkers[358] report that two modifications of trimethylbenzylammonium tetrachlorocuprate(II) exist. An orthorhombic form with $a = 9.54$, $b = 9.18$ and $c = 28.52$ Å is obtained from alcohol solutions of the constituents. The second modification is monoclinic and has the parameters $a = 9.584$, $b = 9.104$, $c = 28.434$ Å and $\beta = 90°$. The space group of this form is $P2_1/n$ and there are four formula units in the cell. The structure, which is shown in Figure 9.23, consists of flattened tetrachlorocuprate(II) tetrahedra, which have S_4 symmetry in one direction, and two crystallographically independent cations. The important internuclear dimensions are given in Table 9.27.

Ammonium tetrachlorocuprate(II) is orthorhombic[340] with $a = 15.46$, $b = 7.20$ and $c = 7.20$ Å. The space group is *Cmca* and there are four formula units in the cell. The structure contains square-planar tetrachlorocuprate(II) anions in which the copper-chlorine bond distances are 2.300 and 2.332 Å. Very long interaction between adjacent anions, which are 2.793 Å apart, gives a pseudo-octahedral environment about each copper atom and forms an infinite two dimensional network.

The unit-cell parameters of some other tetrachlorocuprates(II), which are all orthorhombic, are given in Table 9.28.

The infrared spectra of a number of tetrachlorocuprates(II) have been investigated and the results are given in Table 9.29.

The electronic absorption spectra of the tetrachlorocuprates(II) have been examined in great detail by means of single-crystal, diffuse-reflectance and solution techniques[48,49,334,335,352,356,362–368]. The spectra are characterized by a large number of very intense charge-transfer bands. There is very little difference in the positions of these bands for the anion stereochemistries, but the *d-d* transitions occur in characteristic regions for the different anions. For the square-planar anions the band occurs at about 12,800 cm^{-1} and for the tetrahedral complexes it occurs in the range 5500–9000 cm^{-1}, the actual position being dependent on the degree of flattening of the tetrahedron.

The magnetic susceptibilities of a number of tetrachlorocuprates(II) have been measured over a wide temperature range. The compounds deviate only slightly from Curie-Law behaviour as shown in Table 9.30.

The magnetic anisotropy and electron spin resonance spectra of several tetrachlorocuprates(II) have been investigated in detail[335,340,371–375].

Tetrachlorocuprate(II) dihydrates. Alkali-metal and ammonium tetrachlorocuprate(II) dihydrates have been prepared in aqueous solutions from

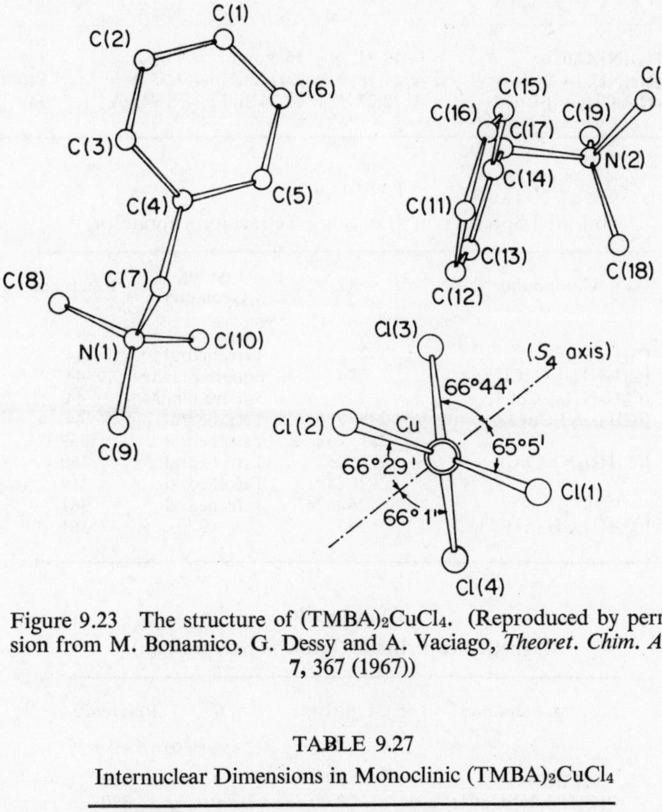

Figure 9.23 The structure of $(TMBA)_2CuCl_4$. (Reproduced by permission from M. Bonamico, G. Dessy and A. Vaciago, *Theoret. Chim. Acta*, 7, 367 (1967))

TABLE 9.27
Internuclear Dimensions in Monoclinic $(TMBA)_2CuCl_4$

Distance	Value (Å)	Angle	Value
Cu—Cl(1)	2.229	Cl(1)—Cu—Cl(2)	100.3°
Cu—Cl(2)	2.263	Cl(1)—Cu—Cl(3)	132.1°
Cu—Cl(3)	2.263	Cl(1)—Cu—Cl(4)	99.1°
Cu—Cl(4)	2.268	Cl(2)—Cu—Cl(3)	98.3°
		Cl(2)—Cu—Cl(4)	132.8°
		Cl(3)—Cu—Cl(4)	99.7°

the constituent chlorides[376–378]. Suga and coworkers[377] have studied the thermal decomposition of potassium and ammonium tetrachlorocuprate(II) dihydrate. Both salts decompose quite readily to give the

TABLE 9.28
Unit-cell Parameters of Some Tetrachlorocuprates(II)

Compound	Parameters	Reference
$[(CH_3)_4N]_2CuCl_4$	$a = 12.11$, $b = 15.5$ and $c = 9.05$ Å	357
$[(C_2H_5)NH_3]_2CuCl_4$	$a = 21.18$, $b = 7.47$ and $c = 7.35$ Å	340
$[(C_6H_5)_3CH_3As]_2CuCl_4$	$a = 32.27$, $b = 25.24$ and $c = 8.943$ Å	349

TABLE 9.29
Infrared Spectra (cm⁻¹) of Some Tetrachlorocuprates(II)

Compound	$\nu_{Cu\text{-}Cl}$	Anion Geometry	Reference
Cs_2CuCl_4	292, 257	Tetrahedral	44
$[CH_3NH_3]_2CuCl_4$	294	Square planar	44
$[C_2H_5NH_3]_2CuCl_4$	279	Square planar	44
$[(CH_3)_4N]_2CuCl_4$	278, 236	Tetrahedral	44
	281, 237	Tetrahedral	359
$[(C_2H_5)_4N]_2CuCl_4$	278, 237	Tetrahedral	360
	267, 248	Tetrahedral	359
	289, 268, 247	Tetrahedral	361
$[(C_6H_5)_3CH_3As]_2CuCl_4$	283	?	361

TABLE 9.30
Magnetic Properties of Tetrachlorocuprates(II)

Compound	μ (BM)	$\theta°$	Reference
Cs_2CuCl_4	2.00	1	369
$[(C_2H_5)_4N]_2CuCl_4$	2.00	0	370
$[(C_6H_5)_3CH_3As]_2CuCl_4$	1.91[a]		344
$[Pt(NH_3)_4]CuCl_4$	1.77	10	340

[a] Room temperature measurement only.

corresponding trichlorocuprate(II), water and the appropriate chloride. The reaction occurs at 93° for the potassium salt and at 138° for the ammonium complex.

The structures of a number of tetrachlorocuprate(II) dihydrates have been investigated. They are all tetragonal and contain the *trans* $[CuCl_4(H_2O)_2]^{2-}$ anion. The available data are given in Table 9.31.

TABLE 9.31

Crystallographic Data for Some Tetrachlorocuprate(II) Dihydrates

Compound	*a* (Å)	*c* (Å)	Cu—O (Å)	Cu—Cl (Å)	Reference
$K_2CuCl_4 \cdot 2H_2O$	7.45	7.88	2.51	2.37, 3.15	376
$Rb_2CuCl_4 \cdot 2H_2O$	7.81	8.00	2.64	2.41, 3.20	376
$Cs_2CuCl_4 \cdot 2H_2O$	7.92	8.24			379
$(NH_4)_2CuCl_4 \cdot 2H_2O$	7.58	7.96	2.55	2.44, 2.31	376
			1.99	2.33, 3.00	380

There is considerable confusion concerning the atomic positions, and hence internuclear distances, in the tetrachlorocuprate(II) dihydrates. Narasimhurty and Premaswarup[381] have discussed this problem in detail from a theoretical point of view.

The copper-chlorine stretching frequencies of the potassium, rubidium and caesium salts occur at 313, 318 and 303 cm^{-1} respectively[44].

Potassium and ammonium tetrachlorocuprate(II) dihydrate obey the Curie-Weiss law down to very low temperatures[382,383]. Below 1°K the magnetic susceptibility increases very sharply and transitions to ferromagnetic behaviour occur at 0.88°K and 0.72°K for the potassium and ammonium compounds respectively[382]. Specific heat maxima also occur at these temperatures[384]. The ESR spectra of the same two salts have been investigated[11,195] and the results of Itoh and coworkers[195] are as follows.

$K_2CuCl_4 \cdot 2H_2O$	$(g_a + g_b)/2 = 2.24$	$g_c = 2.06$
$(NH_4)_2CuCl_4 \cdot 2H_2O$	$(g_a + g_b)/2 = 2.25$	$g_c = 2.05$

Trichlorocuprates(II). Trichlorocuprates(II) with substituted ammonium cations are prepared by direct reaction of stoichiometric amounts of the amine hydrochloride and copper dichloride in water or alcohol[108,345,385–390]. Potassium trichlorocuprate(II) has been prepared in hydrochloric acid solution by interaction of the chlorides using excess copper dichloride[386,387] and the caesium salt can be prepared directly in water, again using excess copper dichloride[391–393]. Silver and caesium trichlorocuprate(II) have been prepared in the melt using stoichiometric amounts of the constituent chlorides[350,394].

The trichlorocuprates(II) are garnet-red crystalline solids. The potassium salt melts[36] at 364° and its heat of formation from the constituent chlorides[354] is −2.7 kcal mole^{-1}. Silver trichlorocuprate(II) melts[394] at 405°.

There are two distinct types of structure in the trichlorocuprates(II).

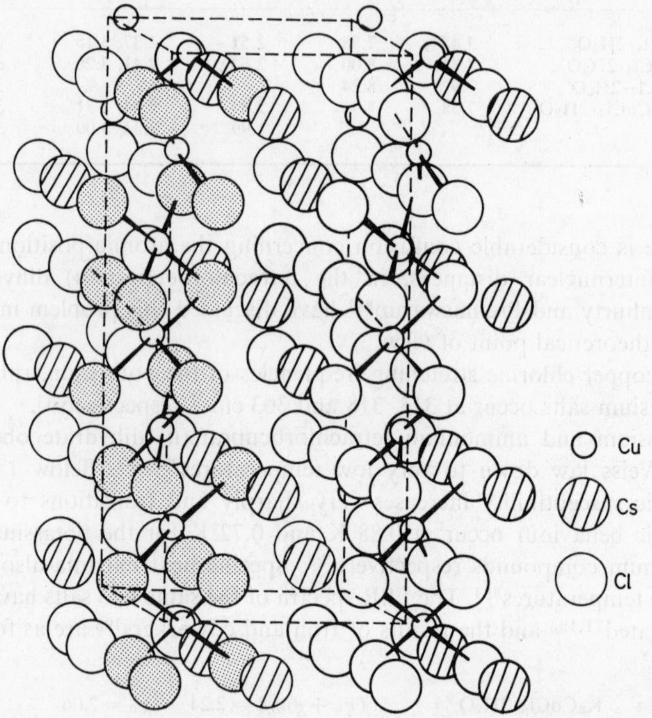

Figure 9.24 The lattice of caesium trichlorocuprate (II). (Reproduced by permission from A. F. Wells, *J. Chem. Soc.*, **1947**, 1661)

Caesium trichlorocuprate(II) is hexagonal[391–393] with $a = 7.216$ and $c = 18.178$ Å. The atomic arrangement in the lattice is shown in Figure 9.24 and shows square-planar $CuCl_4$ units linked by a single bridging atom[392]. Long interaction between the terminal chlorines and the copper atoms resulted in the latter having a distorted octahedral environment.

The structure, as described by Wells[392], may be represented schematically as follows.

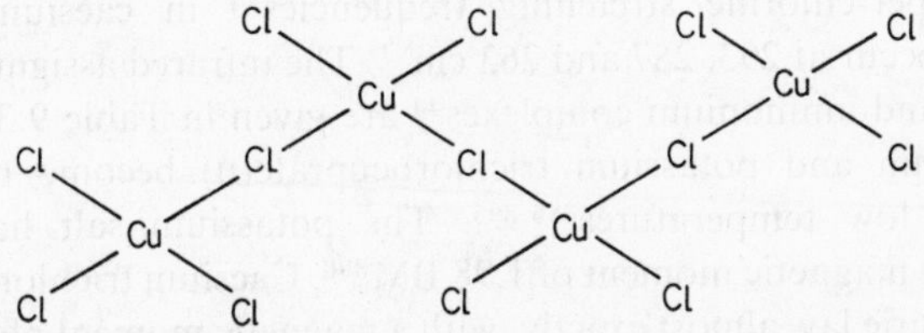

A later refinement of the structure by Rundle and coworkers[391] regarded the coordination about each copper atom to be a distorted octahedron with adjacent octahedra sharing a face with three bridging halogens, which may be represented schematically as

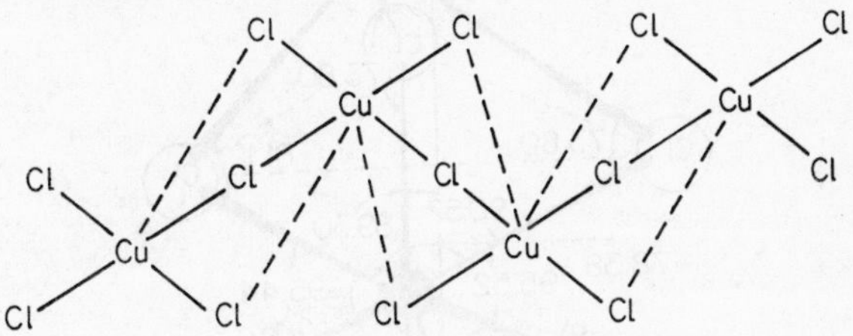

The internuclear dimensions for the refined structure are shown in Figure 9.25.

Potassium and ammonium trichlorocuprates(II) are isostructural, the monoclinic unit cells being of space group $P2_1/c$ and containing four molecules[386]. The unit-cell parameters are:

$KCuCl_3$	$a = 4.029$, $b = 13.785$, $c = 8.736$ Å and $\beta = 97.33°$
NH_4CuCl_3	$a = 4.066$, $b = 14.189$, $c = 9.003$ Å and $\beta = 97.50°$

In these compounds the anions are planar binuclear Cu_2Cl_6 units which are stacked in such a manner as to give tetragonally elongated octahedral coordination about each copper atom. The detailed packing arrangement and the internuclear distances for potassium trichlorocuprate(II) are given in Figure 9.26. The corresponding dimensions for the ammonium salt are given in Table 9.32.

Dimethylammonium trichlorocuprate(II) is monoclinic[390] with $a = 12.09$, $b = 8.63$, $c = 14.49$ Å and $\beta = 97.5°$. There are eight formula units in the cell of space group $I2/a$. The structure consists of isolated Cu_2Cl_6 units which have two bridging halogens but which are unsymmetrical as shown in Figure 9.27. The anions are packed in such a way that long-range interactions give an effective square-pyramidal arrangement about each copper atom as shown in Figure 9.28. The apical copper-chlorine distance is 2.733 Å.

The copper-chlorine stretching frequencies[44] in caesium trichlorocuprate(II) occur at 293, 287 and 263 cm^{-1}. The infrared assignments of the potassium and ammonium complexes[44] are given in Table 9.33.

Ammonium and potassium trichlorocuprate(II) become magnetically ordered at low temperatures[386,395]. The potassium salt has a room-temperature magnetic moment of 1.98 BM[396]. Caesium trichlorocuprate(II) obeys the Curie law almost exactly, with a magnetic moment of 1.95 BM[369].

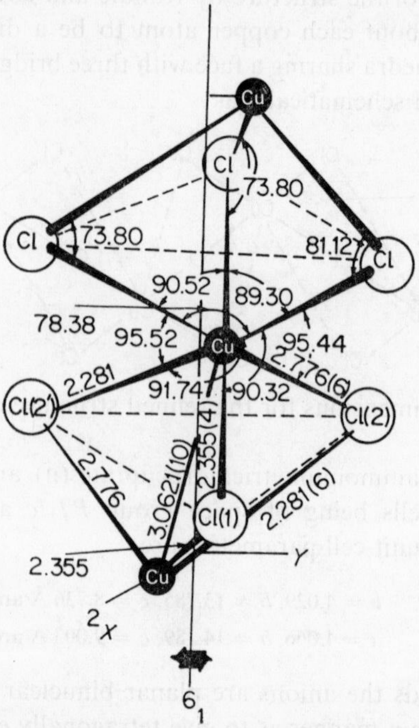

Figure 9.25 The coordination about the copper atom in caesium trichlorocuprate (II). (Reproduced by permission from A. W. Schlueter, R. A. Jaconson and R. E. Rundle, *Inorg. Chem.*, **5**, 277 (1966))

The electronic absorption spectra of several trichlorocuprates(II) have been recorded[48,334,335,397]. A charge-transfer band at about 19,000 cm^{-1} is characteristic of the Cu_2Cl_6 species.

Trichlorocuprate(II) dihydrates. The lithium, potassium and trimethylammonium salts of the trichlorocuprate(II) dihydrate anion are readily obtained by crystallization from aqueous solutions containing stoichiometric quantities of the constituent chlorides[378,387,398].

The crystal structure of red-brown lithium trichlorocuprate(II) dihydrate has been investigated by Rundle and coworkers[399,400] who found the lattice to be monoclinic with $a = 6.078$, $b = 11.145$, $c = 9.145$ Å and $\beta = 108.5°$. There are four molecules in the unit cell of space group $P2_1/c$.

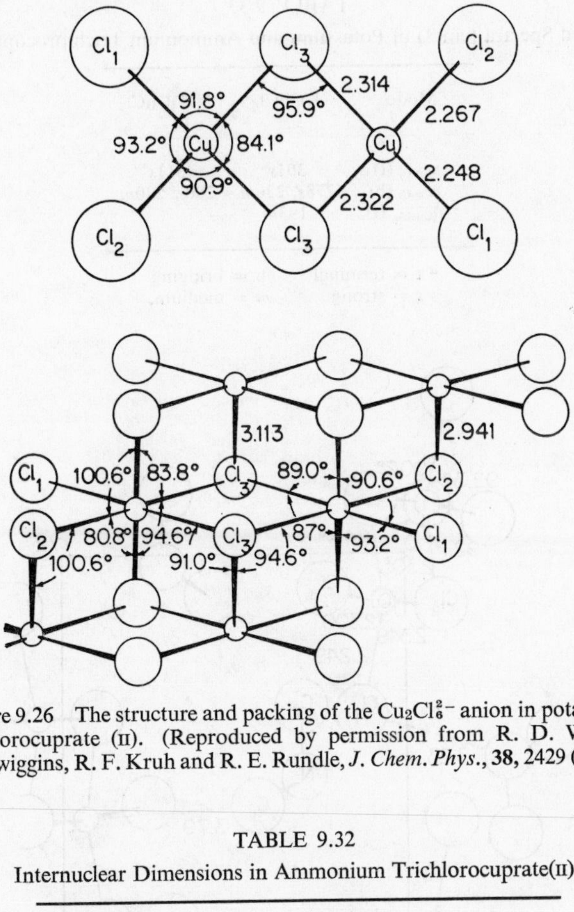

Figure 9.26 The structure and packing of the $Cu_2Cl_6^{2-}$ anion in potassium trichlorocuprate (II). (Reproduced by permission from R. D. Willett, C. Dwiggins, R. F. Kruh and R. E. Rundle, *J. Chem. Phys.*, **38**, 2429 (1963))

TABLE 9.32

Internuclear Dimensions in Ammonium Trichlorocuprate(II)

Distance	Value (Å)	Angle	Value
$Cu—Cl_1$	2.25	$Cl_3—Cu—Cl_3$	84.9°
$Cu—Cl_2$	2.26	$Cl_3—Cu—Cl_2$	90.2°
$Cu—Cl_3$	2.32	$Cl_2—Cu—Cl_1$	94.3°
Cu—Cu	3.42	$Cl_1—Cu—Cl_3$	91.4°
		$Cl_3—Cu—Cl_2$	172.8
		$Cl_3—Cu—Cl_1$	176.2°

The structure of the compound, which is shown in Figure 9.29, consists of nearly planar Cu_2Cl_6 units which are linked together by long copper-chlorine bonds of 2.9 Å, to the groups above and below.

TABLE 9.33

Infrared Spectra (cm^{-1}) of Potassium and Ammonium Trichlorocuprate(II)

Mode[a]	$KCuCl_3$	NH_4CuCl_3
ν_{Cu-Cl} (t)	301*s*	311*s*
ν_{Cu-Cl} (b)	278*s*, 236*m*	280*s*, 230*m*
δ_{Cu-Cl_2} (t)	193*m*	

[a] t = terminal b = bridging
s = strong *m* = medium.

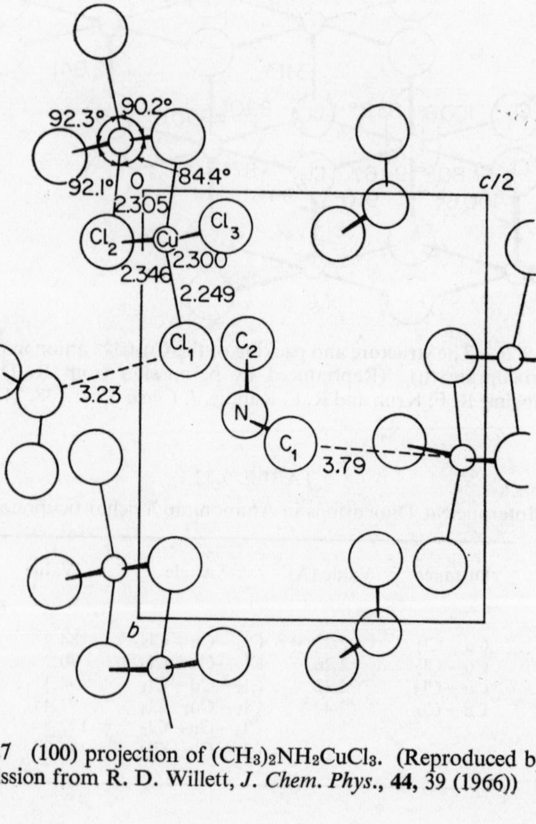

Figure 9.27 (100) projection of $(CH_3)_2NH_2CuCl_3$. (Reproduced by permission from R. D. Willett, *J. Chem. Phys.*, **44**, 39 (1966))

Adams and Lock[44] report a terminal copper-chlorine stretching frequency of 295 cm^{-1} and bridging copper-chlorine stretching frequencies of 272 and 222 cm^{-1} for lithium trichlorocuprate(II) dihydrate.

Magnetic susceptibility, ^1H NMR, ^7Li NMR and neutron-diffraction studies on lithium trichlorocuprate(II) all show that it becomes magnetically

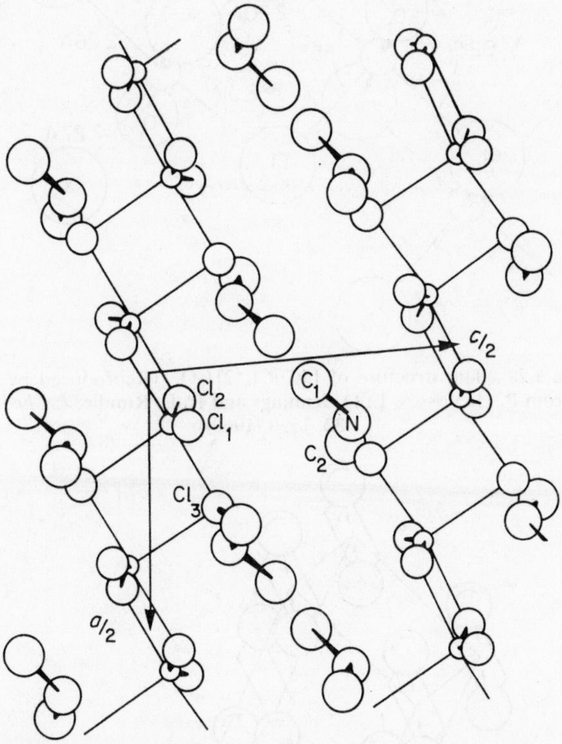

Figure 9.28 (010) projection of $(CH_3)_2NH_2CuCl_3$. (Reproduced by permission from R. D. Willett, *J. Chem. Phys.*, **44**, 39 (1966))

ordered at low temperatures. The reported Neel temperatures range from 4.4 to 6.7°K[400–404]. Over the temperature range 100–300°K the magnetic moment is 1.82 BM while between 20 and 100°K it is 2.04 BM[402]. Neutron-diffraction measurements[402] show that the magnetic spins are coincident with the internuclear dimeric copper-copper linkage as shown in Figure 9.30. As with the anhydrous trichlorocuprates(II) containing the dimeric unit, lithium trichlorocuprate(II) dihydrate shows a strong characteristic charge-transfer band[397] at 19,600 cm^{-1}.

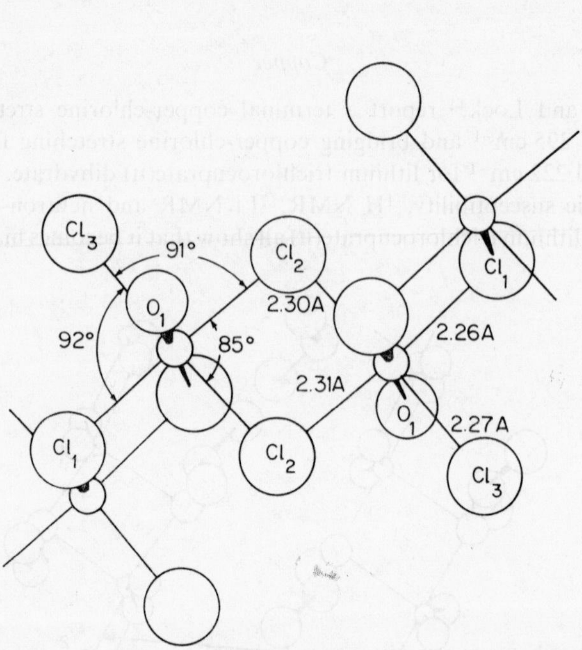

Figure 9.29 The structure of $LiCuCl_3 \cdot 2H_2O$. (Reproduced by permission from P. H. Vossos, L. D. Jennings and R. E. Rundle, *J. Chem. Phys.*, **32**, 1590 (1960))

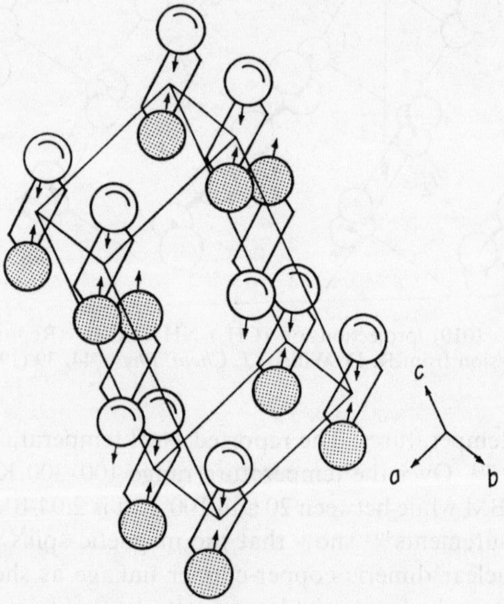

Figure 9.30 The magnetic structure of $LiCuCl_3 \cdot 2H_2O$. (Reproduced by permission from S. C. Abrahams and H. J. Williams, *J. Chem. Phys.*, **39**, 2923 (1963))

Pentabromocuprates(II). A number of compounds reported to contain the pentabromocuprate(II) anion have been prepared[49,300,334]. As with the corresponding chloro complexes, it would appear that more work is necessary to characterize them unambiguously.

Tetrabromocuprates(II). Substituted ammonium and arsonium tetrabromocuprates(II) are normally prepared from the constituent bromides in alcohol[44,344,348,405]. The caesium salt has been prepared by evaporating a saturated aqueous solution of the constituent bromides at room temperature[406].

The tetrabromocuprates(II) are purple to brown crystalline solids and the heat of formation[407] of the tetraethylammonium salt from the constituent bromides is -3.81 kcal mole^{-1}.

Morosin and Lingafalter[406] report that caesium tetrabromocuprate(II) has an orthorhombic unit cell with $a = 10.195$, $b = 7.965$ and $c = 12.936$ Å. There are four formula units in the cell of space group *Pnma*. The tetrahedral tetrabromocuprate(II) anion is considerably distorted and the internuclear dimensions are given in Table 9.34.

TABLE 9.34

Internuclear Dimensions in Caesium Tetrabromocuprate(II)

Distance	Value (Å)	Angle	Value
Cu—Br(1)	2.394	Br(1)—Cu—Br(2)	130.4°
Cu—Br(2)	2.380	Br(1)—Cu—Br(3)	101.9°
Cu—Br(3)	2.354	Br(2)—Cu—Br(3)	99.9°
		Br(3)—Cu—Br(3′)	126.4°

The infrared spectra of a number of tetrabromocuprates(II) have been recorded and they show quite clearly that both square-planar and tetrahedral forms of the anion exist. The observed spectra are summarized in Table 9.35.

The electronic absorption spectra of the tetrabromocuprate(II) complexes resemble those of the corresponding chloro derivatives[49,334,344,348,363–365,405,408–410].

Tetraethylammonium tetrabromocuprate(II) obeys the Curie-Weiss law[370] with a magnetic moment of 1.92 BM and a Weiss constant of 2°, while the triphenylmethylarsonium salt has a room-temperature magnetic moment of 1.96 BM.

Tetrabromocuprate(II) dihydrate. Ammonium tetrabromocuprate(II) dihydrate is readily isolated from aqueous solutions of the constituent bromides[380,411,412]. The compound is isomorphous with the corresponding chloro compound and the tetragonal unit cell[380,411,412] has $a = 7.58$ and $c = 8.41$ Å. The copper atom is surrounded by a distorted octahedron of four bromine atoms and two water molecules, the copper-bromine and copper-oxygen bond distances being 2.46 and 2.20 Å respectively. It has similar magnetic properties to the corresponding chloro compound[384], becoming a ferromagnet at 1.73°K.

TABLE 9.35

Infrared Spectra (cm^{-1}) of Some Tetrabromocuprates(II)

Compound	$\nu_{Cu\text{-}Br}$	Anion Geometry	Reference
Cs_2CuBr_4	224, 189, 172	Tetrahedral	44
$[CH_3NH_3]_2CuBr_4$	227	Square-planar	44
$[C_2H_5NH_3]_2CuBr_4$	219	Square-planar	44
$[(CH_3)_4N]_2CuBr_4$	218, 177	Tetrahedral	44
$[(C_2H_5)_4N]_2CuBr_4$	216, 174	Tetrahedral	359
	248, 222	Tetrahedral	361
$[(C_6H_5)_3CH_3As]_2CuBr_4$	222	?	361

Tribromocuprates(II). Potassium, caesium and substituted ammonium tribromocuprates(II) are readily prepared by evaporating aqueous solutions of the appropriate bromides using excess copper dibromide to prevent formation of the tetrabromocuprates(II)[387,413]. The potassium and caesium salts have also been prepared[414] by evaporating an aqueous solution, containing stoichiometric amounts of the constituent bromides, over concentrated sulphuric acid.

Potassium tribromocuprate(II) has a monoclinic unit cell[414] with $a = 4.28$, $b = 14.43$, $c = 7.19$ Å and $\beta = 108.4°$. The space group is $P2_1/m$ and there are four molecules in the cell.

The magnetic moment of the potassium salt is 1.52 BM at 295°K and falls to 0.83 BM at 78°K[414]. This behaviour has been interpreted as a combination of weak antiferromagnetic interaction between $[Cu_2Br_6]^{2-}$ dimers and stronger spin-spin coupling within the dimer itself. The magnetic properties of caesium tribromocuprate(II) are markedly different in that this compound has only a slight paramagnetism and the susceptibility is constant over the temperature range 80–300°K[414]. This obviously indicates that in this case spin-spin interaction has completely quenched the paramagnetism.

Two *d-d* transitions are observed in the electronic absorption spectra of the tribromocuprates(II)[334] as with the corresponding chloro compounds.

Oxidation State I

Trichlorocuprate(I). Potassium trichlorocuprate(I) has been prepared by adding the appropriate quantity of copper monochloride to a saturated solution of potassium chloride in either water or concentrated hydrochloric acid in the absence of air[415,416]. The compound is colourless, but readily turns green in air through oxidation. It is orthorhombic[415] with $a = 12.00$, $b = 12.55$ and $c = 4.20$ Å. The space group is *Pnam* and the cell contains four formula units. The structure consists of potassium ions

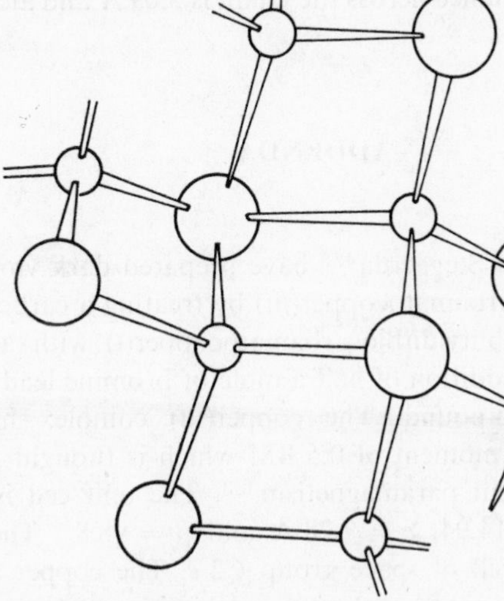

Figure 9.31 The polymeric structure of the $Cu_2Br_3^-$ anion. (Reproduced by permission from C. Romming and K. Waerstad, *Chem. Commun.*, **1965**, 299)

and chains of $CuCl_4$ tetrahedra which are parallel to the c axis and share two corners. The copper-chlorine bond lengths are 2.31, 2.32 and 2.43 Å.

Miscellaneous copper(I) halocomplexes. A large number of complexes containing CuX_2^- or $Cu_2X_3^-$ groups, where X = Cl, Br or I, have been prepared, particularly with large organic cations[417–421]. In general, little is known of the structural properties of these complexes.

Romming and Waerstad[422] have studied the crystal structure of the orange-red compound $(C_5H_5N_2)Cu_2Br_3$ which is isolated as an intermediate in the Sandmeyer reaction. The compound is orthorhombic with $a = 17.69$, $b = 5.72$ and $c = 11.01$ Å and there are four formula units in the cell of space group *Pnma*. The benzenediazonium ions are situated in mirror planes parallel to the (010) plane and each of the 'inner' nitrogen atoms is surrounded by four bromine atoms which are situated in a plane approximately perpendicular to the carbon-nitrogen-nitrogen direction, with a nitrogen-bromine distance of 3.65 Å. These bromine atoms belong to two anionic chains which both run parallel to the *b* axis. In these chains two copper atoms alternate with three bromine atoms as shown in Figure 9.31, giving one third of the bromine atoms the rare stereochemistry of four-coordination. The bromine atom which bridges two copper atoms has copper-bromine distances of 2.45 Å while those that bridge four copper atoms have a copper-bromine bond distance of 2.57 Å. The copper-copper distance across the chain is 3.09 Å and along the chain is 2.86 Å.

ADDENDA

Halides

Beurskens, Cras and Steggerda[423] have prepared dark violet dibromo-*N*,*N*-di-n-butyldithiocarbamatocopper(III) by treating a carbon disulphide solution of *N*,*N*-di-n-butyldithiocarbamatocopper(I) with an equimolar amount of bromine. Addition of half a mole of bromine leads to a monobromo copper(II) compound. The copper(III) complex has a room-temperature magnetic moment of 0.5 BM which is thought to be due to temperature-independent paramagnetism[423]. The unit cell is monoclinic with $a = 13.85$, $b = 14.94$, $c = 7.28$ Å and $\beta = 98.8°$. There are four formula units in the cell of space group $C2/c$. The copper atom is in a square-planar environment, the copper-bromine and copper-sulphur distances being 2.311 Å and 2.193 Å respectively. Both of these distances are considerably shorter than the corresponding dimensions in copper(II) complexes suggesting an appreciable ionic contribution to the copper-ligand bonding[423].

The thermal decomposition of copper difluoride dihydrate and copper(II) hydroxide fluoride have been investigated by Ball and Coultard[424]. In both cases copper(II) oxide is the end product.

Ruthven and Kenny[425] have made a detailed study of the equilibrium chlorine pressure over the copper dichloride–copper monochloride system. The melting points of the di- and monochlorides were given as 633° and 430° respectively. The dichloride-monochloride eutectic point is 378°.

The magnetic properties of copper dichloride have been examined by Adams and coworkers[426]. These workers quote a *g* value of 2.07 and a

Neel temperature of 78°K and give the copper-chlorine stretching frequencies as 329 and 277 cm^{-1}.

Lohr[427] has reported a theoretical treatment of the electronic spectrum and vibronic interactions in gaseous copper dichloride.

The crystal structure of $CuCl_2$(4-vinylpyridine)$_2$ has been determined by Laing and Horsfield[428]. The monoclinic unit cell has $a = 8.68$, $b = 3.91$, $c = 21.11$ Å and $\beta = 90.0°$. There are two formula units in the cell of space group $P2_1/n$. The copper atom has a *trans* square-planar configuration, the copper-chlorine and copper-nitrogen bond distances being 2.38 Å and 2.01 Å respectively. The molecules are stacked parallel to the *b* axis such that the chlorine-copper-chlorine portions are twisted about the nitrogen-copper-nitrogen axis giving a structure similar to that of $CuCl_2(py)_2$. This results in a copper-chlorine contact distance of 3.02 Å.

The infrared spectra of a number of adducts of the type $CuCl_2L_2$, where L is pyridine or a substituted pyridine, have been recorded[429–433]. The effect of the substituent on the pyridine ring on the copper-nitrogen and copper-chlorine stretching frequencies has been discussed in detail. The reported data are given in Table 9.36.

TABLE 9.36
Infrared Spectra (cm^{-1}) of $CuCl_2L_2$

L	ν_{Cu-Cl} (terminal)	ν_{Cu-Cl} (bridge)	ν_{Cu-N}
pyridine	293	235	268
3-C_2H_5pyridine	291		263
3-Clpyridine	292		258
3-CNpyridine	305		256
4-CH_3pyridine	296	259	285
4-C_2H_5pyridine	291		244
4-*i*-C_3H_7pyridine	287		237
4-Clpyridine	295		244
4-CNpyridine	304	236	243
4-CH_2OHpyridine	296	259	285
4-$COCH_3$pyridine	296	241	265
4-$CONH_2$pyridine	301	249	264
4-$COOCH_3$pyridine	301	235	264

The adducts of the type CuX_2L_2 with pyridine and related ligands react[434] with water or alkali to give complexes of the type $Cu_4OX_4L_4$. A single-crystal X-ray diffraction study of the pyridine chloro complex by Kilbourn and Dunitz[435] has shown that the complex consists of a central oxygen atom tetrahedrally coordinated to four copper atoms. The metal atoms are bridged by chlorine atoms. The trigonal-bipyramidal configurations of the copper atoms is completed by the four pyridine molecules. The compound thus has a similar structure to $Cu_4OCl_4L_4$, L = $(C_6H_5)_3PO$ (Figure 9.13, page 506). The compound has a monoclinic unit cell with $a = 11.29$, $b = 21.40$, $c = 11.96$ Å and $\beta = 92.20°$. There are four molecules in the cell of space group $P2_1/n$. The average bond distances and angles are given in Table 9.37.

TABLE 9.37
Internuclear Dimensions in $Cu_4OCl_6(py)_4$

Distance	Value (Å)	Angle	Value
Cu—Cl	2.41	Cl—Cu—Cl	119°
Cu—O	1.90	Cu—Cl—Cu	80°
Cu—N	1.96	O—Cu—Cl	85°

The central Cu_4OCl_6 unit does not have T_d symmetry and the reasons for this are discussed in detail[435].

The infrared and electronic spectra of several derivatives of the above type have been recorded[434] and the copper-oxygen stretching frequencies, together with melting points, are given in Table 9.38.

TABLE 9.38
Properties of $Cu_4OX_6L_4$

L	X = Cl		X = Br	
	M.p.	ν_{Cu-O} (cm^{-1})	M.p.	ν_{Cu-O} (cm^{-1})
Pyridine	275–276°	576	231–232°	533
Isoquinoline	265°	570	217–219°	535
Quinoline	244–245°	573	218–219°	536

Copper dichloride reacts[436] with 2-pyridone (TP) to give orange $CuCl_2L$ or $CuCl_2L_2$, L = (TP), depending upon the relative amounts of the reactants. The room-temperature magnetic moments are 1.80 and 1.82 BM respectively. The 1:1 adduct has a triclinic unit cell with $a = 9.93$, $b = 9.94$, $c = 3.95$ Å, $\alpha = 90.9°$, $\beta = 92.9°$ and $\gamma = 113.2°$. The structure consists of halogen bridged dimeric units which are held together by long copper-chlorine linkages as in $CuCl_2 \cdot CH_3CN$ (Figure 9.7, page 500).

The crystal structure of Cu_3Cl_6[2-methylpyridine *N*-oxide]$_2 \cdot 2H_2O$

has been determined by single-crystal techniques by Sager and Watson[437]. The unit cell is triclinic with $a = 9.732$, $b = 7.380$, $c = 9.596$ A, $\alpha = 104.4°$, $\beta = 119.35°$ and $\gamma = 68.25°$. There is one formula unit in the cell of space group $P\bar{1}$. The structure of the complex is not as suggested previously (page 498), but consists of $Cu_2Cl_4(C_6H_7NO)_2$ dimers bridged by $CuCl_2 \cdot 2H_2O$ units through long copper-chlorine bonds. In this way long chains are formed, the chains being held together by hydrogen bonds. Thus the compound should be formulated as polybis[μ-(2-methylpyridine *N*-oxide)chlorocopper(II)di-μ-chloro]diaquocopper(II). One interesting feature of this structure is that it contains both six- and five coordinate copper atoms. The important internuclear distances are given in Table 9.39. Recrystallization of $Cu_3Cl_6(C_6H_7NO)_2 \cdot 2H_2O$ from methanol yielded green $CuCl_2(C_6H_7NO) \cdot CH_3OH$[437]. This complex looses methanol on standing at room temperature and decomposes at 207–208°. The methanolate has a monoclinic unit cell[437] with $a = 11.81$, $b = 10.47$, $c = 9.02$ Å and $\beta = 103.8°$. There are four molecules in the cell of space group $P2_1/c$.

The crystal structure of chloroaquobis(ethylendiamine)copper(II) chloride has been reported by Waters and coworkers[438]. The purple compound, prepared by adding ethanol to an aqueous solution of copper dichloride containing ethylenediamine, has a monoclinic unit cell with $a = 6.18$, $b = 15.16$, $c = 11.80$ Å and $\beta = 99.0^0$. There are four molecular units in the cell of space group $P2_1/n$. The copper atom is in a tetragonally elongated octahedral environment. It is significantly displaced from the plane of the four nitrogen atoms of the ethylenediamine rings towards the chlorine atom. The copper-chlorine and copper-oxygen distances are 2.81 Å and 2.62 Å respectively, while the copper-nitrogen bonds are in the range 1.98–2.02 Å. The octahedral units are linked together into chains by hydrogen bonding.

Violet dichlorobis(1,2-propanediamine)copper(II) monohydrate[439,440] has a room-temperature magnetic moment of 1.85 BM and an orthorhombic unit cell with $a = 6.411$, $b = 19.75$ and $c = 5.329$ Å. The copper atom has a *trans* octahedral configuration.

Dichlorodimethylnitrosoaminecopper(II) has an orthorhombic unit cell[441] with $a = 6.376$, $b = 14.685$ and $c = 7.265$ Å. There are four formula units in the cell of space group *Pnam*. The copper atom is surrounded by four chlorine atoms and two oxygen atoms in the *trans* positions. The copper atoms are bridged by pairs of chlorine atoms. The copper-oxygen and copper-chlorine bond distances are 2.28 and 2.29 Å respectively. The copper-nitrogen distance is only 3.02 Å indicating also some copper-nitrogen bonding[441].

TABLE 9.39

Internuclear Dimensions in $Cu_3Cl_6(C_6H_7NO)_2 \cdot 2H_2O$[a]

Distance	Value (Å)	Angle	Value	Angle	Value
Cu(1)—Cu(2)	3.581	Cu(2′)—O—Cu(2)	109.3°	Cl(1)—Cu(2)—O′	95.7°
Cu(2)—Cu(2′)	3.255	O—Cu(2)—O′	70.7°	Cl(3)—Cu(2)—O	157.2°
Cu(1)—Cl(1)	2.264	Cu(2)—Cl(1)—Cu(1)	93.1°	Cl(3)—Cu(2)—O′	92.9°
Cu(1)—Cl(2)	2.955	Cu(2′)—Cu(2)—Cu(1)	130.3°	Cl(1)—Cu(1)—OH_2	90.1°
Cu(1)—OH_2	1.99	Cl(1)—Cu(2)—Cl(2)	92.8°	Cu(2)—Cl(2)—Cu(1)	86.3°
Cu(2)—Cl(1)	2.654	C.(2)—Cu(2)—Cl(3)	98.7°	Cl(1)—Cu(1)—Cl(2)	84.4°
Cu(2)—Cl(2)	2.221	Cl(2)—Cu(2)—O	95.4°	Cl(2)—Cu(1)—OH_2	97.4°
Cu(2)—Cl(3)	2.244	Cl(2)—Cu(2)—O′	164.6°	Cu(2′)—O—N	127.1°
Cu(2)—O	2.01	Cl(1)—Cu(2)—Cl(3)	97.7°	Cu(2)—O—N	121.6°
Cu(2)—O′	1.98	Cl(1)—Cu(2)—O	99.3°	Cl(3)—H_2O—Cu(1)	110.2°

[a] Cu(1) and Cu(2) refer to six- and five coordinate copper atoms respectively.

$CuCl_2 \cdot (CH_3)_2SO$ has a monoclinic unit cell[442] with $a = 7.70$, $b = 13.57$, $c = 6.43$ Å and $\beta = 100.63°$. The space group is $P2_1/c$. Square-planar units containing three chlorine atoms and one oxygen atom are linked together into chains by two non-symmetric bridges, Cu—Cl . . . Cu and Cu—O . . . Cl. The copper-chlorine distances are 2.199 Å (*trans* to O), 2.319 and 2.345 Å (*cis* to O, *trans* to each other) and 2.762 Å (axial). The copper-oxygen bond distances are 1.913 and 2.938 Å.

Tomlinson and Hathaway[443] have investigated the ESR, electronic and infrared spectra of the complexes of the type $CuX_2(NH_3)_2$ where X = Cl, Br or I. Some properties of these complexes are given in Table 9.40.

TABLE 9.40
Adducts of the Type $CuX_2(NH_3)_2$

Compound	$g_\perp$	$g_\parallel$	g_i	ν_{Cu-N} (cm^{-1})	δ_{N-Cu-N} (cm^{-1})
α-$CuCl_2(NH_3)_2$	2.030	2.091	2.240	480	
β-$CuCl_2(NH_3)_2$			2.135	512	
α-$CuBr_2(NH_3)_2$			2.126	495	220
β-$CuBr_2(NH_3)_2$	No signal observed			495	228
$CuI_2(NH_3)_2$	2.056	2.36		437	232

The reaction of the copper dihalides with di-2-pyridyl ketone (DPK) has been investigated by Osborne and McWhinnie[444] and by Feller and Robson[445]. Osborne and McWhinnie isolated some of their products as alcoholates and suggested that these had square-pyramidal structures. Feller and Robson have questioned this assignment and suggest that the molecule of alcohol adds across the ketone double bond. The basis for this suggestion is the very large or total disappearance of ν_{CO} in the 'alcoholates'. Osborne and McWhinnie[444] also reacted the dihalides with 2-benzoylpyridine ketone (BPK) and the properties of the complexes isolated with this ligand are given in Table 9.41.

TABLE 9.41
Copper Dihalide–2-Benzoylpyridine Ketone Adducts

Complex	Colour	μ (BM)	ν_{Cu-X} (cm^{-1})	Structure
$CuCl(BPK)_2$	Green	2.10	284	*Trans* octahedral
$CuCl_2(BPK)$	Light green	2.05	315, 292	*Cis* square planar
$CuBr_2(BPK)_2$	Green			*Trans* octahedral
$CuBr_2(BPK)$	Red-brown		249, 233	Tetrahedral

Fowles and coworkers[446] report that copper dichloride and dibromide react with 1,4-dioxan to give adducts of the type $3CuX_2 \cdot 2L$, not simple 1:1 adducts as reported in the earlier literature. The chloro complex has copper-chlorine stretching frequencies of 310 and 263 cm^{-1} which have been assigned to the terminal and bridging modes respectively. The terminal copper-bromine stretching mode is at 237 cm^{-1}. It was suggested that the complexes may have a similar structure to $3CuCl_2 \cdot 2CH_3CN$ (Figure 9.8, page 501), or some other polymeric structure with both four and six coordinate copper atoms.

The properties of a number of other adducts of copper dichloride are summarized in Table 9.42.

Other donor molecules that have been reacted with copper dichloride include pyridine and its derivatives[448], aniline[470], pyridine *N*-oxide[471], acrylonitrile[472], 2,9-dimethyl-1,10-phenanthroline[473], triethanolamine[474], benzophenoneimine[475] and a number of Schiff bases[476].

Uhlig and Staiger[477] have made a further magnetic and spectral study of the previously reported aminoalcohol derivatives of copper dichloride and dibromide (page 498). Adams and coworkers[426] have investigated the magnetic and spectral properties of copper chloro and bromo methoxides. The following properties were reported.

$CuCl(OCH_3)$	$\nu_{Cu-Cl} = 385, 315\ cm^{-1}$	$\mu = 2.0$ BM	$g = 2.01$
$CuBr(OCH_3)$	$\nu_{Cu-Br} = 365, 320\ cm^{-1}$	weakly paramagnetic	

Lindoy and Livingstone[462] report that 2-(2-pyridyl)benzothiazoline rearranges in the presence of copper dichloride to give the complex CuCl(NNS) of the tautomeric Schiff base *N*-2-mercaptophenyl-2'-pyridylmethyleneimine (NNSH). This dark grey compound has a room-temperature magnetic moment of 1.86 BM.

The thermal decomposition of copper dichloride dihydrate has been reinvestigated[424,478]. Loss of water occurs over the temperature range 50–200° to give the anhydrous dichloride which then decomposes at higher temperatures to give the monochloride and chlorine. The translational motion of the water molecules in the hydrate and the corresponding deuterate has been investigated by Biehat and coworkers[479], the following frequencies being reported.

$CuCl_2 \cdot 2H_2O$	227.5, 188.5, 137 cm^{-1}
$CuCl_2 \cdot 2D_2O$	222, 180, 130 cm^{-1}

Copper dibromide looses bromine quite readily over the temperature range 150–300° to give the monobromide[424,478]. It has a Neel temperature

TABLE 9.42

Adducts of the Type $CuCl_2L_n$

Compound and Properties	References
L = 3-methylpyridine, n = 2, light blue, m.p. = 205°, μ = 2.06 BM, tetrahedral	447
L = 4-methylpyridine, n = 2, light blue, m.p. = 190°, μ = 1.92 BM, tetrahedral	447
L = 4-vinylpyridine, n = 2, blue, m.p. = 170°, μ = 2.08 BM, tetrahedral	447
L = 2-chloropyridine, n = 2, blue-green, μ = 1.79 BM, ν_{Cu-N} = 234 cm^{-1}, ν_{Cu-Cl} = 308 cm^{-1}, octahedral	448
L = 3-bromopyridine, n = 2, μ = 1.85 BM, ν_{Cu-N} = 261 cm^{-1}, ν_{Cu-Cl} = 281 cm^{-1}, octahedral	448
L = 4-chloropyridine, n = 2, μ = 1.83 BM, ν_{Cu-N} = 243 cm^{-1}, ν_{Cu-Cl} = 294 cm^{-1}, octahedral	448
L = quinoline, n = 2, ν_{Cu-Cl} = 330 cm^{-1}, octahedral	448
L = 2-methoxyaniline, n = 2, purple, μ = 2.01 BM, octahedral	449
L = *p*-nitroso-*N,N*-dimethylaniline (NODMA), n = 2, μ =1.89 BM, octahedral	450
L = benzene-1,2-diamine, n = 1, green, μ = 1.87 BM, octahedral	451
L = 1-(2-pyridyl)-2-(3-pyridyl)ethylene, n = 1, blue-green, μ = 1.88 BM, octahedral	452
L = 1-(2-pyridyl)-2-(4-pyridyl)ethylene, n = 1, green, μ = 1.83 BM, octahedral	452
L = imidazole, n = 4, blue-purple, μ = 1.84 BM, octahedral	453
L = phenazine, n = 1, brown, μ = 1.68 BM, ν_{Cu-Cl} = 362 cm^{-1}, octahedral	448
L = benzoxazole, n = 2, pale blue, μ = 1.57 BM, ν_{Cu-N} = 239 cm^{-1}, ν_{Cu-Cl} = 318, 314, 301 cm^{-1}, square planar	454
L = 4(5)-bromoimidazole, n = 2, green, μ = 1.88 BM, ν_{Cu-N} = 275 cm^{-1}, ν_{Cu-Cl} = 251 cm^{-1}, octahedral	455
L = pentamethylenetetrazole (PMT), n = 1, light green, m.p. = 250°, μ = 1.55 BM, ν_{Cu-Cl} = 300 cm^{-1}, octahedral	456
L = 1,2,4-triazole, obeys Curie-Weiss law above about 40°K with θ = 19°, ν_{Cu-N} = 320 cm^{-1}, ν_{Cu-Cl} = 275, 246 cm^{-1}, δ_{Cu-Cl} = 198 cm^{-1}	448, 457, 458
L = benzothiazole (BT), n = 2, pale green, μ = 1.78 BM, ν_{Cu-Cl} = 299, 284 cm^{-1}, octahedral with bridging halogens	459
L = benzothiazole (BT), n = 1, golden, μ = 1.89 BM, ν_{Cu-Cl} = 319, 299, 271 cm^{-1}, dimeric, four coordinate with bridging halogens	459
L = ethyl *p*-(2-pyridylmethyleneamino)benzoate (bpda), n = 1, green, μ = 1.88 BM, square planar	460
L = 1-phenyl-2-(2′-pyridylmethyleneamino)propane (bdpa), n = 1, green, μ = 1.98 BM, square planar	460
L = 2,6-pyridylenebis(ethyl *p*-methyleneaminobenzoate) (bcpda), n = 1, orange, μ = 1.98 BM, trigonal-bipyramidal	460
L = 2,6-pyridylenebis(1-phenyl-2′-methyleneaminopropane) (bdpda), n = 1, green, μ = 2.02 BM, trigonalbipyramidal	460
L = 2-(2-pyridyl)benzothiazole (pbt), n = 1, green, μ = 1.81 BM, five or six coordinate	461
L = 2,2′-pyridylmethylthioaniline (NSN), n = 1, dark green, μ = 1.85 BM, five coordinate	462
L = *N*-2-methylthiophenyl-2′-pyridylmethyleneimine (NNSMe), n = 1, green, m.p. = 188°, μ = 1.85 BM, five coordinate	462

TABLE 9.42 (*contd.*)
Adducts of the Type $CuCl_2L_n$

L = biuret, $n = 2$, pale green, $\mu = 1.95$ BM, $\theta = 20°$, octahedral	463–465
L = semicarbazide, $n = 2$, blue, $\mu = 1.83$ BM, $\nu_{Cu-L} = 227$ cm^{-1}, $\nu_{Cu-Cl} = 164$ cm^{-1}, *trans* octahedral	458, 466, 467
L = semicarbazide, $n = 1$, turquoise, $\mu = 1.78$ BM, $\nu_{Cu-L} = 190$ cm^{-1}, $\nu_{Cu-Cl} = 310, 225$ cm^{-1}	458, 466, 467
L = acetone semicarbazone, $n = 2$, green, $\mu = 1.77$ BM, $\nu_{Cu-L} = 284, 210$ cm^{-1}, $\nu_{Cu-Cl} = 166$ cm^{-1}, *trans* octahedral	458, 466, 467
L = thiosemicarbazide, $n = 2$, $\nu_{Cu-L} = 289, 264$ cm^{-1}, $\nu_{Cu-Cl} = 160$ cm^{-1}, *trans* octahedral	458, 466
L = thiosemicarbazide, $n = 1$, $\nu_{Cu-L} = 216$ cm^{-1}, $\nu_{Cu-Cl} = 312, 234$ cm^{-1}, $\delta_{Cu-Cl} = 179$ cm^{-1}, octahedral	458, 466
L = diphenylcyclopropenone (dpcp), $n = 2$, pale olive green, m.p. = 110–111°, $\mu = 2.01$ BM, $\nu_{Cu-Cl} = 336$ cm^{-1}, *trans* square planar	468
L = 1,2-bis(methylsulphonyl)ethane (BMSE), $n = 1$, yellow, m.p. = 174°	469

of 240°K and a g value of 2.03 at room temperature[426]. The copper-bromine stretching frequencies have been given as[426] 254 and 223 cm^{-1}.

The crystallographic, spectroscopic and magnetic properties of $CuBr_2$(N,N'-diethylethylenediamine)$_2 \cdot 2H_2O$[480] and $CuBr_2$(1,2-propanediamine)$_2 \cdot xH_2O$[440,481] have been investigated. Both complexes have a square-planar arrangement of nitrogen atoms about the copper atom with the bromine atoms in the axial positions. The properties of the complexes are given in Table 9.43.

TABLE 9.43
Properties of $CuBr_2$(amine)$_2$ Hydrates

Property	N,N'-diethylethylenediamine	1,2-propanediamine
Symmetry	Monoclinic	Monoclinic
a (Å)	7.248	9.284
b (Å)	10.137	7.332
c (Å)	14.967	12.088
β	109.65°	115.70°
Z	2	2
Space group	$P2_1/c$	$P2_1/c$
μ (BM)	1.83	1.88

Di Vaira and Orioli[482] report that $CuBr_2$(Me_6tren), where Me_6tren is tris(2-dimethylaminoethyl)amine, is isostructural with the corresponding cobalt complex (Figure 7.14, page 360). The cubic cell has $a = 12.137$ Å, contains four formula units and is of space group $P2_13$. The five-coordinate cation $[CuBr(Me_6tren)]^+$ has C_3 symmetry with the dimensions given in Table 9.44.

TABLE 9.44

Internuclear Dimensions in $[CuBr(Me_6tren)]^+$

Distance	Value (Å)	Angle	Value
Cu—Br(1)	2.393	N(1)—Cu—N(2)	84.7°
Cu—N(1)	2.07	N(2)—Cu—Br(1)	95.3°
Cu—N(2)	2.13	N(2)—Cu—N(2′)	119.1°

$Cu[(C_2H_5)_2NC_2H_4NHC_2H_4N(C_2H_5)_2]N_3Br$, Bromoazido-1,1,7,7-tetraethyldiethylenetriaminecopper(II), is a dark green crystalline solid having a triclinic unit cell[483] with $a = 12.95$, $b = 7.65$, $c = 9.87$ Å, $\alpha = 80.7°$, $\beta = 113.2°$ and $\gamma = 100.0°$. There are two molecules in the cell of space group $P\bar{1}$. The copper atom is in a trigonal-bipyramidal environment, with an axial azido group and an equatorial bromine atom.

The properties of a number of other adducts of copper dibromide are given in Table 9.45.

CuI_2(1,2-propanediamine)$_2$, prepared by shaking copper monoiodide with the amine in water containing free iodine[487], has a triclinic unit cell with $a = 7.60$, $b = 8.58$, $c = 6.35$ Å, $\alpha = 103.7°$, $\beta = 105.6°$ and $\gamma = 101.2°$. The copper atom has a *trans* octahedral environment and the room-temperature magnetic moment is 1.99 BM[440].

Addition of the calculated quantity of potassium iodide to a hot ethanolic solution of copper perchlorate and excess imidazole results[453] in the precipitation of violet CuI_2(imidazole)$_4$. The adduct has a magnetic moment of 1.81 BM and a $g_\perp$ value of 2.03 at room temperature[453,488] and its electronic spectrum is consistent with a tetragonally elongated octahedral configuration. This has been confirmed by a single-crystal X-ray diffraction study[488]. The unit cell, of space group *Pna*2, contains four formula units and has $a = 14.55$, $b = 9.48$ and $c = 13.41$ Å. The four imidazole ring nitrogen atoms form a square plane about the copper atom, the copper-nitrogen bond distances being 1.98, 2.00, 2.01 and 2.04 Å. The iodine atoms, in the axial positions, are only very weakly bound to the copper atom, as shown by the very long copper-iodine distances of 3.423 and 3.866 Å. ESR measurements at 80°K show hyperfine splitting due to the nitrogen atoms, indicating extensive delocalization of the unpaired electron onto the imidazole rings[488].

Experimental details for preparing single crystals of copper monochloride have been reported[489]. The melting point of the compound was given[489] as 422°. Nikitine and coworkers[490] have made a further study of the electronic spectrum of the monochloride.

TABLE 9.45
Adducts of the Type $CuBr_2L_n$

Compound and Properties	References
L = 2-chloropyridine, $n = 2$, $\mu = 1.79$ BM, $\nu_{Cu-N} = 234$ cm^{-1}, $\nu_{Cu-Br} = 256$ cm^{-1}, octahedral	448
L = 3-bromopyridine, $n = 2$, $\mu = 1.79$ BM, $\nu_{Cu-N} = 259$ cm^{-1}, $\nu_{Cu-Br} = 230$ cm^{-1}, octahedral	448
L = 4-chloropyridine, $n = 2$, $\mu = 1.84$ BM, $\nu_{Cu-N} = 240$ cm^{-1}, $\nu_{Cu-Br} = 234$ cm^{-1}, octahedral	448
L = benzene-1,2-diamine, $n = 1$, green, $\mu = 1.77$ BM, octahedral	451
L = bis[2-(2-pyridyl)ethyl]amine (dpea), $n = 1$, green, $\mu = 1.89$ BM, five coordinate	484
L = 1-(2-pyridyl)-2-(3-pyridyl)ethylene, $n = 1$, khaki, $\mu = 1.81$ BM, octahedral	452
L = imidazole, $n = 4$, blue-purple, $\mu = 1.84$ BM, octahedral	453
L = benzoxazole, $n = 1$, white, diamagnetic, $\nu_{Cu-N} = 234$ cm^{-1}, probably tetrahedral dimer	454
L = 4(5)-bromoimidazole, $n = 2$, $\mu = 1.82$ BM, $\nu_{Cu-N} = 275$ cm^{-1}, octahedral	455
L = pentamethylenetetrazole (PMT), $n = 1$, rust brown, m.p. = 190°, $\mu = 1.55$ BM, octahedral	456
L = 2-(2-pyridyl)benzothiazole (pbt), $n = 1$, brown-red, $\mu = 1.85$ BM, five or six coordinate	461
L = tris(2-methylthioethyl)amine (TSN), $n = 1$, dark green, $\mu = 1.79$ BM, ionic, five coordinate cation	485
L = 1,2,4-triazole, $n = 1$, $\mu = 1.84$ BM, $\nu_{Cu-N} = 315$ cm^{-1}, $\nu_{Cu-Br} = 243, 226, 206$ cm^{-1}, $\delta_{Cu-Br} = 115$ cm^{-1}, octahedral, isomorphous with corresponding chloride	448, 458, 466
L = benzothiazole, $n = 2$, green, $\mu = 1.63$ BM, $\nu_{Cu-Br} = 257, 246$ cm^{-1}, octahedral with bridging halogens	459
L = semicarbazide, $n = 2$, blue, $\mu = 1.81$ BM, $\nu_{Cu-L} = 209$ cm^{-1}, $\nu_{Cu-Br} = 106$ cm^{-1}, octahedral	448, 466, 467
L = semicarbazide, $n = 1$, olive green, $\mu = 1.53$ BM, $\nu_{Cu-L} = 197$ cm^{-1}, $\nu_{Cu-Br} = 290$ cm^{-1}, $\delta_{Cu-Br} = 172$ cm^{-1}, square planar	448, 466, 469
L = acetone semicarbazone, $n = 2$, green, $\mu = 1.81$ BM, $\nu_{Cu-L} = 282, 205$ cm^{-1}, $\nu_{Cu-Br} = 101$ cm^{-1}, *trans* octahedral	458, 466, 467
L = thiosemicarbazide, $n = 2$, $\nu_{Cu-L} = 286, 240$ cm^{-1}, $\nu_{Cu-Br} = 105$ cm^{-1}, *trans* octahedral	458, 466
L = thiosemicarbazide, $n = 1$, $\nu_{Cu-L} = 196$ cm^{-1}, $\nu_{Cu-Br} = 251, 184$ cm^{-1}, $\delta_{Cu-Br} = 147$ cm^{-1}, octahedral	458, 466
L = diphenylcyclopropenone (dpcp), $n = 2$, dark green, m.p. = 119–120°, $\mu = 2.13$ BM, $\nu_{Cu-Br} = 262$ cm^{-1}, *trans* square planar	468
L = 2-ethylthiocyclohexylamine, $n = 2$, blue-green, $\mu = 1.89$ BM, octahedral	486
L = 2-ethylthiocyclohexylamine, $n = 1$, olive green, $\mu = 1.81$ BM, octahedral	486

Direct interaction of thiourea and copper monochloride leads to the formation[491] of $CuCl[SC(NH_2)_2]_2$. This adduct has a monoclinic cell with $a = 35.81$, $b = 8.24$, $c = 5.81$ Å and $\beta = 95.2°$. There are eight molecules in the cell of space group $P2_1/a$. The structure consists of copper atoms surrounded by a triangular array of sulphur atoms, the triangles sharing vertices to give long chains which spiral in the *c* axis direction. Almost at right angles to the CuS_3 plane are the chlorine atoms. An interesting feature of this structure is, however, that the copper-chlorine distances alternate between 'long' (3.16 Å) and 'short' (2.83 Å). At the same time the copper-sulphur-copper bridge angles vary between 'broad' (138°) and 'sharp' (83°). This results in copper-copper distances of 2.98 and 4.31 Å. Spofford and Amma[491] have discussed the probable bonding models of this compound in detail.

Interaction of copper monochloride and 1,3,5-trithian in ethanolic hydrochloric acid results in the formation of 2:3 and 1:3 adducts depending on the experimental conditions[492]. The 2:3 complex has an orthorhombic unit cell[492] with $a = 6.7628$, $b = 7.3588$ and $c = 31.954$ Å. There are four formula units in the cell of space group $P2_12_12_1$. Each of the three crystallographically independent copper atoms is tetrahedrally bound to two sulphur and two chlorine atoms. The copper-sulphur and copper-chlorine bond distances are in the ranges 2.25–2.41 Å and 2.30–2.43 Å respectively. The ligand molecules are both in the chair conformation. The structure consists of two different types of copper-chlorine chains which are held together by the tridentate ligand.

Copper monochloride and bromide react with tetraethylcyclotetraphosphine $(EtP)_4$ to give[493] pale yellow adducts of the type $(EtP)_4CuX$. These complexes are probably dimeric with halogen bridges, the phosphine acting as a bidentate. Pentamethylcyclopentaphosphine $(MeP)_5$ reacts with copper monobromide to give $(MeP)_5CuBr$, but with the chloride the cyclotetraphosphine complex, $(MeP)_4CuCl$, is formed.

Both the monochloride and monobromide of copper react with allylamine[494] to give 1:1 and 1:3 compounds. These adducts decompose with the evolution of the amine at high temperatures. An infrared study[494] indicates that in the 1:1 complexes the ligand acts as a bidentate, being bound to the copper atom via the lone pair of electrons on the nitrogen atom of the amine *and* the ethylenic carbon-carbon bond. The infrared study revealed that in the 1:3 complexes the second and third ligand molecules are bound to the metal only by the amine nitrogen atom.

Bennett and coworkers[495] report that addition of an ethanolic solution of *o*-allylphenyldiphenylphosphine (AP) to copper monohalide (Cl, Br, I) in a saturated aqueous solution of potassium halide results in the formation of colourless complexes of the type CuX(AP). The melting points of the chloro, bromo and iodo compounds are 257–258°, 214–216° and 175–176° respectively. The compounds are dimeric and non-conducting in chloroform. Infrared spectra of the complexes show that the double bond of the allyl group acts as a coordination site and tetrahedrally coordinated dimers with the ligand as a bidentate and halogen bridges were proposed. Using the same experimental procedure with copper monoiodide and *o*-allylphenyldimethylarsine (AA), Bennett and coworkers[495] isolated colourless $CuI(AA)_2$ which melted at 98–99°. The compound is a non-conductor in nitrobenzene and is monomeric in chloroform. An infrared and ^1H NMR study indicated that only the arsenic atom of the ligand was coordinated to the copper atom, which is thus three coordinate.

Ethanolic solutions of copper dichloride and dibromide are reduced[496] by bis(diphenylphosphino)acetylene (DPPA) to give stable, colourless, diamagnetic complexes of the type 2CuX·3DPPA. Physico-chemical measurements suggest that the compounds are dimeric, the two copper atoms being bridged by the DPPA groups and having terminal halogen atoms. Thus the copper atoms have a tetrahedral stereochemistry. The copper-chlorine and copper-bromine stretching frequencies are 267 and 202 cm^{-1} respectively[496].

Copper monochloride complexes with substituted thioureas have been reported[497].

Reduction of copper dibromide in the presence of acetonitrile results in the formation[498] of $CuBr \cdot CH_3CN$. This adduct has an orthorhombic unit cell with $a = 4.09$, $b = 8.73$ and $c = 13.29$ Å.

An investigation[499] of the high pressure polymorphism of copper monoiodide reveals the presence of three high pressure polymorphs, h_1, h_2 and h_3, which are different from the three phases found at atmospheric pressure, α, β and γ. The high pressure transitions (at 22°) and the triple points are given in Table 9.46.

TABLE 9.46
Polymorphism of Copper Monoiodide

Transition		Triple Point		
$\gamma \rightarrow h_1$	3.15 kbar	h_1—h_2—h_3	216°	9.5 kbar
$h_1 \rightarrow h_2$	4.1 kbar	h_1—h_3—β	280°	7.2 kbar
$h_2 \rightarrow h_3$	14.5 kbar	γ—h_1—β	338°	2.5 kbar
		β—h_3—α	380°	1.8 kbar

Iodo-2-(2-pyridyl)benzothiazolecopper(I) may be prepared as a brown crystalline solid by mixing hot methanolic solutions of the constituents in the presence of excess lithium iodide[461]. A dimeric structure with iodine bridges was postulated for this complex.

Complex Halides

Some of the confusion surrounding the so-called pentachlorocuprates(II) (page 530) has been resolved by two more recent crystallographic studies[442,500]. One of these studies[500] shows that the copper atom has a regular trigonal-bipyramidal configuration, while the other[442] indicates the presence of distorted tetrahedral tetrachlorocuprate(II) groups and lattice chloride ions. Thus it would appear that the stereochemical arrangement adopted by the copper atom depends on the nature of the cation. As yet, there is insufficient crystallographic data available to be able to predict the structure of a particular complex.

Ibers and coworkers[500] report that hexaminechromium(III) pentachlorocuprate(II), $[Cr(NH_3)_6][CuCl_5]$, has a cubic unit cell with $a =$ 22.240 Å. There are thirty-two formula units in the cell of space group *Fd3c*. A very interesting feature of this structure is that the axial bond (2.296 Å) of the trigonal-bipyramidal anion is significantly *shorter* than the equatorial copper-chlorine bond (2.391 Å). The shortening of the axial bond is explained on the basis of the stereochemical activity of the 3*d* electrons[500].

Willett[442] reports that tris(dimethylammonium)monochloride tetrachlorocuprate(II), $[(CH_3)_2NH_2]_3ClCuCl_4$, has an orthorhombic unit cell with $a = 11.304$, $b = 15.638$ and $c = 9.957$ Å. The space group is *Pnma* and the cell contains four formula units. The distortion of the tetrahedral complex anion is pronounced, the chlorine-copper-chlorine angles being 98^0 ($\times 2$) and 136^0 ($\times 2$) although the copper-chlorine bond distances are all 2.230 Å.

Willett[442] has also supplied details of the structure determination of the high temperature form of $(CH_3)_3NHCu_2Cl_5$. At 50° the monoclinic unit cell has $a = 6.16$, $b = 10.30$, $c = 18.96$ Å and $\beta = 90.42°$. The structure consists of planar $Cu_4Cl_{10}^{2-}$ units linked together by long copper-chlorine interactions with adjacent units. The units are linked together in such a way that the terminal atoms have a square-pyramidal configuration whereas the two central copper atoms are in a distorted octahedral environment. Within the planar tetramer the copper-chlorine distances are in the range 2.169–2.410 Å while the chlorine-copper-chlorine angles lie in the range 83–100°.

The preparation of a number of tetrachlorocuprates(II) with organic cations has been reported[474,501,502]. Golden 4-phenyl-1,2-dithiolium tetrachlorocuprate(II)[501], $(C_9H_7S_2)CuCl_4$, has a room-temperature magnetic moment of 1.95 BM and has copper-chlorine stretching frequencies of 285, 272 and 235 cm^{-1}.

The infrared and Raman spectra of several tetrachlorocuprates(II) have been recorded by Avery and coworkers[503] and their results are given in Table 9.47.

TABLE 9.47
Infrared and Raman Spectra (cm^{-1}) of Tetrachlorocuprates(II)

Compound	Method	State of sample	ν_{Cu-Cl}	$\delta_{Cl-Cu-Cl}$
Cs_2CuCl_4	R	Solid	295, 250	134, 116
	IR	Solid	288, 255	
$[(CH_3)_4N]_2CuCl_4$	R	Solid	276, 232	133, 114
$[(C_2H_5)_4N]_2CuCl_4$	R	Solid	277	
	R	CH_3NO_2 solution	274	

TABLE 9.48
Unit-cell Parameters of Tetrabromocuprates(II)

Parameter	$(enH_2)CuBr_4$	$(enH\cdot HBr)_2CuBr_4$
Symmetry	Monoclinic	Monoclinic
a (Å)	8.22	6.78
b (Å)	7.71	20.15
c (Å)	7.42	6.33
β	92.12°	94.92°
Z	2	2
Space group	$P2_1/c$	$P2_1$ or $P2_1/m$

The magnetic moment of caesium tetrachlorocuprate(II) at room temperature is 1.92 BM[504] and the anisotropy is such that $\chi_c > \chi_a > \chi_b$. Spectral studies[505] indicate that there are planar and distorted tetrahedral forms of lithium tetrachlorocuprate(II) in glacial acetic acid.

The magnetic properties of potassium tetrachlorocuprate(II) dihydrate have been discussed from a theoretical point of view[506,507].

Several tetrabromocuprates(II) have been prepared and characterized[442,502,504,505,508]. Interaction[442] of ethylenediamine (en) and copper dibromide in the presence of hydrogen bromide results in the isolation of deep red plate-like crystals of $(enH_2)CuBr_4$ and dark red needles of a compound of empirical formula $(enH_2)_2CuBr_6$. X-ray analysis of the latter compound showed it to be bis(ethylenediammoniumbromide) tetrabromocuprate(II), $(NH_3CH_2CH_2NH_2 \cdot HBr)_2CuBr_4$. The crystallographic parameters of both complexes are given in Table 9.48.

$(enH_2)CuBr_4$ is isomorphous with the corresponding chloro compound and thus contains square-planar tetrabromocuprate(II) anions. Detailed analysis of $(enH \cdot HBr)_2CuBr_4$ indicates the presence of a distorted tetrahedral tetrabromocuprate(II) anion. The copper-bromine bond distances are in the range 2.34–2.45 Å while the bromine-copper-bromine angles are 99.1°, 100.9°, 117.4° and 140.9°.

Ammonium tetrabromocuprate(II) dihydrate is isomorphous with the corresponding chloro complex[509]. The vapour pressure of water above the complex is given by the following equations[509].

$$\log p_{atm} = 8.140 - 3187/T \qquad <95°$$
$$\log p_{atm} = 3.970 - 1650/T \qquad >95°$$

The heat and entropy of decomposition are 14.6 kcal mole^{-1} and 37.2 cal deg^{-1} mole^{-1} respectively[509]. ESR measurements at room temperature give[510] $g_{\parallel} = 2.04$ and $g_{\perp} = 2.196$ for this dihydrate.

REFERENCES

1. H. J. Emeleus and G. L. Hurst, *J. Chem. Soc.*, **1964**, 396.
2. H. M. Haendler, L. H. Towle, E. F. Bennett and W. L. Patterson, *J. Am. Chem. Soc.*, **76**, 2178 (1954).
3. J. M. Crabtree, C. S. Lees and K. Little, *J. Inorg. Nucl. Chem.*, **1**, 213 (1955).
4. R. L. Ritter and H. A. Smith, *J. Phys. Chem.*, **71**, 2036 (1967).
5. R. L. Ritter and H. A. Smith, *J. Phys. Chem.*, **70**, 805 (1966).
6. R. A. Kent, J. D. McDonald and J. L. Margrave, *J. Phys. Chem.*, **70**, 874 (1966).
7. C. Billy and H. M. Haendler, *J. Am. Chem. Soc.*, **79**, 1049 (1957).
8. P. H. Kasai, E. B. Whipple and W. Weltner, *J. Chem. Phys.*, **44**, 2581 (1966).
9. S. G. Salikhov, *Zh. Eksptl. i Theoret. Fiz.*, **34**, 39 (1958).
10. P. Henkel and W. Klemm, *Z. anorg. allgem. Chem.*, **222**, 74 (1935).
11. F. W. Lancaster and W. Gordy, *J. Chem. Phys.*, **19**, 1181 (1951).
12. R. J. Joenk and R. M. Bozorth, *J. Appl. Phys.*, **36**, 1167 (1965).
13. W. J. de Haas, B. H. Shultz and J. Koolhaus, *Physica*, **7**, 57 (1940).
14. E. Konig, Landolt-Bornstein, *Magnetic Properties of Coordination and Organometallic Transition Metal Compounds*, Vol. 2, Springer-Verlag, New York, 1966.
15. D. Oelkrug, *Z. Physik. Chem.* (*Frankfurt*), **56**, 325 (1967).
16. O. Schmitz-DuMont and D. Grimm, *Z. anorg. allgem. Chem.*, **355**, 280 (1967)

17. A. Kurtenacker, W. Finger and F. Hey, *Z. anorg. allgem. Chem.*, **211,** 83 (1933).
18. C. M. Wheeler and H. M. Haendler, *J. Am. Chem. Soc.*, **76,** 263 (1954).
19. R. G. Shulman and B. J. Wyluda, *J. Chem. Phys.*, **35,** 1498 (1961).
20. S. C. Abrahams and E. Prince, *J. Chem. Phys.*, **36,** 50 (1962).
21. R. M. Bozorth and J. W. Nielsen, *Phys. Rev.*, **110,** 879 (1958).
22. M. Le Van, G. Perinet and P. Bianco, *J. Chim. Phys.*, **63,** 719 (1966).
23. S. Geller and W. L. Bond, *J. Chem. Phys.*, **29,** 925 (1958).
24. S. C. Abrahams, *J. Chem. Phys.*, **36,** 56 (1962).
25. A. I. Stekhanov, E. A. Popova and Y. F. Markov, *Optika i Spektroskopiya*, **19,** 583 (1965).
26. P. Escoffier and J. Gauthier, *Compt. Rend.*, **252,** 271 (1961).
27. S. Tazawa, K. Nagata and M. Date, *J. Phys. Soc. Japan*, **20,** 181 (1965).
28. A. R. Pray, *Inorg. Syn.*, **5,** 153 (1957).
29. G. Brauer, *Handbook of Preparative Inorganic Chemistry*, Academic Press, New York, 1963.
30. J. H. Freeman and J. L. Smith, *J. Inorg. Nucl. Chem.*, **7,** 224 (1958).
31. J. C. Fanning and H. B. Jonassen, *J. Inorg. Nucl. Chem.*, **25,** 29 (1963).
32. D. Khristov, S. Karaivanov and V. Kolushki, *Godishnik. Sofiskyia Univ., Khim. Fak.*, **55,** 49 (1960–61).
33. G. W. Watt, P. S. Gentile and E. P. Helvenston, *J. Am. Chem. Soc.*, **77,** 2752 (1955).
34. H. D. Hardt, *Z. anorg. allgem. Chem.*, **301,** 87 (1959).
35. R. R. Hammer and N. W. Gregory, *J. Phys. Chem.*, **68,** 3229 (1964).
36. C. M. Fontana, E. Gorin, G. A. Kidder and C. S. Meridith, *Ind. Eng. Chem.*, **44,** 363 (1952).
37. E. M. Kosower and G. S. Wu, *J. Org. Chem.*, **28,** 633 (1963).
38. E. M. Kosower, W. J. Cole, G. S. Wu, D. E. Cardy and G. Meisters, *J. Org. Chem.*, **28,** 630 (1963).
39. D. C. Nonhebel, *J. Chem. Soc.*, **1963,** 1216.
40. J. C. Ware and E. E. Borchert, *J. Org. Chem.*, **26,** 2263 (1961).
41. G. E. Leroi, T. C. James, J. T. Hougen and W. Klemperer, *J. Chem. Phys.*, **36,** 2879 (1962).
42. S. A. Shchukarev and M. A. Oranskaya, *Zh. Obshchei Khim.*, **24,** 1926 (1954).
43. A. F. Wells, *J. Chem. Soc.*, **1947,** 1670.
44. D. M. Adams and P. J. Lock, *J. Chem. Soc.*, **1967,** A, 620.
45. D. M. Adams, M. Goldstein and E. F. Mooney, *Trans. Faraday Soc.*, **59,** 2228 (1963).
46. N. Perakis, A. Serres and T. Karantassis, *J. Phys. Radium*, **17,** 134 (1956).
47. C. G. Barraclough and C. F. Ng, *Trans. Faraday Soc.*, **60,** 836 (1964).
48. W. E. Hatfield and T. S. Piper, *Inorg. Chem.*, **3,** 841 (1964).
49. R. A. Howald and D. P. Keeton, *Spectrochim. Acta*, **22,** 1211 (1966).
50. J. T. Hougen, G. E. Leroi and T. C. James, *J. Chem. Phys.*, **34,** 1670 (1961).
51. C. W. DeKock and D. M. Gruen, *J. Chem. Phys.*, **44,** 4387 (1966).
52. J. Vymetal and P. Tvaruzek, *Z. anorg. allgem. Chem.*, **351,** 100 (1967).
53. R. Zannetti and R. Serra, *Gazz. Chim. Ital.*, **90,** 328 (1960).
54. J. D. Dunitz, *Acta Cryst.*, **10,** 307 (1957).

55. V. F. Duckworth, D. P. Graddon, N. C. Stephenson and E. C. Watton, *Inorg. Nucl. Chem. Letters*, **3**, 557 (1967).
56. N. S. Gill, R. H. Nuttall, D. E. Scaife and D. W. A. Sharp, *J. Inorg. Nucl. Chem.*, **18**, 79 (1961).
57. R. J. H. Clark and C. S. Williams, *Inorg. Chem.*, **4**, 350 (1965).
58. N. S. Gill, R. S. Nyholm, G. A. Barclay, T. I. Christie and P. J. Pauling, *J. Inorg. Nucl. Chem.*, **18**, 88 (1961).
59. R. J. H. Clark and C. S. Williams, *Chem. Ind.* (*London*), **1964**, 1317.
60. E. Konig and H. L. Schlaffer, *Z. Physik. Chem.* (*Frankfurt*), **26**, 371 (1960).
61. C. W. Frank and L. B. Rogers, *Inorg. Chem.*, **5**, 615 (1966).
62. M. Goldstein, E. F. Mooney, A. Anderson and H. A. Gebbie, *Spectrochim. Acta*, **21**, 105 (1965).
63. W. Libus and I. Uruska, *Inorg. Chem.*, **5**, 256 (1966).
64. J. R. Allan, D. H. Brown, R. H. Nuttall and D. W. A. Sharp, *J. Chem. Soc.*, **1966**, A, 1031.
65. W. R. McWhinnie, *J. Inorg. Nucl. Chem.*, **27**, 1063 (1965).
66. D. P. Graddon, R. Schulz, E. C. Watton and D. G. Weeden, *Nature*, **198**, 1299 (1963).
67. W. R. McWhinnie, *J. Chem. Soc.*, **1964**, 2959.
68. R. N. Patel and D. V. R. Rao, *Indian J. Chem.*, **4**, 314 (1966).
69. N. S. Gill and H. J. Kingdon, *Australian J. Chem.*, **19**, 2197 (1966).
70. M. Goodgame and P. J. Hayward, *J. Chem. Soc.*, **1966**, A, 632.
71. J. Vymetal, *Coll. Czech. Chem. Commun.*, **30**, 2134 (1965).
72. I. S. Ahuja, D. H. Brown, R. H. Nuttall and D. W. A. Sharp, *J. Inorg. Nucl. Chem.*, **27**, 1105 (1965).
73. W. R. McWhinnie, *J. Inorg. Nucl. Chem.*, **27**, 2573 (1965).
74. I. S. Ahuja, D. H. Brown, R. H. Nuttall and D. W. A. Sharp, *J. Inorg. Nucl. Chem.*, **27**, 1625 (1965).
75. D. H. Brown, R. H. Nuttall and D. W. A. Sharp, *J. Inorg. Nucl. Chem.*, **26**, 1151 (1964).
76. E. Hamburg, *Rev. Roumaine Chim.*, **10**, 779 (1965).
77. K. Emerson, American Crystallographic Society Meeting, Tuscon, 1968.
78. G. Morgan and F. H. Burstall, *J. Chem. Soc.*, **1937**, 1649.
79. D. E. C. Corbridge and E. G. Cox, *J. Chem. Soc.*, **1956**, 594.
80. R. Whyman and W. E. Hatfield, *Inorg. Chem.*, **6**, 1859 (1967).
81. M. R. Kidd, R. S. Sager and W. H. Watson, *Inorg. Chem.*, **6**, 946 (1967).
82. R. Whyman, D. B. Copley and H. M. Smith, *J. Am. Chem. Soc.*, **89**, 3135 (1967).
83. H. L. Schafer, J. C. Morrow and H. M. Smith, *J. Chem. Phys.*, **42**, 504 (1965).
84. R. S. Sager, R. J. Williams and W. H. Watson, *Inorg. Chem.*, **6**, 951 (1967).
85. C. M. Harris, E. Kokot, S. L. Lenzer and T. N. Lockyer, *Chem. Ind.* (*London*), **1962**, 651.
86. J. V. Quagliano, J. Fujita, G. Franz, D. J. Phillips, J. A. Walmsley and S. Y. Tyree, *J. Am. Chem. Soc.*, **83**, 3770 (1961).
87. S. Kida, J. V. Quagliano, J. A. Walmsley and S. Y. Tyree, *Spectrochim. Acta*, **19**, 189 (1963).
88. W. E. Hatfield and J. S. Paschal, *J. Am. Chem. Soc.*, **86**, 3888 (1964).
89. Y. Muto and H. B. Jonassen, *Bull. Chem. Soc. Japan*, **39**, 58 (1966).

90. W. E. Hatfield, Y. Muto, H. B. Jonassen and J. S. Paschal, *Inorg. Chem.*, **4,** 97 (1965).
91. W. E. Hatfield, D. B. Copley and R. Whyman, *Inorg. Nucl. Chem. Letters*, **2,** 373 (1966).
92. F. Hein and W. Beerstecher, *Z. anorg. allgem. Chem.*, **282,** 93 (1955).
93. F. Hein and W. Ludwig, *Z. anorg. allgem. Chem.*, **338,** 63 (1965).
94. E. Uhlig and K. Staiger, *Z. anorg. allgem. Chem.*, **346,** 21 (1966).
95. R. D. Willett and R. E. Rundle, *J. Chem. Phys.*, **40,** 838 (1964).
96. R. J. Kern, *J. Inorg. Nucl. Chem.*, **25,** 5 (1963).
97. F. Hanic and I. A. Cakajdova, *Acta Cryst.*, **11,** 610 (1958).
98. R. J. H. Clark and C. S. Williams, *J. Chem. Soc.*, **1966,** A, 1425.
99. J. R. Clifton and J. T. Yoke, *Inorg. Chem.*, **7,** 39 (1968).
100. L. Sacconi and A. Sabatini, *J. Inorg. Nucl. Chem.*, **25,** 1389 (1965).
101. I. de Paolini and C. Goria, *Gazz. Chim. Ital.*, **62,** 1048 (1932).
102. M. Inoue, M. Kishita and M. Kubo, *Inorg. Chem.*, **4,** 626 (1965).
103. E. Sanero, *Period. Miner. Roma*, **7,** 171 (1936).
104. J. A. J. Jarvis, *Acta Cryst.*, **15,** 964 (1962).
105. S. Prasad and P. D. Sharma, *J. Proc. Inst. Chemists (India)*, **30,** 254 (1958).
106. S. Prasad and P. D. Sharma, *J. Proc. Inst. Chemists (India)*, **30,** 249 (1958).
107. J. R. Clifton and J. T. Yoke, *Inorg. Chem.*, **6,** 1258 (1967).
108. V. V. Udovenko and M. V. Artemenko, *Russ. J. Inorg. Chem.*, **4,** 156 (1959).
109. E. J. O'Reiley and R. A. Plowman, *Australian J. Chem.*, **13,** 145 (1960).
110. J. R. Hall, N. K. Marchant and R. A. Plowman, *Australian J. Chem.*, **16,** 34 (1963).
111. J. R. Hall, N. K. Marchant and R. A. Plowman, *Australian J. Chem.*, **15,** 480 (1962).
112. H. S. Preston and C. H. L. Kennard, *Chem. Commun.*, **1967,** 1167.
113. G. J. Sutton, *Australian J. Chem.*, **16,** 371 (1963).
114. S. Utsuno and K. Sone, *J. Inorg. Nucl. Chem.*, **28,** 2647 (1966).
115. K. Issleib and A. Brack, *Z. anorg. allgem. Chem.*, **277,** 258 (1954).
116. D. M. L. Goodgame and F. A. Cotton, *J. Chem. Soc.*, **1961,** 2298.
117. G. A. Rodley, D. M. L. Goodgame and F. A. Cotton, *J. Chem. Soc.*, **1965,** 1499.
118. R. H. Pickard and J. Kenyon, *J. Chem. Soc.*, **89,** 262 (1906).
119. J. A. Bertrand, *Inorg. Chem.*, **6,** 495 (1967).
120. F. A. Cotton, R. Barnes and E. Bannister, *J. Chem. Soc.*, **1960,** 2199.
121. J. C. Sheldon and S. Y. Tyree, *J. Am. Chem. Soc.*, **80,** 4775 (1958).
122. T. F. Dorn, D. E. Campbell, G. R. Cochran, C. M. Hall, J. P. Scheller and J. J. Stuart, *J. Chem. Eng. Data*, **9,** 28 (1964).
123. E. Loyd, C. B. Brown, D. Glynwyn, R. Bonnell and W. J. Jones, *J. Chem. Soc.*, **1928,** 658.
124. J. J. P. Martin, *Compt. Rend.*, **261,** 3622 (1965).
125. G. Booth, *Advances in Inorganic Chemistry and Radiochemistry* (Ed. H. J. Emeleus and A. G. Sharpe), Vol. 6, Academic Press, London, 1964.
126. H. C. Freeman and J. E. W. L. Smith, *Acta Cryst.*, **20,** 153 (1966).
127. A. E. Underhill, *J. Chem. Soc.*, **1965,** 4337.
128. D. E. Billing, A. E. Underhill, D. M. Adams and D. M. Morris, *J. Chem. Soc.*, **1966,** A, 902.
129. J. C. Fanning and L. T. Taylor, *J. Inorg. Nucl. Chem.*, **27,** 2217 (1965).

130. P. S. K. Chia, S. E. Livingstone and T. N. Lockyer, *Australian J. Chem.*, **20,** 239 (1967).
131. L. F. Lindoy, S. E. Livingstone and T. N. Lockyer, *Australian J. Chem.*, **20,** 471 (1967).
132. L. F. Lindoy, S. E. Livingstone and T. N. Lockyer, *Australian J. Chem.*, **19,** 1391 (1966).
133. P. S. K. Chia and S. E. Livingstone, *Australian J. Chem.*, **21,** 339 (1968).
134. P. S. K. Chia, S. E. Livingstone and T. N. Lockyer, *Australian J. Chem.*, **19,** 1835 (1966).
135. I. Bertini and F. Mani, *Inorg. Chem.*, **6,** 2032 (1967).
136. M. Ciampolini and G. P. Speroni, *Inorg. Chem.*, **5,** 45 (1966).
137. M. Ciampolini and N. Nardi, *Inorg. Chem.*, **6,** 445 (1967).
138. C. M. Harris, T. N. Lockyer and H. Waterman, *Nature*, **192,** 424 (1961).
139. H. Elliot, B. J. Hathaway and R. C. Slade, *J. Chem. Soc.*, **1966,** A, 1443.
140. R. J. Berni, R. R. Benerito, W. M. Ayres and H. B. Jonassen, *J. Inorg. Nucl. Chem.*, **25,** 807 (1963).
141. J. F. Geldard and F. Lions, *Inorg. Chem.*, **4,** 414 (1965).
142. J. W. Bourknight and G. McP. Smith, *J. Am. Chem. Soc.*, **61,** 28 (1939).
143. S. Herzog, K. Gustav, E. Kruger, H. Oberenda and R. Gustav, *Z. Chem.*, **3,** 428 (1963).
144. R. Chand, G. S. Handard and K. Lal, *J. Indian Chem. Soc.*, **35,** 28 (1958).
145. E. E. Weaver and W. Keim, *Proc. Indian Acad. Sci.*, **70,** 123 (1960).
146. H. Rheinboldt, A. Luyken and H. Schmittmann, *J. Prakt. Chem.*, **149,** 30 (1937).
147. T. F. Dorn, G. L. Smith and J. C. McKenna, *J. Chem. Eng. Data*, **10,** 195 (1965).
148. C. Reimann and G. Gordon, *Nature*, **205,** 902 (1965).
149. M. V. Artmenko and K. F. Slyusarenko, *Ukr. Khim. Zh.*, **32,** 136 (1966).
150. F. A. Cotton and R. Francis, *J. Am. Chem. Soc.*, **82,** 2986 (1960).
151. S. K. Madan and H. H. Denk, *J. Inorg. Nucl. Chem.*, **27,** 1049 (1965).
152. R. S. Drago and D. Meek, *J. Phys. Chem.*, **65,** 1446 (1961).
153. F. A. Cotton, R. Francis and W. D. Horrocks, *J. Phys. Chem.*, **64,** 1534 (1960).
154. D. W. Meek, D. K. Staub and R. S. Drago, *J. Am. Chem. Soc.*, **82,** 6013 (1960).
155. W. R. McWhinnie, *J. Inorg. Nucl. Chem.*, **27,** 1619 (1965).
156. S. Kirschner, *Inorg. Syn.*, **5,** 14 (1957).
157. I. S. Ahuja, *J. Inorg. Nucl. Chem.*, **29,** 2091 (1967).
158. W. R. McWhinnie, *J. Chem. Soc.*, **1964,** 5165.
159. J. C. Fanning and H. B. Jonassen, *Chem. Ind.* (*London*), **1961,** 1623.
160. M. J. Campbell and R. Grzeskowiak, *J. Chem. Soc.*, **1967,** A, 396.
161. M. Bierley and W. J. Geary, *J. Chem. Soc.*, **1967,** A, 963.
162. W. J. Eilbeck, F. Holmes and A. E. Underhill, *J. Chem. Soc.*, **1967,** A, 757.
163. P. J. Hendra and D. B. Powell, *J. Chem. Soc.*, **1960,** 5105.
164. R. A. Walton, *Inorg. Chem.*, **5,** 643 (1966).
165. C. M. Harris, H. R. H. Patil and E. Sinn, *Inorg. Chem.*, **6,** 1102 (1967).
166. J. D. Curry, M. A. Robinson and D. H. Busch, *Inorg. Chem.*, **6,** 1570 (1967).
167. B. Bosnich, R. D. Gillard, E. D. McKenzie and G. A. Webb, *J. Chem. Soc.*, **1966,** A, 1331.

168. E. Uhlig and G. Heinrich, *Z. anorg. allgem. Chem.*, **330,** 40 (1964).
169. J. R. Hall, M. R. Litzow and R. A. Plowman, *Australian J. Chem.*, **18,** 1331 (1965).
170. M. Goodgame and L. I. B. Haines, *J. Chem. Soc.*, **1966,** A, 174.
171. R. W. Green and M. C. K. Svasti, *Australian J. Chem.*, **16,** 356 (1963).
172. S. Utsuno and K. Sone, *Bull. Chem. Soc. Japan*, **37,** 1038 (1964).
173. G. J. Sutton, *Australian J. Chem.*, **13,** 222 (1960).
174. E. Uhlig and M. Maaser, *Z. anorg. allgem. Chem.*, **322,** 25 (1963).
175. E. Uhlig, P. Schueler and D. Diehlman, *Z. anorg. allgem. Chem.*, **335,** 156 (1965).
176. R. K. Y. Ho and S. E. Livingstone, *Australian J. Chem.*, **18,** 659 (1965).
177. E. Uhlig and H. Schon, *Z. anorg. allgem. Chem.*, **316,** 25 (1962).
178. J. F. Geldard, *Inorg. Chem.*, **4,** 417 (1965).
179. W. J. Eilbeck, F. Holmes, C. E. Taylor and A. E. Underhill, *J. Chem. Soc.*, **1968,** A, 128.
180. A. Doadrio and A. G. Carro, *Anales Real Soc. Espan. Fis. Quim.* (*Madrid*), *Ser. B*, **62,** 329 (1966).
181. R. Perret, *Bull. Soc. Chim. France*, **1966,** 755.
182. D. Harker, *Z. Krist.*, **93,** 136 (1936).
183. J. van den Handel, H. M. Gijsman and N. J. Poulis, *Physica*, **18,** 862 (1952).
184. L. C. van der Marel, J. van den Brock, J. D. Wasscher and C. J. Gorter, *Physica*, **21,** 685 (1955).
185. G. E. G. Hardeman and N. J. Poulis, *Arch. Sci.* (*Geneva*), **9,** 173 (1956).
186. W. Marshall, *J. Phys. Chem. Solids*, **7,** 159 (1958).
187. H. Yamazaki and M. Date, *J. Phys. Soc. Japan*, **21,** 1462 (1966).
188. N. J. Poulis and G. E. G. Hardeman, *Physica*, **18,** 201 (1952).
189. N. J. Poulis and G. E. G. Hardeman, *Physica*, **18,** 429 (1952).
190. G. Shirane, B. C. Frazer and S. A. Friedberg, *Phys. Letters*, **17,** 95 (1965).
191. W. J. O'Sullivan, W. W. Simmons and W. A. Robinson, *Phys. Rev.*, **140A,** 1759 (1965).
192. S. A. Friedberg, *Physica*, **18,** 714 (1952).
193. Y. Ting and D. Williams, *Phys. Rev.*, **82,** 507 (1951).
194. S. M. Rjabchenko and L. A. Shulman, *Soviet Phys.-Solid State*, **8,** 1757 (1967).
195. J. Itoh, M. Fujimoto and H. Ibamoto, *Phys. Rev.*, **83,** 852 (1951).
196. H. J. Geritsen, R. Okkers, B. Bolger and C. J. Gorter, *Physica*, **21,** 629 (1955).
197. R. E. Rundle, K. Nakamoto and J. W. Richardson, *J. Chem. Phys.*, **23,** 2450 (1955).
198. S. W. Peterson and H. A. Levy, *J. Chem. Phys.*, **26,** 220 (1957).
199. R. E. Rundle, *J. Am. Chem. Soc.*, **79,** 3372 (1957).
200. A. C. Hewson, D. Ter Haar and M. E. Lines, *Phys. Rev.*, **137,** 1465 (1965).
201. P. A. van Dalen and P. van der Leeden, *Physica*, **37,** 329 (1967).
202. P. A. van Dalen, *Physica*, **37,** 349 (1967).
203. P. A. van Dalen, *Physica*, **37,** 361 (1967).
204. R. R. Hammer and N. W. Gregory, *J. Phys. Chem.*, **68,** 314 (1964).
205. P. M. Druce, M. F. Lappert and P. N. K. Riley, *Chem. Commun.*, **1967,** 486.
206. K. B. Doifode, *J. Org. Chem.*, **27,** 2665 (1962).

207. E. R. Glazier, *J. Org. Chem.*, **27**, 4397 (1962).
208. R. W. Jemison, *Australian J. Chem.*, **21**, 217 (1968).
209. P. Barrett and N. Guenbaut-Thevenot, *Compt. Rend.*, **242**, 119 (1956).
210. Le-Van-My, G. Perinet and P. Bianco, *Bull. Soc. Chim. France*, **1965**, 3651.
211. P. Barret and P. Perret, *Bull. Soc. Chim. France*, **1957**, 1459.
212. P. Barret and N. Guenbaut-Thevenot, *Bull. Soc. Chim. France*, **1957**, 409.
213. L. Helmoltz, *J. Am. Chem. Soc.*, **69**, 886 (1947).
214. V. Kupcik and S. Durovic, *Czech. J. Phys.*, **10**, 182 (1960).
215. T. I. Torgonskaya and Y. G. Zonov, *Geterogennye Khim. Reaktsii, Inst. Obshch. i Heorgan. Khim. Akad. Nauk Belorussk. SSR*, **1965**, 186.
216. F. Hanic, *Acta Cryst.*, **12**, 739 (1959).
217. C. M. Harris and E. D. McKenzie, *Nature*, **196**, 670 (1962).
218. B. F. Hoskins and F. D. Whillans, *Chem. Commun.*, **1966**, 798.
219. K. Issleib and A. Krebich, *Z. anorg. allgem. Chem.*, **313**, 338 (1961).
220. P. Paoletti and M. Ciampolini, *Inorg. Chem.*, **6**, 64 (1967).
221. M. Ciampolini and N. Nardi, *Inorg. Chem.*, **5**, 41 (1966).
222. S. K. Dhar and F. Basolo, *J. Inorg. Nucl. Chem.*, **25**, 37 (1963).
223. R. Driver and W. R. Walker, *Australian J. Chem.*, **21**, 331 (1968).
224. M. Date, *Phys. Rev.*, **104**, 623 (1956).
225. H. W. Foote and M. Fleischer, *J. Phys. Chem.*, **44**, 647 (1940).
226. N. V. Sidgwick, *The Chemical Elements and Their Compounds*, O.U.P., 1950, p. 153.
227. G. A. Barclay, B. F. Hoskins and C. H. L. Kennard, *J. Chem. Soc.*, **1963**, 5691.
228. G. A. Barclay and C. H. L. Kennard, *Nature*, **192**, 425 (1961).
229. J. W. Mellor, *A Comprehensive Treatise on Inorganic and Theoretical Chemistry*, Vol. 3, Longmans, London, 1923, p. 154.
230. F. Ebert and H. Woitinek, *Z. anorg. allgem. Chem.*, **210**, 269 (1933).
231. T. C. Waddington, *Trans. Faraday Soc.*, **55**, 1531 (1959).
232. J. Jindra and B. Perner, *Chem. Prumysl.*, **15**, 560 (1965).
233. T. Ogura, N. Furano and S. Kawaguchi, *Bull. Chem. Soc., Japan*, **40**, 1171 (1967).
234. R. A. J. Shelton, *Trans. Faraday Soc.*, **57**, 2113 (1961).
235. R. N. Keller and H. D. Wycoff, *Inorg. Syn.*, **2**, 1 (1946).
236. E. C. Stathis, *Chem. Ind. (London)*, **1958**, 633.
237. L. Brewer and N. L. Lofgren, *J. Am. Chem. Soc.*, **72**, 3038 (1950).
238. J. B. Wagner and C. Wagner, *J. Chem. Phys.*, **26**, 1597 (1957).
239. H. M. Rosenstock, J. R. Sites, J. R. Walton and R. Baldock, *J. Chem. Phys.*, **23**, 2442 (1955).
240. C. H. Wong and V. Schomaker, *J. Phys. Chem.*, **61**, 358 (1957).
241. C. G. Maier, *Bur. of Mines, Tech. Paper*, **1925**, 360.
242. M. R. Lorenz and J. S. Prener, *Acta Cryst.*, **9**, 538 (1956).
243. L. Vegard and G. Skofteland, *Arch. Mat. Naturvidenskab.*, **45**, 163 (1942).
244. S. B. Badachhape and A. Goswami, *J. Phys. Soc. Japan*, **17**, 249S (1962).
245. W. Klemperer, S. A. Rice and S. R. Berry, *J. Am. Chem. Soc.*, **79**, 1810 (1957).
246. J. Ringeissen, S. Lewonczuk, A. Coret and S. Nikitine, *Phys. Letters*, **22**, 571 (1966).
247. R. Reiss and S. Nikitine, *Compt. Rend.*, **250**, 2862 (1960).

248. T. L. Davis and P. Ehrlich, *J. Am. Chem. Soc.*, **58,** 2151 (1936).
249. H. J. Worth and H. H. Haendler, *J. Am. Chem. Soc.*, **64,** 1232 (1942).
250. J. T. Yoke, J. F. Weiss and G. Tollin, *Inorg. Chem.*, **2,** 1210 (1963).
251. C. I. Branden, *Acta Chem. Scand.*, **21,** 1000 (1967).
252. I. D. Brown and J. D. Dunitz, *Acta Cryst.*, **13,** 28 (1960).
253. O. Diels and W. Koll, *Ann. Chim. (Paris)*, **443,** 262 (1925).
254. A. B. P. Lever, J. Lewis and R. S. Nyholm, *J. Chem. Soc.*, **1963,** 3156.
255. A. B. P. Lever, J. Lewis and R. S. Nyholm, *Nature*, **189,** 58 (1961).
256. A. U. Malik, *Z. anorg. allgem. Chem.*, **344,** 107 (1966).
257. R. E. Yingst and B. E. Douglas, *Inorg. Chem.*, **3,** 1177 (1964).
258. G. J. Burrows and E. P. Sanford, *J. Proc. Roy. Soc. N.S.W.*, **69,** 182 (1936).
259. D. P. Mellor and D. P. Craig, *J. Proc. Roy. Soc. N.S.W.*, **75,** 27 (1941).
260. L. Cavalca, M. Nardelli and A. Braibanti, *Gazz. Chim. Ital.*, **87,** 146 (1957).
261. K. Cohn and R. W. Parry, *Inorg. Chem.*, **7,** 46 (1968).
262. E. Uhlig and M. Maaser, *Z. anorg. allgem. Chem.*, **344,** 205 (1964).
263. R. G. Wilkins and A. R. Burkin, *J. Chem. Soc.*, **1950,** 127.
264. R. G. Wilkins and A. R. Burkin, *J. Chem. Soc.*, **1950,** 132.
265. J. R. Hall, M. R. Litzow and R. A. Plowman, *Australian J. Chem.*, **18,** 1339 (1965).
266. D. E. Billing and A. E. Underhill, *J. Chem. Soc.*, **1965,** 6639.
267. A. E. Arzubov and V. M. Zorvastrova, *Izv. Akad. Nauk SSSR, Otdel. Khim. Nauk*, **1952,** 809.
268. W. Reppe and W. J. Schweckendiek, *Ann. Chim. (Paris)*, **560,** 104 (1948).
269. J. W. Collier, A. R. Fox, I. G. Hinton and F. G. Mann, *J. Chem. Soc.*, **1964,** 1819.
270. D. G. Hicks and J. A. Dean, *Chem. Commun.*, **1965,** 172.
271. A. Forster, C. S. Cundy, M. Green and F. G. A. Stone, *Inorg. Nucl. Chem. Letters*, **2,** 233 (1966).
272. J. A. W. Dalziel, A. F. Holding and B. E. Watts, *J. Chem. Soc.*, **1967** A, 358.
273. J. H. Hu and H. L. Johnston, *J. Am. Chem. Soc.*, **74,** 4771 (1952).
274. D. G. Peters and R. L. Caldwell, *Inorg. Chem.*, **6,** 1478 (1967).
275. D. B. Briggs, *J. Chem. Soc.*, **127,** 496 (1925).
276. A. G. Galinos and I. K. Kontoyiannakos, *Angew. Chem.*, **70,** 51 (1958).
277. S. Miyake and D. Hoshino, *Rev. Modern Phys.*, **30,** 172 (1958).
278. H. G. F. Winkler, *Z. anorg. allgem. Chem.*, **276,** 169 (1954).
279. J. Krug and L. Sieg, *Z. Naturforsch.*, **7a,** 369 (1952).
280. S. Hoshino, *J. Phys. Soc. Japan*, **7,** 560 (1952).
281. R. N. Kurdyumova and S. A. Semiletov, *Soviet Phys.-Cryst.*, **10,** 529 (1966).
282. S. Prasad and S. R. C. Trivedi, *J. Indian Chem. Soc.*, **42,** 623 (1966).
283. K. Issleib and E. Wenschuh, *Z. anorg. allgem. Chem.*, **305,** 15 (1960).
284. K. Issleib and G. Hohlfeld, *Z. anorg. allgem. Chem.*, **312,** 169 (1961).
285. W. Seidel, *Z. anorg. allgem. Chem.*, **341,** 70 (1965).
286. G. B. Kauffman and R. P. Pinnell, *Inorg. Syn.*, **6,** 3 (1960).
287. M. Chaigneau and M. Chastagnier, *Bull. Soc. Chim. France*, **1958,** 1192.
288. H. D. Hardt and R. Bollig, *Angew. Chem.*, **77,** 860 (1965).
289. W. Jost, H. J. Oel and G. Schneidermann, *Z. Physik. Chem. (Frankfurt)*, **17,** 175 (1958).
290. S. Miyake, S. Hoshino and T. Takenaka, *J. Phys. Soc. Japan*, **7,** 19 (1952).

291. V. V. Pechkovskii and A. V. Sofronova, *Russ. J. Inorg. Chem.*, **10,** 825 (1965).
292. R. W. G. Wychoff and E. Posnjak, *J. Am. Chem. Soc.*, **44,** 30 (1922).
293. R. N. Kurdyamova and R. V. Baranova, *Kristallografiya*, **6,** 402 (1961).
294. G. A. Barclay and R. S. Nyholm, *Chem. Ind. (London)*, **1953,** 378.
295. R. C. Cass, G. E. Coates and R. G. Hayter, *J. Chem. Soc.*, **1955,** 4007.
296. F. G. Mann, D. Purdie and A. F. Wells, *J. Chem. Soc.*, **1936,** 1503.
297. A. U. Malik, *J. Inorg. Nucl. Chem.*, **29,** 2106 (1967).
298. M. G. King and G. P. McQuillan, *J. Chem. Soc.*, **1967,** A, 898.
299. W. Klemm and E. Huss, *Z. anorg. allgem. Chem.*, **258,** 221 (1949).
300. A. G. Sharpe, *Advances in Fluorine Chemistry*, Vol. 1, Butterworths, London, 1960.
301. H. G. Schnering, *Z. anorg. allgem. Chem.*, **353,** 1 (1967).
302. H. G. Schnering, *Z. anorg. allgem. Chem.*, **353,** 13 (1967).
303. W. Rudorff, G. Lincke and D. Babel, *Z. anorg. allgem. Chem.*, **320,** 150 (1963).
304. D. S. Crocket and R. A. Grossman, *Inorg. Chem.*, **3,** 644 (1964).
305. W. G. Bottjer and H. M. Haendler, *Inorg. Chem.*, **4,** 913 (1965).
306. D. Babel, *Z. anorg. allgem. Chem.*, **336,** 200 (1965).
307. W. Rudorff and D. Babel, *Naturwissenschaften*, **49,** 230 (1962).
308. K. Knox, *J. Chem. Phys.*, **30,** 991 (1959).
309. A. D. Liehr and C. J. Ballhausen, *Ann. Phys. (N.Y.)*, **3,** 304 (1958).
310. J. Lecomte, C. Duval and C. Wadier, *Compt. Rend.*, **249,** 1991 (1959).
311. A. J. Edwards and R. D. Peacock, *J. Chem. Soc.*, **1959,** 4126.
312. D. J. Machin, R. L. Martin and R. S. Nyholm, *J. Chem. Soc.*, **1963,** 1490.
313. K. Hirakawa, K. Hirakawa and T. Hashimoto, *J. Phys. Soc. Japan*, **15,** 2063 (1960).
314. S. Kadota, I. Yamada, S. Yoneyama and K. Hirakawa, *J. Phys. Soc. Japan*, **23,** 751 (1967).
315. H. M. Haendler, F. A. Johnson and D. S. Crocket, *J. Am. Chem. Soc.*, **80,** 2662 (1958).
316. D. S. Crocket and H. M. Haendler, *J. Am. Chem. Soc.*, **82,** 4158 (1960).
317. K. Knox, *Acta Cryst.*, **14,** 583 (1961).
318. A. Okazaki, Y. Suemune and T. Fuchikami, *J. Phys. Soc. Japan*, **14,** 1823 (1959).
319. R. Hoppe, *Angew. Chem.*, **71,** 457 (1959).
320. A. Okazaki and Y. Suemune, *J. Phys. Soc. Japan*, **16,** 671 (1961).
321. A. Okazaki and Y. Suemune, *J. Phys. Soc. Japan*, **16,** 176 (1961).
322. D. Babel, *Z. Naturforsch.*, **20a,** 165, (1965).
323. V. Scatturin, L. Corliss, N. Elliot and J. Hastings, *Acta Cryst.*, **14,** 19 (1961).
324. K. Hirakawa and S. Kadota, *J. Phys. Soc. Japan*, **23,** 756 (1967).
325. R. D. Peacock and D. W. A. Sharp, *J. Chem. Soc.*, **1959,** 2762.
326. M. Mori, Y. Saito and T. Watanabe, *Bull. Chem. Soc. Japan*, **34,** 295 (1961).
327. M. Mori, *Bull. Chem. Soc. Japan*, **34,** 454 (1961).
328. M. Mori and S. Fujiwara, *Bull. Chem. Soc. Japan*, **36,** 1636 (1963).
329. M. Mori, *Bull. Chem. Soc. Japan*, **34,** 1249 (1961).
330. M. Mori, *Bull. Chem. Soc. Japan*, **33,** 985 (1960).
331. G. C. Allen and N. S. Hush, *Inorg. Chem.*, **6,** 4 (1967).

332. H. B. Jonassen, T. B. Crumpler and T. D. O'Brien, *J. Am. Chem. Soc.*, **67,** 1709 (1945).
333. W. E. Hatfield, H. D. Bedon and S. M. Horner, *Inorg. Chem.*, **4,** 1181 (1965).
334. P. Day, *Proc. Chem. Soc.*, **1964,** 18.
335. C. Furlani, A. Sgamellotti, F. Magrini and D. Cordischi, *J. Mol. Spectroscopy*, **24,** 270 (1967).
336. B. Zaslow and G. L. Ferguson, *Chem. Commun.*, **1967,** 822.
337. N. V. Fredorenko and T. L. Ivanova, *Russ. J. Inorg. Chem.*, **11,** 414 (1966).
338. L. A. Kazitsyna, N. B. Kapletskaya, V. A. Ptitsyna and O. A. Reutov, *J. Gen. Chem. USSR*, **33,** 3170 (1967).
339. R. D. Whealy, D. H. Bier and B. J. McCormick, *J. Am. Chem. Soc.*, **81,** 5900 (1959).
340. R. D. Willett, *J. Chem. Phys.*, **41,** 2243 (1964).
341. B. Morosin and E. C. Lingafelter, *J. Phys. Chem.*, **65,** 50 (1961).
342. R. Filler and L. Gorelic, *J. Inorg. Nucl. Chem.*, **24,** 1297 (1962).
343. L. A. Il'yukevich and G. A. Shagisultanova, *Russ. J. Inorg. Chem.*, **8,** 1209 (1963).
344. N. S. Gill and R. S. Nyholm, *J. Chem. Soc.*, **1959,** 3997.
345. L. A. Kazitsyna, O. A. Reutov and B. S. Kikot, *J. Gen. Chem. USSR*, **31,** 2751 (1961).
346. L. A. Kazitsyna, O. A. Reutov and Z. F. Buchkovskii, *J. Gen. Chem. USSR*, **31,** 2745 (1961).
347. M. J. Joesten and K. M. Nykerk, *Inorg. Chem.*, **3,** 548 (1964).
348. C. Furlani and G. Morpurgo, *Theoret. Chim. Acta*, **1,** 102 (1963).
349. P. Pauling, *Inorg. Chem.*, **5,** 1498 (1966).
350. H. J. Seifert and K. Klatyk, *Z. anorg. allgem. Chem.*, **334,** 113 (1964).
351. L. Helmholtz and R. F. Kruh, *J. Am. Chem. Soc.*, **74,** 1176 (1952).
352. J. Ferguson, *J. Chem. Phys.*, **40,** 3406 (1964).
353. D. M. Gruen and C. A. Angell, *Inorg. Nucl. Chem. Letters*, **2,** 75 (1966).
354. S. A. Shchukarev, I. V. Vasil'kova and G. M. Barvinok, *Vestn. Leningrad Univ.*, **20,** *Ser. Fiz. i Khim.*, 145 (1965).
355. P. Paoletti and A. Vacca, *Trans. Faraday Soc.*, **60,** 50 (1964).
356. R. D. Willett, O. L. Liles and C. Michelson, *Inorg. Chem.*, **6,** 1885 (1967).
357. D. P. Mellor, *Z. Krist.*, **101,** 160 (1939).
358. M. Bonamico, G. Dessy and A. Vaciago, *Theoret. Chim. Acta*, **7,** 367 (1967).
359. A. Sabatini and L. Sacconi, *J. Am. Chem. Soc.*, **86,** 17 (1964).
360. D. Forster, *Chem. Commun.*, **1967,** 113.
361. R. J. H. Clark and T. M. Dunn, *J. Chem. Soc.*, **1963,** 1198.
362. M. Sharnoff and C. W. Reimann, *J. Chem. Phys.*, **46,** 2634 (1967).
363. B. D. Bird and P. Day, *Chem. Commun.*, **1967,** 741.
364. B. Morosin and K. Lawson, *J. Mol. Spectroscopy*, **12,** 98 (1964).
365. B. Morosin and K. Lawson, *J. Mol. Spectroscopy*, **14,** 397 (1964).
366. P. Ros and G. C. A. Schuit, *Theoret. Chim. Acta*, **4,** 1 (1966).
367. A. van der Avoird and P. Ros, *Theoret. Chim. Acta*, **4,** 13 (1966).
368. C. Furlani, E. Cervone, F. Calzona and B. Baldanza, *Theoret. Chim. Acta*, **7,** 375 (1967).
369. B. N. Figgis and C. M. Harris, *J. Chem. Soc.*, **1959,** 855.
370. L. Sacconi, M. Ciampolini and U. Campighi, *Inorg. Chem.*, **4,** 407 (1965).
371. M. Sharnoff, *J. Chem. Phys.*, **41,** 2203 (1964).

372. M. Sharnoff, *J. Chem. Phys.*, **42,** 3383 (1965).
373. M. Sharnoff and C. W. Reimann, *J. Chem. Phys.*, **43,** 2993 (1965).
374. A. Bose, S. Lahiry and U. S. Ghosh, *J. Phys. Chem. Solids*, **26,** 1747 (1965).
375. S. Mitra, *Indian J. Pure Appl. Phys.*, **2,** 333 (1964).
376. S. B. Hendricks and R. G. Dickinson, *J. Am. Chem. Soc.*, **49,** 2149 (1927).
377. H. Suga, M. Sorai, T. Yamanaka and S. Seki, *Bull. Chem. Soc. Japan*, **38,** 1007 (1965).
378. A. Chretien and R. Weil, *Bull. Soc. Chim. France*, **2,** 1577 (1935).
379. R. Perret, *Bull. Soc. Chim. France*, **1966,** 769.
380. A. Silberstein, *Compt. Rend.*, **202,** 1196 (1936).
381. A. Narasimhurty and D. Premaswarup, *Proc. Phys. Soc.*, **83,** 199 (1964).
382. A. R. Miedema, H. van Kempen and W. J. Huiskamp, *Physica*, **29,** 1266 (1963).
383. H. Abe, H. Moriyaki and K. Koga, *Phys. Rev. Letters*, **9,** 338 (1962).
384. A. R. Miedema, R. F. Wielinga and W. J. Huiskamp, *Physica*, **31,** 1585 (1965).
385. H. B. Jonassen, L. J. Theriot, E. A. Bourdreaux and W. A. Ayres, *J. Inorg. Nucl. Chem.*, **26,** 595 (1964).
386. R. D. Willett, C. Dwiggins, R. F. Kruh and R. E. Rundle, *J. Chem. Phys.*, **38,** 2429 (1963).
387. J. Amiel, *Compt. Rend.*, **205,** 1400 (1937).
388. G. A. Shagisultanova, L. A. Il'yukevich and L. I. Burdyko, *Russ. J. Inorg. Chem.*, **10,** 229 (1965).
389. V. V. Udovenko and M. A. Sherstoboeva, *Ukr. Khim. Zh.*, **32,** 584 (1966).
390. R. D. Willett, *J. Chem. Phys.*, **44,** 39 (1966).
391. A. W. Schuleter, R. A. Jacobson and R. E. Rundle, *Inorg. Chem.*, **5,** 277 (1966).
392. A. F. Wells, *J. Chem. Soc.*, **1947,** 1662.
393. H. P. Klug and G. W. Sears, *J. Am. Chem. Soc.*, **68,** 1133 (1946).
394. M. S. Golubeva and R. D. Sidakova, *Fiz. Khim. Analiz. Solevykh. Sistem. Sb.*, **1962,** 189.
395. G. J. Maass, B. C. Gerstein and R. D. Willett, *J. Chem. Phys.*, **46,** 401 (1967).
396. G. M. Barvinok and V. Vintruff, *Vestn. Leningrad Univ.*, **21,** *Ser. Fiz. i Khim.*, 111 (1966).
397. R. D. Willett and O. L. Liles, *Inorg. Chem.*, **6,** 1666 (1967).
398. V. P. Blindin and V. I. Gordienko, *Dokl. Akad. Nauk. SSSR*, **94,** 1081 (1954).
399. P. H. Vossos, D. R. Fitzwater and R. E. Rundle, *Acta Cryst.*, **16,** 1037 (1963).
400. P. H. Vossos, L. D. Jennings and R. E. Rundle, *J. Chem. Phys.*, **32,** 1590 (1960).
401. R. D. Spence, H. Forstat, C. R. K. Murty and D. R. McNeely, *J. Phys. Soc. Japan*, **17,** 510S (1962).
402. S. C. Abrahams and H. J. Williams, *J. Chem. Phys.*, **39,** 2923 (1963).
403. H. Forstat and D. R. McNeely, *J. Chem. Phys.*, **35,** 1594 (1961).
404. R. D. Spence and C. R. K. Murty, *Physica*, **27,** 850 (1961).
405. P. Ros, *Rec. Trav. Chim.*, **82,** 823 (1963).
406. B. Morosin and E. C. Lingafelter, *Acta Cryst.*, **13,** 807 (1960).

407. P. Paoletti, *Trans. Faraday Soc.*, **61,** 219 (1965).
408. A. G. Karipides and T. S. Piper, *Inorg. Chem.*, **1,** 970 (1962).
409. J. C. Barnes and D. N. Humes, *Inorg. Chem.*, **2,** 444 (1963).
410. P. S. Braterman, *Inorg. Chem.*, **2,** 448 (1963).
411. A. Silberstein, *Compt. Rend.*, **201,** 970 (1935).
412. A. Silberstein, *Bull. Soc. France Minerale*, **59,** 329 (1936).
413. A. Silberstein, *Compt. Rend.*, **209,** 540 (1939).
414. M. Inoue, M. Kishita and M. Kubo, *Inorg. Chem.*, **6,** 900 (1967).
415. C. Brink and C. H. MacGillavary, *Acta Cryst.*, **2,** 158 (1949).
416. W. U. Malik, S. M. F. Rahman and S. A. Ali, *Z. anorg. allgem. Chem.*, **299,** 322 (1959).
417. G. J. Sutton, *Australian J. Chem.*, **17,** 1360 (1964).
418. W. Cochran, F. A. Hart and F. G. Mann, *J. Chem. Soc.*, **1957,** 2816.
419. A. Kabesh and R. S. Nyholm, *J. Chem. Soc.*, **1951,** 38.
420. J. R. Clifton and J. T. Yoke, *Inorg. Chem.*, **5,** 1390 (1966).
421. C. Brink, N. F. Binnendijk and J. van de Linde, *Acta Cryst.*, **7,** 176 (1954).
422. C. Romming and K. Waerstad, *Chem. Commun.*, **1965,** 299.
423. P. T. Beurskens, J. A. Cras and J. J. Steggerda, *Inorg. Chem.*, **7,** 810 (1968).
424. M. C. Ball and R. F. M. Coultard, *J. Chem. Soc.*, **1968,** A, 1417.
425. R. M. Ruthven and C. N. Kenny, *J. Inorg. Nucl. Chem.*, **30,** 931 (1968).
426. R. W. Adams, C. G. Barraclough, R. L. Martin and G. Winter, *Australian J. Chem.*, **20,** 2351 (1967).
427. L. L. Lohr, *Inorg. Chem.*, **7,** 2093 (1968).
428. M. Laing and E. Horsfield, *Chem. Commun.*, **1968,** 735.
429. J. Burgess, *Spectrochim. Acta*, **24A,** 277 (1968).
430. P. T. T. Wong and D. G. Brewer, *Can. J. Chem.*, **46,** 131 (1968).
431. P. T. T. Wong and D. G. Brewer, *Can. J. Chem.*, **46,** 139 (1968).
432. J. Burgess, *Spectrochim. Acta.*, **24A,** 1645 (1968).
433. D. G. Brewer, P. T. T. Wong and M. C. Sears, *Can. J. Chem.*, **46,** 3137 (1968).
434. H. Bock, H. T. Diek, H. Pyttlik and M. Schnoeller, *Z. anorg. allgem. Chem.*, **357,** 54 (1968).
435. B. T. Kilbourn and J. D. Dunitz, *Inorg. Chim. Acta*, **1,** 209 (1967).
436. C. C. Houk and K. Emerson, *J. Inorg. Nucl. Chem.*, **30,** 1493 (1968).
437. R. S. Sager and W. H. Watson, *Inorg. Chem.*, **7,** 2035 (1968).
438. R. D. Ball, D. Hall, C. E. F. Rickard and T. N. Waters, *J. Chem. Soc.*, **1967,** A, 1435.
439. R. Uggla, L. Lemmetti, T. Simonen and S. Lundell, *Suomen Kemistilehti*, **40B,** 265 (1967).
440. R. Uggla, S. Lundell and P. Vaananen, *Suomen Kemistilehti*, **41B,** 250 (1968).
441. U. Klement and A. Schmidpeter, *Angew. Chem., Intern. Ed. Engl.*, **7,** 470 (1968).
442. R. D. Willett, private communication.
443. A. A. G. Tomlinson and B. J. Hathaway, *J. Chem. Soc.*, **1968,** A, 2578.
444. R. R. Osborne and W. R. McWhinnie, *J. Chem. Soc.*, **1967,** A, 2075.
445. M. C. Feller and R. Robson, *Australian J. Chem.*, **21,** 2919 (1968).
446. G. W. A. Fowles, D. A. Rice and R. A. Walton, *J. Chem. Soc.*, **1968,** A, 1842.

447. R. N. Patel and D. V. R. Rao, *Indian J. Chem.*, **6,** 112 (1968).
448. D. E. Billing and A. E. Underhill, *J. Inorg. Nucl. Chem.*, **30,** 2147 (1968).
449. E. J. Duff, *J. Chem. Soc.*, **1968,** A, 1812.
450. C. J. Popp and R. O. Ragsdale, *Inorg. Chem.*, **7,** 1845 (1968).
451. E. J. Duff, *J. Chem. Soc.*, **1968,** A, 434.
452. M. Brierley and W. J. Geary, *J. Chem. Soc.*, **1968,** A, 1641.
453. D. M. L. Goodgame, M. Goodgame, P. J. Hayward and G. W. Rayner-Canham, *Inorg. Chem.*, **7,** 2447 (1968).
454. E. J. Duff and M. N. Hughes, *J. Chem. Soc.*, **1968,** A, 2144.
455. W. J. Eilbeck, F. Holmes, C. E. Taylor and A. E. Underhill, *J. Chem. Soc.*, **1968,** A, 1189.
456. D. M. Bowers and A. I. Popov, *Inorg. Chem.*, **7,** 1594 (1968).
457. M. Inoue, S. Emori and M. Kubo, *Inorg. Chem.*, **7,** 1427 (1968).
458. M. J. Campbell, R. Grzeskowiak and M. Goldstein, *Spectrochim. Acta*, **24A,** 1149 (1968).
459. E. J. Duff, M. N. Hughes and K. J. Rutt, *J. Chem. Soc.*, **1968,** A, 2354.
460. D. StC. Black, *Australian J. Chem.*, **21,** 803 (1968).
461. L. F. Lindoy and S. E. Livingstone, *Inorg. Chim. Acta*, **2,** 119 (1968).
462. L. F. Lindoy and S. E. Livingstone, *Inorg. Chim. Acta*, **2,** 166 (1968).
463. A. W. McLellan and G. A. Melson, *J. Chem. Soc.*, **1967,** A, 137.
464. G. A. Melson, *J. Chem. Soc.*, **1967,** A, 669.
465. B. B. Kedzia, P. X. Armendarez and K. Nakamoto, *J. Inorg. Nucl. Chem.*, **30,** 849 (1968).
466. M. J. Campbell, M. Goldstein and R. Grzeskowiak, *Chem. Commun.*, **1967,** 778.
467. M. J. Campbell and R. Grzeskowiak, *J. Inorg. Nucl. Chem.*, **30,** 1865 (1968).
468. C. W. Bird and E. M. Briggs, *J. Chem. Soc.*, **1967,** A, 1004.
469. J. G. H. du Preez, W. A. Steyn and A. J. Basson, *J. South African Chem. Institute*, **21,** 8 (1968).
470. N. C. Collins and W. K. Glass, *Chem. Commun.*, **1967,** 211.
471. K. E. Hyde, G. Gordon and G. F. Kokoszka, *J. Inorg. Nucl. Chem.*, **30,** 2155 (1968).
472. M. F. Farona and G. R. Tompkins, *Spectrochim. Acta*, **24A,** 788 (1968).
473. H. S. Preston, C. H. L. Kennard and R. A. Plowman, *J. Inorg. Nucl. Chem.*, **30,** 1463 (1968).
474. V. V. Udovenko and M. A. Sherstobpyeva, *Ukr. Khim. Zh.*, **33,** 124 (1967).
475. A. Misono, T. Osa and S. Koda, *Bull. Chem. Soc. Japan*, **41,** 373 (1968).
476. G. E. Batley and D. P. Graddon, *Australian J. Chem.*, **21,** 1473 (1968).
477. E. Uhlig and K. Staiger, *Z. anorg. allgem Chem.*, **360,** 39 (1968).
478. P. Lumme and K. Junkkarinen, *Suomen Kemistilehti*, **41B,** 122 (1968).
479. F. Biehat, B. Wyncke and A. Hadni, *Compt. Rend.*, **267B,** 778 (1968).
480. R. Nasanen and E. Luukkonen, *Suomen Kemistilehti*, **41B,** 27 (1968).
481. R. Uggla and S. Lundell, *Suomen Kemistilehti*, **41B,** 73 (1968).
482. M. Di Vaira and P. L. Orioli, *Acta Cryst.*, **24B,** 595 (1968).
483. Z. Dori, *Chem. Commun.*, **1968,** 714.
484. D. P. Madden and S. M. Nelson, *J. Chem. Soc.*, **1968,** A, 2342.
485. M. Ciampolini, J. Gelsomini and N. Nardi, *Inorg. Chim. Acta*, **2,** 343 (1968).

486. E. Wenschuh, B. Fritzsche and H. Mach, *Z. anorg. allgem. Chem.*, **358,** 233 (1968).
487. R. Uggla, L. Lemmetti, S. Lundell and M. Juvani, *Suomen Kemistilehti*, **41B,** 134 (1968).
488. F. Akhtar, D. M. L. Goodgame, M. Goodgame, G. W. Rayner-Canham and A. C. Skapski, *Chem. Commun.*, **1968,** 1389.
489. M. Soga, R. Imaizumi, Y. Kondo and T. Okabe, *J. Electrochem. Soc.*, **114,** 388 (1967).
490. M. Certier, C. Wexker and S. Nikitine, *Compt. Rend.*, **267B,** 785 (1968).
491. W. A. Spofford and E. L. Amma, *Chem. Commun.*, **1968,** 405.
492. A. Domenicano, R. Spagna and A. Vaciago, *Chem. Commun.*, **1968,** 1291.
493. C. S. Cundy, M. Green, F. G. A. Stone and A. Taunton-Rigby, *J. Chem. Soc.*, **1968,** A, 1776.
494. T. Ogura, T. Hamachi and S. Kawaguchi, *Bull. Chem. Soc. Japan*, **41,** 892 (1968).
495. M. A. Bennett, W. R. Kneen and R. S. Nyholm, *Inorg. Chem.*, **7,** 552 (1968).
496. A. J. Carty and A. Efraty, *Can. J. Chem.*, **46,** 1598 (1968).
497. A. U. Malik, *J. Indian Chem. Soc.*, **45,** 163 (1968).
498. M. J. Bernard and M. Massaux, *Compt. Rend.*, **266C,** 1041 (1968).
499. W. Mao-Chia Yang, L. H. Schwartz and P. N. LaMori, *J. Phys. Chem. Solids*, **29,** 1633 (1968).
500. K. N. Raymond, D. W. Meek and J. A. Ibers, *Inorg. Chem.*, **7,** 1111 (1968).
501. G. A. Heath, R. L. Martin and I. M. Stewart, *Australian J. Chem.*, **22,** 83 (1969).
502. N. S. Bill and F. B. Taylor, *Inorg. Syn.*, **9,** 136 (1967).
503. J. S. Avery, C. D. Burbridge and D. M. L. Goodgame, *Spectrochim. Acta*, **24A,** 1721 (1968).
504. B. N. Figgis, M. Gerloch, J. Lewis and R. C. Slade, *J. Chem. Soc.*, **1968,** A, 2028.
505. R. P. Eswein, E. S. Howald, R. A. Howald and D. P. Keeton, *J. Inorg. Nucl. Chem.*, **29,** 437 (1967).
506. J. W. Halley, *Phys. Rev.*, **168,** 593 (1968).
507. P. D. Loly, *J. Appl. Phys.*, **39,** 1109 (1968).
508. A. Furuhashi, T. Takeuchi and A. Ouchi, *Bull. Chem. Soc., Japan*, **41,** 2049 (1968).
509. R. Perret and H. d'Escrienne, *Bull. Soc. Chim. France*, **1968,** 2379.
510. H. Suzuki and T. Watanabe, *Phys. Letters*, **26A,** 103 (1967).

Index